THE BIOCHEMISTRY OF ARCHAEA (ARCHAEBACTERIA)

New Comprehensive Biochemistry

Volume 26

General Editors

A. NEUBERGER
London

L.L.M. van DEENEN
Utrecht

ELSEVIER
Amsterdam · London · New York · Tokyo

The Biochemistry of Archaea (Archaebacteria)

Editors

M. Kates

*Department of Biochemistry, University of Ottawa,
Ottawa, Ont. K1N 6N5, Canada*

D.J. Kushner

*Department of Microbiology, University of Toronto,
Toronto, Ont. M5S 1A8, Canada*

A.T. Matheson

*Department of Biochemistry and Microbiology,
University of Victoria, Victoria, B.C. V5Z 4H4, Canada*

1993
ELSEVIER
Amsterdam · London · New York · Tokyo

Elsevier Science Publishers B.V.
P.O. Box 211
1000 AE Amsterdam
The Netherlands

Library of Congress Cataloging-in-Publication Data

The Biochemistry of archaea (archaebacteria) / editors, M. Kates, D.J.
Kushner, A.T. Matheson.
    p.   cm. -- (New comprehensive biochemistry ; v. 26)
  Includes bibliographical references and index.
  ISBN 0-444-81713-1 (acid-free paper)
  1. Archaebacteria--Metabolism. 2. Archaebacteria--Molecular
aspects.   I. Kates, Morris. II. Kushner, Donn. III. Matheson, A.
T.   IV. Series
QD415.N48 vol. 26
[QR82.A69]
574.19'2 s--dc20
[589.9]
                                                                           93-33456
                                                                             CIP

ISBN 0 444 81713 1
ISBN 0 444 80303 3 (series)

©1993 Elsevier Science Publishers B.V.   All rights reserved.

No part of this publication may be reproduced, stored in a retrieval system or transmitted in any form or by any means, electronic, mechanical, photocopying, recording or otherwise without the prior written permission of the publisher, Elsevier Science Publishers B.V., Copyright and Permissions Department, P.O. Box 521, 1000 AM Amsterdam, the Netherlands.

No responsibility is assumed by the publisher for any injury and/or damage to persons or property as a matter of products liability, negligence or otherwise, or from any use or operation of any methods, products, instructions or ideas contained in the material herein. Because of the rapid advances in the medical sciences, the publisher recommends that independent verification of diagnoses and drug dosages should be made.

Special regulations for readers in the USA – This publication has been registered with the Copyright Clearance Center Inc. (CCC), Salem, Massachusetts. Information can be obtained from the CCC about conditions under which photocopies of parts of this publication may be made in the USA. All other copyright questions, including photocopying outside the USA, should be referred to the publisher.

Printed on acid-free paper

Printed in the Netherlands

Preface

In the last ten years, considerable information has accumulated on the biochemistry of the archaea. Some aspects of this subject, such as bioenergetics, molecular biology and genetics, membrane lipids, etc., have been dealt with in individual book chapters and review articles in various treatises, but the subject as a whole has not yet been treated in a comprehensive manner.

The present volume attempts to bring together recent knowledge concerning general metabolism, bioenergetics, molecular biology and genetics, membrane lipid and cell-wall structural chemistry and evolutionary relations, of the three major groups of archaea: the extreme halophiles, the extreme thermophiles, and the methanogens. We have called upon a number of experts, all actively involved in research on the above subjects to review their specialized fields.

In the Introduction, C.R. Woese considers the evolutionary relationship of these microorganisms to all other living cells. This is followed by a section on special metabolic features of archaea, covered in Chapters 1 through 4, respectively, by: M.J. Danson on central metabolism in archaea, including carbohydrate metabolism, the citric-acid cycle, and amino-acid and lipid metabolism; V.P. Skulachev on bioenergetics and transport in extreme halophiles; L. Daniels on the biochemistry of methanogenesis; and P. Schönheit dealing with bioenergetics and transport in methanogens and related thermophilic archaea.

A section on protein structural chemistry in archaea includes Chapters 5 through 7, respectively, by: D. Oesterhelt on the structure and function of photoreceptor proteins in the Halobacteriaceae; J. Lanyi on the structure and function of ion-transport rhodopsins in extreme halophiles; and R. Hensel on proteins of extreme thermophiles. In a section on cell envelopes (Chapters 8–10), O. Kandler and H. König discuss the structure and chemistry of archaeal cell walls; M. Kates reviews the chemistry and function of membrane lipids of archaea; and L.I. Hochstein covers membrane-bound proteins (enzymes) in archaea.

Chapters 11 through 14 deal with aspects of molecular biology in archaea and include, respectively, DNA structure and replication by P. Forterre; transcription apparatus by W. Zillig et al.; translation apparatus by R. Amils et al.; and ribosomal structure and function by A. Matheson et al.

The final chapters (15–17) deal with the molecular genetics of archaea: L. Schalkwyk discusses halophilic genes; J. Reeve and J. Palmer describe genes of methanogens; and R. Garrett and J.Z. Dalgaard discuss genes of extreme thermophiles.

In an epilogue, W.F. Doolittle presents an overview of all chapters in the larger context of cellular evolution and our future understanding of this subject.

The editors trust that this volume is sufficiently comprehensive in scope to be of use to researchers actively engaged or interested in various aspects of the biochemistry of archaea. They hope that it will also stimulate further studies of the topics covered, and will open up new areas for investigation.

M. Kates
D.J. Kushner
A.T. Matheson

M. Kates et al. (Eds.), *The Biochemistry of Archaea (Archaebacteria)*
© 1993 Elsevier Science Publishers B.V. All rights reserved

INTRODUCTION

The archaea:
Their history and significance

Carl R. WOESE

Department of Microbiology, University of Illinois, Urbana, IL 61801, USA

1. Introduction

In November of 1977 the president of the National Academy of Sciences of the United States announced to the world that the human growth hormone gene had been cloned. This momentous accomplishment ushered in (or at least symbolized) the era of medical/industrial genetic engineering which so strongly impacts society today. However, the press coverage of this announcement was far less extensive than might have been expected, for a completely unanticipated reason. Coincidentally, NASA had issued a press bulletin announcing the discovery of a "third form of life" (which would become known as the archaebacteria), and it was this that was splashed across the front pages of the New York Times and other major newspapers, mentioned on the evening TV news programs, and even drew a quip from Johnny Carson.

For the press and the public the discovery of the archaea was a highly significant event; it had touched upon that age old basic human concern about where we come from – which interested the layman more than the promise of a brighter tomorrow through biomedical technology. The biology community, on the other hand (though not scientists in general), had a decidedly different reaction. For them the cloning of the somatotropin gene was an important milestone in the ceaseless efforts of biologists to cure disease. As regards the discovery of a "third form of life", however, biologists' attitudes generally ranged from skeptical to intensely negative. Some derisively rejected the claim out of hand. One well-known biologist went so far as to suggest to one of my collaborators that he publicly dissociate himself from the work. Another counseled one of the three major US news magazines not to carry the story, and it didn't. Unfortunately, very little of this negative reaction was expressed in a scientifically proper way, i.e., in the scientific literature. It would have been interesting and instructive to quote today.

Why did biologists react so differently than laymen (and other scientists) to the discovery of the archaea, many even viewing the claim as bogus? Their reaction was nothing new. It (the rejection, its vehemence, and the associated scorn) is a well-recognized sociological phenomenon, which Thomas Kuhn discusses extensively in his now classic work *The Structure of Scientific Revolutions*. What the discovery of the archaea had done was counter the existing paradigm, cross one of biology's deep prejudices:

> All organisms except viruses can be assigned to one of two primary groups [prokaryotes and eukaryotes] ... descriptions of them can be found in the better textbooks of general biology, a sure indication that they have acquired the status of truisms. [1]

> There is little doubt ... that biologists can accept the division of cellular life into two groupings at the highest level expressing the encompassing characters of procaryotic and eucaryotic cellular organization. [2]

To claim that a third primary group existed [3,4] was patently absurd!

As the reader might imagine, those of us involved in the discovery of the archaea had ourselves experienced the same sense of incredulity when first confronted with the data. How could there possibly be something that was neither eukaryotic nor prokaryotic? Yet that seemed the only reasonable interpretation of the data. If so, then what we all took for granted regarding prokaryotes and eukaryotes must be wrong. Once that light dawned, the source of the problem was obvious: A prokaryote had originally been defined as an organism that did *not* possess certain eukaryotic cellular features, e.g., a membrane-bounded nucleus and mitochondria. Defined in this purely negative way "prokaryote" could not be a phylogenetically meaningful grouping. Yet, this is precisely what it had been taken to be – without any supporting evidence. When, decades later, it became possible – through electron microscopy and molecular studies – to redefine "prokaryote" (and "eukaryote") in comparable, positive terms, and so, to test the phylogenetic validity of the prokaryote taxon, the biologist, strangely, felt no need to do so: Prokaryotes were "obviously" all related to one another; therefore, a few representative cases would suffice. As a consequence, the generalizations that came to be known as "prokaryotic characteristics" were all based on very few examples; they were, by and large, merely characteristics of *Escherichia coli* and a few of its relatives. No comprehensive characterization of prokaryotes, with the intent of testing their supposed monophyletic nature, had ever been done!

This was a mistake of major proportions, for, among other things, it almost certainly delayed the discovery of the archaea by at least a decade (see below). Such a logical transgression might be excused among botanists and zoologists on the grounds that they knew and cared little about prokaryotes. But, for the keepers of the prokaryote, the microbiologists, never to have questioned their monophyletic nature would have been unpardonable. As it turns out, this was not actually what happened; but what did happen, nevertheless, had the same effect. Microbiologists of an earlier era were very much concerned with the phylogenetic relationships among bacteria, and were, therefore, skeptical of the idea that "prokaryote" represented a monophyletic grouping. Yet, a few decades later, and for reasons that are hard even in retrospect to understand, this critical scientific attitude was unaccountably replaced by a naive, unscientific acceptance of "prokaryote" as a phylogenetically valid taxon. We will examine this unfortunate transformation, so central to the history of the archaea, in some detail below.

My reason for saying that the prokaryote–eukaryote prejudice delayed the discovery of the archaea by at least a decade is that (anecdotal) evidence for their existence had been in the literature some time before 1977. These bits and pieces in retrospect constituted a prima facie case that something might be wrong with the idea that prokaryote is a monophyletic taxon, and the situation cried for deeper, more comprehensive investigation. Yet, at the time no one came to this conclusion.

The cards were definitely stacked against the discovery of the archaea, for in addition to the hegemony of the prokaryote concept, other factors and prejudices worked to prevent their emergence. The early evidence for the existence of the archaea came in the main from an odd collection of organisms that lived in "extreme" environments. At that time conventional wisdom held that organisms living in extreme environments represent evolutionary adaptations to their environments, and such adaptations required an organism to undergo unusual phenotypic changes. Thus, not only were all bacteria by definition "prokaryotic"; but anything that was atypical and lived in an "extreme" environment was atypical by reason of that circumstance. The case for the archaea was not helped either by the fact that the key organisms in question were very unlike one another in their overall phenotypes, making it particularly difficult for even the best microbiologists to sense their relationship. To make matters worse, idiosyncrasy could also be found scattered among (what turned out to be) the (eu)bacteria as well. And, of course, since the natural relationships among bacteria were not understood, there was no basis for sorting any of this out.

One early piece in the archaeal puzzle was methanogenesis, an unusual biochemistry that involved a variety of new coenzymes [5,6]. Except for this common biochemistry, however, methanogens seemed to have little in common (morphologically, that is) with one another. What did this mean? To those who took morphology as a primary indicator of relationship it meant only that patches of methanogenic metabolism existed scattered across the phylogenetic landscape, a view reflected in the seventh edition of Bergey's Manual [7]. To those who took physiology as a primary indicator of relationship, methanogens constituted a phylogenetically coherent, separate, taxon, the view that prevailed in the Manual's eighth edition [8]. Despite their highly unusual biochemistry, however, methanogens were never perceived as anything but typical "prokaryotes" in the phylogenetic sense.

Another early piece of the puzzle was the unusual isoprenoid ether linked type of lipid found by Morris Kates and his colleagues in the extreme halophiles (refs. [9,10]; see Chapter 9 of this volume]. This sort of lipid was also to be discovered in *Thermoplasma* [11] and in *Sulfolobus* [12,13] before 1977, but was not discovered in the methanogens [13a] until after the archaea were recognized as a group. Yet, no phylogenetic connection among the organisms that possessed them was considered. Conventional wisdom held firm; adaptation to extreme environments had somehow caused those "unrelated" lineages all to independently arrive at the same unusual lipid structure [12].

> The fact that *Sulfolobus* and *Thermoplasma* have similar lipids is of interest, but almost certainly this can be explained by convergent evolution. This hypothesis is strengthened by the fact that *Halobacterium*, another quite different organism, also has lipids similar to those of the two acidophilic thermophiles. [14]

These same organisms shared another peculiar characteristic: cell walls that did not contain peptidoglycan; and this was known to be true even for a methanogen [15–17]. Still, no phylogenetic interpretation followed; but in fairness it should be noted that the walls of certain (eu)bacteria, such as the planctomyces, also contain no peptidoglycan [18].

The extreme halophiles in addition possessed peculiar ribosomes, which contained acidic, rather than basic, proteins [19]. While this fact alone carried no obvious phylogenetic implications, the subsequent finding that sequences of these proteins were remarkably similar to eukaryotic ribosomal proteins – a discovery that occurred at about the same time as, but completely independent of, the rRNA characterizations – certainly did [20].

Still another clue seemed to be the discovery in *Thermoplasma* (a wall-less "prokaryote") of a histone, which was interpreted to mean that this organism represented the ancestry of eukaryotes [21]. Unfortunately, this clue turned out to be a false one, because the histone, upon later sequencing, proved to be of the bacterial, not the eukaryotic, type [22]. True or not, at the time this claim might have been reason to question the prokaryote concept. But, again, nothing of the sort happened – in this case perhaps because we had long been assured that the ancestor of eukaryotic cells was a "prokaryote" that had lost its cell wall [23]. It remained for Sandman et al. [24] to show, much later, that an eukaryotic type of histone does indeed exist in the archaea.

Despite all this tantalizing anecdotal evidence, molecular biologists and microbiologists alike (myself included) remained secure in their belief that a prokaryote is a prokaryote, and all variations from the norm have no phylogenetic significance. It required more compelling evidence, evidence difficult to interpret in equivocal ways, to awaken us from our reverie. And this came in the form of our rRNA oligonucleotide characterizations. [Even then, however, I was occasionally asked in early seminars about archaebacteria why their incredibly unique rRNA sequences weren't merely the result of adaptation to some extreme environment!]

A very few biologists actually did find the initial claim of a third form of life interesting and important. Many of them were microbiologists who worked with one or more of the archaea and found phylogenetic uniqueness a satisfying explanation for the phenotypic uniqueness they were increasingly encountering. Others, such as W. Zillig, who understood some particular molecular system in great detail (RNA polymerase in his case), knew immediately that an archaeal version was no typical bacterial version [25,26]; this was no "prokaryote" with which they were dealing. In my experience Otto Kandler was the only biologist to understand the concept of a third "Urkingdom" (as it was then called) immediately upon encountering it. Realizing the importance of this finding, he began encouraging German biologists to work on the archaea, and soon thereafter gave the field a tremendous boost by convening the first conference on "archaebacteria", in Munich in 1981 [27].

My own involvement with the archaea came about in an unexpected way, having nothing whatever to do with any interest in unusual organisms or phylogeny per se. I had developed a fascination with the genetic code in the 1960's, and soon realized that the problem was not the purely cryptographic one that, in the days of the "comma free code", everyone took it to be [28].

The translation apparatus, an incredibly complex molecular aggregate that today involves of the order of 100 different molecular species, had to have evolved in stages from a far more rudimentary and inaccurate mechanism [29,30]. There seemed no getting around this conclusion, and the corollary that the form of the genetic code was shaped during and by this evolution [30,31]. Not only the degree and type of order in the codon catalog, but probably the actual codon assignments themselves were products of this coevolution [29–31]. Unfortunately, it seemed that none of the crucial tell-tale interactions were evident in the translation apparatus today. If so, then the code's origin must be sought in the long gone rudimentary versions of the translation apparatus – whose nature must be inferred somehow by means of evolutionary reconstruction.

The differences between the eukaryotic and prokaryotic translation mechanisms provided an intriguing clue, for these differences appeared too profound to be trivially explained [32]. A comparative dissection of the problem required a solid phylogenetic framework, and unfortunately, none existed for the prokaryotes. However, Fred Sanger's laboratory, on their way to devising a nucleic acid sequencing technology, had come up with the oligonucleotide cataloging method for partial sequencing of RNAs, which was well suited to the problem at hand. The obvious choice of molecule for comparative oligonucleotide analysis was the small subunit ribosomal RNA: Its function was not only universal but remarkably constant; rRNAs exist in the cell in high copy number, and so, are relatively easy to isolate; and the small subunit rRNA is large enough to give a good deal of information (but not so large as to be unmanageable). What followed is now a matter of record: In screening various diverse prokaryotic (and a few eukaryotic) rRNAs for their similarities and differences, we stumbled across *Methanobacterium thermoautotrophicum,* whose rRNA oligonucleotide catalog was definitely not typical of "prokaryotes" – but neither was it of eukaryotes [3,4,33].

2. Microbiology's changing evolutionary perspective

From a broader biological perspective the microbiology into which the archaea were born was a strange and strangely isolated biological discipline. Its concerns were local, largely microbial biochemistry and physiology in general, taxonomy for purposes of species identification, and the molecular biology of *E. coli.* It functioned as though evolution did not exist. Although it accepted the grand division of the world into eukaryotes and prokaryotes, microbiologists didn't seek to understand the nature of the relationship between them, of their similarities and differences. Neither did the macro-biologists. Evolutionists in particular functioned (and still do) as though prokaryotes didn't exist or don't matter: Their concerns were fossils and mathematical models of evolutionary dynamics and population flow; and prokaryotes offered precious little in the way of the first and didn't fit neatly into the second. Never mind that prokaryotes and eukaryotes shared a common ancestor, and that the question of the nature of that ancestor (answerable or not at the time) is probably the most important question in evolutionary biology. Never mind that most of evolutionary history is written in terms of microorganisms, not multicellular ones. And never mind that the eukaryotes expropriated chlorophyll-based photosynthesis (probably the most important single evolutionary innovation in the history of the organic world) and the capacity to oxidatively produce ATP via the Krebs cycle, from the bacteria (through endosymbioses) – and that without these there would be no plants, indeed no multicellular life at all. I find it particularly telling that some biologists could posit that the entity that became the body of the eukaryotic cell (i.e., the host for the endosymbionts), was derived from a prokaryote that had lost its cell wall [23], yet showed no concern that were this true, then this ancestral mycoplasma would have to have *drastically* altered its biochemistry, the structure of its translation apparatus, its transcription apparatus, etc., in the process of becoming a full-fledged eukaryote. Even today, when it is apparent that the archaea manifest many so-called "eukaryotic traits" and almost certain that the two lineages shared some sort of common ancestry [exclusive of the (eu)bacteria], it is rare to find a microbiologist, macrobiologist, molecular biologist or evolutionist who shows active interest in this relationship. The historical roots of such attitudes cry for examination.

Looking at the microbiology of a few generations ago, one does not sense an intellectual isolation from the rest of biology, nor does one see the almost total lack of interest in evolutionary matters that later came to characterize the field. To earlier microbiologists bacteria represented a primitive stage in the evolutionary flow, and their place in that flow was a matter of considerable interest:

> Perhaps the designation of Schizophytae [bacteria] may recommend itself for this first and simplest division of living beings ... *(F. Cohn, 1875; quoted in translation [1])*

> It is my conviction that ... [microbial ecology] ... is the most necessary and fruitful direction to guide us in organizing our knowledge of that part of nature which deals with the lowest limits of the organic world, and which constantly keeps before our minds the profound problem of the origin of life itself. *(Beijerinck, 1905; quoted in translation [34])*

> ... plants and animals [may have] passed through intermediate [evolutionary] stages of increased complexity which [had] the characteristics of 'bacteria'. [35]

Not only was there genuine concern with the place of bacteria in the natural order of things, but the course of bacterial evolution, evolutionary relationships among the various bacteria, and how best these could be recognized, were all central issues; as can be seen from the following:

> ... the only truly scientific foundation of classification is to be found in appreciation of the available facts from a phylogenetic point of view. Only in this way can the natural interrelationships of the various bacteria be properly understood ... A true reconstruction of the course of evolution is the ideal of every taxonomist. [36]

> It seems acceptable that the diversity of bacterial forms is the outcome of various independent morphological evolutions which have had their startingpoint in the simplest form both existent and conceivable: the sphere. [36]

> ... it is rather naive to believe that in the distribution of their metabolic characters one can discern the trend of physiological evolution. For these reasons, a phylogenetic system based solely or largely on physiological grounds seems unsound. It is our belief that the greatest weight in making the major subdivisions in the Schizomycetes should be laid on morphological characters ... Clearly paramount is the structure of the individual vegetative cell, including such points as the nature of the cell wall, the presence and location of chromatin material, the functional structures (e.g., of locomotion), the method of cell division, and the shape of the cell. [37]

> Orla-Jensen ... regarded the chemosynthetic bacteria as the most primitive group because they can live in the complete absence of organic matter and hence are independent of other living forms. This overlooks the fact that a chemosynthetic metabolism necessarily presupposes a rather highly specialized synthetic ability such as one would not expect to find in metabolically primitive forms. [37]

These microbiologists understood that their natural systems were not perfect, but they also understood that, even so, the search for the true natural bacterial system should never be abandoned:

> ... inasmuch as the course of phylogeny will always remain unknown, the basis of a true phylogenetic system of classification will be very unstable indeed. On the other hand it cannot be denied that the studies in comparative morphology made by botanists and zoologist have made phylogeny a reality. Under these circumstances it seems appropriate to accept the phylogenetic principle also in bacteriological classification. [36]

> ... there is good reason to prefer an admittedly imperfect natural system to a purely empirical one. A phylogenetic system has at least a rational basis, and can be altered and improved as new facts come to light; its very weaknesses will suggest the type of experimental work necessary for improvement. On the other hand, an empirical system is largely unmodifiable because the differential characters employed are arbitrarily chosen and usually cannot be altered to any great extent without disrupting the whole system ... [37]

> ... the mere fact that a particular phylogenetic scheme has been shown to be unsound by later work is not a valid reason for total rejection of the phylogenetic approach. [37]

The culmination of microbiology's efforts to infer a natural bacterial system came with Stanier and van Niel's bold and imaginative global system in 1941, which concluded the following:

> It is true that there are a small number of organisms of whose relationships we are still ignorant, but if it be remembered that these are mostly microbes not yet studied under laboratory conditions, it may be expected that further work will result in an elucidation of their taxonomic positions. All the other bacteria can be readily subdivided into three large groups [the classes Eubacteriae, Myxobacteriae, and Spirochaetae]. [37]

It was not long, however, before doubt as to whether bacterial characteristics had any real phylogenetic significance – doubts that had always lurked in the background – fulminated. What ensued was a rapid devolution, from a simple initial skepticism that a natural bacterial system could be established on the basis of currently definable bacterial characteristics; to the attitude that it will never be possible to determine a natural bacterial system; to the ultimate dismissal of a natural bacterial system as being of little value. This transformation is largely captured in the following series of quotes. In his famous 1946 Cold Spring Harbor Symposium address van Niel commented:

> The only sound conclusion that seems permissible at present is that we cannot yet use physiological or biochemical characters as a sound guide for the development of a 'natural system' of classification of the bacteria ... [However] the search for a basis upon which a 'natural system' can be constructed must continue. [38]

By 1955 the assessment took on a different tone:

> What made Winogradsky (1952) grant that the systematics of plants and animals on the basis of the Linnean system is defensible, while contending that a similar classification of bacteria is out of the question? The answer must be obvious to those who recognize in the former an increasingly successful attempt at reconstructing a phylogenetic history of the higher plants and animals ... and who feel that comparable efforts in the realm of the bacteria (and bluegreen

algae) are doomed to failure because it does not appear likely that criteria of truly phylogenetic significance can be devised for these organisms. [35]

And by 1962 his and Stanier's view of bacterial phylogeny was decidedly jaded:

Any good biologist finds it intellectually distressing to devote his life to the study of a group that cannot be readily and satisfactorily defined in biological terms; and the abiding intellectual scandal of bacteriology has been the absence of a clear concept of a bacterium... Our first joint attempt to deal with this problem... 20 years ago... was framed in an elaborate taxonomic proposal, which neither of us cares any longer to defend. But even though we have become skeptical about the value of developing formal taxonomic systems for bacteria..., the problem of defining these organisms as a group in terms of their biological organization is clearly still of great importance... [39]

an attitude canonized in microbiology's premier text *The Microbial World*:

... any systematic attempt to construct a detailed scheme of natural relationships becomes the purest speculation... The only possible conclusion is, accordingly, that the ultimate scientific goal of biological classification cannot be achieved in the case of bacteria. [40]

the general course of evolution [for bacteria] will probably never be known, and there is simply not enough objective evidence to base their classification on phylogenetic grounds. For these and other reasons, most modern taxonomists have explicitly abandoned the phylogenetic approach... [41]

This retreat from a phylogenetic perspective seems to have left microbiology intellectually stunned and vulnerable. Previously, microbiologists had been highly skeptical of simplistically lumping all bacteria into one grand (monophyletic) taxon. In 1941 Stanier and van Niel had said:

... we believe that the three major groups among the bacteria are of polyphyletic origin. [37]

And Pringsheim had warned:

The entirely negative characteristics upon which this group [prokaryotes] is based should be noted, and the possibility of... convergent evolution... be seriously considered. [42]

Now, in what seemed a desperate search for unifying microbial principles, microbiologists enthusiastically embraced the very notion they had earlier spurned – as the initial two quotes in this article (from Stanier and Murray [1,2]) and the following demonstrate:

It is now clear that among organisms there are two different organizational patterns of cells, which Chatton (1937) called, with singular prescience, the eucaryotic and procaryotic type. The distinctive property of bacteria and bluegreen algae is the procaryotic nature of their cells. [39]

All these organisms share the distinctive structural properties associated with the procaryotic cell..., and we can therefore safely infer a common origin for the whole group in the remote evolutionary past... [40]

What caused the earlier doubting, critical attitude to be replaced by this dogmatic assertion? Granted, it had become possible in the interim to define the prokaryote, through molecular and electron microscopic characterizations, in positive, phylogenetically telling terms. However, as noted above, the phylogenetically important characterizations were confined mainly to *E. coli*. Thus, there had to have been some a priori assumption that all prokaryotes were similar, were related, in order that (i) one or a few examples could be uncritically accepted as representative of all, and (ii) when real discrepancies did surface (see above) they were automatically taken to be the result of adaptations to unusual environments. Very few complained about the danger of so superficial a characterization of the prokaryote. Here is one of the lonely exceptions:

> ... there are remarkably few comparative studies. The result is that the application of the newer adjuncts of morphology for taxonomic purposes entails generalization from limited cases. [43]

As judged by the author's later writings, it was a caution that not even he stressed, and one that clearly was not heeded by other microbiologists or by molecular biologists.

I have no real explanation for this paradigmatic shift, this uncritical turn of mind. It happened, and it happened in the context of the failure to fulfill one of microbiology's major goals, elucidating the natural microbial system. The reasons why it happened, however, appear non-scientific; and are best left to the historian. The consequences certainly were most unfortunate for biology as a whole. Today microbiology, fortunately, has an evolutionary dimension. The question now is how long it will take microbiologists (and evolutionists) to realize this in a meaningful way, to return to the perspective of Beijerinck, Kluyver and (the young) van Niel.

3. A molecular definition of the three domains

As the intellectual wave in microbiology moved away from (a natural) taxonomy and evolutionary considerations in general, an experimental undertow was pulling the field in precisely the opposite direction. The first proteins were sequenced in the 1950s, and the capacity to characterize nucleic acids by oligonucleotide cataloging (as mentioned above) was developed in the 1960s, with full nucleic acid sequencing methods to follow in the 1970s. In the realm of sequences one cannot avoid phylogeny, a point brought home to all by Zuckerkandl and Pauling in their seminal 1965 publication *Molecules as documents of evolutionary history* [44]. Even so, microbiologists, with their counter-evolutionary orientation, largely ignored Zuckerkandl and Pauling's message; and none appeared to appreciate its full implications as regards a natural microbial system. (Granted, a little work was done on bacterial cytochrome c sequences, but it had only minor impact, especially given that those doing the work ultimately concluded, following Joyce Kilmer, that "only God can make a tree" [45], i.e., molecular sequence comparisons cannot determine bacterial relationships [45–47].)

Molecular sequences reveal evolutionary relationships in ways and to an extent that classical phenotypic criteria, and even molecular functions, cannot. What can be seen only dimly, if at all, at higher levels of cellular organization becomes obvious in terms of molecular structures and sequences. This point is nowhere more dramatically illustrated than by the present situation, where at the level of the whole-cell phenotype the archaea and bacteria are, at best, difficult to distinguish, and the differences they do show are impossible to interpret unequivocally. However, the uniqueness of the two groups is blatant and readily interpretable on the molecular level. From the perspective of ribosomal RNA sequence and structure the living world divides into three distinct classes, corresponding to three very distinct rRNA types [48,49]. The same three classes

TABLE 1
Small subunit rRNA sequence signatures defining and distinguishing the archaea, bacteria and eukarya[a]

Position(s) of base or pair[b]	Bacteria		Archaea		Eukarya	
	Composition	% of total	Composition	% of total	Composition	% of total
8	A	99	U	95	Y(C)	98
9:25	G:C*	99	C:G	100[c]	C:G	100
10:24	R(A)[d]:U	100	Y:R	100	U:A	96
33:551	A:U	100	Y(C):R(G)[e,*]	100	A:U	100
52:359	Y:R	96	G:C	98	G:C	96
53:358	A:U*	97	C:G	100	C:G	100
113:314	G:C*	99	C:G	100	C:G	100
121	Y(C)	97	C	100	A	98
292:308	G:C	99	G:C	100	R:U	94
307	Y|A	99	G	95	Y(U)	100
335	C	99	C	100	A*	96
338	A	100	G*	98	A	100
339:350	C:G	100	G:Y(C)*	98	C:G	98
341:348	C:G	99	C:G	100	U:A	96
361	R(G)*	100	C	100	C	98
365	U	100	A	98	A	100
367	U	100	C*	95	U	98
377:386	R(G):Y(C)	99	Y(C):G	100	Y(C):R(G)	96
393	A	100	G*	95	A	98
500:545	G:C	100	G:C	100	U:A*	94
514:537	Y(C):R(G)*	100	G:C	98	G:C	98
549	C	100	U	98	C	94
558	G	99	Y	95	A	96
569:881	Y(C):R(G)	100	Y:R	100	G:C*	100
585:756	R(G):Y(C)*	100	C:G*	98	U:A*	100
675	A	100	U	100	U	98
684:706	U:A	98	G:Y(C)	98	G:Y	98
716	A*	100	C	100	Y(C)	100
867	R(G)	98	Y(C)	95	Y	94
880	C	98	C	100	U	100
884	U	99	U	100	G	98
923:1393	A:U	99	G:C	100	A:U	100
928 1389	G:C	96	G:C	100	A:U	98
930:1387	Y(C):R(G)*	100	A:U	98	G:C	96
931:1386	C:G*	98	G:C	100	G:C	98
933:1384	G:C*	99	A:U	100	A:U	100
962:973	C:G	96	G:C	98	U:G	98
966	G*	98	U	100	U	100
974	A|C	97	G	98	G	96
1098	Y(C)*	97	G	100	G	98
1109	C*	99	A	100	A	100
1110	A*	100	G	100	G	100
1194	U*	100	R(G)	100	R(A)	100
1201	A|C(A)	99	A|C(A)	100	U	96
1211	U	99	G*	100	Y(U)	100
1212	U*	99	A	100	A	94
1381	U*	99	C	100	C	100
1487	G	100	G	100	A*	96
1516	R(G)	98	G	100	U	98

[a] Adapted from ref. [50]. Except where indicated, the compositions shown are invariant in each domain. Analysis based upon approximately 380 bacterial, 40 archaeal, and 50 eucaryal sequences. R = purine; Y = pyrimidine. Those cases in which a signature composition is "pure", i.e., is not seen at all in the other two domains, are marked by an asterisk.   [b] Numbering follows *E. coli* 16S rRNA standard [76].   [c] 100% applies only to cases showing complete invariance; one or a few exceptions (among bacteria) are, therefore, designated as 99%.   [d] If one form dominates, it is given in parentheses.   [e] R:Y as used here does not include A:C pairs.

emerge from the analysis of a variety of other molecular species as well, ribosomal proteins, translation factors, RNA polymerase, etc. [Unfortunately, in these latter cases the sequence collections are far smaller than those that exist for rRNAs.]

The striking differences among the archaeal, bacterial and eukaryal versions of the (small subunit) rRNA manifest themselves in several ways, in terms of homologous sequence characters, non-homologous sequence characters, and in larger secondary structural elements whose form is specific for one (or more) of the domains [50]. Table 1 shows a signature, based upon homologous positions in the small subunit rRNA sequence, that defines and distinguishes the three domains [50]. Fig. 1 shows the positions of various non-homologous elements in the rRNA secondary structure (explained in Table 2) that do likewise [50]. (These various features are discussed in greater detail by Winker and Woese [50].) The unavoidable conclusion from these data, and from similar data from other molecular species, is that at the highest level, life on this planet is organized into three very distinct, monophyletic groupings [4,48–50].

The differences among the rRNA types are relatively profound compared to differences encountered within any given type; so much so that they in aggregate feel qualitatively different. The special nature of these inter-domain differences is made all the more striking by the realization that they had to have evolved over a relatively short time period, less than one billion years; whereas intra-domain changes have been happening over the last three billion years at least [51]. The evolution that transformed the most recent common ancestor of all extant life into the individual ancestors of the three domains would, then, seem of a more rapid and drastic type than that which occurred subsequently within each of the domains. This would be expected if the universal ancestor had been more rudimentary than its descendants, i.e., it had a smaller genome and simpler overall organization than they did. Such a rudimentary entity, the evolution of whose translation function was not yet complete, has been termed a "progenote" [32,51,52]. Although little or nothing can be said with certainty about the universal ancestor at present, it is encouraging to know that sufficient information seems to have been retained in the genomes and molecular structures of modern cells that in the not too distant future biologists will be able to infer a great deal about this most important period in evolutionary history.

The earlier phylogenetic studies involving rRNA sequences that produced the topology of the universal tree did not, and could not, indicate the position of its root, i.e., the location of the universal ancestor – phylogenetic trees based upon a single molecular species are unrootable unless relative evolutionary rates can be determined. However, it is in principle possible to root the universal tree using the Dayhoff strategy [53], which employs pairs of related (paralogous) genes whose common ancestor duplicated (and functionally diverged) in the ancestral lineage *before* the three primary lines of descent came into being. Since both members of such a pair of genes can be present in all descendant lineages, a phylogenetic tree constructed from a combined alignment (of both) will comprise two topologically equivalent halves. One half serves as the outgroup for, determines the root of, the other – in this case the root of the universal tree. To date, two pairs of paralogous genes, the translation factors EFTu and EFG [54] and two related ATPase subunits [54,55], have proven useful for establishing the root of the universal tree. Both give the same result: the root lies between the (eu)bacteria and the

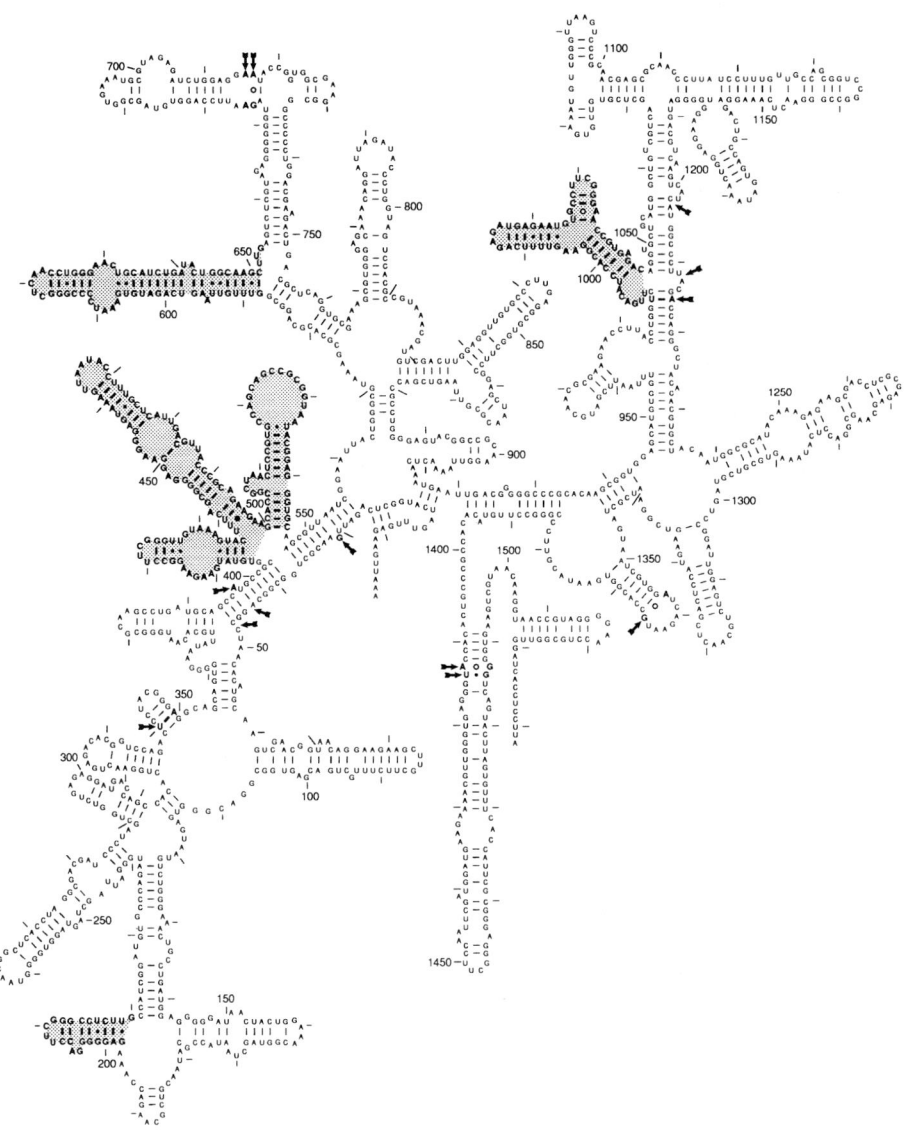

Fig. 1. Secondary structural representation of the non-homologous features in 16S rRNA, i.e., individual nucleotides, pairs, and larger structures that distinguish among the archaea, bacteria and eukarya [50]; underlying secondary structure is that of *E. coli*. Arrows indicate positions of features described in Table 2. Large structures that have a characteristic form in one or more of the domains are shaded.

TABLE 2
Small subunit rRNA sequence idiosyncrasies defining the various domains[a]

Position	Description of unique feature	Found in		
		Bacteria	Archaea	Eukarya
31	bulged base	yes	no	no[b]
44.1:397[c]	paired vs unpaired base	–: A[d]	U:A	–: A[e]
47.1	base added	no	yes	yes
340:349	NCP[f]	U:A 90%	C:G	NCP(A:A)[g] 90%
674:716	G:A pair	G:A 99%	G:C	R:Y(G:C)
675:715	NCP	A:A	U:A	U:A 98%
989:1216	NCP	CP or G:U	CP or G:U	NCP(Y:Y) 90%
1203.1	base added	no	no	yes
1212.1	base added	no	no	yes
1357:1365	R:R pair	A:G or G:A	CP	CP 92%
1413:1487	A:G pair	yes	yes	no, Y:R[h](U:A)
1414:1486	NCP	Y:R	Y:R	NCP(A:A) 92%

[a] Taken from ref. [50]; see the text and Fig. 1 for context.
[b] A bulged nucleotide occurs in a number of eukaryotic sequences at position 29.1, however.
[c] Nucleotides added relative to *E. coli* numbering are designated by decimal additions; e.g., 44.1, 44.2 ... would indicate nucleotides added between *E. coli* positions 44 and 45.
[d] In some cases position 396 is unpaired rather than position 397.
[e] Pairing in helix is largely unproven for eukaryotes; a number of NCP's are present.
[f] "CP" and "NCP" refer to canonical and noncanonical pairs, respectively.
[g] The dominant form is given in parentheses.
[h] "Y:R" does not include C:A pairs.

other two primary lines of descent (which, then, initially share a short common stem); see Fig. 2.

If this result is correct and if the sequences used are actually representative of the eukaryotic cell, i.e., if the basic eukaryotic cell, devoid of genetic contributions from its organelles, is not some kind of phylogenetic chimera [51], then the archaea are specific relatives of the eukaryotes. [Given the importance of these deep evolutionary relationships, one would like to see a great deal more data bearing upon them.] This rooting of the universal tree places the archaea in a different perspective: From long before their discovery, the archaea had been implicitly taken to be bacteria-like, to have the bacterial "Bauplan"; a notion still prevalent [56,57]. This unproductive point of view strongly implies that since the archaea are basically bacterial in nature, there is little point in devoting much effort to their study. A far more productive, and phylogenetically proper, alternative sees the archaea not only as quite distinct from the bacteria in molecular makeup, but, in addition, as specific relatives of the eukaryotes; in which case archaea are not only worthy of significant attention in their own right, but their study should also reveal a great deal about the nature and evolution of the eukaryotic cell.

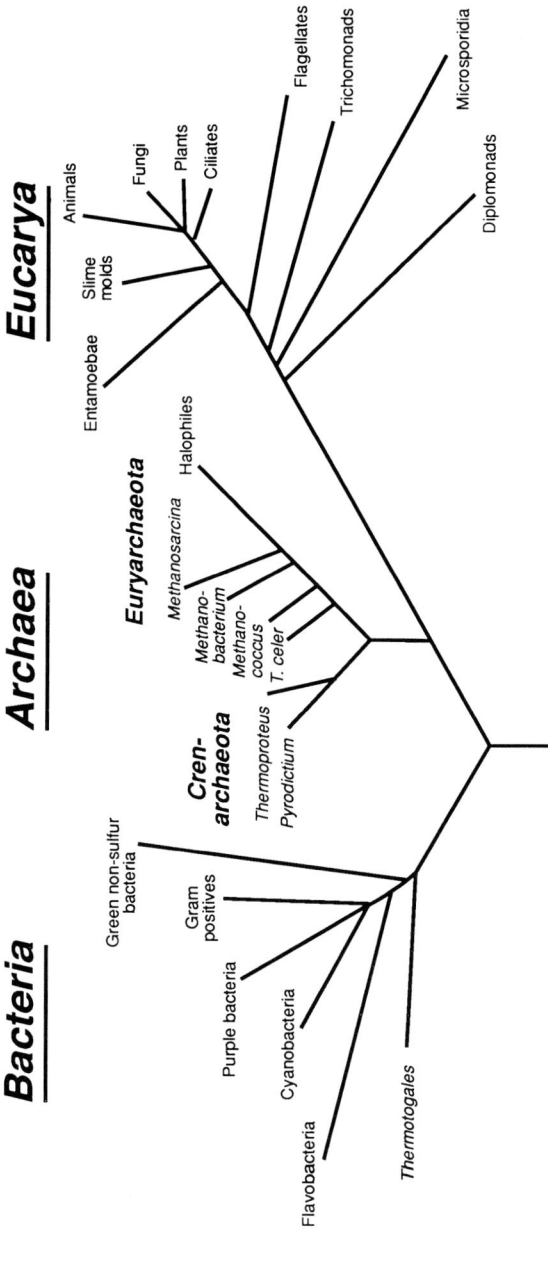

Fig. 2. Universal phylogenetic tree in rooted form, showing the three domains, based upon the corresponding tree in ref. [49] and more recent results concerning eukaryote phylogeny (M.L. Sogin, personal communication). The position of the root was determined by the "Dayhoff strategy", described in the text [54,55].

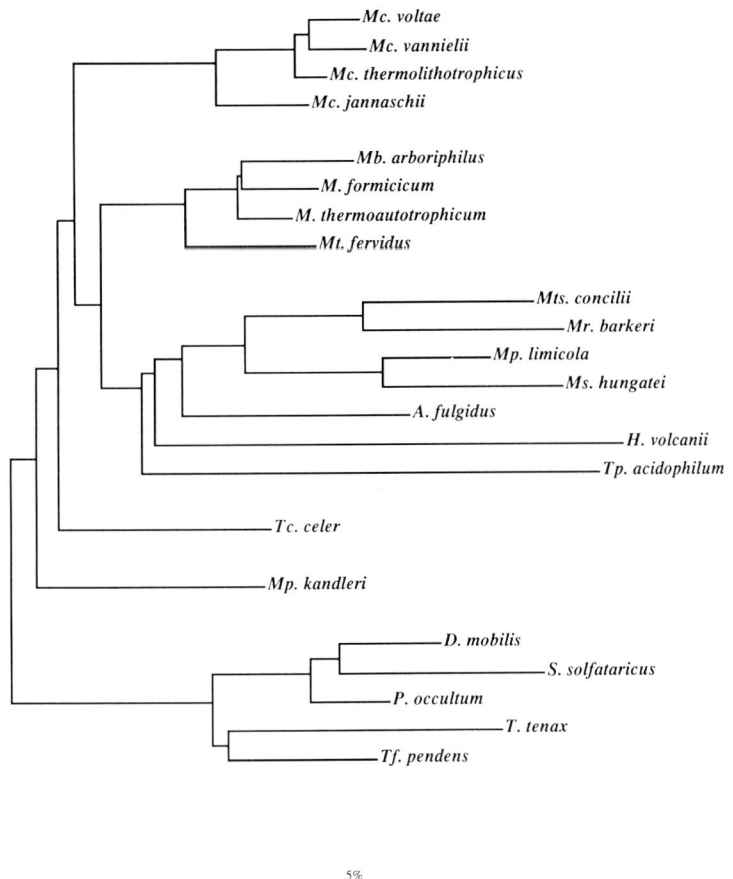

Fig. 3. Representative phylogenetic tree for the archaea, based upon 16S rRNA transversion distances [57]. The scale bar measures 5 nucleotide changes per 100 residues. Generic abbreviations used are: Mc., *Methanococcus*; Mb., *Methanobrevibacter*; M., *Methanobacterium*; Mt., *Methanothermus*; Mts., *Methanosaeta*; Mr., *Methanosarcina*; Mp., *Methanoplanus*; Ms., *Methanospirillum*; H., *Haloferax*; A., *Archaeoglobus*; Tp., *Thermoplasma*; Tc., *Thermococcus*; D., *Desulfurococcus*; S., *Sulfolobus*; P., *Pyrodictium*; T., *Thermoproteus*; Tf., *Thermophilum*.

4. Archaeal phylogenetic relationships

The archaea are known to comprise four quite distinct general phenotypes: the methanogens, the extreme halophiles, a loosely defined thermophilic ("sulfur-dependent") type, and thermophilic sulfate reducers [6]. In that these phenotypes will be thoroughly discussed in other chapters of this book, they will not be detailed here. The four major phenotypes do not correspond to four distinct taxa of equivalent rank, however. Phylogenetic relationships among the four are more complex than this, and suggest particular evolutionary relationships among the phenotypes.

Fig. 3 is a representative phylogenetic tree for the archaea based upon transversion distance analysis of a 16S rRNA sequence alignment [58]. The position of the root of the archaeal tree has been examined in detail, and is considered firmly established [59]. The salient features of the archaeal tree are these: The tree comprises two main branches, now designated the kingdoms Crenarchaeota and Euryarchaeota [48,49,60]. The crenarchaeotes are relatively homogeneous phenotypically, being exclusively of the above thermophilic type. The euryarchaeotes, on the other hand, comprise a potpourri of all the archaeal types. The phylogenetic landscape of the euryarchaeotes is dominated by methanogens, the three main clusters (the Methanococcales, the Methanobacteriales, and the Methanomicrobiales), plus a separate very deeply branching lineage, represented by the genus *Methanopyrus* [48,58,60,61]. Interspersed among these are the extreme halophiles, the thermophilic sulfate reducing archaea, the Thermoplasmales, and the Thermococcales [48,58,60]; all known at present as relatively shallow phylogenetic clusters – whose taxonomic rank would at highest be considered a family among the bacteria. The Thermococcales (and to a lesser extent the Thermoplasmales) exhibit a general crenarchaeal phenotype. A signature analysis leaves little doubt, however, that all the various non-methanogenic phenotypes grouped among the euryarchaeota are definitely more similar to the methanogens than to the crenarchaeota: The 21 positions in the 16S rRNA sequence listed in Table 3 distinguish the three main methanogen clusters (as a whole) from the crenarchaeotes. All of the non-methanogenic euryarchaeotes show predominantly the methanogen signature, each differing from it in no more than four positions; and none shows more than two characters from the crenarchaeal signature. This should be compared to the methanogen *Mp. kandleri,* the deepest branching of euryarchaeal lineage in Fig. 3, whose 16S rRNA shows only 14 methanogen signature characters, the remaining seven being crenarchaeal [61].

Two of the three main methanogen lineages appear phenotypically homogeneous. The third, that leading to the Methanomicrobiales, is not; see Fig. 3. This lineage has also spawned the extreme halophiles, the sulfate reducers and perhaps the Thermoplasmales [48,58,60,62]. All of these relationships to the Methanomicrobiales require further justification, for they are certainly not evident at the phenotypic level [59,62]; and the extreme halophile relationship has been formally questioned [63,64].

Nevertheless, this last mentioned relationship is clearly well established: A specific relationship between the extreme halophiles and the Methanomicrobiales can be demonstrated by a variety of analyses of 16S rRNA sequences – any number of variations on evolutionary distance analysis, parsimony analysis, signature analysis [48,62] and maximum likelihood analysis (G.J. Olsen, personal communication). More importantly, the relationship is also readily given by analysis of 23S rRNA sequences [62].

The placement of the sulfate-reducing archaeon *Archaeoglobus fulgidus* has proven more difficult, but is now considered relatively solid. As have most thermophilic rRNAs, that of *A. fulgidus* has a relatively high G+C content. Because of this, thermophilic lineages tend to be artificially clustered by the usual analyses. (The original placement of *A. fulgidus* in the euryarchaeal tree had been adjacent to *Tc. celer* [65]). The placement of *A. fulgidus* given by transversion distance analysis (Fig. 3) is deemed correct, for this analysis is not subject to the effects of rRNA compositional variation in that total purine content of archaeal rRNAs is almost invariant [58]. The placement of *A. fulgidus* shown in

TABLE 3

Small subunit rRNA signature distinguishing the three groups of methanogens (as a whole) from the Crenarchaeota

Position of nucleotide or base pair	Composition in Methanogens	Composition in Crenarchaeota	Composition[a] in Extreme halophiles	Tp. acidophilum	A. fulgidus	Tc. celer	Mp. kandleri
28:555	G:Y[b]	C:G	–	–	–	–	–
30:553	Y:R[c]	G:C	–	–	–	–	–
34	U	C	–	–	–	–	cr
289:311	C:G	G:C	–	–	–	–	–
501:544	R:Y	C:G	–	–	–	–	cr
503:542	C:G	G:C	–	–	–	–	cr
504:541	Y:R	G:Y	–	–	–	–	–
513:538	C:G	U:A	–	G:C	–	–	–
518	C	U	–	–	G	G	–
658:747	Y:R	G:C	–	–	–	–	–
692	U	C	–	–	–	–	–
939:1344	C:G	G:C	–	–	–	cr	–
965	Y	G	–	–	–	–	–
1074:1083	A:C	G:U	–	–	–	–	cr
1159	C	U	cr	–	cr	–	–
1244:1293	Y:R[d]	R:Y	–	–	–	–	cr
1252	U	C	–	–	–	–	–
1303:1334	C:G	G:C	–	A:U	cr	–	–
1335	C	G	cr	–	–	–	cr
1351	U	C	–	–	–	cr	cr
1408	A	G	–	–	–	–	–

[a] A dash denotes methanogen composition: cr denotes crenarchaeal composition.
[b] Y is C in all cases except some Methanomicrobiales.
[c] Y:R and R:Y pairs include all possibilities except A:C (C:A).
[d] Single exception, U:U.

Fig. 3 has also been confirmed by maximum likelihood analysis of 16S rRNA (G.J. Olsen, personal communication), and by analyses of 23S rRNA sequences [58].

A number of the local relationships among euryarchaeal lineages remain unresolved, however. One is the relative branching order of the extreme halophile and *A. fulgidus* lineages from the Methanomicrobiales stem. Another is the branching order among the three main methanogen lineages (a tight clustering). And, the exact position of the Thermoplasmales remains somewhat variable depending upon analytical method and makeup of the alignment. The deep branching of the Thermococcales and *Mp. kandleri* lineages is given by all methods of analysis, however, and so, is considered highly likely.

Among the crenarchaeota the only branching showing variation is that involving the Sulfolobales [48]. Detailed analysis shows that its correct position is almost certainly as a sister lineage of *Desulfurococcus,* as shown in Fig. [48,58].

Of the three main methanogenic lineages, the Methanomicrobiales lineage is by far the most interesting evolutionarily. This is not merely because of its having spawned the extreme halophiles and sulfate-reducing archaea. Even methanogenesis on this lineage shows unusual variations. Whereas the other two methanogenic lines exhibit a uniform methanogenic biochemistry, varying only in the temperatures at which methanogenesis occurs, various sublines of the Methanomicrobiales produce methane from a variety of sources (acetate, methyl amines) in addition to carbon dioxide, and under a variety of conditions, e.g., halophilic or alkaliphilic [6].

What evolutionary conclusions can one draw from the phylogenetic distribution of phenotypes on the archaeal tree? Unfortunately, very few. It seems safe to infer that halobacterial metabolism arose from methanogenic metabolism, as did archaeal sulfate reduction. When *Tc. celer*, which has the general crenarchaeal phenotype, was thought to be the lowest branching on the euryarchaeal side of the archaeal tree, it was argued that the general crenarchaeal phenotype was the ancestral archaeal phenotype, and so, had given rise to methanogenesis [48]. However, now that a methanogen, *Mp. kandleri*, is known to represent the deepest euryarchaeal branching, this conclusion is no longer defensible. A far better understanding of all types of archaeal metabolism is needed before any further conjectures are made regarding the metabolic history of the archaea.

One conclusion regarding archaeal evolution does seem somewhat safe, however: The archaea have arisen from thermophilic ancestry [48]. Not only are all known crenarchaeal isolates thermophilic, many of them hyperthermophiles, but the two deepest branchings in the euryarchaeota, *Thermococcus* and *Methanopyrus*, are likewise, as are the deepest branchings within two of the three main groups of methanogens and the sulfate-reducing archaea (on the remaining methanogenic lineage, the Methanomicrobiales). Although (eu)bacterial phylogeny is not an issue here, let it be noted that the distribution of phenotypes on the bacterial tree suggests a thermophilic origin there as well [66]. The simplest assumption would then be that the universal ancestor itself lived in a hot environment, as did the ancestor common to the archaea and eukarya. However, this leaves unanswered the question of why there is, then, no indication of thermophilic ancestry in the eukaryal phylogenetic tree.

The archaeal ribosome and the operonal organization of rRNA genes show interesting patterns of phylogenetic variations. Crenarchaeal ribosomes have relatively high ratios of protein to RNA, while the Methanobacteriales, Methanomicrobiales and extreme halophiles show low ratios [67]. However, this cannot be taken as distinguishing the crenarchaeotes from the euryarchaeotes, for the deeply branching euryarchaeon *Thermococcus celer*, as well as the Thermoplasmales and the Methanococcales, all exhibit high protein:RNA ratios [67]. [Protein:rRNA ratios have not been reported for the ribosomes of the sulfate-reducing archaea or for *Mp. kandleri*.] Other ribosomal properties that differ between crenarchaeotes and (some) euryarchaeotes are: (i) the degree of modification of bases in rRNA (and tRNA), much higher in crenarchaeotes than in the euryarchaeotes so far characterized [68]; (ii) the presence of a tRNA gene (for alanine) in the spacer region of the rRNA operon in all euryarchaeotes characterized, but not found so far among the crenarchaeota [69]; and (iii) a 5S rRNA gene terminally linked to the rRNA gene operon, seen in euryarchaeotes (except for *Thermococcus*) but not in crenarchaeotes [70].

The simplest interpretation of these rather striking differences between the crenarchaeal and euryarchaeal ribosomes is that the crenarchaeal type (and its corresponding gene organization) most closely approximates the ancestral archaeal condition. On the euryarchaeal, but not the crenarchaeal branch a rather extensive evolution would then have occurred, which, at various points in euryarchaeal evolution, lowered the ancestral protein:RNA ratio in the ribosome, decreased drastically the degree of base modification in rRNAs, and created an rRNA operon having a tRNA gene in the spacer region and a linked 5S rRNA gene. The reduction in protein:RNA ratio in the ribosome would have occurred only once if those euryarchaeal groups still retaining the high ratio [67] all branched more deeply than those exhibiting the low ratio, which is the case in some phylogenetic analyses of the rRNA sequence data.

5. The new microbiology

Two developments have radically transformed the science of microbiology: (i) the techniques that permitted the determination of a natural microbial system, and (ii) the consequent (but unanticipated) discovery of the archaea. In the past, microbiology was an isolated, excessively anecdotal, and materialistic (in a molecular/biochemical sense) discipline. The prokaryote/eukaryote dichotomy served as an intellectual wall that isolated microbiology from the rest of biology. Now the eukaryote/prokaryote differences serve as a beacon that illumines microbiology's distant evolutionary past, and allows the biologist to begin to glimpse the true relationships between the prokaryotes and the eukaryotes. Microbial taxonomy, which had been largely a collection of scattered facts about a great variety of organisms, is becoming a totally interconnected tree of relationships, on which each species becomes the flowering tip of some particular branch. Microbial metabolic pathways were, in a biological sense, superficially understood, interesting from a molecular mechanistic perspective and because of their impressive variety. Now they will become parts of, and the keys to, a greater evolutionary whole, in which a simple prebiotic chemical seed grows into the myriad complex interconnected biochemical networks of modern living systems.

Microbial ecology too has changed. For the first time it is possible to give a complete accounting of the microbial population in any given environment. Norman Pace, with bold insight, recognized that the sequencing techniques successfully used to characterize cultured microorganisms phylogenetically, can be applied directly to any given niche [71,72]: One can ask the question "Who's there?" [72] of that niche, and answer it by cloning and sequencing particular genes from nucleic acid that is directly extracted from the niche [73]. The construction of specific or class probes from such sequences then permits the direct microscopic identification of those organisms in the environmental sample from which the genes have come [74]. Thus the microbiologist can in principle define every species in any given environment, including those that have eluded culture, and, more importantly, those that have gone completely unrecognized. The combination of these methods with traditional methods of isolation should now permit the isolation, and so, detailed study of whole new groups of microorganisms. Here is the microbial ecology that in the words of Beijerinck (quoted above) "... constantly keeps before our minds the profound problem of the origin of life itself." By adding this evolutionary dimension all aspects of microbiology have been rejuvenated and enriched. Microbiology for the first time has become a fully biological discipline.

It was specifically the discovery of the archaea, which splintered the prokaryote/eukaryote dichotomy, that has given microbiology (and biology in general) true evolutionary perspective. This single event has brought into focus, and allowed us to see in a whole new light, two of the great problems in biology: (i) the origin(s) of the eukaryotic cell; and (ii) the nature and evolution of the universal ancestor. As regards the origin(s) of eukaryotes, phylogenetic analysis shows that the eukaryotic cell is of extremely ancient origin, an origin about as ancient as those of the two prokaryotic groups. Eukaryotes did not arise from within either prokaryotic group, as some versions of the endosymbiosis hypothesis have posited [23]. Given that there are two (unrelated) groups of prokaryotes and that one of them, the archaea, is more closely associated with the eukaryotes than the other, notions of the role of endosymbiosis in the origin of the eukaryotic cell are refining and changing; the possibility of endosymbioses from two very different sources now exists. One wonders whether endosymbioses have been far more instrumental in shaping the eukaryotic cell than can be inferred from the (genetically) small contributions made by the mitochondria and chloroplasts. Were there perhaps hosts of endosymbioses involved in this evolution, which left behind only genetic (not cytostructural) residues? Did the eukaryotic cell receive numerous infusions of smaller parts of prokaryotic chromosomes by means other than endosymbiosis? What fraction of the eukaryotic genome has been contributed by each of the two prokaryotic types? The most extreme view here is represented by W. Zillig and coworkers; based upon their experience with DNA-dependent RNA polymerase sequence analyses, they see the eukaryotic cell as nothing more than a chimeric collection of (eu)bacterial and archaeal genes (ref. [75], and W. Zillig, personal communication). Let this and other possibilities be tested through appropriate genome sequencing studies.

The relatively close phylogenetic relationship between the archaea and the eukaryotes should prove of great value (especially given the small size of archaeal genomes). The archaeal RNA polymerase is closer in sequence to eukaryotic RNA polymerases II and III than these two are to one another [75]. An even clearer and more striking situation holds for the archaeal histone (from *Methanothermus*), whose sequence is closer to the sequences of eukaryotic histones H2a, H2b, H3 and H4 than any of these four are to one another [24]. Are these cases the tip of the iceberg; will a large number of individual archaeal genes turn out to have an ancestral relationship to *families* of eukaryotic genes?

The fact that three primary groups of organisms (rather than two) exist, makes the problem of the nature of the universal ancestor a more tractable one than biologists previously thought. In that nothing can be said with certainty about the ancestor at this point, my mentioning it here serves mainly to underscore its importance. Perhaps the key question concerning the (most recent) universal ancestor is whether it was a progenote (see above) or a full-fledged genote [52]. Given the number of known homologous genes in common between bacteria and eukaryotes, there can be little doubt that this ancestor already contained a fair number of genes. However, as mentioned above, the transition from the rRNA of this ancestor to the rRNAs representative of the three primary lineages involved drastic and relatively rapid change, probably in all lineages; the simplest interpretation of this being that the universal ancestor was a simpler, more rudimentary entity than its descendants (as regards the translation function in particular), which would make it, by definition, a progenote [52,76]. However, the biologist does not have to leave the question of the nature of the universal ancestor in the realm of conjecture. Through a concerted program of sequencing the proper selection of genomes, a great deal can be inferred about the nature of this important entity and its own ancestry. Unfortunately, biology today is in the throes of an internecine struggle between what I would call "technologists", who perceive the science as a discipline whose purpose is to solve practical problems (medical, agricultural, environmental) and "fundamentalists", who perceive it as a basic science, one that tells us at a deep level about the nature of reality. The thrust for genome sequencing (its rationale and choice of genomes to be sequenced) comes now from the technologists. It is time to assert the fundamentalist perspective.

Acknowledgment

The author's work on archaea is supported by NASA grant NSG70440-18.

References

[1] Stanier, R.Y. (1970) Symp. Soc. Gen. Microbiol. 20, 1–38.
[2] Murray, R.G.E. (1974) In: Bergey's Manual of Determinative Bacteriology, 8th edition (Buchanan, R.E. and Gibbons, N.E., Eds.), pp. 4–9, Williams & Wilkins, Baltimore.
[3] Balch, W.E., Magrum, L.J., Fox, G.E., Wolfe, R.S. and Woese, C.R. (1977) J. Mol. Evol. 9, 305–311.
[4] Woese, C.R. and Fox, G.E. (1977) Proc. Natl. Acad. Sci. U.S.A. 74, 5088–5090.
[5] Wolfe, R.S. and Higgins, I.J. (1979) In: Microbial Biochemistry (Quaryl, J.R., Ed.), pp. 267–283. M&P press, Lancaster.
[6] Jones, W.J., Nagle, D.P. and Whitman, W.B. (1987) Microbiol. Rev. 51, 135–177.
[7] Breed, R.S., Murray, E.G.D. and Smith, N.R., Eds. (1957) Bergey's Manual of Determinative Bacteriology, 7th edition, Williams & Wilkins, Baltimore, MD.
[8] Buchanan, R.E. and Gibbons, N.E., Eds., Bergey's Manual of Determinative Bacteriology, 8th edition, Williams & Wilkins, Baltimore, MD.
[9] Seghal, S.N., Kates, M. and Gibbons, N.E. (1962) Can. J. Biochem. Physiol. 40, 69–81.
[10] Kates, M. (1972) In: Ether Lipids, Chemistry and Biology (Snyder, F., Ed.), pp. 351–398, Academic Press, New York.
[11] Langworthy, T.A., Smith, P.F. and Mayberry, W.R. (1972) J. Bacteriol. 112, 1193–1200.
[12] De Rosa, M., Gambacorta, A., Millonig, G. and Bu'Lock, J.D. (1974) Experientia 30, 866–868.
[13] Langworthy, T.A., Smith, P.F. and Mayberry, W.R. (1974) J. Bacteriol. 119, 106–116.
[13a] Makula, R.A. and Singer, M.E. (1978) Biochem. Biophys. Res. Commun. 82, 716–722.
[14] Brock, T.D. (1978) Thermophilic Microorganisms and Life at High Temperatures, p. 174, Springer, New York.
[15] Brown, A.D. and Shorey, C.D. (1963) J. Cell Biol. 18, 681–689.
[16] Weiss, R.L., (1973) J. Bacteriol. 118, 275–284.
[17] Jones, B.J., Bowers, B. and Stadtman, T.C. (1977) J. Bacteriol. 130, 1357–1363.
[18] Konig, W., Schlesner, H. and Hirsch, P. (1984) Arch. Microbiol. 138, 200–205.
[19] Bayley, S.T. and Kushner, D.J. (1974) J. Mol. Biol. 9, 654–669.
[20] Matheson, A.T. (1985) In: The Bacteria, Vol. 8, Archaebacteria (Woese, C.R. and Wolfe, R.S., Eds.), pp. 345–412, Academic Press, New York.

[21] Searcy, D.G., Stein, D.B. and Green, G.R. (1978) Biosystems 10, 19–28.
[22] DeLange, R.J., Williams. L.C. and Searcy, D.G. (1981) J. Biol. Chem. 256, 905–911.
[23] Margulis, L. (1970) Origin of Eucarytoic Cells, Yale University Press, New Haven.
[24] Sandman, K., Krzycki, J.A., Dobrinski, B., Lurz, R. and Reeve, J.N. (1990) Proc. Natl. Acad. Sci. U.S.A. 87, 5788–5791.
[25] Zillig, W., Stetter, K.O. and Janekovic, D. (1979) Eur. J. Biochem. 96, 597–604.
[26] Zillig, W., Stetter, K.O., Schnabel, R. and Thomm, M. (1985) In: The Bacteria, Vol. 8 Archaebacteria (Woese, C.R. and Wolfe, R.S., Eds.), pp. 499–524, Academic Press, New York.
[27] O. Kandler, Ed. (1982) Archaebacteria, Gustav Fischer, Stuttgart.
[28] Woese, C.R. (1969) In: Progress in Molecular and Subcellular Biology, Vol. 1 (Hahn, H., Ed.), pp. 5–46, Springer, Berlin.
[29] Woese C.R. (1965) Proc. Natl. Acad. Sci. U.S.A. 54, 1546–1552.
[30] Woese, C.R. (1967) The Genetic Code: The Molecular Basis of Genetic Expression, Harper and Row, New York.
[31] Woese, C.R., Dugre, D.H., Saxinger, W.C. and Dugre, S.A. (1966) Proc. Natl. Acad. Sci. U.S.A. 55, 966–974.
[32] Woese, C.R. (1970) Symp. Soc. Gen. Microbiol. 20, 39–54.
[33] Balch, W.E., Fox, G.E., Magrum, L.J., Woese, C.R. and Wolfe, R.S. (1979) Microbiol. Rev. 43, 260–296.
[34] van Niel, C.B. (1949) Bacteriol. Rev. 13, 161–174.
[35] van Niel, C.B. (1955) In: A Century of Progress in the Natural Sciences 1853–1953, pp. 89–114, California Academy of Sciences, San Francisco.
[36] Kluyver, A.J. and van Niel, C.B. (1936) Zbl. Bakteriol. Parasit. u. Infekt. II. 94, 369–403.
[37] Stanier, R.Y. and van Niel, C.B. (1941) J. Bacteriol. 42, 437–466.
[38] van Niel, C.B. (1946) Cold Spring Harbor Symp. Quant. Biol. 11, 285–301.
[39] Stanier, R.Y. and van Niel, C.B. (1962) Arch. Mikrobiol. 42, 17–35.
[40] Stanier, R.Y., Doudoroff, M. and Adelberg, E.A. (1963) The Microbial World, 2nd edition, Prentice-Hall, Englewood Cliffs, NJ.
[41] Stanier, R.Y., Doudoroff, M. and Adelberg, E.A. (1970) The Microbial World, 3rd edition, Prentice-Hall, Englewood Cliffs, NJ.
[42] Pringsheim, E.G. (1949) Bacteriol. Rev. 13, 47–98.
[43] Murray, R.G.E. (1962) Symp. Soc. Gen. Microbiol. 12, 199–144.
[44] Zuckerkandl, E. and Pauling, L. (1965) J. Theoret. Biol. 8, 357–366.
[45] Meyer, T.E., Cusanovich, M.A. and Kamen, M.D. (1986) Proc. Natl. Acad. Sci. U.S.A. 83, 217–220.
[46] Ambler, R.P., Daniel, M., Hermoso, J., Meyer, T.E., Bartsch, R.G. and Kamen, M.D. (1979) Nature 278, 659–660.
[47] Ambler, R.P., Meyer, T.E. and Kamen, M.D. (1979) Nature 278, 661–662.
[48] Woese, C.R. (1987) Microbiol. Rev. 51, 221–271.
[49] Woese, C.R., Kandler, O. and Wheelis, M.L. (1990) Proc. Natl. Acad. Sci. U.S.A. 87, 4576–4579.
[50] Winker, S. and Woese, C.R. (1991) System. Appl. Microbiol. 14, 305–310.
[51] Woese, C.R. (1982) Zbl. Bakt. Hyg., I. Abt. Orig. C3, 1–17.
[52] Woese, C.R. and Fox, G.E. (1977) J. Mol. Evol. 10, 1–6.
[53] Schwartz, R.M. and Dayhoff, M.O. (1978) Science 199, 395–403.
[54] Iwabe, N., Kuma, K., Hasegawa, M., Osawa, S. and Miyata, T. (1989) Proc. Natl. Acad. Sci. U.S.A. 86, 9355–9359.
[55] Gogarten, J.P., Kibak, H., Dittrich, P., Taiz, L., Bowman, E.J., Bowmnan, B.J., Manolson, M.F., Poole, R.J., Date, T., Oshima, T., Konishi, J., Denda, K. and Yoshida, M. (1989) Proc. Natl. Acad. Sci. U.S.A. 86, 6661–6665.
[56] Mayr, E. and Ashlock, P.D. (1991) Principles of Systematic Zoology, McGraw-Hill, New York.

[57] Mayr, E. (1990) Nature 348, 491.
[58] Woese, C.R., Achenbach, L., Rouvière, P. and Mandelco, L. (1991) System. Appl. Microbiol. 14, 364–371.
[59] Achenbach-Richter, L., Gupta, R., Zillig, W. and Woese, C.R. (1988) System. Appl. Microbiol. 10, 231–240.
[60] Woese, C.R. and G.J. Olsen. (1986) System. Appl. Microbiol. 7, 161–177.
[61] Burggraf, S., Stetter, K.O., Rouvière, P. and Woese, C.R. (1991) System. Appl. Microbiol. 14, 346–351.
[62] Burggraf, S., Ching, A., Stetter, K.O. and Woese, C.R. (1991) System Appl. Microbiol. 14, 358–363.
[63] Lake, J.A. (1988) Nature 331, 184–186.
[64] Sidow, A. and Wilson, A.C. (1990) J. Mol. Evol. 31, 51–68.
[65] Achenbach-Richter, L., Stetter, K.O. and Woese, C.R. (1987) Nature 327, 348–349.
[66] Achenbach-Richter, L., Gupta, R., Stetter, K.O. and Woese, C.R. (1987) System. Appl. Microbiol. 9, 34–39.
[67] Cammarano, P., Teichner, A. and Londei, P. (1986) System. Appl. Microbiol. 7, 137–145.
[68] Woese, C.R., Gupta, R., Hahn, C.M., Zillig, W. and Tu, J. (1984) System. Appl. Microbiol. 5, 97–105.
[69] Achenbach-Richter, L. and Woese, C.R. (1988) System. Appl. Microbiol. 10, 211–214.
[70] Noll, K. (1989) J. Bacteriol. 171, 6720–6725.
[71] Pace, N.R., Stahl, D.A., Lane, D.J. and Olsen, G.J. (1985) Amer. Soc. Microbiol. News 51, 4–12.
[72] Pace, N.R., Stahl, D.A., Lane, D.J. and Olsen, G.J. (1986) Adv. Microbial Ecol. 9, 1–55.
[73] Schmidt, T.M., DeLong, E.F. and Pace, N.R. (1991) J. Bacteriol. 173, 4371–4378.
[74] DeLong, E.F., Wickham, G.S. and Pace, N.R. (1989) Science 243, 1360–1363.
[75] Puhler, G, Leffers, H., Gropp, F., Palm, P., Klenk, H.-P., Lottspeich, F., Garrett, R.A. and Zillig, W. (1989) Proc. Natl. Acad. Sci. U.S.A. 86, 4569–4573.
[76] Brosius, J., Palmer, J.L., Kennedy, J.P. and Noller, H.F. (1978) Proc. Natl. Acad. Sci. U.S.A. 75, 4801–4805.

List of contributors

R. Amils,
Centro de Biologia Molecular, Universidad Autonoma de Madrid, Canto Blanco, Madrid, Spain

T. Boeckh,
Max-Planck-Institut für Molekulare Genetik, Abt. Wittmann, D-1000 Berlin 33 (Dahlem), Germany

P. Cammarano,
Istituto Pasteur–Fondazione Cenci-Bolognetti; Dipt. Biopatologia Umana, Sez. Biologia Cellulare, Università di Roma "La Sapienza", Policlinico Umberto I, Roma, Italy

J.Z. Dalgaard,
Institute of Molecular Biology, Copenhagen University, Sølvgade 83, DK-1307 Copenhagen K, Denmark

L. Daniels,
Department of Microbiology, The University of Iowa, Iowa City, IA 52242, U.S.A.

M.J. Danson,
School of Biological Sciences, Biochemistry Department, University of Bath, Bath BA2 7AY, England, UK

W. Ford Doolittle,
Department of Biochemistry, Dalhousie University, Halifax, N.S. B3H 4H7, Canada

C. Elie,
Institut de Génétique et Microbiologie, Université Paris-Sud, 91405 Orsay Cedex, France

P. Forterre,
Institut de Génétique et Microbiologie, Université Paris-Sud, 91405 Orsay Cedex, France

R.A. Garrett,
Institute of Molecular Biology, Copenhagen University, Sølvgade 83, DK-1307 Copenhagen K, Denmark

J. Hain,
Max-Planck-Institut für Biochemie, D-82152 Martinsried, Germany

R. Hensel,
FB 9 Mikrobiologie, Universität GHS Essen, Universitätsstr. 5, 45117 Essen 1, Germany
L.I. Hochstein,
Planetary Biology Division, N.A.S.A., Ames Research Center, Moffett Field, CA 94035, U.S.A.
I. Holz,
Max-Planck-Institut für Biochemie, D-82152 Martinsried, Germany
U. Hüdepohl,
Max-Planck-Institut für Biochemie, D-82152 Martinsried, Germany
O. Kandler,
Botanisches Institut der Ludwig-Maximilians-Universität München, 80638 München, Germany
M. Kates,
Department of Biochemistry, University of Ottawa, Ottawa, Ont. K1N 6N5, Canada
H.-P. Klenk,
Max-Planck-Institut für Biochemie, D-82152 Martinsried, Germany
H. König,
Abteilung für Angewandte Mikrobiologie und Mykologie, Universität Ulm, 89069 Ulm, Germany
A.K.E. Köpke,
Max-Planck-Institute for Experimental Medicine, Department of Molecular Neuroendocrinology, 3400 Goettingen, Germany
D. Langer,
Max-Planck-Institut für Biochemie, D-82152 Martinsried, Germany
J.K. Lanyi,
Department of Physiology and Biophysics, University of California, Irvine, CA 92717, U.S.A.
M. Lanzendörfer,
Max-Planck-Institut für Biochemie, D-82152 Martinsried, Germany
P. Londei,
Istituto Pasteur–Fondazione Cenci-Bolognetti; Dipt. Biopatologia Umana, Sez. Biiologia Cellulare, Università di Roma "La Sapienza", Policlinico Umberto I, Roma, Italy
W. Marwan,
Max-Planck-Institut für Biochemie, D-82512 Martinsried, Germany
A.T. Matheson,
Department of Biochemistry and Microbiology, University of British Columbia, Vancouver, B.C. V5Z 4H4, Canada
D. Oesterhelt,
Max-Planck-Institut für Biochemie, D-82152 Martinsried, Germany

P. Palm,
Max-Planck-Institut für Biochemie, D-82152 Martinsried, Germany
J.R. Palmer,
Department of Microbiology, The Ohio State University, Columbus, OH 43210, U.S.A.
C. Ramírez,
Department of Biochemistry and Microbiology, University of British Columbia, Vancouver, B.C., V5Z 4H4, Canada
J.N. Reeve,
Department of Microbiology, The Ohio State University, Columbus, OH 43210, U.S.A.
L.C. Schalkwyk,
Department of Biochemistry, Dalhousie University, Halifax, N.S. B3H 4H7, Canada
P. Schönheit,
Institut für Pflanzenphysiologie und Mikrobiologie, Fachbereich Biologie, Freie Universität Berlin, D-1000 Berlin 33, Germany
V.P. Skulachev,
A.N. Belozersky Laboratorium, Moscow State University, Moscow 119899, Russia
C.R. Woese,
Department of Microbiology, University of Urbana at Urbana-Champaign, Urbana, IL 61801, U.S.A.
D-C. Yang,
Research Center, University of British Columbia, Vancouver, B.C. V5Z 4H4, Canada
W. Zillig,
Max-Planck-Institut für Biochemie, D-82152 Martinsried, Germany

Contents

Preface v

Introduction. The archaea: Their history and significance
Carl R. Woese vii

1. Introduction vii
2. Microbiology's changing evolutionary perspective xi
3. A molecular definition of the three domains xv
4. Archaeal phylogenetic relationships xxi
5. The new microbiology xxv
Acknowledgment xxvii
References xxvii

List of contributors xxxi

Chapter 1. Central metabolism of the archaea
M.J. Danson 1

1. Introduction 1
   1.1. Central metabolism 1
   1.2. Central metabolism in eukaryotes and eubacteria 1
2. Hexose catabolism in the archaea 2
   2.1. The modified Entner–Doudoroff pathway of the halophiles 2
   2.2. The non-phosphorylated pathway of the thermophiles 5
   2.3. The glycolytic pathway of the methanogens 7
   2.4. Gluconeogenesis 7
   2.5. The pentose–phosphate pathway 8
3. Pyruvate oxidation to acetyl–CoA in the archaea 8
   3.1. Pyruvate oxidoreductases 8
   3.2. Comparison with the eukaryotic and eubacterial enzymes 9
   3.3. Dihydrolipoamide dehydrogenase 10
4. The citric acid cycle in the archaea 11
   4.1. The oxidative citric acid cycle 11
   4.2. The reductive citric acid cycle 12
   4.3. Partial citric acid cycles 13
   4.4. Other pathways of acetyl–CoA metabolism 13
5. Amino acid and lipid metabolism in the archaea 14

6. Evolution of central metabolism		15
6.1. Hexose catabolism		15
6.2. Pyruvate oxidation to acetyl–CoA		17
6.3. The citric acid cycle		17
7. Comparative enzymology of central metabolism		18
7.1. The enzymes as molecular chronometers		18
7.2. Citrate synthase		19
7.2.1. The comparative enzymology of citrate synthases		19
7.2.2. Archaebacterial citrate synthases		19
8. Conclusions and perspectives		20
Acknowledgements		20
Note added in proof		21
References		21

Chapter 2. Bioenergetics of extreme halophiles
V.P. Skulachev   25

1. Introduction	25
2. A general scheme of energy transduction in extreme halophiles	25
3. Bacteriorhodopsin and halorhodopsin	27
3.1. Transmembrane charge displacement	27
3.2. Involvement in photoreception	30
4. Respiratory chain	32
5. Arginine fermentation	33
6. H^+-ATP-synthase	33
7. Formation of K^+/Na^+ gradients. $\Delta\bar{\mu}_{H^+}$ buffering	34
8. Na^+, metabolite symports	35
9. A flagellar motor	36
10. Some prospects for future research	37
References	37

Chapter 3. Biochemistry of methanogenesis
L. Daniels   41

1. Introduction	41
2. Novel coenzymes	43
2.1. General	43
2.2. The 5-deazaflavin, F_{420}	45
2.3. Methanofuran (MF)	47
2.4. Tetrahydromethanopterin (H_4MPT)	48
2.5. Coenzyme M	50
2.6. Cobamides	51
2.7. F_{430}	51
2.8. 7-Mercaptoheptanoylthreonine phosphate (HSHTP)	53
3. The pathways and biochemistry of methanogenesis	53
3.1. Methanogenesis from CO_2	53
3.1.1. Reduction of CO_2	53
3.1.2. The RPG effect	57
3.1.3. Source of electrons	57

	3.2.	Methanogenesis from methanol	58
		3.2.1. Reduction of methanol	58
		3.2.2. Oxidation of methanol	59
		3.2.3. Electron transfer	60
		3.2.4. Methanogenesis from methylamines	61
	3.3.	Methanogenesis from acetate	61
		3.3.1. Transport of acetate into the cell	61
		3.3.2. Activation of acetate	61
		3.3.3. Cleavage of acetyl–CoA and CH_3–CoM formation	62
		3.3.4. Reduction of CH_3–CoM to CH_4	65
		3.3.5. Electron transfer	65
4.	Enzymes involved in methanogenesis		66
	4.1.	General	66
	4.2.	Hydrogenase and non-catalytic redox proteins such as ferredoxin and cytochromes	66
		4.2.1. The methylviologen-reducing hydrogenase (MVH)	68
		4.2.2. Redox-active proteins: ferredoxin, cytochromes, and others	69
		4.2.3. The F_{420}-reducing hydrogenase (FRH)	70
	4.3.	Alcohol dehydrogenase (ADH)	72
	4.4.	Formate dehydrogenase	73
	4.5.	Formylmethanofuran dehydrogenase	75
	4.6.	Formylmethanofuran:tetrahydromethanopterin formyltransferase	78
	4.7.	N^5, N^{10}-methenyltetrahydromethanopterin cyclohydrolase	79
	4.8.	Methylenetetrahydromethanopterin dehydrogenase	82
	4.9.	Methylenetetrahydromethanopterin reductase	85
	4.10.	Methyltetrahydromethanopterin:CoM methyltransferase	87
	4.11.	Methyl–Coenzyme M reductase (MR)	88
	4.12.	Heterodisulfide reductase (HR)	92
	4.13.	Methanol methanogenesis-related methyltransferases	93
	4.14.	Carbon monoxide dehydrogenase complex	94
	4.15.	Acetate activating enzymes	98
5.	Key remaining physiological and enzymatic questions		100
Acknowledgements			100
References			100

Chapter 4. Bioenergetics and transport in methanogens and related thermophilic archaea
P. Schönheit 113

Abbreviations			113
1.	Introduction		113
2.	Energy substrates		115
	2.1.	Reduction of CO_2 to CH_4	116
	2.2.	Reduction of a methyl group to CH_4	117
3.	Energetics of methanogenesis from CO_2/H_2		119
	3.1.	Enzymology	119
		3.1.1. CO_2 reduction to formyl–MFR (formate level)	119
		3.1.2. Formyl–MFR reduction to methylene–H_4MPT (formaldehyde level)	123
		3.1.3. Methylene–H_4MPT conversion to methyl–coenzyme M (methanol level)	123
		3.1.4. Methyl–coenzyme M (CH_3–S–CoM) reduction to methane	124

3.2.	Sites of energy coupling	124
	3.2.1. General aspects	124
	3.2.1.1. Growth yields and ATP gains	125
	3.2.1.2. Mechanism of ATP synthesis	125
	3.2.1.3. Thermodynamics of partial reactions	126
	3.2.2. Methyl–coenzyme M reduction to CH_4 – site of primary $\Delta\tilde{\mu}H^+$ generation and of ATP synthesis	127
	3.2.2.1. Heterodisulfide (CoM–S–S–HTP) reduction – coupled to primary H^+ translocation	128
	3.2.2.2. ATP synthase/ATPase	131
	3.2.2.3. Misleading concepts of ATP synthesis	132
	3.2.3. Methylene–H_4MPT conversion to methyl–coenzyme M – site of primary $\Delta\tilde{\mu}Na^+$ generation	133
	3.2.3.1. Methyl–H_4MPT: coenzyme M methyltransferase – coupled to primary Na^+ translocation	133
	3.2.4. CO_2 reduction to methylene–H_4MPT – site of primary $\Delta\tilde{\mu}Na^+$ consumption	135
	3.2.4.1. Formaldehyde oxidation to CO_2 – coupled to primary Na^+ extrusion	135
	3.2.4.2. CO_2 reduction to methylene–H_4MPT – coupled to Na^+ uptake	136
3.3.	Role of the Na^+/H^+ antiporter in CO_2 reduction to CH_4	137
	3.3.1. Na^+/H^+ antiporter	138
	3.3.2. Role of the Na^+/H^+ antiporter	138
	3.3.3. Primary cycles of Na^+ and H^+	139
3.4.	Energetics of CH_4 formation from formate	139
3.5.	Energetics of CH_4 formation from CO_2 reduction by alcohols	139
3.6.	Energetics of CO_2 reduction to CH_4 by methanogens versus CO_2 reduction to acetate by acetogens	141
4. Energetics of methanogenesis from methanol		143
4.1.	Enzymology	143
4.2.	Energetics	144
	4.2.1. Role of the Na^+/H^+ antiporter	145
	4.2.2. Role of methyltransferase	145
	4.2.3. Role of cytochromes	147
	4.2.4. Growth yields	147
4.3.	Energetics of CH_4 formation from methylamines	147
5. Energetics of methanogenesis from acetate		147
5.1.	Enzymology	147
5.2.	Sites of energy coupling	148
	5.2.1. CH_3–S–CoM reduction to CH_4	148
	5.2.2. CO oxidation to CO_2	149
	5.2.3. CH_3–H_4MPT:H–S–CoM methyltransferase	151
5.3.	Acetate fermentation in *Methanothrix soehngenii*	153
6. Energetics of pyruvate catabolism		153
6.1.	Methanogenesis from pyruvate	153
6.2.	Acetate formation from pyruvate in the absence of methanogenesis	154
7. Concluding remarks on energetics		155

8.	Transport in methanogens	156
	8.1. H–S–CoM, CH_3–S–CoM	156
	8.2. Amino acids	157
	8.3. Nickel	157
	8.4. Potassium	157
	8.5. Phosphate	158
9.	Energetics of *Archaeoglobus* and *Pyrococcus* – non-methanogenic thermophilic archaea related to methanogens	158
	9.1. Energetics of *Archaeoglobus fulgidus*	159
	9.1.1. Acetyl–CoA oxidation to CO_2 via a modified acetyl–CoA/carbon monoxide dehydrogenase pathway	159
	9.2. Energetics of *Pyrococcus furiosus*	161
	9.2.1. Novel sugar degradation pathway in *Pyrococcus furiosus*	162
	9.2.2. Sugar degradation to acetate, CO_2 and H_2 via a novel fermentation pathway	162
	9.2.3. Open questions	164
Acknowledgements		164
References		164

Chapter 5. Signal transduction in halobacteria
D. Oesterhelt and W. Marwan   173

1.	Introduction	173
	1.1. General	173
	1.2. Photoreceptors	173
	1.3. Movement	174
	1.4. Signal transduction	174
2.	Flagellar motor and motility	175
	2.1. Mode of movement	175
	2.2. Filament and flagellins	176
	2.3. Motor switching	176
3.	Signal transduction pathway	177
	3.1. Basic observations	177
	3.2. Signal formation	178
	3.3. Identification of a switch factor	180
	3.4. Light-induced release of fumarate	180
	3.5. Methyl-accepting taxis proteins	181
	3.6. Cyclic GMP	182
4.	The photoreceptors	183
	4.1. Spectroscopic and biochemical properties	183
	4.2. The physiology of photoreception	183
Acknowledgement		185
References		185

Chapter 6. Ion transport rhodopsins (bacteriorhodopsin and halorhodopsin): Structure and function
J.K. Lanyi   189

1.	Introduction	189

2. Structure	190
3. Chromophore	193
4. Properties of the Schiff base	195
5. Photoreactions and photocycles	196
6. Transport mechanism	199
7. Energetics and coupling	201
8. Summary	202
References	203

Chapter 7. Proteins of extreme thermophiles
R. Hensel — 209

Abbreviations	209
1. Introduction	209
2. Features of protein thermoadaptation in archaea	212
2.1. How structurally different are proteins from the extreme thermophiles as compared to their mesophilic counterparts?	212
2.2. How are the proteins from thermophiles growing at or above the boiling point of water stabilized towards heat-induced covalent modifications of the peptide chain?	214
2.3. Extrinsic factors stabilizing the native state of proteins at high temperatures	215
3. Proteins with suggested thermoadaptive functions	215
3.1. Proteins which presumably protect DNA	216
3.2. Proteins which presumably protect proteins	216
4. Proteins with biotechnological potential	216
5. Conclusions	218
Acknowledgement	218
References	218

Chapter 8. Cell envelopes of archaea: Structure and chemistry
O. Kandler and H. König — 223

1. Introduction	223
2. Structure and chemistry of cell walls of gram-positive archaea	223
2.1. Methanobacteriales and *Methanopyrus*	223
2.1.1. Morphology	223
2.1.2. Chemical structure and modifications of pseudomurein	224
2.1.3. Secondary and tertiary structure of pseudomurein	228
2.1.4. Lysis of pseudomurein	229
2.1.5. Biosynthesis of the pseudomurein	229
2.1.6. Biological activity of pseudomurein	231
2.1.7. Chemical structure and biosynthesis of the S-layer glycoprotein of *Methanothermus fervidus*	231
2.2. *Methanosarcina*	232
2.2.1. Morphology	232
2.2.2. Chemical structure of methanochondroitin	232
2.2.3. Autolysis	233
2.2.4. Biosynthesis of methanochondroitin	235

		2.3.	*Halococcus*	236
			2.3.1. Morphology	236
			2.3.2. Chemical structure of halococcal heteropolysaccharide	236
	2.4.	*Natronococcus*		237
			2.4.1. Morphology	237
			2.4.2. Chemical composition of the natronococcal "glycosaminoglycan"	237
3.	Structure and chemistry of cell envelopes of gram-negative archaea	239
	3.1.	Proteinaceous sheaths		239
			3.1.1. *Methanospirillum hungatei*	239
			3.1.2. *Methanothrix concilii* (recently renamed *Methanosaeta concilii*)	240
	3.2.	S-layers of gram-negative methanogenic rods and cocci	242
	3.3.	S-layers of gram-negative halobacteria	243
			3.3.1. Chemical structure	243
			3.3.2. Biosynthesis	245
			3.3.3. Halobacterial versus eukaryotic glycoproteins	246
			3.3.4. Three-dimensional structure	246
	3.4.	S-layers of *Archaeoglobus*	247
	3.5.	S-layers of sulfur-metabolizing hyperthermophilic archaea	248
4.	Surface structure of archaea without cell envelopes	252
5.	Concluding remarks	252
References	255

Chapter 9. Membrane lipids of archaea
M. Kates 261

1.	Introduction	261
2.	"Core" archaeal lipids	262
	2.1. Diphytanylglycerol diether (archaeol) and variants	262
	2.2. Dibiphytanyldiglycerol tetraether (caldarchaeol) and variants	265
3.	Polar lipids	265
	3.1. Extreme halophiles	265
	3.1.1. Phospholipids	265
	3.1.2. Glycolipids	266
	3.1.3. Taxonomic relations	267
	3.2. Methanogens	269
	3.2.1. Methanobacteriaceae	270
	3.2.2. Methanomicrobiaceae	272
	3.2.3. Methanococcaceae	272
	3.2.4. Methanosarcinaceae	272
	3.2.5. Taxonomic relations	273
	3.3. Extreme thermophiles	273
	3.3.1. Thermoplasmatales	274
	3.3.2. Sulfolobales	275
	3.3.3. Thermoproteales	276
	3.3.4. Thermococcales	277
	3.3.5. Taxonomic relations	277

4. Biosynthetic pathways 278
   4.1. Archaeol lipid cores 279
   4.2. Archaeol phospholipids 283
        4.2.1. In extreme halophiles 283
        4.2.2. In methanogens 283
   4.3. Archaeol glycolipids 284
   4.4. Caldarchaeol phospholipids, glycolipids and phosphoglycolipids ... 285
        4.4.1. In methanogens 285
        4.4.2. In thermoacidophiles 286
5. Membrane function of archaeal lipids 287
   5.1. Archaeol-derived lipids in extreme halophiles 287
   5.2. Caldarchaeol-derived lipids in methanogens and thermoacidophiles ... 288
6. Evolutionary considerations and conclusions 289
Acknowledgement 291
Note added in proof 291
References 292

Chapter 10. The membrane-bound enzymes of the archaea
L.I. Hochstein 297

1. Introduction 297
2. Methods 298
   2.1. Preparation of membranes 298
   2.2. Isolation of membrane components 298
3. The ATPases 299
   3.1. The ATPases of the methanogens 300
   3.2. The ATPases of *Sulfolobus* 302
   3.3. The ATPases of *Thermoplasma* 304
   3.4. The ATPases of the extreme halophiles 304
4. The electron transport system 308
   4.1. NADH oxidases 308
        4.1.1. The NADH oxidase of *Sulfolobus* 308
        4.1.2. The NADH oxidase of the extreme halophiles 308
   4.2. NADH dehydrogenases 309
        4.2.1. NADH dehydrogenase from *Sulfolobus* 309
        4.2.2. NADH dehydrogenase of extreme halophiles 309
   4.3. Succinic dehydrogenases 311
        4.3.1. Succinic dehydrogenases of *Sulfolobus* 311
        4.3.2. Succinate dehydrogenase from the extreme halophiles 311
   4.4. The cytochromes 312
        4.4.1. The cytochromes of the methanogens 312
        4.4.2. The cytochromes of *Sulfolobus* 312
        4.4.3. The cytochromes of *Thermoplasma* 313
        4.4.4. The cytochromes of the extreme halophiles 314
5. Hydrogenases 316
   5.1. Hydrogenases in methanogens 316
   5.2. Hydrogenase in *Pyrodictium* 316

6.	The enzymes of denitrification	317
	6.1. Nitrate reductase	317
	6.2. Nitrite reductase activity	318
7.	Summary	318
References		319

Chapter 11. Chromosome structure, DNA topoisomerases, and DNA polymerases in archaebacteria (archaea)
P. Forterre and C. Elie 325

1.	Introduction	325
2.	Chromosome structure	326
	2.1. Genome size and organization	326
	2.2. Putative histone-like proteins and nucleosomes	327
	2.2.1. The protein HTa	327
	2.2.2. The protein MC1	329
	2.2.3. The protein HMf	329
	2.2.4. Putative nucleosomal organization	331
	2.3. DNA stability in hyperthermophiles	331
3.	DNA topoisomerases and DNA topology	333
	3.1. Reverse gyrase	336
	3.1.1. Discovery	336
	3.1.2. Biochemical characterization	337
	3.1.3. Mechanistic studies	337
	3.1.4. Primary structure	338
	3.1.5. Mechanism of reverse gyration	338
	3.1.6. Distribution of reverse gyrase in the living world	339
	3.1.7. Putative roles of reverse gyrase	340
	3.2. Other DNA topoisomerases in thermophilic archaebacteria	342
	3.2.1. *Sulfolobus* type II DNA topoisomerase	342
	3.2.2. ATP-independent DNA topoisomerases	343
	3.3. Topological state of the DNA in extremely thermophilic archaebacteria	343
	3.4. DNA topology in halophilic archaebacteria	346
	3.4.1. Sensitivity of halobacteria to DNA topoisomerase II inhibitors	346
	3.4.2. Gene structure and primary sequence of a halobacterial type II DNA topoisomerase	347
	3.4.3. Biological roles of type II DNA topoisomerase in halophilic archaebacteria	349
	3.5. An overview of DNA topology in archaebacteria	349
4.	DNA polymerases	351
	4.1. Sensitivity of archaebacteria to aphidicolin	352
	4.2. DNA polymerases from sulfothermophiles and methanogens	353
	4.2.1. Aphidicolin-sensitive DNA polymerases	353
	4.2.2. Aphidicolin-resistant DNA polymerases	355
	4.3. DNA polymerases from halophiles	356
	4.4. Conclusions	356

5. General discussion 357
   5.1. Future prospects 357
   5.2. Evolutionary considerations 358
        5.2.1. Eukaryotic versus eubacterial features of archaebacteria 358
        5.2.2. The root of the tree of life and other phylogenetic problems 359
        5.2.3. The last common universal ancestor 360
Acknowledgements 360
References 361

Chapter 12. Transcription in archaea
W. Zillig, P. Palm, H.-P. Klenk, D. Langer, U. Hüdepohl, J. Hain, M. Lanzendörfer and I. Holz 367

Abbreviations 367
1. Introduction 367
2. DNA-dependent RNA polymerase 369
   2.1. Composition 369
   2.2. Organization of RNAP component genes 371
   2.3. Similarities between sequences of RNAP components 373
   2.4. The phylogeny of RNAP components 373
   2.5. The structure of the RNAP of *S. acidocaldarius* 377
3. Transcription signals: Promoters and terminators 379
   3.1. Promoters 379
   3.2. Terminators 383
4. In vitro transcription systems 384
5. Control of transcription 384
6. Summary 386
Note added in proof 386
References 386

Chapter 13. Translation in archaea
R. Amils, P. Cammarano and P. Londei 393

1. Introduction 393
2. Structure of translational components 394
   2.1. Transfer RNAs and aminoacyl–tRNA synthetases 394
   2.2. Messenger RNAs 394
   2.3. Messenger RNA–ribosome interaction 395
   2.4. Polypeptide chain initiation 396
   2.5. Elongation factors 396
        2.5.1. Elongation factor sequences 396
        2.5.2. Elongation factor gene order 398
   2.6. Archaeal ribosomes. Halotolerance and heat-stability 400
        2.6.1. Ribosomal subunit interaction 402
        2.6.2. Ribosome mass and composition 402
        2.6.3. Ribosome shape 405
   2.7. In vitro reconstruction of ribosomal subunits 407
        2.7.1. Reconstruction of *Sulfolobus* 50S subunits 407
        2.7.2. Reconstitution of *Haloferax* 50S subunits 408

	2.8.	Protein targeting and signal recognition in archaea	410
3.	In vitro translation systems		411
	3.1.	Translation systems from halophilic archaea	411
	3.2.	Translation systems from methanogenic archaea	412
	3.3.	Translation systems from sulfur-dependent archaea	412
	3.4.	Peptidyltransferase assay systems	413
	3.5.	Distinctness of euryarchaeal and crenarchaeal translation systems	415
4.	Sensitivity of archaea to protein synthesis inhibitors		416
	4.1.	Ribosome-targeted inhibitors: in vivo assays	417
	4.2.	Ribosome-targeted inhibitors: in vitro assays	417
	4.3.	Structural correlates of sensitivity to ribosome-targeted drugs	420
	4.4.	Elongation factor-targeted inhibitors	425
		4.4.1. EF-2- and EF-G-targeted inhibitors	426
		4.4.2. EF-1α- and EF-Tu-targeted inhibitors	427
	4.5.	Phylogeny inference from antibiotic sensitivity spectra	427
5.	Interchangeability of translational components		428
	5.1.	Interchangeability of ribosomal subunits	428
	5.2.	Interchangeability of elongation factors	429
	5.3.	Interchangeability of ribosomal RNAs and proteins	429
		5.3.1. Interchangeability of 5S RNAs	430
		5.3.2. Interchangeability of 50S subunit proteins	430
6.	Conclusions		430
References			432

Chapter 14. The structure, function and evolution of archaeal ribosomes
C. Ramírez, A.K.E. Köpke, D-C. Yang, T. Boeckh and A.T. Matheson . . 439

1.	Introduction		439
2.	The archaeal ribosomes		439
3.	Archaeal ribosomal RNA		441
	3.1.	Gene organization	441
	3.2.	rRNA structure and function	443
		3.2.1. Functional domains	444
		3.2.1.1. The peptidyltransferase center	444
		3.2.1.2. The GTPase center	444
4.	Structure of archaeal ribosomal proteins		446
	4.1.	Nomenclature of ribosomal proteins	446
	4.2.	Comparison of the archaeal r-proteins with those from bacteria and eucarya	446
	4.3.	The L2 r-protein family	451
	4.4.	The stalk protuberance in the large ribosomal subunit (r-proteins L12/L10)	451
	4.5.	Interchangeability of ribosomal components from different organisms	454
5.	Archaeal ribosomal protein genes		454
	5.1.	Gene organization	454
	5.2.	Transcription	455
	5.3.	Translation signals	455
	5.4.	Regulation	457
6.	Evolution of the ribosome		458
Acknowledgements			460
References			460

Chapter 15. Halobacterial genes and genomes
L.C. Schalkwyk 467

Abbreviations 467
1. Introduction 467
2. Halobacterial genomes 468
   2.1. Size 468
   2.2. Plasmids 469
   2.3. Inhomogeneity of composition 470
   2.4. Repeated sequences and instability 471
3. Genetics 473
   3.1. Physical mapping: Introduction 474
   3.2. Clues from comparison of bacterial genetic maps 474
   3.3. The *Haloferax volcanii* map 475
   3.4. Genes and operons 478
      3.4.1. Ribosomal RNA genes 479
      3.4.2. Transfer RNA genes 480
      3.4.3. 7S RNA 480
      3.4.4. RNase P RNA 481
      3.4.5. Bacteriorhodopsin 481
      3.4.6. Halorhodopsin 482
      3.4.7. Sensory rhodopsins 483
      3.4.8. Gas-vesicle proteins 483
      3.4.9. Cell surface glycoprotein 484
      3.4.10. Flagellins 484
      3.4.11. Superoxide dismutase 485
      3.4.12. Dihydrofolate reductase 485
      3.4.13. DNA gyrase 486
      3.4.14. Photolyase 486
      3.4.15. Bacteriophage ΦH 487
      3.4.16. H^+ ATPase 487
      3.4.17. Histidinol-phosphate aminotransferase 487
      3.4.18. 3-Hydroxy-3-methylglutaryl–coenzyme A reductase .. 488
      3.4.19. Tryptophan biosynthesis 488
      3.4.20. Ribosomal proteins 488
      3.4.21. Elongation factors 489
      3.4.22. RNA polymerase 489
4. Future directions 490
References 491

Chapter 16. Structure and function of methanogen genes
J.R. Palmer and J.N. Reeve 497

Abbreviations 497
1. Introduction 497

2.	Genes encoding enzymes involved in methanogenesis	500
	2.1. Methyl–coenzyme M reductase (MR)	500
	2.2. Hydrogenases and ferredoxins	503
	2.3. Formate dehydrogenase (FDH)	504
	2.4. Formylmethanofuran:tetrahydromethanopterin formyltransferase (FTR)	505
	2.5. Carbon-monoxide dehydrogenase (CODH) and acetyl–coenzyme A synthetase (ACS)	506
3.	Amino-acid and purine biosynthetic genes	507
	3.1. Histidine	507
	3.2. Arginine	507
	3.3. Proline and ISM1	508
	3.4. Tryptophan	508
	3.5. Glutamine	509
	3.6. Adenine	510
4.	Transcription and translation machinery genes	510
	4.1. Stable RNA genes	510
	4.1.1. tRNA genes	510
	4.1.2. rRNA genes	511
	4.1.3. 7S RNA genes	512
	4.2. Genes encoding RNA polymerases, ribosomal proteins and elongation factors	512
	4.3. Aminoacyl–tRNA synthetase	514
5.	Nitrogen fixation genes	515
6.	Genes encoding metabolic enzymes	516
	6.1. Glyceraldehyde-3-phosphate dehydrogenase (GAPDH)	516
	6.2. L-Malate dehydrogenase (MDH)	517
	6.3. 3-Phosphoglycerate kinase (PGK)	517
	6.4. ATPases	518
	6.5. Superoxide dismutase	518
	6.6. S-adenosyl-L-methionine:uroporphyrinogen III methyltransferase (SUMT)	519
7.	Chromosomal proteins	520
8.	Surface-layer glycoproteins	520
9.	Flagellins	521
10.	Gene regulation and genetics	521
	10.1. Regulated systems of gene expression	521
	10.2. Transformation systems	522
11.	Summary	523
References		523

Chapter 17. Archaeal hyperthermophile genes
J.Z. Dalgaard and R.A. Garrett 535

1.	Introduction	535
2.	Gene sequences	535
3.	Nucleotide composition and optimal growth temperature	536
4.	Gene organization	543
5.	Transcriptional signals	545
	5.1. Promoter regions	545
	5.2. Terminators	549

6.	Translational signals	551
	6.1. Initiation	551
	6.2. Codon usage	553
	6.3. Termination	553
7.	Phylogenetic considerations	557
8.	Summary	558
	Acknowledgements	559
	References	559

Epilogue
W.F. Doolittle — 565

	Introduction	565
1.	Life's deepest branchings	565
2.	The coherence of the archaea	566
3.	Rooting the universal tree	567
4.	Implications of the root for eucarya	567
5.	Looking for "pre-adaptations" in archaea	568
6.	More courageous scenarios	569
7.	The need for caution and more data	569
8.	Archaea here and now	570
	References	570

Index — 573

M. Kates et al. (Eds.), *The Biochemistry of Archaea (Archaebacteria)*
© 1993 Elsevier Science Publishers B.V. All rights reserved

CHAPTER 1

Central metabolism of the archaea

Michael J. DANSON

*Department of Biochemistry, University of Bath,
Claverton Down, Bath BA2 7AY, England, UK*

1. Introduction

1.1. Central metabolism

Living organisms utilize their nutrients both to supply the precursors of all cell components and to generate the energy necessary for biosynthetic and other endergonic processes. The degradative metabolic pathways by which precursors for cell components are produced are known as 'catabolic' routes, whereas the biosynthetic processes are referred to as 'anabolic' reactions. The exact nature of the catabolic and anabolic pathways will depend on the starting nutrients and the desired cell components; the metabolic link between them is provided by the pathways of central metabolism, the reactions of which also serve as the major routes of energy generation.

It is an obvious, but important, statement to make therefore, that central metabolism is crucial to all activities within a cell. Moreover, because they are central and so important, the pathways are found in all living organisms – the precise details will vary between organisms but the basic patterns span the majority of all species. Therefore, one finds that related organisms have the same or closely similar central metabolic pathways and a comparative study can yield invaluable information on the evolution of organisms and their metabolism.

It is the purpose of this review to describe and discuss the pathways of central metabolism in the archaebacteria; however, for the reasons given above, this will not be done in isolation but as a comparative survey with those found in eubacteria and eukaryotes. Indeed, not only will the pathways be compared, but the comparison will be extended to the enzymes catalysing the reactions of these pathways. For previous reviews the reader is referred to Danson [1,2].

1.2. Central metabolism in eukaryotes and eubacteria

It has been argued that the development of a sugar-based biochemistry might have been an early evolutionary innovation [3] and that sugar catabolism may have been one of the first successful energy-conversion processes. Therefore, I have chosen to consider central metabolism in two parts: (a) the conversion of glucose and other sugars to pyruvate, and

(b) the metabolic fate of pyruvate under both anaerobic and aerobic conditions. It is into these pathways that all nutrients are directed and so they can be regarded as truly central in nature.

The Embden–Meyerhof glycolytic pathway is characteristic of eukaryotic cells and many anaerobic and facultatively anaerobic eubacteria. Two ATP molecules are produced per glucose oxidised to pyruvate. However, phosphofructokinase, a key enzyme of this pathway, is absent in many strictly aerobic eubacteria and, in these, glucose is catabolised via an alternative route, the Entner–Doudoroff pathway, yielding only one ATP per glucose. In both eubacteria and eukaryotes, a third pathway for glucose metabolism is present, namely the pentose–phosphate pathway, its main function being to generate NADPH and pentose and tetrose sugars.

The relationship between these three metabolic routes is illustrated in Fig. 1. Of particular note is the common trunk sequence from triose phosphate to pyruvate that can be regarded as the true central pathway for sugar catabolism in eukaryotes and eubacteria [4].

Under aerobic conditions, pyruvate can be oxidatively decarboxylated via the pyruvate dehydrogenase multienzyme complex to yield acetyl–CoA, which can then be completely oxidised via the citric acid cycle (Fig. 2). In eubacteria growing anaerobically, pyruvate is metabolised fermentatively, thus serving as an electron sink for reducing equivalents generated in its formation from glucose. The diverse array of possible fermentative reactions from pyruvate is reviewed in [5].

2. Hexose catabolism in the archaea

Previous considerations of the pathways of hexose catabolism in archaebacteria have indicated that it is most sensible to consider them with respect to the phenotypes of the organisms, that is the extreme halophiles, the thermophiles and the methanogens [1,2]. This is the strategy to be adopted here, although it is reiterated that these groupings do not necessarily reflect genotypic relationships, a subject comprehensively covered in other chapters.

2.1. The modified Entner–Doudoroff pathway of the halophiles

Halophilic archaebacteria are chemo-organotrophs and many use amino acids and proteins as sole sources of carbon. However, a number of carbohydrate-utilizers have been isolated [6,7] and, from a detailed study of one such halophile, *Halobacterium saccharovorum*, glucose catabolism has been shown to proceed via a modified Entner–Doudoroff pathway [8]. Glucose is oxidised to gluconate and then dehydrated to 2-keto-3-deoxygluconate, which in turn is phosphorylated to 2-keto-3-deoxy-6-phosphogluconate (Fig. 3). Aldol cleavage generates equimolar amounts of pyruvate and glyceraldehyde 3-phosphate, the latter being further metabolised via the common trunk sequence of the glycolytic and normal Entner–Doudoroff pathways found in eukaryotes and eubacteria. In addition to substrate-level phosphorylation of ADP, the NADH generated can be re-oxidised aerobically through an electron transport chain which comprises quinone(s) and various cytochromes, plus a variety of oxidases [9].

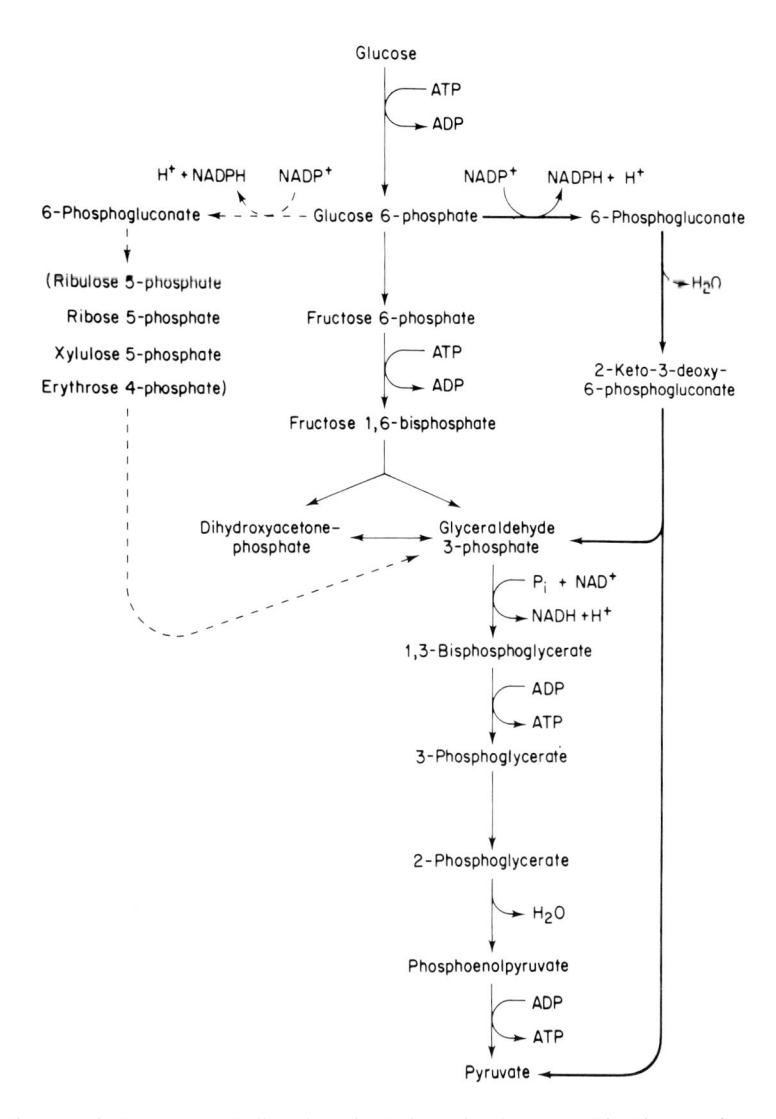

Fig. 1. Pathways of glucose metabolism in eubacteria and eukaryotes. The three major catabolic pathways are the Embden–Meyerhof glycolytic sequence (solid lines), the Entner–Doudoroff pathway (heavy solid lines) and the pentose phosphate pathway (dashed lines). The sequence from glyceraldehyde 3-phosphate to pyruvate is common to all three pathways.

This modified Entner–Doudoroff pathway has been found in other species of *Halobacterium* and in species of *Haloferax* and *Halococcus* [9–11] and may thus be common to the halophiles. Whilst not reported in thermophilic or methanogenic archaebacteria, it is not unique to halophiles, having been found in a few eubacterial genera (see ref. [1] and references therein).

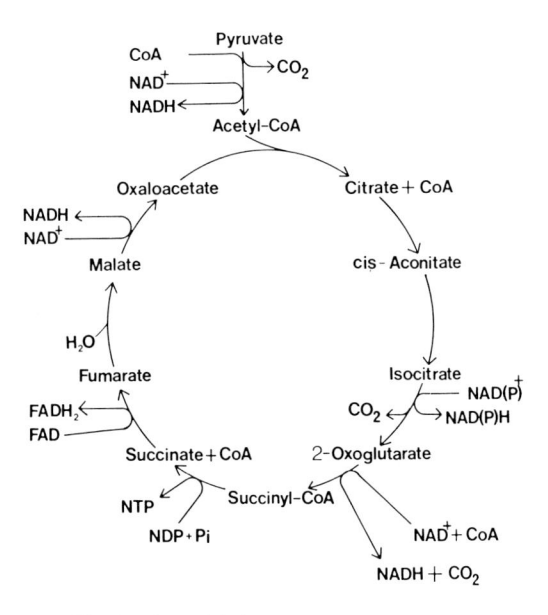

Fig. 2. The oxidative citric acid cycle.

No route for glucose catabolism other than the modified Entner–Doudoroff pathway has been reported in the halophiles. However, Tomlinson et al.[8] have reported that pyruvate production from glucose did not have an absolute requirement for ATP. This might suggest that aldol cleavage does not necessarily require prior phosphorylation and that some of the glucose can be catabolised via a non-phosphorylated pathway. This is an important observation in that such a route occurs in thermoacidophilic archaebacteria (see Fig. 3 and section 2.2) and might suggest a common metabolic link between the halophiles and thermoacidophiles.

Whilst the majority of investigations into halophilic hexose metabolism has been concerned with the catabolism of glucose, it has been recently reported [104,105] that *Haloarcula vallismortis* catabolises fructose via a modified Embden–Meyerhof pathway. Fructose is phosphorylated to fructose 1-phosphate via a ketokinase, and is then converted to fructose 1,6-bisphosphate via 1-phosphofructokinase. Aldol cleavage generates dihydroxyacetone-phosphate and glyceraldehyde 3-phosphate, both of which can be further metabolised via the glycolytic sequence described earlier. It remains to be established whether other halophilic archaebacteria can also catabolise fructose in this manner.

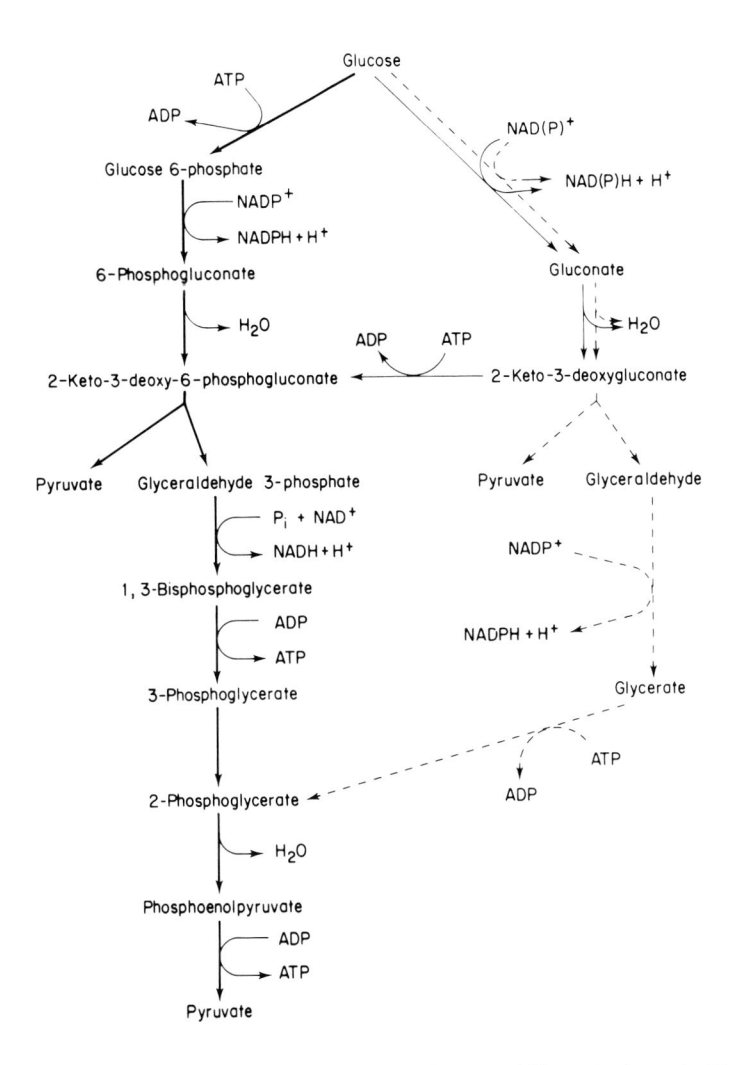

Fig. 3. Pathways of glucose catabolism in halophilic and thermophilic archaebacteria. The modified Entner–Doudoroff pathway of halophiles (solid lines) and the non-phosphorylated Entner–Doudoroff pathway of *Sulfolobus solfataricus* and *Thermoplasma acidophilum* (dashed lines) are shown in comparison with the classical Entner–Doudoroff pathway of eubacteria (heavy solid lines) from Fig. 1.

2.2. The non-phosphorylated pathway of the thermophiles

Hexose catabolism has been studied in detail in two thermophilic archaebacterial genera, *Sulfolobus* and *Thermoplasma*, organisms which are phenotypically close (they are both thermoacidophiles) but phylogenetically distinct. Interestingly, in both genera a further modification of the Entner–Doudoroff pathway has been found.

De Rosa et al. [12] discovered that *Sulfolobus solfataricus* metabolises glucose via an Entner–Doudoroff pathway in which the 2-keto-3-deoxygluconate undergoes direct aldol cleavage to pyruvate and glyceraldehyde; that is, without any prior phosphorylation (Fig. 3). Each step of this non-phosphorylated pathway has been tested at 70°C, the growth temperature of the organism, and the intermediates have been identified. However, the metabolic fate of the glyceraldehyde has not been established. Evidence for the in vivo operation of an Entner–Doudoroff-type pathway has been provided by radiorespirometric assays of glucose oxidation in *Sulfolobus brierleyi* and *Sulfolobus* strain LM (similar to *Sulfolobus acidocaldarius*)[13], although such experiments could not confirm whether the route was via phosphorylated or non-phosphorylated intermediates.

In *Tp. acidophilum* we have also found the production of pyruvate and glyceraldehyde via a non-phosphorylated Entner–Doudoroff pathway[2,14]. Furthermore, we have demonstrated that the glyceraldehyde is oxidised to glycerate, which is then converted to 2-phosphoglycerate by glycerate kinase. Enolase and pyruvate kinase complete the production of a second molecule of pyruvate (Fig. 3). Again, we have characterised the pathway enzymically and by the identification of intermediates[14], and evidence for its in vivo operation has been gained from radiorespirometric studies[2].

To date, this non-phosphorylated Entner–Doudoroff pathway has been found and characterised only in the thermoacidophilic archaebacteria, although as mentioned previously (section 2.1), it may operate to a limited extent in the extreme halophiles[1,2,8]. However, the hyperthermophilic archaebacterium, *Pyrococcus furiosus*, has been found to contain a ferredoxin-linked glucose oxidoreductase (catalysing glucose to gluconate) and glyceraldehyde oxidoreductase (glyceraldehyde to glycerate), and a pathway similar to that in *Sulfolobus* and *Thermoplasma* has been suggested but one in which ferredoxin substitutes for NAD(P)$^+$ [102]. Recently, the other enzymes of the proposed pathway have been detected in cell extracts of *P. furiosus*[107].

From the scheme proposed for *Tp. acidophilum*, there is no net yield of ATP during the oxidation of glucose to pyruvate. As will be discussed later, the citric acid cycle is probably operational and will produce ATP. However, acetate is excreted from this archaebacterium during growth on glucose[15], and we have shown[2,16] that it is generated from acetyl–CoA with the concomitant production of ATP via the enzyme acetyl–CoA synthase (ADP utilising):

$$\text{acetyl–CoA} + \text{ADP} + \text{P}_i \rightarrow \text{acetate} + \text{CoA} + \text{ATP}.$$

With regard to possible mechanisms for the regeneration of NAD(P)$^+$ and the possible concomitant synthesis of ATP in thermoacidophiles, an ATPase involved in oxidative phosphorylation in *S. acidocaldarius* has been found[17–19]. However, no proton-translocating ATPase has been identified in *Tp. acidophilum*, although a membrane-bound enzyme, proposed to function as a sulphate-exporting translocase, has been reported[20].

Finally, the need for further investigations into the metabolism of glucose in thermophilic archaebacteria should be stressed. Firstly, the fate of glyceraldehyde in *Sulfolobus* species needs to be established, and there is still controversy concerning the pathways in *Tp. acidophilum*. That is, Searcy and Whatley[15] have provided evidence from respiratory studies for the operation of glycolysis in *Tp. acidophilum* but we have been unable to detect many of the enzymes of this pathway[14]. Secondly, there is a scarcity of information on metabolism in the other thermophilic archaebacteria, especially the hyperthermophiles (i.e. those growing above 90°C).

2.3. The glycolytic pathway of the methanogens

Methanogens appear to interconvert glucose and pyruvate via the normal glycolytic pathway shown in Fig. 1. However, most methanogens are autotrophs and carbon is fixed into acetyl–CoA; thus, much of the evidence for an Embden–Meyerhof route has been gained in the direction of carbohydrate synthesis and then mainly from the one organism, *Methanobacterium thermoautotrophicum*.

Both radiolabelling studies [21] and the detection of enzymic activities [21,22] support the glycolytic pathway for the synthesis of glucose in *M. thermoautotrophicum*. Fructose 1,6-bisphosphate aldolase was detectable only in the direction of aldol condensation and no 6-phosphofructokinase activity could be found [22]. This would argue against the operation of this pathway for the catabolism of glucose, but ^{13}C-NMR spectroscopy in this same organism suggests that glucose can be converted to pyruvate via this route [23,24]. However, it should be noted that glucose does not serve as an energy or carbon source in *M. thermoautotrophicum*.

This conflict has yet to be resolved. Further investigations are also needed into the presence and function of the metabolite, 2,3-cyclopyrophosphoglycerate, in *M. thermoautotrophicum*. In vivo ^{31}P-NMR spectroscopy identified this metabolite [25,26] and it has been suggested that it is an important gluconeogenic metabolite whose turnover is directly linked to carbohydrate synthesis [24] but in an as yet undefined manner.

2.4. Gluconeogenesis

Whether heterotrophic or autotrophic, all archaebacteria will possess the capacity to carry out gluconeogenesis from C_3 or C_4 compounds.

In both halophiles and methanogens, it appears that gluconeogenesis proceeds via a reversal of the Embden–Meyerhof glycolytic pathway. The evidence supporting this route in methanogens has been previously discussed (section 2.3). In *Halobacterium halobium*, the interconversion of glyceraldehyde and pyruvate has been discussed in the context of the modified Entner–Doudoroff pathway (section 2.1). With the finding in cell extracts of this organism of the enzymes fructose 1,6-bisphosphate aldolase, fructose 1,6-bisphosphatase and phosphoglucose isomerase, the potential for functional reversal of glycolysis is indicated [10,27,28].

The pathways of gluconeogenesis in the thermophilic archaebacteria have not yet been established. We have been unable to detect the enzymes fructose 1,6-bisphosphatase, fructose 1,6-bisphosphate aldolase, glyceraldehyde 3-phosphate dehydrogenase and phosphoglycerate mutase in *Tp. acidophilum* [14], thus appearing to rule out a reversal of the glycolytic pathway. Whether or not gluconeogenesis occurs via a reversal of the non-phosphorylated Entner–Doudoroff pathway remains to be investigated.

2.5. The pentose–phosphate pathway

In the halophilic archaebacteria, *Haloarcula vallismortis* and *Haloferax mediterranei*, Rawal et al. [10] have reported the presence of the non-oxidative pentose–phosphate cycle. Moreover, in species of halophiles accumulating β-hydroxybutyrate, ribulose bisphosphate carboxylase activity has been detected and CO_2-fixation through this reductive pathway has been proposed [10,28]. In other halophiles, CO_2-fixation through a glycine synthase-type reaction has been found [29], although the system may differ from the non-archaebacterial glycine synthase [30] in that NADH and NADPH are apparently not the reductants in the halophilic system. In both cases, the organisms were growing heterotrophically and the function of the CO_2-fixation remains to be established.

Compared with the investigations in the extreme halophiles, there is very little information on the operation of a pentose–phosphate pathway in other archaebacteria. The radiorespirometric analyses of glucose metabolism in *Sulfolobus* species [13], which established the Entner–Doudoroff type pathway (section 2.2), were also consistent with a non-cyclic pentose-phosphate pathway in *S. brierleyi* and a conventional oxidation cycle in *Sulfolobus* strain LM. Similarly, respiratory studies [15] provide evidence for a pentose phosphate cycle capable of glucose oxidation in *Tp. acidophilum*. No data are available for the methanogens.

3. Pyruvate oxidation to acetyl–CoA in the archaea

In contrast to the diversity of pathways leading from glucose to pyruvate, there is a distinct unity in the way archaebacteria oxidatively decarboxylate this C_3-acid to acetyl–CoA. Not only that, but the comparison of the enzymology of this step with the analogous reaction in eubacteria and eukaryotes provides interesting parallels and differences [1].

3.1. Pyruvate oxidoreductases

The conversion of pyruvate to acetyl–CoA is catalysed by pyruvate oxidoreductase in the archaebacteria. The enzyme has been detected and characterised in *Halobacterium halobium* [31,32], *Tp. acidophilum*, *S. acidocaldarius* and *Desulfurococcus mobilis* [33], *Pyrococcus furiosus* [34] and in *Methanobacterium thermoautotrophicum* [35]. In the halophiles and thermophiles, ferredoxin serves as electron acceptor, whereas the methanogens use the deazaflavin derivative F_{420}.

The proposed catalytic mechanism of the ferredoxin oxidoreductase [32] is shown in Fig. 4, a similar mechanism existing for the analogous citric acid cycle enzyme, 2-oxoglutarate oxidoreductase. In outline, the 2-oxoacid is decarboxylated in a TPP-dependent reaction to give an hydroxyalkyl-TPP. From this, one electron is abstracted and transferred to the enzyme-bound iron–sulphur cluster, generating a free-radical-TPP species. This intermediate can then interact direct with coenzyme-A to form acyl-CoA, the iron–cluster receiving the second electron. In each case, ferredoxin serves to re-oxidise the enzyme's redox centre.

The *H. halobium* pyruvate and 2-oxoglutarate ferredoxin oxidoreductases have been shown to possess an a_2b_2 subunit structure [31], each having a total relative molecular

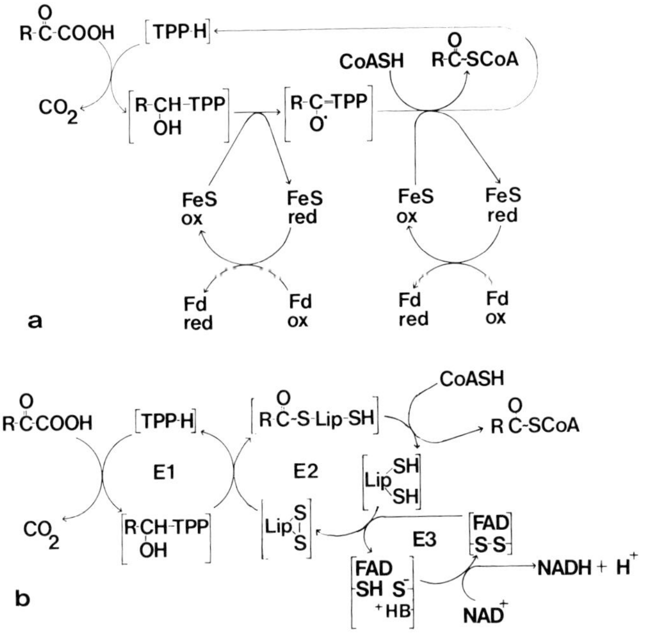

Fig. 4. The enzymic reaction mechanisms for the conversion of pyruvate to acetyl-CoA, yielding the net reaction $R-\overset{O}{C}-COOH + CoASH + NAD^+ \rightarrow R-\overset{O}{C}\sim SCoA + CO_2 + NADH + H^+$. (a) The pyruvate:ferredoxin oxidoreductase of the archaebacteria. (b) The pyruvate dehydrogenase multienzyme complex of eubacteria and eukaryotes (E1: pyruvate decarboxylase; E2: dihydrolipoyl acetyltransferase; E3: dihydrolipoamide dehydrogenase). Symbols: R: CH_3-; Fd: ferredoxin; FeS: an enzyme-bound iron–sulphur cluster; TPP-H: thiamine pyrophosphate; Lip: lipoic acid; B: amino acid base on the dihydrolipoamide dehydrogenase.

mass of approximately 200 000. The genes coding for both subunits of the pyruvate oxidoreductase have now been sequenced and the thiamin-bisphosphate-binding site has been located on the b subunit [109].

3.2. Comparison with the eukaryotic and eubacterial enzymes

In aerobic environments, eukaryotes and many eubacteria oxidatively decarboxylate the 2-oxoacids via pyruvate and 2-oxoglutarate dehydrogenase complexes [36]. For comparison with the archaebacterial oxidoreductases, the catalytic mechanism of these complexes is also shown in Fig. 4. Three enzymic activities are involved, catalysed by three distinct enzymes: a 2-oxoacid decarboxylase (E1), a dihydrolipoyl acyltransferase (E2) and dihydrolipoamide dehydrogenase (E3). Multiple copies of these three enzymes are found in each complex molecule, resulting in relative molecular masses in excess of 2×10^6.

As reviewed previously [1], there are distinct mechanistic similarities between the archaebacterial oxidoreductases and the 2-oxoacid dehydrogenase complexes, despite their obvious structural differences. Thus, both systems form acyl-CoA via a TPP-dependent decarboxylation process, in which a hydroxyl-TPP is formed as an intermediate. The main difference is that the archaebacterial enzyme effects a direct transfer of the C_2-moiety from TPP to coenzyme-A, whereas the dehydrogenase complexes utilise an enzyme-bound lipoic acid as a carrier. In the latter case, an extra enzymic activity, namely dihydrolipoamide dehydrogenase, is required within the complex to re-oxidise the lipoic acid and so transfer the reducing equivalents to NAD^+. Thus, it has been shown [31] that the archaebacterial oxidoreductases do not contain the cofactor lipoic acid and that the transfer of reducing equivalents is via the protein's iron–sulphur centre to ferredoxin.

It should be noted that some anaerobic eubacteria (e.g. *Escherichia coli, Clostridium acidurici, Chlorobium limicola*) possess 2-oxoacid oxidoreductases, either in addition to, or instead of, the dehydrogenase complexes (ref. [37] and references therein). As in the archaebacteria, electrons are transferred to ferredoxin or flavodoxin and the enzymes are small oligomeric proteins. The evolutionary relationships of these enzymes converting pyruvate to acetyl–CoA will be discussed in section 6.2.

3.3. Dihydrolipoamide dehydrogenase

Given the previous discussion on the mechanism of the archaebacterial 2-oxoacid oxidoreductases, and that the only known function of dihydrolipoamide dehydrogenase is as an integral member of the non-archaebacterial 2-oxoacid dehydrogenase complexes, we were surprised to discover the presence of dihydrolipoamide dehydrogenase in the halophilic archaebacteria [38]. The enzyme appears not to be part of any multi-enzyme system in the halophiles, but purification and subsequent characterisation showed it to be very similar to its eubacterial and eukaryotic counterpart in both its dimeric nature and its catalytic mechanism [39]. Using a biological assay [40] and a gas-chromatographic mass-spectrometric procedure [41], the presumed substrate, lipoic acid, was found to be present in the halophiles, although whether or not it is protein-bound has not yet been established.

The function of this archaebacterial dihydrolipoamide dehydrogenase in the absence of its normal multienzyme complexes is unknown [42,43]. Detailed structural studies, beginning with current experiments to clone and sequence the gene [44], may throw light on this. Meanwhile, we have surveyed a number of archaebacterial genera for the presence of the enzyme [43] and have correlated this with the presence of lipoic acid. The data available are summarised in Table 1.

The presence of the enzyme and cofactor are co-incident, indicating that lipoic acid may indeed be the true substrate of the archaebacterial dihydrolipoamide dehydrogenase. Interestingly, their presence can be correlated with the organisms' phylogenetic positions within the archaebacteria. That is, the archaea comprise two main divisions – the methanogens, extreme halophiles, *Thermoplasma* and *Thermococcus* in one, and the remaining sulphur-dependent thermophiles in the other [66,72]. The enzyme and/or the cofactor have been detected in all the phenotypes of the former division but neither have yet been discovered in the latter. Further analyses are required to test this correlation as the data are incomplete.

TABLE 1
Dihydrolipoamide dehydrogenase and lipoic acid in the archaebacteria

Organism	Dihydrolipoamide dehydrogenase[a]	Lipoic acid[a]	References
Halobacterium halobium	+	+	[38,41]
Haloferax volcanii	+	+	[38,40]
Natronobacterium pharaonis	+	nd	[38]
Natronobacterium gregoryi	+	nd	[38]
Natronococcus occultus	+	nd	[38]
Thermoplasma acidophilum	+	+	[100][b]
Thermococcus celer	nd	+	[40]
Methanobacterium thermoautotrophicum	+	+	[40][b]
Sulfolobus acidocaldarius	–	nd	[100]
Sulfolobus solfataricus	nd	–	[40]
Thermoproteus tenax	nd	–	[40]
Pyrodictium occultum	nd	–	[40]

[a] Symbols:
(+) indicates the presence of enzyme and/or lipoic acid;
(–) indicates the absence (or levels at the limits of detection) of enzyme and/or lipoic acid;
(nd) indicates that no data are available.
[b] Also: M.J. Danson, D.W. Hough, S.A. Jackman and K.J. Stevenson, unpublished observations.

4. The citric acid cycle in the archaea

The nature of the pathways of hexose catabolism (section 2) was related to the phenotypic characteristics of the organisms, and it is clear that the form of the citric acid cycle employed can be similarly correlated. The three forms of the cycle (oxidative, reductive and partial) are discussed in turn.

4.1. The oxidative citric acid cycle

The complete citric acid cycle (Fig. 2) is the final pathway for the oxidation of all major nutrients in eukaryotes and eubacteria growing under aerobic conditions. It thus serves to yield reducing equivalents in the form of NADH and $FADH_2$, the re-oxidation of which is coupled to the synthesis of ATP. In addition, the cycle also provides key starting materials for a variety of biosynthetic processes.

This complete oxidative cycle is found in a number of archaebacteria. Halophiles can fulfil their energy requirements by metabolism of amino acids and other nitrogenous compounds, and therefore it is probable that they possess an oxidative citric acid cycle. Aitken and Brown [45] have reported the presence of the cycle's enzymes in *Halobacterium halobium* and we have found the key enzymes, citrate synthase and succinate thiokinase, in a range of classical and alkaliphilic halophiles [46]. Thus, it is probable that the cycle is generally present in this group of archaebacteria, but exhaustive studies have not been carried out.

Thermoplasma acidophilum and *Thermoplasma volcanium* are facultative organotrophs [47], and are phylogenetically related to the halophiles. When growing aerobically, they are also thought to possess a complete, oxidative citric acid cycle. In support of this, many of the constituent enzymes have been reported in *Tp. acidophilum* (ref. [1] and references therein). As noted previously (section 2.2), the conversion of carbohydrates to pyruvate via the non-phosphorylated Entner–Doudoroff pathway in this archaebacterium does not produce a net yield of ATP. Therefore, it may be dependent on the citric acid cycle for its energy supply.

Within the sulphur-dependent thermophilic branch of the archaebacteria, the situation is less clear. *Sulfolobus* species are facultatively autotrophic [13] and many of the citric acid cycle enzymes have been reported (reviewed in ref. [1]). Thus, when growing heterotrophically in the presence of oxygen, an oxidative cycle may be operative, although respirometric analyses suggest that its use may be limited [13].

As for other genera within this branch, there are few metabolic/enzymological studies of the citric acid cycle. A number of organisms are obligate heterotrophs and are able to use proteins and/or amino acids for growth (e.g. *Pyrococcus, Staphylothermus, Thermococcus, Thermophilum* [48,49]); it is possible that the citric acid cycle functions in these organisms, but the evidence is not available yet. The oxidation of pyruvate via pathways other than the citric acid cycle in some of these extreme thermophiles will be described in section 4.4.

4.2. The reductive citric acid cycle

Many of the sulphur-dependent thermophilic archaebacteria can grow autotrophically and, in the two genera where the mechanism of CO_2-fixation has been studied, there is evidence that it is via a reductive citric acid cycle.

Thermoproteus neutrophilus is a strictly anaerobic, sulphur-reducing organism. Autotrophically-grown cells appear to fix CO_2 via a reductive citric acid cycle (Fig. 5): all of the enzymes of this cycle have been detected, except for a citrate-cleavage activity, and radiolabelling studies are consistent with the reductive pathway [50]. Acetate is a preferred carbon source for *T. neutrophilus* and suppresses autotrophic CO_2-fixation. A ^{13}C-NMR study of acetate assimilation [51] demonstrated that it is converted to acetyl–CoA which in turn is reductively carboxylated to pyruvate; phosphoenolpyruvate is formed from pyruvate and a second carboxylation gives rise to oxaloacetate. However, only approximately 15% of 2-oxoglutarate and 5% of pyruvate are generated through the reductive citric acid cycle of the type proposed for autotrophic growth, and a much reduced level of fumarate reductase (converting fumarate to succinate) is observed. The remaining 2-oxoglutarate is formed via an oxidative cycle and, during growth on acetate, a horse-shoe type of citric acid cycle is proposed where the two halves meet at 2-oxoglutarate [51].

Interestingly, autotrophic growth of *S. brierleyi* may also involve CO_2-fixation via a reductive cycle [52]. Wood et al. [13] could provide no further evidence for such a pathway, but suggest that the observed simultaneous use of acetate and CO_2 for cellular biosyntheses indicates that the CO_2-fixing cycle must operate simultaneously with both assimilatory and oxidative pathways.

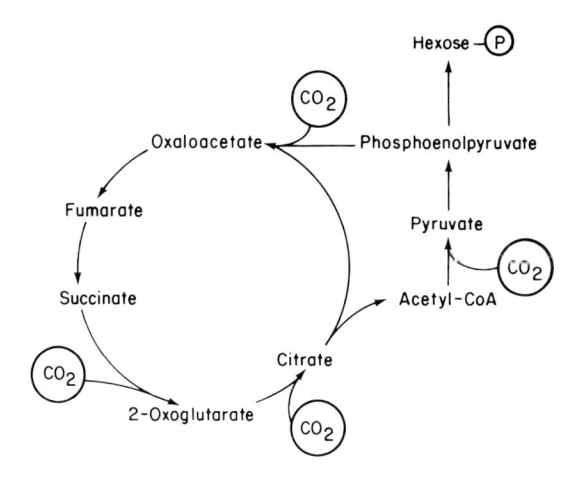

Fig. 5. The reductive citric acid cycle proposed for autotrophic carbon dioxide assimilation in *Thermoproteus neutrophilus*.

4.3. Partial citric acid cycles

In the situation described above for *T. neutrophilus* growing on acetate, the citric acid cycle is complete but is composed of two parts, reductive and oxidative. In the methanogenic archaebacteria, two truly partial versions of the cycle can be observed, both fulfilling an anabolic function (Fig. 6). Thus, in *M. thermoautotrophicum* (reviewed in ref. [53]), *Methanospirillum hungatei* [54] and *Methanococcus voltae* [55], an incomplete, reductive citric acid cycle is present, leading to 2-oxoglutarate via succinate. On the other hand, *Methanosarcina barkeri* possesses an incomplete, oxidative cycle, with 2-oxoglutarate being synthesised via citrate [56,57]. Thus, no methanogen has yet been found with a complete citric acid cycle, whether oxidative or reductive.

4.4. Other pathways of acetyl–CoA metabolism

In several archaebacterial genera, the metabolism of acetyl–CoA may proceed via pathways other than the citric acid cycle. Thus, it has been demonstrated [34] that the hyperthermophilic anaerobe, *P. furiosus*, ferments pyruvate to acetate, CO_2 and H_2 via three enzymes: pyruvate:ferredoxin oxidoreductase (to convert pyruvate to acetyl–CoA), a hydrogenase (to re-oxidise the reduced ferredoxin so produced) and an ADP-utilising acetyl–CoA synthetase (to generate acetate and ATP from acetyl–CoA and ADP). Schäfer and Schönheit [34] suggest that the acetyl–CoA synthetase may be the main energy-conserving site in pyruvate fermentation in this archaebacterium; they have recently reported the presence of the enzyme in *Desulfurococcus amylolyticus*, *Hyperthermus butylicus*, *Thermococcus celer*, and *Halobacterium halobium* [108]. This ATP synthesis via an acetyl–CoA synthetase (ADP) is the same as that reported for *Tp. acidophilum* [2,16], and in all the organisms there is no evidence for phosphate acetyl–transferase or acetate kinase activities. The situation is to be contrasted with the

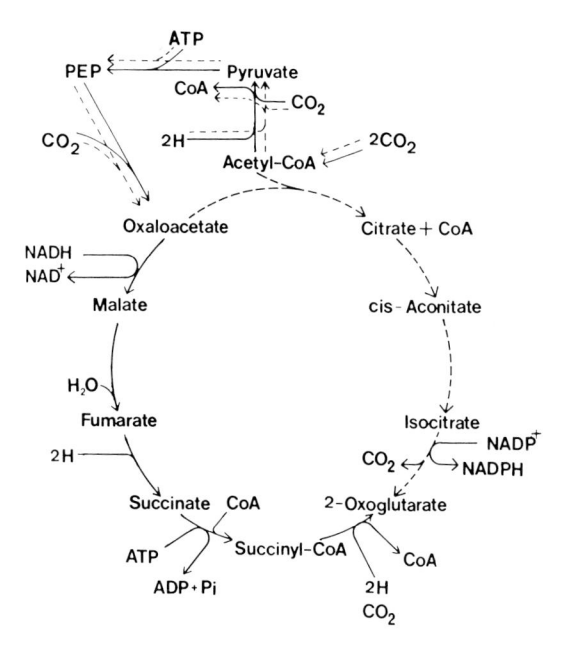

Fig. 6. The partial citric acid cycles in methanogens. The pathways shown are those proposed for *Methanobacterium thermoautotrophicum* (solid lines) [an incomplete, reductive cycle], and for *Methanosarcina barkeri* (dashed lines) [an incomplete, oxidative cycle].

AMP-dependent synthetases of eubacteria, eukaryotes, methanogens and *Thermoproteus neutrophilus*, where the enzyme is thought to function in the activation of acetate to acetyl–CoA (see ref. [1] and references therein).

In the extreme thermophile, *Archeoglobus fulgidus*, pyruvate is converted to acetyl–CoA as in other archaebacteria, but subsequent oxidation to CO_2 proceeds via the carbon monoxide dehydrogenase pathway [58,59]. Several enzymes of the citric acid cycle have been found, but none catalysing the oxidation of 2-oxoglutarate or succinate. Thus the cycle may operate in a biosynthetic mode, the carbon monoxide pathway serving to generate energy.

5. Amino acid and lipid metabolism in the archaea

It is not the purpose of this review to provide detailed descriptions of amino acid and lipid metabolism in the archaea, but rather to see how these compounds feed into and out of the pathways of central metabolism described above.

With regard to amino acid metabolism, the data are scarce but it is probable that these metabolites give rise to, or are derived from, the oxoacids of the citric acid cycle via transamination or analogous reactions. See the literature [60–62] for examples of such enzymes in the halophilic, thermophilic and methanogenic archaebacteria.

In the case of lipids, a detailed description of their structure and biosynthesis is given in Chapter 9. Biosynthesis involves the generation of the glycerol backbone and of the hydrocarbon chain, followed by the formation of the ether linkage. In the halophilic archaebacteria, it is probably glycerol or dihydroxyacetone that accepts the hydrocarbon chain (reviewed in [63] and Chapter 10). Radiolabelling studies indicate that neither of these could arise by aldol-keto isomerisation from the glyceraldehyde 3-phosphate of the modified Entner–Doudoroff pathway; glycerol may therefore be derived from the external medium. Acetyl–CoA and amino acids such as lysine are thought to enter the mevalonate pathway for the synthesis of a C_{20}-prenyl pyrophosphate [101], prior to its condensation with the dihydroxyacetone or glycerol (Fig. 7). In the thermophilic archaebacterium, *S. acidocaldarius*, De Rosa et al. [64,65] have suggested that glycerol is directly alkylated by geranylgeranylpyrophosphate in the formation of ether lipids. However, in the non-phosphorylated Entner–Doudoroff pathway in this organism and in *Tp. acidophilum*, glyceraldehyde is produced (Fig. 3). We have found the enzyme, glycerol:$NADP^+$ oxidoreductase in both these thermophiles [2,16] and have suggested that it might provide the glycerol for lipid biosynthesis according to:

$$\text{DL-glyceraldehyde} + NADPH + H^+ \rightarrow NADP^+ + \text{glycerol}.$$

In cell extracts we have been unable to detect glyceraldehyde 3-phosphate dehydrogenase, glycerol kinase or glyceraldehyde kinase, consistent with a non-phosphorylated glycerol as the glycerolipid precursor [2,16].

Nothing is known about the origin of the glycerol residue of methanogenic lipids. The presence in the Embden–Meyerhof pathway of glyceraldehyde 3-phosphate and dihydroxyacetone phosphate might suggest that these are the precursors as in eubacterial and eukaryotic organisms.

6. Evolution of central metabolism

In order to draw conclusions about the origin and evolution of central metabolism from a study of these pathways in archaebacteria, eubacteria and eukaryotes, a definitive phylogeny of the organisms involved is required. In particular, a knowledge of which organisms are primitive is essential.

Studies on the pathways of central metabolism of the archaebacteria take on special significance when it is realised that, from the universal phylogenetic tree in rooted form, Woese et al. [66] have proposed that the domain of the archaebacteria be known as archaea to denote their apparently primitive nature, especially with respect to the eukaryotes. Furthermore, within the archaea, thermophily is regarded as the ancestral phenotype.

In the following sections, it will be seen that the basic central pathways were established before the divergence of the three domains, but that the small yet notable differences may provide important clues to their evolution.

6.1. Hexose catabolism

From studies of eubacterial and eukaryotic metabolism, it has been previously argued that the Embden–Meyerhof glycolytic pathway is the ancient energy-conserving route of hexose catabolism [3]. However, a key enzyme of this catabolic route,

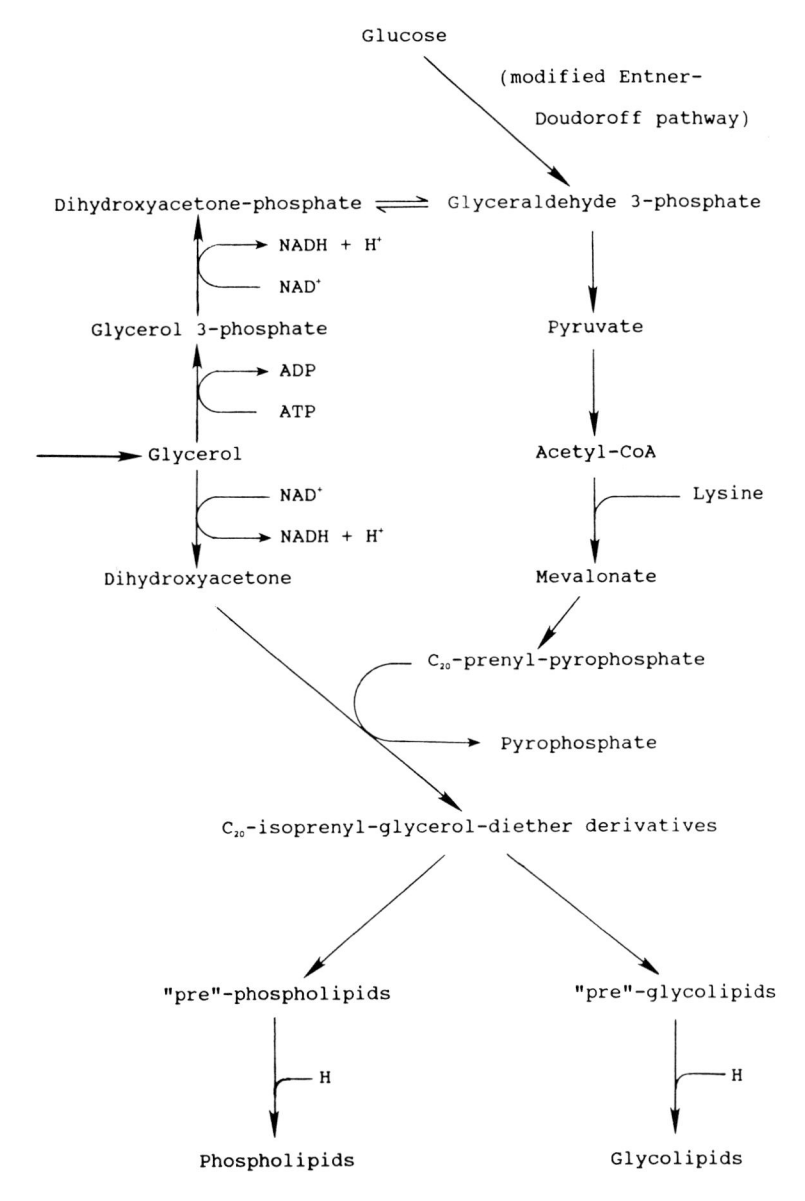

```
                                Glucose
                                      \
                                       \   (modified Entner–
                                        \    Doudoroff pathway)
                                         \
Dihydroxyacetone-phosphate  ⇌  Glyceraldehyde 3-phosphate
          ↑  ⎛→ NADH + H⁺
          │  ⎝— NAD⁺
    Glycerol 3-phosphate                 Pyruvate
          ↑  ⎛→ ADP
          │  ⎝— ATP
    ─────→ Glycerol                    Acetyl-CoA
          │  ⎛— NAD⁺                        │   ⎛— Lysine
          ↓  ⎝→ NADH + H⁺                   ↓
    Dihydroxyacetone                    Mevalonate
                    \                    /
                     \   ⎛— C₂₀-prenyl-pyrophosphate
                      \ /
                       X  ─→ Pyrophosphate
                      / \
                     /   \
         C₂₀-isoprenyl-glycerol-diether derivatives
                    /              \
                   /                \
           "pre"-phospholipids    "pre"-glycolipids
                │  ⎛— H                │  ⎛— H
                ↓                      ↓
            Phospholipids           Glycolipids
```

Fig. 7. Proposed biosynthetic pathway for the formation of diphytanyl glycerol ether lipids from the pathways of central metabolism in halophilic archaebacteria. The scheme outlined is taken from ref. [63] and M. Kates (personal communication).

phosphofructokinase, has not been found in the archaebacteria. Thus, modifications of the Entner–Doudoroff pathway are found in the halophiles and in *Thermoplasma* and *Sulfolobus,* organisms which span both archaebacterial branches. On the other hand, gluconeogenesis via a reversal of the Embden–Meyerhof pathway is found in halophiles and methanogens, no studies having yet been made in the thermophiles. It is therefore a possibility that the Entner–Doudoroff and Embden–Meyerhof pathways were originally for catabolism and anabolism, respectively, with the evolution of phosphofructokinase permitting the latter pathway to operate in a degradative manner.

6.2. Pyruvate oxidation to acetyl–CoA

As discussed in section 3, the conversion of pyruvate to acetyl–CoA is catalysed by pyruvate oxidoreductase in all archaebacteria. A similar pyruvate ferredoxin oxidoreductase is found in anaerobic eubacteria (see ref. [37] and references therein). Thus, Kerscher and Oesterhelt [37] have suggested that these enzymes existed before the divergence of archaebacteria, eubacteria and the ancestral eukaryotes, and that the 2-oxoacid dehydrogenase complexes (found only in respiratory eubacteria and eukaryotes) emerged after the development of oxidative phosphorylation. Archaebacteria and eubacteria that have retained their anaerobic niches still use the 2-oxoacid oxidoreductases and, presumably, eukaryotes gained their dehydrogenase complexes from the eubacterial lineage via the endosymbiotic origin of mitochondria. (In the eukaryotic *Euglena gracilis,* a mitochondrial NADP-dependent pyruvate oxidoreductase has been reported [67], but its catalytic mechanism is thought to differ from the archaebacterial and eubacterial ferredoxin-linked enzymes.)

Using similar arguments, dihydrolipoamide dehydrogenase might also be considered to be an 'ancient' enzyme [42,43] as it is present in archaebacteria, eubacteria and eukaryotes (see section 3.3). Presumably, its original function has been retained in the archaebacteria, but in the respiratory eubacteria and eukaryotes it has been sequestered to function as an integral component of the 2-oxoacid dehydrogenase complexes. Interestingly, in one eukaryote (*Trypanosoma brucei*) which lacks the dehydrogenase complexes, dihydrolipoamide dehydrogenase is still found [68,69]; again, its function is unknown.

6.3. The citric acid cycle

It has been suggested that the reductive conversion of oxaloacetate into succinate via malate and fumarate may have emerged in association with anaerobic hexose metabolism to regenerate NAD^+ [70]. Also, the conversion of pyruvate to acetyl–CoA via an oxidoreductase may have evolved in early anaerobic cells, thus leading to the evolution of the other half of the citric acid cycle whereby 2-oxoglutarate is synthesised from acetyl–CoA and oxaloacetate via citrate. A primitive cell may therefore have possessed two arms of the cycle [71]. The evolution of 2-oxoglutarate oxidoreductase from the pyruvate enzyme would complete the cycle, although it would probably be acting initially in a reductive mode with photoreduced ferredoxin permitting the 2-oxoglutarate oxidoreductase to act as a synthase.

This scenario envisages the reductive citric acid cycle as the primitive pathway, with the oxidative, bioenergetic cycle awaiting the availability of oxygen as a suitable electron acceptor. It may be significant, therefore, that such a reductive cycle is thought to operate in species of *Sulfolobus* and *Thermoproteus,* and incompletely in various methanogens, organisms that are representative of the two main archaebacterial branches.

Once again, the validity of these suggestions awaits the investigations of the pathways of central metabolism in more archaebacterial species, particularly the more-recently isolated extreme thermophiles.

7. Comparative enzymology of central metabolism

7.1. The enzymes as molecular chronometers

The unique phylogenetic position of the archaebacteria was initially proposed on the basis of sequence comparisons of one molecule, namely the 16S/18S rRNA [72]. This proposal has been supported by the comparison of other rRNA and protein sequences, but the evolutionary distance of the domains relative to each other seems to vary depending on the macromolecule chosen (see ref. [73] and references therein).

It is clear that more molecular chronometers need to be analysed. On account of the universality of the core metabolic pathways, a molecular study of the enzymes of central metabolism may prove worthwhile in this context. Moreover, the range of phenotypes within the one domain (e.g. extreme halophilicity and thermophilicity, in addition to mesophilicity) may make a comparative study of these enzymes especially valuable to our understanding of the structural basis for extreme protein stability. For these reasons, a number of laboratories are currently engaged in detailed structure–function investigations of the central metabolic enzymes. For a detailed discussion of the comparative enzymology of these pathways, see ref. [1].

For example, glyceraldehyde 3-phosphate dehydrogenase has been sequenced from mesophilic and thermophilic methanogens [74,75] and from *Pyrococcus woesei* [76], and site-directed mutagenesis studies are in progress [77]. Malate dehydrogenase [78] and 3-phosphoglycerate kinase [79] have also been sequenced from the methanogens. The basis for the salt-stability and -requirement of malate dehydrogenase from *Haloarcula marismortui* has been extensively investigated [80,81], and the enzyme has now been crystallised from this halophile [82] and from the thermoacidophilic archaebacteria, *Sulfolobus acidocaldarius* and *Thermoplasma acidophilum* [83].

In my laboratory, we have cloned and sequenced glucose dehydrogenase [84], the first enzyme of the non-phosphorylated Entner–Doudoroff pathway, and citrate synthase [85], the initial enzyme of the citric acid cycle, from *Thermoplasma acidophilum*. Both genes have been expressed in *Escherichia coli* to give active enzymes, and the glucose dehydrogenase has been purified and crystallised [103]. The cloning and sequencing of dihydrolipoamide dehydrogenase from *Haloferax volcanii* are also in progress [44].

It is not the purpose of this review to provide a comprehensive account of these structure–function studies; however, the case of citrate synthase will be described briefly to illustrate the potential of such investigations. The studies on malate dehydrogenase are reviewed by Eisenberg et al. [81], and those on glyceraldehyde 3-phosphate dehydrogenase are described in detail in Chapter 7.

7.2. Citrate synthase

7.2.1. The comparative enzymology of citrate synthases
Citrate synthase catalyses the condensation of acetyl–CoA and oxaloacetate to form citrate, and so effects the entry of carbon into the citric acid cycle. The enzyme displays a diversity of structure and function which shows a remarkable correlation with the taxonomic status of the source organism (reviewed in refs. [1,86,87]). In summary, citrate synthases from Gram-negative eubacteria are hexameric proteins and are allosterically inhibited by NADH, the primary energetic end-product of the cycle. In the facultatively anaerobic members of this group, the citrate synthase arm of the cycle leads to 2-oxoglutarate; in these cases, this metabolite inhibits the citrate synthase and thus acts as a biosynthetic control in addition to the effect of NADH. In contrast, citrate synthases from Gram-positive eubacteria and eukaryotes are all dimeric enzymes; they are insensitive to NADH but are inhibited by ATP, the ultimate energetic end-product of the cycle.

The sequences of citrate synthases from the eukaryotes pig heart and kidney, *Arabidopsis thaliana* and *Saccharomyces cerevisiae*, and from the eubacteria *Escherichia coli, Rickettsia prowazekii, Acinetobacter anitratum, Acetobacter aceti* and *Pseudomonas aeruginosa* have been determined (see the literature [85,88] for references). In addition, a high-resolution X-ray crystallographic structure is available for the pig heart enzyme [89,90]. This has allowed the identification of 12 residues which are critical for substrate binding and catalysis; multiple sequence alignments [91] have indicated that the majority of these 12 active site residues are conserved between all eukaryotic and eubacterial citrate synthases.

Finally, it should be noted that the subunits of all citrate synthases are approximately the same size (45 000–50 000) and we have evidence that the hexamers and dimers are functionally similar. Thus, proteolysis studies [92] indicate that the hexameric citrate synthase functions as a trimer of dimers, and by homology modelling to the pig heart enzyme, we have shown that the subunit of the *E. coli* hexameric protein has a similar conformation to that in the dimeric enzyme. Further catalytic similarities are indicated by site-directed mutageneses [e.g. 93–95].

7.2.2. Archaebacterial citrate synthases
The wealth of information on eubacterial and eukaryotic citrate synthases, and the correlations of structure, function and taxonomy, have prompted us to extend these studies to the archaebacteria. With the exception of a few methanogens, the enzyme is present in all major groups [46], and thus is a good candidate for comparative enzymology.

Citrate synthases from the thermophiles *Tp. acidophilum* and *S. acidocaldarius* are dimeric enzymes [96,97] with regulatory properties similar to the dimers from Gram-positive eubacteria and eukaryotes [46,98]. The gene encoding the enzyme from *Tp. acidophilum* has been cloned and sequenced and the derived amino acid sequence compared by multiple alignment analysis with other citrate synthases [85]. The sequence identities between the three groups, archaebacteria, eubacteria and eukaryotes, are only 18–28%, consistent with the three separate domains, although the majority of residues implicated in the catalytic action of the enzyme have been conserved across all organisms. Interestingly, the archaebacterial citrate synthase lacks the N-terminal helix (approximately 40 amino acids) of the eubacterial and eukaryotic enzymes.

The *Tp. acidophilum* citrate synthase gene has now been expressed in *E. coli* to give active enzyme and the thermostability used to effect a simple purification scheme [99]. Detailed structural studies are now in progress.

We have also purified the citrate synthase from *Hf. volcanii*, and determined an N-terminal sequence; it aligns with the N-terminus of the *Tp. acidophilum* citrate synthase, and gene cloning and sequencing are now in progress [106]. The enzyme is also dimeric, although it appears to be fairly insensitive to both NADH and ATP [46]. Citrate synthase has been detected in *Ms. barkeri* [46], but no further information is yet available.

With citrate synthase, therefore, we can now generate structural information which, we believe, will help us to fulfil our stated aims; that is, to provide further sequence data for the continued analysis of the phylogeny of the archaebacteria, and to begin to understand how the proteins from these organisms can withstand the extremes of temperature and salinity. As in the cases of glyceraldehyde 3-phosphate dehydrogenase and malate dehydrogenase, it is the wealth of structural data from non-archaebacterial citrate synthases that will permit us to gain the maximum insight from our comparative enzymological studies.

8. Conclusions and perspectives

In primitive organisms, the establishment of metabolic pathways for the production of energy and for biosynthesis would have been the basis for survival and successful competition during early life on earth. Such a belief has been the driving force for the investigation of the central metabolic pathways in the archaea. Coupled with the information on the phylogenetic analysis of these organisms, such investigations are providing important clues to the nature of the earliest organisms and to their subsequent evolution. It is clear, however, that the momentum of these investigations must be maintained, principally by their extension into the recently identified hyperthermophiles, for valid conclusions to be drawn.

From the studies on archaebacterial metabolism, new impetus has been given to the field of comparative enzymology and, in particular, to our investigations of the structural basis of protein stability. Indeed, the fact that the archaebacteria inhabit the extremes of life's environments means that they have a unique contribution to make this area. To realise this potential, the study of archaebacterial enzymes remains a priority.

Acknowledgements

Financial support from SERC (UK), NSERC (Canada), The Royal Society, NATO and ICI Biological Products, Billingham, UK are gratefully acknowledged. Special thanks are also due to Dr D.W. Hough and Dr G.L. Taylor (University of Bath, UK) and Dr K.J. Stevenson (University of Calgary, Canada) with whom I enjoy valuable collaborations, and to past and present archaebacterial students in Bath, N. Budgen, L.D. Smith, K.J. Sutherland, J.R. Bright, M.R.J. Russell, J.M. Muir, M. McCormack, J.J. John and K.D. James, and in Calgary, N. Vettakkorumakankav.

Note added in proof

Since the completion of this manuscript in 1992, Schäfer and Schönheit [110] have demonstrated that gluconeogenesis in *P. furiosus* proceeds via the reactions of the Embden–Meyerhof pathway (see section 2.4).

References

[1] Danson, M.J. (1988) Adv. Micro. Physiol. 29, 165–231.
[2] Danson, M.J. (1989) Can. J. Microbiol. 35, 58–64.
[3] Gest, H. and Schopf, J.W. (1983) In: Earth's Earliest Biosphere (Schopf, J.W., ed.), pp. 135–148, Princeton University Press, Princeton, NJ.
[4] Cooper, R.A. (1986) In: Carbohydrate Metabolism in Cultured Cells (Morgen, M.J., ed.), pp. 461–491, Plenum Press, New York.
[5] Morris, J.G. (1985) In: Comprehensive Biotechnology (Moo-Young, M., ed.), pp. 357–378, Pergamon Press, Oxford.
[6] Tomlinson, G.A. and Hochstein, L.I. (1972) Can. J. Microbiol. 18, 698–701.
[7] Tomlinson, G.A. and Hochstein, L.I. (1972) Can. J. Microbiol. 18, 1973–1976.
[8] Tomlinson, G.A., Koch, T.K. and Hochstein, L.I. (1974) Can. J. Microbiol. 20, 1085–1091.
[9] Hochstein, L.I. (1988) In: Halophilic Bacteria (Rodriguez-Valera, F., ed.), Vol. 2, pp. 67–83, CRC Press, Boca Raton, FL.
[10] Rawal, N., Kelkar, S.M. and Altekar, W. (1988) Ind. J. Biochem. Biophys. 25, 674–686.
[11] Severina, L.O. and Pimenov, N.V. (1988) Mikrobiologiya 57, 907–911.
[12] De Rosa, M., Gambacorta, A., Nicolaus, B., Giardina, P., Poerio, E. and Buonocore, V. (1984) Biochem. J. 224, 407–414.
[13] Wood, A.P., Kelly, D.P. and Norris, P.R. (1987) Arch. Microbiol. 146, 382–389.
[14] Budgen, N. and Danson, M.J. (1986) FEBS Lett. 196, 207–210.
[15] Searcy, D.G. and Whatley, F.R. (1984) Syst. Appl. Microbiol. 5, 30–40.
[16] Budgen, N. (1988) Ph.D. Thesis, University of Bath, Bath, UK.
[17] Wagaki, T.W. and Oshima, T. (1986) Syst. Appl. Microbiol. 7, 342–345.
[18] Lübben, M. and Schäfer, G. (1987) Eur. J. Biochem. 164, 533–540.
[19] Lübben, M., Lündsdorf, H. and Schäfer, G. (1987) Eur. J. Biochem. 167, 211–219.
[20] Searcy, D.G. (1986) System. Appl. Microbiol. 7, 198–201.
[21] Jansen, K., Stupperich, E. and Fuchs, G. (1982) Arch. Microbiol. 132, 355–364.
[22] Fuchs, G., Winter, H., Steiner, I. and Stupperich, E. (1983) Arch. Microbiol. 136, 160–162.
[23] Evans, J.N.S., Tolman, C.J., Kanodia, S. and Roberts, M.F. (1985) Biochemistry 24, 5693–5698.
[24] Evans, J.N.S., Raleigh, D.P., Tolman, C.J. and Roberts, M.F. (1986) J. Biol. Chem. 261, 16323–16331.
[25] Kanodia, S. and Roberts, M.F. (1983) Proc. Natl. Acad. Sci. USA 80, 5217–5221.
[26] Seely, R.J. and Fahrney, D. (1983) J. Biol. Chem. 258, 10835–10838.
[27] D'Souza, S.A. and Altekar, W. (1983) Ind. J. Biochem. Biophys. 20, 29–32.
[28] Altekar, W. and Rajagopalan, R. (1990) Arch. Microbiol. 153, 169–174.
[29] Javor, B.J. (1988) Arch. Microbiol. 172, 756–761.
[30] Kikuchi, G. and Hiraga, K. (1982) Mol. Cell. Biochem. 45, 137–149.
[31] Kerscher, L. and Oesterhelt, D. (1981) Eur. J. Biochem. 116, 587–594.
[32] Kerscher, L. and Oesterhelt, D. (1981) Eur. J. Biochem. 116, 595–600.
[33] Kerscher, L., Nowitzki, S. and Oesterhelt, D. (1982) Eur. J. Biochem. 128, 223–230.
[34] Schäfer, T. and Schönheit, P. (1991) Arch. Microbiol. 155, 366–377.

[35] Zeikus, J.G., Fuchs, G., Kenealy, W. and Thauer, R.K. (1977) J. Bacteriol. 132, 604–613.
[36] Perham, R.N. (1991) Biochemistry 30, 8501–8512.
[37] Kerscher, L. and Oesterhelt, D. (1982) Trends Biochem. Sci. 7, 371–374.
[38] Danson, M.J., Eisenthal, R., Hall, S., Kessell, S.R. and Williams, D.L. (1984) Biochem. J. 218, 811–818.
[39] Danson, M.J., McQuattie, A. and Stevenson, K.J. (1986) Biochemistry 25, 3880–3884.
[40] Noll, K.M. and Barber, T.S. (1988) J. Bacteriol. 170, 4315–4321.
[41] Pratt, K.J., Carles, C., Carne, T.J., Danson, M.J. and Stevenson, K.J. (1989) Biochem. J. 258, 749–754.
[42] Danson, M.J. (1988) Biochem. Soc. Trans. 16, 87–89.
[43] Danson, M.J., Hough, D.W., Vettakkorumakankav, N. and Stevenson, K.J. (1991) In: General and Applied Aspects of Halophilic Microorganisms, NATO ASI Series A (Rodriguez-Valera, F., ed.), Vol. 201, pp. 121–128, Plenum Publishing Corp., New York.
[44] Vettakkorumakankav, N. and Stevenson, K.J. (1992) Biochem. Cell. Biol. 70, 656–663.
[45] Aitken, D.M. and Brown, A.D. (1969) Biochim. Biophys. Acta 177, 351–354.
[46] Danson, M.J., Black, S.C., Woodland, D.L. and Wood, P.A. (1985) FEBS Lett. 179, 120–124.
[47] Segerer, A., Langworthy, T.A. and Stetter, K.O. (1988) System. Appl. Microbiol. 10, 161–171.
[48] Stetter, K.O. (1986) In: Thermophiles, General and Applied Microbiology (Brock, T.D., ed.), pp. 39–74, Wiley, New York.
[49] Kelly, R.M. and Demming, J.W. (1988) Biotech. Prog. 4, 47–62.
[50] Schäfer, S., Barkowski, C. and Fuchs, G. (1986) Arch. Microbiol. 146, 301–308.
[51] Schäfer, S., Paalme, T., Vilu, R. and Fuchs, G. (1989) Arch. Microbiol. 186, 695–700.
[52] Kandler, O. and Stetter, K.O. (1981) Zentralbl. Bakteriol. Hyg. Abt. 1 Orig. C2, 111–121.
[53] Fuchs, G. and Stupperich, E. (1982) In: Archaebacteria (Kandler, O. ed.), pp. 277–288, Gustav Fischer, Stuttgart.
[54] Ekiel, I., Smith, I.C.P. and Sprott, G.D. (1983) J. Bacteriol. 156, 316–326.
[55] Ekiel, I., Jarrell, K.F. and Sprott, G.D. (1985) Eur. J. Biochem. 149, 437–444.
[56] Daniels, L. and Zeikus, J.G. (1978) J. Bacteriol. 136, 75–84.
[57] Weimer, P.J. and Zeikus, J.G. (1979) J. Bacteriol. 137, 332–339.
[58] Thauer, R.K., Moller-Zinkhan, D. and Spormann, A.M. (1989) Ann. Rev. Microbiol. 43, 43–67.
[59] Moller-Zinkhan, D. and Thauer, R.K. (1990) Arch. Microbiol. 153, 215–218.
[60] Bonete, M.J., Camacho, M.L. and Cadenas, E. (1990) Biochim. Biophys. Acta 1041, 305–310.
[61] Consalvi, V., Chiaraluce, R., Politi, L., Gambacorta, A., De Rosa, M. and Scandurra, R. (1991) Eur. J. Biochem. 196, 459–467.
[62] Jones, W.J., Nagle, D.P. and Whitman, W.B. (1987) Microbiol. Rev. 51, 135–177.
[63] Kates, M. and Moldoveanu, N. (1991) In: General and Applied Aspects of Halophilic Microorganisms, NATO ASI Series A (Rodriguez-Valera, F., ed.), Vol. 201, pp. 191–198, Plenum, New York.
[64] De Rosa, M., Trincone, A., Nicolaus, B. and Gambacorta, A. (1991) In: Life Under Extreme Conditions (di Prisco, G., ed.) pp. 61–87, Springer, Berlin.
[65] De Rosa, M., Gambacorta, A. and Gliozzi, A. (1986) Microbiol. Rev. 50, 70–80.
[66] Woese, C.R., Kandler, O. and Wheelis, M.L. (1990) Proc. Natl. Acad. Sci. U.S.A. 87, 4576–4579.
[67] Inui, H., Ono, K., Miyatake, K., Natano, Y. and Kitaoka, S. (1987) J. Biol. Chem. 262, 9130–9135.
[68] Danson, M.J., Conroy, K., McQuattie, A. and Stevenson, K.J. (1987) Biochem. J. 243, 661–666.
[69] Jackman, S.A., Hough, D.W., Danson, M.J., Opperdoes, F.R. and Stevenson, K.J. (1990) Eur. J. Biochem. 193, 91–95.
[70] Gest, H. (1987) Biochem. Soc. Symp. 54, 3–16.

[71] Weitzman, P.D.J. (1985) In: Evolution of Prokaryotes (Schleifer, K.H. and Stackenbrandt, E.), pp. 253–275, Academic Press, London.
[72] Woese, C.R. (1987) Microbiol. Rev. 51, 221–271.
[73] Auer, J., Spicker, G. and Bock, A. (1990) System. Appl. Microbiol. 13, 354–360.
[74] Fabry, S. and Hensel, R. (1988) Gene 64, 189–197.
[75] Fabry, S., Niermann, T., Vingron, M. and Hensel, R. (1989) Eur. J. Biochem. 179, 405–413.
[76] Zwickl, P., Fabry, S., Bogedain, C., Haas, A. and Hensel,R. (1990) J. Bacteriol. 172, 4329–4338.
[77] Biro, J., Fabry, S., Dietmaier, W., Bogedain, C. and Hensel, R. (1990) FEBS Lett. 275, 130–134.
[78] Honka, E., Fabry, S., Niermann, T., Palm, P. and Hensel, R. (1990) Eur. J. Biochem. 189, 623–632.
[79] Fabry, S., Dietmaier, W. and Hensel, R. (1990) Gene 91, 19–25.
[80] Zaccai, G., Cendrin, F., Haik, Y., Borochov, N. and Eisenberg, H. (1989) J. Mol. Biol. 208, 491–500.
[81] Eisenberg, H., Mevarech, M. and Zaccai, G. (1992) Adv. Prot. Chem. 43, 1–62.
[82] Harel, M., Shoham, M., Frolow, F., Eisenberg, H., Mevarech, M., Yonath, A. and Sussman, J.L. (1988) J. Mol. Biol. 200, 609–610.
[83] Stezowski, J.J., Englmaier, R., Galdiga, C., Hartl, T., Rommel, I., Dauter, Z., Görisch, H., Grossebüter, W., Wilson, D. and Musil, D. (1989) J. Mol. Biol. 208, 507–508.
[84] Bright, J.R. (1991) Ph.D. Thesis, University of Bath, UK.
[85] Sutherland, K.J., Henneke, C.M., Towner, P., Hough, D.W. and Danson, M.J. (1990) Eur. J. Biochem. 194, 839–844.
[86] Weitzman, P.D.J. and Danson, M.J. (1976) Curr. Top. Cell. Regul. 10, 161–204.
[87] Weitzman, P.D.J. (1981) Adv. Micro. Physiol. 22, 185–244.
[88] Fukaya, M., Takemura, H., Okumura, H., Kawamura, Y., Horinouchi, S. and Beppu, T. (1990) J. Bacteriol. 172, 2096–2104.
[89] Remington, S., Wiegand, G. and Huber, R. (1982) J. Mol. Biol. 158, 111–152.
[90] Wiegand, G., Remington, S., Diesenhofer, J. and Huber, R. (1984) J. Mol. Biol. 174, 205–219.
[91] Henneke, C.M., Danson, M.J., Hough, D.W. and Osguthorpe, D.J. (1989) Protein Eng. 2, 597–604.
[92] Else, A.J., Danson, M.J. and Weitzman, P.D.J. (1988) Biochem. J. 251, 803–807.
[93] Anderson, D.H. and Duckworth, H.W. (1988) J. Biol. Chem. 263, 2163–2169.
[94] Handford, P.A., Ner, S.S., Bloxham, D.P. and Wilton, D.C. (1988) Biochim. Biophys. Acta 953, 232–240.
[95] Alter, G.M., Casazza, J.P., Zhi, W., Nemeth, P., Srere, P.A. and Evans, C.T. (1990) Biochemistry 29, 7557–7563.
[96] Smith, L.D., Stevenson, K.J., Hough, D.W. and Danson, M.J. (1987) FEBS Lett. 225, 277–281.
[97] Lohlein-Werhahn, G., Goepfert, P. and Eggerer, H. (1988) Biol. Chem. Hoppe-Seyler 369, 109–113.
[98] Grossebüter, W. and Görisch, H. (1985) System. Appl. Microbiol. 6, 119–124.
[99] Sutherland, K.J., Danson, M.J., Hough, D.W. and Towner, P. (1991) FEBS Lett. 282, 132–134.
[100] Smith, L.D., Bungard, S.J., Danson, M.J. and Hough, D.W. (1987) Biochem. Soc. Trans. 15, 1097.
[101] Ekiel, I., Sprott, D.G., and Smith, I.C.P. (1986) J. Bacteriol. 166, 559–564.
[102] Muckund, S. and Adams, M.W.W. (1991) J. Biol. Chem. 266, 14208–14216.
[103] Bright, J.R., Mackness, R., Danson, M.J., Hough, D.W., Taylor, G.L., Towner, P. and Byrom, D. (1991) J. Mol. Biol. 222, 143–144.
[104] Altekar, W. and Rangaswamy, V. (1990) FEMS Microbiol. Lett. 69, 139–144.
[105] Altekar, W. and Rangaswamy, V. (1991) FEMS Microbiol. Lett. 83, 241–246.
[106] James, K.D., Bonete, M.J., Byrom, D., Danson, M.J. and Hough, D.W. (1991) Biochem. Soc. Trans. 20, 12S.

[107] Schäfer, Th. and Schönheit, P. (1992) Arch. Microbiol. 158, 188–202.
[108] Schäfer, Th., Selig, M. and Schönheit, P. (1992) Arch. Microbiol. 159, 72–83.
[109] Plaga, W., Lottspeich, F. and Oesterhelt, D. (1992) Eur. J. Biochem. 205, 391–397.
[110] Schäfer, Th. and Schönheit, P. (1993) Arch. Microbiol. 159, 354–363.

CHAPTER 2

Bioenergetics of extreme halophiles

Vladimir P. SKULACHEV

Department of Bioenergetics, A.N. Belozersky Institute of Physico-Chemical Biology, Moscow State University, Moscow 119899, Russia

1. Introduction

In this chapter, we shall consider a general pattern of energy transduction processes in halophilic archaea. We shall deal above all with membrane-linked phenomena, clearly predominant among bioenergetic events in the extreme halophiles.

Special attention will be paid to the halobacterial light-driven H^+ pump, bacteriorhodopsin, that has proved to be the most prominent and thoroughly studied bioenergetic device found in microorganisms of this type. Bacteriorhodopsin, discovered in 1971 by Oesterhelt and Stoeckenius[1], still remains the simplest and most stable membrane-linked energy transducer. This is why bacteriorhodopsin studies are always very instructive for bioenergeticists investigating other, more complicated ion pumps. The structural and functional relationships in bacteriorhodopsin as well as in halorhodopsin, the light-driven Cl^- pump, will also be discussed in Chapter 6 of this volume, whereas reviews on halobacterial lipids and membrane enzyme proteins other than rhodopsins can be found in Chapters 9 and 10, respectively. As to the halobacterial photoreceptor proteins, i.e. sensory rhodopsins I and II, the reader is referred to Chapter 5 of this volume. Bacteriorhodopsin has proved also to be involved in photoreception but in an indirect way, namely, as the photogenerator of electrochemical H^+ potential difference ($\Delta\bar{\mu}_{H^+}$) which, in some way, is monitored by the cell. Therefore the problem of energy level monitoring will be one of the subjects of this review.

2. A general scheme of energy transduction in extreme halophiles

From the bioenergetic point of view, *Halobacterium halobium* has been studied much more extensively than other extremely halophilic archaea: it is in this species that bacteriorhodopsin was first found. (Later it has also been observed in closely related species: *H. cutirubrum* and *H. salinarium*[2].) In *H. halobium*, three types of energy-supplying processes have been identified, namely, (i) respiration, (ii) light-dependent ion pumping and (iii) arginine fermentation (Fig. 1).

There is some ground to assume that the respiratory chain was the primary energy-producing mechanism of halobacteria, whereas bacteriorhodopsin appeared in the course

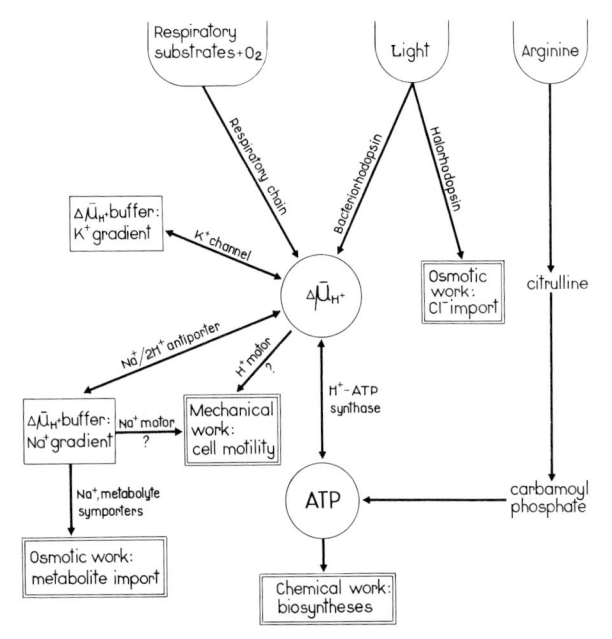

Fig. 1. Energy transduction in *Halobacterium halobium*. Membrane-linked energy transducers are indicated alongside the arrows.

of adaptation to low oxygen tension which is an inevitable consequence of high salinity (O_2 concentration in saturated NaCl solution is about five-fold lower than in seawater).

In *H. halobium*, both the respiratory chain [3] and bacteriorhodopsin [4–6] are shown to operate as $\Delta\bar{\mu}_{H^+}$ generators, extruding H^+ ions from the cell at the expense of the oxidation and light energies, respectively. It is not clear whether halobacteria possess the $\Delta\bar{\mu}_{Na^+}$ motive respiration discovered recently in many marine bacteria (for reviews, see refs. [7,8]).

Both bacteriorhodopsin and the H^+-motive respiratory chain expel H^+ ions from the cell and charge the cell interior negatively. This results in an electrophoretic leakage of intracellular Cl^-. Such a process cannot be avoided because of a very high Cl^- concentration. This is why halobacteria have special systems to return Cl^- to the cytoplasm. These systems are (i) halorhodopsin [9], which was identified by Schobert and Lanyi as a light-driven Cl^- pump [10], and (ii) a Cl^-, nH^+-symporter (or Cl^-/nOH^- antiporter) utilizing $\Delta\bar{\mu}_{H^+}$ [11]. It may be mentioned that halophilic eubacteria do not return the expelled Cl^- to the cell because of the toxicity of high concentrations of this anion [12]. Apparently archaea are more tolerant to high Cl^-. Halorhodopsin, like bacteriorhodopsin, contains retinal. Halorhodopsin pumps Cl^- into the cell whereas bacteriorhodopsin pumps H^+ from the cell so that both proteins generate an electric potential difference ($\Delta\psi$) of one and the same direction. The amount of halorhodopsin in the *H. halobium* cell is much smaller than that of bacteriorhodopsin which is responsible for utilization of the main portion of the light energy available to the halobacterial cell.

$\Delta\bar{\mu}_{H^+}$ formed by bacteriorhodopsin or the respiratory chain is used to perform the cell's main chemical work: ATP synthesis by H^+-ATP-synthase. Moreover, the $\Delta\bar{\mu}_{H^+}$ energy can be invested into K^+ and Na^+ gradients which serve as $\Delta\bar{\mu}_{H^+}$ buffers utilized to form $\Delta\bar{\mu}_{H^+}$ when external energy sources are exhausted. The Na^+ gradient is also used to perform osmotic work, i.e., the uphill uptake of metabolites by means of Na^+,metabolite-symporters. Halobacteria can also perform mechanical work (motility).It is not clear yet whether H^+- or Na^+-motors rotate the halobacterial flagella.

Fermentation of arginine via citrulline and carbamoyl phosphate represents an additional mechanism of energy conservation resulting in ATP synthesis. This substrate-level phosphorylation does not require membrane-bound enzyme systems.

3. Bacteriorhodopsin and halorhodopsin

Bacteriorhodopsin and halorhodopsin are unique energy transducers characteristic of halobacteria and absent in other living cells. The structural and functional aspects of these retinal proteins will be considered in Chapter 6 of this volume. Here we would like to review only two aspects of the problem, namely, the mechanism of the light-induced transmembrane charge displacement by bacteriorhodopsin, and the involvement of bacteriorhodopsin in photoreception, mediated by a $\Delta\bar{\mu}_{H^+}$ sensor.

3.1. Transmembrane charge displacement

Bacteriorhodopsin translocates one H^+ ion per photon which causes the all-*trans* → 13-*cis* photoisomerization of chromophore, retinal. At least two H^+-acceptor groups are shown to be directly involved in the H^+ transfer by bacteriorhodopsin, namely, (a) the Schiff base forming a link between the retinal and the ε-amino group of Lys-216, and (b) the Asp-96 carboxylic group. The involvement of the Schiff base is confirmed by many independent pieces of evidence (e.g., the electrogenic H^+ transfer disappears at a pH below 3.5, i.e., below the pK value of the Schiff base in the M-intermediate of bacteriorhodopsin photocycle; reviewed in ref. [7]). As to Asp-96, its participation in the H^+ transfer relay was recently demonstrated by site-directed mutagenesis studies [13–19].

Three partial electrogenic reactions have been found to contribute to the total electrogenesis directly measured in the proteoliposome–collodion film system:
(i) H^+ transfer from the protonated Schiff base to the exterior.
(ii) Reprotonation of the Schiff base by H^+ belonging to the Asp-96 carboxylic group.
(iii) H^+ transfer from the cytoplasm to the Asp-96 carboxyl [18] (see also [17]).

The exact location of the Schiff base and the Asp-96 carboxyl is yet to be found since the three-dimensional structure of bacteriorhodopsin is not known to atomic resolution. Electron microscopy of two-dimensional bacteriorhodopsin crystals indicates that the Schiff base is localized in the middle of the protein molecule, while Asp-96 is somewhere between the Schiff base and the cytoplasmic surface of the membrane [21]. It has also been shown that the protein regions separating the Schiff base from the outer and cytoplasmic membrane surfaces differ strongly in hydrophobicity, which is low in the former, and high in the latter case. Between the outer membrane surface and the Schiff base, there are four charged amino acids and no valine, leucine and isoleucine. At the same time, between the Schiff base and the cytoplasm, five leucines, valine and only one charged amino acid (Asp-96) seem to be localized [21]. Thus the dielectric

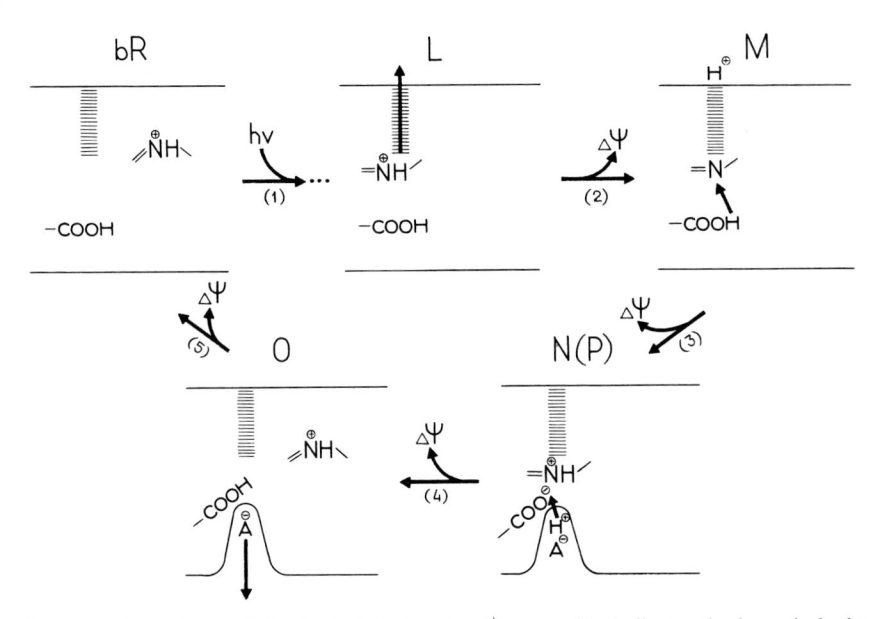

Fig. 2. A tentative scheme of the bacteriorhodopsin H^+ pump. bR indicates the bacteriorhodopsin ground state, and L, M, N(P) and O indicate the corresponding intermediates of the photocycle. $\overset{+}{\text{--NH}}\text{\textbackslash}$, =N′ and $=\overset{+}{\text{NH}}\text{′}$ represent the protonated Schiff base of the all-*trans* retinal residue, the deprotonated and the protonated Schiff bases of 13-*cis* retinal residues, respectively. –COOH and –COO$^-$ are the protonated and the deprotonated Asp-96 carboxylic group, respectively. The outward hydrophilic H^+-conducting pathway (the proton well) is shaded. (From Skulachev [35].)

constant in regions of the outward H^+-conducting pathway is high and that of the inward pathway is low. This may explain why the contribution of the outward H^+ pathway is much lower (20%) than that of the inward one (80%) [22,23].

The outward and inward H^+ pathways have been shown to operate in μs and ms time scales, respectively, so that their contributions to the total $\Delta\psi$ formation can be easily measured. On the other hand, the rate of H^+ transfer from the Asp-96 carboxylic group to the Schiff base is of the same order of magnitude as the rate of reprotonation of this carboxylate by cytoplasmic H^+ ions. To measure the $\Delta\psi$ contributions of these two steps separately, one may specifically decelerate the Asp-96 carboxylate reprotonation by decreasing the H^+ concentration in the medium. Under such conditions, both steps seem to make an almost equal contribution to energy conservation [20].

A tentative scheme of the bacteriorhodopsin H^+ pump is shown in Fig. 2. The following chain of events is assumed.

(i) bR → L. The all-*trans* → 13-*cis* photoisomerization of the retinal residue results in the translocation of the protonated Schiff base from a more hydrophobic to a less hydrophobic environment inside the protein. This causes a strong decline in the pK of the Schiff base, from > 10 to 3.5. In the L state, the Schiff base is supposed to be close to the bottom of the outward H^+ well.

(ii) L → M. The proton leaves the Schiff base, is translocated along the outward H^+ pathway, and appears in the exterior. When moving in this direction, the proton encounters rather low resistance since hydrophilic amino acid residues dominate in this region of the protein.

(iii) M → N. The protein changes its conformation in such a way that (a) the Asp-96 carboxylic group moves in the direction of the Schiff base in order to reprotonate it, and (b) a cleft is formed in the hydrophobic region below the carboxylic group, allowing the cytoplasmic solutes to reach this group. As a result, the N560 intermediate (also known as P, R) is formed. In this intermediate, (a) the Schiff base is protonated, (b) the Asp-96 carboxylate is deprotonated and (c) the retinal is still in the 13-*cis* conformation.

(iv) N → O. Asp-96 carboxylate is reprotonated by the cytoplasmic H^+, and retinal 13-*cis* → all-*trans* isomerization takes place.

(v) O → bR. The cleft disappears and the Asp-96 carboxylic group returns to its starting position. The former event is electrogenic due to an increase in the distance between the protonated Schiff base and a cytoplasmic counter-ion (under natural conditions, Cl^-).

The great dielectric asymmetry of the two half-membrane charge transfer pathways in bacteriorhodopsin is in contrast with the situation in the photosynthetic reaction centre complex. In the latter case, both halves of the membraneous protein are very hydrophobic and equally contribute to $\Delta\psi$ formation [7,24].

This scheme is in agreement with the findings that (i) the N intermediate decomposition (i.e. Asp-96 reprotonation) depends upon the bulk phase pH, whereas the M decomposition does not, and (ii) in Asp-96 → Asn mutant, the M decomposition becomes pH-dependent [13,14,16,17].

An essential feature of the above scheme is that there is practically no inward H^+-conducting pathway in the "resting" bacteriorhodopsin molecule. This pathway is organized only for a short period of time in the "working" protein at N and O stages. In these intermediates, no direct contact of the outward and the inward H^+ pathways is assumed. Such a mechanism excludes a passive H^+ leak via bacteriorhodopsin, a property that seems to be important for this protein which can occupy up to 50% of the membrane in the halobacterial cell.

The scheme shown in Fig. 2 differs from that suggested by Henderson et al. [21] who assumed that two H^+-conducting half-channels, being separated by the protonated Schiff base, pre-exist in the "resting" bacteriorhodopsin. The latter scheme fails to explain how H^+ is translocated through the hydrophobic part of the bacteriorhodopsin molecule separating the Schiff base from the cytoplasmic water phase, a process which is assisted by only one protolytic group, i.e., Asp-96 (the inward H^+ pathway). If this pathway has very low conductivity for H^+, it will not be clear how it can satisfy the high rate of bacteriorhodopsin turnover. If it has high H^+ conductivity, it will not be clear how electrostatic breakdown of the membrane is prevented, since two H^+ wells, connected by a proton-acceptor group, i.e. the Schiff base, oppose each other.

On the other hand, our scheme suggests that the inward H^+ pathway is absent from the bacteriorhodopsin ground state and is organized on a temporary basis when the N intermediate is formed. As a matter of fact, the formation and decomposition of N were found to be accompanied by large light-scattering changes indicating conformational transitions of the protein [20,25]. Moreover, these two stages are characterized by high activation energy, a fact consistent with the assumption that a conformational change occurs in this very part of the photocycle [26].

The idea that a cleft is formed in the hydrophobic region of the protein stems from Mitchell's suggestion about a mobile membrane barrier, meant to explain the mechanism of operation of carriers that translocate hydrophilic metabolites [27,28]. Quite recently Mitchell [29] has presented a concrete version of the mobile membrane barrier principle. He assumes that a slight relative shift of α-helical transmembrane columns (e.g., rotation or tilting of the columns at a small angle) may result in the formation of a cleft allowing solutes to reach the membrane core [1].

The scheme in Fig. 2 may also explain the mechanism of Cl^- pumping by halorhodopsin. Apparently, in this case Arg-200 plays a role similar to that of Asp-96 in bacteriorhodopsin. Some other arginines may also take part in Cl^- transport. It is remarkable that Asp-96 is absent from halorhodopsin (see Chapter 6 of this volume).

It should be noted in this context that (i) there are no additional arginines in the intramembranous part of halorhodopsin in comparison with bacteriorhodopsin [30] and (ii) bacteriorhodopsin is competent in Cl^- binding [31–33] and perhaps even in Cl^- transfer when $[H^+]$ is low and $[Cl^-]$ is high [34].

The specificity of halorhodopsin for the transported anion is not absolute. It can pump not only Cl^- but also NO_3^-. However, the rate of transport of Cl^- is much higher than that of NO_3^-.

Halorhodopsin has also been found in another halophilic microorganism, *Natronobacterium pharaonis*. Here it transports NO_3^- as fast as Cl^- [31].

3.2. Involvement in photoreception

When bacteriorhodopsin was discovered in 1971, its similarity to the visual rhodopsin in the chromophore structure (retinal), the primary light-induced event (retinal isomerization) and the photocycle (short-wavelength intermediate formation) impelled the authors to suggest that this novel retinal protein is somehow involved in photoreception [1]. Such a suggestion seemed reasonable since halobacteria are known to change their direction of swimming in response to a light stimulus.

However, subsequent studies on halobacterial retinal proteins seemed to argue against the photoreceptor function of bacteriorhodopsin. It was found that (i) bacteriorhodopsin operates as a light-driven H^+ pump and (ii) in the same halobacteria there are two other retinal proteins, i.e., sensory rhodopsin I and sensory rhodopsin II (also known as photorhodopsin) which are specialized in photoreception rather than in H^+ pumping, being present in much smaller amounts than bacteriorhodopsin [36–43]. It was shown that

---

[1] When this chapter had already been sent to the publishers, J.E. Draheim, N.J. Gibson and J.Y. Cassim reported (Biophys. J. 60 (1991) 89–100) that in the native purple sheets, the α-helix axes of bacteriorhodopsin are all oriented perpendicular to the membrane whereas the M formation is accompanied by tilting of α-helical columns so that the angle between the membrane normal and the α-helix axes proved to be ~10°. Light bleaching in the presence of hydroxylamine as well as vacuum dehydration of glycerol-treated membranes or their glucose embedding were shown to increase the angle up to ~22–24°. The authors concluded that "the fact that an electron imaging and diffraction analysis of the purple membranes failed to detect any significant structural differences due to the induction of the M_{412} state from the ground bR_{568} state is not surprising in view of the experimental conditions used which include one if not more of the tilt-causing perturbations. The effects of the M_{412} transformation were probably masked by the stronger effects of these perturbations".

the attractant effect of orange and red lights can be mediated by sensory rhodopsin I, whereas the repellent action of light of shorter wavelengths is explained by the involvement of sensory rhodopsin II and/or the 370 nm intermediate of the sensory rhodopsin I photocycle. The inevitable impression was that there is no room for bacteriorhodopsin, nor for halorhodopsin, in the halobacterial photoreceptor system, and one could explain the opposite effects of long- and short-wavelength light on the behaviour of halobacteria by taking into account only two of four retinal proteins, i.e., sensory rhodopsins I and II. This conclusion seemed to be supported by the fact of a very high sensitivity of sensory rhodopsins which, like visual rhodopsin, operate as photon counters (a single photon is sufficient for inducing a behavioural response). However, such maximal sensitivity, while excellent under dim light, is rather inconvenient under bright light.

In 1981 we discovered a specific feature of the response of *Halobacterium halobium* to intensity decreases of bright light [44]. It was shown that cyanide and dicyclohexylcarbodiimide (DCCD) strongly sensitize bacteria to a lowering of the intensity of bright (500–650 nm) light. At the same time, the sensitivity to blue light was not affected by these agents [44]. Later it was found [45] that, under the conditions used, a decrease in light intensity resulted in a lower electric potential difference ($\Delta\Psi$) across the bacterial membrane. The magnitude of this effect was greatly increased by cyanide and DCCD. Sensitization of both the repellent effect and the $\Delta\Psi$ level to decreasing light intensity was also shown to be caused by substituting K^+ for Na^+ in the incubation mixture. Addition of oxidation substrates resulted in some desensitization of both systems.

To explain all these relationships, Glagolev suggested [44–47] that bacteria possess a special device to monitor $\Delta\bar{\mu}_{H^+}$. This $\Delta\bar{\mu}_{H^+}$ receptor, called "protometer", is assumed to produce attractant or repellent signals when $\Delta\bar{\mu}_{H^+}$ increases or decreases, respectively. In light, bacteriorhodopsin generates a $\Delta\bar{\mu}_{H^+}$, an effect which is sensed by the cell as an attractant stimulus. A decrease in light intensity lowers $\Delta\bar{\mu}_{H^+}$ and, hence, causes the repellent effect. The latter is especially strong when light is the only energy source. Apparently, this is the case when respiration and H^+-ATPase are inhibited by cyanide and DCCD, respectively. One more effect of DCCD is directed at the $Na^+/2H^+$ antiporter which is involved in the $\Delta\bar{\mu}_{H^+}$ buffering. Such buffering is absent when extracellular Na^+ is replaced by K^+.

The above reasoning was recently confirmed by several pieces of indirect evidence. It was found that in *H. halobium* (i) protonophorous uncouplers decrease $\Delta\bar{\mu}_{H^+}$ and cause a repellent effect [45], (ii) cyanide and DCCD have no effect on the photoresponse of a mutant which possesses sensory rhodopsins but no bacteriorhodopsin [48], and (iii) in a similar mutant, $\Delta\Psi$ is not involved in photosensing [49].

One could not exclude, however, that $\Delta\bar{\mu}_{H^+}$ somehow modulates the processing of a signal produced by sensory rhodopsin I [49]. If such were the case, a mutant possessing bacteriorhodopsin as the only retinal protein would be as "blind" as those lacking all the retinal proteins. On the other hand, if a protometer does exist after all, some photosensing should be inherent in this mutant. In the latter case, the following features can be predicted. (i) Only the attractant effect of the light must be present; blue light can be, to some degree, attractive like orange and red light because of the broad absorption maximum of bacteriorhodopsin. (ii) Cyanide must sensitize the photosensing. (iii) Low light intensities, sufficient for photoreception by sensory rhodopsins, must be

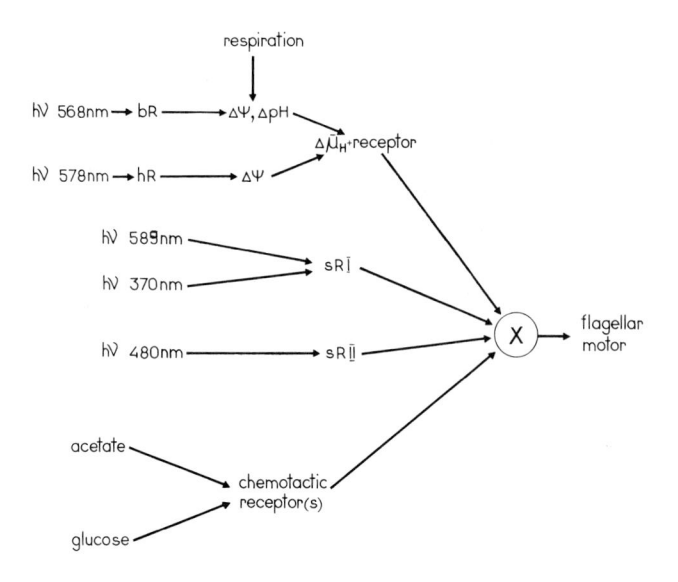

Fig. 3. A taxis system of halobacteria. bR, bacteriorhodopsin; hR, halorhodopsin; sRI, sRII, sensory rhodopsins I and II; X, a flagellar motor-affecting component. (From Bibikov et al. [50].) It is assumed that at low light intensities, the attractant effect of the light is mediated by sRI which operates as a photon counter. At high light intensities when the sRI system is already saturated, bR produces $\Delta\bar{\mu}_{H^+}$ which is monitored by the $\Delta\bar{\mu}_{H^+}$ receptor ('protometer') generating the attractant signal when $\Delta\bar{\mu}_{H^+}$ increases and the repellent signal when it decreases. bR does not participate in the formation of the repellent signal of blue light, which is mediated by sRII and by a short-wavelength intermediate of the sRI photocycle.

ineffective, for in order to increase $\Delta\bar{\mu}_{H^+}$, illumination should be sufficiently strong to support large-scale H$^+$ pumping by bacteriorhodopsin. It is these relationships that have been revealed quite recently in the study performed by our and Oesterhelt's groups [50]. The bacteriorhodopsin gene was introduced to a "blind" H. halobium mutant lacking bacteriorhodopsin, halorhodopsin and sensory rhodopsins I and II. The resulting transformant was shown to acquire the ability to sense light, but in "black-and-white" and not "colour vision", i.e., both orange and blue light caused the attractant effect at high light intensities. Cyanide strongly sensitized the cells to the light stimulus. Thus it was directly proved that bacteriorhodopsin does not require other retinal proteins to be involved in photoreception. It does not mean, however, that bacteriorhodopsin per se is bifunctional ($\Delta\bar{\mu}_{H^+}$ generator and photoreceptor). Rather, in light, bacteriorhodopsin produces $\Delta\bar{\mu}_{H^+}$ which is monitored by the $\Delta\bar{\mu}_{H^+}$ receptor sending an attractant signal when $\Delta\bar{\mu}_{H^+}$ increases (Fig. 3).

4. Respiratory chain

Apparently, there is nothing special about the halobacterial H$^+$ motive respiratory chain except that the enzymes require a high salt concentration to be active. The chain includes flavoproteins, quinone and cytochromes of types a, b, c and o (for references, see [30]). Spectral studies on whole cells and cell-free extracts showed that *H. halobium* [51–53], *H. cutirubrum* [54] and *H. salinarium* [52,55] contain aa$_3$- and o-type oxidases. An attempt to isolate cytochrome aa$_3$ from *H. halobium* [56] resulted in a single two-heme 40 kDa polypeptide containing no Cu. One of the hemes could be reduced by ascorbate and TMPD, whereas the other heme required dithionite for reduction. The isolated enzyme failed to transfer electrons from the reduced cytochrome c or TMPD to oxygen. The membrane fraction from which the cytochrome was purified was found (i) to catalyze such an electron transfer and (ii) to contain Cu in amounts equal to hemes a. The loss of activity occurred immediately on addition of detergents used to fractionate the cytochrome. This resulted, most probably, in the dissociation of Cu or of the Cu-binding subunit.

It is the high sensitivity to detergents as well as the high salt requirement that pose considerable difficulties for isolation of intact respiratory $\Delta\bar{\mu}_{H^+}$ generators and their incorporation into liposomes. This is why respiration-linked $\Delta\bar{\mu}_{H^+}$ generation has been shown only in intact cells [3]. The number and localization of the energy coupling sites in the respiratory chain remain obscure.

5. Arginine fermentation

When neither light nor oxygen are available, *H. halobium* can still grow provided the growth medium is supplemented with substrate amounts of arginine. Under such conditions, ATP is formed from carbamoyl phosphate produced from arginine, with citrulline as an intermediate [57]. ATP, produced by this substrate-level phosphorylation, can be used to support energy-dependent processes in the cytosol or, alternatively, to generate $\Delta\bar{\mu}_{H^+}$ by reversed H$^+$-ATP-synthase (reviewed in [30]).

6. H$^+$-ATP-synthase

H$^+$-ATP-synthase was postulated in view of the evidence that illumination [58,59], respiration [60] or artificially imposed ΔpH of the right direction (inside alkaline) [61] can increase the ATP level in halobacterial cells. This process was shown to be sensitive to N-ethylmaleimide, nitrate, N,N'-dicyclohexylcarbodiimide (DCCD) and 7-chloro-4-nitrobenzo-2-oxa-1,3-diazole (NBD-Cl), and resistant to vanadate and azide. The same inhibitor pattern was found to be inherent in 350 kDa ATPase isolated from *H. halobium* [63] and *H. saccharovorum* [64,65].

The enzyme proved to be composed of four types of subunits (87, 60, 29 and 20 kDa). The subunit ratio was assumed to be equal to 3:3:1:1 [66]. The sequences of major subunits showed significant (more than 50%) homology with the detachable sector of eukaryotic vacuolar (V-type) H$^+$-ATPase. On the other hand, homology with bacterial F$_1$ ATPase proved to be less than 30%. Since the above-mentioned inhibitor pattern is, in fact, identical with those of the vacuolar ATPase, one might assume that halobacterial ATPase corresponds to the V-type. However, the sequence of proteolipid (c subunit) of the membrane sector of H^+-ATP-synthase from another archaea species, *Sulfolobus,* is more related to that of the F_1-type rather than that of the V-type [67].

The membrane sector of halobacterial H^+-ATP-synthase has not yet been studied. As to its detachable sector, 50–65% homology with other archaea (*Sulfolobus* and methanobacteria) has been revealed [66].

Summarizing these observations, Mukohata [68a] concluded that archaea contain a special kind of H^+-ATP-synthase, the so-called A-type, which resembles the V-type and much less the F-type but is not identical with these eukaryotic and bacterial enzymes. On the other hand, Schobert [68b] insists that the regulation pattern of H^+-ATPase of archaea is similar to F_1-ATPase.

It is obvious that the main physiological function of the A-type ATP-synthase is ATP formation rather than hydrolysis, since it should be responsible for the light- and respiration-dependent ATP supply. As to the ATPase activity, it is strongly suppressed in the membrane-linked enzyme because of the inhibition by bound ADP, a feature resembling the F_1-ATPase [30,68b]. Nevertheless, at least under certain in vivo conditions, the enzyme can operate as ATPase, for the discharge of $\Delta\bar{\mu}_{H^+}$ by an uncoupler results in an ATP decrease in cells growing on arginine anaerobically in the dark [30].

7. Formation of K^+/Na^+ gradients. $\Delta\bar{\mu}_{H^+}$ buffering

When bacteriorhodopsin or the respiratory chain pump H^+ from the halobacterial cell, $\Delta\Psi$ is formed and large-scale H^+ efflux ceases. Storage of a large portion of energy appears to be possible if the flux of a penetrating ion, say K^+, discharges $\Delta\Psi$ and converts it to ΔpH. Strong alkalinization of the cytoplasm, which must accompany this effect, can be prevented by exchange of an intracellular cation, say Na^+, for extracellular H^+ [69]. The latter will be equivalent to an increase in the cytoplasmic pH buffer capacity [70]. Assuming that the K^+ import and the Na^+/H^+ antiport are reversible, one may regard the Na^+/H^+ gradients as $\Delta\bar{\mu}_{H^+}$ buffers, ΔpK and ΔpNa being buffers of $\Delta\Psi$ and ΔpH, respectively [69]. A more complicated system may also be involved in the $\Delta\bar{\mu}_{H^+}$ buffering. K^+ import may occur via K^+, H^+ symporter (as seems to be the case in *E. coli* [71]), and the Na^+/H^+ antiporter can substitute one Na^+ ion per two or more H^+ ions. In these instances, (i) the K^+ gradient can stabilize both $\Delta\Psi$ and ΔpH, the effect, however, being stronger for the former, and (ii) the Na^+ gradient buffers not only ΔpH but also (to a lesser degree) $\Delta\Psi$ [7].

In *H. halobium,* an electrophoretic K^+ influx was demonstrated in intact cells [72,73] and cell envelopes [74] (in envelopes, the rate of transport was, however, rather low). It is not clear whether the K^+ influx represents the K^+ uniport or the K^+, H^+ symport.

As to the Na^+ efflux, it is, according to Lanyi and Silverman [75], catalyzed by the $Na^+/2H^+$ antiporter. The antiport could be inhibited by DCCD [76,77]. Konishi and Murakami [78] solubilized the halobacterial membrane by octyl glucoside, and used the mixture obtained, supplemented with phospholipids and bacteriorhodopsin, to reconstitute proteoliposomes which were found to be competent in the DCCD-sensitive Na^+/H^+ antiport. When the proteoliposomes were incubated with [^{14}C]DCCD, radio-labels were detachable on the 50 kDa, but mainly on the 11 kDa component.

Not only ΔpH^+ but also $\Delta\Psi$ were shown to drive the Na^+ export from the *H. halobium* cell. The point is that one Na^+ is substituted by two (or more) H^+. The dependence of the Na^+ efflux rate on the $\Delta\Psi$ level seems to indicate a "gating" effect of $\Delta\Psi$. The rate is low below 100 mV, whereas it increases dramatically above [77].

The $\Delta\bar{\mu}_{H^+}$ decrease results in the reversal of the K^+ and Na^+ fluxes, i.e., K^+ goes out while Na^+ goes in. Assuming that this effect is mediated by the same mechanisms which are involved under high-$\Delta\bar{\mu}_{H^+}$ conditions, one may predict that the dissipation of Na^+/K^+ gradients will stabilize, to some degree, the $\Delta\bar{\mu}_{H^+}$ level and $\Delta\bar{\mu}_{H^+}$-linked functions ($\Delta\bar{\mu}_{H^+}$ buffering). This effect was really observed in our experiments on *H. halobium* cells [79,80] incubated anaerobically in the dark. Two incubation media were tested: (i) high-K^+ (2700 mM KCl^- and 1570 mM NaCl) and (ii) high-Na^+ (27 mM KCl^- and 4270 mM NaCl). It was found that the $\Delta\Psi$ level, [ATP] and the cell motility rate decreased much more in the high-[K^+] than in the high-[Na^+] medium. At high [K^+]$_{out}$ the cells became motionless within 20 min of anaerobic incubation in the dark. At high [Na^+]$_{out}$ they were still motile, although at a lowered rate, at least for 8 h. The motility inhibition in the high-[K^+] medium immediately disappeared in the presence of light or O_2.

As shown in Oesterhelt's laboratory, gradients of monovalent cations stabilize the ATP level in *H. halobium* when other energy sources are absent [81,82]. These authors compared the capacities of the K^+ gradient and ATP as energy pools and found that the former is about 250 times larger than the latter (for discussion, see [69]).

The Na^+ gradient may serve as an additional energy pool. Experimentally this was shown to be the case in *E. coli* and *V. harveyi* where three media were compared, namely KCl, NaCl and Tris-HCl [80,83]. Unfortunately, a Tris-HCl medium cannot be used for halobacteria since the Tris concentration must be as high as 4 M.

At any rate, is seems obvious that halobacteria expend a large portion of energy to expel Na^+. This is hardly due to Na^+ "toxicity" since, under certain conditions, halobacteria can still survive at [Na^+]$_{in}$ as high as 2 M [84]. An attractive possibility is that halobacteria, like many other bacteria, utilize the Na^+ gradient as a $\Delta\bar{\mu}_{H^+}$ buffer [7,69]. For extreme halophiles, this function seems especially effective. Here the amount of energy which can be invested into this gradient should be especially large because of the very high Na^+ concentration outside the cell. In microorganisms such as *H. halobium*, the capacity of Na^+/K^+ gradients is so high that cells can store the light energy available in the daytime and expend it during the night, at least for such a function as motility [7,79].

8. Na^+, metabolite symports

The performance of a specific kind of osmotic work, namely, the uphill import of metabolites, represents yet another function of $\Delta\bar{\mu}_{Na^+}$.

As shown by Lanyi and coworkers [74,85–89], all amino acids, except cysteine, are accumulated by *H. halobium* cells by means of Na^+, amino acid-symporters, with the Na^+:amino acid stoichiometry varying from 1 to 2. In some cases, the symport proved to be electroneutral, being driven by the sodium gradient (ΔpNa). Sometimes it was electrophoretic so that $\Delta\Psi$ was also involved. For instance, $2Na^+$ were shown to be symported with 1 serine when $\Delta\bar{\mu}_{Na^+}$ was in the form of ΔpNa. If it was in the form of $\Delta\Psi$, the 1:1 stoichiometry was observed.

The process of amino acid accumulation is catalyzed by at least five different Na^+, amino acid-symporters transporting (i) asparagine and glutamine, (ii) arginine, lysine and histidine, (iii) alanine, glycine, serine and threonine, (iv) valine, leucine, isoleucine and methionine and (v) phenylalanine, tyrosine and tryptophane. One more symporter seems to be involved in the uptake of glutamate and aspartate. Cysteine is not transported and inhibits the transport of other amino acids (reviewed in [30]).

Plakunov and his colleagues have studied the import of sugars by various halobacteria [90–92b]. They found that *H. saccharovorum*, *H. mediterranei* and *H. volcanii* possess a Na^+, glucose symporter whereas *H. halobium*, *H. salinarium* and *H. morrhuae* lack such a mechanism. The symporter was found to be induced when *H. saccharovorum* grew in glucose-containing media [92b].

Zoratti and Lanyi [93] reported that the inorganic phosphate uptake by *H. halobium* cells is driven by ATP, not by $\Delta\bar{\mu}_{Na^+}$.

$\Delta\bar{\mu}_{H^+}$ often serves as a driving force for the transport of metabolites by bacteria living in fresh water, soil and intestine. On the other hand, $\Delta\bar{\mu}_{H^+}$ was not previously reported to drive metabolite transport in extreme halophiles.

So far none of the halobacterial porters have been isolated and reconstituted into proteoliposomes.

9. A flagellar motor

H. halobium cells swim by using their flagella [94]. During logarithmic and stationary phases of growth, the cells are mono- and bipolarly flagellated. Flagella are organized in bundles consisting of several individual filaments composed of three different proteins [94,95]. Halobacterial flagella were shown to have a right-handed helix in contrast to the left-handed helix found in most bacteria [94]. Experiments on rotation of tethered cells showed that flagella are powered by a rotatory motor [94].

Studies on energetics of *H. halobium* motility were carried out mainly in the early 1980s [44,45,79]. It was assumed then that a H^+ motor, i.e. a device which rotates a flagellum by means of the downhill H^+ influx from exterior to cytosol, represents the only mechanism of motility of the flagellar bacteria. Later it was shown, however, that in some bacteria, influx of Na^+ rather than H^+, is used as a driving force for the rotation of flagella [96–99]. Inhibitor analysis of *H. halobium* motility, supported by the energy of light or respiration, indicates that $\Delta\bar{\mu}_{H^+}$ and/or $\Delta\bar{\mu}_{Na^+}$, not ATP, power the flagellar motor, since motility is DCCD-resistant. At the same time, it is inhibited by protonophores [44,45]. The latter fact seems to be in favour of the H^+ motor. However, the inhibition might be a secondary effect, with the Na^+ motor being operative. The inhibition may well be due to a collapse of $\Delta\bar{\mu}_{H^+}$, which inevitably entails a decrease in $\Delta\bar{\mu}_{Na^+}$ formed by the $\Delta\bar{\mu}_{H^+}$-driven $Na^+/2H^+$ antiporter. It should be stressed that it is the Na^+ motor that rotates flagella in marine and soil alkalotolerant and alkalophilic bacteria utilizing, like halobacteria, $\Delta\bar{\mu}_{Na^+}$ to import all the amino acids and sugars [96,97,100–103]. In both alkalo- and halophiles, the extracellular $[Na^+]/[H^+]$ ratio is high, due to low

[H$^+$] and high [Na$^+$], respectively. Further investigation seems to be necessary to decide which of the two motors is employed by halobacteria.

10. Some prospects for future research

It is thus clear that the nature of the driving force which powers the flagellar motor of halobacteria, remains a problem to be solved. To discriminate between H$^+$- and Na$^+$-motors, one should study motility in a Na$^+$-free medium. If it is the Na$^+$ motor that is inherent in halobacteria, they must be motionless without added Na$^+$, and a pulse of Na$^+$ must activate the motility even in the presence of a protonophore.

Na$^+$-motive respiration is still another moot question of halobacterial energetics. In marine alkalotolerant bacteria, Na$^+$-motive NADH-quinone reductase [104] and Na$^+$-motive terminal oxidase [105,106] have been discovered. Taking into account the above-mentioned considerations relevant to the high [Na$^+$]$_{out}$/[H$^+$]$_{out}$ ratio (section 9), one may hope that Na$^+$ respiration may be present in halobacteria as well. If this is so, one may predict that respiration can support the ΔpNa-driven accumulation of amino acids in the presence of a protonophore and DCCD when bacteriorhodopsin and H$^+$-ATPase fail to do this. At any rate, the presence of Na$^+$-motive respiration seems especially probable in alkalophilic halobacteria (e.g. *Natronobacterium*). In these species, the existence of a Na$^+$-driven ATP-synthase [107] likewise seems to be quite possible. This might be a special enzyme or, alternatively, the same ATP-synthase which catalyzes $\Delta\bar{\mu}_{H^+}$-driven ATP formation. In the latter case, one should assume that, at very low [H$^+$]$_{out}$, Na$^+$ effectively competes with H$^+$ for the gate in the ion channel of the synthase [35]. Competition between Na$^+$ and H$^+$ for the F$_0$ channel has already been shown by Laubinger and Dimroth in *Propionigenium modestum* ATP-synthase [108].

Perhaps it is the Na$^+$ cycle that is responsible for the effect observed by Michel and Oesterhelt [109] (see also [110]) who reported that the protonophorous uncoupler abolishes $\Delta\bar{\mu}_{H^+}$ at much lower concentrations than those affecting the phosphate potential.

Almost nothing is known about the DCCD-sensitive membrane sector of the halobacterial H$^+$, ATP-synthase.

The $\Delta\bar{\mu}_{H^+}$ receptor (protometer) is another intriguing problem of halobacteria. This hypothetical device, which apparently mediates the transmission of light signals from bacteriorhodopsin to the flagellar motor (section 3.2), may also be involved in the adaptation of halobacteria to low-$\Delta\bar{\mu}_{H^+}$ conditions (e.g. to an alkaline medium). Research into the nature of the $\Delta\bar{\mu}_{H^+}$ receptor seems to hold good promise.

Among all the enzymes involved in bioenergetics of halobacteria, only retinal-containing proteins have been isolated in a pure native form and reconstituted into proteoliposomes. However, even for bacteriorhodopsin, the best-studied halobacterial protein, the mechanism of action, i.e. H$^+$ pumping, remains hypothetical. Here, structural studies of on the level of atomic resolution are badly needed.

References

[1] Oesterhelt, D. and Stoeckenius, W. (1971) Nature 233, 149–152.

[2] Kushwaha, S.C., Kates, M. and Stoeckenius, W. (1976) Biochim. Biophys. Acta 426, 703–710.
[3] Bogomolni, R.A., Baker, R.A., Lozier, R.H. and Stoeckenius, W. (1976) Biochim. Biophys. Acta 440, 68–88.
[4] Oesterhelt, D. and Stoeckenius, W. (1973) Proc. Natl. Acad. Sci. U.S.A. 70, 2853–2857.
[5] Kayushin, L.P. and Skulachev, V.P. (1974) FEBS Lett. 39, 39–42.
[6] Drachev, L.A., Kaulen, A.D., Ostroumov, S.A. and Skulachev, V.P. (1974) FEBS Lett. 39, 43–45.
[7] Skulachev, V.P. (1988) Membrane Bioenergetics. Springer, Berlin.
[8] Skulachev, V.P. (1989) J. Bioenerg. Biomembr. 21, 635–647.
[9] Matsuno-Yagi, A. and Mukohata, Y. (1977) Biochem. Biophys. Res. Commun. 78, 237–243.
[10] Schobert, B. and Lanyi, J.K. (1982) J. Biol. Chem. 257, 10306–10313.
[11] Duschl, A. and Wagner, G. (1986) J. Bacteriol. 168, 548–552.
[12] Choquet, C. and Kushner, D.J. (1990) J. Bacteriol. 172, 3462–3468.
[13] Holz, N., Drachev, L.A., Mogi, T., Otto, H., Kaulen, A.D., Heyn, M.P., Skulachev, V.P. and Khorana, H.G. (1989) Proc. Natl. Acad. Sci. U.S.A. 86, 2167–2171.
[14] Drachev, L.A., Kaulen, A.D., Khorana, H.G., Mogi, T., Otto, H., Skulachev, V.P., Heyn, M.P. and Holz, M. (1989) Biokhimiya 54, 1467–1477. In Russian.
[15] Butt, H.J., Fendler, K., Bamberg, E., Tittor, J. and Oesterhelt, D. (1989) EMBO J. 4, 1657–1663.
[16] Gerwert, K., Hess, B., Soppa, J. and Oesterhelt, D. (1989) Proc. Natl. Acad. Sci. U.S.A. 86, 4943–4947.
[17] Otto, H., Marti, T., Holz, M., Mogi, T., Lindau, M., Khorana, H.G. and Heyn, M.P. (1989) Proc. Natl. Acad. Sci. U.S.A. 86, 9228–9232.
[18] Tittor, J., Soell, C., Oesterhelt, D., Butt, H.-J. and Bamberg, E. (1990) EMBO J. 8, 3477–3482.
[19] Engelgard, M., Hess, B., Metz, G., Krentz, N., Siebert, F., Soppa, J. and Oesterhelt, D. (1990) Eur. Biophys. J. 18, 17–24.
[20] Danshina, S.V., Drachev, L.A., Kaulen, A.D. and Skulachev, V.P. (1992) Photochem. Photobiol. 55, 735–740.
[21] Henderson, R., Baldwin, J.M., Ceska, T.A., Zemlin, F., Beckmann, F. and Downing, K.H. (1990) J. Mol. Biol. 213, 899–929.
[22] Drachev, L.A., Kaulen, A.D. and Skulachev, V.P. (1978) FEBS Lett. 87, 161–167.
[23] Drachev, L.A., Kaulen, A.D. and Skulachev, V.P. (1984) FEBS Lett. 178, 331–335.
[24] Dracheva, S.M., Drachev, L.A., Konstantinov, A.A., Semenov, A.Yu., Skulachev, V.P., Arutjunjan, A.M., Shuvalov, V.A. and Zaberezhnaya, S. (1988) Eur. J. Biochem. 171, 253–264.
[25] Drachev, L.A., Kaulen, A.D. and Zorina, V.V. (1989) FEBS Lett. 243, 5–7.
[26] Skulachev, V.P. (1993) Q. Rev. Biophys., accepted for publication.
[27] Mitchell, P. (1957) Nature 180, 134–136.
[28] Mitchell, P. (1959) Biochem. Soc. Symposia 16, 73–93.
[29] Mitchell, P. (1991) In: Annales de l'Institute Pasteur (Saier, M.H., ed.), in press.
[30] Lanyi, J.K. (1991) In: Microbiology of Extreme and Unusual Environments, Vol. 1, Halophiles (Vreeland, R. and Hochstein, L.I., Eds.), Telford Press-CRC/Times Mirror Publishing Co, in press.
[31] Fischer, U. and Oesterhelt, D. (1979) Biophys. J. 28, 211–230.
[32] Drachev, A.L., Drachev, L.A., Kaulen, A.D., Khitrina, L.V., Skulachev, V.P., Lepnev, G.P. and Chekulaeva, L.N. (1989) Biochim. Biophys. Acta 976, 190–195.
[33] Varo, G. and Lanyi, J.K. (1989) Biophys. J. 56, 1143–1151.
[34] Der, A., Toth-Boconadi, R. and Keszthelyi, L. (1989) FEBS Lett. 259, 24–26.
[35] Skulachev, V.P. (1991) Biosci. Rep. 11, 387–444.
[36] Bogomolni, R.A. and Spudich, J.L. (1982) Proc. Natl. Acad. Sci. U.S.A. 79, 6250–6254.
[37] Tsuda, M., Nelson, B., Chang, C.-H., Govindjee, R. and Ebrey, T.G. (1985) Biophys. J. 47, 721–724.

[38] Spudich, J.L. and Bogomolni, R.A. (1984) Nature (London) 312, 509–513.
[39] Sundberg, S.A., Bogomolni, R.A. and Spudich, J.L. (1985) J. Bacteriol. 164, 282–287.
[40] Takahashi, T., Tomioka, H., Kamo, N. and Kobatake, Y. (1985) FEMS Microbiol. Lett. 28, 161–164.
[41] Takahashi, T., Watanabe, M., Kamo, N. and Kobatake Y. (1985) Biophys. J. 48, 235–240.
[42] Wolff, E.H., Bogomolni, R.A., Scherrer, P., Hess, B. and Stoeckenius, W. (1986) Proc. Natl. Acad. Sci. U.S.A. 83, 7272–7276.
[43] Marwan, W. and Oesterhelt, D. (1987) J. Mol. Biol. 195, 333–342.
[44] Baryshev, V.A., Glagolev, A.N. and Skulachev, V.P. (1981) Nature (London) 292, 338–340.
[45] Baryshev, V.A., Glagolev, A.N. and Skulachev, V.P. (1983) J. Gen. Microbiol. 129, 367–373.
[46] Glagolev, A.N. (1980) J. Theor. Biol. 82, 171–185.
[47] Glagolev, A.N. (1984) Trends Biochem. Sci. 9, 397–400.
[48] Bibikov, S.I. and Skulachev, V.P. (1987) FEBS Lett. 243, 303–306.
[49] Oesterhelt, D. and Marwan, W. (1987) J. Bacteriol. 169, 3515–3520.
[50] Bibikov, S.I., Grishanin, R.N., Marwan, W., Oesterhelt, D. and Skulachev, V.P. (1991) FEBS Lett. 295, 223–226.
[51] Lanyi, J.K. (1969) J. Biol. Chem. 244, 2864–2869.
[52] Cheah, K.S. (1970) Biochim. Biophys. Acta 205, 148–160.
[53] Fujiwara, T., Fukumori, Y. and Yamanaka, T. (1987) Plant Cell Physiol. 28, 29–36.
[54] Cheah, K.S. (1969) Biochim. Biophys. Acta 180, 320–333.
[55] Cheah, K.S. (1970) Biochim. Biophys. Acta 197, 84–86.
[56] Fujiwara, T., Fukumori, Y. and Yamanaka, T. (1989) J. Biochem. 105, 287–292.
[57] Hartmann, R., Sickinger, H.-D. and Oesterhelt, D. (1980) Proc. Natl. Acad. Sci. U.S.A. 77, 3821–3825.
[58] Danon, A. and Stoeckenius, W. (1973) Proc. Natl. Acad. Sci. U.S.A. 71, 1234–1238.
[59] Belyakova, T.N., Kadzyauskas, Yu.P., Skulachev, V.P., Smirnova, T.A., Chekulaeva, L.N. and Jasaitis, A.A. (1975) Dokl. Akad. Nauk USSR 223, 483–486. In Russian.
[60] Kushner, D.J. (1978) In: Microbiological Life in Extreme Environments (Kushner, D.J., Ed.), pp. 317–368, Academic Press, London.
[61] Danon, A. and Caplan, S.R. (1976) Biochim. Biophys. Acta 423, 133–140.
[62] Mukohata, Y. and Yoshida, M. (1987) J. Biochem. 101, 311–318.
[63] Nanba, T. and Mukohata, Y. (1987) J. Biochem. 102, 591–598.
[64] Hochstein, L.I., Kristjansson, H. and Altekar, W. (1987) Biochem. Biophys. Res. Commun. 147, 295–300.
[65] Stan-Lotter, H., Bowman, E.J. and Hochstein, L.I. (1991) Arch. Biochim. Biophys. 284, 116–119.
[66] Mukohata, Y., Ihara, K., Yoshida, M. and Sugiyama, Y. (1991) In: New Era of Bioenergetics (Mukohata, Y., Ed.), pp. 169–196, Academic Press, Tokyo.
[67] Denda, K., Konishi, J., Oshima, T., Date, T. and Yoshida, M. (1989) J. Biol. Chem. 264, 7119–7121.
[68a] Mukohata, Y. and Yoshida, M. (1987) J. Biochem. 102, 797–802.
[68b] Schobert, B. (1991) J. Biol. Chem. 266, 8008–8014.
[69] Skulachev, V.P. (1978) FEBS Lett. 87, 171–179.
[70] Mitchell, P. (1968) Chemiosmotic coupling and energy transduction. Glynn Research, Bodmin.
[71] Bakker, E.P. (1980) J. Supramol. Struct. Suppl. 4, 80.
[72] Garty, H. and Caplan, S.R. (1977) Biochim. Biophys. Acta 459, 532–545.
[73] Wagner, G., Hartmann, R. and Oesterhelt, D. (1978) Eur. J. Biochem. 89, 169–179.
[74] Lanyi, J.K., Helgerson, S.L. and Silverman, M.P. (1979) Arch. Biochim. Biophys. 193, 329–339.
[75] Lanyi, J.K. and Silverman, M.P. (1979) J. Biol. Chem. 254, 4750–4755.

[76] Murakami, N. and Konishi, T. (1985) J. Biochem. 98, 897–907.
[77] Murakami, N. and Konishi, T. (1988) J. Biochem. 103, 231–236.
[78] Konishi, T. and Murakami, N. (1990) Biochem. Biophys. Res. Commun. 170, 1339–1345.
[79] Arshavsky, V.Yu., Baryshev, V.A., Brown, I.I., Glagolev, A.N. and Skulachev, V.P. (1981) FEBS Lett. 133, 22–26.
[80] Brown, I.I., Galperin, M.Yu., Glagolev, A.N. and Skulachev, V.P. (1983) Eur. J. Biochem. 134, 345–349.
[81] Wagner, G. and Oesterhelt, D. (1976) Ber. Dtsch. Bot. Ges. 89, 289–292.
[82] Wagner, G., Hartmann, R. and Oesterhelt, D. (1978) Eur. J. Biochem. 89, 169–179.
[83] Brown, I.I. and Kim, Yu.V. (1982) Biokhimiya 47, 137–144. In Russian.
[84] Ginzburg, M., Sachs, L. and Ginzburg, B.Z. (1970) J. Gen. Physiol. 55, 187–207.
[85] MacDonald, R.E. and Lanyi, J.K. (1975) Biochemistry 14, 2882–2889.
[86] MacDonald, R.E., Greene, R.V. and Lanyi, J.K. (1977) Biochemistry 16, 3227–3235.
[87] Lanyi, J.K., Yearwood-Drayton, V. and MacDonald, R.E. (1976) Biochemistry 15, 1595–1603.
[88] Lanyi, J.K., Renthal, R. and MacDonald, R.E. (1976) Biochemistry 15, 1603–1610.
[89] Halgerson, S.L. and Lanyi, J.K. (1978) Biochemistry 17, 1042–1046.
[90] Severina, L.O., Pimenov, N.V. and Plakunov, V.K. (1986) Mikrobiologiya 55, 335–336. In Russian.
[91] Pimenov, N.V., Severina, L.O. and Plakunov, V.K. (1986) Mikrobiologiya 55, 362–367. In Russian.
[92a] Pimenov, N.V. (1987) The study on transport and metabolism of sugars in extremely halophilic archaebacteria, Thesis, Moscow. In Russian.
[92b] Severina, L.O., Pimenov, N.V. and Plakunov, V.K. (1991) Arch. Microbiol. 155, 131–136.
[93] Zoratti, M. and Lanyi, J.K. (1987) J. Bacteriol. 169, 5755–5760.
[94] Alam, M. and Oesterhelt, D. (1984) J. Mol. Biol. 176, 459–475.
[95] Alam, M. and Oesterhelt, D. (1987) J. Mol. Biol. 194, 495–499.
[96] Hirota, N., Kitada, M. and Imae, Y. (1981) FEBS Lett. 132, 278–280.
[97] Chernyak, B.V., Dibrov, P.A., Glagolev, A.N., Sherman, M.Yu. and Skulachev, V.P. (1983) FEBS Lett. 164, 38–42.
[98] Hirota, N. and Imae, Y. (1983) J. Biol. Chem. 258, 10577–10581.
[99] Bakeeva, L.E., Chumakov, K.M., Drachev, A.L., Metlina, A.L. and Skulachev, V.P. (1986) Biochim. Biophys. Acta 850, 466–472.
[100] Tokuda, H., Sugasawa, M. and Unemoto, T. (1982) J. Biol. Chem. 257, 788–794.
[101] Kakinuma, Y. and Unemoto, T. (1985) J. Bacteriol. 163, 1293–1295.
[102] Koyama, N., Kiyomiya, A. and Nosoh, Y. (1976) FEBS Lett. 72, 77–78.
[103] Krulwich, T.A., Guffanti, A.A., Bornstein, R.E. and Hoffstein, J. (1982) J. Biol. Chem. 257, 1885–1889.
[104] Tokuda, H. and Unemoto, T. (1982) J. Biol. Chem. 257, 10007–10014.
[105] Semeykina, A.L., Skulachev, V.P., Verkhovskaya, M.L., Bulygina, E.S. and Chumakov, K.M. (1989) Eur. J. Biochem. 183, 671–678.
[106] Kostyrko, V.A., Semeykina, A.L., Skulachev, V.P., Smirnova, I.A., Vaghina, M.L. and Verkhovskaya, M.L. (1991) Eur. J. Biochem. 198, 527–534.
[107] Dimroth, P. (1987) Microbiol. Rev. 51, 320–340.
[108] Laubinger, W. and Dimroth, P. (1989) Biochemistry 28, 7194–7198.
[109] Michel, H. and Oesterhelt, D. (1980) Biochemistry 19, 4615–4619.
[110] Helgerson, S.L., Requadt, C. and Stoeckenius, W. (1983) Biochemistry 22, 5746–5753.

CHAPTER 3

Biochemistry of methanogenesis

Lacy DANIELS

Department of Microbiology, University of Iowa, Iowa City, IA 52242, USA

1. Introduction

This review will mostly focus on research during the past 15 years; older material is covered in a variety of other reviews [1–5]. We discuss microbiological, biosynthetic, and bioenergetic aspects of methanogens only when useful in understanding the biochemistry. More information on the microbiology of methanogens is available in several recent reviews [6–11]. Detailed coverage of bioenergetic and biosynthetic aspects is found in other chapters of this book.

Methanogens are strict anaerobes, but vary in their oxygen sensitivity [12–14]. Specialized equipment, supplies, and skills needed to work with them are well-described [15–22]. Although many methanogens are thermophiles, most isolates are mesophiles; no psychrophiles have been reported, although methanogenesis occurs in nature at temperatures as low as 3°C (refs. [23,24,447], and Daniels, L. and Jung, K.Y., unpublished data). Methanogen coenzymes and enzymes have a wide range of oxygen and heat sensitivities, and do not necessarily follow patterns expected from properties of the microbes.

Most methanogens are capable of using a variety of one-carbon substrates for methane production and as major substrates for growth, and a few can use acetate for these purposes. Also, a few can use multiple-carbon alcohols as a source of electrons, but cannot use the carbon atoms for methane or growth [25–29]. In general, other-than-acetate compounds with carbon–carbon bonds are not broken down by methanogens, but are either oxidized in a limited fashion, or assimilated intact into biosynthetic pathways. Table 1 describes the substrates and key properties of a number of methanogens that have been studied from the viewpoint of their methanogenic pathway. This table is not comprehensive, but is intended to guide the reader as we discuss coenzymes, pathways, and enzymes studied in these diverse methanogens. One genus needing special comment is *"Methanothrix"*, methanogens that use only acetate as a methanogenic substrate; the name of this ecologically important genus is currently being contested, with Patel and colleagues proposing that the name be *Methanosaeta* [30,31]. We have chosen to place quotes around the more familiar original name.

Table 2 provides balanced equations for the use of methanogenic substrates. Values of free energy change for each reaction ($\Delta G^{o'}$) are in kJ per mole methane; a negative value indicates favorability, and conversion of one mole ATP to ADP+P_i releases energy equivalent to −32 kJ. However, in natural methanogenic environments, and in laboratory

TABLE 1
Key methanogens in the study of the biochemistry of methane production

Organism	Substrates[a]	Typical growth temperature (°C)	References
Methanobacterium thermoautotrophicum strains Marburg and ΔH	HC	60–65	34,357
Methanobacterium formicicum JF-1	HC, F	37	436
Methanococcus voltae	HC, F	30–37	36
Methanococcus thermolithotrophicus	HC, F	60–65	35
Methanococcus jannaschii	HC	80–85	437
Methanosphaera stadtmanii	HCM	37	98
Methanolobus tindarius	M	25	438
Methanosarcina barkeri	HC, M, A	37	439–441
Methanosarcina Göl	HC, M, A	37	442
Methanosarcina thermophila	HC, M, A	50	284,443
"*Methanothrix*" *soehngenii*	A	37	30,31,285,286
"*Methanothrix*" CALS-1	A	60	288,444

[a] Abbreviations: HC, H_2–CO_2; F, formate; M, methanol or methylamines; A, acetate; HCM, H_2–CO_2 plus methanol.

culture, substrate and product concentrations modify these values. A good thermodynamic calculation reference is the review by Thauer et al. [32].

Methanogens use methanogenic substrates for biosynthetic needs. Also, mixotrophic growth (e.g. with methanol + H_2–CO_2) is possible with some organisms. Table 1 and the reviews mentioned above provide key references on use of these substrates.

Microbiological features important for the study of methanogenesis include growth rates, ability to grow to a high density in fermentors, and ease of application of molecular biology techniques. For example, *Methanobacterium thermoautotrophicum* strains Marburg and ΔH grow moderately fast, and to high cell density (> 1 g dry cell weight per liter) in fermentors [17,33,34], making their use for coenzyme and enzyme purification attractive. If test tube or bottle growth is all that is needed, *Methanococcus thermolithotrophicus* or *Methanococcus voltae* are attractive due to their rapid growth and in some cases their accessibility to molecular biology techniques [35,36]; however, their growth in fermentors has not been uniformly successful, and growth is not to a high cell density (<0.5 g dry cell weight per liter) (ref. [37], and Mukhopadhyay, B. and Daniels, L., unpublished observations), and thus are not good subjects for coenzyme production. *Methanosarcina* cells grow slowly, but to high density (> 10 g dry cell weight per liter) [21]; however, they are more likely to become contaminated.

Fig. 1 provides a detailed overview of the methanogenic pathways. Below we discuss the coenzymes and enzymes involved, and summarize how the paths operate as a unit.

TABLE 2
Reactions for methanogenesis by methanogenic bacteria

Reaction	$\Delta G^{o\prime}$ [a] (kJ/mol CH$_4$)	Reaction number [b]
$4H_2(g) + CO_2(g) \rightarrow CH_4(g) + 2H_2O$	−130.8	(1)
$4H_2(g) + HCO_3^- + H^+ \rightarrow CH_4(g) + 3H_2O$	−135.6	(1A)
$4HCOO^-(aq) + 4H^+ \rightarrow CH_4(g) + 3CO_2(g) + 2H_2O$	−144.5	(2)
$4CH_3CHOHCH_3(aq) + CO_2(g) \rightarrow CH_4(g) + 4CH_3COCH_3(aq) + 2H_2O$	−31.7	(3)
$2CH_3CH_2OH(aq) + CO_2(g) \rightarrow CH_4(g) + 2CH_3COO^-(aq) + 2H^+$	−111.5	(4)
$4CH_3OH(aq) \rightarrow 3CH_4(g) + CO_2(g) + 2H_2O$	−106.5	(5)
$4CH_3OH(aq) \rightarrow 3CH_4(g) + HCO_3^- + H^+ + H_2O$	−104.9	(5A)
$4CH_3NH_3^+(aq) + 2H_2O \rightarrow 3CH_4(g) + CO_2(g) + 4NH_4^+(aq)$	−76.6	(6)
$H_2(g) + CH_3OH(aq) \rightarrow CH_4(g) + H_2O$	−112.5	(7)
$CH_3COO^-(aq) + H^+ \rightarrow CH_4(g) + CO_2(g)$	−35.8	(8)
$CH_3COO^-(aq) + H_2O \rightarrow CH_4(g) + HCO_3^-$	−31.0	(8A)

[a] $\Delta G^{o\prime}$, standard free energy change at pH 7.0, calculated from ref. [184].
[b] These reaction numbers have been referred to in the text.

2. Novel coenzymes

2.1. General

Many unusual features of methanogens arise from the fact that they are members of the archaea (archaebacteria). It is now clear that their metabolism, structure, and biochemistry are in some cases very different from bacteria (typical bacteria or eubacteria); at the same time there are similarities where least expected, emphasizing that generalities cannot be made about the novelty of biochemical features in methanogens. However, they use an array of coenzymes in their methanogenic pathway that are sometimes not found in other archaea, and that are either without any similarity to coenzymes in non-archaea, or that are structural analogs to coenzymes found in the bacteria and eucarya. The reader should consult several recent reviews of methanogen coenzymes for greater detail [2,4,5,38].

Unusual coenzymes in methanogens do not replace many commonly used coenzymes found in bacteria, since most of the traditional water-soluble vitamins are found in methanogens, including biotin, riboflavin, pantothenic acid, nicotinic acid, pyridoxine, and lipoic acid, although their levels are sometimes low [39,40]. Folic acid may be absent; using a variety of bioassay methods, Worrell and Nagle [41] concluded "cell extracts of methanogens ... contained little or no folic acid (pteroylglutamate) or pteroylpolyglutamate activity (<0.1 nmol/g dry weight)", which corresponds to <0.0002 nmol per mg protein. As shown in Table 3, this is far below the normal range of ∼1–100 nmoles coenzyme per mg protein.

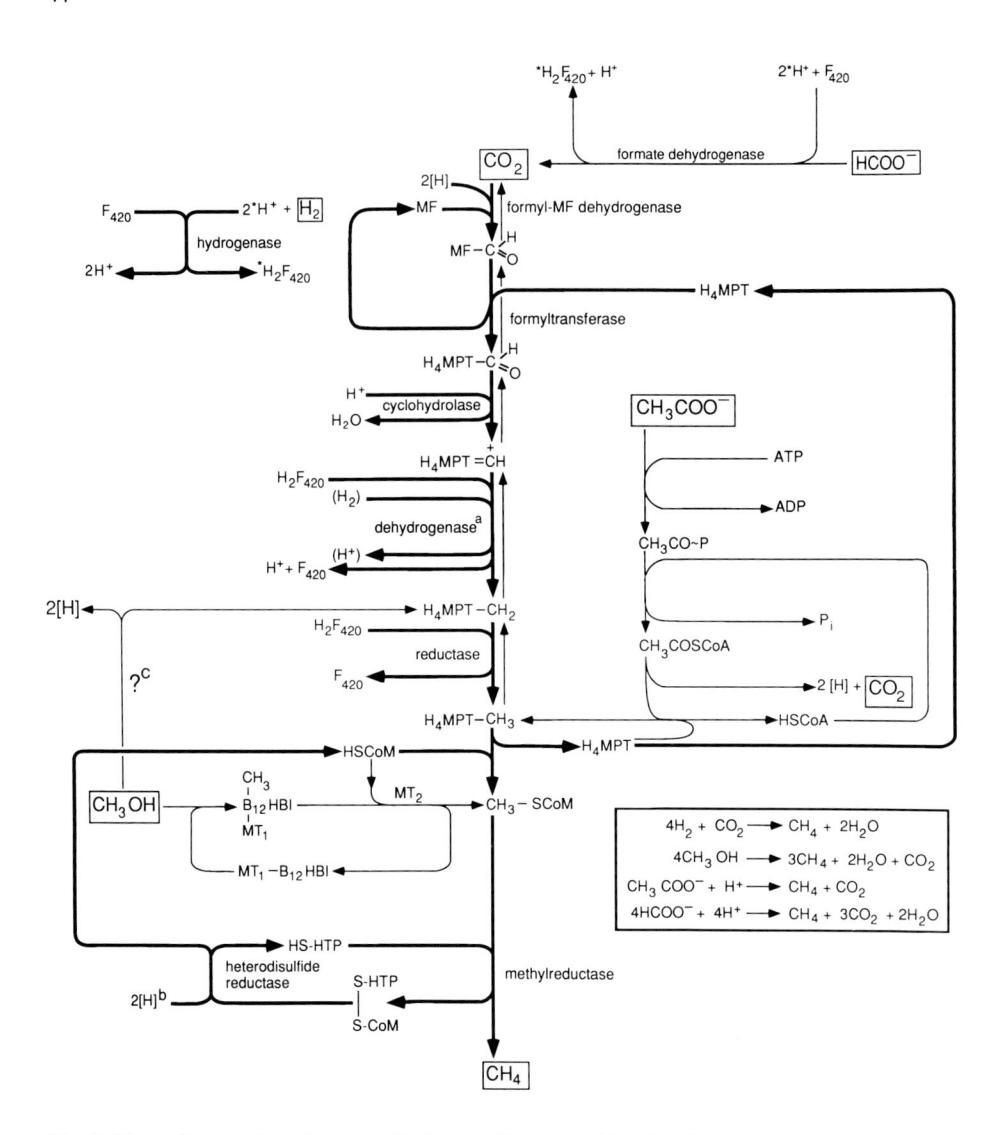

Fig. 1. The pathways of methanogenesis. Intermediates are abbreviated as in the text. The thick lines indicate the pathway for H_2–CO_2 methanogenesis, which is also in common to some extent with methanogenesis from one or more other substrates. The thin lines indicate specialized portions of pathways of methanogenesis from methanol, formate, and acetate. (a) Two different dehydrogenases have been reported, one dependent on H_2F_{420}, and one dependent on H_2. (b) The source of these electrons may be H_2F_{420} in some cases, but in other cases it is unknown: see the text. (c) This is a possible alternative for methanol oxidation to the methylene–H_4MPT level; see the text for details.

TABLE 3
Coenzyme content of methanogenic bacteria

Coenzyme	Mol.wt.[a]	Commonly found levels (nmol/mg protein)	Stability
F_{420}	774	0.2–3.4	Air- and heat-stable; somewhat acid- and base-sensitive; reduced form air-stable for hours in absence of flavins
Methanofuran	749	1–6	Stable to air and heat
Methanopterin	772	100–200	Oxidized form stable in air for short periods; reduced form very air-sensitive
CoM	142	3–12	Sensitive to air (forms disulfide that is air-stable); heat-, acid- and base-stable
F_{430}	905[b]	1–4	Sensitive to air and heat, especially both at same time, but can be used in air with care
HSHTP	343	~1[c]	Sensitive to air (forms disulfide that is air-stable); heat-stable

[a] The originally described protonated and anhydrous form is given; alternate structures and salts, and differing degrees of hydration will vary in molecular weight. See the text for references.
[b] The molecular weight of the penta-acid form that is thought to be present in vivo is given. The originally described methanolysis product has a molecular weight of 975.
[c] Based on mg of pure material obtained per kg of cell wet weight, assuming 25% recovery [178].

2.2. The 5-deazaflavin, F_{420}

Methanogenic bacteria use the 5-deazaflavin F_{420} (the N-(N-L-lactyl-γ-L-glutamyl)-L-glutamic acid phosphodiester of 7,8-didemethyl-8-hydroxy-5-deazariboflavin 5'-phosphate), originally discovered by Cheeseman et al. [42], and shown in Fig. 2, as a major electron-transferring coenzyme. The structure was determined first in *Methanobacterium thermoautotrophicum* strain ΔH by Eirich et al. [43,44], but in methanogens several forms exist, differing only in the number of glutamates; in *Methanosarcina*, three glutamates (or more) are observed [45,46]. F_{420} is involved in several reactions in the methanogenic pathway: hydrogenase [47], formate dehydrogenase [48,49], methylene tetrahydromethanopterin (H_4MPT) dehydrogenase [50–52], methylene H_4MPT reductase [53–55], and heterodisulfide reductase involved in the terminal step of the methanogenic pathway, possibly via the H_2F_{420}-dehydrogenase [56]. (See Fig. 1.) It is likely to play a role in biosynthetic pathways, directly as well as via NADP-F_{420}-oxidoreductase [57,58]. It has a role in DNA photorepair, where thymine dimers are cut by an F_{420}-containing enzyme in *Streptomyces* [59], and methanogens [60]. Table 3 describes the range of F_{420} levels in methanogens. Levels vary during growth, and with the substrate.

Although the structure of F_{420} was first determined from a methanogen [43], it (or Fo) is present in all *Streptomyces* species examined, and in a variety of related genera, e.g. *Nocardia* and *Mycobacterium* [59,61–64]; it is also present in one genus of eucarya (and is probably in several [65,66]), and in some other archaea [46,67–70].

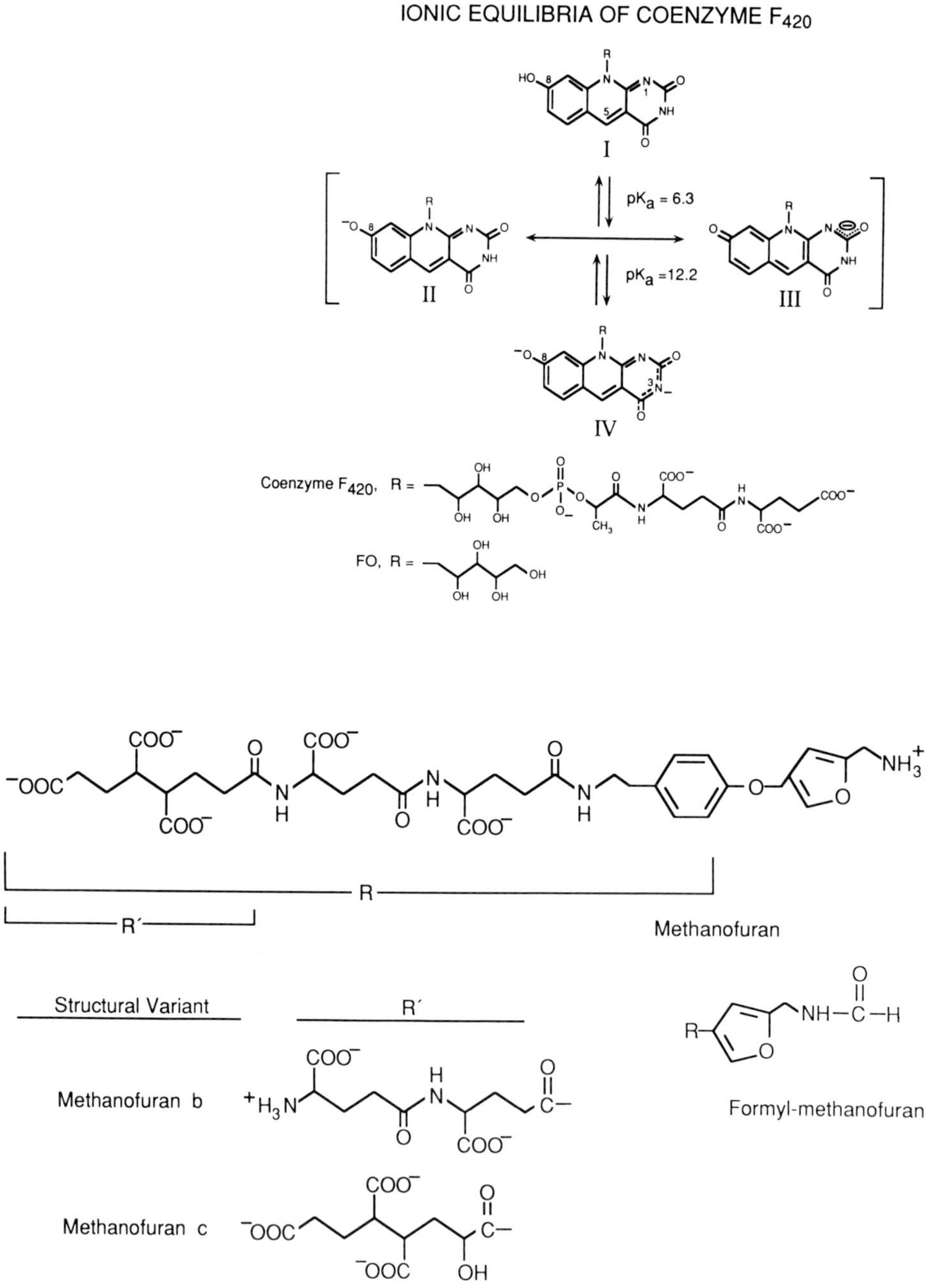

Fig. 2. Structures of F_{420} and methanofuran.

Fo is excreted by methanogens into their media [71], and is catalytically active with all known F_{420}-dependent methanogen enzymes [38,44,72]. Some *Streptomyces* work has dealt with 5-deazaflavins excreted into the medium, as Fo, but probably F_{420} is the normal coenzyme [62]. Compared to methanogens, ca. 10-fold lower levels (or less) are found in *Streptomyces,* and in *Archaeoglobus fulgidus* levels are similar to methanogens. The role of this 5-deazaflavin in most *Streptomyces* and all the *Mycobacterium* species is not known, but in specific *Streptomyces* it is essential for synthesis of chlortetracycline, oxytetracycline, and lincolnomycin [63,73–75], and is involved in DNA light repair [59,76].

Oxidized F_{420} is an obligate two-electron acceptor (under physiological non-photoreduction conditions). The oxidized form is bright yellow, with an absorbance maximum at 401–420 nm; as the pH drops below 7, the yellow color fades as the anion indicated in Fig. 2 is protonated, with a pKa at pH 6.3 for the hydroxyl proton at the 8 position [38,44]. F_{420} reduction is monitored by decrease in absorbance at 401 or 420 nm. The extinction coefficient of F_{420} at 420 nm changes as a function of temperature, as a result of the temperature effect on the pKa of the 8-OH– group [77]. This difference is significant in enzyme assays monitored at 420 nm and at pH $<$ 7, when the reaction cuvette is heated to thermophilic temperatures; appropriate corrections can be made.

Although F_{420} structurally resembles flavins, its chemistry is more like a nicotinamide [38,72]. Properties that make it similar to nicotinamides include: ability to only accept two electrons (under physiological conditions); redox potential of -350 mV; high stability of the reduced form to air; rapid reaction of the reduced form with oxidized flavin; structural and chemical similarity of the central pyridine ring of F_{420} to the pyridine ring in NAD(P). Nonetheless, 5-deazaflavins are useful biochemical tools to study flavoprotein catalysis [38,78]. F_{420} (oxidized) is stable to air, and to boiling at near neutral pH, but is degraded by light at high pH, and the side chain is cleaved by exposure to low pH [42–44,79]. It is unknown whether, except in the photolyase, F_{420} is ever enzyme-bound. So far, it appears to act as a soluble electron-transfer cofactor, but it has not been investigated fully for its ability to bind proteins.

An interesting, puzzling, derivative of F_{420} known as F_{390} is enzymatically synthesized when methanogen cells or extracts are exposed to oxygen under proper conditions. Although Schönheit et al. [80] originally observed the disappearance of F_{420} when cells were exposed to oxygen, Hausinger and others [81–86] documented the transformation into F_{390} by adenylation or quanylation of the 5-deazaflavin at the 8-OH position. When cells are briefly exposed to air, and then made anaerobic again, there is a conversion back to F_{420}. However, recovery efficiencies of F_{420} preparations can be greatly influenced by conversion to F_{390} [79]. Thus far, the function of F_{390} is unknown, although it is interesting to speculate a role in regulation, or protection from oxygen.

2.3. Methanofuran (MF)

The structure of methanofuran (MF), (4-[N-(4, 5, 7-tricarboxyheptanoyl-γ-L-glutamyl-γ-L-glutamyl-)-p-(β-aminoethyl)phenoxymethyl]-2-(aminomethyl)furan), determined by Leigh et al. using *M. thermoautotrophicum* ΔH [87,88], is shown in Fig. 2. This coenzyme was originally called CDR (carbon dioxide reduction) factor, for its role in the initial steps of CO_2 reduction [89,90]. There are several structural variations in different methanogens, all with changes in the carboxy terminal portion of the molecule [91,92]; two of these are shown in Fig. 2: methanofuran b (from *Methanosarcina barkeri*) and methanofuran c (from *Methanobrevibacter smithii*). All varieties are catalytically active in methanogens. Also, an abbreviated and commercially available chemical (see Fig. 2, where R=H), *N*-furfurylformamide, functions as a substrate, but with a high K_m and low V_{max} [93]. The formyl derivative of MF can be chemically produced by the reaction of MF with nitrophenylformate [94]. Methanofuran participates only in the initial steps of the methanogenic pathway, where carbon dioxide is bound to the furan at the primary amine, is reduced to the formyl level, and then transferred to the next coenzyme, H$_4$MPT [88,89]. Fig. 2 also shows the structure of the formylmethanofuran, the first stable intermediate in the methanogenic pathway. The initial enzyme in the reduction of carbon dioxide is formylmethanofuran dehydrogenase [95,96]. This enzyme can also oxidize the formylmethanofuran, and it is thought to play a role in the oxidation of methanol, and similar substrates, to CO$_2$. A second enzyme, formylmethanofuran:H$_4$MPT formyltransferase, transfers the carbon group to H$_4$MPT to create 5-formyl–H$_4$MPT [94].

Methanofuran is found in methanogens at ~1–6 nmoles per mg protein, as indicated in Table 3 [91,97]. *M. barkeri* grown on H$_2$–CO$_2$, methanol, or acetate contained methanofuran in this range [97]. However, due to its inability to use CO$_2$ in methanogenesis, or to oxidize methanol, *Methanosphaera stadtmanii* does not contain MF [98,99], and possibly only low levels may be found in some acetate-grown methanogens. It has thus far been found (as methanofuran b) in one non-methanogen, *Archaeoglobus* [69,92]. It has not been found in bacteria or eucarya.

The UV-visible spectra of methanofuran, and its formyl form, are very similar, with an absorbance peak at 273 nm, and a shoulder at 280 nm, with a maximum at <240 nm [87–89,100]. This coenzyme is stable to both boiling and air exposure.

2.4. Tetrahydromethanopterin (H$_4$MPT)

Methanopterin is a carbon-carrying coenzyme structurally related to folate; it exists in methanogens mainly as tetrahydromethanopterin (H$_4$MPT) or one of its one-carbon-carrying forms. H$_4$MPT participates in the methanogenic pathway as four different species, and with the carbon in three different oxidation stages. Fig. 3 provides the structure of H$_4$MPT and, for comparison, the structure of tetrahydrofolate (H$_4$folate); note that the more traditional numbering scheme is used in our H$_4$MPT structure (e.g. with methyl groups attached at carbons 7 and 9), in contrast to that by Van Beelen et al. [101]. Methanopterin was first described in one of its forms as YFC [102], and later as FAF [103,104]; it was determined that these were related pterins, distinct from folate, and capable of carrying a one-carbon group. Soon thereafter, work from the laboratories of both Vogels and Wolfe led to the structural identification of the parent methanopterin, its four-electron reduced form H$_4$MPT, and the various species (5-formyl–, 5,10-methenyl–, methylene–, and methyl–H$_4$MPT) [101,104–109]. In addition to the originally described methanopterin structures, studied principally in *M. thermoautotrophicum* ΔH and screened for in a variety of other methanogens, we now know that at least two structural variations occur. *M. barkeri* and some *Methanococcus* species contain (in addition to methanopterin) sarcinapterin, differing only in the additional presence of a glutamyl moiety attached by an amide linkage to the α-carboxylic acid portion of

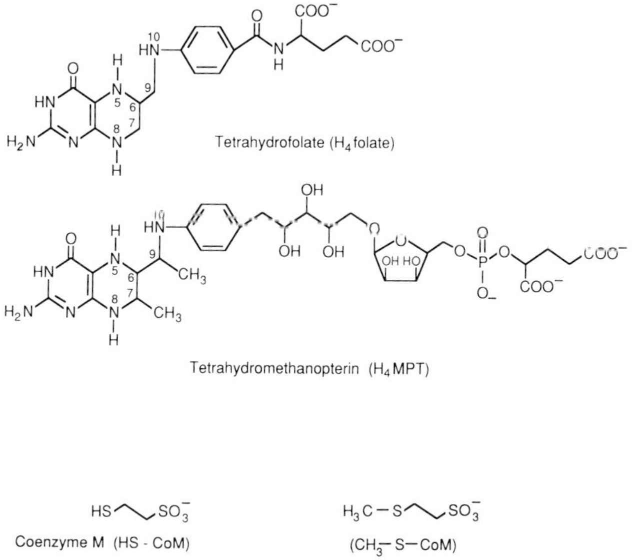

Fig. 3. Structures of tetrahydromethanopterin (and for comparison, tetrahydrofolate), and Coenzyme M and methyl–CoM.

α-hydroxyglutarate via an amide linkage [108]. Methanopterin and sarcinapterin both differ from folate by the presence of methyl groups at the 7 and 9 positions of the molecule, and in the nature of the non-pterin groups. Other variations are two types of "tatiopterin" (in two *Methanogenium* species), where the 7-methyl is absent, and where in addition to the second glutamyl group, an aspartyl group is attached, or where only one glutamyl group remains, and two aspartyl groups are present [110–112]. The variants work interchangeably in methanogen enzyme reactions, but detailed kinetic experiments have not been conducted [111]. Methanogen extracts cannot catalyze methanogenic transformations with folate coenzymes and bacterial extracts cannot catalyze reactions using methanopterin coenzymes [50,103,111].

H_4MPT participates in carbon transfer steps in the pathway of methane production from CO_2, and probably in the path by which methanol is oxidized to produce CO_2 and electrons. The following enzymes act in series, in the reductive direction: formylmethanofuran:H_4MPT formyltransferase [94,113], methenyl-H_4MPT cyclohydrolase [100], methylene H_4MPT dehydrogenase [50–52,114], methylene H_4MPT reductase [115,116], and methyl H_4MPT:CoM methyltransferase [117].

Methanopterin (or a variant) is found in all methanogens, except possibly at very low levels in *Methanosphaera stadtmanii* [46,118,119]. This latter organism produces methane solely by the reduction of methanol to methane, with no methanol oxidation, and thus needs no methanopterin in the methanogenic pathway (however, it may need a low level for biosynthetic needs, e.g. in synthesis of serine or glycine [120]).

Analysis of this coenzyme in cells is more complicated than with other coenzymes. Estimates (of total 7-methyl pterin) vary considerably in the literature, from about

238 nmol/mg protein down to 2 nmol/mg. Higher values come from recent HPLC assay techniques [45,46,119], whereas the lower numbers for a given organism generally come from spectrophotometric assays [97], and estimates accompanying the original purification of the cofactor [101,104,110,112]. Analysis is complicated further by the difference between H$_4$MPT and MPT: the latter is probably not detected by the spectrophotometric assay, and the fully reduced species must be oxidized to MPT without destruction for proper measurement in the HPLC assay (e.g., Raemakers-Franken et al. [112] incubated originally anaerobic cells under an atmosphere of N$_2$–CO$_2$–O$_2$ (80/15/5, v/v) to "convert pterin derivatives to the oxidized form"). The total pool of "methanopterin" includes a collection of species bearing a one-carbon unit. An example of the range of values in one organism is with *M. thermoautotrophicum*: Van Beelen et al. [101] reported 2 nmol/mg protein (assuming 25% yield), and Jones et al. [97] reported 6 nmol/mg, both using enzymatic assays; recently Gorris et al. [46] reported 238 nmol/mg using HPLC methods. No bacteria contain any of the methanopterin structures, but several non-methanogen archaea contain unusual pterins, e.g. the 7-methyl pterins in *Archaeoglobus* [46], and phosphohalopterin and solfapterin, respectively, in *Halobacterium* and *Sulfolobus* [121].

Laboratories studying H$_4$MPT enzymes isolate this reduced form from methanogens, using strictly anaerobic chromatography of boiled cell extracts [51,103–105]. Methanopterin itself, unlike H$_4$MPT, is air stable, and is thus technically easier to purify. However, even though it has been reported that methanopterin can be converted enzymatically to H$_4$MPT by methanogen extracts [104], a suitable procedure has not been described for large-scale, preparative use; such a procedure should take into account "conditioning" the cells, so that the oxidized methanopterin predominates, without destroying the oxygen-sensitive H$_4$MPT. Once H$_4$MPT is obtained, all other one-carbon-carrying derivatives can be made in the laboratory. Methylene H$_4$MPT is synthesized by a simple chemical reaction of H$_4$MPT with formaldehyde [51,103–105]; this species can be reduced chemically or enzymatically to methyl–H$_4$MPT [105,119]. Methenyl H$_4$MPT can be produced enzymatically from methylene H$_4$MPT, and is the likely identity of the YFC initially described as a carbon-carrying intermediate in methanogens [102].

Both methenyl- and 5-formyl derivatives are reasonably stable in air [109]; under acidic conditions, 5-formyl–H$_4$MPT is converted back to methenyl-H$_4$MPT, and under very basic conditions methenyl-H$_4$MPT is converted to the oxygen-sensitive 10-formyl derivative [122]. The various one-carbon-H$_4$MPT derivatives can be assayed by spectral changes in some cases, or by HPLC analysis [2,109,122]. No H$_4$MPT coenzymes discussed above have been reported to be bound to enzymes.

An enzyme-bound molybdopterin is found in methanogen formate dehydrogenase, and formylmethanofuran dehydrogenase [123,124]. It is distinct from the H$_4$MPT coenzymes used in the methanogenic pathway (see the enzyme sections for detail).

2.5. Coenzyme M

Coenzyme M (CoM), 2-mercaptoethanesulfonic acid, was the first novel methanogen coenzyme described, and is the smallest organic coenzyme found in biological systems. The structure (Fig. 3) arose from work by Taylor, McBride, and Wolfe on methane production by extracts of *Methanobacterium* strain MOH, where a component was identified that was essential for methanogenesis from methyl-cobalamin [125,126]. Purification of the oxidized dimer, 2, 2'-dithiodiethanesulfonic acid, allowed the structural determination of the active monomer, and further work showed that methane arose by reduction of a methyl group on the mercapto portion of the molecule. Thus CoM acts as a methyl-carrying coenzyme in the last step of the methanogenic pathway. This step is present in methanogens using any substrate. CoM interacts with at least three enzymes: methyl–H_4MPT:CoM methyltransferase [117], methyl CoM reductase [127,128], and heterodisulfide reductase [129–131]. Also, an enzyme of relatively low specific activity, the "reduced nicotinamide dependent 2, 2'-dithioethanesulfonic acid reductase", has been described [132].

CoM is found only in methanogens, not in bacteria or eucarya; it is present in the range of 3–12 nmol/mg cell protein (Table 3). An analog of CoM, BrCoM (with bromine replacing the mercapto group), inhibits the methyl reductase [133], and is a specific inhibitor of methanogens [134]. A stronger inhibitor, Br-propanesulfonate, has also been reported [135]. Several alkyl-mercapto analogs have been synthesized as potential alternate substrates or inhibitors [133,136,137]; of these, only ethyl–CoM is an active substrate, although at lower rates than with methyl–CoM. CoM is stable to heat and air; although it is oxidized to its dimer by oxygen exposure, it can be reduced to the original monomer without destruction. At one time it was proposed that CoM was bound to methyl–CoM reductase [138], but it is clear now that it occurs in methyl reductase in a non-covalent fashion, and enzyme assays treat it as a soluble coenzyme [139–142].

2.6. Cobamides

Two types of cobamides (or corrinoids) are found in methanogens. Methanogens do not contain the normal, bacterial, dimethylbenzimidazole (DMBI) B_{12} cobamides, but instead use either a 5-hydroxybenzimidazole (HBI) to make B_{12}-HBI, or use Coα[α-7-adenyl)]cobamide to make "pseudo-B_{12}"; abbreviated structures of these compounds are compared in Fig. 4. *M. barkeri*, *M. thermoautotrophicum*, and *Methanobrevibacter ruminantium* use B_{12}-HBI, whereas pseudo-B_{12} is used in *Methanococcus voltae* and *Methanococcus thermolithotrophicus*. *M. barkeri* has ca. 10 times more corrinoid than organisms not using methanol or acetate [143–145]. Interestingly, if DMBI-B_{12} is provided in media, it is taken up and inserted into methanogen proteins [146].

Cobamides or corrinoids are involved in methyl-transfer reactions in the methanogenic pathways, especially from methyl substrates [143–156]. At least one cobamide is involved in methanogenesis from H_2-CO_2 by *M. thermoautotrophicum*, where it is found in the methyl–H_4MPT:CoM methyltransferase [157]. The majority are either associated with membranes or bound to soluble proteins [143,158]. Methanogens are inhibited by corrinoid antagonists, suggesting an important metabolic role [159].

2.7. F_{430}

Factor F_{430} is a nickel-containing tetrapyrrole (or hydrocorphinoid) found only in methanogens. It was originally observed by Jean LeGall [2], and purified and partly characterized by Gunsalus and Wolfe [160] as a brightly yellow factor with maximal visible absorbance at 430 nm. It was demonstrated to contain nickel in two independent

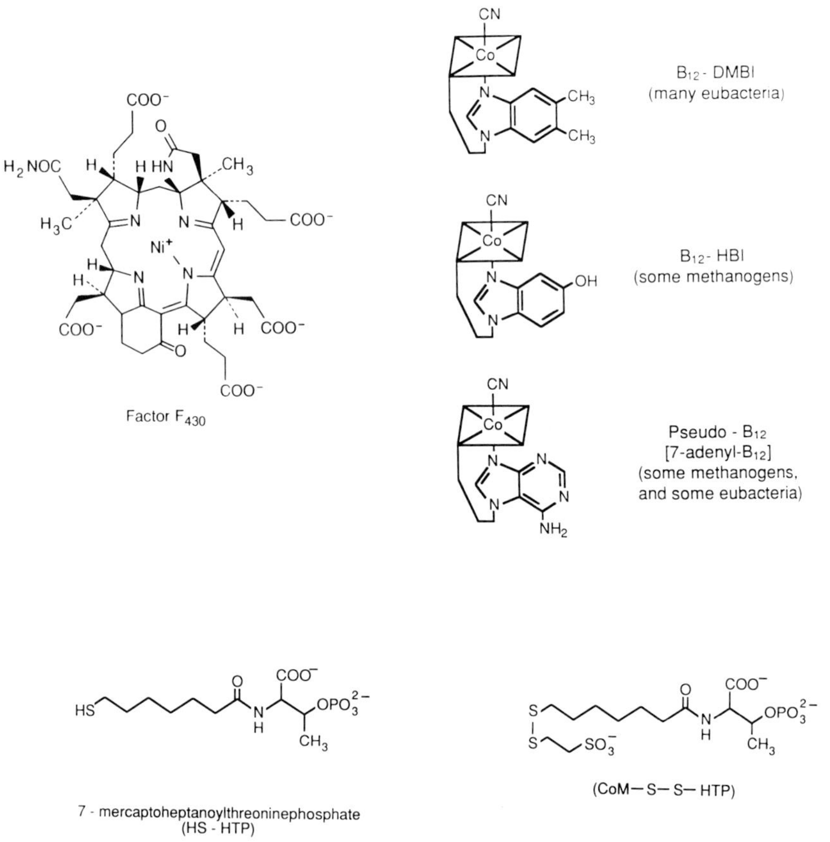

Fig. 4. Partial structures of B_{12}–dimethylbenzimidazole, B_{12}–hydroxybenzimidazole, and Pseudo-B_{12}, and the complete structures of F_{430}, HSHTP, and its heterodimer with CoM.

studies [161,162], and its structure was determined by Eshenmoser, Thauer, and colleagues [142,163] (see Fig. 4). It was later examined with respect to its physiological form [140], and with respect to structural heterogeneity in F_{430} preparations [164]. It does not contain covalently-bound CoM or other coenzymes [139,141]. F_{430} is found in all methanogens at about 1–4 nmol/mg cell protein (see Table 3), and is not found in non-methanogens [46,165]. It is bound to methyl reductase, although some also appears in the unbound form [140,166,167]. Its only role appears to be as a coenzyme in the methyl reductase reaction, via an unknown mechanism; it has been hypothesized that the Ni in F_{430} is transiently methylated during catalysis to form methane [168,169]. Evidence on the nature of its catalytic involvement is given in the enzymatic section on methyl reductase. F_{430} is relatively unstable to heat and acid, but is oxygen-stable for short periods; thermal degradation produces two chromatographically separable isomers, and eventually a variety of colored oxidation and degradation products [140,164,165,170,171].

2.8. 7-Mercaptoheptanoylthreonine phosphate (HSHTP)

One of the most elusive of the methanogen coenzymes was N-(7-mercaptoheptanoyl)threonine-O-phosphate (HSHTP), which was originally known as Component B [128,172,173]. The reduced form is required as an electron donor for methyl reductase [174–177]. The reduced form transfers 2 electrons to methyl–CoM, producing methane and a heterodisulfide of HSHTP and CoM; it may also play a role, via its heterodisulfide with CoM, in the activation of the first, methanofuran-requiring, reaction in methanogenesis from carbon dioxide. The heterodisulfide is reduced by heterodisulfide reductase, regenerating CoM and HSHTP (see Fig. 1). The structure of HSHTP, determined by Noll et al. [178], is shown in Fig. 4, along with the heterodisulfide. Enantiomeric specificity is for the L-threonine form [179]. It is one of the few methanogen coenzymes synthesized chemically [177]. A structural analog of HSHTP (with a hexanoyl rather than a heptanoyl chain) inhibits methanogenesis in cell extracts [136]. HSHTP is produced by virtually all methanogens, roughly at the levels shown in Table 3; an exception is *Methanomicrobium mobile*, where it is required as a growth factor [180]. It has not been reported in non-methanogens. It is non-covalently bound to methyl reductase [181]. The compound has no visible or near-UV absorbance.

3. The pathways and biochemistry of methanogenesis

The pathways for methanogenesis from CO_2, $HCOO^-$, CH_3OH, and CH_3COO^- are shown in Fig. 1. Generally, the H_4MPT-mediated steps in these pathways show close homology to the H_4folate system [182].

3.1. Methanogenesis from CO_2

Almost all methanogen species can grow on H_2+CO_2 carrying out autotrophic cell carbon biosynthesis [183]. Due to simplicity of the catabolites, CO_2-reduction to methane has been extensively used as a model system to study the biochemistry of methanogenesis. Most of this work has used *M. thermoautotrophicum*, and to a lesser extent, *M. barkeri*.

3.1.1. Reduction of CO_2
Reduction of CO_2 to CH_4 is a multi-step process involving intermediates at formyl- (methenyl-), methylene-, and methyl-stages of oxidation (Fig. 1) [2,184]. Formyl–MF is the first stable intermediate in the CO_2-reduction pathway [88]. Purified formyl–MF dehydrogenase can oxidize the formyl group to CO_2, using methylviologen as electron acceptor [95]. It has been hypothesized that the first step traps CO_2 as a carbamate, MF–COO^- (Reaction 9) [184]. However, formation of MF–COO^- has not been demonstrated. In water, furfuralamine carbamate is spontaneously decarboxylated (Reaction 10); a similar situation is expected with MF–COO^-.

$$CO_2 + MF \rightarrow MF-COO^- + H^+ \quad (\Delta G^{o\prime} = +8.1 \text{ kJ/mol}), \tag{9}$$

$$R-CH_2N-COO^- + H^+ \rightarrow R-CH_2-NH_2 + CO_2. \tag{10}$$

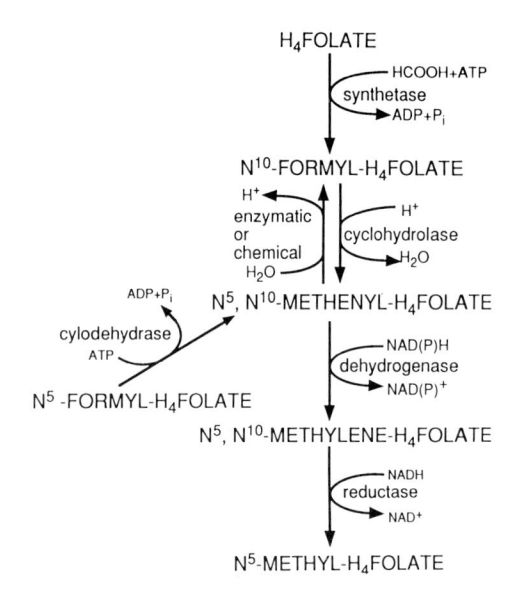

```
                    H₄FOLATE
                         │ ┌─ HCOOH+ATP
                         │ │ synthetase
                         │ └─► ADP+Pᵢ
                         ▼
              N¹⁰-FORMYL-H₄FOLATE
              H⁺ ◄─┐  ┌─ H⁺
              enzymatic │ cyclohydrolase
              or        │ ►H₂O
              chemical ─┘
              H₂O ─┘
     ADP+Pᵢ    
 cylodehydrase     N⁵, N¹⁰-METHENYL-H₄FOLATE
   ATP                   │ ┌─ NAD(P)H
                         │ │ dehydrogenase
 N⁵-FORMYL-H₄FOLATE      │ └─► NAD(P)⁺
                         ▼
              N⁵, N¹⁰-METHYLENE-H₄FOLATE
                         │ ┌─ NADH
                         │ │ reductase
                         │ └─► NAD⁺
                         ▼
              N⁵-METHYL-H₄FOLATE
```

Fig. 5. The clostridial path for methyl–H₄folate synthesis.

But, possibly, decarboxylation of MF–COO⁻ could be prevented by formation of a transient enzyme-bound intermediate, to be reduced to formyl–MF by formyl–MF dehydrogenase (Reaction 11). The $\Delta G^{o\prime}$ for Reaction (11), assuming H_2 as the direct reductant, is +7.9 kJ/mol [184].

$$\text{MF-COO}^- + \text{H}^+ + 2[\text{H}] \rightarrow \text{MF-CHO} + \text{H}_2\text{O}. \tag{11}$$

Thus, overall (Reactions 9 + 11), formation of formyl–MF requires an energy input of 16 kJ/mol. The source of this energy, electron carriers involved, and the cellular site of CO_2 activation are unknown, but work with *M. thermoautotrophicum* ΔH suggests CO_2 activation is energetically coupled to the last step of the methanogenic pathway (see section 3.1.2).

In the H₄folate system, formate is the formylating agent for N^{10}-formyl–H₄folate synthesis; this reaction requires ATP (Fig. 5):

$$\text{HCOO}^- + \text{MgATP}^{-2} + \text{H}_4\text{folate} \rightarrow \text{CHO-H}_4\text{folate} + \text{MgADP}^- + \text{HPO}_4^{-2}. \tag{12}$$

In methanogenic CO_2 reduction, H₄MPT is formylated by formyl–MF:H₄MPT formyltransferase (Reaction 13) [94,113,185], producing N^5-formyl–H₄MPT. This step does not require ATP. However, energy input has already occurred in the formation of formyl–MF (Reactions 9 and 11):

$$\text{MF-CHO} + \text{H}_4\text{MPT} \rightarrow \text{CHO} - \text{H}_4\text{MPT} + \text{MF} \quad (\Delta G^{o\prime} = -4.4 \text{ kJ/mol}). \tag{13}$$

The enzyme N^5, N^{10}-methenyl-H₄MPT cyclohydrolase dehydrates N^5-formyl–H₄MPT to N^5, N^{10}-methenyl-H₄MPT [100,113,122,186]):

$$\text{CHO-H}_4\text{MPT} + \text{H}^+ \rightarrow \text{HC}^+=\text{H}_4\text{MPT} + \text{H}_2\text{O} \quad (\Delta G^{o\prime} = -4.6 \text{ kJ/mol}). \tag{14}$$

In methanogenesis, N^5-formyl–H_4MPT is the biologically active species, and N^{10}-formyl–H_4MPT is produced only by chemical hydrolysis of methenyl-H_4MPT [122]. On the other hand, both chemical and cyclohydrolase-mediated hydrolysis of methenyl-H_4folate yields N^{10}-formyl–H_4folate (Fig. 5; [182]). N^{10}-formyl–H_4MPT is not an energetically favorable structural form, since when it is produced by chemical hydrolysis of methenyl-H_4MPT, it spontaneously converts to the N^5 form at ambient temperature, whereas the corresponding conversion for N^{10}-formyl–H_4folate requires incubation at 100°C [187].

A two-electron reduction of methenyl-H_4MPT yields methylene–H_4MPT; the source of electrons can either be H_2F_{420} or H_2 [50–52,114]):

$$HC^+=H_4MPT + H_2F_{420} \rightarrow H_2C=H_4MPT + H^+ + F_{420} \quad (15)$$
$$(\Delta G^{o\prime} = +6.5 \text{ kJ/mol}),$$

$$HC^+=H_4MPT + H_2 \rightarrow H_2C=H_4MPT + H^+ \quad (\Delta G^{o\prime} = -5.5 \text{ kJ/mol}). \quad (16)$$

Reaction (15) is catalyzed by F_{420}-dependent N^5,N^{10}-methylene–H_4MPT dehydrogenase [50–52,55,188,189] and Reaction (16) by F_{420}-independent (H_2ase-type) N^5,N^{10}-methylene–H_4MPT dehydrogenase [55,114,190]. If energetics (under standard conditions) is the only consideration, it could be suggested that Reaction (16) and not Reaction (15) participates in CO_2-methangenesis. On the other hand, kinetic data show that the H_2ase-type enzyme functions only at high H_2-partial pressures (K_m=0.6 atm= 60 000 Pa) [191], and not at levels (1–10 Pa) found in natural methanogen habitats [192]. Thus, from ecological and physiological standpoints a combination of Reaction (15) and the hydrogenase discussed below (combined $\Delta G^{o\prime} = -5.5$ kJ/mol), is probably functional in CO_2 reduction under normal conditions. For oxidation of methylene groups, as in methanol methanogenesis (see below), Reaction (15) should be favored. The parallel step in the folate system uses NAD(P)H [243].

Methylene–H_4MPT is reduced by F_{420}-dependent methylene–H_4MPT reductase to methyl–H_4MPT [115,116,189,190,193]:

$$H_2C=H_4MPT + H_2F_{420} \rightarrow CH_3-H_4MPT + F_{420} \quad (\Delta G^{o\prime} = -5.2 \text{ kJ/mol}). \quad (17)$$

The parallel step in the H_4folate system is dependent on NAD(P)H, ferredoxin or $FADH_2$.

The involvement of H_4MPT in CO_2-reduction ends with the transfer of the methyl group from methyl–H_4MPT to CoM to form methyl–CoM [157,184]:

$$CH_3-H_4MPT + CoM \rightarrow CH_3-CoM + H_4MPT \quad (\Delta G^{o\prime} = -29.7 \text{ kJ/mol}). \quad (18)$$

Two models have been proposed for this methyl-transfer reaction. (i) Recent evidence from Kengen et al. suggests that the methyl group of CH_3–H_4MPT is first transferred to a corrinoid protein called methyltransferase a (MT_a) yielding protein-bound CH_3–B_{12} HBI (Reaction 19) which acts as the methyl-donor to CoM in a reaction catalyzed by methyltransferase MT_2, a non-corrinoid protein (Reaction 20) [157]:

$$CH_3-H_4MPT + MT_a-B_{12}HBI \rightarrow MT_a-\overset{CH_3}{\underset{|}{B}}_{12}HBI + H_4MPT, \quad (19)$$

$$\mathrm{MT_a-\overset{CH_3}{\underset{|}{B}}_{12}HBI + CoM \xrightarrow{MT_2} MT_a-B_{12}HBI + CH_3-CoM.} \qquad (20)$$

The oxygen-insensitive MT_2 protein was previously purified by Taylor and Wolfe as methylcobalamin:CoM methyltransferase from *Methanobacterium bryantii* [194]. Reactions (19) and (20) are analogous to steps in methyl transfer from methanol to CoM.

(ii) Recently Fischer et al. [117] showed that membrane fractions of H_2-CO_2-grown *M. thermoautotrophicum* Marburg, and H_2-CO_2-, methanol-, and acetate-grown *M. barkeri*, are devoid of MT_2-like methylcobalamin:CoM transferase activity, yet they catalyze methyl transfer from CH_3-H_4MPT to CoM; from kinetic data they proposed the involvement of only one membrane-bound corrinoid protein with MT_a-type activity in Reaction (18).

In vitro, Reaction (18) proceeds irreversibly; the analogous reaction in methionine biosynthesis is also highly exergonic [184]. In *Methanosarcina* strain Göl excess energy from Reaction (18) is conserved [195]. Working with washed everted vesicles of this organism Becker et al. [195] showed that the methyl transfer from CH_3-H_4 MPT to CoM generates a primary sodium motive force.

Both models (i) and (ii) above suggest that $B_{12}HBI$ is the methyl carrier during methylation of CoM irrespective of the methanogenic substrate catabolized. Thus, CH_3-cobamide probably is the first common intermediate, albeit in an enzyme-bound form, during methanogenesis from various substrates; this is similar to the conclusion of Blaylock [196], although she in addition proposed CH_3-B_{12} to be the terminal substrate of methanogenesis, which was later identified as CH_3-CoM.

Reduction of CH_3-CoM to CH_4 (Reaction 21) is the common terminal step in methanogenesis. The methylreductase reaction was originally described [126] as CH_3-CoM + H_2 → CH_4 + CoM, but is now resolved into two reactions [174–176,197,198]:

$$\mathrm{CH_3-CoM + HSHTP \rightarrow CH_4 + \underset{\underset{S-CoM}{|}}{S-HTP}} \qquad (\Delta G^{o\prime} = -45\,\mathrm{kJ/mol}), \qquad (21)$$

$$\mathrm{CoM-S-S-HTP + 2[H] \rightarrow CoM + HSHTP}$$
$$(\Delta G^{o\prime} = -40\,\mathrm{kJ/mol\ if\ H_2\ reductant}). \qquad (22)$$

Reaction (21), the most exergonic in the pathway [184], is catalyzed by methyl–CoM methyl reductase [199]. The source of electrons for CH_3-CoM reduction is HSHTP [174,176]. Heterodisulfide reductase reductively cleaves the disulfide bond of the heterodisulfide CoM–S–S–HTP, a product of the methylreductase reaction (21), and regenerates HSHTP and CoM (Reaction 22) [130,131,184,198,200]. Although in crude preparations (soluble extracts or washed vesicles) H_2 and H_2F_{420} can supply electrons for heterodisulfide reduction [129,130,201,202], the direct electron carrier is unknown [131,201]. It has been speculated that in methylotrophic methanogens, electrons are funneled to heterodisulfide reductase via a membrane transport system, and that CoM–S–S–HTP acts as the terminal electron acceptor [201–203]; evidence suggests that during chemolithotropic growth on H_2-CO_2 these electrons are derived from H_2 via an F_{420}-nonreducing hydrogenase [201,204], and in cells growing on methanol the source of electrons is H_2F_{420} [202] generated from the oxidation of methyl groups. Work with vesicle preparations of methylotrophic methanogens *Methanosarcina* strain Göl and *Methanolobus tindarius* shows that the proton motive force generated by such electron transport systems drives ATP synthesis from ADP and P_i. Thus, in these methanogens reduction of heterodisulfide is associated with energy production. However, such a role for heterodisulfide reductase in energy generation in other H_2–CO_2 methanogenic organisms, if any, is not clear.

3.1.2. The RPG effect

In 1977 Gunsalus and Wolfe [205] reported that the rate of methanogenesis from CO_2 in cell extracts of *Methanobacterium* is greatly enhanced if CH_3–CoM is added to the reaction mixture; this phenomenon of the dependence of the first step of methanogenesis on the last step is known as the RPG effect. The RPG effect is exhibited by all methanogenic intermediates, most likely because they are ultimately converted to CH_3–CoM [105,206]. Leigh et al. [88] showed that in a cell extract system formyl–MF accumulates in the absence of H_4MPT only if CH_3–CoM is added; this indicates a dependence on the methylreductase reaction for the fixation of CO_2 as formyl–MF. Bobik and Wolfe [185,198] showed that CoM–S–S–HTP, a product of the methylreductase reaction, activates formyl–MF synthesis. This specific requirement for CoM–S–S–HTP in the activation of CO_2 is further confirmed by the following observations: (i) Reduction of fumarate in *M. thermoautotrophicum* is dependent on CoM and HSHTP, which are oxidized to form CoM–S–S–HTP [207]. Addition of fumarate to cell extracts induces a large RPG effect [208] and activates CO_2 fixation to form formyl–MF [207]. (ii) Addition of reductants that cleave CoM–S–S–HTP to CoM and HSHTP abolishes the activation. Evidence suggests that CoM–S–S–HTP activates a low potential electron carrier that participates in the formation of formyl–MF; formyl–MF dehydrogenase is a possible candidate for this task.

3.1.3. Source of electrons

During methanogenesis from $H_2 + CO_2$ the H-atoms in CH_4 originate from H_2O, and H_2 serves as the source of electrons [209,210]. Although H_2 is the primary reductant, the direct electron donors in Reactions (11) and (22) remain unknown (see above and Fig. 1). Methanogenic bacteria possess two kinds of hydrogenase [184]: an F_{420}-reducing hydrogenase (Reaction 23), and an F_{420}-non reducing hydrogenase (Reaction 24):

$$H_2 + F_{420} + 2H^+ \rightarrow H_2F_{420} + 2H^+ \quad (\Delta G^{o\prime} = -12 \text{ kJ/mol}), \tag{23}$$

$$H_2 \rightarrow 2[\text{II}]. \tag{24}$$

Thus, for H_2F_{420}-requiring steps needs could be met by F_{420}-reducing hydrogenase. The physiological electron carrier for the other hydrogenase has not been identified, although ferredoxin and a unique polyferredoxin have been discussed as possibilities (see section 4.2. below). The F_{420}-nonreactive hydrogenase may not have a direct role in the CO_2 reduction pathway, but rather may be involved in a membrane transport system that supplies electrons for the cleavage of CoM–S–S–HTP and the reductive activation of CO_2 to form formyl–MF, and that generates a proton motive force [201,204,211].

A few methanogens can also use some non-methanol alcohols as a primary electron source for CO_2-methanogenesis (Table 2 [25–28,212]) by expressing an alcohol dehydrogenase; e.g., an F_{420}-reducing alcohol dehydrogenase has been purified from *Methanogenium thermophilum* [27],

$$CH_3CHOHCH_3 + F_{420} \rightarrow CH_3COCH_3 + H_2F_{420}, \qquad (25)$$

and a $NADP^+$-specific activity is present in extracts of *Methanobacterium palustre* [212]:

$$CH_3CHOHCH_3 + NADP^+ \rightarrow CH_3COCH_3 + NADPH + H^+. \qquad (26)$$

The H_2F_{420} or NADPH obtained could supply electrons for CO_2 reduction either directly, or via carriers or H_2; electron flow from NADPH may proceed via F_{420} by the action of the NADPH:F_{420} oxidoreductase described in several methanogens [57,58].

Electrons for CO_2 reduction can also come from the oxidation of CO to CO_2:

$$CO + H_2O \rightarrow CO_2(g) + H_2 \quad (\Delta G^{o\prime} = -20 \, \text{kJ/reaction}). \qquad (27)$$

This arrangement is the basis for methanogenesis and/or growth of methanogens on CO [213–215], although it may not be significant in natural systems.

3.2. Methanogenesis from methanol

Methanogenesis from methanol occurs by its disproportionation to methane and carbon dioxide (Reactions (5) and (5A), Table 2). Reducing equivalents derived from the oxidation of one methanol to CO_2 are used to reduce three methanols to CH_4.

3.2.1. Reduction of methanol

The first evidence for cobalamin involvement in the conversion of methanol to methane was provided by Blaylock and Stadtman [196,216–218]; with extracts of methanol-grown *M. barkeri* they demonstrated enzymatic formation of methylcobalamin from methanol, and subsequent reduction of methylcobalamin to methane. Later Blaylock [196] showed that conversion of methanol to methylcobalamin requires a heat-stable cofactor and at least three proteins, a 100–200 kDa B_{12}-enzyme (methyltransferase), a ferredoxin, and an unidentified protein. Blaylock speculated that the role of hydrogen and ferredoxin in the conversion of methanol to methylcobalamin was in the reduction of the B_{12}-protein. This work led to the proposal that methylcobalamin was the direct precursor of methane in methanogenesis from various substrates [196,218].

In 1971 McBride and Wolfe showed that for methane formation from methylcobalamin by extracts of H_2–CO_2-grown *Methanobacterium*, an additional coenzyme is required; this coenzyme was identified as CoM [125,126]. Later Taylor and Wolfe [194] described the methylcobalamin:CoM methyltransferase, but its role in methanogenesis remained unknown. In 1978 Daniels and Zeikus [102] reported that [^{14}C]–CH_3–CoM is a major radioactive fixation product isolated from *M. barkeri* cells exposed to [^{14}C]–CH_3OH. These findings, in conjunction with the observation that CoM is present in all methanogens [15], described CH_3–CoM as the direct precursor of methane in methanogenesis from methanol. Using a B_{12}-protein from *M. barkeri* (first purified by Blaylock [196]) labelled with a [^{14}C]-methyl group, Wood et al. [219] showed that in

M. barkeri extracts methanogenesis from this labelled protein is at least 100 times faster than the non-enzymatic reaction between CoM and methylcobalamin; they postulated that this protein acts as a methyltransferase in methanogenesis from methanol. Finally, Van der Meijden et al. [152–155] resolved the *M. barkeri* methyl transfer from methanol to CoM into two steps involving methyltransferase 1 and 2, or MT_1 and MT_2. The oxygen-sensitive corrinoid protein MT_1 in its most reduced state (Co^{+1}) accepts the methyl group from methanol [153,154]:

$$CH_3OH + MT_1-B_{12}HBI \rightarrow MT_1-\overset{\overset{\displaystyle CH_3}{|}}{B_{12}HBI} + H_2O. \qquad (28)$$

In the next step,

$$MT_1-\overset{\overset{\displaystyle CH_3}{|}}{B_{12}HBI} + CoM \xrightarrow{MT_2} MT_1-B_{12}HBI + CH_3-CoM, \qquad (29)$$

catalyzed by the non-corrinoid oxygen-stable MT_2 protein, methylated MT_1 serves as the methyl donor to CoM; the resulting CH_3–CoM is then reduced to CH_4 by the methyl reductase. The sum of Reactions (28) and (29) has $\Delta G^{o\prime} = -27.5$ kJ/mol [184]. MT_1 is inactivated by oxidation of its reduced corrinoid (to the Co^{+2} or Co^{+3} state); reactivation by reduction to Co^{+1} requires appropriate reductants and ATP; in vivo reduction is probably via H_2ase and ferredoxin [153,154]. Thus, the scheme first proposed by Blaylock [196] has found validation.

3.2.2. Oxidation of methanol

A variety of evidence suggests that methanol oxidation occurs via the steps between CO_2 and CH_3–CoM of the CO_2-reduction pathway (Fig. 1) but operating in the reverse direction: (i) If *M. barkeri* grown on methanol is exposed to [^{14}C]–CH_3OH for a short time, methenyl-H_4MPT (formerly called YFC; [102]) is labelled. (ii) Methanol-grown cells of *M. barkeri* produce methane from H_2–CO_2 or formaldehyde (in the presence or absence of H_2) [220]. (iii) Methanol-grown cells of *M. barkeri* produce H_2 and CO_2 from methanol if methylreductase is inhibited with BES [221], and the corresponding cell extracts oxidize HCHO to CO_2 only if MF and H_4MPT are present (see Fig. 1; [222]). (iv) Methanol-grown cells of *Methanosarcina* strain Göl, when treated with BES, produce H_2 and CO_2 from CH_3OH or HCHO [223]. (v) Equivalent levels of formylmethanofuran dehydrogenase, methenyl-H_4MPT cyclohydrolase, methylene–H_4MPT dehydrogenase, and methylene–H_4MPT reductase are present in both H_2–CO_2- and methanol-grown *M. barkeri* [55,224].

However, it is not clear at what point of this pathway the methyl groups of methanol that are destined for oxidation to CO_2 enter. Entry as CH_3–CoM is possible, but would mean a reversal of Reaction (18), which has not been demonstrated [117,184]. Hypothetically, methanol could be directly activated to the formaldehyde level of oxidation as methylene–H_4MPT. Both possibilities would include at least one endergonic step. Results with whole cells of *M. barkeri* and *Methanosarcina* strain Göl indicate the requirement of an electrochemical gradient for the activation of methanol destined for oxidation to CO_2: (i) If methylreductase is inhibited by BES, methanol and HCHO are oxidized to H_2 and CO_2 [223]. (ii) But, if in addition to BES, tetrachlorosalicylanilide (TCS, a protonophore) is added, only HCHO but not methanol is oxidized to CO_2 and H_2; methanol oxidation could be temporarily re-established by a pulse of sodium ions [225]. These results indicate that methanol oxidation is driven by Na^+-influx, and possibly a proton pump associated with the methylreductase system helps generate the necessary Na^+ gradient via a Na^+/H^+ antiporter (assuming low-level methylreductase activity in the presence of BES can activate methanol for oxidation). Contrary to the situation with methanol, HCHO is oxidized without involvement of the membrane transport process mentioned above. These observations, and the energetics of the following reactions, suggest that in methanol oxidation to $CO_2 + 3H_2$ (Reaction 30), one or more steps in the activation of methanol to the methylene–H_4MPT stage (Reaction 32) would require energy input, and steps beyond the methylene stage are exergonic (Reaction 31):

$$CH_3OH + H_2O \rightarrow CO_2(aq) + 3H_2 \quad (\Delta G^{o\prime} = +26.55 \text{ kJ/reaction}), \quad (30)$$

$$HCHO + H_2O \rightarrow CO_2(aq) + 2H_2 \quad (\Delta G^{o\prime} = -18.3 \text{ kJ/reaction}), \quad (31)$$

$$CH_3OH \rightarrow HCHO + H_2 \quad (\Delta G^{o\prime} = +44.85 \text{ kJ/reaction}). \quad (32)$$

(However, note that Reactions 30–32 do not exactly represent the energy changes of coenzyme-bound intermediates.) It is not clear whether these membrane-dependent processes are for pushing the conversion of CH_3–CoM to CH_3–H_4MPT uphill (reverse of Reaction 18), or to activate methanol directly to $H_2C=H_4$MPT or CH_3–H_4MPT. Work with *M. barkeri* and everted vesicles of *Methanosarcina* Göl suggests that methylation of CoM with CH_3–H_4MPT leads to extrusion of Na^+ from cells [182,184,195,220,226]. Hence, it is possible that Reaction (18) is reversed using a Na^+-motive force set up directly or indirectly, and entry of methanol molecules is as CH_3–CoM to the central pathway for oxidation to CO_2.

3.2.3. Electron transfer

It is not clear how electrons from methanol oxidation are used for heterodisulfide reduction (Reaction 22). Indirect evidence suggests that H_2F_{420} generated from oxidation of methyl and methylene carbons (reverse of Reactions 15 and 17) donates electrons to a membrane-bound electron-transport system that uses CoM–S–S–HTP as the terminal electron acceptor. In *Methanosarcina* Göl, electrons from H_2F_{420} (generated from methanol oxidation) are probably transferred to a cytochrome via a membrane-bound H_2F_{420} dehydrogenase [56,201], and the reduced cytochrome supplies electrons (directly or via unknown carriers) for heterodisulfide reduction [211]. Hydrogen is required for methanogenesis from methanol in extracts which suggests membrane involvement in electron transfer. In extract methanogenesis, H_2 supplies electrons for those steps (cleavage of CoM–S–S–HTP, reductive activation of components such as MT_1, etc.) which in whole cells may receive electrons from intact membrane-bound electron transport systems.

3.2.4. Methanogenesis from methylamines

Methanogenesis from methylamines (Reaction 6, Table 2) follows almost the same route as methanol to CH_4 and CO_2 [227], although each system probably requires specific components for substrate transport, and to form a common methylated product. For example, extracts of *M. barkeri* grown on trimethylamine (TMA) produce methane from TMA + H_2, but extracts from methanol- or dimethylamine-grown cells do not [227].

3.3. Methanogenesis from acetate

About two-thirds of the methane produced in nature originates from acetate, but only *Methanosarcina* and *"Methanothrix"* species use this substrate for methanogenesis. Acetate is the least energy-yielding of all methanogenic substrates (Reactions 8 and 8A, Table 2). *Methanosarcina*, which produce methane from the widest variety of substrates, use acetate only when all other sources of energy are depleted; K_s for acetate is 3–5 mM and the threshold concentration is 0.4–1.2 mM [228]. Alternatively, for *"Methanothrix"* acetate is the only source of energy. It grows more slowly than *Methanosarcina*, but in sewage digestors where acetate levels are 0.3–5.8 mM [229,230], *"Methanothrix"* species grow efficiently; their K_s and threshold concentrations for acetate (as low as 0.1 mM and 0.01 mM, respectively) are much lower than those of *Methanosarcina*.

Experiments with ^{14}C- and ^2H-acetate showed that the methyl group of acetate is the major precursor of methane and the carboxyl-carbon is oxidized to CO_2, providing electrons for reduction of the methyl-carbon to CH_4 [7,231,232]. However, Krzycki et al. [233] showed that ~14% of the methane produced by whole cells of an acetate-adapted strain of *M. barkeri* originated from the carboxyl-carbon, and an equivalent amount of CO_2 came from methyl-group oxidation; the reduction of carboxyl-carbon to methane has also been observed in extracts [7,234].

With all organisms, methanogenesis from acetate involves the following steps in sequence: 1. acetate transport into the cell, 2. activation of acetate to acetyl–CoA, 3. cleavage of the C–S and C–C bonds of CH_3CO–SCoA, generating CO_2, 2[H] and a CH_3, 4. transfer of the methyl group to CoM to form CH_3 CoM, and 5. reduction of CH_3–CoM to CH_4. There are major differences between *Methanosarcina* and *"Methanothrix"* acetate catabolism, but *M. barkeri* and *M. thermophila* use similar reactions.

3.3.1. Transport of acetate into the cell

Acetate anion may be taken up by *Methanosarcina* via a bicarbonate/acetate antiport, based on the discovery of carbonic anhydrase in *M. barkeri* [235]:

$$HCO_3^- + H^+ \leftrightarrow CO_2(g) + H_2O \quad (\Delta G^{o\prime} = -4.8 \, kJ/mol). \tag{33}$$

3.3.2. Activation of acetate

In *Methanosarcina*, acetate is activated by acetate kinase (Reaction 34) and phosphotransacetylase (Reaction 35) before cleavage. These enzymes are produced by *Methanosarcina* grown on acetate [236–240]. The energetics of Reaction (8) (Table 2) suggest that at most 0.5 mol ATP can be generated from 1 mol of acetate [184]. Since activation of 1 mol acetate (Reaction 34+35) requires 1 mol ATP, steps beyond the formation of acetyl–CoA must generate more than 1 mol of ATP:

$$CH_3COO^- + ATP \rightarrow CH_3CO\sim P + ADP \quad (\Delta G^{o\prime} = +13\,kJ/mol), \quad (34)$$

$$CH_3CO\sim P + HS-CoA \rightarrow CH_3CO-SCoA + P_i \quad (\Delta G^{o\prime} = -9\,kJ/mol). \quad (35)$$

"*Methanothrix*" activates acetate differently than *Methanosarcina*, using only one enzyme, acetyl–CoA synthetase:

$$CH_3COO^- + ATP + HS-CoA \rightarrow CH_3CO-SCoA + AMP + PP_i$$
$$(\Delta G^{o\prime} = -5.9\,kJ/mol). \quad (36)$$

M. soehngenii lacks acetate kinase, phosphotransacetylase, and PP_i:AMP- or PP_i:ADP-phosphotransferase [241], but possesses high levels of adenylate kinase (Reaction 37) and pyrophosphatase (Reaction 38) activities:

$$ATP + AMP \rightarrow 2ADP \quad (\Delta G^{o\prime} = 0\,kJ/mol), \quad (37)$$

$$PP_i + H_2O \rightarrow 2P_i \quad (\Delta G^{o\prime} = -21.9\,kJ/mol). \quad (38)$$

Thus, activation of 1 mol acetate in *M. soehngenii* could require 2 mol of ATP (Reaction 39, sum of Reactions 36–38) [241]:

$$CH_3COO^- + HSCoA + 2ATP + H_2O \rightarrow CH_3CO-SCoA + 2ADP + 2P_i$$
$$(\Delta G^{o\prime} = -27.8\,kJ/mol). \quad (39)$$

It is surprising that an organism growing on an energy-poor substrate like acetate would use such an expensive means of activation; the organism must have to generate more than 2 mol of ATP per mol of acetyl–CoA converted to CH_4 and CO_2. These energetics may explain the poor growth yields and growth rates of "*Methanothrix*" compared to *Methanosarcina*. Evidence suggests that, in addition to providing highly favorable thermodynamics to acetyl–CoA formation at low acetate concentrations (by PP_i hydrolysis), the pyrophosphatase in a membrane-associated configuration may help generate a proton-motive force and conserve energy [228].

3.3.3. Cleavage of acetyl–CoA and CH_3–CoM formation

In *Methanosarcina* and "*Methanothrix*" CH_3–CoM, an intermediate of acetoclastic methanogenesis, is formed by transfer of the methyl group of acetyl–CoA to CoM [231,237,239,242]. This conversion,

$$CH_3CO-SCoA + CoM \rightarrow CH_3-CoM + HS-CoA + 2[H] + CO_2, \quad (40)$$

is mechanistically analogous (but opposite in direction) to acetyl–CoA formation in acetogenic bacteria [243] and in *M. thermoautotrophicum* [183]. Clostridial acetogenesis has been studied in detail using *Clostridium thermoaceticum* as the model organism, with emphasis on its carbon monoxide dehydrogenase, and results from this work have provided excellent guidance for work on acetyl–CoA cleavage in methanogens; readers are referred to recent reviews on this topic [243,244]. For comparison, a brief description of the acetate synthesizing machinery of *C. thermoaceticum* is presented below while describing the acetyl–CoA cleavage system of *M. thermophila*.

Carbon monoxide dehydrogenase (CODH) is the key enzyme for acetyl–CoA cleavage in *Methanosarcina*. A CODH complex, produced in *M. thermophila* when grown on acetate, catalyzes oxidation of CO, and synthesizes (in vitro) acetyl–CoA from CH_3I, CO and coenzyme A [245,246]. It catalyzes isotope exchange between the carbonyl group of CH_3CO–SCoA and CO or between HS–CoA and CH_3CO–SCoA, indicating the acetoclastic capability of the enzyme and the reversible nature of C–S and C–C bond cleavage [247]. The CODH complex of *M. thermophila* is designed more for acetate cleavage than for acetate synthesis. Electron paramagnetic resonance (EPR) and isotope exchange studies with the purified enzyme complex have established CH_3CO–SCoA as the physiological substrate and CO as an acetoclastic intermediate in *M. thermophila* [247,248]. Ferredoxin acts as an electron acceptor for the CODH complex, and thus it could participate in the acetoclastic reaction in vivo to collect electrons generated by carboxyl-carbon oxidation [249].

In *C. thermoaceticum* the methyl group of CH_3–H_4folate (derived from CO_2 via the pathway in Fig. 5) is transferred to the C/Fe–SP (corrinid Fe–sulfur protein) by a methyltransferase [243] to yield an enzyme-bound CH_3–Cob(III)-amide. Only the Co^{+1} form of C/Fe–SP is an active methyl acceptor; thus, conversion of Co^{+2} to Co^{+1} (probably by electrons from reduced CODH or ferredoxin) [243,250] is a prerequisite for methylation. CODH is the site for acetyl–CoA synthesis, and thus is properly called acetyl–CoA synthase [243,244]. Transfer of the methyl group from CH_3–Cob(III)-amide of C/Fe–SP methylates CODH. CODH catalyzes the reduction of CO_2 to CO and forms an enzyme-bound one-carbon organometallic intermediate [243]; this one-carbon species can also be formed directly from CO. Thus, CO is an intermediate in acetogenesis from CO_2 [243]. Electrons for reduction of CO_2 to CO are supplied from H_2 by hydrogenase possibly via a ferredoxin [243], although for thermodynamic reasons this role of ferredoxin in vivo remains uncertain [251]. From the CODH-bound methyl and carbonyl groups an acetyl–metal intermediate is proposed to be formed by methyl migration or CO insertion at the Ni–Fe site [243]. Next, HSCoA binds to CODH. Formation of acetyl–CoA is probably by the thiolytic cleavage of an acetyl–metal intermediate with –SH of HSCoA as the nucleophile.

In analogy to the clostridial system the following reaction sequence describing the functions of Ni/Fe–SP (Ni/Fe sulfur protein) and C/Fe–SP (corrinoid/Fe–sulfur protein) of *M. thermophila* has been proposed [252,253]:

$$\text{Ni/Fe–SP} + CH_3CO\text{–SCoA} + H^+ \leftrightarrow \text{Ni/Fe–SP–CO–}CH_3 + \text{HS–CoA}, \tag{41}$$

$$\begin{aligned}&\text{Ni/Fe–SP–CO–}CH_3 + [Co^{+1}]\text{–C/Fe–SP} + H_2O \\ &\leftrightarrow \text{Ni/Fe–SP} + CH_3\text{–}[Co^{+3}]\text{–C/Fe–SP} + CO_2 + 2[H].\end{aligned} \tag{42}$$

In vitro, in the absence of ferredoxin, the Ni/Fe–SP component transfers electrons from CO to the C/Fe–SP component to reduce from Co^{+2} to Co^{+1} [252]; this may activate C/Fe–SP. Also, EPR spectroscopy shows that similar to the cobamide in C/Fe–SP of *C. thermoaceticum*, the B_{12}HBI in the *M. thermophila* C/Fe–SP is in a base-off configuration;

corrinoids in this base off configuration have a mid-point potential of about -500 mV (compared to -600 mV for base-on configuration) for the Co^{+1}/Co^{+2} couple, which makes reduction of Co^{+2} by physiological electron donors possible [253]. Thus, the C/Fe–SP of *M. thermophila* has the potential to accept the methyl group of acetyl–CoA from the Ni/Fe–SP, but the fate of CH_3–C/Fe–SP or the specific steps involved in methyl transfer to CoM in *M. thermophila* remain unknown [253].

In *M. barkeri* CH_3–H_4MPT has been identified as an intermediate in the formation of CH_3–CoM [254] from acetyl–CoA and thus, for this organism, Reaction (40) can be divided into two steps, Reaction (43) and Reaction (18),

$$CH_3CO-SCoA + H_4MPT \rightarrow CH_3-H_4MPT + CO_2 + 2[H] + HS-CoA, \qquad (43)$$

where the participation of H_4MPT is analogous to the role of H_4folate in clostridial acetate synthesis [243]. Work with extracts and purified CODH (or CODH complex) suggests that also in *M. barkeri* CODH cleaves the C–C and C–S bonds of CH_3CO–SCoA [233,239,255–259], and indirect evidence suggests that, as in *C. thermoaceticum* and *M. thermophila*, a ferredoxin is the electron carrier [260]. However, early work showed that the *M. barkeri* CODH purifies as an $\alpha_2\beta_2$ protein without associated corrinoid-containing components [255,257], and unlike in *M. thermophila* [247,248], free CO was not an intermediate in acetate cleavage [261]. Recently Grahame [258] purified CODH in association with a corrinoid protein, and this complex catalyzes cleavage of CH_3CO–SCoA yielding CO_2 and CH_3-H_4sarcinopterin. Evidence consistent with corrinoid involvement, e.g. appearance of CH_3–B_{12}HBI as an intermediate in CH_3–CoM formation from acetyl–CoA in *M. barkeri*, is also available [156]. Fischer and Thauer [262] showed that corrinoids are involved in steps after acetyl–CoA cleavage that lead to CH_3–H_4MPT. They proposed the following scheme for CH_3–H_4MPT formation: Reaction (41) in combination with

$$CH_3-CO-CODH + corrinoid-protein + H_2O$$
$$\leftrightarrow CH_3-corrinoid-protein + CO_2 + 2[H] + CODH, \qquad (44)$$

$$CH_3-corrinoid-protein + H_4MPT \leftrightarrow CH_3-H_4MPT + corrinoid-protein \qquad (45)$$

could describe a mechanism for Reaction (43) in *M. barkeri*; the corrinoid–protein of this scheme is analogous to the C/Fe–SP in clostridial acetate synthesis. Recently Cao and Krzycki [147] demonstrated that acetate-grown *M. barkeri* contain two electrophoretically distinct corrinoid–proteins (480 kDa and 29 kDa) that are methylated by acetate in the presence of BES; the 480 kDa protein is methylated at the onset of methanogenesis from acetate and demethylated as methane formation ceases. Further work with these proteins may sharpen understanding of the mechanism described in Reactions (44) and (45).

Conversion of CH_3–H_4MPT to CH_3–CoM in acetate methanogenesis is expected to be mechanistically similar to the CO_2-reduction pathway. Acetate-grown *M. barkeri* has a methylcobalamin:CoM methyltransferase distinct from this enzyme in methanol-grown cells [263], suggesting that a specific version of the two-step mechanism given by Reactions (19) and (20) operates during acetate methanogenesis. However, recent work with membrane preparations of acetate-grown cells suggests that methylation of CoM with CH_3-H_4MPT may be catalyzed by a single membrane-bound corrinoid enzyme [117]. From related work [195] it could be speculated that this methyl transfer generates a Na^+-motive force required for certain steps in acetate methanogenesis [264,265].

Purified CODH from *"Methanothrix" soehngenii* catalyzes acetyl–CoA/CO exchange, and EPR data shows that acetyl–CoA and CO interact with certain metal centers of this enzyme [266,267]. Thus, in *"Methanothrix"* CODH is the likely site for acetate cleavage, and acetyl–CoA is the substrate for this reaction; however, nothing is known about the steps beyond the cleavage of acetyl–CoA to form CH_3–CoM.

3.3.4. Reduction of CH_3–CoM to CH_4

Methane formation from CH_3–CoM in acetate methanogenesis follows the same course described for H_2–CO_2 methanogenesis; methyl–coenzyme M reductases have been purified from acetate-grown *Methanosarcina* and *"Methanothrix"* [242,268].

3.3.5. Electron transfer

The route for transport of electrons from carboxyl-group oxidation to the site of heterodisulfide reduction is unknown. Methanogenesis from acetate in extracts of *M. barkeri* requires H_2, whereas such conversion by cells is inhibited by H_2 [234,259,262]. Ti(III)-citrate can substitute for H_2 in methanogenesis from acetyl–CoA by extracts, but the rate of methanogenesis is lower; high rates are obtained only if H_2 is present in addition. This means the role of H_2 in extract methanogenesis from acetate is not limited to activation. But, H_2 may not have a physiologically significant role in vivo. Membrane-dependent electron transfer systems to shuttle electrons from the carboxyl oxidation site to the heterodisulfide reduction site are probably not available in extract systems, and thus H_2 may replace this in vivo role of membrane components.

The role of F_{420} in acetate methanogenesis remains unclear. Acetate-grown *M. barkeri* contains F_{420} at a 3- to 4-fold lower concentration (0.2 mmol/g dry cell) compared to H_2–CO_2- or methanol-grown cells [269], and methanogenesis from acetyl–CoA or CH_3–CoM in cell extracts is independent of F_{420} addition [254]. Compared to H_2–CO_2- or methanol-grown cells of *M. barkeri* strain Fusaro, acetate- or H_2–CO_2 + methanol grown cells contain 2- to 3-fold lower levels of F_{420}-reducing hydrogenase [224]; it could be speculated that if the steps from CO_2 to CH_3–CoM (in Fig. 1) are not operative in either direction, F_{420}-reducing hydrogenase level is down-regulated. No information on the role of F_{420}-nonreactive hydrogenase in acetate-methanogenesis is available; this enzyme may act as the conduit for electrons needed for heterodisulfide reduction.

Methanogenesis from acetate in extracts of *Methanosarcina* does not require membrane addition [260]. However, this does not exclude a function for cytochromes in acetoclastic methanogenesis by whole cells. Rather, the role of H_2 in cell extracts, the ability of cytochrome b from *Methanosarcina* species to react with CO, and the observation that membrane-bound cytochrome b of *M. barkeri* is reduced by H_2, and is oxidized by CH_3–CoM + ATP or CH_3–CoM + acetyl-phosphate, all point to the participation of cytochromes in *Methanosarcina*. A role of cytochromes in transport of electrons generated from carboxyl-group oxidation to the heterodisulfide reductase is a logical hypothesis.

TABLE 4

Methylviologen-reducing hydrogenases (MVH; non-F_{420}-reducing) from methanogenic bacteria

Organism	Native molecular weight (kDa)	Minimal active size (kDa)	Subunits (kDa) [a]	Components (per min. size)				Ref.
				Fe	Ni	Flavin	Se	
M. thermoautotrophicum Marburg	nr[b]	60	52,38	nr	0.8	none	nr	272,291
M. thermoautotrophicum ΔH	nr	nr	52,40; 57,42	+	+	none	nr	47,275
M. voltae[c]								277,294
M. jannaschii	475	nr	43,31	nr	nr	nr	nr	279
M. formicicum MF	70	70	48,38[d]	10	1	nr	nr	270,276
Methanosarcina Göl	78	78	60,40	15	0.8	none	nr	204

[a] From protein, not sequence.
[b] nr = not reported.
[c] Genetic studies indicate its presence, but it has not yet been purified.
[d] Data from Laemmli gels; Weber–Osborne gels show two bands of 50 and 31 kDa.

4. Enzymes involved in methanogenesis

4.1. General

A wide range of novel enzymes are found in the methanogenic pathway, many deriving their novelty from the unusual coenzymes used. However, examples of parallel reactions in bacteria and methanogens can be found, e.g. the strong chemical and biochemical similarities of the methanopterin and folate reactions. Although methanogens are strict anaerobes, many of their enzymes are oxygen stable. We provide here a description of each enzyme known in the methanogenic pathways, as shown in Fig. 1. Their reactions have already been described in the previous section.

4.2. Hydrogenase and non-catalytic redox proteins such as ferredoxin and cytochromes

Hydrogenase (H_2ase) in methanogens catalyzes several reactions, and may be involved in H_2 uptake or production, depending on the circumstance or organism. The most commonly studied cell-free reactions are uptake of H_2 using either F_{420} or methylviologen (MV) as electron acceptors, as described in Reactions (23) and (46).

$$H_2 + 2MV^{++} \rightarrow 2H^+ + 2MV^{+\bullet}. \qquad (46)$$

All H_2ase-plus methanogens examined so far contain the F_{420}-reducing hydrogenase (FRH), which catalyzes both Reaction (23) and Reaction (46), and the non-F_{420}-reducing enzyme, also called the MV-only H_2ase (MVH); see Tables 4 and 5 [47,270–281]. In vivo, both H_2ases are probably of importance, with the MVH donating electrons to an electron carrier other than F_{420}.

TABLE 5
F_{420}-reducing hydrogenases from methanogenic bacteria

Organism	Native mol.wt (kDa)	~Minimal active size (kDa)	Subunits (kDa) [a]	Components (per min. size)				Ref.
				Fe	Ni	Flavin	Se	
M. thermoautotrophicum Marburg[b]								
M. thermoautotrophicum ΔH	~800	115	47,31,26[c][1:1:1]	14	0.7	0.9 FAD	nr[d]	47, 275
M. vannielii	~1300	340	56,42,35	nr	nr	nr	3–4	281
M. voltae	~745	105	55,45,37,27	5	0.6	1 FAD	0.7[e]	277
M. jannaschii	~990	115	48,32,25	nr	nr	nr	nr	279
M. formicicum MF	~600	380	74,43,34,24[f]	nr	nr	nr	nr	276
M. formicicum JF-1	~1000 or 600	109	44,37,29	13	1	1 FAD	none	271, 278
M. barkeri MS	~800	nr	200,130,60	9	0.7[g]	1 FMN[g]	nr	273
M. barkeri Fusaro	~845	198	48,33,30 [2:2:1]	nr	nr	nr	nr	274
Methanosarcina Göl[h]								204
M. hungatei	~720	144	51,31 [1:3]	nr	nr	nr	nr	280

[a] From protein, not sequence. Stoichiometry given in brackets.
[b] Not yet reported, almost certainly present.
[c] Original molecular weight reported as 40,31,26 kDa.
[d] nr, not reported.
[e] Another form of FRH not containing Se may also be produced [277].
[f] Data from Laemmli gels; Weber–Osborne gels showed three bands of 41, 32, and 21 kDa.
[g] Per 60 kDa subunit.
[h] Present, but not yet purified.

Since most methanogens can use H_2 as a source of electrons, this enzyme is found in the vast majority of them. In several cases examined with species able to use an alternate substrate, H_2ase has been consistently present, regardless of the substrate used at the time [221,224,282,283]. One example to the contrary is the adaptation time required by M. thermophila prior to growth on H_2–CO_2, which is probably due to low hydrogenase levels [284]. In methanogens unable to use $H_2 + CO_2$ for growth under any circumstances, e.g. "Methanothrix" [30,285–287], Methanoccoides methylutens, or Methanosarcina acetivorans, an uptake hydrogenase is not likely to be present in high levels; however, low levels (0.05 μmol/min mg) of benzylviologen-reducing H_2ase have been reported in "Methanothrix" CALS-1 [288]. In some cases, a methanogen may be isolated, and originally found unable to use H_2, but after long-term incubation with H_2–CO_2, it will develop the ability to do so [289]. In some organisms growing on non-H_2 substrates, e.g. methanol, acetate, or CO, an H_2-evolution function may be important [221,290], and production of low levels of H_2 is seen when cells are grown on acetate [288]. It has not been established in any methanogen whether different proteins are involved in H_2 uptake as opposed to production.

As shown in Fig. 1, FRH can provide electrons to several points of the methanogenic pathway via F_{420}, but in some cases F_{420} is not the direct reductant. Whenever investigated, an MVH which carries out only Reaction (46) has also been observed in cells containing FRH (see Table 4); since MV is an artificial electron acceptor, another physiologically relevant acceptor must be used by this H_2ase, e.g. possibly a ferredoxin. H_2ase is a major cellular protein, and either of the two types can account for about 2% of the total extracted protein [47,274,275,291]. A newly discovered type of "hydrogenase" which reacts directly with methylene–H_4MPT has been described, and is discussed separately in section 4.8 on methylene–H_4MPT dehydrogenase [114].

Purification of methanogen H_2ase is often done aerobically, since it is unstable under reduced conditions, but stable under air in the absence of reducing agents [47,270,271,274–276]. However, the assay must be under anaerobic and very reduced conditions, and requires some form of reductive activation for reproducible and accurate analysis. H_2ase has been purified from a variety of methanogens, as shown in Tables 4 and 5.

4.2.1. The methylviologen-reducing hydrogenase (MVH)
The MVH has been isolated and studied [47,270,272,275,276,279,291–293] from several methanogens; see Table 4. In one instance (*M. voltae*) it was originally reported as absent [277], but genetic approaches from the same lab have revealed genes corresponding to what should be an MVH [294]. The MVH is composed of two subunits; in one case (unpublished work described in a review [295]) the MVH from *M. thermoautotrophicum* was stated to contain four subunits, but this has not been confirmed. The MVH cannot reduce F_{420}, but can reduce MV. It contains both non-heme iron–sulfur groups and one nickel per molecule, but it contains no flavin. Subunit sizes, polypeptide mapping, and sequence analysis have proven that the two types of H_2ase are not coded by the same genes [276,293,294,296–298]; this may suggest that not only are they different structurally and catalytically, but they are used for different metabolic purposes, and could be regulated separately in some cases. However, as described below, there is similarity in the sequence of homologous subunits.

The MVH of *M. thermoautotrophicum* strains Winter and ΔH have been cloned and sequenced [293]. Accompanied by data on the *M. thermoautotrophicum* ΔH enzyme subunit composition [292,295,299], and data from *M. thermoautotrophicum* Marburg [272], the sequence data give a good description of MVH in methanogens. The MVH from *M. thermoautotrophicum* ΔH was originally reported to contain two subunits of 52 and 40 kDa molecular weight [292]. The MVH gene cluster mvhDGAB contains four open reading frames. The hydrogenase is composed of two subunits coded for by the mvhA and mvhG genes; calculated molecular weights for these subunits are 57 and 42 kDa, corresponding reasonably well to the original protein data. The β subunit is lost during purification, and is not essential for MVH activity. Although the protein arising from mvhD copurifies to some extent with the MVH, its role in hydrogenase activity is unknown.

The β subunit corresponds to the mvhB gene, which codes for an unusual protein termed a polyferredoxin that contains six tandemly repeated bacterial ferredoxin regions, with as many as 12 ferredoxin iron–sulfur clusters [293,300,301]. This protein is the first example of such a tandem repeat molecule. It may play a role in electron transfer from hydrogenase, perhaps to the methanogenic pathway, or, alternatively [301], it may be used for iron storage. It accepts electrons from hydrogenase, under the conditions examined, at a rate of 90 nmol/min mg, which is 30- to 40-fold slower than the rate at which MV is reduced by the hydrogenase [301]. The authors conclude that it is "questionable whether this interesting protein is the physiological electron acceptor although this cannot be excluded". It is not processed to form smaller molecules, and is a soluble protein.

The MVH from *Methanothermus fervidus* has also been cloned and sequenced [298], and is similar to the corresponding enzyme from *M. thermoautotrophicum* in that its genes are located in a cluster (mvhDGAB) which gives rise to the small and large hydrogenase subunits (mvhG and mvhA), and to a polyferredoxin (mvhB). The sequences for the mvhG, mvhA, and mvhB genes are, respectively 73, 70, and 63% identical between *M. thermoautotrophicum* ΔH and *M. fervidus*. The role of mvhD is unknown.

The sequence and organization of the mvh A and G genes suggest a common evolutionary relationship with bacterial hydrogenases [293,297,298]. The mvh sequence data from *M. thermoautotrophicum* strain Winter shows 88% identical bases with *M. thermoautotrophicum* ΔH, which encode 90% identical amino-acid residues. In a recent study of hydrogenases in *Methanococcus voltae,* Halboth and Klein [294] have shown that *M. voltae* uses a similar gene organization mvhDGAB; moreover, two gene clusters were found, one where the Ni-containing subunit gene coded for a selenocysteine, and the other where the same subunit gene coded for a normal cysteine. The hydrogenase Ni is thought to be bound by the cysteine or selenocysteine coded for by these mvhA genes; however, neither of the MVH proteins have been purified or characterized. As well, there is a parallel duplication in the FRH in *M. voltae*, as described below.

4.2.2. Redox-active proteins: ferredoxin, cytochromes, and others
Concerning the general topic of ferredoxin and redox proteins, it is important to note the existence of several of these in various methanogens. Ferredoxin has been reported in *M. barkeri* [302], *M. thermophila* [249,303], and *Methanococcus thermolithotrophicus* [304,305]. In one case, ferredoxin is required in electron transfer from carbon monoxide dehydrogenase to a membrane-bound hydrogenase in *M. thermophila* [306]. Another redox protein, referred to as a "glutaredoxin-like protein", has been isolated from *M. thermoautotrophicum* [307], but has no known function. As mentioned above, MVH gene clusters contain a sequence for a polyferredoxin of unknown function.

Cytochromes are reported principally in the family Methanosarcinaceae [308–310], which include the genera *Methanosarcina,* "*Methanothrix*", and *Methanolobus*. Although cytochromes can be reduced by hydrogen [311], such reactions are slower than when H_2F_{420} [211] is used. Reduced cytochromes are also oxidized by the heterodisulfide, leading Kamlage and Blaut [211] to conclude, on the basis of work with methanol-grown *Methanosarcina* Göl, that electrons from methyl-group oxidation are transferred by one or more cytochromes to the heterodisulfide oxidoreductase system, functioning in the oxidation of reduced F_{420} [211]. However, details of this reaction remain to be described. They also comment that since acetotrophic methanogens also possess cytochromes, they serve a different function there. Indeed, experiments with acetate-grown *M. thermophila* provide evidence for a linkage between cytochrome b and hydrogenase [306].

4.2.3. The F_{420}-reducing hydrogenase (FRH)

Several features are seen with most FRHs [47,271,273–281,292,312]. All use both F_{420} and MV as electron acceptors. The native enzyme is large, generally with several sizes ranging from 100 to 1000 kDa, and a minimal size of 100–200 kDa. Three subunits are usually present with the largest subunit being ~50 kDa, and the other two being ~25–45 kDa. See Table 5. In *Methanospirillum*, only two subunits were seen [280], but it is possible that the smaller 31 kDa protein band could result from unresolved electrophoresis of two subunits of very similar size, given the small difference between the small subunits shown in Table 5. The *Methanococcus voltae* FRH has 4 subunits of ca. 55, 45, 37, and 27 kDa. All the FRHs examined for the specific components contain non-heme iron and sulfur, most likely as Fe–S clusters, and contain ca. one Ni and one flavin per minimal enzyme size [47,271,273,275,277,278,281].

The FRH from *M. thermoautotrophicum* ΔH has been cloned and sequenced [296]. The three subunits α, β, and γ have protein molecular weights of ca. 47, 31, and 26 kDa, and the sizes predicted for these subunits from their respective sequences are 45, 31, and 26 kDa. The three subunits are coded by an apparent transcriptional unit, frhADGB, with the D gene coding for an ORF of unknown function; the gene product of frhD is not copurified with the hydrogenase.

Cloning and sequence analysis of the FRH from *M. voltae* has proven interesting, since there are two FRH genes: one coding for a selenocysteine, and the other coding cysteine at the same sequence position [294]. Both are arranged in a similar fashion to the *M. thermoautotrophicum* genes, as fruADGB (coding the selenium enzyme) and frcADGB (coding the cysteine enzyme), and are analogous to the frhADGB region of *M. thermoautotrophicum*. However, of the proteins produced by these genes (and the analogous pair of proteins arising from the MVH genes discussed above), only one has been purified, the selenium-containing FRH [277]. Interestingly, this protein has four subunits, which may come from the four genes in the putative transcriptional unit. Despite the presence of this fourth subunit, it remains of unknown function. It is not clear if and when the four different *M. voltae* hydrogenases are produced and used, but as put by Halboth and Klein, "It is tempting to consider an alternative expression of selenium-containing or selenium-free hydrogenases in *M. voltae* depending on the availability of this trace element" [294].

The frhA-derived peptide from *M. thermoautotrophicum* is clearly related to the large subunits of other nickel-containing hydrogenases, both bacterial, and to the large subunit of the MVH from the same methanogen [293,296,297]; similar homology is found with the A gene of *M. voltae* [294]. Unfortunately, separation of intact *M. thermoautotrophicum* subunits causes major losses of both Ni and flavin, making knowledge of cofactor distribution tentative at this point. Currently, data suggest, but do not prove, that Ni is bound to the α subunit in the active enzyme. Fe is also present in this subunit, as indicated by iron assays following electrophoretic separation of the subunits [275]. Methylviologen activity staining of gels also indicate the α subunit alone can use hydrogen to reduce MV [275]. The sequence of the γ subunit suggests that this protein is the site of two [4Fe–4S] clusters. The β subunit may be the site where the flavin is bound, an hypothesis supported by the fact that it has 26% identity with the β subunit of the formate dehydrogenase from *M. formicicum,* since this formate dehydrogenase contains a loosely bound flavin.

The enzymatic detail of the FRH has been most thoroughly studied in *M. thermoautotrophicum*. Physical biochemical methods have focused on the properties of the metals and flavins. Early observations by Lancaster[313,314] demonstrated an unusual EPR signal in *Methanobacterium bryantii* extracts which he correctly proposed as arising from nickel. This was confirmed in a series of nickel hyperfine EPR studies on *M. thermoautotrophicum* ΔH [275,292], and was in agreement with parallel studies on the MVH of *M. thermoautotrophicum* Marburg [315]; both FRH and MVH show very similar EPR spectra, which are similar to the signals originally observed by Lancaster. The original EPR signals observed with aerobic enzyme arise from Ni(III); upon reduction with dithionite or H_2, the Ni signal disappears, consistent with a change to the Ni(II) state [292]. Upon anaerobic oxidation (by replacing H_2 with argon), the flavin semiquinone contributes an EPR signal. Iron EXAFS data indicate the presence of iron–sulfur clusters, and nickel EXAFS suggest that three sulfur molecules are at $\sim$0.2 nm distance from Ni in the protein [316]. Electron spin echo spectroscopy suggests that a nitrogen atom is $\sim$0.35 nm or more from the Ni site [317], but it is unclear if this is a protein or flavin nitrogen. Despite the information obtained by these studies, and although it undergoes redox transformation, Ni participation in catalysis is not certain.

The catalytic mechanism of the *M. thermoautotrophicum* FRH has been more heavily studied than others, although it is still unclear what the structure of the active site looks like. As suggested earlier with whole-cell studies using D_2 and D_2O [209,210], the cleavage of H_2 yields an H_2F_{420} product molecule containing very little "H" that was in the H_2, but rather the C-5 position carries an "H" that arises from H^+ in water [312]. The F_{420} is reduced on the si face [318]. Kinetic studies suggest a two-site hybrid ping-pong mechanism [312]. Also, no steps involving D transfer are rate limiting; both substrate and solvent kinetic isotope effects were <1.1.

A characteristic of virtually all methanogen hydrogenases is their very large native size (although they are soluble enzymes), typically 600–1000 kDa. This is often explained as a result of hydrophobic interactions with other proteins, and as an indication that it is a loosely-bound membrane protein. H_2ase is often associated with membranes early during its purification [271,280]. These properties have been investigated in several ways by electron microscopy.

Examination of purified FRH from *M. thermoautotrophicum* ΔH by negative staining and shadow-casting electron microscopy has revealed a 800 kDa multisubunit complex [319]. The proposed model depicts the FRH as two stacked rings, with an inner "channel", an outer diameter of $\sim$9 nm, and a height of 16 nm. Sprott et al. [280] also isolated, in an undisturbed way, the FRH from *Methanospirillum hungatei* as a 750 kDa complex; electron microscopy revealed a "coin-shaped" structure of 16 nm diameter, and a central depression. Muth et al. [277] showed similar electron microscopic pictures of hydrogenase from *M. voltae*, with diameters of 12 or 18 nm. Wackett et al. [319] conclude that the large FRH "supramolecular assemblies may have relevance in vivo in the construction of multiprotein arrays that function in methane biogenesis".

Using immunogold electron microscopy, several labs have examined the location of FRH in cells. The enzyme is membrane-associated in *M. vannielii* [320] and *M. barkeri* [321]. The membrane location of H$_2$ase suggests its role in membrane-bound electron transport processes, or in creation of a proton gradient; as yet these possible roles have not been clearly described in methanogens (see also Chapter 4 of this volume), and several other proteins have been proposed as generators of the proton gradient, e.g. the methyl reductase or the heterodisulfide reductase.

4.3. Alcohol dehydrogenase (ADH)

Several groups have reported the unusual and unexpected ability of some methanogens (new isolates and known strains) to oxidize alcohols other than methanol, and to use the electrons obtained to reduce CO$_2$ to methane [25–29,212,322]. Alcohols are oxidized to either the ketone or acid form, depending on the substrate; the methanogens do not make methane from the alcohol carbon, but simply use the alcohols as an electron donor (Reactions 3 and 4 in Table 2). Alcohols used by at least one strain include 2-propanol, 2-butanol, ethanol, 1-propanol, cyclopentanol, 2-pentanol, and 2,3-butanediol; growth is typically slight, and evidence for the presence of an alcohol dehydrogenase (ADH) is commonly obtained by observation of methane production dependent on the target alcohol. Despite the uncomfortable similarities of this phenomenon to "*Methanobacillus omelianskii*", the mixed culture grown using ethanol [323,324], most researchers have rigorously established culture purity.

Thus far, most work on this phenomenon has been with whole cells or crude extract activities, and thus little enzymatic detail is available. The ADH activity is not present in cells grown in abundant H$_2$ without the alcohol [26,27], suggesting regulation; H$_2$ starvation in the absence of alcohol may also induce the ADH. Several alcohols can cause cells to produce the active enzyme.

The ADH in *Methanobacterium palustre* (grown with 2-propanol as the electron donor) was assayed with 2-propanol in crude extracts and found to use NADP, but not NAD, F$_{420}$, FMN, or FAD as electron acceptors (Reaction 26) [212]; V_{max} was about 0.1 µmol/min mg. 2-butanol, 2,3-butanediol, and cyclopentanol were also substrates, but ethanol, 1-propanol, and 1-butanol were not. This suggests this organism has a secondary ADH. Several ketones could also be reduced to their corresponding alcohols using NADPH as the electron donor. Crude extract ADH activity dependent on NADP was also observed with ethanol-grown *Methanogenium organophilum*. With ethanol and 2-propanol the specific activities were, respectively, 5 and 10 µmol/min mg; NAD served as a poor electron acceptor (0.1 µmol/min mg), and F$_{420}$ did not work at all [25].

In contrast, an F$_{420}$-dependent ADH was purified from *Methanogenium thermophilum* [27]. Extract activity was about 15 µmol/min mg with 2-propanol and F$_{420}$ as substrates (Reaction 25); no activity was seen with NAD(P), FMN, or FAD. Anaerobic purification yielded a 12-fold purified enzyme with 26% recovery; exposure to air was not harmful for up to 30 min, but longer exposures caused activity loss. The native enzyme was a 65 kDa protein, composed of two subunits of 39.5 kDa. The pure enzyme oxidized 2-propanol at about 176 µmol/min mg, and ethanol or 1-propanol at about 0.2 µmol/min mg, using F$_{420}$ as the electron acceptor. With reduced F$_{420}$ as the electron donor, acetone, acetaldehyde, and propionaldehyde were reduced, suggesting that ADH could also serve as an aldehyde reductase, and that it was responsible for the dismutation of aldehydes seen in whole cells. Purification of this F_{420}-dependent enzyme provides reassuring evidence that the methanogen is actually responsible for the ADH, due to the novelty of F_{420} itself.

A secondary ADH dependent on F_{420} was also detected with 2-propanol using crude extracts of *M. hungatei,* with an activity of ca. 0.2 μmol/min mg [27]. Bleicher et al. [322] conducted a survey using extracts of nine secondary alcohol-using methanogens to determine substrate and coenzyme specificities of the ADH activities. With 2-propanol as the electron donor, specific activities varied from 0.05 to 8.4 μmol/min mg. Four strains used NADP as electron acceptor, but not F_{420}; the other five strains used F_{420}, but not NAD(P). A wide but variable range of primary and secondary alcohols were used as electron donors.

4.4. Formate dehydrogenase

About half of all methanogen species can use formate as a methanogenic and growth substrate, including some species of *Methanococcus, Methanobacterium,* and *Methanospirillum* [8]. It is not used for growth by any *Methanosarcina*. Formate is thought to be oxidized to produce CO_2 and a reduced electron carrier; if free H_2 is produced, it can be oxidized by hydrogenase. The enzyme formate dehydrogenase (FDH) catalyzes the following reaction:

$$H^+ + HCOO^- + F_{420} \rightarrow H_2F_{420} + CO_2. \qquad (47)$$

In enzyme assays, methylviologen is often used as an artificial electron acceptor.

In work with *M. thermolithotrophicus*, using D_2O and ^{13}C-formate, the hydrogen in the methane molecule was shown to originate from the H^+ in the medium, not via the "H" originally on formate [325], in contrast to direct reduction reported based on experiments with rumen fluid enrichments [326]. Work with purified FDH from *M. formicicum* agrees with the conclusion that the "H" in methane arises from the H^+ in water [327].

Tzeng et al. [328] first reported the use of F_{420} as an electron acceptor for formate oxidation in extracts of *Methanobrevibacter smithii*, but proof of the role of F_{420} in FDH catalysis was obtained by purification of this enzyme from *Methanococcus vannielii* [48,49,329] and *Methanobacterium formicicum* [327,330–332]. Most enzymatic detail of the methanogen FDH comes from study of these two organisms. A review on FDH has recently been published by Ferry [253].

There are two chromatographically separable FDHs expressed in *M. vannielii*, depending on the levels of selenium and tungsten in the medium [48,49]. Both are very oxygen sensitive. If abundant selenium is available, growth is faster than under Se-deficient conditions, and a high molecular weight selenoprotein complex contains FDH activity; the Se is located in Se-cysteine. When both Se and tungsten (W) are provided in the medium, this selenoprotein is the predominant FDH in the cell, and some replacement of molybdenum (Mo) by W appears to occur. Under Se-starved conditions only one smaller FDH is produced; after purification, it was found to be a 105 kDa enzyme with 6 Fe per molecule, but no Se was present. Little information is available on similarities at the molecular or genetic level of these two FDHs, e.g. subunit composition similarities, or if they arise from different genes. It is of interest that *E. coli* also produces a selenium-containing FDH [333].

The FDH from *M. formicicum* has been studied in more detail. Cells grow well without added Se or W, and produce an FDH with a native molecular weight of 177 kDa, composed of two subunits with 85 and 50 kDa sizes in an $\alpha_1\beta_1$ arrangement [331,332]. This enzyme contains 1 Mo, 1 FAD, 2 Zn, and ca. 23 Fe and S per molecule, and no Se or W. During purification or manipulation, FAD is lost, but the apo-enzyme can be reconstituted with FAD; FMN or reduced-FAD cannot substitute. The FDH also contains a molybdopterin guanine dinucleotide similar to the pterin cofactors in the molybdoproteins nitrate reductase and xanthine oxidase [123,124,334]. The extracted cofactor (although originally reported to be unable to do so [124]), is able to complement the molybdopterin-deficient nitrate reductase of *Neurospora crassa* like those from most molybdoenzymes [123]. The pterin is similar to the folate-related pterins since it has a 6-alkyl side-chain (not the 7-alkyl group expected in methanopterin), and its fluorescent properties are also similar to those of folates, not a methanopterin. A structure has been proposed for the extracted cofactor [123]. EPR analysis of the reduced enzyme revealed Mo-derived EPR signals [335]; exposure to cyanide caused both irreversible loss of activity and a change in the EPR signals. The potential role of Se in FDH catalysis was examined by the introduction of a selenocysteine, replacing the normal cysteine, into the FDH gene from *M. formicicum* [336]; although immunoreactive material was produced in *E. coli* transformed by this gene, no catalytic activity was observed. It will be interesting to look for activity when this modified gene is transferred to an FDH-negative *M. formicicum* strain, since the environment is more suitable for an active methanogen FDH.

During catalysis, formate is most likely oxidized by the FeS or molybdopterin portions of the molecule, and the electrons transferred to FAD; the electrons are then transferred two electrons at a time to the F_{420} [124,327,332]. FAD is essential for F_{420} reduction, but the artificial electron acceptor methylviologen is reduced with or without the flavin, most likely by one-electron transfer from the Fe–S centers to the oxidized viologen. Like the previously studied F_{420}-dependent enzymes NADP:F_{420}-oxidoreductase [58,337] and FRH [318], the F_{420} is reduced on the si face [327]. Schauer et al. [327] conclude that hydride transfer from formate probably generates "an EH_2 species and then by hydride transfer back out to F_{420}, the formate-derived hydrogen exchanged with solvent protons before transfer back out to F_{420}" [327]. The lack of significant deuterium solvent or substrate isotope effects suggest that transfers of substrate-derived D is not rate imiting in catalysis [327]. The presence of Mo in the enzyme is essential for activity [334].

Although *M. formicicum* appears to always produce FDH, variation in levels occur, and the molecular details of FDH synthesis have been studied [334,338,339]. Mo depletion in the medium yields 35- and 100-fold drops in the MV-and F_{420}-dependent activities, respectively, and a 15-fold drop in FDH protein as assayed by ELISA methods. Although a drop in Mo levels leads to increased mRNA production from the fdh genes, Mo is needed for protein synthesis and FDH activity [340]. Addition of excess W leads to loss of FDH activity, which can be reversed competitively by addition of more Mo [334]. The FDH from *M. formicicum* has been cloned and sequenced. The two subunits overlap by one base pair, and are in different reading frames. The genes have been expressed in *E. coli* [339].

Variation of FDH activity has been studied in *M. thermolithotrophicus*, where two different types of strains have been described [283]. In one strain, initial adaptation and growth on formate was preceded by a very long lag, but once growth was observed, formate growth and FDH levels became reproducibly similar at moderately high levels, even after many transfers on H_2–CO_2 medium, and FDH levels varied over a two-fold range. In another strain, FDH activity was modulated; in the absence of H_2 and presence of formate, FDH levels were ca. 10–40-fold higher than when cells were grown on H_2–CO_2. In all strains, two- to three-fold fluctuations were observed during growth, with peak activities in the exponential phase. The molecular approaches used with the mostly unmodulated FDH activity from *M. formicicum* have not yet been used with the clearly modulated *M. thermolithotrophicus* strain.

Formate–hydrogen lyase and formate–NADP reductase systems have been reconstituted from FDH and, respectively, hydrogenase [341] and NADP-F_{420} oxidoreductase [48]; this demonstrates that it is possible for formate to produce free H_2, and to provide NADPH for biosynthetic needs. As well, using the formate hydrogen lyase in the reverse direction [341], or with whole cells [448], formate can be synthesized from H_2–CO_2. Cells of an uncharacterized formate-utilizing methanogen can also produce high levels of formate when H_2–CO_2 and methylviologen are present with whole cells [342].

In several instances, FDH activity has been described in methanogens unable to grow on formate. One *Methanosarcina* strain has low FDH activity in crude extracts [343], but cannot grow on formate. Nelson and Ferry [344] reported FDH activity in *M. thermophila*, with a specific activity of 0.3 μmol/min mg in an assay with methylviologen as an electron acceptor [344], which compares to ca. 7.8 μmol MV reduced/min mg in *M. formicicum* grown on formate [332], and 80 μmol MV reduced/min mg with formate-grown *M. thermolithotrophicus* [283]. Extract of *"Methanothrix"* species strain CALS-1 reduces benzylviologen with formate at a rate of 5.8 μmol/min mg [288]. *M. thermoautotrophicum* Marburg also has low levels of NAD- or NADP-dependent FDH activities, ca. 0.008 μmol/min mg [345]; this is ~0.3–0.7% of the F_{420}-dependent activity observed in *M. formicicum* [345]. Most interestingly, an FDH-minus mutant has been obtained which has lower FDH levels (0.003 μmol/min mg using NADP, and no activity with NAD as electron acceptor), and which requires low levels of formate in the growth medium, suggesting a possible biosynthetic need for formate in methanogens not making methane from formate. The physiological role of these diverse FDH activities is unknown.

4.5. Formylmethanofuran dehydrogenase

This enzyme catalyzes reduction of CO_2 to formyl–MF (Reaction 48) during methanogenesis from CO_2, and the oxidation of formyl–MF (the reverse of Reaction 48) during methylotrophic methanogenesis [184]:

$$MF + CO_2 + 2[H] \rightarrow MF-CHO + H_2O \quad (48)$$
(if H_2 is reductant, $\Delta G^{o\prime} = +16 \text{ kJ/mol}$).

It is present (with varying specific activities) in H_2–CO_2-grown *M. thermoautotrophicum*, *Methanobacterium wolfei*, and *Methanobrevibacter arboriphilus*, H_2–

CO_2, methanol- and acetate-grown *M. barkeri*, methanol-and acetate-grown *M. thermophila*, and acetate-grown *"Methanothrix" soehngenii* [55,346]. H_2–CO_2-grown *M. hungatei*, *Methanogenium organophilum* and *M. voltae* and methanol–H_2-grown *Methanosphaera* have very low levels of this enzyme [55]; except for *Methanosphaera* this observation might be due to the use of inappropriate types of MF (see section 2.3) and electron acceptor (methylviologen). *M. thermophila* grown on acetate has less activity than methanol-grown cells [346]. *Methanosphaera stadtmanii* may not need formyl–MF dehydrogenase since it does not reduce CO_2, and reduces methanol only when H_2 is present [98]; it also does not possess MF [99]. This is consistent with the observation that formyl–MF dehydrogenase is down-modulated in *Methanosarcina* during growth on acetate [55,346]. Formyl–MF dehydrogenase is also present in *Archaeoglobus fulgidus* [69].

The $E^{0\prime}$ of the MF + CO_2/formyl–MF couple is −497 mV and that of H^+/H_2 is −414 mV [184]. Thus, in vitro H_2 is not a good electron donor for Reaction (48). However, an HPLC-based assay in the direction of CO_2-reduction, using Ti(III)-citrate ($E^{0\prime} = -480$ mV) as an artificial electron donor, has been described [185]. In the opposite direction, the enzyme is easily assayed spectrophotometrically using the artificial electron acceptor methylviologen ($E^{0\prime} = -446$ mV):

$$\text{MF-CHO} + 2\text{MV}^{++} + H_2O \rightarrow \text{MF-H} + CO_2 + 2\text{MV}^{+\bullet} + 2H^+. \tag{49}$$

Work on the purification and characterization of this enzyme is based on the MV-dependent assay [95,96,347–350].

Formyl–MF dehydrogenase has been purified from *M. thermoautotrophicum* Marburg [348], *M. wolfei* [350], and methanol-grown *M. barkeri* [95]. Properties of these purified enzymes are summarized in Table 6. The enzyme from different methanogens differs in size and the number of subunits. Formyl–MF dehydrogenases are iron–sulfur proteins that possess tightly bound metallopterins [95,348,350]. *M. wolfei* contains two types of formyl–MF dehydrogenases: a molybdoenzyme (the predominant form in cells grown with Mo) and a tungsten-containing enzyme (predominant if Na_2WO_4 but not Mo is added to the medium) [350]. Schmitz et al. [350] suggest that W is required for the activity of the tungsten-enzyme, and this enzyme is not a molybdoenzyme with Mo replaced by W; however, it remains to be demonstrated that these enzymes correspond to different apoproteins [350].

The tungsten-containing formyl–MF dehydrogenase is extremely oxygen sensitive, whereas the corresponding molybdoenzymes are less sensitive [350]; such differences have also been observed with other molybdoenzymes and their W-containing counterparts [350], and unlike molybdoenzymes, tungsten enzymes have so far been found only in strictly anaerobic bacteria and archaea [221]. Tungsten-containing formyl–MF dehydrogenase is one of only four W enzymes purified so far, others being formate dehydrogenase, carboxylic acid reductase, and aldehyde:ferredoxin oxidoreductase from several organisms [49,350]. During purification the tungsten enzyme from *M. wolfei* undergoes rapid inactivation, and the purified enzyme loses half of its activity in less than 12 hours [350]; such lability is probably due to extreme sensitivity to O_2. The molybdenum-containing formyl–MF dehydrogenases from *M. thermoautotrophicum* and *M. barkeri* are relatively stable [95,348].

TABLE 6
Properties of formylmethanofuran dehydrogenase

Property	M. thermoautotroph-icum strain Marburg [96,348]	M. wolfei [a] W-enzyme [350]	M. barkeri [95,349]
Apparent molecular mass (kDa)			
Native	110	130	220
Subunit	60,45[b]	65,51,35	65,50,37
			34,29,17
Metal content (mol/mol protein)			
Non-heme iron	4	5	28
Acid-labile sulfur	4	nr [c]	28
Molybdenum	0.3–0.6	non-detectable	0.5–0.8
Tungsten	nr	0.3–0.4	nr
Pterin content (mol/mol protein)	0.5–0.7	0.4	0.6–0.9
Type of pterin(s)	guanine, adenine and hypoxanthine dinucleotide (1:0.4:0.1 ratio)	guanine dinucleotide	guanine dinucleotide
Kinetic constants			
V_{max} (U/mg)	70	nr	175
Apparent K_m (mM) (formyl–MF)	0.03	0.013	0.02
Apparent K_m (MV, mM)[d]	0.10	0.4	<0.02
k_{cat} (s^{-1})	128	nr	640

[a] This organism also contains a Mo-containing enzyme; see text.
[b] An additional 38 kDa band is seen when a large amount of protein is applied to SDS-PAGE.
[c] nr, not reported.
[d] Non-physiological artificial electron carrier; physiological electron carrier is unknown.

Formyl–MF dehydrogenase of M. thermoautotrophicum and M. barkeri are associated with the 160 000 × g pellet of extracts, and a membrane-bound structure for these enzymes has been proposed. Evidence suggests that in M. barkeri, formyl–MF oxidation (during methanol methanogenesis) is coupled to generation of a electrochemical sodium potential that helps activate methanol to the methylene level [220,223,226,351]; as mentioned by Karrasch et al. [95] the M. barkeri enzyme might be suitable for such activities [95]. Although a role for formyl–MF dehydrogenase in Na$^+$ translocation has been hypothesized, purified formyl–MF dehydrogenase from M. wolfei is not stimulated by NaCl [350]. Extracts of both M. thermoautotrophicum Marburg and M. barkeri reduce F_{420}, but not NAD$^+$ and flavins, with formyl–MF [347], although purified enzymes are devoid of the F_{420}-reducing activity [95,96,348,350].

The energetics discussed above suggest that electrons from H_2 cannot be used directly for formyl–MF formation during H_2–CO_2 methanogenesis. At partial pressures of H_2 prevailing in the natural habitats (10 Pa; [192]) the energetics become even poorer, since under this condition $E^{0\prime}$ for the H^+/H_2 couple is -300 mV [184]. Thus, electrons from the primary donor (H_2) must be transported uphill ($E^{0\prime}$ for $CO_2 +$ MF/formyl–MF is -497 mV [184]) via unidentified (membrane-bound?) electron carriers. It has been suggested that the endergonic reduction of CO_2 to formaldehyde is driven by a transmembrane electrochemical potential of Na^+ [226]; of all steps, the most likely one to receive help is the reduction of CO_2 to formyl–MF. See also section 3.1.2 on the RPG effect for more on the activation of formyl–MF synthesis.

All Mo-containing formyl–MF dehydrogenases described above contain tightly-bound molybdopterin (it is likely that the W in a non-Mo enzyme is in a "tungstopterin"). Two types of molybdopterins are known in non-archaeal domains: molybdopterins of eukaryotic molybdoenzymes and bactopterins [352]. Molybdopterins of formyl–MF dehydrogenase (and also of formate dehydrogenase from *M. formicicum*) are bactopterins [123,348–350]. The enzyme from *M. thermoautotrophicum* contains three pterins, of which two are novel in having adenine– or hypoxanthine–dinucleotide groups [348]; the presence of a guanine–dinucleotide group is common [123,348–350,353].

N-furfurylformamide, which contains the furfuralamine moiety of MF but not the side chain (see Fig. 2), is a substrate for formyl–MF dehydrogenase from *M. barkeri* [93]; this enzyme also oxidizes formamide and formate [350]. The apparent K_m for *N*-furfurylformamide is 20 000 times higher than for formyl–MF, and the V_{max} with the artificial substrate is 8-fold lower [93], indicating the importance of the MF side chain. Such pseudosubstrate activities are not found with the enzyme from *M. thermoautotrophicum* and *M. wolfei* (W-enzyme) [93,350].

4.6. Formylmethanofuran:tetrahydromethanopterin formyltransferase

In reduction of CO_2 to CH_4, N^5-formyl H_4MPT is produced by formyl–MF:H_4MPT formyltransferase with formyl–MF as the formylating agent (Reaction 13) [184]. The reverse reaction occurs during the oxidation of methyl groups from methylotrophic substrates. Formyl–MF:H_4MPT formyltransferase has been purified from H_2–CO_2-grown *M. thermoautotrophicum* ΔH [94] and methanol-grown *M. barkeri* [113]; the gene has been cloned from *M. thermoautotrophicum* ΔH and sequenced [354]. The transferase is in methanol- and acetate-grown *M. thermophila* at equivalent levels, but the role of this enzyme in acetate catabolism is unknown [346]. This enzyme is also involved in oxidation of lactate-derived methyl groups in *Archaeoglobus fulgidus* [69].

The enzyme is typically assayed in the direction shown in Reaction (13) [94,113]; the conversion of H_4MPT to N^5-formyl–H_4MPT gives a rise in absorbance at 282 nm ($\Delta\varepsilon = 5.1$ mM^{-1} cm^{-1}; [94,113]). The formyltransferase reaction can be coupled to the methenyl–H_4MPT cyclohydrolase reaction yielding methenyl–H_4MPT [94,113], demonstrating the roles of these enzymes and N^5-formyl–H_4MPT in two consecutive steps of the CO_2-methanogenesis pathway (Fig. 1).

Purified enzymes from *M. thermoautotrophicum* and *M. barkeri* are similar in structural and kinetic properties, as shown in Table 7. The *M. barkeri* enzyme is a

TABLE 7

Properties of formylmethanofuran:tetrahydromethanopterin formyltransferase

Properties	*Methanobacterium thermoautotrophicum*		*Methanosarcina barkeri* strain Fusaro [113]
	strain ΔH [94]	strain Marburg [113]	
Apparent molecular mass (kDa)			
Native	160	–	35
Subunit	41	39.5	32
Catalytic properties			
K_m for H_4MPT (mM)	–	0.6 [a]	0.4 [b]
K_m for formyl–MF (mM)	–	0.5 [a]	0.4 [b]
V_{max} (U/mg)	–	4200 [a]	3700 [b]
k_{cat} (s^{-1})	–	2765 [a]	1973 [b]

[a] Determined at 65°C.
[b] Determined at 37°C.

monomer [113], whereas that from *M. thermoautotrophicum* ΔH is a homotetramer [94]. The purified formyl–MF:H_4MPT transferases are stable in air [94,113,354]; the *M. barkeri* enzyme is stable at 65°C (the temperature for its maximum activity) for >4 h. *N*-furfurylformamide, an analog of formyl–MF, is used by formyltransferases, although the K_m for this pseudosubstrate is about 1000-fold higher than that for the natural substrate [93]; formate, formamide and *N*-methylformamide are not used.

The gene for formyltransferase from *M. thermoautotrophicum* seems to be a part of an operon [354]. The cloned gene is functional in *E. coli*, indicating correct folding of this enzyme from a thermophilic organism in a mesophilic host [354].

4.7. N^5, N^{10}-methenyltetrahydromethanopterin cyclohydrolase

Methenyl–H_4MPT cyclohydrolase, first detected by Donnelly et al. [122], catalyzes hydrolysis of N^5, N^{10}-methenyl–H_4MPT to N^5-formyl–H_4MPT (Reaction 14). See also Fig. 6. The corresponding alkaline chemical hydrolysis yields N^{10}-formyl–H_4MPT [122]. These two reactions (enzymatic and chemical) can be distinguished by characteristic spectra and oxygen sensitivities of the products; N^5-formyl–H_4MPT is oxygen-stable whereas N^{10}-formyl–H_4MPT is oxygen-sensitive [94,122]. Although an early report on *M. barkeri* cyclohydrolase identified N^{10}-formyl–H_4MPT as the product [186], it is actually N^5-formyl–H_4MPT [113,187]. In the H_4folate systems of bacteria and eucarya N^{10}-formyl–H_4folate is the product of both chemical and enzymatic hydrolysis of methenyl–H_4folate [182]. However, a N^5-formyl–H_4folate cyclohydrolase that converts N^5-formyl–H_4folate to N^5, N^{10}-methenyl–H_4folate has been described [355]; this conversion requires ATP–hydrolysis [182], whereas dehydration of N^5-formyl–H_4MPT to N^5, N^{10}-methenyl–H_4MPT is ATP independent [100,113,122,356]. See section 3 for a discussion of differences between H_4MPT and H_4folate systems.

Methenyl–H_4MPT cyclohydrolase is usually assayed in the direction of hydrolysis (Reaction 14), although the dehydration reaction has been demonstrated [100,113, 122,186,356]. While assaying cyclohydrolase in the direction of hydrolysis, the rates

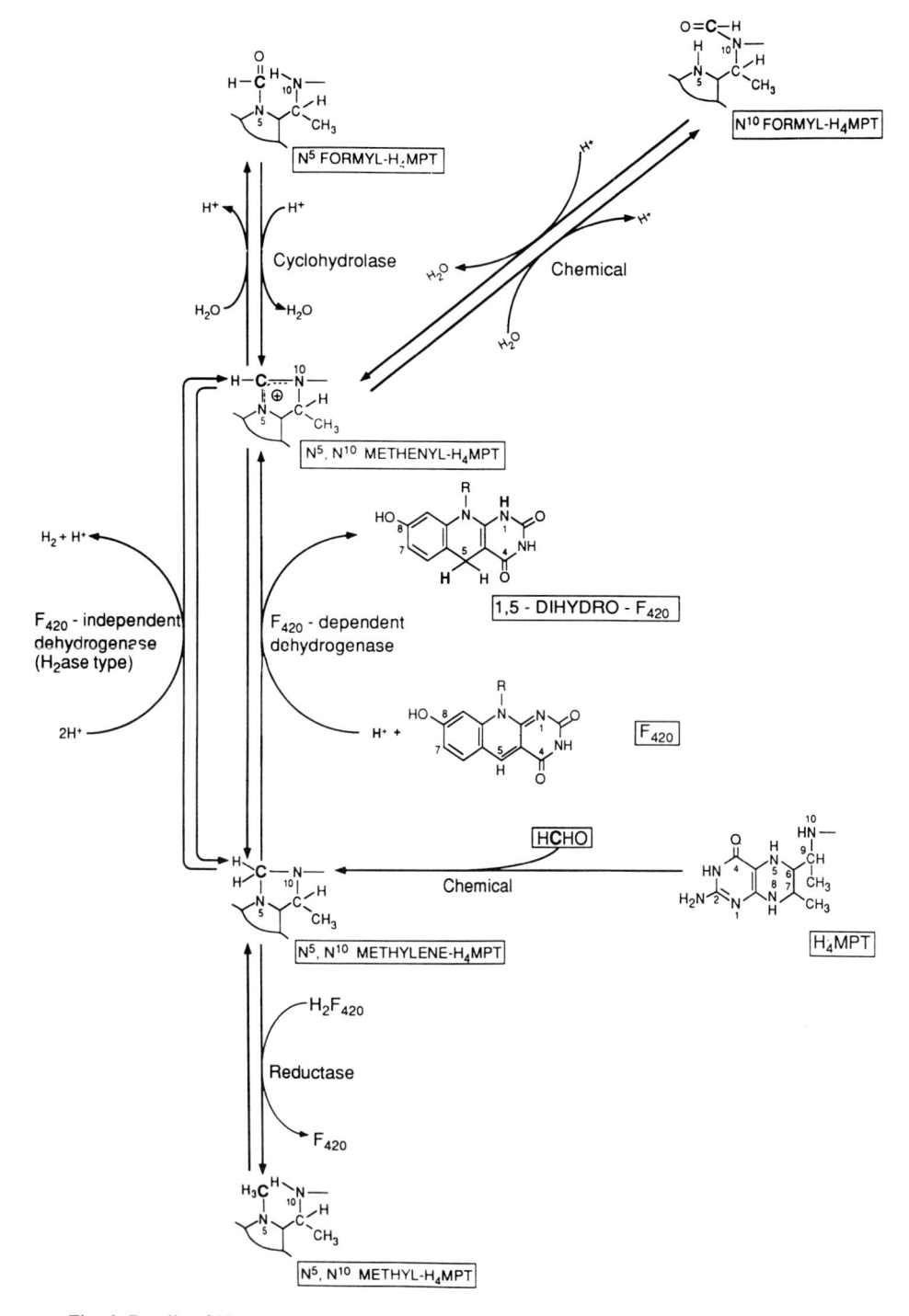

Fig. 6. Details of H₄MPT reactions in methanogens, and relevant chemical transformations.

TABLE 8
Properties of N^5, N^{10}-methenyltetrahydromethanopterin cyclohydrolase from methanogens [a]

Property	Methanobacterium thermoautotroph- icum strain ΔH [100]	Methanosarcina bark- eri strain MS [186]	Methanopyrus kand- leri strain MS [356]
Apparent molecular mass (kDa)			
Native enzyme	82	82	42
Subunit	41	41	41.5
Chomophoric prosthetic group	nr[b]	nr	absent
Catalytic properties			
Apparent K_m for $HC^+=H_4MPT$ (mM)	nr	0.57	0.04[c]; 2.0[d]
Specific activity at 30 μM $HC^+=H_4MPT$ (U/mg)	130	470	5026[c]
k_{cat} (s^{-1})	nr	nr	9200[c]; 1385[d]
V_{max} (U/mg)	nr	nr	13 300[c]; 2000[d]

[a] This table has been adapted mostly from Breitung et al. [356].
[b] nr, not reported.
[c] Determined in the presence of 1.5 M K_2HPO_4
[d] Determined in the absence of K_2HPO_4

must be corrected for simultaneous chemical hydrolysis. However, chemical hydrolysis is not observed at 20°C [356], and thus characterization of enzymatic hydrolysis can be done at this temperature without interference.

Methenyl–H$_4$MPT cyclohydrolase has been purified from *M. thermoautotrophicum* (ΔH and Marburg) [100,113], *Methanopyrus kandleri* [356], and methanol-grown *M. barkeri* [186]. Properties of these purified enzymes are given in Table 8.

The optimum pH values for these enzymes are in the alkaline range, not surprisingly, since assays are in the direction of methenyl hydrolysis, which releases a proton (Reaction 14). On the other hand, dehydration of N^5-formyl–H$_4$MPT occurs only when pH < 7 [113,122]. The optimum temperature for the *M. barkeri* enzyme is substantially higher than its optimum growth temperature, and the enzyme is thermostable. Unexpectedly high temperature optima have also been noted for formyl–MF dehydrogenase, formyl–MF:H$_4$MPT formyltransferase, methylene–H$_4$MPT dehydrogenase, and methylene–H$_4$MPT reductase, from *M. barkeri*. The cyclohydrolase from *M. kandleri* has several notable features [356]: this enzyme is a monomer, is stable at near boiling temperature in the presence of high levels of salts, and salts also stimulate activity up to 200-fold. Both K_m for $HC^+=H_4MPT$ and V_{max} are affected by salt concentrations; the catalytic efficiency of *M. kandleri* cyclohydrolase is 330 times higher in 1.5 M K_2HPO_4 than that in its absence. Activation by salts, although of lower magnitude, is also observed with *M. thermoautotrophicum* Marburg and *M. barkeri* enzymes (Table 8). The extent of stimulation by salts for these enzymes is related to the optimum growth temperature, and their intracellular cyclic 2,3-diphosphoglycerate (cDPG) contents [356]. The optimum growth temperatures of

M. thermoautotrophicum and *M. kandleri* are 65°C and 98°C [357,358], respectively, and the corresponding cDPG contents are 0.065 M and 1.1 M [356,359,360].

Cyclohydrolase activity in cell extract of *M. kandleri* is sensitive to air, but the purified enzyme is fairly air-stable [356]. Purified enzymes from *M. thermoautotrophicum* and *M. barkeri* are air-stable [51,100,186]. However, the *M. barkeri* enzyme requires non-ionic detergents and ethylene glycol for stability [186]. On the other hand, cyclohydrolase activity in cell extracts of *M. thermoautotrophicum* Marburg is stimulated 7-fold upon anaerobic incubation at 30°C in the presence of 2-mercaptoethanol or dithiothreitol, and 5-fold under the corresponding aerobic incubation [51]; the mechanism of this stimulation is unknown.

All methenyl–H_4MPT cyclohydrolases purified so far are monofunctional. In bacteria the analogous enzyme, methenyl–H_4folate cyclohydrolase, is found in both mono- and bifunctional forms (with methylene–H_4folate dehydrogenase activity) [182]. In eucarya, methenyl–H_4folate cyclohydrolases exist in three different forms [182]: mono-, bi- (with methylene–H_4folate dehydrogenase), and trifunctional (with dehydrogenase and formyl–H_4folate synthetase).

Methenyl–H_4MPT cyclohydrolase is essential for H_2–CO_2 methanogenesis. Pathway analysis and enzyme level studies suggest that this enzyme is also involved in methanol oxidation [224,346]. In acetate catabolism it does not have an obvious role, and in *Methanosarcina* growing on acetate this enzyme is down-modulated [224,346].

4.8. Methylenetetrahydromethanopterin dehydrogenase

This enzyme activity is responsible for the interconversion of methenyl–H_4MPT and methylene–H_4MPT. Early work with *M. thermoautotrophicum* revealed two types of dehydrogenase activities, one requiring F_{420} and the other F_{420}-independent [50,51]. The F_{420}-dependent enzyme has been purified from *M. thermoautotrophicum* strains Marburg and ΔH [50,51,361] and *M. barkeri* strains MS and Fusaro [188,189]; it catalyzes Reaction (15), as described further in Fig. 6. The F_{420}-independent dehydrogenase activity in *M. thermoautotrophicum* Marburg was ascribed to a H_2ase-type enzyme by Zirngibl et al. [114], who showed the purified enzyme catalyzes Reaction (16); such an H_2-evolving methylene–H_4MPT oxidizing activity is also present in *M. thermoautotrophicum* ΔH [105], and has been purified from *M. kandleri* and *M. wolfei* [190,191]. In anaerobic cell extracts the H_2ase-type enzyme out-competes the F_{420}-dependent enzyme [52], and thus the oxidation of methylene–H_4MPT does not accompany the reduction of F_{420}. This explains early failures to detect the F_{420}-dependent enzyme in "methanobacterium type" organisms [52,114].

From early results it was speculated that in *M. thermoautotrophicum* Marburg the F_{420}-dependent enzyme could be a processed form (chemical or enzymatic) of the H_2ase-type enzyme, whose subunit size is ~10 kDa larger than that of the former (ref. [52], and Mukhophadhyay, B., unpublished data), but cloning and sequencing (ref. [52], and Mukhophadhyay, B. and Purwantini, E., unpublished data) has shown that the two enzymes are genetically distinct, and expression does not involve cleavage of a precursor protein. A survey by Schworer and Thauer [55], and purification of these two types of enzymes from several organisms, suggest that both enzymes are present in the more "primitive" methanogens (most of which possess peptidoglycan cell walls), and the late-evolving methanogens (the order *Methanomicrobiales*) possess only the F_{420}-dependent enzyme.

Both types of methylene–H_4MPT dehydrogenases are routinely assayed in the direction of methylene–H_4MPT oxidation, by following the appearance of methenyl–H_4MPT (a rise in absorbance at 340 nm) [50,51,189]; for the F_{420}-dependent enzyme, assay at 340 nm requires consideration of spectral changes from both methenyl–H_4MPT formation and F_{420} reduction [50,51,189,361]. In addition, the F_{420}-dependent enzyme can be assayed by following the reduction of F_{420} at 400 or 420 nm [188,189,361], and the H_2ase-type enzyme by measuring H_2 evolution [114,190]. For both enzymes, reduction of methenyl–H_4MPT (with H_2 for H_2ase-type enzyme and with H_2F_{420} for the other) has been demonstrated [50,114,189–191,361].

All purified F_{420}-dependent methylene–H_4MPT dehydrogenases are homo-hexamers of 30–36 kDa subunits. All but the enzyme from *M. barkeri* show maximal activity at around the optimum growth temperatures of the source organisms; like many other enzymes of *M. barkeri*, the F_{420}-dependent enzyme from this mesophile is fastest at 60°C [188,361]. The F_{420}-dependent enzyme from *M. thermoautotrophicum* Marburg shows maximum activity at pH 4.0, whereas all others function best at pH 6.0. The purified enzymes absolutely require coenzyme F_{420}. Neither the nicotinamides (electron carriers for the analogous enzymes in the H_4folate system) [182], FAD, nor artificial electron acceptors such as methylviologen, participate in the dehydrogenase reaction (refs. [188,189,361], and Mukhophadhyay, B., unpublished data). However, F_{420}-dependent methylene–H_4MPT dehydrogenases resemble the methylene–H_4folate dehydrogenase in that both are devoid of chromophoric prosthetic groups [188,189,361] and use electron carriers of similar redox chemistry (see section 2.2 on coenzyme F_{420} for its similarities to nicotinamides). Purified methylene–H_4MPT dehydrogenase (F_{420}-dependent) from *M. barkeri* strain Fusaro and *M. thermoautotrophicum* ΔH require monovalent cations for maximal activity, but do not show a specific requirement for Na^+ for activity [188,189,361]. The purified enzymes exhibit similar kinetic characteristics with few exceptions. The enzymes from *M. thermoautotrophicum* strain ΔH and *M. barkeri* strain MS show a sequential bi–bi mechanism [189,361]. The same kinetic mechanism is observed with all methylene–H_4folate dehydrogenases [182,362], except for the enzyme from *A. woodii* [243], which shows a ping-pong mechanism; however, work by Wohlfarth et al. [362] shows a need to re-evaluate this exception.

The H_2ase-type methylene–H_4MPT dehydrogenase has proven one of the most interesting discoveries in work with methanogens. There is no parallel to this enzyme in the H_4folate system [182]. Several characteristics make this dehydrogenase an unusual member in the hydrogenase family. All other hydrogenases contain nickel and/or iron–sulfur clusters, and due to these resident electron carriers they catalyze an isotope exchange with H_2 and H^+ in the absence of an external electron acceptor, reduce one-electron acceptors such as viologen dyes, and exhibit a ping-pong mechanism [191]; these characteristics are absent in the H_2ase-type methylene–H_4MPT dehydrogenases from *M. thermoautotrophicum* Marburg and *M. wolfei*, and also most likely the *M. kandleri* enzyme [190]. The uv-visible spectra of the H_2ase-type enzymes do not suggest the presence of any prosthetic groups, and the enzyme from *M. thermoautotrophicum* Marburg does not contain FAD, FMN, or F_{420}. Purified enzymes from *M. thermoautotrophicum* Marburg and *M. wolfei* do not contain Ni, Fe or other transition metals, except Zn, which is not required for activity, but is present in substoichiometric and often varying amounts in the *M. thermoautotrophicum* enzyme [191]. Also, CO, acetylene, nitrite, cyanide or azide, which are traditional transition-metal-directed inhibitors of hydrogenases, do not inhibit the H_2ase-type enzyme from *M. thermoautotrophicum*; the enzyme is EPR silent under both N_2 and H_2, indicating the absence of Ni or iron–sulfur clusters. This enzyme is rapidly inactivated by oxygen; inactivation is enhanced by dithiothreitol or 2-mercaptoethanol [190,191].

Unlike observations in the absence of an electron acceptor or donor, the H_2ase-type enzyme from *M. thermoautotrophicum* Marburg catalyzes an isotope exchange between H_2 and H^+ in the presence of either $H_2C=H_4MPT$ or $HC^+=H_4MPT$; the enzyme also forms ^{3}H-labelled $H_2C=H_4MPT$ (the label belonging to the $=CH_2$ group) from $^3H_2$ and $HC^+=H_4MPT$. These observations are consistent with a hydride transfer mechanism; also, the enzyme removes only one of the hydrogens from the methylene group of $H_2C=H_4MPT$ during its oxidation to $HC^+=H_4MPT$ [191]. The hydride transfer mechanism is consistent with the 2-electron-redox nature of this enzyme. In all, the F_{420}-independent methylene–H_4MPT dehydrogenase is a novel type of hydrogenase whose catalytic site for hydrogenation seems to be unique.

The subunit sizes of the H_2ase-type dehydrogenases are 43–44 kDa; the enzyme from *M. thermoautotrophicum* Marburg is a monomer [114], the *M. kandleri* enzyme is a homotetramer [190], and that of *M. wolfei* a homodimer [191]. The amino acid sequences of *M. thermoautotrophicum* and *M. kandleri* enzymes (derived from the DNA sequence of the cloned genes; [52,191]) are 57% identical. A remarkable conservation is observed in the N-terminal sequences of the enzymes. Thus, the H_2ase-type enzyme has been conserved over a considerable phylogenetic distance. The H_2ase-type enzymes of *M. thermoautotrophicum* Marburg and *M. kandleri* possess four cysteine residues, but their locations in the primary structure are not characteristic of those in Ni/Fe hydrogenase or iron–sulfur proteins [191].

The H_2ase-type enzyme from *M. kandleri* shows increased activity (methylene–H_4MPT oxidation) at low pH. Its apparent K_m for H^+ is $2\,\mu M$ (pH 5.7) and that for $H_2C=H_4MPT$ is $50\,\mu M$ (pH 5–6.5) [190]. Between pH 4.5 and pH 7.0 the enzyme exhibits a sequential mechanism. The *M. thermoautotrophicum* enzyme shows maximum activity at pH 6.5 [191]. The K_m for H_2 for all purified H_2ase-type dehydrogenases is about 60% H_2 ($pH_2 = 0.6$ atm) in the gas phase [190,191]. Thus, in natural methanogen habitats ($pH_2 = 1$–$10\,Pa = 10^{-4}$–10^{-5} atm); [184]) this enzyme most likely would not participate in the CO_2 reduction pathway. It would be interesting to see if this enzyme is expressed by methanogens in nature; the enzyme has so far been purified from cells grown with ample H_2 [114,190,191]. The temperatures for maximum activity of the H_2ase-type enzymes correspond well to the optimum growth temperatures of the source organisms; these enzymes require salts (KCl or NaCl) for full activity, although the requirement seems to be non-specific [190,191].

All methylene–H_4MPT dehydrogenases (both types) purified so far are monofunctional. In the H_4folate system, the analogous enzyme can be monofunctional, or may have methenyl–H_4folate cyclohydrolase and/or formyl–H_4folate synthetase activity [362–371].

4.9. Methylenetetrahydromethanopterin reductase

This enzyme catalyzes reduction of $H_2C=H_4MPT$ to CH_3-H_4MPT (Reaction 17) during CO_2-methanogenesis, and presumably the reverse reaction is used in methyl-group oxidation during methylotrophic methanogenesis. See also Fig. 6. The electron carrier is F_{420}; thus, activity can be assayed in either direction at 400 or 420 nm [53-55,115,116,189,193,372,373]. In extracts, however, it is difficult to estimate methylene-H_4MPT reductase activity due to interferences from F_{420}-dependent and F_{420}-independent (H_2-forming) methylene-H_4MPT dehydrogenase, and F_{420}-reactive hydrogenase. The F_{420} formed in Reaction (17) could be used as the electron acceptor for F_{420}-dependent dehydrogenase (Reaction 15); also, F_{420}-independent dehydrogenase and F_{420}-reactive hydrogenase, in combination, can consume $H_2C=H_4MPT$ and F_{420}. If assayed in the reverse direction, similar interferences come into play.

Methylene–H_4MPT reductase is present in all methanogens examined [53–55], but the activity in *Methanosphaera stadtmanii* is very low [55], consistent with this organism's inability to reduce CO_2 or oxidize methyl groups [98]. Methylene–H_4MPT reductase is also present in *A. fulgidus*, where it participates in oxidation of lactate-derived methyl groups to CO_2 [69]). Methylene–H_4MPT reductase has been purified from five methanogens and from *A. fulgidus* (see Table 9). Except for the *M. thermoautotrophicum* ΔH enzyme, all are multimers of identical subunits. The N-terminal sequences of these subunits show a high degree of homology [54].

The optimum temperatures and thermostabilities of methylene–H_4MPT reductases correspond well to the optimum growth temperatures of the host organisms, except the high temperature optimum of the enzyme from *M. barkeri*. The methylene–H_4MPT reductases of *M. thermoautotrophicum* Marburg and *M. kandleri* have low isoelectric points (pI), and are stimulated by salts [373]. On the other hand, salts inhibit the reductases with high pI. Salts and the components (both low and high molecular weight) of cell extracts provide thermal stability to the *M. kandleri* enzyme which is not otherwise heat stable [54]; K_2HPO_4 and KCl are the best stabilizing agents. The stimulatory and stabilizing effects of these salts are due to anions rather than cations. As for methenyl–H_4MPT cyclohydrolase [356], stimulation of methylene–H_4MPT reductases parallel the cDPG contents of the source organisms [356,359,360,374]. Also, the stimulatory effect of salts on the *M. kandleri* reductase is via an increase in its apparent V_{max}, but its K_m for methylene–H_4MPT is unaffected [356].

It was proposed from whole-cell methanogenesis studies that Reaction (17) is involved in generation of an electrochemical Na^+ gradient [220,226]. But, methylene–H_4MPT reductases do not require Na^+ for activity [54,115,116,189,193,372,373], and can be purified from soluble cell fractions. The effects of aggressive cell-breakage procedures on the properties and localization of enzymes may account for this, but Ma and Thauer [116] have also described thermodynamic grounds for discounting a role of methylene–H_4MPT reductase in energy conservation during H_2–CO_2 methanogenesis. Also, Becker et al. [195] have shown that washed everted vesicles of *Methanosarcina* Göl do not translocate Na^+ while reducing methylene–H_4MPT.

All methylene–H_4MPT reductases purified are air stable, to some extent [54,115,116, 189,193,372,373]. None contain flavins or Fe–S centers as prosthetic groups, and they are incapable of reacting with one-electron carriers such as viologen

TABLE 9
Properties of N^5,N^{10}-methylene-tetrahydromethanopterin reductases and N^5,N^{10}-methylene-H$_4$folate reductases [a]

Organisms	Apparent molecular mass[b] (kDa)	Prosth. group	Electr. donor	K_m (mM)[c]	V_{max} (U/mg)	k_{cat} (s^{-1})	Ref.
Methylene-H$_4$MPT reductase							
Methanobacterium thermoautotrophicum strain Marburg	150 (36)	none	H$_2$F$_{420}$	0.30 / 0.003	6000	3600	54,116
Methanobacterium thermoautotrophicum strain ΔH	35 (35)	none	H$_2$F$_{420}$	nr[d]	nr	nr	193
Methanosarcina barkeri strain Fusaro[e]	130 (36)	none	H$_2$F$_{420}$	0.015 / 0.012	2200	1320	54,115
Methanopyrus kandleri	300 (38)	none	H$_2$F$_{420}$	0.006 / 0.004	435	275	54,373
Archaeoglobus fulgidus	200 (35)	none	H$_2$F$_{420}$	0.016 / 0.004	450	265	54
Methylene–H$_4$folate reductase							
Clostridium formicoaceticum	237 (26, 35)	2 FAD + Fe/S	ferredoxin?	–	140[f]	–	116
Peptostreptococcus productus	250 (32)	4 FAD	NADH	–	380[f]	–	116
Escherichia coli	–	FAD	FADH$_2$	–	0.2[f]	–	116
pig liver	150 (75)	2 FAD	NADPH	–	1600	–	116

[a] Mostly adapted from tables compiled by Ma and Thauer [116], Te Brommelstroet et al. [189], and Schmitz et al. [54].
[b] For native enzyme; values for subunit(s) are given in parentheses.
[c] First value is for H$_2$C=H$_4$MPT, second for H$_2$F$_{420}$
[d] nr, not reported.
[e] With *M. barkeri* strain MS, the purified enzyme catalyzes the oxidation of CH$_3$–H$_4$MPT with a V_{max} of 200 μmol/min mg, and with a k_{cat} of 127 s^{-1} [189]. In this example, apparent K_m for CH$_3$–H$_4$MPT was 0.25 mM and the K_m for F$_{420}$ was 0.04 mM.
[f] Specific activities and not V_{max}.

dyes [54,115,116,189,193,372,373]. In contrast, methylene–H$_4$folate reductases contain prosthetic groups (flavins, and/or Fe–S centers) capable of carrying single electrons, and reduce viologen dyes (see Table 9). These differences also explain why methylene–H$_4$MPT reductases use a ternary complex mechanism whereas the methylene–H$_4$folate reductases follow a ping-pong mechanism for catalysis [375,376].

4.10. Methyltetrahydromethanopterin:CoM methyltransferase

This enzyme, or enzyme system, is responsible for the transfer of the methyl group of CH_3–H_4MPT to CoM to yield CH_3–CoM (Reaction 18; [184]). Current evidence suggests involvement of this enzyme system in both H_2–CO_2 and acetate methanogenesis (see section 3 above). It may also play a role in methanol oxidation during methanol methanogenesis by catalyzing the reverse of Reaction (18); this endergonic process may be coupled to a Na^+-motive force [184,195].

Most reports pertinent to Reaction (18) describe experiments with extracts or membrane preparations [117,377–380]. Results indicate involvement of corrinoid as an intermediate methyl carrier [379], and an oxygen-labile enzyme [377], similar to the methanol:CoM methyltransferase system (MT_1) of *M. barkeri* [154], that requires ATP-dependent reductive activation for activity. Since *M. bryantii*, *M. formicicum* [380], and *M. thermoautotrophicum* possess cobalamin:CoM methyltransferase activities that resemble the analogous protein MT_2 in the methanol to CH_3–CoM conversion pathway (see sections 3 and 4.13) [152], a scheme similar to that shown in Reactions (19) and (20) has been envisaged, where CH_3–H_4MPT replaces methanol in Reaction (28). Three recent reports [117,157,195] provide more information about this system, and are reviewed below.

Kengen et al. [157] purified an oxygen-sensitive methyltransferase (MT_a from *M. thermoautotrophicum* ΔH) that is required for conversion of HCHO to CH_3–CoM under H_2 when 55% ammonium sulfate-saturated cell extract supernatant provides H_4MPT, F_{420}, methylene–H_4MPT reductase and hydrogenase. The 55% supernatant can convert HCHO to CH_3–H_4MPT using H_2F_{420} generated from F_{420} by hydrogenase; thus, MT_a's suggested role is in the CH_3–$H_4MPT \rightarrow CH_3$–CoM conversion. The 100 kDa MT_a protein is composed of 35, 33, and 31 kDa subunits in an α–β–γ configuration [157]. The protein forms high molecular weight complexes of up to 2000 kDa; centrifugation experiments suggest that MT_a is either membrane bound or forms large aggregates. MT_a requires the presence of CHAPS, a detergent, to maintain high activity, and contains 0.2 mol B_{12}HBI per mol enzyme (100 kDa). The corrinoid in MT_a is methylated by CH_3–H_4MPT only if Ti(III)-citrate is present, and MT_a also methylates externally supplied B_{12}HBI and B_{12}DMBI, using CH_3–H_4MPT as a methyl donor. The role of Ti(III)-citrate could be to convert inactive Co(II) corrinoid to the active methyl acceptor Co(I). To validate the two-component model, Kengen et al. attempted to purify the protein that could methylate CoM with CH_3–B_{12}HBI. Their results suggest that a non-corrinoid oxygen-stable 35 kDa protein in *M. thermoautotrophicum*, which they call MT_2, methylates CoM with CH_3–B_{12}DMBI. However, the ability of MT_a and MT_2, in combination, to catalyze Reaction (18) has not been demonstrated [157].

The 33 kDa subunit of a corrinoid-containing membrane protein from *M. thermoautotrophicum* Marburg [158,381] immunologically cross-reacts with the 33 and 31 kDa subunits of MT_a and a 33 kDa membrane protein from *M. thermoautotrophicum* ΔH [157,382], suggesting MT_a is membrane-bound in *M. thermoautotrophicum*. The cross-reacting membrane-bound corrinoid protein [382] could itself be the MT_a in *M. thermoautotrophicum* Marburg.

Recently, Fischer et al. [117] demonstrated that membrane fractions of *M. barkeri* grown on acetate, methanol, or H_2–CO_2, and of H_2–CO_2-grown *M. thermo-*

autotrophicum Marburg can transfer the methyl group of CH_3–H_4MPT to CoM but cannot carry out the reverse reaction [117]. Membrane fractions of *M. barkeri* were devoid of methylcobalamin:CoM methyltransferase activity, indicating this activity has no role in methylation of CoM by CH_3–H_4MPT, and contradicting the two-step model for Reaction (18) described above [157] (methylcobalamin:CoM methyltransferase was found in the supernatant, a fraction containing little CH_3–H_4MPT activity). Involvement of only one protein in the CH_3–$H_4MPT \rightarrow CH_3$–CoM conversion was also supported by the linear relationship between the membrane protein concentration and the reaction rate [117]. In the absence of CoM, corrinoids in these membranes were methylated by CH_3–H_4MPT; this reaction required catalytic amounts of ATP and Ti(III), for activation. Methylated membranes methylated CoM without ATP and Ti(III). In the presence of CoM (which can continually demethylate CH_3–Co(III) of the membranes) methylation of membranes by CH_3–H_4MPT does not require ATP and Ti(III), which indicated (in analogy to known corrinoid methylation systems) [250,383–385]) that demethylation of CH_3–Co(III) generates an active methyl acceptor. Fischer et al. [117] have proposed the following one-enzyme model for Reaction (18) in which the role of ATP is in the reductive activation of the bound corrinoid from Co(II) to Co(I):

$$CH_3-H_4MPT + MT-(Co^I)B_{12}HBI \rightarrow H_4MPT + MT-(Co^{III})B_{12}HBI, \quad (50)$$
$$\phantom{CH_3-H_4MPT + MT-(Co^I)B_{12}HBI \rightarrow H_4MPT + MT-(}|$$
$$\phantom{CH_3-H_4MPT + MT-(Co^I)B_{12}HBI \rightarrow H_4MPT + MT-}CH_3$$

$$MT-(Co^{III})B_{12}HBI + CoM \rightarrow CH_3-CoM + MT-(Co^I)B_{12}HBI. \quad (51)$$
$$|$$
$$CH_3$$

However, it is possible that in a complete membrane form, methylcobalamin has no access to the enzyme, and/or that in a membrane form, a multicomponent enzyme system looks as if it were one enzyme. Even if CH_3–B_{12} did serve as a reactant, it would not mean that it was the normal intermediate.

Recent work by Becher et al. [195] shows that during CO_2 reduction to methane, energy from the highly exergonic Reaction (18) is conserved. Washed everted vesicles of *Methanosarcina* Göl translocate Na^+ into their lumen as membrane-bound CH_3–H_4MPT:CoM methyltransferase activity methylates CoM [195].

4.11. Methyl–Coenzyme M reductase (MR)

The methyl–Coenzyme M reductase (MR) reaction, whereby the methyl group of methyl–CoM is reduced by two electrons to form methane, is given in Reaction (21). It is likely catalyzed by one enzyme in vivo, but may be physically associated with one or more enzymes in what has been termed a methanoreductosome. The reaction involves three coenzymes: methyl–CoM, F_{430}, and HSHTP. The picture is more complex in vitro, since various activation and reduction components have been used in attempts to obtain high specific activity enzyme preparations, although even in vitro the reaction clearly proceeds by MR catalysis alone. The MR is also known as Component C [127,128,173,175,386,387].

TABLE 10
Methyl–Coenzyme M reductase from methanogenic bacteria

Organism	Apparent Mol. Weight (kDa)		K_m, CH_3CoM (mM)	K_m, HSHTP (mM)	Max. specific activity (μmol min^{-1} mg^{-1})		Ref.
	Native	Subunits			Pure[a]	Other[a]	
M. thermoautotrophicum strain ΔH[b]	300	68,45,39	0.1[c]	0.45	nr[d]	0.14	[137,199, 386,387]
MCRI	300	66,48,38	4	0.08	0.5	nr	[391]
MCRII	300	66,48,33	4	0.08	1.1	nr	[391]
M. thermoautotrophicum strain Marburg[b]	300	66,48,37	4	0.08	2.5[b]	20.1	[135,175, 392]
MCRI	300	66,48,38	4	0.08	0.4	nr	[135,391]
MCRII	300	66,48,33	4	0.08	1.6	nr	[135,391]
M. voltae [b]	297	66,44,35	nr	nr	nr	0.14	[386]
M. jannaschii [b]	300	65,44,36	nr	nr	nr	0.16	[386]
M. barkeri [b]	305	73,42,33	nr	nr	nr	0.04	[386]
M. thermophila [b]	137[e]	69,42,33	3.3	0.06	0.2	nr	[268]

[a] Pure, rigorously pure assay; Other, other protein in assay.
[b] Work described before discovery of two MR isozymes in some methanogens, and which may reflect a mixture of isozymes.
[c] Assays conducted with crude extract [137].
[d] nr, not reported.
[e] Average of two estimate methods [268].

Most MR enzyme data comes from work with M. thermoautotrophicum strains Marburg and ΔH [135,137,139,140,173–176,181,199,319,387–394], but some work is reported on a variety of other organisms [36,268,386,395–397]. Table 10 describes the properties of several purified MR. Actually, there are two isozymes of the MR (MCRI and MCRII) in M. thermoautotrophicum strains ΔH and Marburg [391,398,399]; both are similar, but are genetically distinct, and not the result of protein processing. However, MCRII has a higher specific activity, compared to the previously studied MCRI [391]. Levels of the two forms vary with culture conditions, and are modulated by hydrogen abundance and medium pH. Presence of these forms has not been investigated in a wide variety of methanogens.

The MR makes up roughly 7–10% of the total cellular protein in methanogens [166, 172,175,199,388]. It is composed of three subunits of ca. 65, 45, and 35 kDa size, and has a native molecular weight of ca. 140 or 300 kDa, and an α_2–β_2–γ_2 structure in most cases; see Table 10. The MR was originally thought to be oxygen-stable, but it is likely that massive loss of enzyme activity occurs upon exposure to air [392]. Also, several associated components are oxygen-sensitive. Purifications are routinely conducted under anaerobic conditions in a glove bag. The MR contains 2 molecules each of F_{430} and CoM per 300 kDa protein [138,139,166,167,175,268,389]. HSHTP is also present in non-dialyzable form, but in unknown amounts [181]. These cofactors are extractable by solvent or heat denaturation, and are not covalently bound; CoM in the enzyme is in equilibrium with free methyl–CoM [139,141]. F_{430} is present only in the enzyme under low-Ni growth conditions, but in addition some is freely soluble in the cytoplasm when excess Ni is in the medium [165]. The previous report [400] that a lumazine derivative was present in the enzyme is not true for the pure enzyme [140,141].

A series of other proteins are not likely involved in the physiological MR reaction, but have been used in the laboratory to obtain increased specific activities and activation of the MR; some of these can be replaced by alternate reductants or other components. The most complex set of such proteins have been described in *M. thermoautotrophicum* ΔH by Wolfe et al. [173,199]. Component A1 is a partly pure component with MV- and F_{420}-hydrogenase activity; it can reduce the heterodisulfide (CoM–S–S–HTP), via a second protein in the component, heterodisulfide reductase [173,199]. Component A2 is a pure, colorless, air-stable protein thought to be an ATP-binding protein important in MR activation [393]. Component A3a is an air-sensitive iron–sulfur protein which appears to transfer electrons to the MR as part of its activation; component A3b, not pure, contains MVH, but no FRH activity, and transfers electrons from H_2 to activate the MR [394]. Thus, parts of Component A1 may serve physiologically in an electron transfer role from H_2 to the heterodisulfide; however, Components A2 and A3, although important in activation of MR inactivated during purification in the lab, may play only an occasional role in the organism's response to oxygen exposure.

In studies by Wolfe et al. [199,387,394], some of these proteins were replaceable. Ti(III)-citrate reductively activated the MR via A3a, eliminating the need for A3b, and when Ti(III) and CN–B_{12} were both present, the heterodisulfide was chemically reduced, eliminating the need for A1 [199,387,394]. Thus, *M. thermoautotrophicum* ΔH required as a minimum for conversion of methyl–CoM to methane: MR, A2 + ATP/Mg, A3a, HSHTP, Ti(III), and CN–B_{12}; if A3b, A1, and H_2 were present, then Ti(III) and CN–B_{12} could be omitted.

In the past few years, there has been uncertainty over the requirement for various A proteins in both *M. thermoautotrophicum* strains ΔH and Marburg. Work from Thauer's laboratory, originally and still mainly with *M. thermoautotrophicum* Marburg, has demonstrated that MR is active without any A proteins: low activity was observed with HSHTP alone; this activity was increased by the presence of DTT, and greatly stimulated by DTT + B_{12a} + Ti(III) [135,175,388]. The purity of MR from both labs has been rigorously established [135,175,199,387], and it is unlikely that small amounts of contaminating material cause the difference. However, an explanation for the difference may be the details of purification: e.g., time, buffer, or temperature.

Another catalytic issue, related to requirements for activation, is that activities in extracts and purified systems have been significantly lower than in whole cells. In *M. thermoautotrophicum* strains ΔH and Marburg the maximal observed specific activities for purified MR are routinely 1–2.5 μmol/min mg [135,175,199], but based on methanogenic rates of whole cells, a specific activity of ~30–50 μmol/min mg is predicted; thus, observed activity is ~2–8% of that expected. As put by Ellerman et al. [175], "either correct assay conditions ... have not yet been found or ... upon breakage of the cells the enzyme system disaggregates to a less active form". Both factors may be responsible, but one approach to obtaining high specific activities has recently been successful in

M. thermoautotrophicum Marburg: preincubation of cells prior to harvest with an excess of hydrogen + carbon dioxide resulted in crude extract activities of 2 μmol/min mg compared to 0.2 when cells were gassed with nitrogen + carbon dioxide [392]; the partly purified enzyme from these cells had an activity of 20 μmol/min mg, and EPR analysis revealed a much higher level of the signals thought to represent the Ni(I) form of the F_{430}. The EPR data have been confirmed by Brenner et al. [399]. The enzyme (a mixture of both MCRI and MRCII) was about 90% pure, but clearly non-MR bands were visible; unfortunately, efforts "to further purify the reductase resulted in a 90% loss in specific activity" [392]. The work by Rospert et al. [392] to develop a highly active MR by use of excess hydrogen also describes the ability of methyl–CoM and CoM to stabilize the activity. In summary, using the approach of Rospert et al., the MR in *M. thermoautotrophicum* can be assayed in vitro with high specific activity (20 μmol/min mg), using only HSHTP as an electron donor, CH_3–CoM as electron acceptor and source of methane, MR as catalyst, and DTT + cobalamin as "activation" factors, but not with a rigorously purified MR [392]. Alternatively, a rigorously pure MR can catalyze the reaction without additional proteins or "activation factors" at a slow rate of 0.08 μmol/min mg, which can be stimulated to ~1 μmol/min mg by addition of DTT and cobalamin [135].

The MR from *M. thermophila* does not require A proteins in vitro [268]. The enzyme is reductively activated by Ti(III) (a process stimulated by, but not requiring, ATP), to give a specific activity of 0.22 μmol/min mg. The minimal components needed for activity are MR, Ti(III) and HSHTP, but physiologically it is likely that ferredoxin, cytochrome b, and heterodisulfide reductase are involved. The MR from *M. voltae*, *M. jannaschii*, and *M. barkeri* 227 have also been purified [386], and it was shown with these enzymes that crude A components were active when used with MR from other methanogens.

As a catalyst, MR is active with the alternate substrate ethyl–CoM, and is inhibited by several CoM analogs, including Br–ethanesulfonate, Br–methanesulfonate, Br–propanesulfonate, allyl–ethanesulfonate, and azidoethanesulfonate, and the methane analogs chloroform and carbon tetrachloride [133,135–137,401]. An HSHTP analog, 6(methylthio)hexanoyl-L-threonine phosphate, also is inhibitory [136]. Little is known about the actual mechanism of reduction of methyl–CoM to methane at the active site, or of the specific role F_{430} plays. It is thought the Ni in F_{430} is in the Ni^{+1} or Ni^{+3} form [402,403], and it has been proposed that the methyl group from methyl–CoM is transferred to the Ni in the F_{430}, and that methane arises via protonolysis [168]; using labelled ethyl–CoM, evidence was obtained on the steric course of the reaction which was consistent with this concept [404]. Lin and Jaun [169] have presented NMR evidence for the existence of a nickel-bound methyl group in CH_3–F_{430}MII; although their work used material synthesized via CH_3I treatment of F_{430}, it suggests that the idea of nickel involvement in methyl transfer and reduction in the MR is not unreasonable. Work on the conformation of F_{430} also supports this possible role of nickel in F_{430} [405]. Efforts to make crystals for X-ray studies have thus far been unsuccessful.

Genes for the three MR subunits from several methanogens have been cloned and sequenced [406–411]. In all cases, the three proteins are coded in a transcription unit, mcrBDCGA, containing two additional open reading frames (ORF) of unknown function. The proteins coded by these ORFs have been produced by cloning into *E. coli*, and do not purify with the MR, or stimulate the MR reaction [135], but are present in crude methanogen extract in low levels. Subunit order is the same in all organisms examined, and subunit sizes are similar. Sequence homologies are strong at both DNA and protein levels.

The location of MR is thought by some to be soluble, but most suggest it is membrane associated, based on whole-cell immunolabelling and electron microscopy studies. Data from *M. voltae* [396,412], *M. thermoautotrophicum* [319,412,413] and *Methanosarcina* Göl [414] support a membrane location, but in contrast Thomas et al. [415] report that "immunogold labelling showed that the Component C [methyl reductase] was located randomly in the cytoplasm in *Methanosarcina* species and in *Methanothrix soehngenii*". Biochemical work suggests it participates in proton translocation in *Methanosarcina* Göl [225,416–421]. The term methanoreductosome has been coined to describe this large molecular weight entity [414].

It has been suggested that instead of the MR, the heterodisulfide reductase is "coupled with the phosphorylation of ADP" [175]. One of the most interesting experimental investigations of the role of MR and the heterodisulfide reductase, both involved in the terminal step of methanogenesis, has been the study by Gottschalk's laboratory of vesicle systems of Göl [201,203,420–422]. They have demonstrated that both the MR and heterodisulfide reductase reactions are coupled in some way to ATP synthesis, and that the heterodisulfide is "the terminal electron acceptor of a membrane-bound electron transport chain" [421]. The synthesis of ATP in acetate-grown *Methanosarcina* and H_2–CO_2-grown methanogens is not so clear, but useful heuristic models have been proposed.

4.12. Heterodisulfide reductase (HR)

Elucidation of the structure of component B (HSHTP) and the discovery of heterodisulfide reductase are two recent successes in work on the biochemistry of methanogens [131,174,178,197,423]. The enzyme heterodisulfide reductase (HR) carries out Reaction (22), regenerating CoM and HSHTP [131,174,423]. HR is found in all methanogens examined [55], and has been proposed to be involved in energy conservation. Two approaches have been used to study this enzyme: characterization of the purified enzyme, and studies with membranes and everted vesicles.

The HR of *M. thermoautotrophicum* Marburg has been purified from a soluble cell fraction [131]. The enzyme aggregates with MVH, and even after several steps of purification, still contains minor amounts of hydrogenase. Whether this association has physiological significance is unknown [131]. Purified HR does not use F_{420} or NAD(P) as an electron carrier [131]. However, extracts and partially purified enzyme from *M. thermoautotrophicum* Marburg catalyze reduction of CoM–S–S–HTP with H_2 [130]. Thus, in vivo, electrons for heterodisulfide reduction may come from H_2 via hydrogenase and intermediate electron carrier(s). For purification and characterization, HR has been assayed with dithionite-reduced benzylviologen as electron donor for CoM–S–S–HTP reduction and methylene blue as electron acceptor for oxidation of HSHTP and CoM [131].

The *M. thermoautotrophicum* enzyme has an apparent molecular mass of 550 kDa [131]. It contains 4 mol FAD and 72 mol Fe/S per mol enzyme. The purified enzyme, as isolated, is partially reduced; complete reduction can be achieved with CoM + HSHTP, or dithionite. The reduced enzyme can be oxidized by CoM–S–S–HTP [131]. In the direction of CoM–S–S–HTP reduction the k_{cat} is 68 s^{-1}; K_m for CoM–S–S–HTP is 0.1 mM. In the direction of HSHTP and CoM oxidation, the V_{max}

is 15 U/mg (k_{cat} = 34 s^{-1}); the apparent K_m values for CoM and HSHTP are 0.2 mM and <0.05 mM, respectively. 6-Mercaptohexanoyl-L-threonine phosphate, a mild inhibitor of HR in the direction of CoM–S–S–HTP reduction, can replace HSHTP in the reverse direction, although its K_m is 4 times higher and the corresponding V_{max} is about 4 times lower than with HSHTP [131].

The *M. thermoautotrophicum* Marburg HR is similar to other known disulfide reductases in some ways [131], probably including the use of FAD in catalysis. But unlike others, the HR does not use a nicotinamide (or its functional analog, coenzyme F_{420}) as electron carrier, and it possesses Fe/S centers [131].

The role of HR in energy generation has been established from work with everted vesicles of *Methanosarcina* Göl and *Methanolobus tindarius* [129,201–204]. *Methanosarcina* Göl grows on H_2–CO_2, methylotrophic substrates, and acetate, whereas *Methanolobus* cannot use H_2–CO_2 or acetate, but uses methanol. Vesicles of methanol-grown *Methanosarcina* Göl prepared via protoplast formation and gentle lysis show H_2F_{420}-dependent HR activity which is absent in the corresponding cytoplasmic fraction [129]. Everted vesicles of *Methanosarcina* Göl also catalyze a H_2-dependent, F_{420}-independent, reduction of CoM–S–S–HTP [201]. Thus, *Methanosarcina* Göl possesses two HR systems: one H_2F_{420}-dependent and the other H_2-dependent; both couple heterodisulfide reduction to proton translocation and ATP synthesis [201–203]. Evidence suggests that the H_2F_{420}-dependent system operates during methylotrophic methanogenesis using H_2F_{420} generated from methyl-group oxidation [201]; electrons are probably channeled from H_2F_{420} to HR through a membrane transport system involving H_2F_{420} dehydrogenase and cytochrome b [56,211]. The possibility that the F_{420}-dependent HR activity is a combination of the F_{420}-dependent hydrogenase and a H_2-dependent heterodisulfide reductase (system) is unlikely since membranes of methanol-grown *M. tindarius* are devoid of F_{420}-dependent hydrogenase, but they reduce CoM–S–S–HTP with H_2F_{420} [129]. Isolation of a H_2F_{420} dehydrogenase from *Methanolobus* [56] strengthens the above hypothesis. The H_2-dependent heterodisulfide system of *Methanosarcina* Göl is suggested to participate during H_2–CO_2 methanogenesis [56]; electrons from H_2 are proposed to enter a membrane-bound electron transport chain directed towards the HR via an F_{420}-nonreactive hydrogenase. Purification of a membrane-bound F_{420}-nonreactive hydrogenase from *Methanosarcina* Göl [204] is consistent with this hypothesis.

4.13. Methanol methanogenesis-related methyltransferases

As discussed above in the pathway section (3.2.1) on reduction of methanol, it is likely that the B_{12}-containing enzyme reported by Blaylock [196] is directly relevant to the methanol-to-methyl–CoM methyl transfers. The original Blaylock–Stadtman papers [196,216–218], and further work by Wood et al. [219], provide evidence for B_{12} involvement in methanol metabolism, and for the importance of a 100–200 kDa protein originally purified from *M. barkeri*. Recent information on enzymes involved in methanol entry into the methanogenic pathway includes work published in the early 1980s by Vogels' group, using *M. barkeri* [152–155,424]. They partially purified two enzymes, methanol:HBI methyltransferase (named MT$_1$) and methyl–HBI:CoM methyltransferase (MT$_2$) (HBI is the cobamide cofactor used in *Methanosarcina* as a B_{12} analog [149,150]; see section 2.6). The two enzymes act in sequence as shown in

Reactions (28) and (29). The first enzyme, MT_1, is a cobalamin-containing oxygen-sensitive protein; it was purified 6-fold to $\sim$90% homogeneity with 44% yield, and required ATP-dependent reductive activation [154,155]. The native molecular weight was $\sim$122 kDa, and was composed of subunits of 34 and 53 kDa. MT_1 is probably the same B_{12}-protein described by Blaylock [196]; in vivo the H_2-reduced ferredoxin may maintain the enzyme in a reduced, active Co(I) state, ready to accept a methyl group. As shown in Reaction (28), it is methylated by methanol. A different enzyme is needed for methylamine entry into the pathway, since extracts from trimethylamine-grown cells methylate CoM with trimethylamine, but methanol-grown cells cannot [227].

MT_1 appears to be central in the entry of methanol in the path leading to methane; however, it is not clear that in cells grown on methanol without H_2 the methanol molecules destined to become CO_2 enter via the same reaction. There is little evidence on this topic, but note the methyl from acetate appears to enter the pathway as methyl–H_4MPT (Fig. 1).

The second enzyme, MT_2, is an oxygen-stable protein purified 86-fold with 27% yield; it transfers the methyl group, bound to the Co in the MT_1–B_{12}, to CoM, and can use methylcobalamin as a substrate [153,424]. It is composed of one polypeptide with a molecular weight of $\sim$40 kDa. MT_2 is similar to the "methylcobalamin:CoM methyltransferase" from *Methanobacterium*, isolated and described by Taylor and Wolfe [194], and might be the same enzyme; their organism was grown on H_2–CO_2, and cannot grow on methanol, and thus this enzyme, if physiologically relevant, is likely to be involved in methanogenesis from CO_2. Support for the idea of a two-step transfer via the two enzymes in Reactions (28) and (29) was obtained by Zydowsky et al. [425] in their study of the stereochemical course of methyl transfer in *M. barkeri*.

A complicating factor is the recent results of Fischer et al. [117], who reported that membrane fractions of H_2–CO_2-grown *M. thermoautotrophicum* Marburg, and H_2–CO_2-, acetate-, and methanol-grown *M. barkeri* contain very little MT_2-like methylcobalamin:CoM transferase activity in fractions that could transfer a methyl from methyl–H_4MPT to CoM. A direct transfer from methyl–H_4MPT to CoM occurs, probably via only one enzyme [117]. This suggests that: (i) the Taylor–Wolfe enzyme mentioned above [194] may not be the enzyme involved in the normal progress of the CO_2-to-methane pathway, via methyl–H_4MPT; (ii) the Taylor–Wolfe enzyme preparation has unmasked an ability to interact with methyl–B_{12}; or (iii) the newly-reported membrane fraction has lost ability to catalyze that part of the normal reaction. Further work with both membrane fractions and pure enzymes is needed to resolve this issue; one approach would be to use the stereochemical methods described by Zydowsky et al. [425] with the Fischer et al. [117] membrane preparations, to see if there is evidence for one or two methyl transfers [117,425].

4.14. Carbon monoxide dehydrogenase complex

In methanogens, cleavage of the C–C and C–S bonds of CH_3CO–S–CoA and oxidation of the carboxyl-carbon to CO_2 (Reaction 43) is catalyzed by a single enzyme called carbon monoxide dehydrogenase or CODH [239,241,245–248,252,255,257–259,261,267,426,427]. In *C. thermoaceticum* CODH catalyzes the reverse reaction forming C–C and C–S bonds and synthesizing acetyl–CoA [243]; thus, it is called acetyl–CoA synthase [243]. Terminologies used with the clostridial acetate synthesizing machinery (Ni/Fe–SP, C/Fe–SP, etc.; see acetate pathway section 3.3) are used below in reviewing methanogen CODH. Properties of purified CODH from methanogens are summarized in Table 11.

A nickel-containing CODH composed of 92 and 18 kDa subunits in an $\alpha_2\beta_2$ configuration has been purified from acetate-grown *M. barkeri* [255]. When CODH in *M. barkeri* cell extracts is precipitated with antisera raised against this purified protein, extracts cannot convert acetate to methane [255]. This observation confirms the involvement of this CODH in acetate cleavage, consistent with its role in methanogenesis rather than acetate biosynthesis. It should be noted that *M. bryantii* possesses a nickel-containing CODH whose function is probably in acetate synthesis, and an analogous enzyme is expected to be present in all acetate-synthesizing methanogens for biosynthetic purposes.

Cell extracts of acetate- or methanol-grown *M. thermophila* contain two chromatographically separable CODHs, one expressed at 5-fold higher levels in acetate-grown cells than in methanol-grown cells [246]. This predominant form has been characterized. The purified CODH complex of *M. thermophila* is composed of 5 subunits (Table 11) [245,246]. This complex oxidizes CO to CO_2, and catalyzes isotope exchange between the carbonyl group of CH_3CO–S–CoA and CO or between HS–CoA and CH_3CO–S–CoA, demonstrating the acetoclastic ability of the enzyme and reversible nature of the C–C and C–S bond cleavage [245–247]. The HS–CoA/acetyl–CoA exchange rate is about 6 times higher than that of the *C. thermoaceticum* enzyme [247], consistent with an acetoclastic role for the *M. thermophila* enzyme and an acetate-synthesizing role of the clostridial enzyme. Like the *C. thermoaceticum* enzyme (in conjunction with the corresponding C/Fe–SP; [428]) the *M. thermophila* CODH complex can also synthesize acetyl–CoA from CH_3I, CO, and HS–CoA [245]; this activity would be physiologically relevant only if this CODH complex also participates in acetate biosynthesis in vivo. The CO-oxidizing activity of the *M. thermophila* CODH complex is inhibited by cyanide, and the inhibition is reversed by CO; thus, the binding sites for CO and CN– might be either the same or close to each other.

Anion exchange chromatography in the presence of a detergent resolves the CODH complex of *M. thermophila* into two components, one containing the 89 and 19 kDa subunits and the other the 60 and 58 kDa subunits [252] that are, respectively, reminiscent of the Ni/Fe–SP and C/Fe–SP of the clostridial acetate synthesis machinery [243]; the 71 kDa subunit of the *M. thermophila* CODH complex is not recovered [252]. The Ni/Fe–SP of the *M. thermophila* CODH complex contains Ni, Zn, Fe, acid-labile sulfur, but no corrinoid [252]. Ferredoxin from *M. thermophila* accepts electrons from the CO-reduced CODH complex; in the resolved system, the Ni/Fe–SP catalyzes CO dependent reduction of ferredoxin and methylviologen [249,252,306]. Thus, the Ni/Fe–SP of *M. thermophila* is the CODH. Also, from its subunit composition ($\alpha_1\varepsilon_1$, 89 and 19 kDa) the Ni/Fe–SP of *M. thermophila* is easily recognized as a counterpart of the $\alpha_2\beta_2$ (92 and 18 kDa) CODH of *M. barkeri* [255,257].

EPR spectra of the CO-reduced CODH complex from *M. thermophila* show features characteristic of the spin-coupled Ni–Fe–C center in *C. thermoaceticum* CODH (acetate synthase) [243,248]. Treatment of CO-reduced CODH complex of *M. thermophila* under N_2 with acetyl–CoA perturbs these EPR features, an effect also observed with the clostridial enzyme [243,248]. This perturbation is not observed with *M. thermophila* CODH complex if the gas phase is CO or if HS–CoA is used in place of acetyl–CoA [248]; such is not the case with the *C. thermoaceticum* CODH [243].

TABLE 11
Properties of carbon monoxide dehydrogenase of *Methanosarcina* and *"Methanothrix"* species

Property	*Methanosarcina barkeri* [255,257,258,427]	*Methanosarcina thermophila* [246,247,252]	*"Methanothrix" soehngenii* [266,267,431]
Apparent molecular mass (kDa)			
Native	232[a]	250 (CODH complex)	220[b]
Subunit	92,18	89,71,60,58,19	89,21
Subunits in complex	–	Ni/Fe–SP (89,19), C/Fe–SP (60,58)	–
Metal and cofactor content (mol/mol protein)			
Ni	2	CODH complex, 2.3; Ni/Fe–SP, 0.21; C/Fe–SP, 0	2
Fe	30	CODH complex, 21; Ni/Fe–SP, 7.7; C/Fe–SP, 3	25
Acid-labile sulfur	nr[c]	CODH complex, 19.7; Ni/Fe–SP, 13.2; C/Fe–SP, 2.9	24
Corrinoid (B_{12}HBI)	0	CODH complex, 0.8; Ni/Fe–SP, 0; C/Fe–SP, 0.7	nr
Zn	1	CODH complex, 1.7; Ni/Fe–SP, 2.7; C/Fe–SP, 0	nr
Cu	1	nr	nr
CO-oxidizing activity			
V_{max} (U/mg)	1300 (37°C)	6.4 (27°C)	140 (35°C)
K_m for CO (mM)	5	nr	0.7
k_{cat} (s^{-1}, based on native mol. wt.)	5027	27	513
Acetyl–CoA/CO exchange activity			
K_m for acetyl–CoA (mM)	nr	0.2	nr
V_{max} (U/mg)	nr	0.05 (55°C)	0.035 (35°C)
k_{cat} (s^{-1}, based on native mol. wt.)	nr	0.2	0.1

[a] Not a complex; complex contains 0.9 mol corrinoid per mol minimal size of native enzyme; complex includes proteins of 85, 63, 53, 51, and 20 kDa in a native molecule of 1600 kDa.
[b] Not a complex.
[c] nr, not reported.

These results indicate that acetyl–CoA, but not HS–CoA, interacts with the Ni–Fe–C of the *M. thermophila* CODH complex, and they are consistent with an acetoclastic role for this enzyme. Jablonski et al. [429] have reported EPR data on the resolved components of the *M. thermophila* CODH complex. The Ni/Fe–SP, which harbors the CO-oxidizing ability, has three types of Fe/S clusters, two with EPR spectra characteristic of 4Fe–4S centers (the other remains uncharacterized) [429]. The Ni/Fe–SP, like the unresolved complex [248], exhibits EPR spectra characteristic of a spin-coupled Ni–Fe–C center [429]. Since the Ni–Fe–C center of the *C. thermoaceticum* enzyme has been proposed as the site for acetyl–CoA assembly [243], the corresponding site on the Ni/Fe–SP (or CODH) of *M. thermophila* is suggested as the acetate cleavage site [253].

The EPR spectra of the as-isolated C/Fe–SP show that its corrinoid is in the Co^{+2} state, and in a base-off configuration; thus, the mid-point potential of the Co^{+2}/Co^{+} couple would be about 100 mV less negative than that in a base-on configuration, making the Co^{+2} more accessible to physiological electron donors for reductive activation [253,429,445]. As mentioned above, the C/Fe–SP of the *M. thermophila* complex is the proposed primary methyl acceptor after the cleavage of acetyl–CoA [253], but the resolved system is yet to meet the requirement of catalyzing the partial reactions of acetate cleavage. The Ni/Fe–SP of *M. thermophila* has so far been unable to catalyze either the acetyl–CoA/CO exchange or acetate synthesis (in combination with the C/Fe–SP) from CH_3I, CO, and HS–CoA [252], properties exhibited by the unresolved complex [245,247].

The metal centers of *M. barkeri* CODH have also been studied [426,427]. The enzyme is very oxygen-sensitive, and inhibited by cyanide [255]; since CO provides partial protection from cyanide inactivation, cyanide may bind at or near the CO binding site [257]. Core extrusion experiments show that this $\alpha_2\beta_2$ enzyme contains six 4Fe–4S clusters [427]. EPR spectroscopy with dithionite-reduced enzyme shows one 4Fe–4S cluster with a midpoint potential of -390 mV and an uncharacterized center (termed center 2 [426]) with a midpoint potential of -35 mV; the latter interacts with CO with a more pronounced shift in EPR signals that the former [427]. Whole-cell studies show that during active methanogenesis from acetate the signals from center 2 of CODH exhibit changes similar to those seen with the purified enzyme upon interaction with CO [426]. These results suggest that cleavage of acetate yields a moiety that CODH recognizes as CO [426]. EPR spectra of dithionite-reduced CODH before or after treatment with CO do not show a signal attributed to a mononuclear nickel site or a spin-coupled Ni–Fe–C center [427]. Failure to detect an EPR-active nickel center in *M. barkeri* CODH could be due to inactivation of the enzyme by dithionite or oxygen (J.A. Krzycki, personal communication), or due to a novel nickel center not detectable under conditions used. Little is known of the C/Fe–SP in *M. barkeri*. Recently, a CODH from *M. barkeri* was isolated in association with a corrinoid protein which catalyzes all reactions analogous to, but reverse of, the clostridial acetate synthesis system [243]. Resolution of this complex, study of its components, and reconstitution experiments are needed to fully understand the mechanism of acetate cleavage in *M. barkeri*.

A carbon monoxide dehydrogenase with two subunits ($\alpha_2\beta_2$; 89 and 21 kDa) has been purified from *"Methanothrix" soehngenii* [266,267]. It contains nickel and iron, and constitutes 4% of the soluble protein in the cell. Unlike CODH from *Methanosarcina* and most anaerobic bacteria, the CO-oxidizing activity is air-stable and only slightly affected by cyanide. However, acetyl–CoA/CO exchange activity is extremely oxygen-sensitive, and EPR spectra of aerobically purified (unreduced or reduced) and anaerobically purified enzyme are different. The anaerobically purified enzyme is also $\alpha_2\beta_2$ in structure. The CO-oxidizing activity of *"Methanothrix"* CODH is 4000-fold higher than the acetyl–CoA/CO exchange rate. This could contradict an acetoclastic role, or indicate changes in catalytic capabilities during purification; the latter suggestion is more likely. But, even for extremely air-sensitive CODHs, a difference in stabilities of the CO-oxidizing and exchange activates have been observed [430]; with respect to oxygen and storage, the

CO-oxidizing activity of the CODH from *C. thermoaceticum* is much more stable than the exchange and acetyl–CoA synthase activities.

EPR spectra of aerobically purified *M. soehngenii* CODH are characteristic of 3Fe–4S clusters [267]; a similar signal is seen with thionine oxidized *M. barkeri* CODH [427]. Reduction of aerobically purified *M. soehngenii* CODH with dithionite gives complex EPR spectra characteristic of reduced bacterial eight-iron ferredoxins (two [4Fe–4S] clusters) [267]. Anaerobically purified CODH, after reduction, exhibits two major EPR signals: (i) One characteristic of [4Fe–4S] centers with a midpoint potential of -410 mV that is typical of bacterial [4Fe–4S] clusters, and also observed with *M. barkeri* and *M. thermophila* enzymes (see above); this potential makes these clusters suitable to accept electrons from CO (the $E^{0\prime}$ for CO/CO_2 couple is -520 mV; [427]). (ii) The other signal, in substoichiometric amounts, is not characteristic of typical bacterial [4Fe–4S] clusters, but may arise from [6Fe–6S] prismane-like clusters [267,431]. The midpoint potential associated with this signal is -260 to -280 mV [431], and it changes notably upon incubation of the enzyme with CO [267]; similar midpoint potentials are also seen with the *M. barkeri* enzyme (see above; [427]). Like the *M. barkeri* enzyme, the CODH of *M. soehngenii* does not show CO-dependent EPR signals characteristic of a spin-coupled Ni–Fe–C complex. Similar to the observation with the *M. thermophila* CODH complex [248], acetyl–CoA and not HS–CoA perturbs EPR signals of CO-treated *M. soehngenii* CODH. Since acetyl–CoA and CO interact with certain metal centers of its CODH, in *M. soehngenii* this enzyme is the likely site for acetate cleavage, and acetyl–CoA is the substrate for this reaction. Nothing is known about the C/Fe–SP of *M. soehngenii*, and its CODH has not been purified as a complex with a corrinoid protein.

The genes for α and β subunits of *M. soehngenii* CODH, cdhA and cdhB, respectively, have been cloned, sequenced, and expressed in *E. coli*, albeit without conferring CODH activity to the recombinant [432]; these genes are organized in an operon-like arrangement. The cdhA contains a stretch of 64 amino acid residues with the characteristics to form an archaeal-type ferredoxin domain harboring two [4Fe–4S] clusters. The NH_2-terminal sequence of the cdhA protein could also provide a fixation site for the [6Fe–4S] prismane like structure mentioned above. No region of these genes shows amino acid sequences similar to the proposed consensus Ni binding sequence [432].

4.15. Acetate activating enzymes

To convert acetate to acetyl–CoA, *Methanosarcina* requires two enzymes, acetate kinase (ACK; Reaction 34) and phosphotransacetylase (PTA; Reaction 35) [236,237,239,240, 256]; only one enzyme, acetyl–CoA synthetase (AS; Reaction 36) is required in *"Methanothrix" soehngenii* [241]. The ACK and PTA have been purified from *M. thermophila* [236,240] and AS from *M. soehngenii* [241] and *"Methanothrix"* CALS-1 [446]. Properties of these purified enzymes are given in Table 12. A pyrophosphatase is involved in acetate activation in *M. soehngenii* (Reaction 38); it pulls the AS reaction thermodynamically by hydrolyzing the energy-rich PP_i, and may also play a role in energy conservation [228]. Table 12 also provides properties of purified pyrophosphatase from *M. soehngenii*. All these enzymes are air-stable [228,236,240,241].

The ACK and AS, two ATP-requiring enzymes, are homodimers, and interact directly with acetate. The affinity of AS for acetate is ca. 20-fold higher than that for ACK, consis-

TABLE 12
Properties of acetate activating enzymes

Property	Acetate kinase (ACK) [236]	Phosphotransacetylase (PTA) [240]	Acetyl–CoA synthetase [241][a]	Pyrophosphatase [228]
Source	*Methanosarcina thermophila*	*Methanosarcina thermophila*	*"Methanothrix" soehngenii*	*"Methanothrix" soehngenii*
Apparent molecular mass (kDa)				
Native	94	52	148	139
Subunit	53	43	73	35,33
Kinetic characteristics				
Apparent K_m (mM)				
acetate	22	–	0.86	–
acetylphosphate	–	0.17	–	–
ATP	2.8	–	1	–
HS–CoA	–	0.09	0.048	–
pyrophosphate	–	–	–	0.1
V_{max} (U/mg)	668	5517	55	590
k_{cat} (s^{-1})	1047[b]	4781[b]	136[b]	1367[b]

[a] This enzyme has also been purified from *"Methanothrix"* CALS-1 by Teh and Zindler [446] as a 165 kDa native protein with a 78 kDa subunit. [b] Based on native molecular weight.

tent with the ability of *"Methanothrix"* to grow at very low acetate concentrations where *Methanosarcina* cannot (see pathway section 3.3). Use of a kinase–phosphotransacetylase system at high acetate concentrations and an acetyl–CoA synthetase activity at low acetate concentrations for forming acetyl–CoA has also been described in *E. coli* [433]. The initial velocity versus ATP concentration plot for AS from *M. soehngenii* is slightly sigmoidal, whereas that for the ACK of *M. thermophila* is hyperbolic [236,241]. Thus, in *M. soehngenii* AS may be allosterically regulated by ATP; since *"Methanothrix"* needs twice the ATP for acetate activation than *Methanosarcina*, such modulation is logical. Kinetic studies show two interacting ATP binding sites on the AS [241]. The gene for AS from *M. soehngenii* has been cloned and sequenced [432]; analysis shows two putative ATP binding sequences per AS subunit [432]. The cloned AS gene has been expressed in *E. coli* with a specific activity higher than that in *M. soehngenii* [432].

PTA from *M. thermophila* has much lower K_m values for its substrates than the ACK (Table 12; [236,240]), and lower K_m values than other phosphotransacetylases [240]. Activity of *M. thermophila* PTA in the physiological direction of acetyl–CoA formation is 10-fold greater than in the direction of acetyl-phosphate formation [240]. Potassium and ammonium ions stimulate the activity of PTA about 7-fold, but phosphate, arsenate, and Na$^+$ are inhibitory [240]; phosphate probably acts as an end product inhibitor. Sodium also inhibits other phosphotransacetylases.

The pyrophosphatase of *M. soehngenii* differs from other pyrophosphatases in some aspects. The *M. soehngenii* enzyme is an $\alpha_2\beta_2$ enzyme of 139 kDa [228], whereas bacterial and yeast enzymes have only one kind of subunit [434]. It is insensitive to fluorides, an inhibitor of other pyrophosphatases [228,434]. Membrane fractions of *M. soehngenii* recovered from gentle disruption of cells contain 5% of the total cellular activity; thus, a role in energy conservation has been speculated [228]. As pointed out by Jetten et al. [228], chromatophores of phototrophic bacteria contain a proton-translocating pyrophosphatase that maintains proton-motive force under energy-poor situations [435].

5. Key remaining physiological and enzymatic questions

Understanding of the overall pathway of methane production from H_2–CO_2 appears to be largely complete; as a result of the massive research efforts by a small number of labs, a great deal is known about the biochemistry of methanogenesis, including pathways for the use of other methanogenic substrates. There remain several unclear areas, and in many cases reactions have not been studied in mechanistic or structural detail. Regulation is mostly unstudied; *Methanosarcina,* with its array of potential substrates, is particularly promising in this regard.

Several portions of the pathway are still unclear. How the methyl of methanol enters the reversed methanogenic pathway is unknown, since some suggest it may not proceed via methyl–CoM; nonetheless, evidence is clear that most of the reversed H_2–CO_2 path is used. Also, our understanding of methyl-transfer reactions at several portions of the pathways remains incomplete, e.g. the way in which methanol is reduced to methane, the enzymes involved in methyl transfer in CO_2 methanogenesis, and routes of non-methanol methyl-substrate entry into the path. In several cases, the source of electrons for a reductive step is unknown, e.g. the heterodisulfide reductase and formyl–methanofuran dehydrogenase steps.

One of the most important issues is how ATP is made. This topic is dealt with in detail in Chapter 4 of this volume. A variety of possible reactions have been identified as energetically favorable steps that could be coupled to ATP generation, and several enzymes have been suggested as playing a role in proton or ion gradient formation. The potential sites for energy conservation may differ, depending on the organism and substrate. A particularly challenging task is to find how *"Methanothrix"* can possibly make a living growing on acetate, given the price it seems to have to pay to get access to very low substrate levels. It appears that resolution of these topics will be the major task of the coming decade.

Acknowledgements

The author thanks Biswarup Mukhopadhyay for his generous help in the collection of information for this review, and in the preparation of figures, tables and text; much of the material in the figures was prepared during the compilation of his thesis [449], and is presented therein. The author also thanks B.K. Kim for word processing assistance in the final preparation of the manuscript.

References

[1] Jones, W.J., Nagle Jr., D.P. and Whitman, W.B. (1987) Microbiol. Rev. 51, 135–177.

[2] DiMarco, A.A., Bobik, T.A. and Wolfe, R.S. (1990) Annu. Rev. Biochem. 59, 335–394.
[3] Daniels, L., Sparling, R. and Sprott, G.D. (1984) Biochim. Biophys. Acta 768, 113–163.
[4] Keltjens, J.T. and Vogels, G.D. (1981) In: Microbial Growth on C_1 Compounds (Dalton, H., Ed.), pp. 152–158, Heyden, London.
[5] Keltjens, J.T. and Van der Drift, C. (1986) FEMS Microbiol. Rev. 39, 259–302.
[6] Balch, W.E., Fox, G.E., Magrum, L.J., Woese, C.R. and Wolfe, R.S. (1979) Microbiol. Rev. 43, 260–296.
[7] Bhatnagar, L., Jain, M.K. and Zeikus, J.G. (1991) In: Variations in Autotrophic Life (Shively, J.M. and Barton, L.L., Eds.), pp. 251–270, Academic Press, San Diego.
[8] Jarrell, K.F. and Kalmokoff, M.L. (1988) Can. J. Microbiol. 34, 557–576.
[9] Mah, R.A. and Smith, M.R. (1981) In: The Prokaryotes (Starr, M.P., Stolp, H., Truper, H.G., Balows, A. and Schlegel, H.G., Eds.), pp. 948–977, Springer-Verlag, Heidelberg.
[10] Oremland, R.S. (1987) In: Environmental Microbiology of Anaerobes (Zehnder, A., Ed.), pp. 641–705, Wiley, New York.
[11] Winfrey, M.R. (1984) In: Petroleum Microbiology (Atlas, R.M., Ed.), pp. 153–219, MacMillan, New York.
[12] Jarrell, K.F. (1985) Bioscience 35, 298.
[13] Keiner, A. and Leisinger, T. (1983) Syst. Appl. Microbiol. 4, 305–312.
[14] Patel, G.B., Roth, L.A. and Agnew, B.J. (1984) Can. J. Microbiol. 30, 228–235.
[15] Balch, W.E. and Wolfe, R.S. (1976) Appl. Environ. Microbiol. 32, 781–791.
[16] Bryant, M.P., McBride, B.C. and Wolfe, R.S. (1966) J. Bacteriol. 95, 1118–1123.
[17] Daniels, L., Belay, N. and Mukhopadhyay, B. (1984) Bioeng. Biotechnol. Sympos. 14, 199–213.
[18] Edwards, T. and McBride, B.C. (1975) Appl. Microbiol. 29, 540–545.
[19] Gunsalus, R.P., Tandon, S.M. and Wolfe, R.S. (1980) Anal. Chem. 101, 327–331.
[20] Jones, W.J., Whitman, W.B., Fields, R.D. and Wolfe, R.S. (1983) Appl. Environ. Microbiol. 46, 220–226.
[21] Nishio, N., Kakizono, T., Silveira, R.G., Takemoto, S. and Nagai, S. (1992) J. Ferm. Bioeng. 73, 481–485.
[22] Nishimura, N., Kitaura, S., Mimura, A. and Takahara, Y. (1992) J. Ferm. Bioengin. 73, 477–480.
[23] Macario, A.J.L., Bardulet, M., Conway de Macario, E. and Paris, J.M. (1991) Syst. Appl. Microbiol. 14, 85–92.
[24] Conrad, R., Bak, F., Seitz, H.J., Thebrath, B., Mayer, H.P. and Schutz, H. (1989) FEMS Microbiol. Ecol. 62, 285–294.
[25] Frimmer, U. and Widdel, F. (1989) Arch. Microbiol. 152, 479–483.
[26] Widdel, F. (1986) Appl. Environ. Microbiol. 51, 1056–1062.
[27] Widdel, F. and Wolfe, R.S. (1989) Arch. Microbiol. 152, 322–328.
[28] Widdel, F., Rouviere, P.E. and Wolfe, R.S. (1988) Arch. Microbiol. 150, 477–481.
[29] Zellner, G. and Winter, J. (1987) FEMS Microbiol. Lett. 44, 323–328.
[30] Patel, G.B. (1990) Int. J. Syst. Bacteriol. 40, 79–82.
[31] Patel, G.B. (1992) Int. J. Syst. Bacteriol. 42, 324–326.
[32] Thauer, R.K., Jungermann, K. and Decker, K. (1977) Bacteriol. Rev. 41, 100–180.
[33] Schönheit, P., Moll, J. and Thauer, R.K. (1980) Arch. Microbiol. 127, 59–65.
[34] Zeikus, J.G. and Wolfe, R.S. (1972) J. Bacteriol. 109, 707–713.
[35] Huber, H., Thomm, M., Konig, H., Thies, G. and Stetter, K.O. (1982) Arch. Microbiol. 132, 47–50.
[36] Whitman, W.B., Ankwanda, E. and Wolfe, R.S. (1982) J. Bacteriol. 149, 852–863.
[37] Fardeau, M.-L. and Belaich, J.P. (1986) Arch. Microbiol. 144, 381–385.
[38] Walsh, C. (1986) Acc. Chem. Res. 19, 216–221.
[39] Leigh, J.A. (1983) Appl. Environ. Microbiol. 45, 800–803.
[40] Noll, K.M. and Barber, T.S. (1988) J. Bacteriol. 170, 4315–4321.

[41] Worrell, V.E. and Nagle Jr., D.P. (1988) J. Bacteriol. 170, 4420–4423.
[42] Cheeseman, P., Toms-Wood, A. and Wolfe, R.S. (1972) J. Bacteriol. 112, 527–531.
[43] Eirich, L.D., Vogels, G.D. and Wolfe, R.S. (1978) Biochemistry 17, 4583–4593.
[44] Eirich, L.D., Vogels, G.D. and Wolfe, R.S. (1979) J. Bacteriol. 140, 20–27.
[45] Gorris, L.G.M., Van der Drift, C. and Vogels, G.D. (1988) J. Microbiol. Methods 8, 175–190.
[46] Gorris, L.G.M., Voet, A.C.W.A. and Van der Drift, C. (1991) BioFactors 3, 29–35.
[47] Jacobson, F.S., Daniels, L., Fox, J.A., Walsh, C.T. and Orme-Johnson, W.H. (1982) J. Biol. Chem. 257, 3385–3388.
[48] Jones, J.B. and Stadtman, T.C. (1980) J. Biol. Chem. 255, 1049–1053.
[49] Jones, J.B. and Stadtman, T.C. (1981) J. Biol. Chem. 256, 656–663.
[50] Hartzell, P.L., Zvilius, G., Escalante-Semerena, J.C. and Donnelly, M.I. (1985) Biochem. Biophys. Res. Commun. 133, 884–890.
[51] Mukhopadhyay, B. and Daniels, L. (1989) Can. J. Microbiol. 35, 499–507.
[52] Von Bunau, R., Zirngibl, C., Thauer, R.K. and Klein, A. (1991) Eur. J. Biochem. 202, 1205–1208.
[53] Rospert, S., Breitung, J., Ma, K., Schworer, B., Zirngibl, C., Thauer, R.K., Huber, R. and Stetter, K.O. (1991) Arch. Microbiol. 156, 49–55.
[54] Schmitz, R.A., Linder, D., Stetter, K.O. and Thauer, R.K. (1991) Arch. Microbiol. 156, 427–434.
[55] Schworer, B. and Thauer, R.K. (1991) Arch. Microbiol. 155, 459–465.
[56] Haase, P., Deppenmeier, U., Blaut, M. and Gottschalk, G. (1992) Eur. J. Biochem. 203, 527–531.
[57] Eirich, L.D. and Dugger, R.S. (1984) Biochim. Biophys. Acta 802, 454–458.
[58] Yamazaki, S. and Tsai, L. (1980) J. Biol. Chem. 255, 6462–6465.
[59] Eker, A.P.M., Dekker, R.H. and Berends, W. (1981) Photochem. Photobiol. 33, 65–72.
[60] Kiener, A., Gall, R., Rechsteiner, T. and Leisinger, T. (1985) Arch. Microbiol. 143, 147–150.
[61] Daniels, L., Bakhiet, N. and Harmon, K. (1985) Syst. Appl. Microbiol. 6, 12–17.
[62] Eker, A.P.M., Pol, A., Van der Meijden, P. and Vogels, G.D. (1980) FEMS Microbiol. Lett. 8, 161–165.
[63] McCormick, J.R.D. and Morton, G.O. (1982) J. Am. Chem. Soc. 104, 4014–4015.
[64] Naraoka, T., Momoi, K., Fukasawa, K. and Goto, M. (1984) Biochim. Biophys. Acta 797, 377–380.
[65] Eker, A.P.M., Hessels, J.K.C. and Van de Velde, J. (1988) Biochemistry 27, 1758–1765.
[66] Halldal, P. (1961) Phisiol. Plant. 14, 558–575.
[67] Dewit, L.E.A. and Eker, A.P.M. (1987) FEMS Microbiol. Lett. 48, 121–125.
[68] Lin, X. and White, R.H. (1986) J. Bacteriol. 168, 444–448.
[69] Moller-Zinkhan, D., Borner, G. and Thauer, R.K. (1989) Arch. Microbiol. 152, 362–368.
[70] Moller-Zinkhan, D. and Thauer, R.K. (1990) Arch. Microbiol. 153, 215–218.
[71] Kern, R., Keller, P.J., Schmidt, G. and Bacher, A. (1983) Arch. Microbiol. 136, 191–193.
[72] Jacobson, F.S. and Walsh, C.T. (1984) Biochemistry 23, 979–988.
[73] Kuo, M.S.T., Yurek, D.A., Coats, J.H. and Li, G.P. (1989) J. Antibiotics 42, 475–478.
[74] Miller, P.A., Sjolander, N.O., Nalesnyk, S., Arnold, N., Johnson, S., Doerschuk, A.P. and McCormick, J.R.D. (1960) J. Amer. Chem. Soc. 82, 5002–5003.
[75] Rhodes, P.M., Winskill, N., Friend, E.J. and Warren, M. (1981) J. Gen. Microbiol. 124, 329–338.
[76] Rokita, S.E. and Walsh, C.T. (1984) J. Am. Chem. Soc. 106, 4589–4595.
[77] Purwantini, E., Mukhopadhyay, B., Spencer, R.W. and Daniels, L. (1992) Anal. Biochem. 205, 342–350.
[78] Walsh, C., Fisher, J., Spencer, R., Graham, D.W., Ashton, W.T., Brown, J.E., Brown, R.D. and Rogers, E.F. (1978) Biochemistry 17, 1942–1951.
[79] Purwantini, E. (1992) Coenzyme F_{420}, Masters Thesis, University of Iowa.
[80] Schönheit, P., Keweloh, H. and Thauer, R.K. (1981) FEMS Microbiol. Lett. 12, 347–349.

[81] Gloss, L.M. and Hausinger, R.P. (1987) FEMS Microbiol. Lett. 48, 143–145.
[82] Gloss, L.M. and Hausinger, R.P. (1988) BioFactors 1, 237–240.
[83] Hausinger, R.P., Orme-Johnson, W.H. and Walsh, C. (1985) Biochemistry 24, 1629–1633.
[84] Kengen, S.W.M., Keltjens, J.T. and Vogels, G.D. (1989) FEMS Microbiol. Lett. 60, 5–10.
[85] Kiener, A., Orme-Johnson, W.H. and Walsh, C.T. (1988) Arch. Microbiol. 150, 249–253.
[86] Van de Wijngaard, W.M.H., Vermey, P. and Van der Drift, C. (1991) J. Bacteriol. 173, 2710–2711.
[87] Leigh, J.A., Rinehart Jr., K.L. and Wolfe, R.S. (1984) J. Am. Chem. Soc. 106, 3636–3640.
[88] Leigh, J.A., Rinehart Jr., K.L. and Wolfe, R.S. (1985) Biochemistry 24, 995–999.
[89] Leigh, J.A. and Wolfe, R.S. (1983) J. Biol. Chem. 258, 7536–7540.
[90] Romesser, J.A. and Wolfe, R.S. (1982) Zentralbl. Bakteriol. Hyg. I. Abt., Orig. C. 3, 271–276.
[91] Bobik, T.A., Donnelly, M.I., Rinehart Jr., K.L. and Wolfe, R.S. (1987) Arch. Biochem. Biophys. 254, 430–436.
[92] White, R.H. (1988) J. Bacteriol. 170, 4594–4597.
[93] Breitung, J., Borner, G., Karrasch, M., Berkessel, A. and Thauer, R.K. (1990) FEBS Lett. 268, 257–260.
[94] Donnelly, M.I. and Wolfe, R.S. (1986) J. Biol. Chem. 261, 16653–16659.
[95] Karrasch, M., Borner, G., Enssle, M. and Thauer, R.K. (1990) Eur. J. Biochem. 194, 367–372.
[96] Karrasch, M., Borner, G., Enssle, M. and Thauer, R.K. (1989) FEBS Lett. 253, 226–230.
[97] Jones, W.J., Donnelly, M.I. and Wolfe, R.S. (1985) J. Bacteriol. 163, 126–131.
[98] Miller, T.L. and Wolin, M.J. (1985) Arch. Microbiol. 141, 116–122.
[99] Van de Wijngaard, W.M.H., Creemers, J., Vogels, G.D. and Van der Drift, C. (1991) FEMS Microbiol. Lett. 80, 207–212.
[100] DiMarco, A.A., Donnelly, M.I. and Wolfe, R.S. (1986) J. Bacteriol. 168, 1372–1377.
[101] Van Beelen, P., Stassen, P.M., Bosch, W.G., Vogels, G.D., Guijt, W. and Haasnoot, A.G. (1984) Eur. J. Biochem. 138, 563–571.
[102] Daniels, L. and Zeikus, J.G. (1978) J. Bacteriol. 136, 75–84.
[103] Escalante-Semerena, J.C. and Wolfe, R.S. (1984) J. Bacteriol. 158, 721–726.
[104] Escalante-Semerena, J.C., Leigh, J.A., Rinehart Jr., K.L. and Wolfe, R.S. (1984) Proc. Natl. Acad. Sci. U.S.A. 81, 1976–1980.
[105] Escalante-Semerena, J.C., Rinehart Jr., K.L. and Wolfe, R.S. (1984) J. Biol. Chem. 259, 9447–9455.
[106] Escalante-Semerena, J.C. and Wolfe, R.S. (1985) J. Bacteriol. 161, 696–701.
[107] Van Beelen, P., Thiemessen, H.L., De Cock, R.M. and Vogels, G.D. (1983) FEMS Microbiol. Lett. 18, 135–138.
[108] Van Beelen, P., Labro, F.A., Keltjens, J.T., Geerts, W.J., Vogels, G.D., Laarhoven, W.H., Guijt, W. and Haasnoot, A.G. (1984) Eur. J. Biochem. 139, 359–365.
[109] Van Beelen, P., Van Neck, J.W., De Cock, R.M., Vogels, G.D., Guijt, W. and Haasnoot, C.A.G. (1984) Biochemistry 23, 4448–4454.
[110] Raemakers-Franken, P., Voncken, F.G.J., Korteland, J., Keltjens, J.T., Van der Drift, C. and Vogels, G.D. (1989) BioFactors 2, 117–122.
[111] Raemakers-Franken, P.C., Kortstee, A.J., Van der Drift, C. and Vogels, G.D. (1990) J. Bacteriol. 172, 1157–1159.
[112] Raemakers-Franken, P.C., Van Elderen, H.M., Van der Drift, C. and Vogels, G.D. (1991) BioFactors 3, 127–130.
[113] Breitung, J. and Thauer, R.K. (1990) FEBS Lett. 275, 226–230.
[114] Zirngibl, C., Hedderich, R. and Thauer, R.K. (1990) FEBS Lett. 261, 112–116.
[115] Ma, K. and Thauer, R.K. (1990) FEMS Microbiol. Lett. 70, 119–124.
[116] Ma, K. and Thauer, R.K. (1990) Eur. J. Biochem. 191, 187–193.
[117] Fischer, R., Gartner, P., Yeliseev, A. and Thauer, R.K. (1992) Arch. Microbiol. 158, 208–217.

[118] Gorris, L.G.M., Van der Drift, C. and Vogels, G.D. (1988) BioFactors 1, 105–109.
[119] Keltjens, J.T. and Vogels, G.D. (1988) BioFactors 1, 95–103.
[120] Hoyt, J.C., Oren, A., Escalante-Semerena, J.C. and Wolfe, R.S. (1986) Arch. Microbiol. 145, 153–158.
[121] Lin, X. and White, R.H. (1988) Arch. Microbiol. 150, 541–546.
[122] Donnelly, M.I., Escalante-Semerena, J.C., Rinehart Jr., K.L. and Wolfe, R.S. (1985) Arch. Biochem. Biophys. 242, 430–439.
[123] Johnson, J.L., Bastian, N.R., Schauer, N.L., Ferry, J.G. and Rajagopalan, K.V. (1991) FEMS Microbiol. Lett. 77, 213–216.
[124] May, H.D., Schauer, N.L. and Ferry, J.G. (1986) J. Bacteriol. 166, 500–504.
[125] McBride, B.C. and Wolfe, R.S. (1971) Biochemistry 10, 2317–2324.
[126] Taylor, C.D. and Wolfe, R.S. (1974) J. Biol. Chem. 249, 4879–4885.
[127] Ellefson, W.L. and Wolfe, R.S. (1980) J. Biol. Chem. 255, 8388–8389.
[128] Gunsalus, R.P. and Wolfe, R.S. (1980) J. Biol. Chem. 255, 1891–1895.
[129] Deppenmeier, U., Blaut, M., Mahlmann, A. and Gottschalk, G. (1990) FEBS Lett. 261, 199–203.
[130] Hedderich, R. and Thauer, R.K. (1988) FEBS Lett. 234, 223–227.
[131] Hedderich, R., Berkessel, A. and Thauer, R.K. (1990) Eur. J. Biochem. 193, 255–261.
[132] Smith, S.G. and Rouviere, P. (1990) J. Bacteriol. 172, 6435–6441.
[133] Gunsalus, R.P., Romesser, J.A. and Wolfe, R.S. (1978) Biochemistry 17, 2374–2377.
[134] Sparling, R. and Daniels, L. (1988) Can. J. Microbiol. 33, 1132–1136.
[135] Ellermann, J., Rospert, S., Thauer, R.K., Bokranz, M., Klein, A., Voges, M. and Berkessel, A. (1989) Eur. J. Biochem. 184, 63–68.
[136] Olson, K.D., Chmurkowska-Cichowlas, L., McMahon, C.W. and Wolfe, R.S. (1992) J. Bacteriol. 174, 1007–1012.
[137] Wackett, L.P., Honek, J.F., Begley, T.P., Wallace, V., Orme-Johnson, W.H. and Walsh, C.T. (1987) Biochemistry 26, 6012–6018.
[138] Keltjens, J.T., Whitman, W.B., Caerteling, C.G., Kooten, A.M.V., Wolfe, R.S. and Vogels, G.D. (1982) Biochem. Biophys. Res. Commun. 108, 495–503.
[139] Hartzell, P.L., Donnelly, M.I. and Wolfe, R.S. (1987) J. Biol. Chem. 262, 5581–5586.
[140] Hausinger, R.P., Orme-Johnson, W.H. and Walsh, C. (1984) Biochemistry 23, 801–804.
[141] Huster, R., Gilles, H.H. and Thauer, R.K. (1985) Eur. J. Biochem. 148, 107–111.
[142] Livingston, D.A., Pfaltz, A., Schreiber, J., Eschenmoser, A., Ankel-Fuchs, D., Moll, J., Jaenchen, R. and Thauer, R.K. (1984) Helv. Chim. Acta. 67, 334–351.
[143] Dangel, W., Schulz, H., Diekert, G., Konig, H. and Fuchs, G. (1987) Arch. Microbiol. 148, 52–56.
[144] Hollriegl, V., Scherer, P. and Renz, P. (1983) FEBS Lett. 151, 156–158.
[145] Krzycki, J.A. and Zeikus, J.G. (1980) Curr. Microbiol. 3, 243–245.
[146] Stupperich, E., Steiner, I. and Eisinger, H.J. (1987) J. Bacteriol. 169, 3076–3081.
[147] Cao, X. and Krzycki, J. (1991) J. Bacteriol. 173, 5439–5448.
[148] Kohler, E.H. (1988) Arch. Microbiol. 150, 219–223.
[149] Pol, A., Van der Drift, C. and Vogels, G.D. (1982) Biochem. Biophys. Res. Commun. 108, 731–737.
[150] Pol, A., Gage, R.A., Neis, J.M., Reijnen, J.W.M., Van Der Drift, C. and Vogels, G.D. (1984) Biochim. Biophys. Acta 797, 83–93.
[151] Stupperich, E. and Krautler, B. (1988) Arch. Microbiol. 149, 268–271.
[152] Van der Meijden, P., Heythuysen, H.J., Pouwels, A., Houwen, F., Van der Drift, C. and Vogels, G.D. (1983) Arch. Microbiol. 134, 238–242.
[153] Van der Meijden, P., Heythuysen, H.J., Sliepenbeek, H.T., Houwen, F.P., Van der Drift, C. and Vogels, G.D. (1983) J. Bacteriol. 153, 6–11.

[154] Van der Meijden, P., Te Brommelstroet, B.W., Poirot, C.M., Van der Drift, C. and Vogels, G.D. (1984) J. Bacteriol. 160, 629–635.
[155] Van der Meijden, P., Van der Lest, C., Van der Drift, C. and Vogels, G.D. (1984) Biochem. Biophys. Res. Commun. 118, 760–766.
[156] Van de Wijngaard, W.M.H., Van de Drift, C. and Vogels, G.D. (1988) FEMS Microbiol. Lett. 52, 165–172.
[157] Kengen, S.W.M., Daas, P.J.H., Duits, E.F.G., Keltjens, J.T., Van der Drift, C. and Vogels, G.D. (1992) Biochim. Biophys. Acta 1118, 249–260.
[158] Schulz, H., Albracht, S.P.J., Coremans, J.M.C.C. and Fuchs, G. (1988) Eur. J. Biochem. 171, 589–597.
[159] Kenealy, W.R. and Zeikus, J.G. (1981) J. Bacteriol. 146, 133–140.
[160] Gunsalus, R.P. and Wolfe, R.S. (1978) FEMS Microbiol. Lett. 3, 191–193.
[161] Diekert, G., Klee, B. and Thauer, R.K. (1980) Arch. Microbiol. 124, 103–106.
[162] Whitman, W.B. and Wolfe, R.S. (1980) Biochem. Biophys. Res. Commun. 92, 1196–1201.
[163] Pfaltz, A., Jaun, B., Fassler, A., Eschenmoser, A., Jaenchen, R., Gilles, H.-H., Diekert, G. and Thauer, R.K. (1982) Helv. Chim. Acta. 65, 828–865.
[164] Shiemke, A.K., Hamilton, C.L. and Scott, R.A. (1988) J. Biol. Chem. 263, 5611–5616.
[165] Diekert, G., Konheiser, U., Piechulla, K. and Thauer, R.K. (1981) J. Bacteriol. 148, 459–464.
[166] Ankel-Fuchs, D., Jaenchen, R., Gebhardt, N.A. and Thauer, R.K. (1984) Arch. Microbiol. 139, 332–337.
[167] Ellefson, W.L., Whitman, W.B. and Wolfe, R.S. (1982) Proc. Natl. Acad. Sci. U.S.A. 79, 3707–3710.
[168] Jaun, B. and Pfaltz, A. (1988) J. Chem. Soc. Chem. Commun., p. 293.
[169] Lin, S.-K. and Jaun, B. (1991) Helvetica Chim. Acta 74, 1725–1738.
[170] Keltjens, J.T., Caerteling, C.G., Van Kooten, A.M., Van Dijk, H.F. and Vogels, G.D. (1983) Arch. Biochem. Biophys. 223, 235–253.
[171] Pfaltz, A., Livingston, D.A., Jaun, B., Diekert, G. and Thauer, R.K. (1985) Helv. Chim. Acta. 68, 1338–1358.
[172] Ellefson, W.L. and Wolfe, R.S. (1981) J. Biol. Chem. 256, 4259–4262.
[173] Nagle Jr., D.P. and Wolfe, R.S. (1983) Proc. Natl. Acad. Sci. U.S.A. 80, 2151–2155.
[174] Ellermann, J., Kobelt, A., Pfaltz, A. and Thauer, R.K. (1987) FEBS Lett. 220, 358–362.
[175] Ellermann, J., Hedderich, R., Bocher, R. and Thauer, R.K. (1988) Eur. J. Biochem. 172, 669–677.
[176] Noll, K.M. and Wolfe, R.S. (1987) Biochem. Biophys. Res. Commun. 145, 204–210.
[177] Noll, K.M., Donnelly, M.I. and Wolfe, R.S. (1987) J. Biol. Chem. 262, 513–515.
[178] Noll, K.M., Rinehart Jr., K.L., Tanner, R.S. and Wolfe, R.S. (1986) Proc. Natl. Acad. Sci. U.S.A. 83, 4238–4242.
[179] Kobelt, A., Pfaltz, A., Ankel-Fuchs, D. and Thauer, R.K. (1987) FEBS Lett. 214, 265–268.
[180] Kuhner, C.H., Smith, S.S., Noll, K.M., Tanner, R.S. and Wolfe, R.S. (1991) Appl. Environ. Microbiol. 57, 2891–2895.
[181] Noll, K.M. and Wolfe, R.S. (1986) Biochem. Biophys. Res. Commun. 139, 889–895.
[182] MacKenzie, R.E. (1984) In: Folates and Pterins (Blakely, R.L. and Benkovic, S., Eds.), pp. 255–306, Wiley, New York.
[183] Fuchs, G. (1986) FEMS Microbiol. Rev. 39, 181–213.
[184] Thauer, R.K. (1990) Biochim. Biophys. Acta 1018, 256–259.
[185] Bobik, T.A. and Wolfe, R. (1989) J. Bacteriol. 171, 1423–1427.
[186] Te Brommelstroet, B.W., Hensgens, C.M.H., Geerts, W.J., Keltjens, J.T., Van der Drift, C. and Vogels, G.D. (1990) J. Bacteriol. 172, 564–571.
[187] Keltjens, J.T., Brugman, A.J.A.M., Kesseleer, J.M.A., Te Brommelstroet, B.W.J., Van der Drift, C. and Vogels, G.D. (1992) BioFactors 3, 249–255.

[188] Enssle, M., Zirngibl, C., Linder, D. and Thauer, R.K. (1991) Arch. Microbiol. 155, 483–490.
[189] Te Brommelstroet, B.W.J., Geerts, W.J., Keltjens, J.T., Van der Drift, C. and Vogels, G.D. (1991) Biochim. Biophys. Acta 1079, 293–302.
[190] Ma, K., Zirngibl, C., Linder, D., Stetter, K.O. and Thauer, R.K. (1991) Arch. Microbiol. 156, 43–48.
[191] Zirngibl, C., Van Dongen, W., Schworer, B., Van Bunau, R., Richter, M., Klein, A. and Thauer, R.K. (1992) Eur. J. Biochem. 208, 511–520.
[192] Conrad, R., Schink, B. and Phelps, T.J. (1986) FEMS Microbiol. Ecol. 38, 353–360.
[193] Te Brommelstroet, B.W., Hensgens, C.M.H., Keltjens, J.T., Van der Drift, C. and Vogels, G.D. (1990) J. Biol. Chem. 265, 1852–1857.
[194] Taylor, C.D. and Wolfe, R.S. (1974) J. Biol. Chem. 249, 4886–4890.
[195] Becher, B., Miller, V. and Gottschalk, G. (1992) FEMS Microbiol. Lett. 91, 239–244.
[196] Blaylock, B.A. (1968) Arch. Biochem. Biophys. 124, 314–324.
[197] Ankel-Fuchs, D., Bocher, R., Thauer, R.K., Noll, K.M. and Wolfe, R.S. (1987) FEBS Lett. 213, 123–127.
[198] Bobik, T.A. and Wolfe, R.S. (1987) Proc. Natl. Acad. Sci. U.S.A. 85, 60–63.
[199] Rouviere, P.E., Bobik, T.A. and Wolfe, R.S. (1988) J. Bacteriol. 170, 3946–3952.
[200] Hedderich, R., Berkessel, A. and Thauer, R.K. (1989) FEBS Lett. 255, 67–71.
[201] Deppenmeier, U., Blaut, M. and Gottschalk, G. (1991) Arch. Microbiol. 155, 272–277.
[202] Deppenmeier, U., Blaut, M., Mahlmann, A. and Gottschalk, G. (1990) Biochemistry 87, 9449–9453.
[203] Peinemann, S., Hedderich, R., Blaut, M., Thauer, R.K. and Gottschalk, G. (1990) FEBS Lett. 263, 57–60.
[204] Deppenmeier, U., Blaut, M., Schmidt, B. and Gottschalk, G. (1992) Arch. Microbiol. 157, 505–511.
[205] Gunsalus, R.P. and Wolfe, R.S. (1977) Biochem. Biophys. Res. Commun. 76, 790–795.
[206] Romesser, J.A. (1978) The activation and reduction of carbon dioxide to methane in *Methanobacterium thermoautotrophicum*, Ph.D. thesis, University of Illinois.
[207] Bobik, T.A. and Wolfe, R. (1989) J. Biol. Chem. 264, 18714–18718.
[208] Romesser, J.A. and Wolfe, R.S. (1982) J. Bacteriol. 152, 840–847.
[209] Daniels, L., Fulton, G., Spencer, R.W. and Orme-Johnson, W.H. (1980) J. Bacteriol. 141, 694–698.
[210] Spencer, R.W., Daniels, L., Fulton, G. and Orme-Johnson, W.H. (1980) Biochemistry 19, 3678–3683.
[211] Kamlage, B. and Blaut, M. (1992) J. Bacteriol. 174, 3921–3927.
[212] Zellner, G., Bleicher, K., Braun, E., Kneifel, H., Tindall, B.J., Conway de Macario, E. and Winter, J. (1989) Arch. Microbiol. 151, 1–9.
[213] Daniels, L., Fuchs, G., Thauer, R.K. and Zeikus, J.G. (1977) J. Bacteriol. 132, 118–126.
[214] Kluyver, A.J. and Schnellen, C.G.T.P. (1947) Arch. Biochem. 14, 57–70.
[215] O'Brien, J.M., Wolkin, R.H., Moench, T.T., Morgan, J.B. and Zeikus, J.G. (1984) J. Bacteriol. 158, 373–375.
[216] Blaylock, B.A. and Stadtman, T.C. (1963) Biochem. Biophys. Res. Commun. 11, 34–38.
[217] Blaylock, B.A. and Stadtman, T.C. (1964) Biochem. Biophys. Res. Commun. 17, 475–480.
[218] Blaylock, B.A. and Stadtman, T.C. (1966) Arch. Biochem. Biophys. 116, 138–152.
[219] Wood, J.M., Moura, I., Moura, J.J.G., Santos, M.H., Xavier, A.V., LeGall, J. and Scandellari, M. (1982) Science 216, 303–305.
[220] Muller, V., Winner, C. and Gottschalk, G. (1988) Eur. J. Biochem. 178, 519–525.
[221] Bhatnagar, L., Krzycki, J.A. and Zeikus, J.G. (1987) FEMS Microbiol. Lett. 41, 337–343.
[222] Mahlmann, A., Deppenmeier, U. and Gottschalk, G. (1989) FEMS Microbiol. Lett. 61, 115–120.

[223] Winner, C. and Gottschalk, G. (1989) FEMS Microbiol. Lett. 65, 259–264.
[224] Mukhopadhyay, B., Purwantini, E. and Daniels, L. (1993) Arch. Microbiol. 159, 141–146.
[225] Blaut, M., Muller, V., Fiebig, K. and Gottschalk, G. (1985) J. Bacteriol. 164, 95–101.
[226] Kaesler, B. and Schonheit, P. (1989) Eur. J. Biochem. 186, 309–316.
[227] Naumann, E., Fahlbusch, K. and Gottschalk, G. (1984) Arch. Microbiol. 138, 79–83.
[228] Jetten, M.S.M. (1991) Acetate metabolism in *Methanothrix soehngenii*, Ph.D. Thesis, Agricultural University, Wageningen.
[229] Kaspar, H.F. and Wuhrman, K. (1978) Appl. Environ. Microbiol. 36, 1–7.
[230] Smith, P.H. and Mah, R.A. (1966) Appl. Microbiol. 14, 368–371.
[231] Lovely, D.R., White, R.H. and Ferry, J.G. (1984) J. Bacteriol. 160, 521–525.
[232] Pine, M.J. and Barker, H.A. (1956) J. Bacteriol. 71, 644–648.
[233] Krzycki, J.A., Wolkin, R.H. and Zeikus, J.G. (1982) J. Bacteriol. 149, 247–254.
[234] Krzycki, J.A. and Zeikus, J.G. (1984) FEMS Microbiol. Lett. 25, 27–32.
[235] Karrasch, M., Bott, M. and Thauer, R.K. (1989) Arch. Microbiol. 151, 137–142.
[236] Aceti, D.J. and Ferry, J.G. (1988) J. Biol. Chem. 263, 15444–15448.
[237] Grahame, D.A. and Stadtman, T.C. (1987) Biochem. Biophys. Res. Commun. 147, 254–258.
[238] Kenealy, W.R. and Zeikus, J.G. (1982) J. Bacteriol. 151, 932–941.
[239] Krzycki, J.A., Lehman, L.J. and Zeikus, J.G. (1985) J. Bacteriol. 163, 1000–1006.
[240] Lundie, L.L.J. and Ferry, J.G. (1989) J. Biol. Chem. 264, 18392–18396.
[241] Jetten, M.S.M., Stams, A.J.M. and Zehnder, A.J.B. (1989) J. Bacteriol. 171, 5430–5435.
[242] Jetten, M.S.M., Stams, A.J.M. and Zehnder, A.J.B. (1990) FEMS Microbiol. Lett. 66, 183–186.
[243] Ragsdale, S.W. (1991) Crit. Rev. Biochem. Mol. Biol. 26, 261–300.
[244] Wood, H.G. (1991) FASEB J. 5, 156–163.
[245] Abbanat, D.R. and Ferry, J.G. (1990) J. Bacteriol. 172, 7145–7150.
[246] Terlesky, K.C., Nelson, M.J. and Ferry, J.G. (1986) J. Bacteriol. 168, 1053–1058.
[247] Raybuck, S.A., Ramer, S.E., Abbanat, D.R., Peters, J.W., Orme-Johnson, W.H., Ferry, J.G. and Walsh, C.T. (1991) J. Bacteriol. 173, 929–932.
[248] Terlesky, K.C., Barber, M.J., Aceti, D.J. and Ferry, J.G. (1987) J. Biol. Chem. 262, 15392–15395.
[249] Terlesky, K.C. and Ferry, J.G. (1988) J. Biol. Chem. 263, 4075–4079.
[250] Harder, S.R., Lu, W.P., Feinberg, B.A. and Ragsdale, S.W. (1989) Biochemistry 28, 9080–9087.
[251] Diekert, G. (1988) In: The Bioinorganic Chemistry of Nickel (Lancaster, J.R., Ed.), pp. 299–309, VCH Publishers, New York.
[252] Abbanat, D.R. and Ferry, J.G. (1991) Proc. Natl. Acad. Sci. U.S.A. 88, 3272–3276.
[253] Ferry, J.G. (1992) J. Bacteriol. 174, 5489–5495.
[254] Fischer, R. and Thauer, R.K. (1989) Arch. Microbiol. 151, 459–465.
[255] Krzycki, J.A. and Zeikus, J.G. (1984) J. Bacteriol. 158, 231–237.
[256] Fischer, R. and Thauer, R.K. (1988) FEBS Lett. 228, 249–253.
[257] Grahame, D.A. and Stadtman, T.C. (1987) J. Biol. Chem. 262, 3706–3712.
[258] Grahame, D.A. (1991) J. Biol. Chem. 266, 22227–22233.
[259] Eikmanns, B. and Thauer, R.K. (1984) Arch. Microbiol. 138, 365–370.
[260] Fischer, R. and Thauer, R.K. (1990) FEBS Lett. 269, 368–372.
[261] Laufer, K., Eikmanns, B., Frimmer, U. and Thauer, R.K. (1986) Z. Naturfor. 42C, 360–372.
[262] Fischer, R. and Thauer, R.K. (1990) Arch. Microbiol. 153, 156–162.
[263] Grahame, D.A. (1989) J. Biol. Chem. 264, 12890–12894.
[264] Peinemann, S., Muller, V., Blaut, M. and Gottshalk, G. (1988) J. Bacteriol. 170, 1369–1372.
[265] Perski, H.J., Schonheit, P. and Thauer, R.K. (1982) FEBS Lett. 143, 323–326.
[266] Jetten, M.S.M., Stams, A.J.M. and Zehnder, A.J.B. (1989) Eur. J. Biochem. 181, 437–441.
[267] Jetten, M.S M., Hagen, W.R., Pierik, A.J., Stams, A.J.M. and Zehnder, A.J.B. (1991) Eur. J. Biochem. 195, 385–391.
[268] Jablonski, P.E. and Ferry, J.G. (1991) J. Bacteriol. 173, 2481–2487.

[269] Gorris, L.G.M. and Van der Drift, C. (1986) In: Biology of Anaerobic Bacteria (Dubourguier, H.C., Ed.), pp. 144–150, Elsevier Science Publishers, Amsterdam.
[270] Adams, M.W.W., Jin, S.L.C., Chen, J.-C. and Mortenson, L.E. (1986) Biochim. Biophys. Acta 869, 37–47.
[271] Baron, S.F. and Ferry, J.G. (1989) J. Bacteriol. 171, 3846–3853.
[272] Coremans, J.M.C.C., Van der Zwaan, J.W. and Albracht, S.P.J. (1989) Biochim. Biophys. Acta 997, 256–267.
[273] Fauque, G., Teixeira, M., Moura, I., Lespinat, P.A., Xavier, A.V., Der Vartanian, D.V., Peck Jr., H.D., LeGall, J. and Moura, J.G. (1984) Eur. J. Biochem. 142, 21–28.
[274] Fiebig, K. and Friedrich, B. (1989) Eur. J. Biochem. 184, 79–88.
[275] Fox, J.A., Livingston, D.J., Orme-Johnson, W.H. and Walsh, C.T. (1987) Biochemistry 26, 4219–4227.
[276] Jin, S.L.C., Blanchard, D.K. and Chen, J.S. (1983) Biochim. Biophys. Acta 784, 8–20.
[277] Muth, E., Morschel, E. and Klein, A. (1987) Eur. J. Biochem. 169, 571–577.
[278] Nelson, M.J.K., Brown, D.P. and Ferry, J.G. (1984) Biochem. Biophys. Res. Commun. 120, 775–781.
[279] Shah, N.N. and Clark, D. (1990) Appl. Environ. Microbiol. 56, 858–863.
[280] Sprott, G.D., Shaw, K.M. and Beveridge, T.J. (1987) Can. J. Microbiol. 33, 896–904.
[281] Yamazaki, S. (1982) J. Biol. Chem. 257, 7926–7929.
[282] Baresi, L. and Wolfe, R.S. (1981) Appl. Environ. Microbiol. 41, 388–391.
[283] Sparling, R. and Daniels, L. (1990) J. Bacteriol. 172, 1464–1469.
[284] Murray, P.A. and Zinder, S.H. (1985) Appl. Environ. Microbiol. 50, 49–55.
[285] Huser, B.A., Wuhrmann, K. and Zehnder, A.J.B. (1982) Arch. Microbiol. 132, 1–9.
[286] Patel, G.B. (1984) Can. J. Microbiol. 30, 1383–1396.
[287] Zehnder, A.J.B., Huser, B.A., Brock, T.D. and Wuhrmann, K. (1980) Arch. Microbiol. 124, 1–11.
[288] Zinder, S.H. and Anguish, T. (1992) Appl. Environ. Microbiol. 58, 3323–3329.
[289] Mukhopadhyay, B., Purwantini, E., Conway de Macario, E. and Daniels, L. (1991) Curr. Microbiol. 23, 165–173.
[290] Phelps, T.J., Conrad, R. and Zeikus, J.G. (1985) Appl. Environ. Microbiol. 50, 589–594.
[291] Graf, E.G. and Thauer, R.K. (1981) FEBS Lett. 136, 165–169.
[292] Kojima, N., Fox, J.A., Hausinger, R.P., Daniels, L., Orme-Johnson, W.H. and Walsh, C. (1983) Proc. Natl. Acad. Sci. U.S.A. 80, 378–382.
[293] Reeve, J.N., Beckler, G.S., Cram, D.S., Hamilton, P.T., Brown, J.W., Krzycki, J.A., Kolodziej, A.F., Alex, L. and Orme-Johnson, W.H. (1989) Proc. Natl. Acad. Sci. U.S.A. 86, 3031–3035.
[294] Halboth, S. and Klein, A. (1992) Mol. Gen. Genet. 233, 217–224.
[295] Walsh, C.T. and Orme-Johnson, W.H. (1987) Biochemistry 26, 4901–4906.
[296] Alex, L.A., Reeve, J.N., Orme-Johnson, W.H. and Walsh, C.T. (1990) Biochemistry 29, 7237–7244.
[297] Reeve, J.N. and Beckler, G.S. (1990) FEMS Microbiol. Rev. 87, 419–424.
[298] Steigerwald, V.J., Beckler, G.S. and Reeve, J.N. (1990) J. Bacteriol. 172, 4715–4718.
[299] Bastian, N.R., Wink, D.A., Wackett, L.P., Livingston, D.J., Jordan, L.M., Fox, J.L., Orme-Johnson, W.H. and Walsh, C.T. (1988) In: The Bioinorganic Chemistry of Nickel (Lancaster, J.R., Ed.), VCH Publishers, New York.
[300] Steigerwald, V.J., Phil, T.D. and Reeve, J. (1992) Proc. Natl. Acad. Sci. U.S.A. 89, 6929–6933.
[301] Hedderich, R., Albracht, S.P.J., Linder, D. and Thauer, R.K. (1992) FEBS Lett. 298, 65–68.
[302] Hatchikian, E.C., Bruschi, M., Forget, N. and Scandellari, M. (1982) Biochem. Biophys. Res. Commun. 109, 1316–1323.
[303] Clements, A.P. and Ferry, J.G. (1992) J. Bacteriol. 174, 5244–5250.

[304] Bruschi, M., Bonicel, J., Hatchikian, E.C., Fardeau, M.L., Belaich, J.P. and Frey, M. (1991) Biochim. Biophys. Acta 1076, 79–85.
[305] Hatchikian, E.C., Fardeau, M.L., Bruschi, M., Belaich, J.P., Chapman, A. and Cammack, R. (1989) J. Bacteriol. 171, 2384–2390.
[306] Terlesky, K.C. and Ferry, J.G. (1988) J. Biol. Chem. 263, 4075–4079.
[307] McFarlan, S.C., Terrell, C.A. and Hogenkamp, H.P.C. (1992) J. Biol. Chem. 267, 10561–10569.
[308] Kuhn, W., Fiebig, K., Walther, R. and Gottschalk, G. (1979) FEBS Lett. 105, 271–274.
[309] Kuhn, W., Fiebig, K., Hippe, H., Mah, R.A., Huser, B.A. and Gottschalk, G. (1983) FEMS Microbiol. Lett. 20, 407–410.
[310] Kuhn, W. and Gottschalk, G. (1983) Eur. J. Biochem. 135, 89–94.
[311] Kemner, J.M., Krzycki, J.A., Prince, R.C. and Zeikus, J.G. (1987) FEMS Microbiol. Lett. 48, 267–272.
[312] Livingston, D.J., Fox, J.A., Orme-Johnson, W.H. and Walsh, C.T. (1987) Biochemistry 26, 4228–4237.
[313] Lancaster Jr., J.R. (1980) FEBS Lett. 115, 285–288.
[314] Lancaster Jr., J.R. (1982) Science 216, 1324–1325.
[315] Albracht, S.P.J., Graf, E.-G. and Thauer, R.K. (1982) FEBS Lett. 140, 311–313.
[316] Lindahl, P.A., Kojima, N., Hausinger, R.P., Fox, J.A., Teo, B.K., Walsh, C.T. and Orme-Johnson, W.H. (1984) J. Am. Chem. Soc. 106, 3062–3064.
[317] Tan, S.L., Fox, J.A., Kojima, N., Walsh, C.T. and Orme-Johnson, W.H. (1984) J. Am. Chem. Soc. 106, 3064–3066.
[318] Teshima, T., Nakaji, A., Shiba, T., Tsai, L. and Yamazaki, S. (1985) Tetrahedron Lett. 26, 351–354.
[319] Wackett, L.P., Hartwieg, E.A., King, J.A., Orme-Johnson, W.H. and Walsh, C.T. (1987) J. Bacteriol. 169, 718–727.
[320] Muth, E. (1988) Arch. Microbiol. 150, 205–207.
[321] Lunsdorf, H., Niedrig, M. and Fiebig, K. (1991) J. Bacteriol. 173, 978–984.
[322] Bleicher, K., Zellner, G. and Winter, J. (1989) FEMS Microbiol. Lett. 59, 307–312.
[323] Barker, H.A. (1941) J. Biol. Chem. 137, 153–167.
[324] Bryant, M.P., Wolin, E.A. and Wolfe, R.S. (1967) Arch. Mikrobiol. 59, 20–31.
[325] Sparling, R. and Daniels, L. (1986) J. Bacteriol. 168, 1402–1407.
[326] Fina, L.R., Sincher, H.J. and De Cou, D.F. (1960) Arch. Biochem. Biophys. 91, 159–162.
[327] Schauer, N.L., Honek, J.F., Orme-Johnson, W.H., Walsh, C. and Ferry, J.G. (1986) Biochemistry 25, 7163–7168.
[328] Tzeng, S.F., Bryant, M.P. and Wolfe, R.S. (1975) J. Bacteriol. 121, 192–196.
[329] Jones, J.B., Dilworth, G.L. and Stadtman, T.C. (1979) Arch. Biochem. Biophys. 195, 255–260.
[330] Schauer, N.L. and Ferry, J.G. (1982.) J. Bacteriol. 150, 1–7.
[331] Schauer, N.L. and Ferry, J.G. (1983) J. Bacteriol. 155, 467–472.
[332] Schauer, N.L. and Ferry, J.G. (1986) J. Bacteriol. 165, 405–411.
[333] Zinoni, F., Heider, J. and Bock, A. (1990) Proc. Natl. Acad. Sci. U.S.A. 87, 4660–4664.
[334] May, H.D., Patel, P.S. and Ferry, J.G. (1988) J. Bacteriol. 170, 3384–3389.
[335] Barber, M.J., May, H.D. and Ferry, J.G. (1986) Biochemistry 25, 8150–8155.
[336] Heider, J. and Bock, A. (1992) J. Bacteriol. 174, 659–663.
[337] Yamazaki, S., Tsai, L., Stadtman, T.C., Teshima, T., Nakaji, A. and Shiba, T. (1985) Proc. Natl. Acad. Sci. U.S.A. 82, 1364–1366.
[338] Patel, P.S. and Ferry, J.G. (1988) J. Bacteriol. 170, 3390–3395.
[339] Shuber, A.P., Orr, E.C., Recny, M.A., Schendel, P.F., May, H.D., Schauer, N.L. and Ferry, J.G. (1986) J. Biol. Chem. 261, 12942–12947.
[340] White, W.B. and Ferry, J.G. (1992) J. Bacteriol. 174, 4997–5004.
[341] Baron, S.F. and Ferry, J.G. (1989) J. Bacteriol. 171, 3854–3859.

[342] Eguchi, S.Y., Nishio, N. and Nagai, S. (1985) Appl. Microbiol. Biotechnol. 22, 148–151.
[343] Nishio, N., Eguchi, S.Y., Kawashima, H. and Nagai, S. (1983) J. Ferm. Technol. 61, 557–561.
[344] Nelson, M.J.K. and Ferry, J.G. (1984) J. Bacteriol. 160, 526–532.
[345] Tanner, R.S., McInerney, M.J. and Nagle Jr., D.P. (1989) J. Bacteriol. 171, 6534–6538.
[346] Jablonski, P.E., DiMarco, A.A., Bobik, T.A., Cabell, M.C. and Ferry, J.G. (1990) J. Bacteriol. 172, 1271–1275.
[347] Borner, G., Karrasch, M. and Thauer, R.K. (1989) FEBS Lett. 244, 21–25.
[348] Borner, G., Karrasch, M. and Thauer, R.K. (1991) FEBS Lett. 290, 31–34.
[349] Karrasch, M., Borner, G. and Thauer, R.K. (1990) FEBS Lett. 274, 48–52.
[350] Schmitz, R.A., Richter, M., Linder, D. and Thauer, R.K. (1992) Eur. J. Biochem. 207, 559–565.
[351] Kaesler, B. and Schonheit, P. (1989) Eur. J. Biochem. 184, 223–232.
[352] Kruger, B. and Mayer, O. (1987) Biochim. Biophys. Acta 912, 357–364.
[353] Johnson, J.L., Bastian, N.R. and Rajagopalan, K.V. (1990) Proc. Natl. Acad. Sci. U.S.A. 87, 3190–3194.
[354] DiMarco, A.A., Sment, K.A., Konisky, J. and Wolfe, R.S. (1990) J. Biol. Chem. 265, 472–476.
[355] Grimshaw (1984) J. Biol. Chem. 259, 2728–2733.
[356] Breitung, J., Schmitz, R.A., Stetter, K.O. and Thauer, R.K. (1991) Arch. Microbiol. 156, 517–524.
[357] Brandis, A., Thauer, R.K. and Stetter, K.O. (1981) Zentralbl. Bakteriol. Mikrobiol. Hyg. 1 Abt. Orig. C2, 311–317.
[358] Kurr, M., Huber, R., Konig, H., Jannasch, H.W., Fricke, H., Trincone, A., Kristjansson, J.K. and Stetter, K.O. (1991) Arch. Microbiol. 156, 239–247.
[359] Kanodia, S. and Roberts, M.F. (1983) Biochemistry 80, 5217–5221.
[360] Hensel, R. and Konig, H. (1988) FEMS Microbiol. Lett. 49, 75–79.
[361] Te Brommelstroet, B.W., Hensgens, C.M.H., Keltjens, J.T., Van der Drift, C. and Vogels, G.D. (1991) Biochim. Biophys. Acta 1073, 77–84.
[362] Wohlfarth, G., Geerligs, G. and Diekert, G. (1991) J. Bacteriol. 173, 1414–1419.
[363] Barlowe, C.K. and Appling, D.R. (1990) Biochemistry 29, 7089–7094.
[364] Dev, I.K. and Harvey, R.J. (1978) J. Biol. Chem. 253, 4245–4253.
[365] Mejia, N.R. and MacKenzie, R.E. (1988) Biochem. Biophys. Res. Com. 155, 1–6.
[366] Mejia, N.R. and MacKenzie, R.E. (1985) J. Biol. Chem. 260, 14616–14620.
[367] Mejia, N.R., Rios-Orlandi, E.M. and MacKenzie, R.E. (1986) J. Biol. Chem. 261, 9509–9513.
[368] Moore, M.R., O'Brien, W.E. and Ljungdahl, L.G. (1974) J. Biol. Chem. 249, 5250–5253.
[369] O'Brien, W.E., Brewer, J.M. and Ljungdahl, L.G. (1973) J. Biol. Chem. 248, 403–408.
[370] Scrigmour, K.G. and Huennekens, F.M. (1960) Biochem. Biophys. Res. Com. 2, 230–233.
[371] Uyeda, K. and Rabinowitz, J.C. (1967) J. Biol. Chem. 242, 4378–4385.
[372] Ma, K. and Thauer, R.K. (1990) FEBS Lett. 268, 59–62.
[373] Ma, K., Linder, D., Stetter, K.O. and Thauer, R.K. (1991) Arch. Microbiol. 155, 593–600.
[374] Rudnick, H., Hendrich, S., Pilatus, U. and Blotevogel, K.H. (1990) Arch. Microbiol. 154, 584–588.
[375] Vanoni, M.A., Daubner, S.C., Ballou, D.P. and Matthews, R.G. (1983) J. Biol. Chem. 258, 11510–11514.
[376] Wohlfarth, G., Geerligs, G. and Diekert, G. (1990) Eur. J. Biochem. 192, 411–417.
[377] Kengen, S.W.M., Daas, P.J.H., Keltjens, J.T., Van der Drift, C. and Vogels, G.D. (1990) Arch. Microbiol. 154, 156–161.
[378] Sauer, F.D. (1986) Biochem. Biophys. Res. Commun. 136, 542–547.
[379] Poirot, C.M., Kengen, S.W.M., Valk, E., Keltjens, J.T., Van der Drift, C. and Vogels, G.D. (1987) FEMS Microbiol. Lett. 40, 7–13.
[380] Van der Meijden, P., Van der Drift, C. and Vogels, G.D. (1984) Arch. Microbiol. 138, 360–364.
[381] Schulz, H. and Fuchs, G. (1986) FEBS Lett. 198, 279–282.

[382] Stupperich, E., Jusa, A., Eckerskorn, C. and Edelman, L. (1990) Arch. Microbiol. 155, 28–34.
[383] Banarjee, R.V., Frasca, V., Ballou, D.P. and Matthews, R.G. (1990) Biochemistry 29, 11101–11109.
[384] Saveant, J.M., De Tacconi, N., Lexa, D. and Zickler, J. (1979) In: Vitamin B_{12} (Zagalak, B. and Friedrich, W., Eds.), pp. 203–212, Walter de Gruiter, Berlin.
[385] Schweiger, G., Dutcho, R. and Buckel, W. (1987) Eur. J. Biochem. 169, 441–448.
[386] Hartzell, P.L. and Wolfe, R.S. (1986) Syst. Appl. Microbiol. 7, 376–382.
[387] Hartzell, P.L., Escalante-Semerena, J.C., Bobik, T.S. and Wolfe, R.S. (1988) J. Bacteriol. 170, 2711–2715.
[388] Ankel-Fuchs, D. and Thauer, R.K. (1986) Eur. J. Biochem. 156, 171–177.
[389] Ankel-Fuchs, D., Huster, R., Morschel, E., Albracht, S.P.J. and Thauer, R.K. (1986) Syst. Appl. Microbiol. 7, 383–387.
[390] Hamilton, C.L., Scott, R.A. and Johnson, M.K. (1989) J. Biol. Chem. 264, 11605–11613.
[391] Rospert, S., Linder, D., Ellermann, J. and Thauer, R.K. (1990) Eur. J. Biochem. 194, 871–877.
[392] Rospert, S., Bocher, R., Albracht, S.P.J. and Thauer, R.K. (1991) FEBS Lett. 291, 371–375.
[393] Rouviere, P.E., Escalante-Semerena, J.C. and Wolfe, R.S. (1985) J. Bacteriol. 162, 61–66.
[394] Rouviere, P.E. and Wolfe, R.S. (1989) J. Bacteriol. 171, 4556–4562.
[395] Deppenmeier, U., Blaut, M., Jussofie, A. and Gottschalk, G. (1988) FEBS Lett. 241, 60–64.
[396] Hoppert, M. and Mayer, F. (1990) FEBS Lett. 267, 33–37.
[397] Whitman, W.B. (1985) In: The Bacteria (Woese, C.R. and Wolfe, R.S., Eds.), pp. 3–84, Academic Press, New York.
[398] Bonacker, L.G., Baudner, S. and Thauer, R.K. (1992) Eur. J. Biochem. 206, 87–92.
[399] Brenner, M.C., Ma, L., Johnson, K.K. and Scott, R.A. (1992) Biochim. Biophys. Acta 1120, 160–166.
[400] Keltjens, J.T., Caerteling, C.G., Van Kooten, A.M., Van Dijk, H.F. and Vogels, G.D. (1983) Biochim. Biophys. Acta 743, 351–358.
[401] Hutten, T.J., De Jong, M.H., Peeters, B.P.H., Van der Drift, C. and Vogels, G.D. (1981) J. Bacteriol. 145, 27–34.
[402] Albracht, S.P.J., Ankel-Fuchs, D., Van der Zwaan, J.W., Fontijn, R.D. and Thauer, R.K. (1986) Biochim. Biophys. Acta 870, 50–57.
[403] Jaun, B. and Pfaltz, A. (1986) J. Chem. Soc. Chem. Commun., pp. 1327–1328.
[404] Ahn, Y., Krzycki, J.A. and Floss, H.G. (1991) J. Amer. Chem. Soc. 113, 4700–4701.
[405] Zimmer, M. and Crabtree, R.H. (1990) J. Am. Chem. Soc. 112, 1062–1066.
[406] Bokranz, M., Baumer, G., Allmansberger, R., Ankel-Fuchs, D. and Klein, A. (1988) J. Bacteriol. 170, 568–577.
[407] Bokranz, M. and Klein, A. (1987) Nucleic Acids Res. 15, 4350–4351.
[408] Cram, D.S., Sherf, B.A., Libby, R.T., Mattaliano, R.J., Ramachandaran, K.L. and Reeve, J.N. (1987) Proc. Natl. Acad. Sci. U.S.A. 84, 3992–3996.
[409] Klein, A., Allmansberger, R., Bokranz, M., Knaub, S., Muller, B. and Muth, E. (1988) Mol. Gen. Genet. 213, 409–420.
[410] Weil, C.F., Cram, D.S., Sherf, B.A. and Reeve, J.N. (1988) J. Bacteriol. 170, 4718–4726.
[411] Weil, C.F., Sherf, B.A. and Reeve, J.N. (1989) Can. J. Microbiol. 35, 101–108.
[412] Ossmer, R., Mund, T., Hartzell, P.L., Konheiser, U., Kohring, G.W., Klein, A., Wolfe, R.S., Gottschalk, G. and Mayer, F. (1986) Proc. Natl. Acad. Sci. U.S.A. 83, 5789–5792.
[413] Aldrich, H.C., Beimborn, D.B., Bokranz, M. and Schonheit, P. (1987) Arch. Microbiol. 147, 190–194.
[414] Mayer, F., Rohde, M., Salzmann, M., Jussofie, A. and Gottschalk, G. (1988) J. Bacteriol. 170, 1438–1444.
[415] Thomas, I., Dubourguier, H.C., Prensier, G., Debeire, P. and Albagnac, G. (1987) Arch. Microbiol. 148, 148.

[416] Blaut, M. and Gottschalk, G. (1984) Eur. J. Biochem. 141, 217–222.
[417] Blaut, M. and Gottschalk, G. (1984) FEMS Microbiol. Lett. 24, 103–107.
[418] Blaut, M. and Gottschalk, G. (1985) TIBS. 10, 486–489.
[419] Blaut, M., Muller, V. and Gottschalk, G. (1986) Syst. Appl. Microbiol. 7, 354–357.
[420] Blaut, M., Muller, V. and Gottschalk, G. (1987) FEBS Lett. 215, 53–57.
[421] Gottschalk, G. and Blaut, M. (1990) Biochim. Biophys. Acta 1018, 263–266.
[422] Peinemann, S., Blaut, M. and Gottschalk, G. (1989) Eur. J. Biochem. 186, 175–180.
[423] Bobik, T.A., Olson, K.D., Noll, K.M. and Wolfe, R.S. (1987) Biochem. Biophys. Res. Commun. 149, 455–460.
[424] Van der Meijden, P., Jansen, L.P.J.M., Van der Drift, C. and Vogels, G.D. (1983) FEMS Microbiol. Lett. 19, 247–251.
[425] Zydowsky, L.D., Zydowsky, T.M., Haas, E.S., Brown, J.W., Reeve, J.N. and Floss, H.G. (1987) J. Am. Chem. Soc. 109, 7922–7923.
[426] Krzycki, J.A. and Prince, R.C. (1989) Biochim. Biophys. Acta 1015, 53–60.
[427] Krzycki, J.A., Mortenson, L.E. and Prince, R.C. (1989) J. Biol. Chem. 264, 7217–7221.
[428] Lu, W.P., Harder, S.R. and Ragsdale, S.W. (1990) J. Biol. Chem. 265, 3124–3133.
[429] Jablonski, P.E., Lu, W.-P., Ragsdale, S.W. and Ferry, J.G. (1991). Abstr. K-136, Abstr. 91st Gen. Meet. Am. Soc. Microbiol., Dallas, TX, p. 237, ASM.
[430] Raybuck, S.A., Bastian, N.R., Orme-Johnson, W.H. and Walsh, C.T. (1988) Biochemistry 27, 7698–7702.
[431] Jetten, M.S.M., Pierik, A.J. and Hagen, W.R. (1991) Eur. J. Biochem. 202, 1291–1297.
[432] Eggen, R.I.L., Geerling, A.C.M., Jetten, M.S.M. and De Vos, W.M. (1991) J. Biol. Chem. 266, 6883–6887.
[433] Brown, T.D.K., Jones-Mortimer, M.C. and Kornberg, H.L. (1977) J. Gen. Microbiol. 102, 327–336.
[434] Lahti, R. (1983) Microbiol. Rev. 47, 169–179.
[435] Nyren, P. and Strid, A. (1991) FEMS Microbiol. Lett. 77, 265–270.
[436] Schauer, N.L. and Ferry, J.G. (1980) J. Bacteriol. 142, 800–807.
[437] Jones, W., Leigh, J.A., Mayer, F., Woese, C.R. and Wolfe, R.S. (1983) Arch. Microbiol. 136, 254–261.
[438] Konig, H. and Setter, K.O. (1982) Zentralbl. Bakteriol. Mikrobiol. Hyg. 1 Abt. Orig. C3, 478–490.
[439] Hippe, H., Caspari, D., Fiebig, K. and Gottschalk, G. (1979) Microbiol. 76, 494–498.
[440] Smith, M.R. and Mah, R.A. (1980) Appl. Environ. Microbiol. 39, 993–999.
[441] Weimer, P.J. and Zeikus, J.G. (1978) Arch. Microbiol. 119, 175–182.
[442] Jussofie, A., Mayer, F. and Gottschalk, G. (1986) Arch. Microbiol. 146, 245–249.
[443] Zinder, S.H., Sowers, K.R. and Ferry, J.G. (1985) Int. J. Syst. Bacteriol. 35, 522–523.
[444] Zinder, S.H., Anguish, T. and Lobo, A.L. (1987) Arch. Microbiol. 146, 315–322.
[445] Jablonski, P.E., Lu, W.P., Ragsdale, S.W. and Ferry, J.G. (1993) J. Biol. Chem. 268, 325–329.
[446] Teh, Y.L. and Zinder, S. (1992) FEMS Microbiol. Lett. 98, 1–8.
[447] Franzmann, P.D., Springer, N., Ludwig, W., Conway de Macario, E. and Rohde, M. (1992) Syst. Appl. Microbiol. 15, 573–581.
[448] Wu, W.M., Hickey, R.F., Jain, M.K. and Zeikus, J.G. (1993) Arch. Microbiol. 159, 57–65.
[449] Mukhopadhyay, B. (1993) Purification and characterization of F_{420}-dependent methylenetetrahydromethanopterin dehydrogenase of *Methanobacterium thermoautotrophicum* strain Marburg, and cloning and expression in *E. coli* and sequencing of the corresponding gene and effects of catabolites on the levels of three catabolic enzymes in *Methanosarcina barkeri*, Ph.D. Thesis, University of Iowa.

M. Kates et al. (Eds.), *The Biochemistry of Archaea (Archaebacteria)*
© 1993 Elsevier Science Publishers B.V. All rights reserved

CHAPTER 4

Bioenergetics and transport in methanogens and related thermophilic archaea

Peter SCHÖNHEIT

Institut für Pflanzenphysiologie und Mikrobiologie, Fachbereich Biologie, Freie Universität Berlin

Abbreviations

MFR	methanofuran	SLP	substrate level phosphorylation
H_4MPT	5,6,7,8-tetrahydromethanopterin	ETP	electron transport phosphorylation
H–S–CoM	coenzyme M (2-mercaptoethanesulfonate)	$\Delta\tilde{\mu}H^+$	transmembrane electrochemical potential of H^+
H–S–HTP	7-mercaptoheptanoylthreonine phosphate	$\Delta\tilde{\mu}Na^+$	transmembrane electrochemical potential of Na^+
CoM–S–S–HTP	heterodisulfide of H–S–CoM and H–S–HTP	$\Delta\psi$	transmembrane electrical gradient
F_{420}	the N-(N-L-lactyl-γ-L-glutamyl)-L-glutamic acid phosphodiester of 7,8-didemethyl-8-hydroxy-5-deazariboflavin	ΔpH, ΔpNa	transmembrane concentration differences of H^+ and Na^+
		F_1F_0-type ATPase	H^+-translocating ATPase
DCCD	N,N'-dicyclohexylcarbodiimide	P-type ATPase	ion-translocating ATPase involving a phosphorylated intermediate in the catalytic cycle
TCS	3,5,3',4'-tetrachlorosalicyl-anilide		
amiloride	3,5-diamino-6-chloropyrazinoylguanidine		

1. Introduction

Methanogenic bacteria are strictly anaerobic organisms that are defined by their ability to form methane as end product of their energy metabolism. This group of organisms has attracted the interest of microbiologists, molecular biologists and biochemists, especially in the last two decades. At the beginning of this period (1970) pure cultures of methanogens had become available and techniques for anaerobic handling and mass culturing had been established [1]. During these studies the phylogenetic position of methanogens and their ecological role in the anaerobic carbon cycle was elucidated and the unique biochemical features of methanogenesis were discovered: methanogens are important representatives of the archaea (archaebacteria) [2–5]. Many features leading to the development of the concept of archaea were derived from studies with methanogens (see the Introduction by Woese in this volume). Methanogens play an important quantitative role in the anaerobic degradation of organic matter. In anoxic habitats with low sulfate concentrations, e.g. in freshwater sediments, rice fields, swamps and ruminants, methanogens catalyze the terminal reactions in the complete degradation of organic compounds to CH_4 and CO_2. About 10^9 tons of methane are formed per year by the metabolic activity of methanogens (for a recent review see Oremland [6]). The study of methane formation is of particular interest because of the participation of several unique types of cofactors [7] and a number of unusual types of enzymes (for detailed literature references see Chapter 3 by Daniels in this volume).

This article deals with the energetics of methanogenesis. Because of the unique biochemistry involved it was not possible to deduce the bioenergetic principles of methanogenesis on the basis of well-established energy transducing systems. Methanogens gain energy by coupling the exergonic formation of methane with the synthesis of ATP. Furthermore the process of methanogenesis was specifically dependent on low concentrations of Na^+ ions. Thus the major energetic problems were the mode of coupling of ATP formation with methanogenesis, and the role of Na^+ in energy coupling.

In this chapter the present knowledge of the bioenergetics of methanogenesis is discussed. This topic has been previously reviewed [8–18]. The main items discussed here are: (i) the sites and the mechanism of ATP synthesis coupled to the conversion of various substrates to methane; (ii) the role of Na^+ in energy transduction associated with the various methanogenic pathways; (iii) transport in methanogens. The review includes a short treatment of the energetics of the extremely thermophilic, non-methanogenic *Archaeoglobus* and *Pyrococcus* [19]. These anaerobic euryarchaeota are phylogenetically closely related to methanogens [4,5,20].

Most biochemical and energetic studies in this field have been performed with a few methanogenic genera: e.g. *Methanosarcina, Methanobacterium* and *Methanococcus*. Since these genera are phylogenetically somewhat distantly related (Fig. 1), the results obtained may be considered to be representative for all methanogens. Based on 16S rRNA sequence data methanogens have been divided into three major phylogenetic groups: the Methanococcales (e.g. *Methanococcus*), the Methanobacteriales (e.g. *Methanobacterium*), and the Methanomicrobiales (e.g. *Methanosarcina*), all of which are distributed within the euryarchaeotal branch of the archaea (Fig. 1). The very recently isolated *Methanopyrus kandleri*, an extremely thermophilic methanogen growing at temperatures up to 110°C [21,22] was found not to be related to any of the three methanogen groups: the organism seems to branch very deeply on the euryarchaeotal side of the archaeal tree [23]. Included in Fig. 1 are the other groups of the euryarchaeota, *Thermoplasma* and the extreme halophiles, the metabolism of which is discussed in other chapters of this book.

The phylogenetic diversity of the methanogens is also reflected by a great diversity in morphology, cell wall structure and metabolic and physiological properties (see Chapters 3 and 8 by Daniels and Kandler/König in this volume). Most methanogens can be cultivated on mineral salt media containing various energy substrates (see

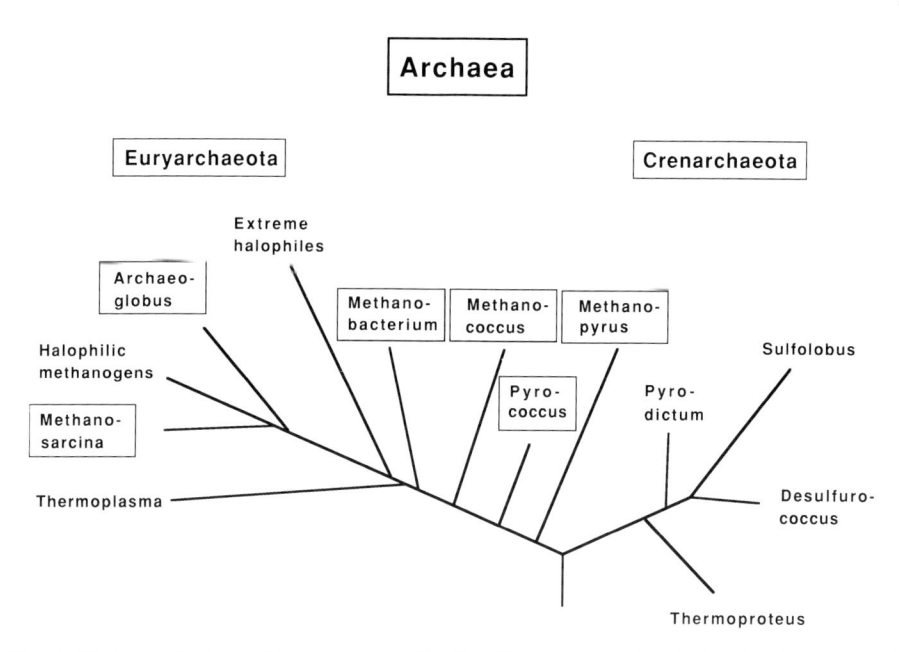

Fig. 1. Phylogenetic tree of the archaea indicating the phylogenetic relationship between various methanogenic genera, *Archaeoglobus* and *Pyrococcus*. The branching orders are based upon rRNA sequence comparisons according to Woese et al. [5,20] and Burggraf et al. [23]. The line lengths do not correspond to the phylogenetic distances.

below). Ammonia and sulfide serve generally as nitrogen and sulfur sources; however molecular nitrogen [24–26] and oxidized sulfur compounds, such as sulfate [27], can also be used by some species. Various trace elements proved to be essential for growth of methanogens, most notably nickel, cobalt, and molybdenum [28–30], which are components of various enzymes and coenzymes involved in methanogenesis, and Na^+, which is involved in energy coupling (see below). Most methanogens are mesophilic, non-halophilic, neutrophilic organisms. However, extreme thermophilic methanogens, e.g. *Methanopyrus kandleri* growing at up to 110°C [21,22] and halophilic alkaliphilic organisms, e.g. *Methanohalophilus zhilinae* [31], have also recently been isolated. Details about the physiology as well as the biochemistry of anabolic and catabolic pathways of methanogens are given in various reviews (e.g., refs. [2,13,32–34]) and in Chapter 3 by Daniels in this volume.

2. Energy substrates

Methanogens can use only a limited number of C_1 compounds and acetate as substrates for growth and methane formation (Reactions 1–11 of Table 1). Very recently pyruvate was shown to be a methanogenic substrate [35]. The substrates can be divided into two groups: CH_4 is formed either by the reduction of CO_2 (Reactions 1–4) or by the reduction of a methyl-group (Reactions 5–11) with different electron donors. For a distribution of

TABLE 1
Energy substrates for growth of methanogenic bacteria

#	Reaction	$\Delta G^{o\prime}$ (kJ/mol CH$_4$) [a]
1.	$4H_2 + CO_2 \rightarrow CH_4 + 2H_2O$	-131
2.	$4HCOO^- + 4H^+ \rightarrow CH_4 + 3CO_2 + 2H_2O$	-145
3.	$2CH_3CH_2OH\ ^b + CO_2 \rightarrow CH_4 + 2CH_3COO^- + 2H^+$	-112
4.	$4CH_3CHOHCH_3\ ^b + CO_2 \rightarrow CH_4 + 4CH_3COCH_3 + 2H_2O$	-32
5.	$4CH_3OH \rightarrow 3CH_4 + CO_2 + 2H_2O$	-107
6.	$H_2 + CH_3OH \rightarrow CH_4 + H_2O$	-112
7.	$4CH_3NH_3^+ + 2H_2O \rightarrow 3CH_4 + CO_2 + 4NH_4^+$	-77
8.	$2(CH_3)_2NH_2^+ + 2H_2O \rightarrow 3CH_4 + CO_2 + 2NH_4^+$	-75
9.	$4(CH_3)_3NH^+ + 6H_2O \rightarrow 9CH_4 + 3CO_2 + 4NH_4^+$	-76
10.	$CH_3COO^- + H^+ \rightarrow CH_4 + CO_2$	-36
11.	$CH_3COCOO^- + H^+ + 0.5H_2O \rightarrow 1.25CH_4 + 1.75CO_2$	-96

[a] $\Delta G^{o\prime}$ values are calculated from ΔG_f° values as given by Thauer et al. [8]; CO_2, gaseous state.
[b] Other alcohols shown to be electron donors for CO_2 reduction to CH_4 are: 1-propanol, 2-butanol and cyclopentanol.

substrate utilization by methanogenic species and for the ecological importance of the various substrates see recent reviews [2,13,36,37].

2.1. Reduction of CO_2 to CH_4

CO_2 can be reduced to CH_4 with H_2, formate and primary and secondary alcohols as electron donors.

Most methanogens form methane by the reduction of CO_2 with molecular hydrogen as the electron donor:

$$4H_2 + CO_2 \rightarrow CH_4 + 2H_2O. \tag{1}$$

About 50% of these species can use formate as substrate. Formate conversion to CH_4 (Reaction 2) involves the oxidation of formate to CO_2 by formate dehydrogenase, generating reducing equivalents which are subsequently used to reduce CO_2 to CH_4:

$$4HCOO^- + 4H^+ \rightarrow 4CO_2 + 8[H], \tag{2a}$$

$$8[H] + 1CO_2 \rightarrow CH_4 + 2H_2O \tag{2b}$$

$$4HCOO^- + 4H^+ \rightarrow CH_4 + 3CO_2 + 2H_2O \tag{2}$$

A few methanogens have recently been isolated which can use primary and secondary alcohols as electron donors for CO_2 reduction to CH_4 [38–40]. Ethanol and 1-propanol or 2-propanol, 2-butanol and cyclopentanol have been shown to serve as hydrogen donors for

CO_2 reduction to CH_4. Primary alcohols (e.g. ethanol) are oxidized to their corresponding acids:

$$2CH_3CH_2OH + 2H_2O \rightarrow 2CH_3COO^- + 2H^+ + 8[H], \tag{3a}$$

$$CO_2 + 8[H] \rightarrow CH_4 + 2H_2O \tag{3b}$$

$$2CH_3CH_2OH + CO_2 \rightarrow 2CH_3COO^- + 2H^+ + CH_4 \tag{3}$$

Secondary alcohols (e.g. 2-propanol) are oxidized to their corresponding ketones:

$$2(CH_3)_2CHOH + 2H_2O \rightarrow 4CH_3COCH_3 + 8[H], \tag{4a}$$

$$CO_2 + 8[H] \rightarrow CH_4 + 2H_2O \tag{4b}$$

$$2(CH_3)_2CHOH + CO_2 \rightarrow 2CH_3COCH_3 + CH_4 \tag{4}$$

Almost all of the alcohol-utilizing methanogens can in addition use H_2/CO_2 and formate as methanogenic substrates.

2.2. Reduction of a methyl group to CH_4

The second group of methanogenic substrates have in common that CH_4 is formed by the reduction of a methyl group with electrons derived either from methyl-group oxidation, from molecular hydrogen or from the oxidation of an enzyme-bound carbon monoxide.

Several methanogens disproportionate methanol to CH_4 and CO_2. Disproportionation involves oxidation of CH_3OH to CO_2, generating 6 reducing equivalents which reduce 3 mol of CH_3OH to methane according to the following reactions:

$$1CH_3OH + H_2O \rightarrow 1CO_2 + 6[H], \tag{5a}$$

$$3CH_3OH + 6[H] \rightarrow 3CH_4 + 3H_2O \tag{5b}$$

$$4CH_3OH \rightarrow 3CH_4 + 1CO_2 + 2H_2O \tag{5}$$

Those methanogens which possess a hydrogenase can in addition form methane by the reduction of methanol with molecular hydrogen:

$$H_2 + CH_3OH \rightarrow CH_4 + H_2O \tag{6}$$

The reduction of methanol with H_2 to methane represents the simplest form of methane fermentation, and turned out to be extremely useful for studying the mechanism of ATP synthesis in cell suspensions [41] (see below). Growth at the expense of CH_3OH reduction with H_2 has been demonstrated so far only with *Methanosphaera stadtmanae* [42] and

Methanosarcina barkeri [43]. While *Methanosphaera stadtmanae* is restricted to this kind of methanogenesis, *Methanosarcina barkeri* in addition can disproportionate methanol. *Methanolobus tindarius* being devoid of a hydrogenase can only disproportionate methanol rather than reduce it with molecular hydrogen [44].

As was first shown for *Methanosarcina barkeri* [45], methanogens can grow on methylamine, dimethylamine and trimethylamine as sole carbon and energy sources. The various methylamines are disproportionated analogously to methanol; in addition ammonia is formed (Reactions 7–9 of Table 1).

Acetate is the most important methanogenic substrate in nature; about 70% of the methane formed biologically is produced from this acid [33,46]. However, only a few species of the genera *Methanosarcina* and *Methanothrix* have been shown to convert this compound to CO_2 and CH_4 (Reaction 10). Acetate conversion involves carbon–carbon cleavage of acetyl–CoA to yield enzyme-bound methanol [CH_3OH] and enzyme-bound carbon monoxide [CO]. Reduction of [CH_3OH] to CH_4 is accomplished with reducing equivalents derived from [CO] oxidation to CO_2:

$$CH_3COOH \rightarrow [CH_3OH] + [CO], \tag{10a}$$

$$[CO] + H_2O \rightarrow CO_2 + 2[H], \tag{10b}$$

$$[CH_3OH] + 2[H] \rightarrow CH_4 + H_2O, \tag{10c}$$

$$CH_3COOH \rightarrow CH_4 + CO_2. \tag{10}$$

Recently we found [35] that *Methanosarcina barkeri* can grow on pyruvate as sole carbon and energy source after an "adaptation period" of several weeks in the presence of pyruvate. During growth only CO_2 and CH_4 were found as end products. H_2 was not formed. The fermentation balance during growth on pyruvate can best be described by the following partial reactions:

$$CH_3COCOOH + H_2O \rightarrow 2[H] + CO_2 + [CH_3COOH], \tag{11a}$$

$$[CH_3COOH] \rightarrow CO_2 + CH_4, \tag{11b}$$

$$2[H] + 0.25CO_2 \rightarrow 0.25CH_4 + 0.5H_2O, \tag{11c}$$

$$CH_3COCOOH + 0.5H_2O \rightarrow 1.25CH_4 + 1.75CO_2. \tag{11}$$

Pyruvate is oxidized to acetyl–CoA ([CH_3COOH]), CO_2 and 2[H]; acetyl–CoA is converted to CH_4 and CO_2, and simultaneously CH_4 is formed by the reduction of CO_2 with the reducing equivalents generated. This is the first report of the growth of a methanogenic bacterium on pyruvate as sole carbon and energy source, i.e. a substrate more complex than acetate.

The utilization of carbon monoxide [47,48] and dimethyl sulfide [49] as growth substrates by methanogens has been reported.

Formaldehyde has been shown to be a methanogenic substrate for cell suspensions of methanogens, rather than a growth substrate. This is probably due to the extreme toxicity of this compound. In cell suspensions formaldehyde can either be reduced to methane with molecular H_2 (HCHO + $2H_2$ → CH_4 + H_2O) or, under a N_2 gas phase, be disproportionated to CH_4 and CO_2 (2HCHO → CH_4 + CO_2); upon inhibition of methanogenesis by specific inhibitors formaldehyde is oxidized to CO_2 and H_2 (HCHO + H_2O → CO_2 + $2H_2$). As will be seen below, studies on the energetics of the various modes of formaldehyde conversion in cell suspensions of methanogens gave significant insight into the mechanism of energy transduction of partial reactions of methanogenesis from CO_2, methanol or acetate.

The following sections describe the energetics of methanogenesis from the various substrates given in Table 1. Most data are from studies with methanogenic cell suspensions that form methane from CO_2/H_2, methanol, methanol/H_2, formaldehyde or acetate. The following topics are discussed in particular: (i) enzymology and thermodynamics of energetic relevant partial reactions; (ii) the sites of energy coupling of the different methanogenic pathways; (iii) the mechanism of ATP synthesis coupled to methanogenesis; (iv) the sites of Na^+ dependence of the various methanogenic pathways; (v) the function of Na^+ in energy transduction. Also, brief remarks will be made on the energetics of methanogenesis from other substrates.

3. Energetics of methanogenesis from CO_2/H_2

Almost all methanogens form methane by the reduction of CO_2 with H_2 as electron donor and couple this exergonic reaction with the synthesis of ATP:

$$CO_2 + 4H_2 \rightarrow CH_4 + 2H_2O, \qquad \Delta G^{o\prime} = 131 \text{ kJ/mol } CH_4.$$

3.1. Enzymology

The reduction of CO_2 to CH_4 proceeds in 4 two-electron steps via the formal redox states of formate, formaldehyde and methanol (see below). As early as 1956 Barker [50] proposed that the C_1-intermediates are not free but that they are all coenzyme-bound. The C_1-carrying coenzymes have been identified as methanofuran (MFR), tetrahydromethanopterin (H_4MPT) and coenzyme M (H–S–CoM) (for recent reviews see refs. [7,51–53]). The structures of the C_1-carrying coenzymes, of C_1-intermediates, of electron carriers and of the coenzyme F_{430}, the prosthetic group of methyl–coenzyme M reductase are given in Figs. 2A–C (for structural derivatives of MFR and H_4MPT and their distribution within methanogenic species see refs. [7,51,53]). Fig. 3 shows partial reactions, intermediates and enzymes involved in CH_4 formation from CO_2, formaldehyde and from methanol. In Table 2 the energetics of partial reactions (Reactions 1–8) of CO_2 reduction to CH_4 are given.

3.1.1. CO_2 reduction to formyl–MFR (formate level)
The reduction of CO_2 to formyl–MFR (CHO–MFR) is catalyzed by formyl–MFR dehydrogenase (Reaction 1). It is assumed that first a N-substituted

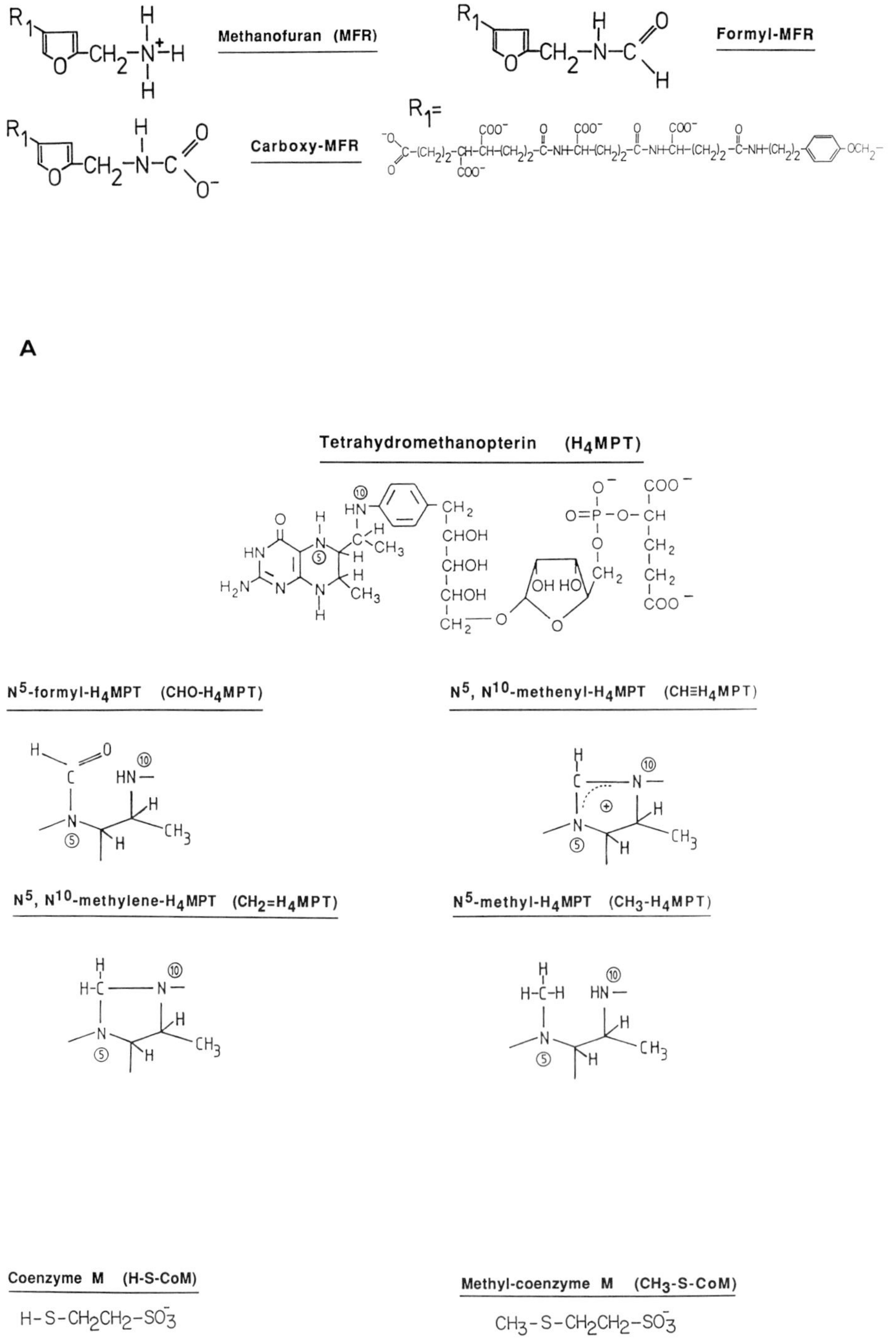

Fig. 2. Structures of coenzymes and electron carriers involved in methanogenesis. (A) Structures of C_1-carrying coenzymes and of C_1-intermediates.

B Coenzyme F$_{420}$

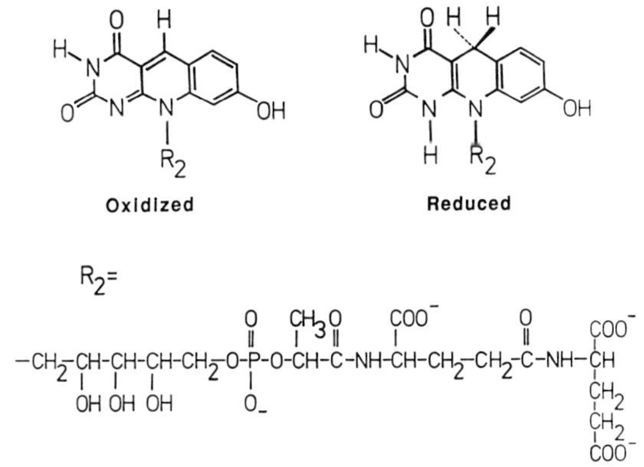

Oxidized   Reduced

R$_2$ =

N-7-mercaptoheptanoylthreonine phosphate (H-S-HTP)

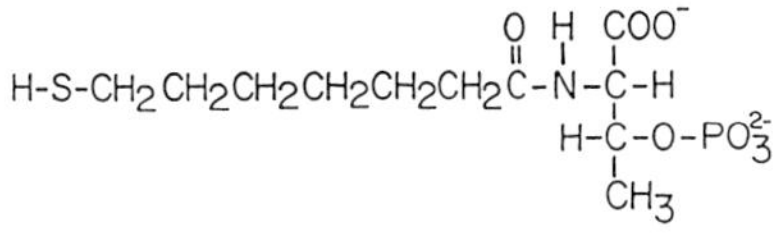

C Coenzyme F$_{430}$

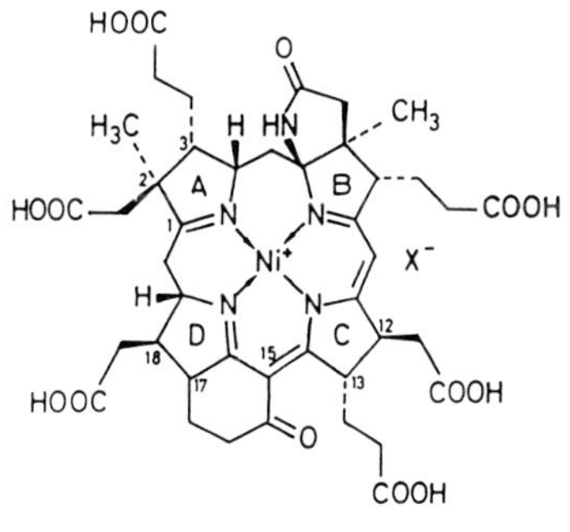

Fig. 2(BC). (B) structures of electron carriers; (C) structure of coenzyme F$_{430}$, the prosthetic group of methyl–coenzyme M reductase.

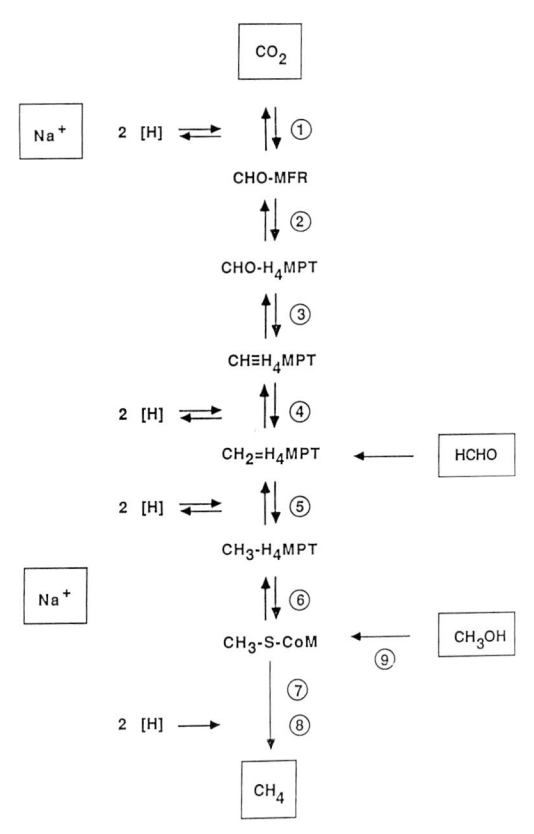

Fig. 3. Proposed pathway of methanogenesis from CO_2/H_2 or from methanol: intermediates, enzymes and sites of Na^+ dependence. It is assumed that methanol binds first to coenzyme M prior to oxidation and that oxidation proceeds via the reversal of the reactions involved in CO_2 reduction to methyl–coenzyme M. MFR, methanofuran; CHO–MFR, formyl–MFR; H_4MPT, tetrahydromethanopterin; CHO–H_4MPT, formyl–H_4MPT; CH≡H_4MPT, methenyl–H_4MPT; CH_2=H_4MPT, methylene–H_4MPT; CH_3–H_4MPT, methyl–H_4MPT; H–S–CoM, coenzyme M; CH_3–S–CoM, methyl–coenzyme M. Numbers in circles refer to enzymes involved: (1) formyl–MFR dehydrogenase; (2) formyl–MFR:H_4MPT formyltransferase; (3) methenyl–H_4MPT cyclohydrolase; (4) methylene–H_4MPT dehydrogenase; (5) methylene–H_4MPT reductase; (6) methyl–H_4MPT:H–S–CoM methyltransferase; (7) methyl–coenzyme M reductase; (8) heterodisulfide reductase; (9) methanol:H–S–CoM methyltransferase(s). The non-enzymatic formation of CH_2=H_4MPT from formaldehyde (HCHO) is also shown.

carbamate (carboxy–MFR, see Fig. 2A) is non-enzymatically formed from CO_2 and MFR, which is then reduced to formyl–MFR by formyl–MFR dehydrogenase. The enzyme has been purified from *Methanosarcina barkeri* [54,55] and from *Methanobacterium thermoautotrophicum* [56,57]. Both enzymes are iron–sulfur proteins containing different molybdopterin derivatives: The enzyme of *Methanosarcina barkeri* contains molybdopterin guanine dinucleotide [55], that of *Methanobacterium*

TABLE 2

Free energy changes ($\Delta G^{o\prime}$) and redox potentials ($E^{o\prime}$) of partial reactions involved in CH_4 formation from CO_2 and H_2 [a]

#	Reaction [b]	$\Delta G^{o\prime}$ (kJ/react.)	$E^{o\prime}$ (mV)
1.	$CO_2 + MFR + H_2 \rightarrow$ formyl–MFR $+ H_2O$	+16	−497
2.	formyl–MFR $+ H_4MPT \rightarrow N^5$-formyl–$H_4MPT + MFR$	−5	
3.	N^5-methyl–$H_4MPT + H^+ \rightarrow N^5, N^{10}$-methenyl–$H_4MPT^+ + H_2O$	−2	
4.	N^5, N^{10}-methenyl–$H_4MPT^+ + H_2 \rightarrow N^5, N^{10}$-methylene–$H_4MPT + H^+$	−5	−386
5.	N^5, N^{10}-methylene–$H_4MPT + H_2 \rightarrow N^5$-methyl–$H_4MPT$	−20	−323
6.	N^5-methyl–$H_4MPT + $ H–S–CoM $\rightarrow$ methyl–S–CoM $+ H_4MPT$	−30	
7.	methyl–S–CoM + H–S–HTP $\rightarrow CH_4 + $ CoM–S–S–HTP	−45	
8.	CoM–S–S–HTP $+ H_2 \rightarrow$ H–S–CoM + H–S–HTP	−40	−210
1–8.	$CO_2 + 4H_2 \rightarrow CH_4 + 2H_2O$	−131	−245

[a] $\Delta G^{o\prime}$ and $E^{o\prime}$ values are taken from references [12,18,51].
[b] Abbreviations: MFR, methanofuran; H_4MPT, tetrahydromethanopterin; H–S–CoM, coenzyme M; H–S–HTP, N-7-mercaptoheptanoylthreonin phosphate.

thermoautotrophicum contains molybdopterin guanine dinucleotide, molybdopterin adenine dinucleotide and molybdopterin hypoxanthine dinucleotide [57]. The physiological electron donor for formyl–MFR dehydrogenase is not known.

3.1.2. Formyl–MFR reduction to methylene–H_4MPT (formaldehyde level)

The reduction of formyl–MFR to N^5, N^{10}-methylene-H_4MPT involves N^5-formyl–H_4MPT and N^5, N^{10}-methenyl–H_4MPT as intermediates [7]. The enzymes catalyzing Reaction (2), formyl–MFR: H_4MPT formyltransferase, and Reaction (3), N^5, N^{10}-methenyl–H_4MPT cyclohydrolase, have been purified [58–60]. Reduction of N^5, N^{10}-methenyl–H_4MPT to N^5, N^{10}-methylene–H_4MPT (Reaction 4, Table 2) is catalyzed in *Methanosarcina barkeri* by a coenzyme F_{420}-dependent methylene–H_4MPT dehydrogenase activity and a coenzyme F_{420}-reducing hydrogenase [61] (for the structure of coenzyme F_{420} see Fig. 2B). In *Methanobacterium thermoautotrophicum* and other Methanobacteriales the reduction is catalyzed by an F_{420}-dependent enzyme [62–64] and an F_{420}-independent enzyme. The latter protein constitutes a novel type of hydrogenase that reduces protons to H_2 with methylene–H_4MPT as electron donor [65]. For a distribution of factor F_{420}-reducing and of H^+-reducing methylene–H_4MPT dehydrogenase within methanogens see ref. [66].

3.1.3. Methylene–H_4MPT conversion to methyl–coenzyme M (methanol level)

Reduction of N^5, N^{10}-methylene–H_4MPT to N^5-methyl–H_4MPT is catalyzed by methylene–H_4MPT reductase (Reaction 5, Table 2). Reduced coenzyme factor F_{420} is the physiological electron donor. The enzyme appears to be soluble and not to contain a prosthetic group [67–69a]. The subsequent methyl-group transfer from

H₄MPT to coenzyme M (H–S–CoM) (Reaction 6, Table 2) is catalyzed by a methyl–H₄MPT:H–S–CoM methyltransferase. The enzyme has been partially purified as a tightly membrane-bound protein from *Methanosarcina barkeri* and *Methanobacterium thermoautotrophicum* [69b].

3.1.4. Methyl–coenzyme M (CH₃–S–CoM) reduction to methane

The final step in methanogenesis is the reductive demethylation of CH_3–S–CoM to CH_4. This reduction involves two reactions: CH_3–S–CoM is reduced with *N*-7-mercaptoheptanoylthreonine phosphate (H–S–HTP) (Fig. 2B) as electron donor to yield CH_4 and a heterodisulfide of H–S–CoM and H–S–HTP (CoM–S–S–HTP) (Reaction 7, Table 2). This reaction is catalyzed by CH_3–S–CoM reductase [70–72] which contains a nickel porphinoid, factor F_{430}, as prosthetic group (Fig. 2C) (for a recent review see Friedmann et al. [73]). The subsequent reduction of the heterodisulfide with H_2 to yield H–S–HTP and H–S–CoM (Reaction 8, Table 2) is catalyzed by CoM–S–S–HTP-dependent heterodisulfide reductase. The enzyme is an iron–sulfur protein containing FAD as prosthetic group [74]. The physiological electron donor for the heterodisulfide reductase is not known.

The activation of molecular H_2 (in Reactions 1, 2, 5 and 8 of Table 2) is catalyzed in methanogens by three types of hydrogenases: one type reduces coenzyme F_{420} as physiological electron acceptor and, in addition, viologen dyes as artificial electron acceptors [75–80]; a second type reduces only viologen dyes. The physiological electron acceptor of this enzyme is not known [81–84]; a third type of hydrogenase (methylene–H₄MPT dehydrogenase) specifically reduces methenyl–H₄MPT with H_2 [65]. The F_{420}-reducing hydrogenase has been purified from various species; it belongs to the group of Ni–Fe hydrogenases and contains FAD [75–80]. Methylene–H₄MPT dehydrogenase has been purified from *Methanobacterium thermoautotrophicum*. It appears not to contain metals [65].

Most of the enzymes involved in CO_2 reduction to CH_4 have been purified from the phylogenetically distantly related *Methanosarcina barkeri* and *Methanobacterium thermoautotrophicum* strains Marburg and ΔH, and, very recently, from the extreme thermophile *Methanopyrus kandleri* [85–88]. This suggests that CO_2 reduction to methane proceeds by the same set of enzymes and thus by the same mechanism in all methanogens. The enzyme activities of formyl–MFR dehydrogenase, methylene–H₄MPT dehydrogenase, methylene–H₄MPT reductase and heterodisulfide reductase have been determined in various methanogenic species after growth on H_2/CO_2 or on other substrates [66].

3.2. Sites of energy coupling

3.2.1. General aspects

The reduction of CO_2 to CH_4 is an exergonic process that is coupled with the synthesis of ATP as evidenced by growth of methanogens on CO_2 and H_2 as sole energy and carbon sources:

$$CO_2 + 4H_2 \rightarrow CH_4 + 2H_2O, \quad \Delta G^{o\prime} = -131 \text{ kJ/mol}. \tag{12}$$

3.2.1.1. Growth yields and ATP gains. The free energy change of CO_2 reduction to CH_4 under standard conditions is -131 kJ/mol. However, in natural habitats methanogens grow at a H_2 partial pressure of 10^{-4}–10^{-5} atm (1–10 Pa). Under these conditions the free energy change is about -30 to -40 kJ/mol CH_4. The energy requirement for the synthesis of ATP from ADP and P_i under cellular conditions is about 60–80 kJ/mol assuming a thermodynamic efficiency of energy conversion of 60–70% [8,89,90]. Thus, in vivo less than 1 mol ATP/mol CH_4 can be formed. In accordance with the thermodynamics, the experimentally determined molar growth yields (Y_{CH_4}, grams of cell dry mass formed per mol CH_4 produced) during growth on H_2/CO_2 were low. Values of 1.6–6 g/mol CH_4 have been reported for various methanogens growing in batch and continuous cultures [91–97]. Assuming that about 10 g dry weight cells were formed per mol ATP (Y_{ATP}) during heterotrophic growth on glucose, Y_{ATP} during autotrophic growth on H_2/CO_2 has to be lower than 10 g/mol [8,98,99] (see also ref. [13]). From the growth yields, ATP gains of 0.3–0.7 mol ATP/mol CH_4 can be deduced. The growth yield (Y_{CH_4}) was found to increase, for example in *Methanobacterium thermoautotrophicum*, from 1.6 g/mol up to values of 3–6 g/mol [94,100] when the hydrogen partial pressure in the culture medium was lowered. This indicates that CH_4 formation and ATP synthesis are coupled more efficiently when the H_2 concentration is low, as is the case in natural habitats, or, vice versa, high H_2 concentrations appear to uncouple CH_4 formation and ATP synthesis. Uncoupling by high H_2 concentrations was also observed during coupling of CH_4 formation with ATP synthesis in non-growing cell suspensions. An apparent ATP/CH_4 stoichiometry of 0.2–0.3 was found when the H_2 concentration in the gas phase was low (gas phase: 0.2% H_2, 20% CO_2, 80% N_2); at high H_2 concentration in the gas phase (80% H_2/20% CO_2) the coupling ratio ATP/CH_4 decreased to <0.01 [100]. The rationale behind this uncoupling effect of high H_2 concentrations is not understood.

3.2.1.2. Mechanism of ATP synthesis. The mechanism by which ATP synthesis is coupled with methanogenesis from H_2 and CO_2 has been discussed controversially (see below). ATP may be formed either by electron transport phosphorylation (ETP) or by substrate level phosphorylation (SLP). Since H_2 is the electron donor for CO_2 reduction it seemed most likely that ETP is the mechanism for ATP synthesis; furthermore, the free energy changes ($\Delta G^{\alpha\prime}$) associated with methanogenesis from H_2/CO_2 from acetate (see below) are too low (-30 to -40 kJ/mol) to be coupled with stoichiometric ATP formation. This also suggested ATP synthesis to occur via ETP, since ETP allows fractional numbers of ATP molecules to be formed (for a discussion see ref. [89]). The mechanism of ETP according to the Mitchell hypothesis [101] implies that methane formation from CO_2/H_2 generates an electrochemical potential of protons ($\Delta\tilde{\mu}H^+$) across the cytoplasmic membrane, which constitutes the driving force for the synthesis of ATP via a membrane-bound H^+-translocating ATP synthase. To determine whether methanogenesis, i.e. electron transport, primarily generates $\Delta\tilde{\mu}H^+$, as is indicative for a chemiosmotic mechanism, the effects of various ionophores, which dissipate the electrochemical potential, and of inhibitors of H^+-ATP synthases were tested. However, this approach was not useful in studying ATP coupling with methanogenesis from H_2/CO_2: addition of ionophores, which dissipated the electrochemical proton potential, caused an inhibition of both ATP synthesis and methane formation, i.e. electron transport. This inhibition could be explained later (see below), following an understanding of the enzymology and the energetics of the partial reactions involved in the complex CO_2 reduction pathway: the

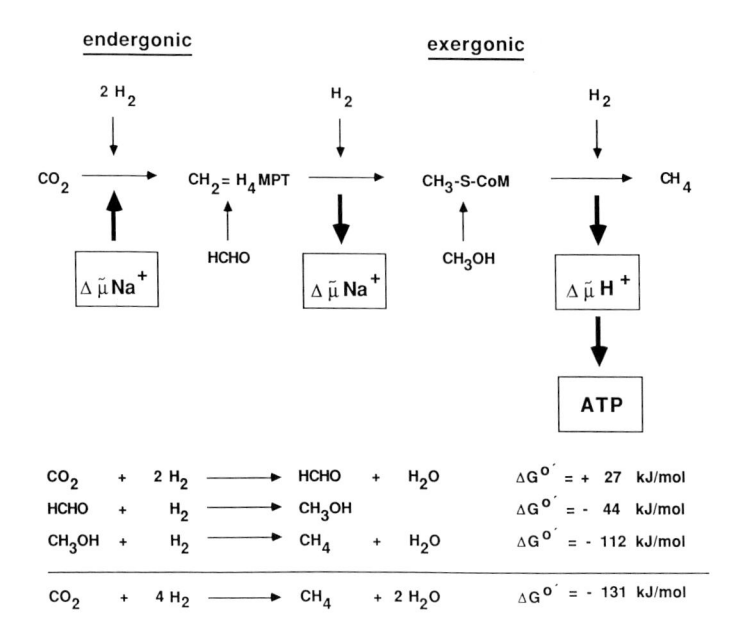

Fig. 4. Energetics and modes of energy coupling of endergonic and exergonic partial reactions involved in methane formation from CO_2 and H_2. $CH_2=H_4MPT$, methylene–H_4MPT; CH_3–S–CoM, methyl–coenzyme M; $\Delta\tilde{\mu}H^+$, transmembrane electrochemical potential of H^+, $\Delta\tilde{\mu}Na^+$, transmembrane electrochemical potential of Na^+. The free energy changes given for the partial reactions are calculated using the thermodynamic data for HCHO and CH_3OH rather than for the coenzyme-bound intermediates.

first step of CO_2 reduction, the synthesis of formyl–MFR, is an endergonic reaction that is energy driven in vivo. It became possible to elucidate the mechanism of ATP synthesis by studying only the exergonic partial reactions of CO_2 reduction pathway.

3.2.1.3. Thermodynamics of partial reactions. On the basis of thermodynamic data (Table 2) the energetics of the reduction of CO_2 to CH_4 can be divided into three parts, the energy coupling of which can be tested experimentally with cell suspensions: the reduction of CO_2 to methylene–H_4MPT ($CH_2=H_4MPT$, formaldehyde level), the conversion of methylene–H_4MPT to methyl–coenzyme M (CH_3–S–CoM, methanol level), and the reduction of CH_3–S–CoM to CH_4 (Fig. 4).

The first part, the reduction of CO_2 to methylene–H_4MPT (Reactions 1–4 of Table 2) is endergonic. It involves the strongly endergonic formation of formyl–MFR from CO_2 and H_2 (Reaction 1). This endergonic part is followed by two exergonic reactions: the conversion of methylene–H_4MPT to methyl–coenzyme M (Reactions 5, 6), and the reduction of methyl–coenzyme M to methane (Reactions 7, 8). It was possible to study the energetics of these reactions since methanogens have been shown to form CH_4 from H_2 and formaldehyde [102,103]; in addition, several methanogens, including *Methanosarcina* species, are able to form methane by the reduction of methanol with H_2 [41]. Formaldehyde binds non-enzymatically to H_4MPT to form N^5, N^{10}-methylene–H_4MPT, $CH_2=H_4MPT$ [104]; methanol is transferred to H–S–CoM by two specific methyltransferases (for a review see ref. [13]) to yield CH_3–S–CoM. Thus, formaldehyde and methanol can be used as substitutes for the coenzyme-bound intermediates. The modes of energy coupling of the exergonic parts had been studied by comparing the reduction of formaldehyde and of methanol by H_2 to CH_4 in different methanogens (Fig. 4):

$$HCHO + 2H_2 \rightarrow CH_4 + H_2O, \quad \Delta G^{o\prime} = -158 \text{ kJ/mol } CH_4,$$

$$CH_3OH + H_2 \rightarrow CH_4 + H_2O, \quad \Delta G^{o\prime} = -112 \text{ kJ/mol } CH_4.$$

The energetics of the endergonic part of the pathway, the reduction of CO_2 to methylene–H_4MPT, has been studied by analyzing the reverse reaction, the exergonic conversion of formaldehyde to CO_2 and H_2:

$$HCHO + H_2O \rightarrow CO_2 + 2H_2, \quad \Delta G^{o\prime} = -27 \text{ kJ/mol}.$$

Methanogens catalyze this reaction if methanogenesis was blocked by inhibitors of the CH_3–S–CoM reductase [105,106].

3.2.2. Methyl–coenzyme M reduction to CH_4 – site of primary $\Delta\tilde{\mu}H^+$ generation and of ATP synthesis

The most exergonic part of the CO_2 reduction pathway is the reduction of CH_3–S–CoM to CH_4 (CH_3–S–CoM + H_2 → CH_4 + H–S–CoM, $\Delta G^{o\prime} = -85$ kJ/mol)(Reactions 7, 8 of Table 2). Evidence that this reduction is a coupling site for ATP synthesis came from the finding that *Methanosphaera stadtmanae* [42] and *Methanosarcina barkeri* [43] grow on CH_3OH and H_2 as sole energy source. The mechanism of ATP synthesis coupled to CH_4 formation from H_2 and CH_3OH was studied with cell suspensions of *Methanosarcina barkeri* by Gottschalk and coworkers [41]. They measured the electrochemical proton potential ($\Delta\tilde{\mu}H^+$), the ATP pools and the rates of methanogenesis of the cells both in the absence and in the presence of protonophores and of dicyclohexylcarbodiimide (DCCD), a specific inhibitor of the H^+-translocating F_1/F_0-ATP synthases: (i) protonophores caused dissipation of $\Delta\tilde{\mu}H^+$, a decrease of the ATP content and a stimulation of methanol reduction (electron transport); (ii) DCCD slightly increased $\Delta\tilde{\mu}H^+$, decreased the ATP pool and partially inhibited methane formation. Inhibition of methane formation by DCCD was reversed by protonophores; (iii) methanol reduction to CH_4 was coupled with H^+ extrusion (3–4 mol H^+/mol CH_3OH). H^+ translocation could be inhibited by protonophores rather than by DCCD [107]. These data clearly indicate a chemiosmotic mechanism of ATP synthesis [11]: CH_4 formation from H_2 and CH_3OH is associated with the build-up of an electrochemical proton potential as a result of primary electrogenic proton translocation. The electrochemical proton potential then drives the synthesis of ATP via a H^+-translocating ATP synthase.

It was shown later that ATP synthesis can be coupled to methanogenesis at very low $\Delta\tilde{\mu}H^+$ values (−90 to −100 mV) [103]: Proton potentials of defined magnitude, adjusted with K^+ gradient in the presence of valinomycin, were applied to cells of *Methanosarcina barkeri* and *Methanobacterium thermoautotrophicum*, and the ATP pool and the rates of methane formation from H_2/methanol and from H_2/CO_2 were followed as a function of

$\Delta\tilde{\mu}H^+$. It was found that $\Delta\tilde{\mu}H^+$ could be lowered to $-90\,mV$ without affecting ATP synthesis and methanogenesis. Further decrease of $\Delta\tilde{\mu}H^+$ resulted in an inhibition of ATP synthesis and in methane formation from H_2/CO_2 but in a stimulation of methanogenesis from H_2/methanol. The energy of an electrogenic proton at $\Delta\tilde{\mu}H^+ = -100\,mV$ is $10\,kJ/mol\,H^+$. Chemiosmotic coupling of ATP synthesis at this proton potential would implicate the H^+/ATP stoichiometry to be greater than 5 assuming ATP formation from ADP and P_i in methanogens under reversible conditions to require $50\,kJ/mol$ [8]. H^+/ATP stoichiometries of other chemiosmotic coupling systems have been reported to be about 3 (see ref. [89]). The H^+/ATP stoichiometry of ATP synthase of methanogens has not yet been determined.

CH_4 formation from H_2/CO_2 is dependent on Na^+ ions [109,110]. Since CH_4 formation from H_2/CH_3OH was found not to be dependent on Na^+ ions, a role for this cation in the terminal part of the CO_2 reduction pathway, namely $CH_3-S-CoM$ reduction to CH_4, and in ATP synthesis could be excluded [111]. CH_4 formation from H_2/CH_3OH involves three steps: (i) binding of CH_3OH to coenzyme M via two methyltransferases (see ref. [13]):

$$CH_3OH + H-S-CoM \rightarrow CH_3-S-CoM + H_2O, \qquad \Delta G^{o\prime} = -27\,kJ/mol,$$

(ii) the reduction of methyl–CoM with H–S–HTP as electron donor (Reaction 7 of Table 2):

$$CH_3-S-CoM + H-S-HTP \rightarrow CH_4 + CoM-S-S-HTP, \qquad \Delta G^{o\prime} = -40\,kJ/mol,$$

and (iii) the reduction of the heterodisulfide with H_2 (Reaction 8 of Table 2):

$$CoM-S-S-HTP + H_2 \rightarrow H-S-CoM + H-S-HTP, \qquad \Delta G^{o\prime} = -45\,kJ/mol.$$

3.2.2.1. Heterodisulfide (CoM–S–S–HTP) reduction – coupled to primary H^+ translocation. The reaction in which ATP is synthesized during methanol reduction to CH_4 could be identified with energetically competent membrane vesicles of the methanogenic strain Göl. These vesicles, which are orientated more than 90% inside-out, catalyzed CH_4 formation from $CH_3-S-CoM$ by reduction with H_2 (Reactions 7, 8) and coupled this process with the synthesis of ATP [112]. $CH_3-S-CoM$ reduction generated a ΔpH (inside acidic) as monitored by acridine dye quenching; protonophores and ATP synthase inhibitors exerted their effects in accordance with a chemiosmotic type of ATP synthesis [113].

After it became clear that the reduction of $CH_3-S-CoM$ to CH_4 consists of two reactions, one of these, the reduction of the heterodisulfide of CoM–S–H and H–S–HTP with H_2 (Reaction 7 of Table 2), was considered to be the coupling site for ATP synthesis [14,71]. Indeed, it was shown that everted vesicles of Göl also catalyzed the reduction of the heterodisulfide with H_2 or with chemically reduced factor F_{420}, $F_{420}H_2$, to H–S–CoM and H–S–HTP and that this reaction was coupled with the synthesis of ATP via the mechanism of electron transport phosphorylation [114–117] (Fig. 5): (i) the reduction of the heterodisulfide was associated with primary proton translocation at a ratio of up to $2H^+$/CoM–S–S–HTP; proton translocation was inhibited by protonophores rather than by DCCD; (ii) reduction of the heterodisulfide was stimulated by protonophores and inhibited by DCCD; inhibition by DCCD could be relieved by protonophores; (iii) ADP stimulated heterodisulfide reduction, indicating "respiratory control" and thus a tight coupling of electron transport and ATP synthesis; (iv) ATP/CoM–S–S–HTP stoichiometries of up to 2 were obtained, from which a H^+/ATP ratio of 4 was calculated.

The findings clearly indicate that the last step of the CO_2 reduction pathway, the reduction of CoM–S–S–HTP by H_2, is a coupling site for ATP synthesis. It is concluded that CoM–S–S–HTP is the terminal electron acceptor ($E^{o\prime} = -200$ mV) of a membrane-bound electron transport chain, with molecular H_2 ($E^{o\prime} = -414$ mV) being the electron donor in hydrogenotrophic methanogens [115]. The physiological electron donor for CoM–S–S–HTP reduction is not known (Fig. 5).

The electrons coming from H_2 may be channeled into the chain either via F_{420} using an F_{420}-dependent hydrogenase or via a yet unknown electron carrier using an F_{420}-nonreducing hydrogenase. Both F_{420}-dependent ($E^{o\prime} = -350$ mV) and H_2-dependent (F_{420}-independent) heterodisulfide reduction via electron transport chains have been demonstrated in the vesicle system of strain Göl, both processes being coupled with ATP synthesis [115,117]. Chemiosmotic ATP synthesis in washed vesicles indicate that $F_{420}H_2$ dehydrogenase, hydrogenase, heterodisulfide reductase and ATPase are membrane-associated enzymes (Fig. 5). The heterodisulfide reductase from *Methanobacterium thermoautotrophicum* has recently been purified from the soluble fraction, indicating that the enzyme is not an integral membrane protein [74]. Factor F_{420}-reducing hydrogenase [77,79,118–121] and CH_3–S–CoM reductase [122,123] have also been purified from various species as soluble proteins but were found to be membrane-associated in vivo as shown by the immunogold labeling technique. Furthermore, a membrane-associated supercomplex named "methanoreductosome", consisting of methyl–CoM reductase, and other proteins have been detected by electron microscopy in the methanogenic strain Göl [124].

Methane formation from CO_2, methanol and acetate proceeds via the same terminal reaction sequence, CH_3–S–CoM reduction to CH_4 involving the reduction of the heterodisulfide (Tables 2, 3, 4; Figs. 3, 9, 10). It can therefore be concluded that reduction of CoM–S–S–HTP is a coupling site for ATP synthesis common in methanogenesis from all substrates. The hydrogen donors for heterodisulfide reduction are generated in various dehydrogenase reactions of the different pathways. It is not known whether, in addition to the heterodisulfide reduction, the exergonic CH_3–S–CoM reductase reaction (-40 kJ/mol) or the methanol:H–S–CoM–methyltransferase reaction (-45 kJ/mol) are also involved in proton translocation. Stoichiometries of $4H^+/2e^-$ have been reported for the reduction of methanol to methane in whole cells, and of $2H^+/2e^-$ for reduction of heterodisulfide in a tightly coupled vesicle system. The higher $H^+/2e^-$ ratio for methanol reduction might indicate another coupling site of ATP synthesis beside the heterodisulfide reduction.

3.2.2.1.1. Electron carriers. Little information is available about the electron carriers involved in electron transfer from H_2 or $F_{420}H_2$ to the heterodisulfide. Membrane-bound iron–sulfur proteins have been implicated in methanogenic electron transfer in *Methanobacterium thermoautotrophicum* [125]. Several methanogens have been shown to contain a membrane-bound corrinoid protein, and a role of this protein in electron transfer has been suggested [126,127]. Quinones (menaquinone, ubiquinone) appear to be absent in methanogens. Cytochromes have been found only in those methanogens capable of oxidizing methyl groups [128] (see below). Participation of

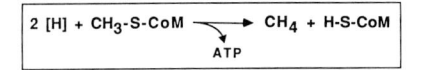

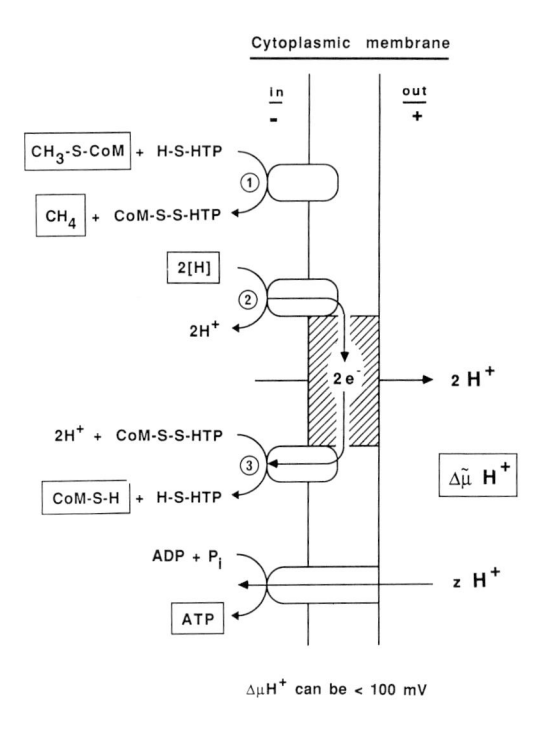

Fig. 5. Proposed mechanism of ATP synthesis coupled to methyl–coenzyme M (CH_3–S–CoM) reduction to CH_4: The reduction of the heterodisulfide (CoM–S–S–HTP) as a site for primary H^+ translocation. ATP is synthesized via membrane-bound H^+-translocating ATP synthase. CoM–S–S–HTP, heterodisulfide of coenzyme M (H–S–CoM) and 7-mercaptoheptanoylthreonine phosphate (H–S–HTP); numbers in circles, membrane-associated enzymes: (1) CH_3–S–CoM reductase; (2) dehydrogenase; (3) heterodisulfide reductase; 2[H] can be either H_2, reduced coenzyme F_{420} ($F_{420}H_2$) or carbon monoxide; the hatched box indicates an electron transport chain catalyzing primary H^+ translocation; the stoichiometry of H^+ translocation ($2H^+/2e^-$, determined in everted vesicles) was taken from ref. [117]; z is the unknown H^+/ATP stoichiometry; $\Delta\tilde{\mu}H^+$, transmembrane electrochemical potential of H^+.

cytochromes in electron transport to the heterodisulfide in obligate hydrogenotrophic species can therefore be excluded. Ferredoxin has been isolated from *Methanosarcina barkeri*, *Methanosarcina thermophila* and *Methanococcus thermolithotrophicus* [129–131]. A role for ferredoxin in CH_4 formation from CO_2/H_2 is not obvious; however, a function in acetate catabolism has been indicated [132,133] (see below). Recently, polyferredoxin-encoding genes tightly linked to genes of methylviologen-reducing hydrogenase of *Methanobacterium thermoautotrophicum* [82], *Methanothermus fervidus* [83] and *Methanococcus voltae* [78] have been identified. The polyferredoxin

Errata

Chapter 3 – Biochemistry of methanogenesis

Page 56, line 14:	Becker should be Becher.
Page 74, line 14:	insert the words in bold – '6-alkyl side-chain (**but** not the **additional** 7-alkyl group ...'.
Page 82, line 9:	change the reference from our published paper to Biswarup's MS thesis, a new reference [356a].
Page 95, line 16:	*arboriphilicus* should replace *bryantii*, and a new reference [427a] should be added.
Page 101, ref. [30]:	insert the name of the second author, G.D. Sprott.
Page 110:	insert a new reference – [356a] Mukhopadhyay, B. (1987) MS thesis, University of Iowa, Iowa City, Iowa.
Page 112:	insert a new reference – [427a] Hammel, K.E., Cornwell, K.L., Diekert, G.B., and Thauer, R.K. (1984) J. Bacteriol. 157, 975–978.

from *Methanobacterium thermoautotrophicum* has been purified [134]. This protein might be involved in H_2-dependent electron transport.

3.2.2.2. ATP synthase/ATPase.

3.2.2.2.1. F_1F_0-type ATPase. ATP synthesis driven by an electrochemical proton potential ($\Delta\tilde{\mu}H^+$) requires a membrane-bound H^+-translocating ATP synthase. The DCCD sensitivity of ATP synthesis driven either by an artificially imposed pH gradient [135] or by methanogenic electron transport [41] suggested the presence of an F_1F_0-like H^+-translocating ATP synthase in methanogens. Electron microscopy studies demonstrated the presence of membrane-bound F_1-like particles in *Methanobacterium thermoautotrophicum* [136] and in the methanogenic strain Göl [137]. A DCCD-sensitive ATPase activity has been purified from membranes of *Methanosarcina barkeri* [138] and of *Methanolobus tindarius* [139]. SDS gel electrophoresis of the soluble part revealed two subunits (α, β) of the *Methanosarcina barkeri* ATPase [138] and four subunits ($\alpha, \beta, \gamma, \delta$) of the *Methanolobus tindarius* ATPase [139]. The subunits had similar molecular weights to those of the $\alpha, \beta, \gamma, \delta$ subunits of bacterial F_1F_0-ATPases, e.g. of the *Escherichia coli* enzyme (for reviews on various ATPases, see refs. [140–142]). The genes coding for the α and β subunits of *Methanosarcina barkeri* ATPase have been cloned and sequenced [143]. Alignment of the deduced amino acid sequences revealed 50–60% homology with the corresponding subunits of the archaeon *Sulfolobus acidocaldarius* and with the A and B subunits of the vacuolar ATPases of various eucaryotes. Less (25%) homology was found with the α, β subunits of F_1F_0-ATPase from *Escherichia coli*. Membrane fractions of various methanogens show immunological cross-reactivity with antibodies directed against the β subunit of *Sulfolobus acidocaldarius* or *Escherichia coli* ATPases [144]. Furthermore, a DCCD-binding membrane protein of 6000 D has been detected in *Methanolobus tindarius* [139] and purified from *Methanosarcina barkeri* [145]. The presence of this "proteo-lipid"-like protein suggests that the membrane integral part of the methanogen ATPase/synthase is also similar to that of F_1F_0-ATPases [140]. Speculations about the evolution of H^+-ATPases are given in ref. [146].

3.2.2.2.2. P-type ATPase. *Methanococcus voltae*, a marine organism growing on H_2/CO_2, contains high activities of a membrane-bound ATPase, which was sensitive towards vanadate rather than to DCCD [147], suggesting the presence of a P-type ATPase (see ref. [141]). Accordingly the purified enzyme, composed of one 74 kD subunit, could be phosphorylated in a vanadate-sensitive fashion [148], a characteristic property for P-type ATPases, which involve a phosphoprotein as intermediate in the catalytic cycle.

The physiological function of this ATPase is not known. The enzyme may be involved in Na^+ extrusion, functioning in maintaining internal ion homeostasis, which might be of importance in this moderate halophilic organism [149]. A Na^+-translocating ATPase, acting in the direction of ATP synthesis, could also explain the observed Na^+-stimulated formation of ATP in *Methanococcus voltae* when driven by a diffusion potential of protons in the presence of protonophores [150]. The presence of a Na^+-translocating ATPase has also been proposed for *Methanobacterium thermoautotrophicum* [151,152]. So far a DCCD-sensitive H^+ ATPase has not been demonstrated in *Methanococcus voltae*. However, membranes of the organism showed immunological cross-reactivity with antibodies directed against the β subunit of F_1F_0-type *Sulfolobus* ATPase [Lübben and Schäfer, personal communication], indicating the presence of a similar enzyme in *Methanococcus voltae*. Thus, the mechanism of ATP synthesis driven by $\Delta\tilde{\mu}H^+$ via a

H⁺-translocating ATP synthase is most probably the same for all methanogens (for a controversial discussion of energy coupling in *Methanococcus voltae* see ref. [15]).

3.2.2.3. Misleading concepts of ATP synthesis.

3.2.2.3.1. Substrate level phosphorylation versus electron transport phosphorylation.

The finding that in *Methanococcus voltae* protonophores did not affect ATP synthesis coupled to methanogenesis from H_2 and CO_2 was in apparent contrast to a chemiosmotic type of ATP synthesis [150]. Since the organism apparently does not contain a DCCD-sensitive ATP synthase, Lancaster [15,151] postulated ATP to be synthesized in Methanococcales by a direct mechanism, substrate level phosphorylation, without a transmembrane intermediacy of ion gradients. An apparent insensitivity to uncoupling was also described for *Methanobacterium thermoautotrophicum* [154]. Furthermore, it has been reported that in a cell-free system of *Methanobacterium thermoautotrophicum* 1 mol of pyrophosphate is formed at the expense of 1 mol CH_3–S–CoM reduced to CH_4 [155], an observation which was interpreted to support substrate level phosphorylation for ATP synthesis in methanogens.

However, the apparent uncoupler insensivity of ATP synthesis in *Methanobacterium thermoautotrophicum* could later be explained by the findings that methanogens couple ATP synthesis and methanogenesis at low $\Delta\tilde{\mu}H^+$, -100 mV, and that most protonophores in *Methanobacterium thermoautotrophicum*, even when used in high concentrations, did not dissipate $\Delta\tilde{\mu}H^+$ below this value [103]. A similar situation might be true for *Methanococcus voltae*. Since $\Delta\tilde{\mu}H^+$ was not measured in the experiments [150] and only low protonophore concentrations were used, it is most likely that $\Delta\tilde{\mu}H^+$ in *Methanococcus voltae* was not completely dissipated under the experimental conditions. Furthermore, pyrophosphate formation in the cell extracts of *Methanobacterium thermoautotrophicum* could not be confirmed by other groups [156]. Thus, the data do not argue against electron transport phosphorylation as the mechanism for ATP synthesis in methanogens.

3.2.2.3.2. "Methanochondrion" concept. Several authors have proposed that methanogenesis and ATP synthesis in methanogenic bacteria are located on specialized intracytoplasmic organelles ("methanochondrions") rather than on the cytoplasmic membrane [9]. This concept was based mainly on the following published findings: (i) structural studies by electron microscopy indicate the presence of internal membranes in various methanogens. These membranes were often arranged in concentric circles, suggesting the presence of closed compartments [157–159]; (ii) experiments on the transport of adenine nucleotides and inhibitor studies suggest the presence of a mitochondrion-like ATP/ADP translocator [160]; (iii) ATP synthesis did not show a sensitivity towards uncouplers [150,154,161].

According to the "methanochondrion" concept ATP is synthesized by $\Delta\tilde{\mu}H^+$ across internal membranes and ATP is transported from the organelle into the cytoplasm via an ATP/ADP translocator (for a cartoon see ref. [162]). Thus, the observed uncoupler insensitivity of ATP synthesis might be explained on the assumption that the internal membranes are not accessible to these compounds. However, upon reinvestigation of the electron microscopic data and the adenine nucleotide transport this explanation could be ruled out: the internal membranes which had been described in the literature for *Methanobacterium thermoautotrophicum* were found to be artefacts of fixation [163] (see also ref. [164]). The adenine nucleotide "transport" could be explained by a tight and specific binding of ADP and ATP to internal binding sites, presumably ATPases, and no ADP/ATP translocator could be detected [165].

3.2.3. Methylene–H$_4$MPT conversion to methyl–coenzyme M – site of primary $\Delta\tilde{\mu}Na^+$ generation

The second exergonic site of the CO$_2$ reduction pathway is the conversion of methylene–H$_4$MPT to methyl-CoM, involving methylene–H$_4$MPT reductase and methyl–H$_4$MPT:H–S–CoM methyltransferase (Reactions 5, 6 of Table 2). The mode of energy transduction coupled to these exergonic reactions was studied by comparing the energetics of CH$_4$ formation from formaldehyde/H$_2$ with that from methanol/H$_2$ in cell suspensions of *Methanosarcina barkeri*. Formaldehyde binds to H$_4$MPT, methanol to H–S–CoM. A role of Na$^+$ ions in energy coupling became apparent. It was found that formaldehyde reduction to CH$_4$ rather than methanol reduction to CH$_4$ was dependent on Na$^+$ ions, indicating that the conversion of methylene–H$_4$MPT to CH$_3$–S–CoM is a site of Na$^+$ dependency [111]. It was further demonstrated that formaldehyde reduction to methane is associated with primary electrogenic Na$^+$ translocation, generating an electrochemical Na$^+$ potential ($\Delta\tilde{\mu}Na^+$) across the cytoplasmic membrane. This conclusion was based on the following findings [166]: (i) the reduction of formaldehyde with H$_2$ was coupled with the extrusion of Na$^+$ ions, which could not be inhibited by protonophores, by inhibitors of the Na$^+$/H$^+$ antiporter or by the H$^+$-ATP synthase inhibitor DCCD; the stoichiometry of primary Na$^+$ translocation, calculated from the initial rates of Na$^+$ extrusion and formaldehyde reduction, was 3–4 mol Na$^+$/mol formaldehyde [167]. (ii) Na$^+$ efflux generates a protonophore-insensitive membrane potential indicating electrogenic Na$^+$ transport [166].

Thus, the free energy change associated with the conversion of formaldehyde, via methylene–H$_4$MPT, to CH$_3$–S–CoM is conserved in a primary electrochemical Na$^+$ potential (Fig. 6). This conversion represents a second coupling site in methane formation from CO$_2$.

3.2.3.1. Methyl–H$_4$MPT: coenzyme M methyltransferase – coupled to primary Na$^+$ translocation.
Methylene–H$_4$MPT conversion to methyl–CoM involves two reactions (Table 2): methylene–H$_4$MPT reductase (Reaction 5) and CH$_3$–H$_4$MPT:H–S–CoM methyltransferase (Reaction 6).

Methylene–H$_4$MPT reductase catalyzes the reduction of methylene–H$_4$MPT to methyl–H$_4$MPT with factor F$_{420}$ as the physiological electron donor:

$$N^5, N^{10}\text{-methylene–H}_4\text{MPT} + F_{420}H_2 \rightarrow N^5, N^{10}\text{-methyl–H}_4\text{MPT} + F_{420},$$
$$\Delta G^{o\prime} = -5.2 \text{ kJ/mol}.$$

Assuming a redox potential $E^{o\prime}$ of the methyl–H$_4$MPT/methylene–H$_4$MPT couple of -323 mV [18,51] the free energy change associated with the reduction using reduced F$_{420}$ as physiological electron donor ($E^{o\prime} = -350$ mV) would be too small (-5.2 kJ/mol) to be coupled with electrogenic Na$^+$ translocation. The methylene–H$_4$MPT reductase has been purified as a soluble protein from *Methanobacterium thermoautotrophicum* and *Methanosarcina barkeri*; it does not show a Na$^+$ stimulation and appears not to contain a prosthetic group [67–69a]; these properties argue against an involvement of the reductase in Na$^+$ translocation.

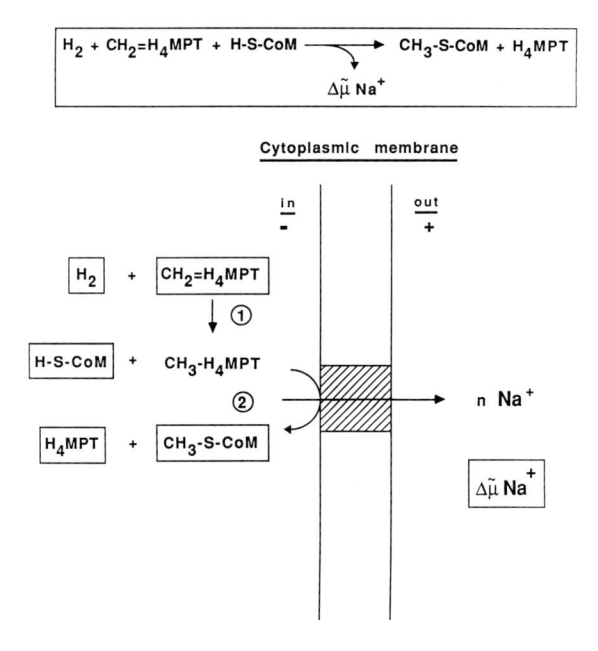

Fig. 6. Proposed mechanism of the generation of a primary electrochemical sodium potential ($\Delta\tilde{\mu}Na^+$) coupled to methylene–H$_4$MPT (CH$_2$=H$_4$MPT) conversion to methyl–coenzyme M (CH$_3$–S–CoM): Methyl–H$_4$MPT:coenzyme M (H–S–CoM) methyltransferase as a primary Na$^+$ pump. Numbers in circles: (1) methylene–H$_4$MPT reductase; (2) CH$_3$–H$_4$MPT: H–S–CoM methyltransferase; the hatched box indicates membrane-bound methyltransferase; n is an unknown stoichiometric factor.

Methyl–H$_4$MPT:H–S–CoM methyltransferase catalyzes methyl-group transfer from methyl–tetrahydromethanopterin to coenzyme M:

N^5-methyl–H$_4$MPT + H–S–CoM → CH$_3$–S–CoM + H$_4$MPT,

$\Delta G^{o\prime} = -29.7$ kJ/mol.

The free energy change associated with the reaction (−29.7 kJ/mol) is assumed to be identical to that of the methyl transfer from methyl–tetrahydrofolate to homocysteine forming tetrahydrofolate and methionine [12].

The following data indicate that CH$_3$–H$_4$MPT:H–S–CoM methyltransferase is the site of primary Na$^+$ translocation (see Figs. 6 and 12): (i) the enzyme has been partially purified from *Methanosarcina barkeri* and *Methanobacterium thermoautotrophicum* and found to be tightly membrane bound [69b]; (ii) inverted vesicles of the methanogenic strain Göl catalyzed methyl transfer from CH$_3$–H$_4$MPT to H–S–CoM. This reaction was stimulated by Na$^+$ ions and was coupled with the accumulation of Na$^+$ into the vesicles. Na$^+$ uptake was inhibited by Na$^+$ ionophores rather than by protonophores indicating primary Na$^+$ translocation [168].

The mechanism by which the exergonic methyl-transfer reaction is coupled with vectorial electrogenic Na$^+$ translocation across the membrane is not known. An electron transport chain appears not to be involved in Na$^+$ transport. Sodium ion transport decarboxylases of anaerobic bacteria catalyze primary electrogenic Na^+ translocation at the expense of a decarboxylation reaction. These enzymes have been characterized as primary Na^+ pumps (see ref. [246]). In analogy to these decarboxylases, $CH_3-H_4MPT:H-S-CoM$ methyltransferase of methanogens might catalyze primary Na^+ translocation by a conformational pump mechanism.

3.2.4. CO_2 reduction to methylene–H_4MPT – site of primary $\Delta\tilde{\mu}Na^+$ consumption
The first reaction of the CO_2 reduction pathway, the formation of formyl–MFR from CO_2, H_2 and MFR (Reaction 1 of Table 2) is an endergonic reaction for which a $\Delta G^{o\prime}$ value of $+16$ kJ/mol was calculated. Under conditions of low H_2 partial pressure (10 Pa), as found in natural ecosystems, this reaction is even more endergonic ($\Delta G^{o\prime} = +41$ kJ/mol). Since the subsequent reactions, catalyzed by formyltransferase, cyclohydrolase and methenyl–H_4MPT dehydrogenase (Reactions 2–4 of Table 2) are only slightly exergonic, the overall reduction of CO_2 to methylene–H_4MPT (formaldehyde level) is endergonic. CH_4 formation from CO_2/H_2, rather than from formaldehyde/H_2, was found to be sensitive towards uncouplers. This result suggested that CO_2 reduction to the formaldehyde level (methylene–H_4MPT) ("CO_2 activation") is also endergonic in vivo and can proceed only when coupled to the exergonic formaldehyde reduction to CH_4 [9,102].

In principle, two mechanisms of coupling can be envisaged: (i) activation of CO_2 occurs at the level of the substrate at the expense of ATP hydrolysis ("substrate activation"), or (ii) the redox potentials ($E^{o\prime}$) of the electron required for CO_2 reduction are pushed towards more negative values at the expense of electrochemical potentials of either H^+ or Na^+ by the mechanism of reversed electron transport ("redox activation"). Since ATP-consuming synthetases are not involved in CO_2 reduction to methylene–H_4MPT (Reactions 1–4 of Table 2) the latter mechanism is more likely.

In the following subsections experiments are described which indicate that CO_2 reduction to methylene–H_4MPT is driven by a primary electrochemical Na^+ potential generated by formaldehyde reduction to CH_4. These experiments include: (1) studies of the mode of energy transduction of the reverse reaction, the exergonic formaldehyde oxidation to CO_2 and $2H_2$; (2) experiments on the effects of ionophores and inhibitors on CH_4 formation from CO_2/H_2 and CH_4 formation from formaldehyde/H_2, and the determination of stoichiometries of primary Na^+ translocation.

3.2.4.1. Formaldehyde oxidation to CO_2 – coupled to primary Na^+ extrusion. Cell suspensions of methanogens catalyze the oxidation of formaldehyde to CO_2 and $2H_2$ ($HCHO + H_2O \rightarrow CO_2 + 2H_2$, $\Delta G^{o\prime} = -27$ kJ/mol CO_2), if methanogenesis was inhibited by 2-bromoethanesulfonate [105,106]. It was found that the conversion of formaldehyde to CO_2 and $2H_2$ in *Methanosarcina barkeri* was coupled with primary electrogenic Na^+ extrusion resulting in the formation of an electrochemical Na^+ potential ($\Delta\tilde{\mu}Na^+$) [105]: (i) formaldehyde oxidation was coupled with the extrusion of Na^+ ions. Na^+ extrusion was a primary process since it was inhibited by Na^+ ionophores but was not affected by protonophores and Na^+/H^+ antiporter inhibitors; (ii) formaldehyde oxidation was associated with the generation of a membrane potential of the order of 100 mV (inside negative). The membrane potential could be dissipated by sodium ionophores rather than by protonophores; (iii) formaldehyde oxidation was coupled with the synthesis of ATP in an indirect process, since it was inhibited by Na^+/H^+ antiporter inhibitors, by protonophores and by the H^+-ATPase inhibitor dicyclohexylcarbodiimide. These findings indicate the following sequences of reactions: formaldehyde oxidation is coupled with the generation of primary Na^+ potential, $\Delta\tilde{\mu}Na^+$, which is converted into a transmembrane proton potential, $\Delta\tilde{\mu}H^+$, via Na^+/H^+ antiporter. Subsequently, $\Delta\tilde{\mu}H^+$ drives the synthesis of ATP via H^+-translocating ATP synthase.

The possibility that primary Na^+ extrusion was driven by a Na^+-translocating ATPase was excluded: The protonophore tetrachlorosalicylanilide was found to uncouple formaldehyde oxidation from ATP synthesis without affecting Na^+ extrusion. Thus, ATP cannot be the driving force for Na^+ extrusion in *Methanosarcina barkeri*. This situation may be different in *Methanococcus voltae* which appears to contain a Na^+-translocating ATPase (see above).

The site of Na^+ translocation during formaldehyde conversion to CO_2 and $2H_2$ is not known. After formaldehyde has reacted with tetrahydromethanopterin (H_4MPT) to yield methylene–H_4MPT, it is probably oxidized to CO_2 and $2H_2$ most likely via the reversal of Reactions (1)–(4) of Table 2, since e.g. the oxidation of formaldehyde to CO_2 in cell extracts requires both H_4MPT and methanofuran [169,222]. Since the formyl–MFR dehydrogenase reaction is the most exergonic step of the sequence it is assumed that electron transport from formyl–MFR ($E^{o'} = -497$ mV) as electron donor to protons ($E^{o'} = -414$ mV) as electron acceptors yielding CO_2 and H_2 is coupled with electrogenic Na^+ translocation. Formyl–MFR dehydrogenase is apparently membrane associated [56]. The components of this electron transport chain are not known.

Thus, in methanogens, formaldehyde oxidation to CO_2 represents a second site of a primary $\Delta\tilde{\mu}Na^+$ generation. The energy stored in the gradient can be used, via $\Delta\tilde{\mu}H^+$, for the synthesis of ATP (see above) or $\Delta\tilde{\mu}Na^+$ can drive endergonic reactions directly (see below).

3.2.4.2. CO_2 reduction to methylene–H_4MPT – coupled to Na^+ uptake. If Reactions (1)–(4) of Fig. 3 and Table 2 are reversible (see Reactions 6–9 of Table 3, below), it is likely that CO_2 reduction to the formaldehyde level is driven by an electrochemical Na^+ potential. Evidence for this notion was obtained from the following experiments with cell suspensions of *Methanosarcina barkeri* [167].

3.2.4.2.1. Ionophore and inhibitor studies. (i) CO_2 reduction to CH_4 was insensitive towards protonophores when the Na^+/H^+ antiporter was inhibited; under these conditions the electrochemical Na^+ potential ($\Delta\tilde{\mu}Na^+$) was -120 mV, and was composed of a membrane potential of -80 mV and a chemical Na^+ gradient of -40 mV. The electrochemical proton potential ($\Delta\tilde{\mu}H^+$) was almost absent (-10 mV) since an inverse pH gradient (inside acidic) of $+70$ mV was present which counteracts the inwardly directed membrane potential. Thus, no driving force for H^+ ions exists under these conditions and accordingly the cellular ATP content was low (<1 nmol/mg). This result clearly indicates that $\Delta\tilde{\mu}Na^+$ rather than $\Delta\tilde{\mu}H^+$ or ATP hydrolysis is the driving force for CO_2 reduction to methylene–H_4MPT; (ii) CO_2 reduction to CH_4, rather than formaldehyde reduction to CH_4, was sensitive towards Na^+ ionophores, which dissipated $\Delta\tilde{\mu}Na^+$. This result also suggests CO_2 activation driven by $\Delta\tilde{\mu}Na^+$.

3.2.4.2.2. Stoichiometries of primary Na^+ translocation. Further support for the conclusion that CO_2 activation is driven by $\Delta\tilde{\mu}Na^+$ came from the determination of stoichiometries of primary Na^+ transport coupled to CH_4 formation from CO_2/H_2 and from formaldehyde/H_2. Na^+ transport experiments were performed with whole cells of *Methanosarcina barkeri* equilibrated with $^{22}Na^+$; the stoichiometry of Na^+ export was determined from the initial rate of Na$^+$ efflux and the corresponding electron transport rate, i.e. methanogenesis. CO$_2$ reduction to CH$_4$, in the presence of protonophores and Na$^+$/H$^+$ antiport inhibitors, was coupled with the extrusion of 1–2 mol Na$^+$/mol CH$_4$; formaldehyde reduction to CH$_4$ was coupled with the extrusion of 3–4 mol Na$^+$/mol CH$_4$. Thus, during CO$_2$ reduction to the formaldehyde level 2–3 mol Na$^+$ ions must have been taken up. In accordance, the reverse reaction, formaldehyde oxidation to CO$_2$ and H$_2$, was coupled with the extrusion of 2–3 mol Na$^+$/mol CO$_2$ [105]. Furthermore the disproportionation of formaldehyde to CH$_4$ and CO$_2$ was coupled with primary translocation of 5–7 mol Na$^+$/2 mol formaldehyde, which equals the sum of the amount of Na$^+$ ions extruded by formaldehyde reduction to CH$_4$ and that extruded by formaldehyde oxidation to CO$_2$. These stoichiometries were determined at low, but equal, rates of CH$_4$ or CO$_2$ formation. During preincubation of the cells, required for Na$^+$ equilibration across the cytoplasmic membrane, the rates of catabolic reactions decreased by more than 90%. It is assumed that the stoichiometries of Na$^+$ translocation at high rates of CH$_4$ or CO$_2$ formation are the same as at low reaction rates.

In summary, the data strongly suggest that the driving force for the endergonic reduction of CO$_2$ to methylene–H$_4$MPT is the primary electrochemical sodium potential, which is generated during exergonic formaldehyde reduction to CH$_4$. The mechanism of coupling between the exergonic and endergonic reactions is that of a reversed electron flow: Electrons coming from H$_2$ [$E^{o\prime}$ = −414 mV or E'(pH$_2$ = 10 Pa) = −300 mV] have to be pushed to more negative values for the reduction of CO$_2$ to formyl–MFR ($E^{o\prime}$ = −497 mV). The energy stems from the electrochemical sodium potential (see Fig. 8, below). The subsequent reduction of formyl–MFR to methylene–H$_4$MPT (formaldehyde level) is slightly exergonic.

These conclusions have been drawn from whole-cell studies. Evidently, conclusive evidence for primary Na$^+$ translocation coupled to CO$_2$ reduction to formyl–MFR will require the purification and reconstitution of the components involved and demonstration of a direct role of Na$^+$ in the reactions catalyzed by the reconstituted system.

3.2.4.2.3. CO$_2$ reduction to formyl–MFR in cell extracts. Cell extracts of *Methanobacterium thermoautotrophicum* catalyze the formation of formyl–MFR from H$_2$, CO$_2$ and MFR. The rates of formation of formyl–MFR were stimulated by CH$_3$–S–CoM [170,171] or CoM–S–S–HTP [172], suggesting that CO$_2$ activation is coupled with the terminal steps of methane formation ("RPG effect" [170], for literature see [173]). The mechanism of that coupling is far from clear. The heterodisulfide may act as an allosteric effector in the activation of low-potential electron carriers in electron flow from H$_2$ to formyl–MFR [173]. It should be pointed out that the "RPG effect" was not observed in cell extracts of *Methanosarcina barkeri* [174].

3.3. Role of the Na$^+$/H$^+$ antiporter in CO$_2$ reduction to CH$_4$

Perski et al. [109,110] found that growth and CH$_4$ formation from CO$_2$ and H$_2$ in various methanogens were dependent on Na$^+$ ions. Subsequent studies [110] showed that CH$_4$ formation from other substrates, methanol and acetate (see below), also requires Na$^+$ ions. Thus a specific role of the cation in the coupling mechanism of ATP synthesis was envisaged. Later it was found that ATP synthesis driven by a potassium diffusion potential in *Methanobacterium thermoautotrophicum* was stimulated by Na$^+$ [175]. However, a direct role of this cation in the mechanism of ATP synthesis could be excluded since CH_4 formation from H_2 and CH_3OH, as well as ATP synthesis coupled to this reaction, were not stimulated by Na^+ ions [111]. Two Na^+-dependent sites involved in CO_2 reduction to CH_4 have been identified: methyl–H_4MPT:H–S–CoM methyltransferase and the reduction of CO_2 to formyl–MFR (see above).

3.3.1. Na^+/H^+ antiporter
In order to explain the Na^+ stimulation of ATP synthesis driven by a diffusion potential the presence of a Na^+/H^+ antiporter was proposed [175]. In this artificial system the acidification of the cytoplasm, which occurs in response to electrogenic potassium efflux, could be prevented by the antiporter. Subsequently, Na^+/H^+ antiporter activity has been demonstrated in both *Methanobacterium thermoautotrophicum* [176] and in *Methanosarcina barkeri* [108]. An important result of these studies was that the Na^+/H^+ antiporter could be inhibited by amiloride and harmaline, which have been described as inhibitors of eucaryotic Na^+/H^+ antiporters [177]. Using these inhibitors it has been shown that an active antiporter is essential for methanogenesis from H_2/CO_2 [176,178]. The antiporter also accepts Li^+ instead of Na^+, since Li^+ stimulates CH_4 formation from H_2/CO_2 in the absence of Na^+ [176]. In subsequent studies the use of amiloride and the more potent derivative ethyl–isopropylamiloride permitted the discrimination of primary and secondary Na^+ potentials generated in partial reactions of the CO_2 reduction pathway.

The Na^+/H^+ exchange by the methanogenic antiporter is probably electrogenic. A Na^+/H^+ stoichiometry has been calculated in whole cells of *Methanosarcina barkeri* by measuring the membrane potential ($\Delta\psi$) and the transmembrane chemical gradients of Na^+ and H^+ in the steady state of CH_4 formation from H_2/CO_2 [179]: the membrane potential ($\Delta\psi$) was about -120 mV, a ΔpH could not be detected, the measured Na^+ gradient ($[Na^+]_{outside}/[Na^+]_{inside}$) was at least 10 (higher gradients could not be quantitated correctly). The stoichiometry of an electrogenic Na^+/H^+ antiport (yNa^+/xH^+) can be estimated from the following equation [160]:
$-Z\log([Na^+]_{in}/[Na^+]_{out}) = -[(x-y)\Delta\psi - xZ\Delta pH]; \quad Z = 60\,\text{mV}.$
On the basis of the measured values a stoichiometry of about $1Na^+/1.5H^+$ was calculated.

3.3.2. Role of the Na^+/H^+ antiporter
In principle, a Na^+/H^+ antiporter converts transmembrane potentials of protons ($\Delta\tilde{\mu}H^+$) into those of Na^+ ions ($\Delta\tilde{\mu}H^+$), and vice versa. The direction of the exchange is determined by the magnitude of the prevailing ion gradients. During methane formation from H_2/CO_2 the antiporter is involved in the conversion of a $\Delta\tilde{\mu}Na^+$ into $\Delta\tilde{\mu}H^+$ as indicated by the following findings obtained with *Methanosarcina barkeri* [167,179]: (i) in the steady state of CH_4 formation from H_2/CO_2 the value for $\Delta\tilde{\mu}Na^+$ was higher (-180 mV) than $\Delta\tilde{\mu}H^+$ (-120 mV); (ii) Na^+/H^+ antiporter inhibitors inhibited all reactions that are coupled with primary Na^+ extrusion, the degree of the inhibition being roughly proportional to the amount of Na^+ ions translocated; for example, formaldehyde oxidation to CO_2 and $2H_2$ (2–3 mol Na^+/mol CO_2), formaldehyde reduction to CH_4 (3–4 mol Na^+/mol CH_4) or formaldehyde disproportionation to CO_2 and CH_4 (5–7 mol Na^+/2 mol formaldehyde) were inhibited by 30%, 50% and 80%, respectively. This inhibition could be reversed by Na^+ ionophores. These data indicate that the Na^+/H^+ antiporter is involved in catalyzing Na^+ backflow into the cell, in exchange for protons, thereby closing the Na^+ cycle.

Accordingly CO_2 reduction to CH_4, which is coupled with the net primary extrusion of 1–2 mol Na^+/mol CH_4, is partially inhibited ($\sim$30%) by Na^+/H^+ antiporter inhibitors.

3.3.3. Primary cycles of Na^+ and H^+

Summarizing the published data, a scheme is proposed showing how transmembrane H^+ and Na^+ cycles couple exergonic and endergonic reactions to each other during CH_4 formation from CO_2/H_2 (Fig. 7): The exergonic reactions are the conversion of methylene H_4MPT ($CH_2=H_4MPT$, formaldehyde-level) to methyl–CoM (CH_3–S–CoM, methanol level), and the reduction of CH_3–S–CoM to CH_4. The former reaction is coupled with primary extrusion of Na^+ ions, the latter with primary extrusion of protons. During methylene–H_4MPT conversion to methyl–CoM more Na^+ ions are translocated than are consumed by CO_2 reduction to methylene–H_4MPT. The "extra" Na^+ are exchanged against H^+ via the Na^+/H^+ antiporter. The resulting $\Delta\tilde{\mu}H^+$ can be used for ATP synthesis. Thus, the necessity of a Na^+/H^+ antiporter in CH_4 formation from H_2/CO_2 follows directly from the imbalance of the different Na^+ stoichiometries of the partial reactions involved. In summary, coupling of exergonic and endergonic reactions in CH_4 formation from H_2 and CO_2 involves both primary Na^+ and primary H^+ cycles, and these are linked by the electrogenic Na^+/H^+ antiporter.

According to this scheme the antiporter has a stoichiometric function in CO_2 reduction to CH_4 and ATP synthesis. Besides this function a role of the methanogenic Na^+/H^+ antiporter in pH regulation (for reviews see refs. [177,180]) has also been proposed [176] to explain Na^+-dependent ΔpH formation (inside alkaline at external pH 5, inside acidic at external pH 9) in cell suspensions of *Methanobacterium thermoautotrophicum*. However, the data are difficult to interpret because of interference of Na^+-dependent methanogenesis from H_2/CO_2 (for a discussion of pH regulation in methanogens see ref. [16]).

3.4. Energetics of CH_4 formation from formate

CH_4 formation from formate involves the oxidation of formate to CO_2 and the reduction of CO_2 to CH_4 by the reducing equivalents generated. There is no evidence that free formate binds directly to methanofuran or tetrahydromethanopterin [181]. Thus, formate ($E^{o\prime}$, $CO_2/HCOO^-$ = -431 mV) acts as the electron donor for CO_2 reduction to CH_4 which is energetically almost equivalent to CO_2 reduction by H_2 ($E^{o\prime} = -414$ mV) (Reactions 1,2 of Table 1). Accordingly, the molar growth yields Y_{CH_4} of *Methanobacterium formicicum* on formate (4.8 g cells/mol CH_4) and on H_2/CO_2 (3.8 g cells/mol CH_4) were similar [95] (see also ref. [182]). Formate oxidation is catalyzed by formate dehydrogenase. The enzyme has been purified from various methanogens [183]. Formate dehydrogenase from *Methanobacterium formicicum* is a Fe/S protein containing FAD and a molydopterin guanine dinucleotide [184]. The physiological electron acceptor of formate dehydrogenase is factor F_{420}. The enzyme is membrane associated as shown by immunogold labeling [121].

3.5. Energetics of CH_4 formation from CO_2 reduction by alcohols

Several methanogenic species have been described that use ethanol, propanol and 2-propanol, 2-butanol and cyclopentanol as electron donors for CO_2 reduction to

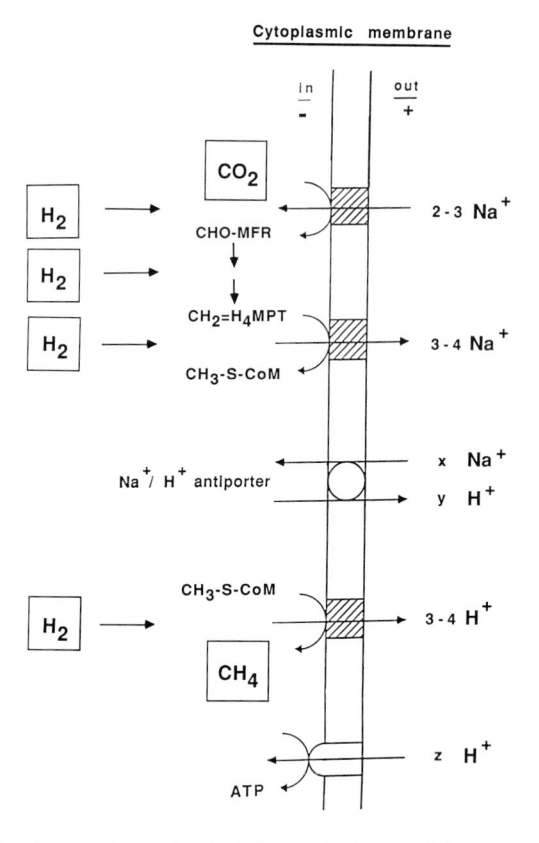

Fig. 7. Proposed function of electrochemical H^+ and Na^+ potentials in energy conservation coupled to CH_4 formation from CO_2/H_2. The Na^+/H^+ antiporter is involved in the generation of $\Delta\tilde{\mu}H^+$ from $\Delta\tilde{\mu}Na^+$. CHO–MFR, formyl–methanofuran; CH_2=H_4MPT, methylene–tetrahydromethanopterin; CH_3–H_4MPT, methyl–tetrahydromethanopterin; CH_3–S–CoM, methyl–coenzyme M. The hatched boxes indicate membrane-bound electron transport chains or membrane-bound methyltransferase catalyzing either Na^+ or H^+ translocation (see Figs. 5, 6 and 12). ATP is synthesized via membrane-bound H^+-translocating ATP synthase. The stoichiometries of Na^+ and H^+ translocation were taken from refs. [105,107,167]. x, y and z are unknown stoichiometric factors.

CH_4 [38–40]. In addition, most species can use H_2 or formate as electron donors. The energetics of methane formation from primary alcohols/CO_2, H_2/CO_2 and formate are comparable in terms of their free energy changes under standard conditions (see Table 1). Methanogenesis from secondary alcohols/CO_2 is thermodynamically less efficient. This is obvious by comparing the redox potentials of the redox couples involved: The average redox potential $E^{o'}$ of the CO_2/CH_4 couple is -244 mV, and the $E^{o'}$ values of the hydrogen donating reactions are -414 mV (H^+/H_2), -431 mV ($CO_2/HCOO^-$), -390 mV (acetate/ethanol), and -286 mV (acetone/2-propanol). Thus, secondary alcohols are the weakest reductants for CO_2 reduction to methane.

In *Methanogenium thermophilum*, which can use both H_2 and ethanol as electron donor for CO_2 reduction, the energetics and the mechanism of ethanol oxidation was studied. Growth yields per mol CH_4 on ethanol and H_2 were 7.6 g cells or 2.2 g cells, respectively [185]. The high growth yield on ethanol suggested additional ATP formation coupled to the oxidation of ethanol oxidation to acetate by the mechanism of substrate level phosphorylation. Ethanol oxidation is catalyzed by alcohol dehydrogenase and by acetaldehyde dehydrogenase. The ADH from *Methanogenium organophilum* has been purified [186]. Acetaldehyde conversion to acetate was studied in cell extracts and found to be independent of coenzyme A, furthermore neither phosphate acetyltransferase nor acetate kinase activities were detected. These data exclude acetyl–coenzyme A as an intermediate of ethanol oxidation to acetate and argue against additional ATP formation by the mechanism of substrate level phosphorylation [186]. The higher growth yield on ethanol compared to H_2 might be explained by a better coupling of ATP synthesis and methanogenesis from CO_2 with ethanol as electron donor. The growth yield on H_2/CO_2 was determined at high H_2 partial pressure, i.e. under conditions for which uncoupling has been indicated (see above).

Oxidation of secondary alcohols to the corresponding ketones is catalyzed by inducible secondary alcohol dehydrogenases (secADH). The physiological electron acceptor of these enzymes from various methanogens was found to be either $NADP^+$ or coenzyme F_{420} [187,188]. Both F_{420}-dependent and $NADP^+$-dependent secADHs have been purified [186,187,189].

3.6. Energetics of CO_2 reduction to CH_4 by methanogens versus CO_2 reduction to acetate by acetogens

Several reactions of CO_2 reduction to CH_4 by methanogens resemble those of CO_2 reduction to by acetogens. Most acetogens, which are anaerobic bacteria (eubacteria), catalyze in their energy metabolism the formation of acetate from CO_2 and H_2 according to the following equation:

$$2CO_2 + 4H_2 \rightarrow CH_3COO^- + H^+ + 2H_2O, \qquad \Delta G^{o\prime} = -95 \text{ kJ/mol acetate}.$$

Acetate is synthesized via the acetyl–CoA pathway (for recent reviews see refs. [190–192]): one molecule CO_2 is reduced via formyl–tetrahydrofolate (formyl–H_4F) and methylene–H_4F to methyl–H_4F, a second one is reduced to a enzyme-bound carbonyl. The methyl group is condensed with the carbonyl group and coenzyme A (CoA) to acetyl–CoA. Acetyl–CoA is then converted to acetate, and ATP is formed in the acetate kinase reaction.

CO_2 reduction to acetate and CO_2 reduction to CH_4 involve two energetically analogous sites: (1) the endergonic reduction of CO_2 to the formate level ("CO_2 activation"), and (2) the exergonic conversion of a coenzyme-bound formaldehyde to the methanol level.

(1) In acetogens CO_2 is converted to formyl–H_4F (formate level), which is energetically similar to CO_2 conversion to formyl–H_4MPT or formyl–MFR in methanogens. However, the mode of energy coupling is different. In acetogens this endergonic reaction is driven at the expense of ATP hydrolysis: CO_2 is reduced to free formate via formate dehydrogenase, formate is then activated in the ATP-consuming formyl–H_4F synthetase reaction to yield formyl–H_4F. Thus in acetogens CO_2 activation is accomplished by the mechanism of "substrate activation" rather than by reversed electron flow ("redox activation") as shown

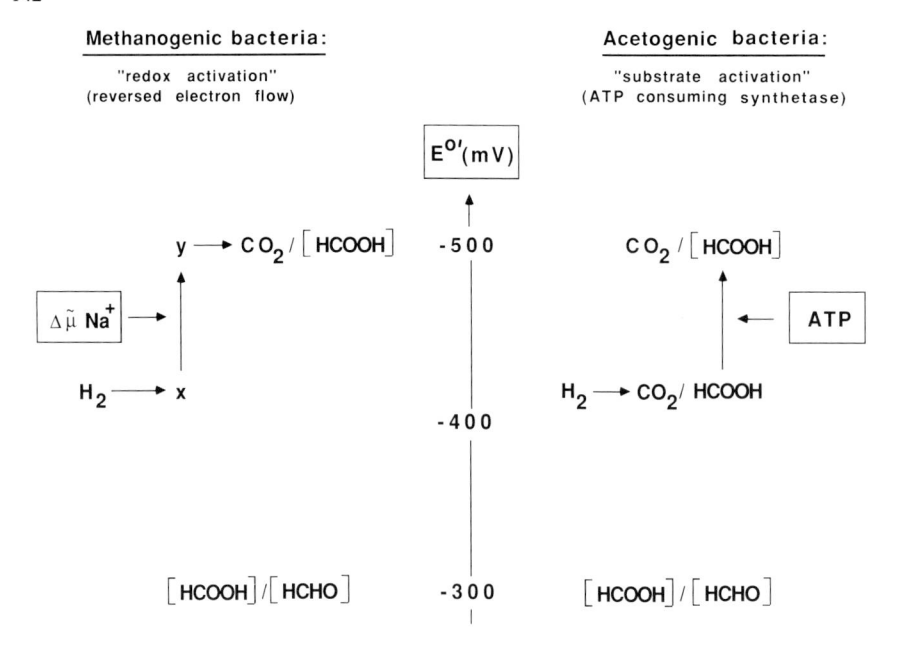

Fig. 8. Different mechanisms of endergonic CO_2 activation in methanogens versus acetogens. [HCOOH], coenzyme-bound formate (in methanogens, formyl–methanofuran or formyl–tetrahydromethanopterin; in acetogens, formyl–tetrahydrofolate); [HCHO], coenzyme-bound formaldehyde (in methanogens, methylene–tetrahydromethanopterin; in acetogens, methylene–tetrahydrofolate). It is assumed that both the redox potentials of the CO_2/[HCOOH] couple and of the [HCOOH]/[HCHO] couple in methanogens and acetogens are equal. $\Delta\tilde{\mu}Na^+$ is the electrochemical potential of Na^+; x and y are unknown electron carriers involved in $\Delta\tilde{\mu}Na^+$-driven reversed electron flow.

for methanogens (Fig. 8). Why these two completely different pathways and mechanisms of CO_2 activation have evolved is presently not understood.

(2) Acetate formation from CO_2/H_2, like CH_4 formation from CO_2/H_2, is a Na^+-dependent process [193,194]. There is evidence that the Na^+-dependent site of acetogenesis is located between methylene–H_4F (formaldehyde level) and acetyl–CoA (methanol level). The conversion of methylene–H_4F to acetyl–CoA involves the following reactions (enzymes in parentheses): reduction of methylene–H_4F to methyl–H_4F (methylene–H_4F reductase), transfer of the methyl group to a corrinoid protein (specific methyltransferase) and condensation of a methyl group with CO and coenzyme A to acetyl–CoA (carbon monoxide dehydrogenase). In methanogens the conversion of methylene–H_4MPT to methyl–coenzyme M involves methylene–H_4MPT reductase and CH_3–H_4MPT:H–S–CoM methyltransferase; the latter enzyme has been shown to generate a primary Na^+ potential. In analogy to the methanogenic system, it has been proposed that also methylene–H_4F conversion to acetyl–CoA is coupled with the generation of a primary Na^+ potential ($\Delta\tilde{\mu}Na^+$) and that either methylene–H_4F reductase and/or the methyltransferase is the site of Na^+ translocation (for discussion see refs. [195,196]). Methylene–H_4F reductase appears to be membrane associated [197,198].

The $\Delta\tilde{\mu}Na^+$ formed may be converted via a Na^+/H^+ antiporter into a $\Delta\tilde{\mu}H^+$ which then drives the synthesis of ATP via a DCCD-sensitive H^+-translocating ATP synthase. This ATP formation explains net ATP synthesis coupled to acetate formation from H_2/CO_2 [192,195,199]. Alternatively, $\Delta\tilde{\mu}Na^+$ could drive ATP synthesis directly via Na^+-translocating ATP synthase. A Na^+-stimulated ATP-synthase activity has recently been reported for *Acetobacterium woodii* [200].

4. Energetics of methanogenesis from methanol

Several methanogens can grow at the expense of methanol disproportionation to CH_4 and CO_2 according to the following equation:

$$4CH_3OH \rightarrow CO_2 + 3CH_4 + 2H_2O, \qquad \Delta G^{o\prime} = -107 \text{ kJ/mol } CH_4.$$

4.1. Enzymology

Methanol disproportionation involves the reduction of 3 mol CH_3OH to CH_4 with 6[H] derived from the oxidation of 1 mol CH_3OH to CO_2:

$$CH_3OH + H_2O \rightarrow CO_2 + 6[H],$$

$$3CH_3OH + 6[H] \rightarrow 3CH_4 + 3H_2O.$$

Methanol-grown cells of *Methanosarcina barkeri* and *Methanosarcina thermophila* have been shown to contain all enzymes that are present in H_2/CO_2-grown cells, indicating that the reduction of methanol to CH_4 and the oxidation of methanol to CO_2 are catalyzed by the same set of enzymes involved in CO_2 reduction to CH_4 [66,201]. The pathway of methanol disproportionation to CH_4 and CO_2, as well as thermodynamic data of the partial reactions involved, are shown in Fig. 3 and Table 3 (Reactions 1–9). The reduction of methanol to CH_4 starts with binding of methanol to H–S–CoM via two specific methyltransferases (Reaction 1 of Table 3) to form CH_3–S–CoM. The subsequent reduction of CH_3–S–CoM to CH_4 involves CH_3–S–CoM reductase and heterodisulfide reductase (Reaction 2, 3, Table 3). The oxidation of methanol to CO_2 proceeds most probably via the reverse sequence (Reactions 4–9, Table 3) involved in CO_2 reduction to the methanol level (Reactions 1–6, Table 2): all enzymes catalyzing these reactions have been purified from methanol-grown *Methanosarcina barkeri* (strain Fusaro): Formyl–MFR dehydrogenase [54,55], formyl–MFR:H_4MPT formyltransferase [59], formyl–H_4MPT cyclohydrolase [202], methylene–H_4MPT dehydrogenase [61], methylene–H_4MPT reductase [67], and methyl–H_4MPT:H–S–CoM methyltransferase [69b]. It is assumed that methanol binds first to H–S–CoM prior to oxidation and that methyltransferase operates reversibly in vivo, which has yet to be demonstrated.

TABLE 3
Free energy changes ($\Delta G^{o\prime}$) and redox potentials ($E^{o\prime}$) of partial reactions involved in CH_3OH disproportionation to CH_4 and CO_2 [a]

# Reaction [b]	$\Delta G^{o\prime}$ (kJ/reaction)	$E^{o\prime}$ (mV)
1. $CH_3OH + H-S-CoM \rightarrow$ methyl$-S-CoM + H_2O$	−27	
Reductive part		
2. methyl$-S-CoM + H-S-HTP \rightarrow CH_4 + CoM-S-S-HTP$	−45	
3. $CoM-S-S-HTP + 2[H] \rightarrow H-S-CoM + H-S-HTP$	−40	−210
Oxidative part		
4. methyl$-S-CoM + H_4MPT \rightarrow N^5$-methyl$-H_4MPT + H-S-CoM$	+30	
5. N^5-methyl$-H_4MPT \rightarrow N^5, N^{10}$-methylene$-H_4MPT + 2[H]$	+20	−323
6. N^5, N^{10}-methylene$-H_4MPT + H^+ \rightarrow N^5, N^{10}$-methenyl$-H_4MPT^+ + 2[H]$	−5	−386
7. N^5, N^{10}-methenyl$-H_4MPT^+ + H_2O \rightarrow N^5$-formyl$-H_4MPT + H^+$	+2	
8. N^5-formyl$-H_4MPT + MFR \rightarrow$ formyl$-MFR + H_4MPT$	+5	
9. formyl$-MFR \rightarrow CO_2 + 2[H] + MFR$	−16	−497

[a] $\Delta G^{o\prime}$ and $E^{o\prime}$ values are taken from references [12, 18, 51]; 2[H] are considered to be H_2.
[b] Abbreviations: MFR, methanofuran; H_4MPT, tetrahydromethanopterin; H–S–CoM, coenzyme M; H–S–HTP, N-7-mercaptoheptanoylthreonine phosphate.

4.2. Energetics

ATP is formed in the reductive part of methanol disproportionation involving the heterodisulfide reductase system as a coupling site (see above). Reducing equivalents required for heterodisulfide reduction are generated in the oxidative pathway. Factor F_{420} has been shown to be the electron acceptor of Reactions (5) and (6) of Table 3. The electron acceptor of formyl–MFR dehydrogenase (Reaction 9, Table 3) is not known. In organisms containing a hydrogenase, molecular hydrogen can serve as the electron donor for the reduction of the heterodisulfide.

The oxidation of methanol to CO_2 is assumed to start from CH_3–S–CoM and to proceed energetically via the reversal of CO_2 reduction to the methanol level. The oxidation involves an endergonic site, the oxidation of methyl–CoM to methylene–H_4MPT, and an exergonic site, the oxidation of formyl–MFR to CO_2 (Table 3). The mode of energy coupling of formyl–MFR oxidation to CO_2 has been studied for the conversion of formaldehyde to CO_2 and $2H_2$ (see above) in cell suspensions of *Methanosarcina barkeri*. This reaction was shown to generate a primary Na^+ potential, which can be converted into $\Delta \tilde{\mu} H^+$ via Na^+/H^+ antiporter. $\Delta \tilde{\mu} H^+$ then drives the synthesis of ATP. Based on these findings one could definitely exclude ATP formation by the mechanism of substrate level phosphorylation coupled to the conversion of formyl–H_4MPT to CO_2 via formyl phosphate. Such a mechanism has been proposed in methanogens [13] by analogy to ATP formation coupled to the conversion of formyl–tetrahydrofolate to CO_2 in several anaerobic bacteria.

The energetics of the endergonic reaction, i.e. the oxidation of CH_3–S–CoM to the level of formaldehyde, has been studied with *Methanosarcina barkeri*. The disproportionation of methanol rather than the reduction of methanol with H_2 has been shown to be sensitive towards uncouplers [41] and to be dependent on Na^+ ions [110]. Thus, it was proposed that the oxidation of methanol to CO_2 involves an endergonic reaction that is energy driven in vivo and can proceed only when coupled with the exergonic reactions of the pathway. Furthermore, Na^+ ions were somehow implicated in the coupling mechanism [111]. Experiments with cell suspensions of *Methanosarcina barkeri* indicate that the oxidation of methanol to the redox level of formaldehyde (methylene–H_4MPT) is the endergonic step and that the driving force for this reaction is an electrochemical potential of Na^+ ions [203]: for example, (i) methanogenesis from methanol rather than from methanol and H_2 was inhibited by an artificially imposed inversed Na^+ gradient ($[Na^+]_{in} > [Na^+]_{out}$); (ii) an influx of Na^+ ions coupled to methanol disproportionation, rather than to methanol reduction with H_2, was observed when the Na^+/H^+ antiporter was inhibited; Na^+ influx was electrogenic as indicated by a decrease in the membrane potential.

Methanol disproportionation includes two sites of $\Delta\tilde{\mu}Na^+$ generation: (1) Methanol reduction generates a secondary $\Delta\tilde{\mu}Na^+$ from a primary $\Delta\tilde{\mu}H^+$ by the activity of the Na^+/H^+ antiporter. This was concluded from the finding that CH_4 formation from H_2/CH_3OH in *Methanosarcina barkeri* was coupled with Na^+ extrusion, which was sensitive to Na^+/H^+ antiporter inhibitors and protonophores [108]. (2) The oxidation of methylene–H_4MPT to CO_2 and 4[H] is coupled with the generation of a primary $\Delta\tilde{\mu}Na^+$ as indicated from the fact that Na^+ translocation associated with formaldehyde conversion to CO_2 and $2H_2$ was not sensitive towards Na^+/H^+ antiporter inhibitors and protonophores [105].

4.2.1. Role of the Na^+/H^+ antiporter

Methanol disproportionation was found to be completely inhibited by Na^+/H^+ antiporter inhibitors [167,203], indicating that a secondary Na^+ potential is required for methanol oxidation. This can be explained assuming that methanol oxidation to CO_2 is energetically the reversal of CO_2 reduction to the methanol level. Accordingly, during methanol oxidation to CO_2 the Na^+ translocating reactions and the Na^+/H^+ antiporter operate in the opposite direction as described for the CO_2 reduction pathway. The stoichiometries of electrogenic Na^+ translocation can explain why methanol oxidation requires secondary transport of Na^+ ions: during methanol oxidation to the formaldehyde level (methylene–H_4MPT) 3–4 Na^+ ions are taken up; formaldehyde oxidation to CO_2 and $2H_2$ is coupled with the extrusion of only 2–3 Na^+. Thus, the "missing" Na^+ ions have to be provided by the Na^+/H^+ antiporter via transformation of $\Delta\tilde{\mu}H^+$ into a secondary $\Delta\tilde{\mu}Na^+$ (Fig. 9). The endergonic methanol oxidation is therefore energetically linked to CH_3OH reduction, which generates a primary $\Delta\tilde{\mu}H^+$. Thus, the role of the Na^+/H^+ antiporter during methanol disproportionation is to generate $\Delta\tilde{\mu}Na^+$ from $\Delta\tilde{\mu}H^+$ by mediating Na^+ extrusion.

4.2.2. Role of methyltransferase

The mechanism of $\Delta\tilde{\mu}Na^+$-driven methanol oxidation to the level of formaldehyde is not known. CH_3–H_4MPT:H–S–CoM methyltransferase has been identified to be the site of primary $\Delta\tilde{\mu}Na^+$ generation during formaldehyde conversion to the methanol level

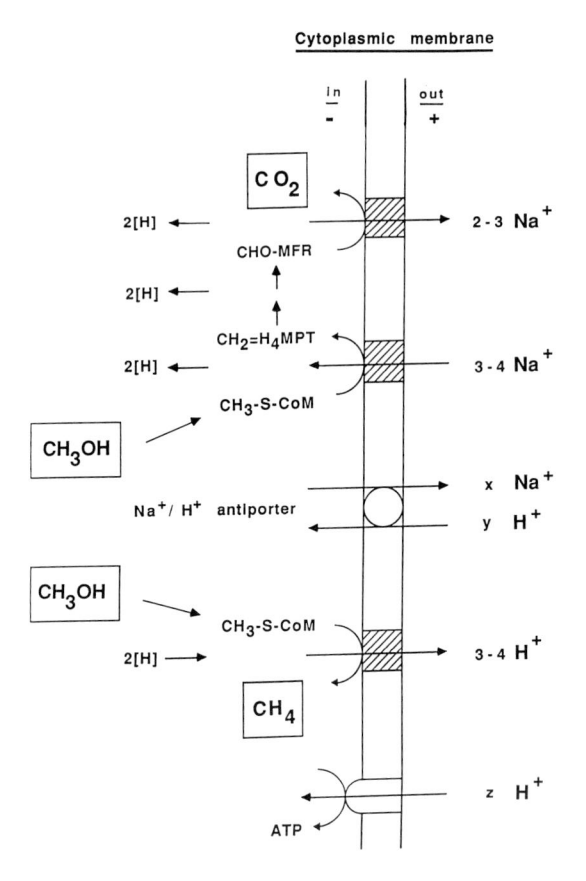

Fig. 9. Proposed function of electrochemical H^+ and Na^+ potentials in energy conservation coupled to methanol disproportionation to CH_4 and CO_2. It is assumed that prior to oxidation methanol binds first to coenzyme M and that the oxidation is mechanistically and energetically the reversal of CO_2 reduction to methyl–coenzyme M. The Na^+/H^+ antiporter is involved in the generation of $\Delta\tilde{\mu}Na^+$ from $\Delta\tilde{\mu}H^+$. CHO–MFR, formyl–methanofuran; CH_2=H_4MPT, methylene–tetrahydromethanopterin; CH_3–H_4MPT, methyl–tetrahydromethanopterin; CH_3–S–CoM, methyl–coenzyme M. The hatched boxes indicate membrane-bound electron transport chains or membrane-bound methyltransferase catalyzing either Na^+ or H^+ translocation (see Figs. 5, 6 and 12). ATP is synthesized via membrane-bound H^+-translocating ATP synthase. The stoichiometries of Na^+ and of H^+ translocation were taken from refs. [105,107,167]. x, y and z are unknown stoichiometric factors.

(methyl–CoM). Assuming methyltransferase to operate reversibly in vivo it is likely that the $\Delta\tilde{\mu}Na^+$-driven reaction during methanol oxidation to the formaldehyde level is the endergonic methyl transfer from methyl–CoM to methyl–H_4MPT (Fig. 12, below). Figure 9 summarizes the proposed role of Na^+ and H^+ cycles involved in coupling exergonic and endergonic reactions during methanol disproportionation to CH_4 and CO_2.

4.2.3. Role of cytochromes

Methanogens growing at the expense of methanol disproportionation have been shown to contain cytochromes [128,204,205]. A role of these electron carriers in methanol oxidation rather than in methanol reduction has been postulated. Accordingly, *Methanosphaera stadtmanae*, which forms CH_4 exclusively by methanol reduction with H_2, does not contain cytochromes [42]. The cytochromes have been implicated in the endergonic oxidation of methanol to the level of formaldehyde. In cell suspensions of *Methanolobus tindarius* the redox spectra of membrane-bound cytochromes changed in response to artificially imposed Na^+ gradients [206], indicating a role in the postulated $\Delta\tilde{\mu}Na^+$ driven methanol oxidation.

4.2.4. Growth yields

Molar growth yields (Y_{CH_4}) were determined for *Methanosarcina barkeri* growing on a mineral salt medium with either methanol (gas phase N_2) or with methanol/H_2 as energy source and acetate as carbon source: The Y_{CH_4} values on methanol/N_2 (3.2 g cells/mol CH_4) and on methanol/H_2 (2.8 g cells/mol CH_4) were almost the same [43] and can therefore not be used to calculate the energy requirement for methanol oxidation.

4.3. Energetics of CH_4 formation from methylamines

Methanosarcina barkeri disproportionates methylamine, dimethylamine and trimethylamine to CH_4, CO_2 and NH_4^+ according to the Equations (7)–(9) of Table 1. The growth yields Y per mol CH_4 of *Methanosarcina barkeri* after growth on methanol and on the different methylamines are similar [45], indicating energetic equivalence of the pathways of methyl-group disproportionation. In analogy to the H_2/methanol system it was shown that cell suspension of *Methanosarcina barkeri* couple methanogenesis from trimethylamine and H_2 with ATP synthesis by a chemiosmotic mechanism [207]. Furthermore the oxidation of trimethylamine to the level of formaldehyde has been shown to require energy [208].

5. Energetics of methanogenesis from acetate

The utilization of acetate by methanogens has so far been shown only for species of the genera *Methanosarcina* and *Methanothrix*. Most studies concerning the enzymatics and energetics of acetate fermentation to CH_4 and CO_2 have been performed with *Methanosarcina barkeri*, *Methanosarcina thermophila* and *Methanothrix soehngenii* (for a recent review see [209]. *Methanosarcina* species are the most versatile methanogens since they can utilize CO_2/H_2, methanol, methanol/H_2, methylamines and acetate as substrates. *Methanothrix* species are restricted to acetate fermentation [210].

5.1. Enzymology

The reactions involved in acetate conversion to CH_4 and CO_2 as well as the thermodynamic data are given in Fig. 10 and Table 4. Acetate is activated in *Methanosarcina* to acetyl–CoA via acetate kinase and via phosphate acetyl transferase (Reactions 1, 2 of Table 4) [211–217]. Acetyl–CoA is then converted to methyl–H_4MPT, CoA, CO_2 and 2[H] (Reactions 3, 3a, 3b). This complex reaction involves several enzymes, a carbon monoxide dehydrogenase (CO–DH), a corrinoid protein and a methyltransferase [218–221] and is not completely understood in its details. Acetyl–CoA is the substrate of the C–C cleavage reaction catalyzed by CO–DH to generate a methyl group, and carbon monoxide [CO] both bound to the enzyme (Reaction 3a). CO–DH also catalyzes the oxidation of [CO] to CO_2 with the formation of two reducing equivalents (Reaction 3b). The physiological electron acceptor of CO–DH appears to be ferredoxin [132,133]. The methyl group is transferred to H_4MPT [222] via a specific methyltransferase to give CH_3–H_4MPT. CH_3–H_4MPT is converted to CH_3–S–CoM via CH_3–H_4MPT:H–S–CoM methyltransferase. The enzyme has been partially purified from acetate-grown *Methanosarcina barkeri* and found to be essentially membrane associated [69b]. Both methyltransferases appear to contain corrinoids [69b,223,224]. CH_3–S–CoM [225] is then reduced to CH_4 involving CH_3–S–CoM reductase [218,226] and heterodisulfide reductase [66].

5.2. Sites of energy coupling

Acetate is fermented to CO_2 and CH_4 according to the equation

$$CH_3COO^- + H^+ \rightarrow CH_4 + CO_2, \quad \Delta G^{o\prime} = -36\,kJ/mol.$$

The small free energy change (−36 kJ/mol) associated with the reaction is sufficient to allow the synthesis of about 0.5 mol ATP/mol CH_4. In accordance, the growth yields Y_{CH_4} of *Methanosarcina* and *Methanothrix* on acetate as sole carbon and energy source are low, of the order of 2–4 g cells/mol CH_4 [35,227–230].

After uptake of acetate into the cells (for a proposed transport mechanism see [231]) acetate is activated in *Methanosarcina* species to acetyl–CoA via acetyl-phosphate by stoichiometric consumption of 1 ATP in the acetate kinase reaction. This indicates that during acetyl–CoA conversion to CH_4 more than 1 mol ATP (about 1.5) has to be formed.

Three sites of energy coupling have been proposed for CH_4 formation from acetate (Fig. 11): (1) CH_3–S–CoM reduction to CH_4, (2) oxidation of enzyme-bound CO to CO_2 and H_2, and (3) CH_3–H_4MPT:H–S–CoM methyltransferase reaction.

5.2.1. CH_3–S–CoM reduction to CH_4
CH_3–S–CoM is reduced to methane via the heterodisulfide of H–S–CoM and H–S–HTP. The reduction of the heterodisulfide has been shown to be coupled with ATP synthesis according to a chemiosmotic mechanism (see above). The electrons required for the reduction are derived from the oxidation of enzyme-bound CO ([CO]) which is oxidized to CO_2 via CO–DH. It is assumed that electron transport from [CO] to the heterodisulfide is coupled with the generation of an electrochemical proton potential which then drives ATP synthesis. Possible electron transport components, a cytochrome b and a membrane-bound hydrogenase, have been identified [232]. Probably two H^+-translocating sites are present in electron transport from CO to the heterodisulfide: the oxidation of CO to CO_2 and H_2, and the reduction of the heterodisulfide (or methyl–CoM) by H_2. Both H_2 and

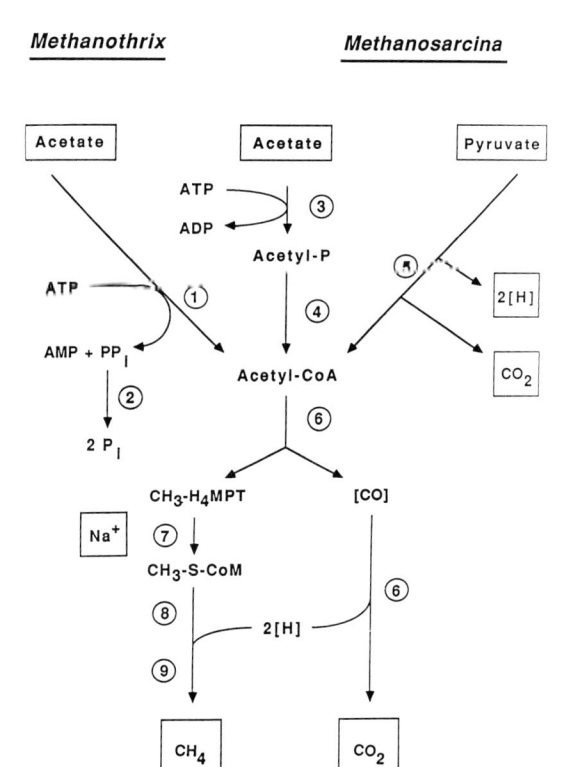

Fig. 10. Proposed pathway of methanogenesis from acetate (*Methanosarcina*, *Methanothrix*) and from pyruvate (*Methanosarcina*): Intermediates, enzymes and a site for Na^+ dependence. CoA, coenzyme A; PP, pyrophosphate; CH_3–H_4MPT, methyl–tetrahydromethanopterin, CH_3–S–CoM, methyl–coenzyme M; [CO], CO bound to carbon monoxide dehydrogenase. Numbers in circles refer to enzymes involved: (1) acetyl–CoA synthetase; (2) pyrophosphatase; (3) acetate kinase; (4) phosphate acetyltransferase; (5) pyruvate:ferredoxin oxidoreductase; (6) carbon monoxide dehydrogenase; (7) CH_3–H_4MPT:H–S–CoM methyltransferase; (8) methyl–coenzyme M reductase; (9) heterodisulfide reductase.

CO have been shown to reduce the heterodisulfide in cell extracts of *Methanosarcina barkeri* [133] (see also ref. [233]).

5.2.2. CO oxidation to CO_2

A second site for ATP synthesis in acetate fermentation appears be the oxidation of the enzyme-bound CO to CO_2. This was deduced from the following findings: cell suspensions of *Methanosarcina barkeri* catalyze the oxidation of free CO to CO_2 with H^+ as electron acceptor, when methanogenesis is inhibited by bromoethanesulfonate [233]:

$$CO + H_2O \rightarrow CO_2 + H_2, \quad \Delta G^{o\prime} = -20 \, kJ/mol.$$

TABLE 4

Free energy changes ($\Delta G^{o\prime}$) and redox potentials ($E^{o\prime}$) of partial reactions involved in acetate fermentation to CH_4 and CO_2 in *Methanosarcina barkeri* [a]

#	Reaction [b]	$\Delta G^{o\prime}$ (kJ/react.)	$E^{o\prime}$ (mV)
1.	Acetate + P_i → Acetyl–P	+47	
2.	Acetyl–P + H–S–CoA → Acetyl–S–CoA + P_i	−9	
3.	Acetyl–CoA + H_4MPT → N^5-methyl–H_4MPT + H–S–CoA + CO_2 + 2[H]	+41	−200
3a.	Acetyl–S–CoA + H_4MPT → N^5-methyl–H_4MPT + H–S–CoA + [CO]	+61	
3b.	[CO] + H_2O → CO_2 + 2[H]	−20	−524
4.	N^5-methyl–H_4MPT + H–S–CoM → methyl–S–CoM + H_4MPT	−30	
5.	methyl–S–CoM + H–S–HTP → CH_4 + CoM–S–S–HTP	−45	
6.	CoM–S–S–HTP + 2[H] → H–S–CoM + H–S–HTP	−40	−210
1–6.	Acetate$^-$ + H$^+$ → CH_4 + CO_2	−36	

[a] $\Delta G^{o\prime}$ and $E^{o\prime}$ values are taken from references [12, 18, 51]; 2[H] are considered to be H_2.
[b] Abbreviations: P_i, inorganic phosphate; H–S–CoA, coenzyme A; H_4MPT, tetrahydromethanopterin; [CO], enzyme-bound carbon monoxide; H–S–CoM, coenzyme M; H–S–HTP, N-7-mercaptoheptanoyl-threonin phosphate.

This reaction was found to be coupled with the synthesis of ATP. The effects of protonophores and the H$^+$-ATP synthase inhibitor DCCD on the rate of CO oxidation, on the electrochemical proton potential and on the ATP content of the cells indicate $\Delta\tilde{\mu}H^+$-driven ATP synthesis by a H$^+$-translocating DCCD-sensitive ATP synthase [233]: (i) protonophores dissipated $\Delta\tilde{\mu}H^+$, inhibited ATP synthesis and stimulated CO oxidation; (ii) DCCD inhibited ATP synthesis and slowed down the CO oxidation rate without affecting $\Delta\tilde{\mu}H^+$. Protonophores relieved the inhibition of CO oxidation rate by DCCD; (iii) CO oxidation to CO_2 was coupled with primary electrogenic H$^+$ translocation at a stoichiometry of 2 mol H$^+$ per mol of CO oxidized to CO_2 and H_2 [234]; (iv) the reverse reaction, the reduction of CO_2 to CO, was found to be driven by a $\Delta\tilde{\mu}H^+$ [235]. The findings indicate that in *Methanosarcina barkeri* electron transport from CO ($E^{o\prime}$ = −525 mV) as electron donor to protons ($E^{o\prime}$ = −414 mV) as terminal electron acceptors is coupled with primary H$^+$ translocation. Assuming that the thermodynamics of enzyme-bound CO is similar to that of free CO these findings indicate that the oxidation of bound CO to CO_2 is a site of H$^+$ translocation and ATP synthesis in methanogenesis from acetate.

Coupling of CO oxidation to CO_2 and H_2 with energy conservation has also been demonstrated in acetogenic bacteria: CO oxidation drives the endergonic uptake of histidine in *Acetobacterium woodii* [236]. Furthermore the phototroph *Rhodocyclus gelatinosus* (formerly *Rhodopseudomonas gelatinosa*) can grow in the dark at the expense of CO oxidation to CO_2 and H_2 [237,238]. The CO–DH in the latter organism has been shown to be membrane associated [239]; there are also indications for membrane association of CO–DH in homoacetogenic bacteria [197].

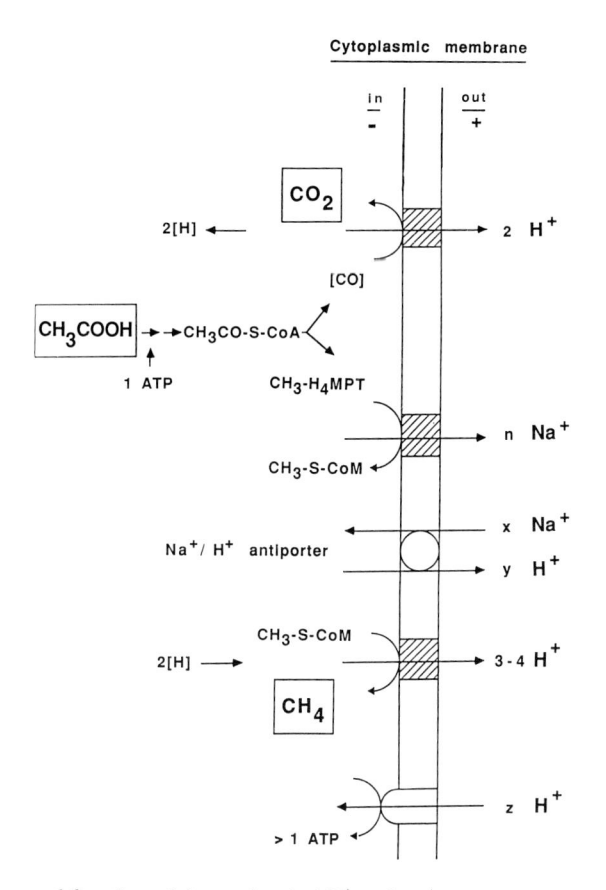

Fig. 11. Proposed function of electrochemical H$^+$ and Na$^+$ potentials in energy conservation coupled to acetate fermentation to CH$_4$ and CO$_2$. The Na$^+$/H$^+$ antiporter is involved in the generation of $\Delta\tilde{\mu}$H$^+$ from $\Delta\tilde{\mu}$Na$^+$. CH$_3$CO–S–CoA, acetyl–coenzyme A; [CO], CO bound to carbon monoxide dehydrogenase; CH$_3$–H$_4$MPT, methyl–tetrahydromethanopterin; CH$_3$–S–CoM, methyl–coenzyme M. The hatched boxes indicate membrane-bound electron transport chains or membrane-bound methyltransferase catalyzing either H$^+$ or Na$^+$ translocation (see Figs. 5, 6 and 12). It is assumed that enzyme-bound [CO] is energetically equal to free CO. ATP is synthesized via membrane-bound H$^+$-translocating ATP synthase. The stoichiometries of H$^+$ translocation were taken from refs. [107,234]; n, x, y and z are unknown stoichiometric factors.

5.2.3. CH$_3$–H$_4$MPT:H–S–CoM methyltransferase

Acetate conversion to CH$_4$ and CO$_2$ is strictly dependent on Na$^+$ ions [110]. Cell suspensions of *Methanosarcina barkeri* have been shown to generate a Na$^+$ gradient ([Na$^+$]$_{out}$/[Na$^+$]$_{in}$ ≈ 5) across the cytoplasmic membrane [230]. The sites of Na$^+$ dependence and of Na$^+$ translocation in acetate conversion to CH$_4$ and CO$_2$ are not known. Both the reduction of CH$_3$–S–CoM to CH$_4$ and the oxidation of CO to CO$_2$ are not dependent on Na$^+$ [230,235]. Recent evidence indicates that CH$_3$–H$_4$MPT:H–S–CoM methyltransferase is a Na$^+$-dependent membrane protein, which catalyzes primary Na$^+$

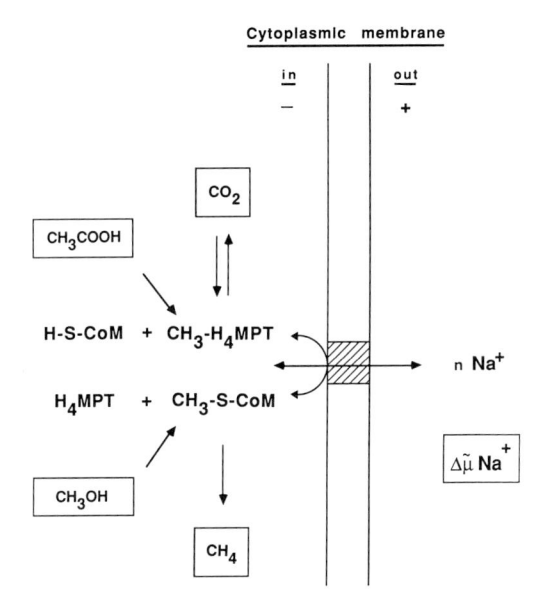

Fig. 12. Proposed role of methyl–tetrahydromethanopterin (CH_3–H_4MPT):coenzyme M (H–S–CoM) methyltransferase in methanogenesis from CO_2, acetate and methanol: methyltransferase as a reversible Na^+-translocating membrane protein. During CH_4 formation from CO_2 and from acetate, methyltransferase is involved in primary Na^+ extrusion generating $\Delta\tilde{\mu}Na^+$; during methanol disproportionation to CH_4 and CO_2 the enzyme is involved in $\Delta\tilde{\mu}Na^+$-driven methanol oxidation. n is an unknown stoichiometric factor.

extrusion during methane formation from CO_2 [69b,168] (Fig. 12). Since *Methanosarcina barkeri* grown on acetate also contain such a membrane-bound methyltransferase, it is likely that this enzyme is responsible for the Na^+ dependence of acetate fermentation to CH_4 and CO_2 and that it is involved in the generation of a primary electrochemical Na^+ potential, $\Delta\tilde{\mu}Na^+$. Thus, CH_3–H_4MPT:H–S–CoM methyltransferase appears to be a third site of energy conservation in CH_4 formation from acetate.

The proposed functions of electrochemical potentials of H^+ and Na^+ in energy conservation coupled to acetate fermentation to CH_4 and CO_2 are summarized in Fig. 11: CH_3–S–CoM reduction, including heterodisulfide reduction, and the oxidation of enzyme-bound CO are coupled with the primary translocation of H^+ generating $\Delta\tilde{\mu}H^+$, which drives the synthesis of ATP via H^+-translocating ATPase. Methyltransferase is coupled with the primary translocation of Na^+ ions across the membrane. The $\Delta\tilde{\mu}Na^+$ generated can be converted by the Na^+/H^+ antiporter into a $\Delta\tilde{\mu}H^+$ which then drives ATP synthesis.

Acetyl–CoA conversion to CH_3–H_4MPT, CoA, CO_2 and 2[H] is strongly endergonic (+41 kJ/mol) under standard conditions, e.g. with ferredoxin ($E^{o\prime} = -420$ mV) as physiological electron acceptor (Reaction 3 of Table 4). This might indicate that acetyl–CoA cleavage is energy driven in vivo, e.g. by a Na^+ gradient [230]. However, $\Delta G^{o\prime}$ is highly concentration dependent because four products are formed in the reaction; assuming, e.g., 0.1 mM concentrations of all substrates and products under physiological conditions the reaction is almost in equilibrium [18]. This is in accordance with the finding that in cell-free extracts of *Methanosarcina barkeri,* acetyl phosphate, or acetate plus ATP, is converted via acetyl–CoA to CH_4 and CO_2 in significant rates, indicating that acetyl–CoA cleavage can proceed in vivo in the absence of transmembrane ion gradients [213,215,224].

5.3. Acetate fermentation in Methanothrix soehngenii

Cell extracts of *Methanothrix soehngenii* contain high activities of acetyl–CoA synthetase rather than acetate kinase and phosphate acetyltransferase [240,241] (Fig. 10). This indicates that in this organism acetate is activated to acetyl–CoA by acetyl–CoA synthase (AS). Since AMP is converted to ADP via adenylate kinase (AK), and pyrophosphate (PP_i) is completely hydrolyzed via pyrophosphatase (PP_i-ase), acetate activation by acetyl–CoA synthetase requires the input of 2 ATP equivalents:

acetate + ATP + CoA → acetyl–CoA + AMP + PP_i	(AS)
AMP + ATP → 2ADP,	(AK)
PP_i → $2P_i$	(PP_i ase)

acetate + 2ATP + CoA → acetyl–CoA + 2ADP + 2Pi.

The presence of both adenylate kinase and pyrophosphatase has been demonstrated in cell extracts of *Methanothrix soehngenii* [241]. Furthermore, the organism contains CO–DH [242], methyl–CoM reductase [243] and heterodisulfide reductase [66], indicating the same mechanism of acetyl–CoA conversion to CH_4 and CO_2 as shown for *Methanosarcina* (Fig. 10, above). Pyrophosphate:AMP- or pyrophosphate:ADP-phosphotransferase could not be detected [241], indicating that the energy-rich anhydride bond of pyrophosphate is not directly used to regenerate ADP or ATP. Activation of acetate at the expense of 2ATP poses an energetic problem since for thermodynamic reasons acetyl–CoA conversion to CH_4 and CO_2 can be coupled with the synthesis of at most 1.5 ATP (see above). One possibility to solve this problem is that pyrophosphate hydrolysis is somehow coupled with energy conservation, e.g. via the formation of $\Delta\tilde{\mu}H^+$ generated by a membrane-bound H^+-translocating pyrophosphatase. Such an enzyme has been demonstrated in the phototrophic bacteria, e.g. *Rhodospirillum rubrum* (see ref. [244]). Pyrophosphatase from *Methanotrix soehngenii* [245] might have a similar energetic function as the enzyme from phototrophs.

6. Energetics of pyruvate catabolism

6.1. Methanogenesis from pyruvate

Methanosarcina barkeri has been shown to grow on pyruvate as sole carbon and energy source provided the organism has been incubated in the presence of high pyruvate concentrations (100 mM) for several weeks. Evidence was presented that during incubation a mutant of *M. barkeri* was selected wich was able to utilize pyruvate [35]. Growing cultures of pyruvate-utilizing *Methanosarcina barkeri* converted pyruvate to CO^2 and CH^4 according to

$$CH_3COCOO^- + H^+ + 0.5H_2O \rightarrow 1.25CH_4 + 1.75CO_2,$$
$$\Delta G^{o\prime} = -120 \text{ kJ/mol pyruvate}.$$

The specific growth rate of *M. barkeri* was linearly dependent on the pyruvate concentration up to 100 mM, indicating that pyruvate was taken up by passive diffusion. Cell extracts of pyruvate-grown *M. barkeri* catalyzed the coenzyme A-dependent oxidative decarboxylation of pyruvate to acetyl–CoA with ferredoxin as electron aceptor indicating a pyruvate:ferredoxin oxidoreductase to be operative in acetyl–CoA formation from pyruvate. Coenzyme F_{420} did not serve as primary electron acceptor of pyruvate oxidation [35]. Furthermore, the cells contained an active carbon-monoxide dehydrogenase. These findings indicate that pyruvate is converted to acetyl–CoA, which is further split into CO_2 and CH_4 by the mechanism described for acetate fermentation (Fig. 10). In addition, CH_4 is formed by the reduction of CO_2 with reducing equivalents generated by pyruvate oxidation. The molar growth yields Y_{CH_4} (g cell dry weight/mol CH_4) were up to 15 g cells/mol CH_4, which are the highest yields obtained in methanogens. Assuming Y_{ATP} with pyruvate as anabolic substrate to be about 10 g cells/mol ATP the yield data indicate that up to 1.5 mol ATP were formed during methane formation from pyruvate.

The growth yield Y_{CH_4} of *Methanosarcina barkeri* on acetate as carbon and energy source was determined to about 2–4 g/mol CH_4 indicating an ATP gain of about 0.3 mol ATP during methane formation from acetate. The higher cell yield and ATP gain on pyruvate compared acetate can be explained by the different energetics of acetyl–CoA formation: acetyl–CoA formation from pyruvate does not consume energy, whereas acetyl–CoA formation from acetate in *Methanosarcina barkeri* requires the energy input of 1 ATP equivalent (about 10 g cells/mol). In addition, about 0.1 mol ATP/mol pyruvate was formed by the reduction of CO_2 with 2[H]. Thus, the growth yield per mol substrate of *Methanosarcina barkeri* on pyruvate is more than 10 g higher than on acetate as carbon and energy source.

6.2. Acetate formation from pyruvate in the absence of methanogenesis

It is interesting to note that pyruvate-utilizing *Methanosarcina barkeri* also grew on pyruvate when methanogenesis was inhibited completely by bromoethanesulfonate (BES). Under these conditions the cells fermented 1 mol pyruvate to about 0.8 mol of each, acetate and CO_2, and minor amounts of alanine, acetolactate and valine [35a]. The growth yield per mol acetate formed was about 10 g cells/mol acetate, indicating that about 1 mol ATP is formed in this fermentation per mol acetate generated in the catabolism. Since CH_4 was not formed during growth on pyruvate in the presence of BES these findings indicate that ATP was synthesized almost completely in the acetate-kinase reaction via the mechanism of substrate level phosphorylation. Cell extracts of *M. barkeri* grown on pyruvate in the presence of BES contained high activities of both acetate kinase and phosphate acetyltransferase and of pyruvate:ferredoxin oxidoreductase, indicating that these enzymes are involved in acetate formation from pyruvate. This is the first demonstration of growth of a methanogen at the expense of an energy metabolism different from methanogenesis, in which energy is conserved by the mechanism of substrate level phosphorylation.

7. Concluding remarks on energetics

The review up to this point has dealt with the mechanism of ATP synthesis coupled to methanogenesis and with the role of Na^+ ions. The study of these two topics revealed certain features that were found only in methanogens. Of particular interest was the discovery that electrochemical potentials of both H^+ and Na^+ are involved in coupling of exergonic and endergonic reactions of methanogenesis from various substrates. It is now accepted that ATP is formed by the mechanism of electron transport phosphorylation with $\Delta\tilde{\mu}H^+$ as driving force. A unique electron transport chain is involved in the generation of a primary $\Delta\tilde{\mu}H^+$: a heterodisulfide (formed by the typical methanogenic coenzymes H–S–CoM and H–S–HTP) ($E^{o\prime} = -200$ mV) is reduced with reducing equivalents, 2[H], at redox potentials ($E^{o\prime}$) ranging from -350 mV ($F_{420}H_2$) to -525 mV (CO). This reaction, which is a partial reaction of CH_3–S–CoM reduction to CH_4, is the common site for H^+ translocation and ATP synthesis in methanogenesis from all substrates. In addition, the oxidation of CO ($E^{o\prime} = -525$ mV) to CO_2 and H_2 ($E^{o\prime} = -414$ mV) is coupled with primary H^+ translocation and ATP synthesis; this reaction appears to be a coupling site in methane formation from acetate.

Two reversible primary Na^+ translocating reactions have been identified in methanogenesis from H_2/CO_2, from methanol and from acetate, which explain the Na^+ dependence of methanogenesis from these substrates: CH_3–H_4MPT:H–S–CoM methyltransferase and the conversion of formyl–MFR to CO_2. Both reactions represent novel modes of energy transduction : (1) CH_3–H_4MPT:H–S–CoM methyltransferase couples a methyl-group transfer with the primary vectorial translocation of Na^+ ions; (2) the conversion of formyl–MFR to CO_2 involves the oxidation of a formamide to CO_2, which also appears to be coupled with the generation of a primary Na^+ potential.

A unique Na^+ cycle is operative in methanogenesis from CO_2: the exergonic conversion of methylene–H_4MPT to methyl–CoM via CH_3–H_4MPT:H–S–CoM methyltransferase, generates a primary $\Delta\tilde{\mu}Na^+$, which drives the endergonic reduction of CO_2 to the formamide formyl–MFR. This novel mechanism of CO_2 activation is different from that in acetogens, in which CO_2 is activated at the expense of ATP hydrolysis. Methanol oxidation to CO_2 as part of methanol disproportionation involves the reverse reactions: the exergonic formyl–methanofuran oxidation to CO_2 generates a primary $\Delta\tilde{\mu}Na^+$ and the endergonic conversion of methyl–CoM to methylene–H_4MPT is driven by a $\Delta\tilde{\mu}Na^+$. Primary Na^+-translocating reactions are rare in bacterial bioenergetics [246–248]. Thus, it is of interest that two of those reactions are operative in the process of methanogenesis.

An electrogenic Na^+/H^+ antiporter turns out to have an important function in the energetics of methanogens. This cation exchanger has been shown to be stoichiometrically involved in the interconversion of primary potentials of Na^+ and H^+ in the various methanogenic pathways. During methane formation from H_2/CO_2 and probably also from acetate the antiporter generates a secondary $\Delta\tilde{\mu}H^+$ from a primary $\Delta\tilde{\mu}Na^+$ and is therefore indirectly involved in ATP synthesis. Thus, the Na^+/H^+ antiporter in combination with the H^+-translocating ATP synthase allows the complete conservation of free energy of the electrochemical Na^+ potentials in the form of ATP via the mechanism of electron transport phosphorylation. This is important since less than 1 mol ATP/mol CH_4 is formed during methane formation from CO_2 and from acetate. During methanogenesis from methanol the antiporter has a different function; it converts a primary $\Delta\tilde{\mu}H^+$ into a secondary $\Delta\tilde{\mu}Na^+$ which is involved in the endergonic $\Delta\tilde{\mu}Na^+$-driven methanol oxidation to the level of formaldehyde.

One can only speculate about the question why two different cation gradients, of H^+ and of Na^+, are operative in driving endergonic reactions. During CH_4 formation from H_2 and CO_2 such a situation would allow an independent regulation of $\Delta\tilde{\mu}Na^+$-driven electron transport (methanogenesis) and of $\Delta\tilde{\mu}H^+$-driven ATP synthesis. This might be important in certain metabolic situations in nature, e.g. when growth of the bacteria is initiated after periods of energy starvation.

8. Transport in methanogens

Only a few transport studies have been performed with methanogens. Since most methanogens are chemolithoautotrophs they probably do not contain transport systems for substrate uptake or product excretion: H_2, CO_2 and CH_4 are considered to permeate freely the cytoplasmic membrane. This may probably also be the case for other energy substrates such as methanol, ethanol, and 2-butanol. Nothing is known about the transport of formate ($pK_s = 3.7$), acetate ($pK_s = 4.8$) and methylamine ($pK_s = 9.2$). Being weak acids and bases they are assumed to permeate the membrane in the undissociated form. It has not been calculated whether the rate of uptake of these compounds is sufficient to explain the observed growth rates. The presence of transport systems for all three compounds cannot be excluded. Evidence for acetate transport systems in bacteria has been presented [249,250]. *Methanosarcina barkeri* has been shown to contain an active carbonic anhydrase after growth on acetate rather than on H_2/CO_2 or on methanol [231], indicating that HCO_3^- rather than CO_2 is the product of acetate fermentation. It has been suggested that in *M. barkeri* an acetate/HCO_3^- antiport is involved in acetate transport [231].

The uptake of essential trace elements, e.g. Na^+, K^+, Ni^{2+}, Co^{2+}, Mg^{2+}, PO_3^{3-}, MoO_4^{2-} and, in the case of auxotrophic mutants, of coenzymes, vitamins, and amino acids requires transport systems. In addition to Na^+ transport (see previous sections), active transport systems for Ni^{2+}, K^+, PO_3^{3-}, coenzyme M (H–S–CoM), methyl–coenzyme M (CH_3–S–CoM), isoleucine and other branched-chain amino acids have been described in methanogens (for recent literature on ion transport in prokaryotes see refs. [251,252]).

8.1. H–S–CoM, CH₃–S–CoM

Growth of *Methanobrevibacter* (formerly *Methanobacterium ruminantium*) on H_2/CO_2 is dependent on exogenous coenzyme M (H–S–CoM) as a vitamin [253]. The uptake of

H–S–CoM was studied in cell suspensions [254]. The data indicate the presence of a high-affinity (apparent K_m = 70 nM) active transport system: H–S–CoM accumulates against a concentration gradient. Uptake requires H_2/CO_2 and is inhibited by O_2 and inhibitors of methanogenesis. The driving force for HS–CoM uptake was not determined. An active transport system for H–S–CoM and CH_3–S–CoM has also been described in *Methanococcus voltae* [255]. These compounds are transported against concentration gradients and uptake was dependent on methanogenesis. The transport system also mediates the uptake of the H–S–CoM analogue bromoethanesulfonate (BES). BES-resistant mutants showed a decreased uptake of the inhibitor and of CH_3–S–CoM rather than of H–S–CoM, suggesting the presence of specific transport systems for H–S–CoM and CH_3–S–CoM.

8.2. Amino acids

Methanococcus voltae requires the branched-chain amino acids leucine and isoleucine for growth on H_2/CO_2 [30]. Isoleucine transport was studied in cell suspensions of this organism [256]. The amino acid was accumulated against a concentration gradient of 100; transport was stimulated by Na^+ ions (optimal concentration 50–100 mM). Uptake was inhibited by protonophores and sodium ionophores. The data suggest that in analogy to other Na^+-dependent amino acid transport systems [257] a positively charged Na^+/isoleucine symport system is operative in *Methanococcus voltae*. The transport of several other amino acids in *Methanococcus voltae* has also been shown to be Na^+ stimulated [258].

8.3. Nickel

Nickel is an essential trace element for growth of methanogenic bacteria [28–30,259]. The cation has been shown to be a component of factor F_{430}, of two hydrogenases and of carbon monoxide dehydrogenase (for reviews see refs. [260,261]). Ni^{2+} uptake was studied in cell suspension of *Methanobacterium bryantii* [262]: the cation was transported against a concentration gradient via a high-affinity (apparent K_m = 3.1 µM) system, which was specific for Ni^{2+}. Mg^{2+} did not compete for Ni^{2+} uptake indicating that Ni^{2+} was not transported by a Mg^{2+} transport system. Uptake of Ni^{2+} was dependent on a H_2/CO_2 gas phase reflecting energy dependence. Under a N_2 gas phase an artificially imposed pH gradient (inside alkaline) induced transient Ni^{2+} accumulation. The mode of energy coupling has not been elucidated. Ionophore and inhibitor studies were not conclusive. A proton-coupled Ni^{2+} uptake system was suggested [262]. In *Methanothrix concilii* an apparently very rapid Ni^{2+} uptake was identified as an energy-independent semispecific adsorption of the cation to the cell surface [264]. For literature on Ni^{2+} transport in bacteria see refs. [260,263].

8.4. Potassium

Methanogenic bacteria contain high internal K^+ concentrations, ranging from 150 mM in some mesophilic strains, to about 1 M in some thermophilic strains, up to 3 M in the extreme thermophile *Methanopyrus* [22,175,265,266]. The anion of K^+ in thermophilic methanogens is the trianionic cyclic 2,3-diphosphoglycerate [266–268]; this compound has been found in *Methanothermus fervidus* ($[K^+]_{in} \approx 1$ M) at a concentration of 300 mM [266] and in *Methanopyrus kandleri* ($[K^+]_{in} \approx 3$ M) at a concentration of about 1 M [21,22]. The K^+ salt of the 2,3-diphosphoglycerate has been implicated as thermostabilizer of proteins in these extremely thermophilic methanogens [266] (see also refs. [85,88]). Furthermore, a correlation of the K^+ content of the cells and the presence of acidic ribosomal proteins was suggested [265].

The mechanism of active K^+ transport has been studied in cultures of *Methanobacterium thermoautotrophicum* growing on media low in K^+ (<1 mM) [269]. The organism accumulated the cation up to a concentration gradient (in/out) of 60 000. Under these conditions the membrane potential was found to be -180 mV. Addition of valinomycin caused the K^+ gradient to collapse by a factor of 500–1000, which is the accumulation factor expected from the existing membrane potential (-180 mV). The data indicate that at high external K^+ concentration (>1 mM) K^+ accumulation can be explained by an electrogenic uniport mechanism in response to the membrane potential. At low external K^+ concentrations (<1 mM K^+) a different mechanism for K^+ uptake must be operative. Two possible mechanism are: (i) K^+ is accumulated in a H^+ (cation) symport system causing electrogenic K^+ uptake to higher values at a given membrane potential; (ii) K^+ transport is coupled directly to ATP hydrolysis. A K^+-translocating ATPase, induced at low K^+ concentration, has been studied in detail in *Escherichia coli* [270,271]. Preliminary experiments argue against the presence of an *Escherichia-coli* type K^+-ATPase in *Methanobacterium thermoautotrophicum*: No K^+ stimulation of ATPase activity could be detected in membranes of cells after growth on low K^+. Furthermore, immunoblotting of solubilized membrane proteins does not show a specific reaction against the purified A, B, and C subunits of *Escheria coli* K^+-ATPase. Instead, a prominent membrane protein ($\sim$30 kD) was induced after growth on K^+-limited medium [271].

Rubidium uptake via the K^+ transport system was measured in K^+ depleted cells of *Methanospirillum hungatei* [272]. At low external Rb^+ concentration the accumulation gradient was significantly larger (2300-fold) than could be explained by the existing membrane potential (~ -120 mV). The driving force for Rb^+ uptake has not been elucidated.

8.5. Phosphate

Phosphate uptake was studied in phosphate-limited chemostat cultures of *Methanobacterium thermoautotrophicum*. In cells adapted to nmolar phosphate concentrations the presence of a high-affinity transport system ($K_m = 25$ nM; $V_{max} = 60$ nmol/min × mg) was demonstrated [273].

9. Energetics of Archaeoglobus and Pyrococcus – non-methanogenic thermophilic archaea related to methanogens

Based upon rRNA sequence comparison, the genera *Archaeoglobus* and *Pyrococcus* are closely related to the methanogens. *Archaeoglobus* is related to *Methanosarcina*, and *Pyrococcus* branches off between the extreme thermophiles *Methanococcus* and

Methanopyrus (Fig. 1). In this section of the review we report recent data on the energy metabolism of *Archaeoglobus fulgidus* and *Pyrococcus furiosus*.

9.1. Energetics of Archaeoglobus fulgidus

In 1987 Stetter and coworkers isolated the thermophilic archaebacterium *Archaeoglobus fulgidus*, which gains its energy by sulfate reduction [274,275] (see also refs. [276,277]). Surprisingly, the organism showed fluorescence under an ultraviolet microscope, which is typical for the methanogenic coenzyme factor F_{420}. Further studies established the presence of the "methanogenic" coenzymes factor F_{420}, tetrahydromethanopterin (H_4MPT) and methanofuran (MFR) in *Archaeoglobus fulgidus* [53,274,278]. Coenzyme M, 7-mercaptoheptanoylthreonine phosphate (H–S–HTP) and factor F_{430} could not be detected, indicating that the organism is not able to gain energy by methanogenesis. *Archaeoglobus fulgidus* grows at 83°C (temperature optimum) with lactate and sulfate forming CO_2 and H_2S as sole products:

$$CH_3CHOHCOO^- + 1.5SO_4^{2-} + 4H^+ \rightarrow 3CO_2 + 1.5H_2S + 3H_2O,$$
$$\Delta G^{o\prime} = -150 \text{ kJ/mol lactate}.$$

The fermentation involves complete oxidation of lactate to CO_2, generating reducing equivalents, which reduce sulfate to H_2S:

$$CH_3CHOHCOO^- + 4H^+ \rightarrow 3CO_2 + 12[H], \quad 1.5SO_4^{2-} + 12[H] \rightarrow 1.5H_2S + H_2O$$

9.1.1. Acetyl–CoA oxidation to CO_2 via a modified acetyl–CoA/carbon monoxide dehydrogenase pathway

The pathway of lactate oxidation was elucidated by Thauer and coworkers [279–281]: lactate is oxidized to CO_2 via pyruvate and acetyl–CoA involving a membrane-bound lactate dehydrogenase and a pyruvate:ferredoxin oxidoreductase. Oxidation of acetyl–CoA to $2CO_2$ proceeds via a modified acetyl–CoA/carbon monoxide dehydrogenase (CO–DH) pathway rather than by the citric acid cycle (for literature see ref. [209]) as indicated by the enzyme activities found in cell-free extracts [279–281]: acetyl–CoA is cleaved to a coenzyme-bound methanol (methyl–H_4MPT) and an enzyme-bound carbon monoxide [CO]; [CO] is oxidized to CO_2. Both the C–C cleavage reaction and the oxidation of [CO] are catalyzed by CO–DH, which shows high activities in *Archaeoglobus fulgidus* [281]. CO–DH is membrane associated. Methyl–H_4MPT is further oxidized to CO_2 via reactions shown for methanol oxidation to CO_2 in methanogens. The following enzyme activities have been detected [279–281] (see Fig. 13): methylene–H_4MPT reductase and methylene–H_4MPT dehydrogenase, which both use factor F_{420} as electron acceptor, methenyl–H_4MPT cyclohydrolase, formyl–H_4MPT:MFR formyltransferase and formyl–MFR dehydrogenase. Furthermore the C–C cleavage reaction of acetyl–CoA has been demonstrated [281]. Coenzyme F_{420}-dependent methylene–H_4MPT reductase and formyl–MFR dehydrogenase have been purified from *Archaeoglobus fulgidus* [280]. The N-terminal amino-acid sequence of the former enzyme

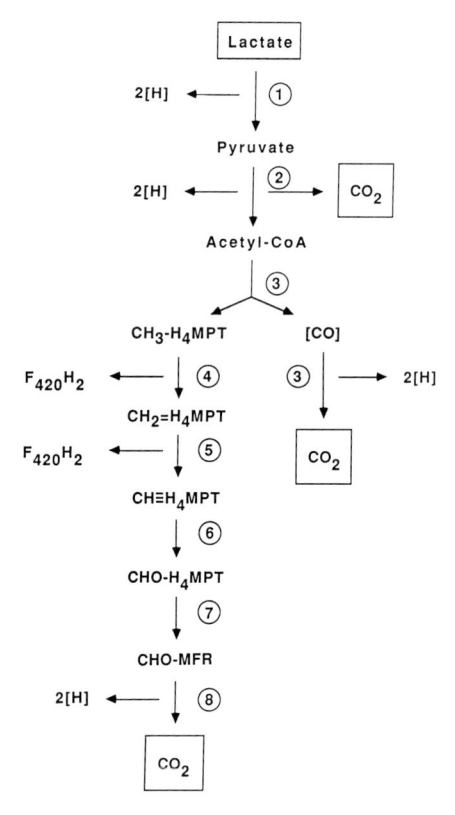

Fig. 13. Proposed pathway of lactate oxidation to 3CO$_2$ in *Archaeoglobus fulgidus*. CoA, coenzyme A; H$_4$MPT, tetrahydromethanopterin; MFR, methanofuran; CH$_3$–H$_4$MPT, methyl–H$_4$MPT; CH$_2$=H$_4$MPT, methylene–H$_4$MPT; CH≡H$_4$MPT, methenyl–H$_4$MPT; CHO–H$_4$MPT, formyl–H$_4$MPT, CHO–MFR, formyl–MFR; [CO], CO bound to carbon monoxide dehydrogenase; F$_{420}$H$_2$, reduced coenzyme factor F$_{420}$. Numbers in circles refer to enzymes involved: (1) lactate dehydrogenase; (2) pyruvate:ferredoxin oxidoreductase; (3) carbon monoxide dehydrogenase; (4) methylene–H$_4$MPT reductase; (5) methylene–H$_4$MPT dehydrogenase; (6) methenyl–H$_4$MPT cyclohydrolase; (7) formyl–H$_4$MPT:MFR formyltransferase; (8) formyl–MFR dehydrogenase.

was determined and found to be similar to that of methylene–H$_4$MPT reductase from various methanogens [280].

Thus, acetyl–CoA oxidation via the acetyl–CoA pathway in *Archaeoglobus fulgidus* differs from that of eubacterial sulfate reducers in several respects: It involves tetrahydromethanopterin rather than tetrahydrofolate (H$_4$F) as C$_1$ carrier, and formyl–methanofuran rather than free formate as an intermediate. Furthermore, coenzyme F$_{420}$ serves as electron acceptor of two dehydrogenases. In eubacterial sulfate reducers the oxidation of acetyl–CoA to CO$_2$ involves the exergonic formyl–H$_4$F conversion to formate and H$_4$F, which is catalyzed by formyl–H$_4$F synthetase; this reaction is coupled with ATP synthesis by the mechanism of substrate level phosphorylation (for literature see refs. [90,209]). The different mechanism of formyl–H$_4$MPT conversion to

CO$_2$ in *Archaeoglobus fulgidus* and the absence of a formyl–H$_4$MPT synthetase exclude ATP formation in this organism via substrate level phosphorylation in this pathway. *Archaeoglobus fulgidus* has been shown to form small amounts of methane during growth. This "mini methane formation" may be explained by unspecific reduction of coenzyme-bound CH$_3$ groups generated in the pathway as has been proposed for eubacterial sulfate reducers [282]. *Archaeoglobus fulgidus* does not use acetate instead of lactate, which is probably due to the lack of acetate activating enzymes [281].

The reducing equivalents generated during lactate oxidation are used to reduce sulfate to hydrogen sulfide. *Archaeoglobus fulgidus* contains ATP sulfurylase (sulfate adenylyltransferase), pyrophosphatase, adenylylsulfate (APS) reductase and bisulfite reductase indicating that sulfate reduction involves adenosine phosphosulfate (APS) and sulfite as intermediates [283,284]. These compounds serve as terminal electron acceptors. The redox potentials of the APS/SO$_3^{2-}$ couple ($E^{o\prime}$ = -60 mV) and the SO$_3^{2-}$/H$_2$S couple ($E^{o\prime}$ = -105 mV) are positive enough to allow the oxidation of pyruvate ($E^{o\prime}$ = -500 mV), of reduced F$_{420}$ ($E^{o\prime}$ = -350 mV), of formyl–methanofuran ($E^{o\prime}$ = -497 mV) and of carbon monoxide ($E^{o\prime}$ = -524 mV). The redox potential differences of the various redox reactions are high enough to allow ATP formation via the mechanism of electron transport phosphorylation (ETP). A lipophilic menaquinone, an indicator for ATP synthesis via ETP in anaerobic bacteria, has been isolated from membranes of *Archaeoglobus fulgidus* [285]. In summary, both the mechanism of dissimilatory sulfate reduction and that of ATP synthesis in this archaeon appear to be very similar to those operative in eubacterial (bacterial) sulfate reducers (for a recent review see ref. [286]).

It has been proposed that *Archaeoglobus fulgidus* is a biochemical missing link between the sulfur-dependent, extremely thermophilic archaea, e.g. Thermococcales, and the thermophilic methanogens [287]. However, very recent results on the phylogenetic position of *Archaeoglobus fulgidus* indicate that the organism is more related to *Methanosarcina* [20] and that the extreme thermophile *Methanopyrus* rather than the Thermococcales represents the deepest branch of the euryarchaeota [23] (Fig. 1).

9.2. Energetics of Pyrococcus furiosus

The genus *Pyrococcus* (species: *Pyrococcus furiosus* [288], *Pyrococcus woesei* [289]) and the genus *Thermococcus* (species: *Thermococcus celer* [290,291], *Thermococcus stetteri* [292], *Thermococcus litoralis* [293]) belong to the order of Thermococcales. They are strictly anaerobic, extremely thermophilic organisms [19] with optimal growth temperatures at 100°C (*Pyrococcus furiosus*) or at 87°C (*Thermococcus celer*). The organisms grow on complex media containing tryptone, yeast extract or casein. Also, growth of several organisms on starch and maltose has been described. Growth of the organisms in closed bottles was found to be strongly stimulated by elemental sulfur, which serves as an electron acceptor (see below). With the exception of *Pyrococcus furiosus* the mechanism of energy conservation has not been elucidated. ATP formation by sulfur respiration or by unknown modes of fermentation have been proposed.

Within the Thermococcales the metabolism of *Pyrococcus furiosus* has been studied in detail. The organism has been isolated in 1986 by Fiala and Stetter and has been shown to grow on tryptone, casein and yeast extract and on maltose and starch, respectively [288]. Recently the metabolism of *Pyrococcus furiosus* on the defined energy and carbon substrates maltose and pyruvate was studied. Fermentation balances and enzyme activities in cell extracts were determined [294,295]. It was found that both maltose and pyruvate were fermented to acetate, CO_2 and H_2. In growing cultures one mol of maltose was converted to 3–4 mol acetate, 3–4 mol CO_2 and 6–7 mol H_2 [295].

9.2.1. Novel sugar degradation pathway in Pyrococcus furiosus

Based on enzyme activities detected in cell extracts, evidence was presented that maltose is degraded by *Pyrococcus furiosus* via a novel sugar fermentation pathway (Fig. 14): Maltose is most likely split into two glucose molecules by an active cyctoplasmatic α-glucosidase, which has been purified from both *Pyrococcus furiosus* [296] and *Pyrococcus woesei* [297]. The conversion of glucose to pyruvate is catalyzed by reactions of a modified non-phosphorylated Entner–Doudoroff pathway similar to that described for glucose catabolism of the aerobic thermophilic archaea *Sulfolobus acidocaldarius* and *Thermoplasma acidophilum* [298–300] (see also Chapter 1 by Danson in this volume). Cell extracts of *Pyrococcus furiosus* contained glucose:ferredoxin (methylviologen) oxidoreductase, 2-keto-3-deoxygluconate (KDG) alolase, glyceraldehyde:ferredoxin (benzylviologen) oxidoreductase, 2-phosphoglycerate-forming glycerate kinase, enolase and pyruvate kinase [295,301] (Fig. 14). Gluconate dehydratase has not been detected so far in *Pyrococcus*. Glyceraldehyde:ferredoxin oxidoreductase has been purified from *Pyrococcus furiosus*. It is a novel tungsten–iron–sulfur protein [301,302] (see also ref. [303]).

The pathway of glucose oxidation to pyruvate found in *Pyrococcus furiosus* differs from that of *Thermoplasma* and *Sulfolobus* in that the dehydrogenation of both glucose and of glyceraldehyde are coupled with the reduction of ferredoxin rather than of pyridine nucleotides [301]. Since the subsequent dehydrogenation of pyruvate to acetyl–CoA is catalyzed in *Pyrococcus furiosus* by a pyruvate:ferredoxin oxidoreductase [294] these findings indicate that all reducing equivalents generated during glucose oxidation are released as molecular hydrogen via ferredoxin-dependent hydrogenase [304]. In accordance, $\sim$7 mol H_2/mol maltose were formed during growth [295].

9.2.2. Sugar degradation to acetate, CO_2 and H_2 via a novel fermentation pathway

Further conversion of pyruvate to acetate, CO_2 and H_2 is catalyzed by three enzymes: pyruvate:ferredoxin oxidoreductase, hydrogenase and ADP-forming acetyl–CoA synthetase. These enzyme activities were found in both maltose- and pyruvate-grown cells of *Pyrococcus furiosus* [294,295]. Phosphate acetyltransferase and acetate kinase activity could not be detected. These findings indicate that acetate formation from acetyl–CoA and the synthesis of ATP from ADP and phosphate (P_i) in *Pyrococcus furiosus* are catalyzed by a single enzyme, an ADP-forming acetyl–CoA synthetase:

$$\text{acetyl–CoA} + \text{ADP} + P_i \rightarrow \text{acetate} + \text{ATP} + \text{CoA–SH}.$$

From the postulated fermentation pathway (Fig. 14) it is probable that glucose degradation to pyruvate is not coupled with net ATP synthesis. Thus, ADP-forming acetyl–CoA synthetase appears to be the only energy-conserving site during maltose (and pyruvate) fermentation, which is in accordance with the growth yield data [294,295]. The enzyme

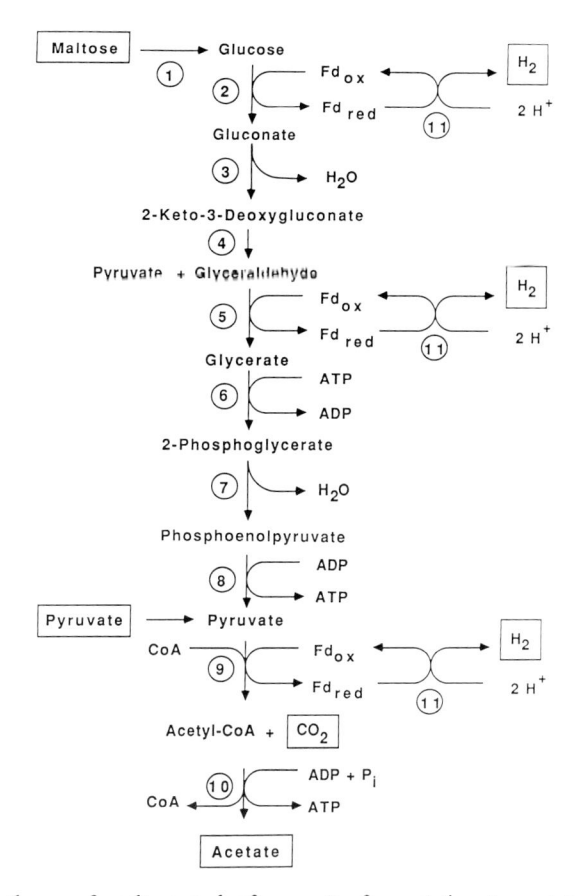

Fig. 14. Proposed pathway of maltose and of pyruvate fermentation to acetate, H_2 and CO_2 in *Pyrococcus furiosus*. Fd_{ox}, oxidized ferredoxin; Fd_{red}, reduced ferredoxin; CoA, coenzyme A. Numbers in circles refer to enzymes involved: (1) α-glucosidase [296]; (2) glucose:ferredoxin oxidoreductase; (3) gluconate dehydratase (this enzyme has not been detected so far in *Pyrococcus furiosus*); (4)· 2-keto-3-deoxygluconate aldolase; (5) glyceraldehyde:ferredoxin oxidoreductase; (6) glycerate kinase (2-phosphoglycerate forming); (7) enolase; (8) pyruvate kinase; (9) pyruvate:ferredoxin oxidoreductase; (10) ADP-forming acetyl–CoA synthetase; (11) hydrogenase [311].

therefore represents a novel prokaryotic enzyme catalyzing ATP synthesis from ADP and P_i via the mechanism of substrate level phosphorylation.

Further studies indicate that an ADP-forming acetyl–CoA synthetase is also operative in other extremely thermophilic archaea (*Pyrococcus woesei, Thermococcus celer, Hyperthermus butylicus, Desulfurococcus amylolyticus*), which form acetate as end product of their fermentation [305]. In contrast, in acetate forming (eu)bacteria, acetate formation from acetyl–CoA and the synthesis of ATP from ADP and P_i are catalyzed by two enzymes: phosphate acetyltransferase and acetate kinase. This holds true for the extremely thermophilic (eu)bacterium, *Thermotoga maritima* [305], which ferments glucose to lactate and acetate as major products [306]. These findings indicate that ADP-forming acetyl–CoA synthetase represents a typical archaeal enzyme rather than an enzyme adapted to high temperatures. In accordance we could show that the aerobic mesophilic archaeon *Halobacterium saccharovorum,* which forms significant amounts of acetate during growth on glucose [307], also contained ADP-dependent acetyl–CoA synthetase rather than phosphate acetyltransferase and acetate kinase. The presence of ADP-forming acetyl–CoA synthetase has also been postulated for the aerobic archaeon *Thermoplasma acidophilum* (see ref. [300]).

9.2.3. Open questions

Cell extracts of *Pyrococcus* contain low activities of fructose-1,6-bisphosphate aldolase [295] and NAD(P)-dependent glyceraldehyde-3-phosphate dehydrogenase [301, 308a], which probably have anabolic functions (see ref. [308b]). Glyceraldehyde-3-phosphate dehydrogenase has been purified from *Pyrococcus woesei* [308a] (see Chapter 7 by Hensel in this volume).

Sulfur respiration in the metabolism of *Pyrococcus* has been discussed in order to explain the higher growth yields on various substrates in the presence of elemental sulfur. However, as suggested first by Fiala and Stetter [288], sulfur stimulates growth most likely via removal of the inhibitory fermentation product H_2 by forming H_2S rather than by an increased formation of ATP via sulfur respiration: the stimulatory effect of sulfur during growth on peptone or pyruvate could almost be overcome by gassing the cultures in open fermenter systems with N_2 in order to remove H_2 [288,294]. The same result was obtained when *Pyrococcus furiosus* was co-cultured with the thermophilic methanogen *Methanococcus jannaschii,* which consumes H_2 and CO_2 to form methane [309]. However, more studies are necessary to understand the role of sulfur [310] in the energy metabolism of Thermococcales.

Acknowledgements

The work performed in the author's laboratory both in Marburg (FRG) and in Berlin (FRG) was supported by grants from the Deutsche Forschungsgemeinschaft, from the Bundesministerium für Forschung und Technologie, and from the Fonds der Chemischen Industrie. The author thanks Prof. Thauer (Marburg) for helpful discussions and Prof. Friedmann (Chicago) for reading the manuscript.

References

[1] Wolfe, R.S. (1971) Adv. Microbial Physiol. 6, 107–146.
[2] Balch, W.E., Fox, G.E., Magrum, L.J., Woese, C.R. and Wolfe, R.S. (1979) Microbiol. Rev. 43, 260–296.
[3] Jones, W.J., Nagle, D.P. and Whitman, W.B. (1987) Microbiol. Rev. 51, 135–177.
[4] Woese, C.R. (1987) Microbiol. Rev. 51, 221–271.
[5] Woese, C.R., Kandler, O. and Wheelis, M.L. (1990) Proc. Natl. Acad. Sci. U.S.A. 87, 4576–4579.

[6] Oremland, R.S. (1988) In: Biology of Anaerobic Microorganisms (Zehnder, A.J.B., Ed.), pp. 641–705, Wiley, New York.
[7] DiMarco, A.A., Bobik, T.A. and Wolfe, R.S. (1990) Annu. Rev. Biochem. 59, 355–394.
[8] Thauer, R.K., Jungermann, K. and Decker, K. (1977) Bacteriol. Rev. 41, 100–180.
[9] Kell, D.B., Doddema, H.J., Morris, J.G. and Vogels, G.D. (1981) In: Proc. 3rd Int. Symp. on Microbial Growth on C-1 Compounds (Dalton, H., Ed.), pp. 159–170, Hyden, London.
[10] Daniels, L., Sparling, R. and Sprott, G.D. (1984) Biochim. Biophys. Acta 768, 113–163.
[11] Blaut, M. and Gottschalk, G. (1985) Trends Biochem. Sci. 10, 486–489.
[12] Keltjens, J.T. and van der Drift, C. (1986) FEMS Microbiol. Rev. 39, 259–303.
[13] Vogels, G.D., Keltjens, J.T. and van der Drift, C. (1988) In: Biology of Anaerobic Microoranisms (Zehnder, A.J.B., Ed.), pp. 707–770, Wiley, New York.
[14] Hauska, G. (1988) Trends Biochem. Sci. 13, 2–4.
[15] Lancaster Jr., J.R. (1989) J. Bioenerg. Biomembr. 21, 717–740.
[16] Blaut, M., Müller, V. and Gottschalk, G. (1990) In: The Bacteria, Vol. 12 (Krulwich, T.A., Ed.), pp. 505–537, Academic Press, New York.
[17] Gottschalk, G. and Blaut, M. (1990) Biochim. Biophys. Acta 1018, 263–266.
[18] Thauer, R.K. (1990) Biochim. Biophys. Acta 1018, 256–259.
[19] Stetter, K.O., Fiala, G., Huber, G., Huber, R. and Segerer, A. (1990) FEMS Microbiol. Rev. 75, 117–124.
[20] Woese, C.R., Achenbach, L., Rouvière, P. and Mandelco, L. (1991) System. Appl. Microbiol. 14, 364–371.
[21] Huber, R., Kurr, M., Jannasch, H.W. and Stetter, K.O. (1989) Nature 342, 833–834.
[22] Kurr, M., Huber, R., König, H., Jannasch, H.W., Fricke, H., Trincone, A., Kristjansson, J.K. and Stetter, K.O. (1991) Arch. Microbiol. 156, 239–247.
[23] Burggraf, S., Stetter, K.O., Rouvière, P. and Woese, C.R. (1991) System. Appl. Microbiol. 14, 346–351.
[24] Murray, P.A. and Zinder, S.H. (1984) Nature 312, 284–288.
[25] Belay, N., Sparling, R. and Daniels, L. (1984) Nature 312, 286–288.
[26] Bomar, M., Knoll, K. and Widdel, F. (1985) FEMS Microbiol. Ecol. 31, 47–55.
[27] Daniels, L., Belay, N. and Rajagopal, B.S. (1986) Appl. Environ. Microbiol. 51, 703–709.
[28] Schönheit, P., Moll J. and Thauer, R.K. (1979) Arch. Microbiol. 123, 105–107.
[29] Scherer, P. and Sahm, H. (1981) Acta Biotechnol. 1, 57–65.
[30] Whitman, W.B., Ankwanda, E. and Wolfe, R.S. (1982) J. Bacteriol. 149, 852–863.
[31] Mathrani, I.M., Boone, D.R., Mah, R.A., Fox, G.E. and Lau, P.P. (1988) Int. J. System. Bacteriol. 38, 139–142.
[32] Zeikus, J.G. (1977) Bacteriol. Rev. 41, 514–541.
[33] Mah, R.A., Ward, D.M., Baresi, L. and Glass, T.L. (1977) Annu. Rev. Microbiol. 31, 309–341.
[34] Fuchs, G. and Stupperich, E. (1986) System. Appl. Microbiol. 7, 364–369.
[35] Bock, A.-K., Prieger-Kraft, A. and Schönheit, P. (1993) Arch. Microbiol. 160, in press.
[35a] Bock, A.-K. (1991) Diploma Thesis, Free University, Berlin.
[36] Dubach, A.C. and Bachofen, R. (1985) Experientia 41, 441–446.
[37] Garcia, J.L. (1990) FEMS Microbiol. Rev. 87, 297–308.
[38] Widdel, F. (1986) Appl. Environ. Microbiol. 51, 1056–1062.
[39] Widdel, F., Rouvière, P. and Wolfe, R.S. (1988) Arch. Microbiol. 150, 477–481.
[40] Zellner, G. and Winter, J. (1987) FEMS Microbiol. Lett. 44, 323–328.
[41] Blaut, M. and Gottschalk, G. (1984) Eur. J. Biochem. 141, 217–222.
[42] Miller, T.L. and Wolin, M.J. (1985) Arch. Microbiol. 141, 116–122.
[43] Müller, V., Blaut, M. and Gottschalk, G. (1986) Appl. Environ. Microbiol. 52, 269–274.
[44] König, H. and Stetter, K.O. (1982) Zbl. Bakt. Hyg. I Abt. Orig. C3, 478–490.

[45] Hippe, H., Caspari, D., Fiebig, K. and Gottschalk, G. (1979) Proc. Natl. Acad. Sci. U.S.A. 76, 494–498.
[46] Zinder, S.H. (1990) FEMS Microbiol. Rev. 75, 125–138.
[47] Daniels, L., Fuchs, G., Thauer, R.K. and Zeikus, J.G. (1977) J. Bacteriol. 132, 118–126.
[48] O'Brien, J.M., Wolkin, R.H., Moench, T.T., Morgan, J.B. and Zeikus, J.G. (1984) J. Bacteriol. 158, 373–375.
[49] Kiene, R.P., Oremland, R.S., Catena, A., Miller, L.G. and Capone, D.G. (1986) Appl. Environ. Microbiol. 52, 1037–1045.
[50] Barker, H.A. (1956) Bacterial Fermentations, Wiley, New York.
[51] Keltjens, J.T. and Vogels, G.D. (1988) BioFactors 1, 95–103.
[52] Keltjens, J.T., te Brömmelstroet, B.W., Kengen, S.W.M., van der Drift, C. and Vogels, G.D. (1990) FEMS Microbiol. Rev. 87, 327–332.
[53] White, R.H. (1988) J. Bacteriol. 170, 4594–4597.
[54] Karrasch, M., Börner, G., Enssle, M. and Thauer, R.K. (1990) Eur. J. Biochem. 194, 367–372.
[55] Karrasch, M., Börner, G. and Thauer, R.K. (1990) FEBS Lett. 274, 48–52.
[56] Börner, G., Karrasch, M. and Thauer, R.K. (1989) FEBS Lett. 244, 21–25.
[57] Börner, G., Karrasch, M. and Thauer, R.K. (1991) FEBS Lett. 290, 31–34.
[58] Donnelly, M.I. and Wolfe, R.S. (1986) J. Biol. Chem. 261, 16653–16659.
[59] Breitung, J. and Thauer, R.K. (1990) FEBS Lett. 275, 226–230.
[60] DiMarco A.A., Donnelly, M.I. and Wolfe, R.S. (1986) J. Bacteriol. 168, 1372–1377.
[61] Enßle, M., Zirngibl, C., Linder, D. and Thauer, R.K. (1991) Arch. Microbiol. 155, 483–490.
[62] Hartzell, P.L., Zvilius, G., Escalante-Semarena, J.C. and Donelly, M.I. (1985) Biochem. Biophys. Res. Commun. 133, 884–890.
[63] Mukhopadhyay, B. and Daniels, L. (1989) Can. J. Microbiol. 35, 499–507.
[64] te Brömmelstroet, B.W., Hensgens, C.H.M., Keltjens, J.T., van der Drift, C. and Vogels, G.D. (1991) Biochim. Biophys. Acta 1073, 77–84.
[65] Zirngibl, C., Hedderich, R. and Thauer, R.K. (1990) FEBS Lett. 261, 112–116.
[66] Schwörer, B. and Thauer, R.K. (1991) Arch. Microbiol. 155, 459–465.
[67] Ma, K. and Thauer, R.K. (1990) FEMS Microbiol. Lett. 70, 119–124.
[68] Ma, K. and Thauer, R.K. (1990) Eur. J. Biochem. 191, 187–193.
[69a] te Brömmelstroet, B.W., Hensgens, C.M.H., Keltjens, J.T., van der Drift, C. and Vogels, G.D. (1990) J. Biol. Chem. 265, 1852–1857.
[69b] Fischer, R., Gärtner, P., Yeliseev, A. and Thauer, R.K. (1992) Arch. Microbiol. 158, 208–217.
[70] Ellefson, W.L., Whitman, W.B. and Wolfe, R.S. (1982) Proc. Natl. Acad. Sci. U.S.A. 79, 3707–3710.
[71] Ellermann, J., Hedderich, R., Böcher, R. and Thauer, R.K. (1988) Eur. J. Biochem. 172, 669–677.
[72] Rospert, S., Linder, D., Ellermann, J. and Thauer, R.K. (1990) Eur. J. Biochem. 194, 871–877.
[73] Friedmann, H.C., Klein, A. and Thauer, R.K. (1990) FEMS Microbiol. Rev. 87, 339–348.
[74] Hedderich, R., Berkessel, A. and Thauer, R.K. (1990) Eur. J. Biochem. 193, 255–261.
[75] Graf, E.G. and Thauer, R.K. (1981) FEBS Lett. 136, 165–169.
[76a] Fox, J.A., Livingston, D.J., Orme-Johnson, W.H. and Walsh, C.T. (1987) Biochemistry 26, 4219–4227.
[76b] Alex, L.A., Reeve, J.N., Orme-Johnson, W.H. and Walsh, C.T. (1990) Biochemistry 29, 7237–7244.
[77] Muth, E., Mörschel, E. and Klein, A. (1987) Eur. J. Biochem. 169, 571–577.
[78] Halboth, S. and Klein, A. (1992) Mol. Gen. Genet. 233, 217–224.
[79] Fiebig, K. and Friedrich, B. (1989) Eur. J. Biochem. 184, 79–88.
[80] Baron, S.F. and Ferry, J.G. (1989) J. Bacteriol. 171, 3846–3853.

[81] Coremans, J.M.C.C., van der Zwaan, J.W. and Albracht, S.P.J. (1989) Biochim. Biophys. Acta 997, 256–267.
[82] Reeve, J.N., Beckler, G.S., Cram, D.S., Hamilton, P.T., Brown, J.W., Krzycki, J.A., Kolodziej, A.F., Alex, L., Orme-Johnson, W.H. and Walsh, C.T. (1989) Proc. Natl. Acad. Sci. U.S.A. 86, 3031–3035.
[83] Steigerwald, V.J., Beckler, G.S and Reeve, J.N. (1990) J. Bacteriol. 172, 4715–4718.
[84] Reeve, J.N. and Beckler, G.S. (1990) FEMS Microbiol. Rev. 87, 419–424.
[85] Ma, K., Linder, D., Stetter, K.O. and Thauer, R.K. (1991) Arch. Microbiol. 155, 593–600.
[86] Ma, K., Zirngibl, C., Linder, D., Stetter, K.O. and Thauer, R.K (1991) Arch. Microbiol. 156, 43–48.
[87] Rospert, S., Breitung, J., Ma, K., Schwörer, B., Zirngibl, C., Thauer, R.K., Linder, D., Huber, R. and Stetter, K.O. (1991) Arch. Microbiol. 156, 49–55.
[88] Breitung, J., Schmitz, R.A., Stetter, K.O. and Thauer, R.K. (1991) Arch. Microbiol. 156, 517–524.
[89] Thauer, R.K. and Morris, J.G. (1984) In: The Microbe: Part II. Prokaryotes and Eukaryotes (Kelly, D.P. and Carr, N.G., Eds.), pp. 123–168, Cambridge University Press, Cambridge.
[90] Thauer, R.K. (1988) Eur. J. Biochem. 176, 497–508.
[91] Roberton, A.M. and Wolfe, R.S. (1970) J. Bacteriol. 102, 43–51.
[92] Zehnder, A.J.B. and Wuhrmann, K. (1977) Arch. Microbiol. 111, 199–205.
[93] Weimer, P.J. and Zeikus, J.G. (1978) Arch. Microbiol. 119, 49–57.
[94] Schönheit, P., Moll, J. and Thauer, R. K. (1980) Arch. Microbiol. 127, 59–65.
[95] Schauer, N.L. and Ferry, J.G. (1980) J. Bacteriol. 142, 800–807.
[96] Fardeau, M.-L. and Belaich, J.-P. (1986) Arch. Microbiol. 144, 381–385.
[97] Fardeau, M.-L., Peillex, J.-P. and Belaich, J.-P. (1987) Arch. Microbiol. 148, 128–131.
[98] Decker, K., Jungermann, K. and Thauer, R.K. (1970) Angew. Chemie (Int. Edn. Engl.) 9, 138–158.
[99] Stouthamer, A.H. (1979) In: Microbial Biochemistry (Quayle, J.R., Ed.), pp. 1–47, University Park Press, Baltimore, MD.
[100] Kaesler, B. (1986) Diploma Thesis, Philipps-University, Marburg.
[101] Mitchell, P. (1966) Biol. Rev. 41, 445–502.
[102] Blaut, M. and Gottschalk, G. (1984) FEMS Microbiol. Lett. 24, 103–107.
[103] Kaesler, B. and Schönheit, P. (1988) Eur. J. Biochem. 174, 189–197.
[104] Escalante-Semerena, J.C. and Wolfe, R.S. (1984) J. Bacteriol. 158, 721–726.
[105] Kaesler, B. and Schönheit, P. (1989) Eur. J. Biochem. 184, 223–232.
[106] Winner, C. and Gottschalk, G. (1989) FEMS Microbiol. Lett. 65, 259–264.
[107] Blaut, M., Müller, V. and Gottschalk, G. (1987) FEBS Lett. 215, 53–57.
[108] Müller, V., Blaut, M. and Gottschalk, G. (1987) Eur. J. Biochem. 162, 461–466.
[109] Perski, H.J., Moll, J. and Thauer, R.K. (1981) Arch. Microbiol. 130, 319–321.
[110] Perski, H.J., Schönheit, P. and Thauer, R.K. (1982) FEBS Lett. 143, 323–326.
[111] Blaut, M., Müller, V., Fiebig, K. and Gottschalk, G. (1985) J. Bacteriol. 164, 95–101.
[112] Peinemann, S., Blaut, M. and Gottschalk, G. (1989) Eur. J. Biochem. 186, 175–180.
[113] Blaut, M., Peinemann, S., Deppenmeier, U. and Gottschalk, G. (1990) FEMS Microbiol. Rev. 87, 367–372.
[114] Peinemann, S., Hedderich, R., Blaut, M., Thauer, R.K. and Gottschalk, G. (1990) FEBS Lett. 263, 57–60.
[115] Deppenmeier, U., Blaut, M., Mahlmann, A. and Gottschalk, G. (1990) Proc. Natl. Acad. Sci. U.S.A. 87, 9449–9453.
[116] Deppenmeier, U., Blaut, M., Mahlmann, A. and Gottschalk, G. (1990) FEBS Lett. 261, 199–203.
[117] Deppenmeier, U., Blaut, M. and Gottschalk, G. (1991) Arch. Microbiol. 155, 272–277.

[118] Muth, E. (1988) Arch. Microbiol. 150, 205–207.
[119] Lünsdorf, H., Niedrig, M. and Fiebig, K. (1991) J. Bacteriol. 173, 978–984.
[120] Baron, S.F. and Ferry, J.G. (1989) J. Bacteriol. 171, 3846–3853.
[121] Baron, S.F., Williams, D.S., May, H.D., Patel, P.S., Aldrich, H.C. and Ferry, J.G. (1989) Arch. Microbiol. 151, 307–313.
[122] Ossmer, R., Mund, T., Hartzell, P.L., Konheiser, U., Kohring, G.W., Klein, A., Wolfe, R.S., Gottschalk, G. and Mayer, F. (1986) Proc. Natl. Acad. Sci. U.S.A. 83, 5789–5792.
[123] Aldrich, H.C., Beimborn, D.B., Bokranz, M. and Schönheit, P. (1987) Arch. Microbiol. 147, 190–194.
[124] Mayer, F., Rohde, M., Salzmann, M., Jussofie, A. and Gottschalk, G. (1988) J. Bacteriol. 170, 1438–1444.
[125] Rogers, K.R., Gillies, K. and Lancaster Jr., J.R. (1988) Biochem. Biophys. Res. Commun. 153, 87–95.
[126] Dangel, W., Schulz, H., Diekert, G., König, H. and Fuchs, G. (1987) Arch. Microbiol. 148, 52–56.
[127] Schulz, H., Albracht, S.P.J., Coremans, J.M.C.C. and Fuchs, G. (1988) Eur. J. Biochem. 171, 589–597.
[128] Kühn, W., Fiebig, K., Hippe, H., Mah, R.A., Huser, B.A. and Gottschalk, G. (1983) FEMS Microbiol. Lett. 20, 407–410.
[129] Hausinger, R.P., Moura, I., Moura, J.J.G., Xavier, A.V., Santos, M.H., LeGall, J. and Howard, J.B. (1982) J. Biol. Chem. 257, 14192–14197.
[130] Terlesky, K.C. and Ferry, J.G. (1988) J. Biol. Chem. 263, 4080–4082.
[131] Bruschi, M., Bonicel, J., Hatchikian, E.C., Fardeau, M.L., Belaich, J.P. and Frey, M. (1991) Biochim. Biophys. Acta 1076, 79–85.
[132] Terlesky, K.C. and Ferry, J.G. (1988) J. Biol. Chem. 263, 4075–4079.
[133] Fischer, R. and Thauer, R.K. (1990) FEBS Lett. 269, 368–372.
[134] Hedderich, R., Albracht, S.P.J., Linder, D., Koch, J. and Thauer, R.K. (1992) FEBS Lett. 298, 65–68.
[135] Mountfort, D.O. (1978) Biochem. Biophys. Res. Commun. 85, 1346–1351.
[136] Mountfort, D.O., Mörschel, E., Beimborn, D.B. and Schönheit, P. (1986) J. Bacteriol. 168, 892–900.
[137] Mayer, F., Jussofie, A., Salzmann, M., Lübben, M., Rohde, M. and Gottschalk, G. (1987) J. Bacteriol. 169, 2307–2309.
[138] Inatomi, K.-I. (1986) J. Bacteriol. 167, 837–841.
[139] Scheel, E. and Schäfer, G. (1990) Eur. J. Biochem. 187, 727–735.
[140] Schneider, E. and Altendorf, K. (1987) Microbiol. Rev. 51, 477–497.
[141] Pedersen, P.L. and Carafoli, E. (1987) Trends Biochem. Sci. 12, 146–150.
[142] Pedersen, P.L. and Carafoli, E. (1987) Trends Biochem. Sci. 12, 181–186.
[143] Inatomi, K.-I., Eya, S., Maeda, M. and Futai, M. (1989) J. Biol. Chem. 19, 10954–10959.
[144] Lübben, M., Lünsdorf, H. and Schäfer, G. (1987) Eur. J. Biochem. 167, 211–219.
[145] Inatomi, K.-I., Maeda, M. and Futai, M. (1989) Biophys. Res. Commun. 162, 1585–1590.
[146] Nelson, N. and Taiz, L. (1989) Trends Biochem. Sci. 14, 113–116.
[147] Dharmavaram, R.M. and Koniski, J. (1987) J. Bacteriol. 169, 3921–3925.
[148] Dharmavaram, R.M. and Konisky, J. (1989) J. Biol. Chem. 264, 14085–14089.
[149] Carper, S.W. and Lancaster Jr., J.R. (1986) FEBS Lett. 200, 177–180.
[150] Crider, B.P., Carper, S.W. and Lancaster Jr., J.R. (1985) Proc. Natl. Acad. Sci. U.S.A. 82, 6793–6796.
[151] Al-Mahrouq, H.A., Carper, S.W. and Lancaster Jr., J.R. (1986) FEBS Lett. 207, 262–265.
[152] Šmigáň, P., Horovská, L. and Greksák, M. (1988) FEBS Lett. 242, 85–88.
[153] Lancaster Jr., J.R. (1986) FEBS Lett. 199, 12–18.

[154] Schönheit, P. and Beimborn, D.B. (1985) Eur. J. Biochem. 148, 545–550.
[155] Keltjens, J.T., van der Erp, E., Mooijaart, R.J., van der Drift, C. and Vogels, G.D. (1988) Eur. J. Biochem. 172, 471–476.
[156] Ellermann, J., Rospert, S., Thauer, R.K., Bokranz, M., Klein, A., Voges, M., and Berkessel, A. (1989) Eur. J. Biochem. 184, 63–68.
[157] Zeikus, J.G. and Wolfe, R.S. (1973) J. Bacteriol. 123, 461–467.
[158] Doddema, H.J., van der Drift, C., Vogels, G.D. and Veenhuis, M. (1979) J. Bacteriol. 140, 1081–1089.
[159] Sauer, F.D., Erfle, J.D. and Mahadevan, S. (1981) J. Biol. Chem. 256, 9843–9848.
[160] Doddema, H.J., Claesen, C.A., Kell, D.B., van der Drift, C. and Vogels, G.D. (1980) Biochem. Biophys. Res. Commun. 95, 1288–1293.
[161] Doddema, H.J., Hutten, T.J., van der Drift, C., and Vogels, G.D. (1978) J. Bacteriol. 136, 19–23.
[162] Harold, F. (1986) The Vital Force: A Study of Bioenergetics, Freeman, San Francisco.
[163] Aldrich, H.C., Beimborn, D.B. and Schönheit, P. (1987) Can. J. Microbiol. 33, 844–849.
[164] Sprott, G.D., Sowden, L.C., Colvin, J.R., Jarrell, K.F. and Beveridge, T.J. (1984) Can. J. Microbiol. 30, 394–604.
[165] Krämer, R. and Schönheit, P. (1987) Arch. Microbiol. 146, 370–376.
[166] Müller, V., Winner, C. and Gottschalk, G. (1988) Eur. J. Biochem. 178, 519–525.
[167] Kaesler, B. and Schönheit, P. (1989) Eur. J. Biochem. 186, 309–316.
[168] Becher, B., Müller, V. and Gottschalk, G. (1992) FEMS Microbiol. Lett. 91, 239–244.
[169] Mahlmann, A., Deppenmeier, U. and Gottschalk, G. (1989) FEMS Microbiol. Lett. 115–120.
[170] Gunsalus, R.G. and Wolfe, R.S. (1977) Biochem. Biophys. Res. Commun. 76, 790–795.
[171] Leigh, J.A., Rinehart Jr., K.L. and Wolfe, R.S. (1985) Biochemistry 24, 995–999.
[172] Bobik, T.A. and Wolfe, R.S. (1988) Proc. Natl. Acad. Sci. U.S.A. 85, 60–63.
[173] Bobik, T.A., DiMarco, A.A. and Wolfe, R.S. (1990) FEMS Microbiol. Rev. 87, 323–326.
[174] Romesser, J.A. and Wolfe, R.S. (1982) J. Bacteriol. 152, 840–847.
[175] Schönheit, P. and Perski, H.J. (1983) FEMS Microbiol. Lett. 20, 263–267.
[176] Schönheit, P. and Beimborn, D.B. (1985) Arch. Microbiol. 142, 354–361.
[177] Krulwich, T.A. (1983) Biochim. Biophys. Acta 726, 245–264.
[178] Schönheit, P. and Beimborn, D.B. (1986) Arch. Microbiol. 146, 181–185.
[179] Kaesler, B. (1989) Ph.D. Thesis, Philipps-University, Marburg.
[180] Booth, I.R. (1985) Microbiol. Rev. 49, 359–378.
[181] Sparling, R. and Daniels, L. (1986) J. Bacteriol. 168, 1402–1407.
[182] Chua, H.B. and Robinson, J.P. (1983) Arch. Microbiol. 135, 158–160.
[183] Ferry, J. G. (1990) FEMS Microbiol. Rev. 87, 377–382.
[184] Johnson, J.L., Bastian, N.R., Schauer, N.L., Ferry, J.G. and Rajagopalan K.V. (1991) FEMS Microbiol. Lett. 77, 213–216.
[185] Frimmer, U. and Widdel, F. (1989) Arch. Microbiol. 152, 479–483.
[186] Widdel, F. and Frimmer, U. (1993) In: Protocols for Archaeal Research (Robb, F.F., et al., Eds.), Cold Spring Harbor Laboratory Press, in press.
[187] Widdel, F. and Wolfe, R.S. (1989) Arch. Microbiol. 152, 322–328.
[188] Bleicher, K., Zellner, G. and Winter, J. (1989) FEMS Microbiol. Lett. 59, 307–312.
[189] Bleicher, K. and Winter, J. (1991) Eur. J. Biochem. 200, 43–51.
[190] Wood, H.G., Ragsdale, S.W. and Pezacka, E. (1986) FEMS Microbiol. Rev. 39, 345–362.
[192] Fuchs, G. (1986) FEMS Microbiol. Rev. 39, 181–213.
[192] Diekert, G. (1990) FEMS Microbiol. Rev. 87, 391–396.
[193] Geerligs, G., Schönheit, P. and Diekert, G. (1989) FEMS Microbiol. Lett. 57, 253–258.
[194] Heise, R., Müller, V. and Gottschalk, G. (1989) J. Bacteriol. 171, 5473–5478.
[195] Müller, V., Blaut, M., Heise, R., Winner, C. and Gottschalk, G. (1990) FEMS Microbiol. Rev. 87, 373–376.

[196] Wohlfahrt, G. and Diekert, G. (1991) Arch. Microbiol. 155, 378–381.
[197] Hugenholtz, J., Ivey, D.M. and Ljungdahl, L.G. (1987) J. Bacteriol. 169, 5845–5847.
[198] Wohlfarth, G., Geerligs, G. and Diekert, G. (1990) Eur. J. Biochem. 192, 411–417.
[199] Hugenholtz, J. and Ljungdahl, L.G. (1990) FEMS Microbiol. Rev. 87, 383–390.
[200] Heise, R., Reidlinger, J., Müller, V. and Gottschalk, G. (1991) FEBS Lett. 295, 119–122.
[201] Jablonski, P.E., Di Marco, A.A., Bobik, T.A., Cabell, M.C. and Ferry, J.G. (1990) J. Bacteriol. 172, 1271–1275.
[202] te Brömmelstroet, B.W., Hensgens, C.M.H., Geerts, W.J., Keltjens, J.T., van der Drift, C. and Vogels, G.D. (1990) J. Bacteriol. 172, 564–571.
[203] Müller, V., Blaut, M. and Gottschalk, G. (1988) Eur. J. Biochem. 172, 601–606.
[204] Kühn, W. and Gottschalk, G. (1983) Eur. J. Biochem. 135, 89–94.
[205] Jussofie, A. and Gottschalk, G. (1986) FEMS Microbiol. Lett. 37, 15–18.
[206] Kuchenbecker, R. and Hauska, G. (1988) In: EBEC Report, Vol. 5, p. 157, Aberystwyth, U.K.
[207] Müller, V., Kozianowski, G., Blaut, M. and Gottschalk, G. (1987) Biochim. Biophys. Acta 892, 207–212.
[208] Müller, V., Blaut, M. and Gottschalk, G. (1987) FEMS Microbiol. Lett. 43, 183–186.
[209] Thauer, R.K., Möller-Zinkhan, D. and Spormann, A. (1989) Annu. Rev. Microbiol. 43, 43–67.
[210] Huser, B.A., Wuhrmann, K. and Zehnder, A.J.B. (1982) Arch. Microbiol. 132, 1–9.
[211] Kenealy, W.R. and Zeikus, J.G. (1982) J. Bacteriol. 151, 932–941.
[212] Krzycki, J.A. and Zeikus, J.G. (1984) FEMS Microbiol. Lett. 25, 27–32.
[213] Krzycki, J.A., Lehman, L.J. and Zeikus, J.G. (1985) J. Bacteriol. 163, 1000–1006.
[214] Grahame, D.A. and Stadtman, T.C. (1987) Biochem. Biophys. Res. Commun. 147, 254–258.
[215] Fischer, R. and Thauer, R.K. (1988) FEBS Lett. 228, 249–253.
[216] Aceti, D.J. and Ferry, J.G. (1988) J. Biol. Chem. 263, 15444–15448.
[217] Lundie Jr., L.L. and Ferry, J.G. (1989) J. Biol. Chem. 264, 18392–18396.
[218] Krzycki, J.A. and Zeikus, J.G. (1984) J. Bacteriol. 158, 231–237.
[219] Abbanat, D.R. and Ferry, J.G. (1991) Proc. Natl. Acad. Sci. U.S.A. 88, 3272–3276.
[220] Terlesky, K.C., Nelson, M.J.K. and Ferry, J.G. (1986) J. Bacteriol. 168, 1053–1058.
[221] Fischer, R. and Thauer, R.K. (1990) Arch. Microbiol. 153, 156–162.
[222] Fischer, R. and Thauer, R.K. (1989) Arch. Microbiol. 151, 459–465.
[223] Eikmanns, B. and Thauer, R.K. (1985) Arch. Microbiol. 142, 175–179.
[224] van de Wijngaard, W.M.H., van der Drift, C. and Vogels, G.D. (1988) FEMS Microbiol. Lett. 52, 165–172.
[225] Lovley, D.R., White, R.H. and Ferry, J.G. (1984) J. Bacteriol. 160, 521–525.
[226] Nelson, M.J.K. and Ferry, J.G (1984) J. Bacteriol. 160, 526–532.
[227] Weimer, P.J. and Zeikus, J.G. (1978) Arch. Microbiol. 119, 175–182.
[228] Smith, M.R. and Mah, R.A. (1980) Appl. Environ. Microbiol. 39, 993–999.
[229] Zinder, S.H. and Elias, A.F. (1985) J. Bacteriol. 163, 317–323.
[230] Peinemann, S., Müller, V., Blaut, M. and Gottschalk, G. (1988) J. Bacteriol. 170, 1369–1372.
[231] Karrasch, M., Bott, M. and Thauer, R.K. (1989) Arch. Microbiol. 151, 137–142.
[232] Kemner, J.M., Krzycki, J.A., Prince, R.C. and Zeikus, J.G. (1987) FEMS Microbiol. Lett. 48, 267–272.
[233] Bott, M., Eikmanns, B. and Thauer, R.K. (1986) Eur. J. Biochem. 159, 393–398.
[234] Bott, M. and Thauer, R.K. (1989) Eur. J. Biochem. 179, 469–472.
[235] Bott, M. and Thauer, R.K. (1987) Eur. J. Biochem. 168, 407–412.
[236] Diekert, G., Schrader, E. and Harder, W. (1986) Arch. Microbiol. 144, 386–392.
[237] Uffen, R.L. (1976) Proc. Natl. Acad. Sci. U.S.A. 73, 3298–3302.
[238] Uffen, R.L. (1983) J. Bacteriol. 155, 956–965.
[239] Wakim, B.T. and Uffen, R.L. (1983) J. Bacteriol. 153, 571–573.
[240] Kohler, H.-P.E. and Zehnder, A.J.B. (1984) FEMS Microbiol. Lett. 21, 287–292.

[241] Jetten, M.S.M., Stams, A.J.M. and Zehnder, A.J.B. (1989) J. Bacteriol. 171, 5430–5435.
[242] Jetten, M.S.M., Stams, A.J.M. and Zehnder, A.J.B. (1989) Eur. J. Biochem. 181, 437–441.
[243] Jetten, M.S.M., Stams, A.J.M. and Zehnder, A.J.B. (1990) FEMS Microbiol. Lett. 66, 183–186.
[244] Nyren, P. and Strid, A. (1991) FEMS Microbiol. Lett. 77, 265–270.
[245] Jetten, M.S.M., Fluit, T.J., Stams, A.J.M. and Zehnder, A.J.B. (1992) Arch. Microbiol. 157, 284–289.
[246] Dimroth, P. (1987) Microbiol. Rev. 51, 320–340.
[247] Skulachev, V. (1988) Membrane Bioenergetics. Springer, New York.
[248] Unemoto, T., Tokuda, H. and Hayashi, M. (1990) In: The Bacteria, Vol. 12 (Krulwich, T.A., Ed.), pp. 33–54, Academic Press, New York.
[249] Boenigk, R., Dürre, P. and Gottschalk, G. (1989) Arch. Microbiol. 152, 589–593.
[250] Ebbighausen, H., Weil, B. and Krämer, R. (1991) Arch. Microbiol. 155, 505–510.
[251] Rosen, B.P. and Silver, S., Eds. (1987) Ion Transport in Prokaryotes, Academic Press, New York.
[252] Krulwich, T.A., Ed. (1990) The Bacteria, Vol. 12, Bacterial Energetics, Academic Press, New York.
[253] Taylor, C.D., McBride, B.C., Wolfe, R.S. and Bryant, M.P. (1974) J. Bacteriol. 120, 974–975.
[254] Balch, W.E. and Wolfe, R.S. (1979) J. Bacteriol. 137, 264–273.
[255] Dybas, M. and Konisky, J. (1989) J. Bacteriol. 171, 5866–5871.
[256] Jarrell, K.F., Bird, S.E. and Sprott, G.D. (1984) FEBS Lett. 166, 357–361.
[257] Maloy, S.R. (1990) In: The Bacteria, Vol. 12 (Krulwich, T.A., Ed.) pp. 203–224, Academic Press, New York.
[258] Ekiel, I., Jarrell, K.F. and Sprott, G.D. (1985) Eur. J. Biochem. 149, 437–444.
[259] Diekert, G., Konheiser, U., Piechulla, K. and Thauer, R.K. (1981) J. Bacteriol. 148, 459–464.
[260] Hausinger, R.P. (1987) Microbiol. Rev. 51, 22–42.
[261] Ankel-Fuchs, D. and Thauer, R.K. (1988) In: Bioinorganic Chemistry of Nickel (Lancaster Jr., J.R., Ed.), pp. 93–110, VHC Publishers, Deerfield Beach, USA.
[262] Jarrell, K.F. and Sprott, G.D. (1982) J. Bacteriol. 151, 1195–1203.
[263] Lohmeyer, M. and Friedrich, C.G. (1987) Arch. Microbiol. 149, 130–135.
[264] Baudet, C., Sprott, G.D. and Patel, G.B. (1988) Arch. Microbiol. 150, 338–342.
[265] Jarrell, K.F., Sprott, G.D. and Matheson, A.T. (1984) Can. J. Microbiol. 30, 663–668.
[266] Hensel, R. and König, H. (1988) FEMS Microbiol. Lett. 49, 75–79.
[267] Seeley, R.J. and Fahrney, D.E. (1983) J. Biol. Chem. 258, 10835–10838.
[268] Kanodia, S. and Roberts, M.F. (1983) Proc. Natl. Acad. Sci. U.S.A. 80, 5217–5221.
[269] Schönheit, P., Beimborn, D.B. and Perski, H.J. (1984) Arch. Microbiol. 140, 247–251.
[270] Epstein, W. (1990) In: The Bacteria, Vol. 12 (Krulwich, T.A., Ed.), pp. 87–110, Academic Press, New York.
[271] Siebers, A. and Altendorf, K.H. (1992) In: Alkali Cation Transport Systems in Prokaryotes (Bakker, E.P., Ed.), pp. 225–252, CRC press, Boca Raton, FL.
[272] Sprott, G.D., Shaw, K.M. and Jarrell, K.F. (1985) J. Biol. Chem. 260, 9244–9250.
[273] Krueger, R.D., Harper, S.H., Campbell, J.W. and Fahrney, D.E. (1986) J. Bacteriol. 167, 49–56.
[274] Stetter, K.O., Lauerer, G., Thomm, M. and Neuner, A. (1987) Science 236, 822–824.
[275] Stetter, K.O. (1988) System. Appl. Microbiol. 10, 172–173.
[276] Zellner, G., Stackebrandt, E., Kneifel, H., Messner, P., Sleytr, U.B., Conway de Macario, E., Zabel, H.P., Stetter, K.O. and Winter, J. (1989) System. Appl. Microbiol. 11, 151–160.
[277] Burggraf, S., Jannasch, H.W., Nicolaus, B. and Stetter, K.O. (1990) System. Appl. Microbiol. 13, 24–28.
[278] Gorris, L.G.M., Voet, A.C.W.A. and van der Drift, C. (1991) BioFactors 3, 29–35.
[279] Möller-Zinkhan, D., Börner, G. and Thauer, R.K. (1989) Arch. Microbiol. 152, 362–368.

[280] Schmitz, R.A., Linder, D., Stetter, K.O. and Thauer, R.K. (1991) Arch. Microbiol. 156, 427–434.
[281] Möller-Zinkhan, D. and Thauer, R.K. (1990) Arch. Microbiol. 153, 215–218.
[282] Schauder, R., Eikmanns, B., Thauer, R.K., Widdel, F. and Fuchs, G. (1986) Arch. Microbiol. 145, 162–172.
[283] Speich, N. and Trüper, H.G. (1988) J. Gen. Microbiol. 134, 1419–1425.
[284] Dahl, C., Koch, H.-G., Keuken, O. and Trüper, H.G. (1990) FEMS Microbiol. Lett. 67, 27–32.
[285] Tindall, B.J., Stetter, K.O. and Collins, M.D. (1989) J. Gen. Microbiol. 135, 693–696.
[286] Thauer, R.K. (1989) In: Autotrophic Bacteria (Schlegel, H.G. and Bowien, B., Eds.), pp. 397–413, Science Tech Publishers, Madison, WI.
[287] Achenbach-Richter, L., Stetter, K.O. and Woese, C.R. (1987) Nature 327, 348–349.
[288] Fiala, G. and Stetter, K.O. (1986) Arch. Microbiol. 145, 56–61.
[289] Zillig, W., Holz, I., Klenk, H.-P., Trent, J., Wunderl, S., Janekovic, D., Imsel, E. and Haas, B. (1987) System. Appl. Microbiol. 9, 62–70.
[290] Zillig, W., Holz, I., Janekovic, D., Schäfer, W. and Reiter, W.D. (1983) System. Appl. Microbiol. 4, 88–94.
[291] Achenbach-Richter, L., Gupta, R., Zillig, W. and Woese, C.R. (1988) System. Appl. Microbiol. 10, 231–240.
[292] Miroshnichenko, M.L., Bonch-Osmolovskaya, E.A., Neuner, A., Kostrikina, N.A., Chernych, N.A. and Alekseev, V.A. (1989) System. Appl. Microbiol. 12, 257–262.
[293] Neuner, A., Jannasch, H.W., Belkin, S. and Stetter, K.O. (1990) Arch. Microbiol. 153, 205–207.
[294] Schäfer, T. and Schönheit, P. (1991) Arch. Microbiol. 155, 366–377.
[295] Schäfer, T. and Schönheit, P. (1992) Arch. Microbiol. 158, 188–202.
[296] Costantino, H.R., Brown, S.H. and Kelly, R.M. (1990) J. Bacteriol. 172, 3654–3660.
[297] Koch, R., Spreinat, A., Lemke, K. and Antranikian, G. (1991) Arch. Microbiol. 155, 572–578.
[298] De Rosa, M., Gambacorta, A., Nicolaus, B., Giardina, P., Poerio, E. and Buonocore, V. (1984) Biochem. J. 224, 407–414.
[299] Budgen, N. and Danson, M.J. (1986) FEBS Lett. 196, 207–210.
[300] Danson, M.J. (1988) Adv. Microbial Physiol. 29, 165–231.
[301] Mukund, S. and Adams, M.W.W. (1991) J. Biol. Chem. 266, 14208–14216.
[302] Mukund, S. and Adams, M.W.W. (1990) J. Biol. Chem. 265, 11508–11516.
[303] White, H., Strobl, G., Feicht, R. and Simon, H. (1989) Eur. J. Biochem. 184, 89–96.
[304] Adams, M.W.W. (1990) FEMS Microbiol. Rev. 75, 219–238.
[305] Schäfer, T., Selig, M. and Schönheit, P. (1993) Arch. Microbiol. 159, 72–83.
[306] Huber, R., Langworthy, T.A., König, H., Thomm, M., Woese, C.R., Sleytr, U.B. and Stetter, K.O. (1986) Arch. Microbiol. 144, 324–333.
[307] Tomlinson, G.A., Koch, T.K. and Hochstein, L.I. (1974) Can. J. Microbiol. 20, 1085–1091.
[308a] Zwickl, P., Fabry, S., Bogedain, C., Haas, A. and Hensel, R. (1990) J. Bacteriol. 172, 4329–4338.
[308b] Schäfer, T., and Schönheit, P. (1993) Arch. Microbiol. 159, 354–363.
[309] Bonch-Osmolovskaja, E.A. and Stetter, K.O. (1991) System. Appl. Microbiol. 14, 205–208.
[310] Blumentals, I.I., Itoh, M., Olson, G.J. and Kelly, R.M. (1990) Appl. Environ. Microbiol. 56, 1255–1262.
[311] Bryant, F.O. and Adams, M.W.W. (1989) J. Biol. Chem. 264, 5070–5079.

CHAPTER 5

Signal transduction in halobacteria

Dieter OESTERHELT and Wolfgang MARWAN

Max-Planck-Institut für Biochemie, D-82152 Martinsried, Germany

1. Introduction

1.1. General

"Halobacteria" is a commonly used term comprising all species of the order Halobacteriales. Members of this group prefer extreme environments, which is typical of archaebacteria in general. Halobacteria are extremely halophilic and facultatively anaerobic organisms. In the presence of oxygen the cells grow by respiration. Under anaerobic conditions they can produce ATP either by fermentation of arginine [1] or by photophosphorylation [2]. More recently, anaerobic respiration with fumarate as electron acceptor has also been observed [3]. The capability for anaerobic growth is important since the oxygen tension in the natural habitat of brines and salt ponds can be very low.

1.2. Photoreceptors

In addition to the bioenergetic functions of fermentation and electron transport, the halobacteria possess a unique system of photosynthesis not found in other archaebacteria or in eukaryotes. In contrast to the chlorophyll-dependent "green" photosynthesis based on a light-driven electron transport chain, this alternative photosynthetic pathway uses a retinal protein which directly drives ion translocation upon absorption of light. This chromoprotein, resembling in some respect the visual pigments, has been termed bacteriorhodopsin. Even more surprising, the entire photobiochemistry of halobacteria resides in the action of retinal proteins. Two light–energy converters, bacteriorhodopsin (BR) and halorhodopsin (HR) power all energy-driven processes of the cell (see Chapter 6 by Lanyi, and Chapter 2 by Skulachev in this volume). Bacteriorhodopsin as an outwardly directed proton pump supplies the ATP synthase with proton motive force (pmf) and thus allows photophosphorylation. In addition, the BR-generated pmf is used for sodium–potassium exchange in the molar range, thus creating an enormous electroneutral energy store available to the cell during dark periods [4]. Halorhodopsin is an inwardly directed light-driven chloride pump which, in cooperation with a potassium uniport, mediates the net salt uptake by the cells, which is required for growth. Under conditions where the halobacterial cell depends on pmf created only by light, bacteriorhodopsin can act not only as an energy converter but also as a light sensor [5]. This capacity of "black and white vision" will be dealt with below in more detail. At lower irradiation of the environment two additional retinal proteins, sensory rhodopsin-I and -II (SR-I, SR-II), mediate a kind of "colour vision" that aids the cell in finding the optimal photosynthetic microenvironment (for a review see ref. [6]).

The primary photochemical process in all four proteins is the isomerization of all-*trans* retinal to its 13-*cis* state which is followed by a conformational change of the protein. The light-independent re-isomerization to the all-*trans* state is catalysed by the protein and thermodynamically driven by part of the photon's energy which has been stored in the protein conformation. The major differences between the pumps and the sensors are a longer lifetime of the intermediate 13-*cis* state of the sensors and the connection of ion release and binding to *cis*–*trans* isomerization in the pumps. In other words, retinal acts as a light-triggered switch in all four proteins and causes the translocation of an ion, in the case of bacteriorhodopsin and halorhodopsin, or the formation and decay of a signalling state in the sensory rhodopsins. Several reviews on the physiology, biochemistry and biophysics of the four retinal proteins have been published [7–9].

1.3. Movement

Halobacterium halobium is a rod-shaped archaebacterium that swims by means of a polarly inserted flagellar bundle. The flagella are driven by rotary motors at their bases. The cells can swim back and forth simply by switching the rotational sense of the bundle. Switching occurs spontaneously causing the cells to move at random in their suspending medium. Light and some chemical stimuli can induce or suppress motor switching, thereby introducing a bias in the random walk which ultimately results in an accumulation in a preferential region of the biotope.

Photomovement as a tool to locate optimal conditions for photosynthesis is not the only tactic behaviour halobacteria can master. The cells respond to many more positive and negative stimuli than light alone. Examples found are chemotaxis, aerotaxis and, presumably, thermotaxis. All the signal chains initiated by the activation of various receptors of the cell merge into one and the same final target, the flagellar motor.

1.4. Signal transduction

The halobacterial signal chain is based on the same basic principles as are known for other sensory systems. These are: reception of the stimulus, amplification of the signal input, integration of signals caused by different types of sensory stimuli, and adaptation to the stimulus background. Since halobacteria integrate, for example, light and chemical stimuli for a common control of the flagellar motor switch, they provide a unique system for studying a branched signal chain and, if mutants defective in BR and HR are used, there will be no interference from photosynthetic processes. A previous disadvantage was the lack of a gene transfer but this has now been overcome by the development of a transformation system [9]. *H. halobium* can be considered to be currently the best understood system in prokaryotic photobiology with respect to identification of molecular components involved, and analysis of their action in the cell.

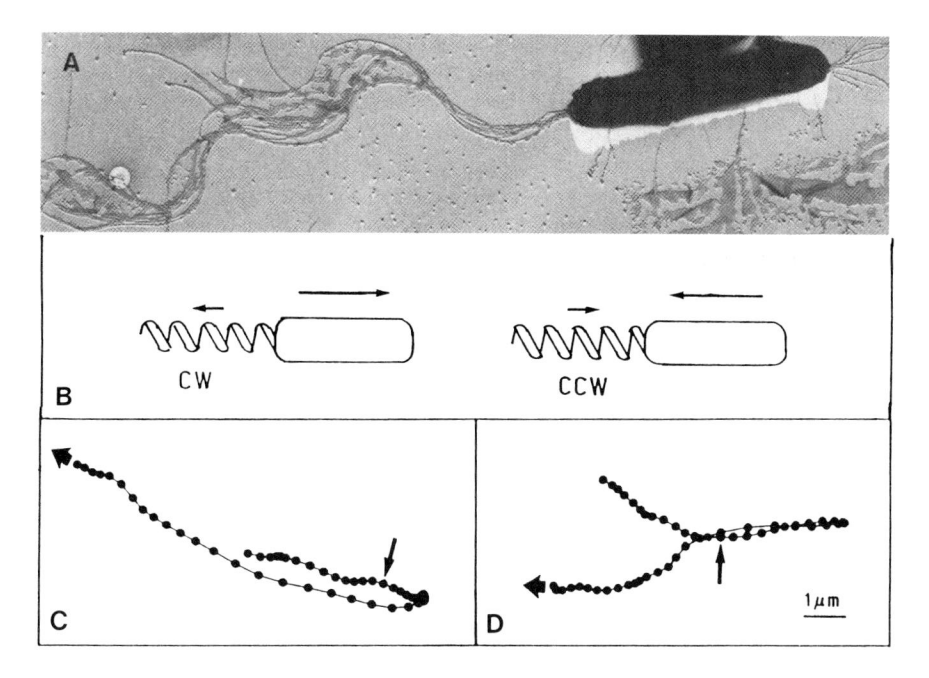

Fig. 1. Flagellation and swimming behaviour of *Halobacterium halobium*. (A) Electron micrograph of a polarly flagellated cell. The cells were adsorbed to the grid from a suspension in 4 M salt and shadowed with carbon/platinum. (B) Modes of rotation of the flagellar bundle in a monopolarly flagellated cell (CW, clockwise; CCW, counterclockwise). The swimming direction resulting from either rotational sense is indicated by the long arrows. Note that in a growing culture most cells are monopolarly flagellated. If a second flagellar bundle is formed during the course of cell division, one bundle is usually much shorter than the other one and thus contributes less to the overall movement of the cell. (C,D) Swimming patterns of cells during light-induced motor switching (photophobic response) as recorded with a computerized cell tracking system. The successive centers of the cell are displayed in 100 ms time intervals. The thin arrows mark the onset of light, the thick ones indicate the swimming direction of the cell. The period of transient acceleration followed by motor switching is clearly seen. The photograph was taken from Alam and Oesterhelt [10] and the cell tracks from Marwan et al. [22].

2. Flagellar motor and motility

2.1. Mode of movement

Halobacterium halobium (according to the newest edition of Bergey's Manual now called *H. salinarium*) cells are propelled by a monopolarly inserted flagellar bundle (Fig. 1A). In the stationary growth phase, bipolarly flagellated cells also occur, but generally one of the flagellar bundles is much shorter than the other and thus contributes less to the overall movement of the cell [10]. Each flagellar bundle consists of five to ten flagellar filaments [11] which form a right-handed semi-rigid helix. The filaments are connected to the cell by a rotatory link called the motor. Bacterial flagella are too thin to be visible in the light microscope under standard conditions. Fortunately, they can be visualized by high-intensity dark-field microscopy [12,13] so that their motion can be directly observed in vivo. Alam and Oesterhelt [10] used this method to study the flagellation and swimming mechanism of *H. halobium*. Their experiments revealed that clockwise (CW) rotation of the right-handed flagellar bundle exerts a pushing force on the cell and propels it in the forward direction. In the counterclockwise (CCW) rotational mode, on the other hand, the bundle pulls the cell which then swims with the flagellated pole at the front (Fig. 1B). Thus, in both rotational modes, a translational movement results. This is quite different from the motion of a peritrichously flagellated *Escherichia coli* cell which only swims when the flagella are rotating CCW, whereas CW rotation produces tumbling [14,15].

2.2. Filament and flagellins

Flagellar filaments detached from the cells spontaneously aggregate into thick superflagella which can grow in length to sizes of about 10–20 times that of an individual cell. These superflagella are easily isolated from stationary cultures of overproducing strains by differential centrifugation of the nutrient broth [10]. Polymorphic transitions from normal to a curly, a ring, or a straight form are induced by different pH values and heat treatments [16].

Superflagella contain three protein species, Fla I (23.5 kDa), Fla II (26.5 kDa) and Fla III (31.5 kDa) [10] which were characterized as sulphated glycoproteins with N-glycosidically linked oligosaccharides [17,18]. The glycoprotein nature of the flagellins is apparently instrumental in the flagellar function, because a mutant strain producing superflagella, as a result of the loss of the filaments from the cell, has a changed glycosylation pattern. Halobacterial glycoproteins, including the flagellins, are glycosylated at the extracellular surface of the cell membrane [18], therefore the flagellin polypeptides must be translocated across the cell membrane before glycosylation [19]. This implies an export mechanism different from that found in eubacteria (for a review see ref. [20]).

Gerl and Sumper [19] isolated a gene fragment of one of the flagellins in an expression vector using immunological probes. With this fragment as a probe, five highly homologous flagellin genes were identified and sequenced. The genes are arranged tandemly in a group of two and three, respectively, at two different loci on the chromosome [19]. It remains to be shown whether all halobacteria or even other families of the archaebacteria share the basic features found for flagellation and motility of *H. halobium*. So far, the same basic mechanism for motility, including superflagella formation, has also been described for peritrichously flagellated halophilic square bacteria. A major difference was found in the molecular weight of the component flagellins, being about 73 kDa [21].

2.3. Motor switching

In the absence of stimuli, halobacterial cells spontaneously switch the rotational sense of their flagellar bundle. The time course of the probability for spontaneous motor switching was calculated from frequency distributions and was shown to be independent of the rotational sense [22]. Switching occurs with a constant probability per unit of time except in the early phase immediately after a switching event. Kinetically, spontaneous switching can satisfactorily be described by stochastic transitions of four states of the motor [23,24]. The transition rates depend highly on temperature and in addition are modulated by the sensory input from various receptors.

In order to get more insight into the mechanism of motor switching, the swimming speed of the cells and the switching process of the bundle were investigated with a computer-assisted motion analysis system. The cells swim faster by CW rotation than by CCW rotation of the bundle. From the small relative magnitude of speed fluctuation, it is concluded that the majority, if not all, of the individual flagellar motors of a cell rotate in the same direction at any given time [22]. Whether this functional coordination of the individual flagellar motors is purely mechanical coupling by friction between the individual filaments or is produced by a more complicated regulatory process remains to be elucidated.

After stimulation by light the cells continue swimming with almost constant speed for some seconds until they slow down about 100 ms before reversing the direction. This behaviour was independent of the photosystem activated since a blue light pulse or an orange step-down stimulus produced the same result [22]. These observations strongly suggest that the flagellar motors switch after stimulation of the cell in an all-or-none response, just as observed for unstimulated behaviour.

The process of reversing the direction of swimming is not an immediate one but the cells pass through a pausing state of several seconds during the change of the rotational sense of the flagellar bundle. After that the cells exhibit a transient acceleration (Figs. 1C, D). Both the average length of the pausing period and the transient acceleration were shown to be independent of the stimulus intensity and thus to represent intrinsic properties of the flagellar motor assembly [22].

3. Signal transduction pathway

3.1. Basic observations

Spontaneous switching in the direction of flagellar rotation in the absence of stimuli results in a random walk around fixed spatial coordinates of the cell since due to collisional events the cells often do not take the original path after turns. Attractant stimuli suppress spontaneous switching and thus lengthen swimming intervals. This causes a shift in the spatial coordinates towards the source of an attractant. Repellent stimuli, on the other hand, activate the switch leading to a stimulus-induced reversal which is also called a phobic response. Repellents and attractants (light and chemicals) can both act in the two alternative modes, depending on whether they are applied or removed. For example, if a repellent stimulus is continuously present the cell adapts after the initial phobic response, i.e., it returns to its pre-stimulus behaviour. Removal of the repellent, after the cell has adapted, then acts as an attractant stimulus. The analogous situation is found for attractants [25]. The result of these behavioural responses is a net accumulation of cells in a green to orange light spot and a depletion of cells from a blue light spot [26,27]. Cells entering the blue light spot reverse and cells leaving the spot swim on. Fig. 2 summarizes our current knowledge of the signal transduction pathway.

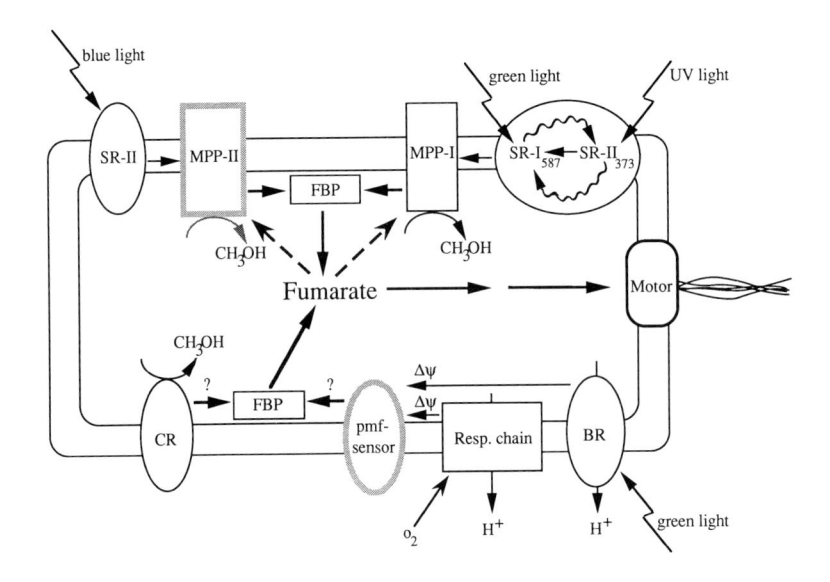

Fig. 2. Schematic presentation of a minimal model of sensory reception and signal transduction in *Halobacterium*. Elements drawn in thin lines have been detected biochemically; the molecular existence of others, i.e., methyl-accepting phototaxis proteins (MPP-II) [46] and pmf sensor are postulated from physiological studies. The arrows indicate the flow of information but do not imply direct physical interaction of the components. SR-I and SR-II are photoreceptors. Upon light excitation they undergo a conformational change which might be transmitted via methyl-accepting phototaxis proteins (MPP-I, MPP-II). These pass the information on to the fumarate binding protein which upon activation releases fumarate to cause the flagellar motor to switch the sense of rotation. A feedback from the integrated signal which modulates the activities of the MPP's by changing their methylation level has to be postulated on the basis of physiological experiments. A pmf sensor is activated upon a drop in the membrane potential generated by either bacteriorhodopsin or by the respiratory chain. Whether the signal relay from the pmf sensor and the chemoreceptors (CR) to the motor is mediated by the fumarate binding protein is currently under investigation.

3.2. Signal formation

Blue light and near- and far-UV light act as repellents with maxima in their action spectrum at 280, 373 and 480 nm [28,29,23,30]. Green to orange light is an attractant (maximal effectiveness at 565 and 590 nm [28,31,32]). Response to orange and near-UV light is mediated by SR-I, the response to blue light by SR-II [33,34,29,23,35]. The phobic response to far-UV light at 280 nm, which all proteins absorb, cannot easily be linked to a specific receptor. After light absorption by the receptor, signalling means the transfer of information to the final target, e.g. the halobacterial flagellar motor. This process, also called "signal transduction pathway", could involve physical phenomena such as changes in membrane potential or ionic gradients and molecular interaction including small and macromolecules. Because different stimuli can cause the same response of the motor a branched signal transduction pathway must exist, and in the following we call the signal a substance which either interacts directly with the motor or is transformed into such a substance. In any case, its formation must occur after the merging of the various signalling pathways of the cells.

An obvious candidate for the signal is a change in membrane potential which could directly trigger the flagellar response. The fact that the sensory rhodopsins are non-electrogenic constitutes an argument against direct electric coupling between the receptor and the motor switch [36]. Rigorous exclusion of a membrane potential change involved in the sensory rhodopsin-triggered transduction pathway was demonstrated in the following way. Long and bipolarly flagellated cells were grown in the presence of a DNA-polymerase inhibitor and irradiated under microscopic observation with a microbeam of light. The flagellar bundles of these cells only switched their sense of rotation when the outermost pole of the cell was irradiated. The distal, non-irradiated bundle showed no response although the irradiated one reacted. This demonstrated that there is no long-range signalling in halobacteria. In addition, the change of membrane potential as a causal element of the SR-dependent signal chain was excluded since an electrical signal would not decay to a significant extent over the length of a cell [37]. Hence, the flagellar motor switch is controlled by a chemical signalling pathway, as is the case in *Escherichia coli* (for a review see ref. [38] and references therein).

Although a change in membrane potential is not involved in the signal transduction pathway, halobacteria can sense changes in proton motive force. This has been shown by expression of bacteriorhodopsin in a strain which is totally defective in halorhodopsin and the sensory rhodopsins. Under anaerobic conditions in the absence of a fermentative energy source the cells swim only when exposed to light. This phenomenon, called photokinesis, indicates that the proton motive force of the cell is generated exclusively by light under this condition. If the cells are exposed to a step-down in irradiation they reply with a photophobic response. However, if the cells would be able also to swim in the absence of light no response to light changes would occur. From these experiments and similar observations it was concluded that a pmf-sensor exists which triggers the flagellar motor switch after merging into the general signal transduction pathway of the cell [5].

Quantitation of the relationship between size of stimulus and response of the target can be obtained by statistical or stochastic methods. Both have been applied to halobacterial behaviour. The SR-II-mediated photophobic response was analyzed statistically by measuring the response time of the cells as a function of blue light stimulus input. The analysis revealed that the period of time an activated photoreceptor molecule interacts with the signal chain defines its effectiveness in signal formation. This strongly suggests a photocatalytic amplification step as an element of the signal chain [23]. Amplification indeed has to be expected, since it was shown by stochastic analysis that one activated SR-II molecule can be sufficient to switch the rotation of all the flagellar motors of a cell [39]. Amplification has now been demonstrated biochemically since one activated sensory rhodopsin causes the release of a large number of fumarate molecules (see below).

Independently of their nature, attractant and repellent stimuli are integrated by the cell if applied simultaneously [25,40]. For example, a blue step-up (repellent) stimulus can be repressed in its repellent action by a green step-up (attractant) stimulus, provided an appropriate intensity ratio of both light colours is applied. Integration was also demonstrated for a symmetrically inverted stimulus program, i.e. a blue step-down (attractant) stimulus and a green step-down (repellent) stimulus. If, on the other hand, two attractant stimuli are applied, e.g. blue step-down and green step-up, the attractant effect is additive. Integration also occurs between chemical stimuli and light [25].

3.3. Identification of a switch factor

The smooth swimming mutant strain M415 is defective in spontaneous and stimulus-induced flagellar motor switching. Cells of this strain can be cured after permeabilization by mild ultrasonication in the presence of a cell extract from wild-type cells. This somatic complementation was used to identify the active compound, called the switch factor, and this was purified from extracts of wild-type cells [41]. Chemical group analysis suggested that the switch factor was fumarate. Indeed, fumaric acid, when used in the complementation assay, replaces the switch factor and enables the cell to respond to blue light by motor switching. Fumaric acid is active at the level of only a few molecules per cell, suggesting that fumarate itself or a close derivative is the active compound. Quantitative mass-spectrometric analysis with deuterated fumaric acid as internal standard showed that the amount of fumarate present in "switch factor" preparations correlated well with the minimal content of the active compound as determined by the complementation assay. It therefore seems probable that a major part of the activity is due to fumaric acid. It cannot, however, be excluded that a derivative of fumaric acid is also involved. The switch factor is bound to the membrane fraction of the cell, and it is released upon heat treatment, presumably by denaturation of a membrane-associated protein (fumarate-binding protein, FBP) which binds the factor in vivo.

If the switch factor or fumarate are applied to cells of the mutant M415 in higher concentrations, not only stimulated responses but also spontaneous stopping responses of the flagellar motor are observed [41]. This suggests two things: (i) that the factor is involved in the mechanism of motor switching and (ii) that the factor acts in the part of the signal chain located beyond the point of stimulus integration.

3.4. Light-induced release of fumarate

Identification of fumarate as a switch factor raises the question whether it acts as a cofactor required for the switching process itself or as a cellular response regulator. The latter assumption would be correct if the cellular concentration of fumarate is under receptor control.

A suspension of wild-type cells was exposed to blue light stimulation and after a defined period of time the cells were lysed by rapid mixing with lysis buffer of low ionic strength. After ultrafiltration which removed the high-molecular-weight compounds, the switch factor activity in the filtrate was measured by its ability to reconstitute motor switching in mutant cells. Blue light, received by SR-II, or a decrease in orange light (sensed by SR-I) caused a release of the switch factor (fumarate) from a protein-bound pool [42].

By using an enzymatic assay for fumarate which detects the compound in the lower pmol range the kinetics of fumarate release after activation of sensory rhodopsin-II could be measured. The concentration of fumarate released from the protein-bound pool increased transiently after a blue light pulse. The kinetics correlate consistently well with the probability of motor switching in intact cells as measured by computer-assisted motion analysis. Cells defective in retinal biosynthesis and therefore lacking functional photoreceptors showed only light-induced release of fumarate when the sensory opsins were reconstituted in the cell by addition of retinal. This experimental result leads to the conclusion that fumarate release is caused by receptor activation rather than by an independent light-induced change in the activity of metabolic enzymes. Under the experimental conditions used, at least 200 molecules of fumarate were released per activated SR-II molecule. This strongly suggests that amplification occurs and is consistent with the model of photocatalytic signal formation derived from behavioural measurements [23]. These results confirm the role of fumarate as a photoreceptor-controlled response regulator of the flagellar motor. As a working hypothesis we assume that release of fumarate upon stimulation occurs from the fumarate-binding protein (FBP) pool of the membrane (Fig. 2). This pool is linked to the concentration of fumarate in the cytoplasm, which might account for the stochastic fluctuations of the steady state. Upon light activation of the photoreceptors the cytoplasm is flushed with a fumarate pulse which would immediately cause the flagellar motor to switch.

So far all stimulatory processes of the halobacterial cell depend on fumarate as seen by the fact that smooth swimming mutant cells of strain M415 do not respond to any stimulus. This also holds true for the step-down photophobic response mediated by bacteriorhodopsin [5]. This points to the role of a central metabolite as an integrative tool for measurement of the metabolic state of the cell. Especially fumarate would allow us to measure and compare activity states of electron transport, fermentation and photosynthesis (see Fig. 2).

Fumarate was also demonstrated to be a switch factor in eubacteria. Fumarate and the CheY protein are the only cytoplasmic components required for spontaneous motor switching in *Escherichia coli* and *Salmonella typhimurium* [43].

3.5. Methyl-accepting taxis proteins

It has been shown by various groups that methyl-accepting proteins occur in *H. halobium* [44–47]. Proteins which receive their methyl groups from S-adenosylmethionine can be radiolabelled specifically in vivo by supplying methyl-^{3}H methionine to cells in which protein synthesis is blocked by, e.g., puromycin. Analysis by electrophoresis and fluorography revealed methyl-^{3}H-labelled protein species between 90 and 135 kDa [47] and between 65 and 150 kDa [46]. The label released from the protein by mild alkali treatment was shown to be a volatile compound as expected for methanol, the product of carboxymethylester hydrolysis. Chemostimuli caused distinct changes on specific bands whereas no changes could be detected upon photostimulation. The release of labelled volatile methyl groups in intact labelled cells could, however, be measured upon photostimulation in a flow apparatus where cells on a filter were washed continuously with buffer containing an excess of nonradioactive methionine [48].

Chemo- and photostimulation of halobacteria both caused a transient increase in the net rate of demethylation [47]. This increased rate is observed independently of whether the stimulus is an attractant or a repellent. In contrast, *Escherichia coli* exhibits responses to positive and negative stimuli which consist of a transient decrease and increase in methylesterase activity, respectively [48]. The same demethylation phenomenon as in

Halobacterium has also been seen in *Bacillus subtilis* [49] and can be explained on the basis of specific regulatory properties of the enzymes involved.

Several arguments support the view that methyl-accepting proteins are involved in the sensory transduction machinery of *Halobacterium* [46,47]: (1) Demethylation is transiently increased by photo- and chemostimuli which are known to control behaviour. (2) Mutant strains (Pho 71, Pho 72 and M 402) [50,46,47] which show neither photo- nor chemoresponses lack almost all the methyl-labelled bands in the 70–135 kDa region. (3) In cells of strain Pho 81, which are deleted in SR-I and SR-II [50] but perform normal chemotaxis, no light-induced release of volatile methyl groups can be observed, whereas the demethylation response to chemostimuli is unaltered [47].

One of the methyl-accepting proteins seems to be specifically involved in phototransduction by SR-I. This 94 kDa protein was shown to be present in increased amounts in an overproducer of SR-I but absent in a strain which contained only SR-II and lacked SR-I. It is also absent in a mutant lacking both photoreceptors [46].

Currently, it is unclear whether the methylation system is involved in adaptation, as is the case in enteric bacteria [51], or alternatively in the excitation system. Treatment of cells with ethionine, which should lower intracellular pools of the methyl donor S-adenosyl methionine, causes increased response times to both 360 nm step-up (repellent) and 360 nm step-down (attractant) stimuli. The repellent response, therefore, is weakened and the attractant response is enhanced by ethionine although the frequency of spontaneous motor switching remains constant [52]. These results suggest that ethionine suppresses stimulus-induced motor switching in general. Not necessarily, however, does ethionine influence adaptation. The observation that stimulus-induced demethylation persists over periods roughly equivalent to adaptation times is currently the only argument for assigning methylation to adaptation.

Two antagonistic stimuli, i.e. attractant and repellent, delivered at the same time cancel demethylation, therefore a feedback loop from the integrated signal to the demethylation system has to be postulated [46].

Methyl-accepting taxis proteins were first discovered in eubacteria. In *E. coli* they function as chemoreceptors for specific ligands and are deactivated by methylation of several glutamyl residues which causes sensory adaptation (for a review see ref. [53]). Antisera raised to methyl-accepting proteins of *E. coli* cross react with methyl-accepting proteins from *H. halobium* as shown by Western blotting [54]. The authors conclude that the genes for the contemporary methyl-accepting proteins are related through an ancestral gene that existed before the divergence of archaebacteria and eubacteria.

3.6. Cyclic GMP

A second small molecule invoked in signalling is cyclic GMP (cGMP). Addition of cGMP to crude preparations of membranes causes a slow hydrolysis to GMP which is partially inhibited by orange light [55]. In intact cells, dibutyryl-cyclic GMP and 3-isobutyl-1-methylxanthine (IBMX), both inhibitors of eukaryotic phosphodiesterase, were found to suppress motor switching [52]. Aluminium tetrafluoride (AlF_4^-), which activates phosphodiesterase via a G-protein in eukaryotic cells and thus reduces the cGMP level, shortens spontaneous and attractant intervals. In addition, an immunological cross-reaction of a halobacterial protein with a G_a antiserum was found. This led to the conclusion that a

G-protein-controlled phosphodiesterase may be involved in signal transduction [55], but the exact localization remains to be established.

4. The photoreceptors

4.1. Spectroscopic and biochemical properties

Two sensory rhodopsins, SR-I and SR-II, were detected spectroscopically in the cell membrane of strains defective in BR and HR [29,56–58]. In wild-type cells the copy number of BR is 100 times that of SR-I, and that of HR 10 times that of SR-I. This is the reason why spectroscopic resolution of the sensory rhodopsins only became possible when mutants defective in the two ion pumps were available. SR-II is constitutively present in a copy number of about 400 per cell. SR-I in parallel with BR and HR is induced during growth of the culture, presumably by a drop in oxygen tension; SR-I then reaches a level of expression at about 4000 copies per cell [35]. The differential expression of the two receptors results in the cell avoiding blue light during heterotrophic growth, when oxygen is abundant, but actively seeking green light during phototrophic growth.

SR-I has been isolated from halobacterial membranes after solubilization in laurylmaltoside. In the presence of retinal, detergent and salt, the native protein was obtained in pure form by sucrose density gradient centrifugation, hydroxyapatite chromatography and gel filtration [59]. The apparent molecular weight of the molecule is 24 kDa as analysed by SDS gel electrophoresis and 49 kDa by sedimentation and size-exclusion chromatography, suggesting that the native protein occurs as a dimer. The purified SR-I has an absorption maximum of 580 nm, which is 7 nm blue-shifted compared to the membrane-bound state ($SR-I_{587}$). The action spectrum for the formation of the longest living intermediate, $SR-I_{373}$, coincided with the absorption spectrum.

The primary structure of SR-I has been determined by protein sequencing and by cloning and sequencing the gene [60]. The protein consists of 239 amino acids which span the membrane by seven predicted transmembrane helices. A sequence identity of 14% to bacteriorhodopsin and halorhodopsin, the light-driven proton pump and chloride pump, respectively, was found. Since there exists no homology to bovine rhodopsin and the family of seven helix receptors which trigger G-protein cascades in eukaryotes, the halobacterial light sensors form a new family of photoreceptor proteins. A striking feature of the predicted secondary structure of SR-I is the absence of the large cytoplasmic domains found in animal rhodopsins as well as in eubacterial chemotaxis receptors, suggesting that the signalling state might be transduced within the membrane.

4.2. The physiology of photoreception

SR-I is a photochromic protein discovered by Bogomolni and Spudich [56], who proposed its function as a photoreceptor. Upon absorption of orange light by $SR-I_{587}$, a metastable intermediate ($SR-I_{373}$) is formed which sends an attractant signal to the flagellar motor.

The intermediate can relax either thermally to the original state $SR-I_{587}$ or photochemically by absorption of near-UV light via an alternative pathway. This photochemical pathway includes the formation of a repellent signalling state that induces motor switching. Therefore a single photoreceptor protein provides the molecular basis for the colour discrimination mechanism by triggering two antagonistic responses.

A general problem in the photobiology of sensing and response is the definitive assignment of photoreceptor function, as seen by the biological effect, to the spectroscopically and biochemically defined chromoprotein that is believed to be the receptor. Because of the fact that the sensitivity of cells to UV is greatly enhanced by orange light, Spudich and Bogomolni[61] have proposed SR-I_{373} to be the near-UV receptor in halobacteria. Since one or two alternative explanations would hold for this experimental result we made a kinetic fingerprint of the functional near-UV (UV-A) receptor in situ and compared it with the known photocycle of SR-I. The photophobic response of the cells to near-UV light pulses was recorded as a function of pulse intensity. Evaluation of stimulus response curves taken at various orange background light irradiation revealed the following properties of the near-UV receptor: (1) Active UV-receptor molecules are formed by orange light. The alternative possibility that orange light might increase the UV-sensitivity by tuning the gain of the signal chain, e.g. through additional photoreceptors, was excluded. (2) The number of UV-receptor molecules that are active under photostationary conditions depends on the orange background light intensity as expected from the kinetic scheme for the photocycle of SR-I. (3) The half-lifetime of the near-UV receptor corresponds to the value found spectroscopically for SR-I_{373}. These results confirmed SR-I_{373} to be the halobacterial near-UV receptor[62]. SR-I turned out to be the far-UV (UV-B) photoreceptor as well[63]. Intramolecular energy transfer from aromatic amino acid residues of the SR-I apoprotein to the holopigment appears possible, as was shown for bacteriorhodopsin in intact cells[30,64].

By dichromatic action spectroscopy, Wagner et al.[65] measured peak responsivity of photoattraction in *Halobacterium halobium* cells as a function of wavelength and irradiation. Peak responsivity of photoattraction showed a steady hypsochromic shift from 590 nm wavelength under low irradiance conditions to 560 nm under high irradiance conditions. Inversion of the photoattractant response, as dependent on blue versus red background light, is compatible with the known properties of SR-I with maximum absorption of the original state at 587 nm. An antagonistic pigment or intermediate state, different from the original state SR-I, with peak sensitivity at 620 nm or even above is suggested. The less sensitive photoattractant response at 560 nm persists without inactivation in light and represents the peak responsivity under high irradiation conditions. Thus, a long-standing controversy has been resolved[65,66], in favour of a predominant function of SR-I, both in visible and in ultraviolet light[67].

It is an important experimental advantage that the halobacterial photoreceptor apoproteins (so-called sensory opsins, SO) are expressed by the cell even in absence of retinal. Retinal biosynthesis can be blocked either genetically or simply by addition of nicotine to a growing culture. Upon addition of external retinal, active photoreceptors spontaneously reconstitute in vivo. This kind of experiment showed that photoreception is strictly retinal-dependent. By incorporation of artificially synthesized retinal analogues into sensory opsins, John Spudich and coworkers obtained valuable information on the activation process of the sensory rhodopsins.

Introduction of an analogue of all-*trans* retinal in which all-*trans*/13-*cis* isomerization is blocked by a carbon bridge from C13 to C14 into SO-I or SO-II in vivo did not restore photophobic responses through any of the three known photosensory systems (SR-I attractant, SR-I repellent, or SR-II repellent) although analogue pigments of SR-I and SR-II with absorption spectra similar to those of the native pigments were found. However, the formation of the long-lived intermediate of the SR-I photocycle (SR-I$_{373}$) and those of the SR-II photocycle (S-II$_{360}$ and S-II$_{530}$) was blocked by all-*trans*-locked retinal [67]. In conclusion, specific all-*trans*/13-*cis* double-bond isomerization of retinal is essential for generating signalling states in SR-I and SR-II.

By reconstitution with other retinal analogues it was shown that a steric interaction between the retinal 13-methyl group and the protein is required for SR-I activation but not for that of BR. These results reveal a key difference between SR-I and BR that is likely to be the initial diverging point in their photoactivation pathways [68]. Obviously the 13-methyl group–protein interaction functions as a trigger for SR-I activation – i.e., converts photoabsorption by the chromophore into protein conformational changes. All-*trans*/13-*cis* isomerization also occurs during photoactivation of *Chlamydomonas* rhodopsin [69,70]. A similar steric trigger, i.e. interaction of the protein and the 9-methyl group of retinal, is essential for activation of mammalian rhodopsin [71], indicating a common mechanism for receptor activation in archaebacterial, vertebrate and plant retinylidene photosensors.

Acknowledgement

The scheme in Fig. 2 was designed by Barbara Nordmann.

References

[1] Hartmann, R., Sickinger, H.-D. and Oesterhelt, D. (1980) Proc. Natl. Acad. Sci. U.S.A. 77, 3821–3825.
[2] Oesterhelt, D. and Krippahl, G. (1983) Ann. Microbiol. (Inst. Pasteur) B 134, 137–150.
[3] Oren, A. (1991) J. Gen. Microbiol. 137, 1387–1390.
[4] Wagner, G., Hartmann, R. and Oesterhelt, D. (1978) Eur. J. Biochem. 89, 169–179.
[5] Bibikov, S.I., Grishanin, R.N., Marwan, W., Oesterhelt, D. and Skulachev, V.P. (1991) FEBS Lett. 295, 223–226.
[6] Spudich, J.L. and Bogomolni, R.A. (1988) Annu. Rev. Biophys. Biophys. Chem. 17, 193–215.
[7] Oesterhelt, D. and Tittor, J. (1989) TIBS 14, 57–61.
[8] Lanyi, J. (1984) In: Comparative biochemistry and bioenergetics (Ernster, L., Ed.), pp. 315–335, Elsevier, Amsterdam.
[9] Cline, S.W., Schalkwyk, L.C. and Doolittle, W.F. (1989) J. Bacteriol. 171, 4987–4991.
[10] Alam, M. and Oesterhelt, D. (1984) J. Mol. Biol. 176, 459–475.
[11] Houwink, A.L. (1956) J. Gen. Microbiol. 15, 146–150.
[12] Reichert, K. (1909) Centralbl. f. Bakt. Abt. I, 51.
[13] Macnab, R.M. (1976) J. Clin. Microbiol. 4, 258–265.
[14] Larsen, S.H., Reader, R.W., Kort, E.N., Tso, W.-W. and Adler, J. (1974) Nature (London) 249, 74–77.
[15] Macnab, R.M. and Ornston, M.K. (1977) J. Mol. Biol. 112, 1–30.
[16] Alam, M. and Oesterhelt, D. (1987) J. Mol. Biol. 194, 495–499.

[17] Wieland, F., Paul, G. and Sumper, M. (1985) J. Biol. Chem. 260, 15180–15185.
[18] Sumper, M. (1987) Biochim. Biophys. Acta 906, 69–79.
[19] Gerl, L. and Sumper, M. (1988) J. Biol. Chem. 263, 13246–13251.
[20] Macnab, R.M. (1990) In: Biology of the Chemotactic Response (Armitage, J.P. and Lackie, J.M., Eds.), pp. 79–106, Cambridge Univ. Press.
[21] Alam, M., Claviez, M., Oesterhelt, D. and Kessel, M. (1984) EMBO Journal 3, 2899–2903.
[22] Marwan, W. and Oesterhelt, D. (1991) Naturwiss. 78, 127–129.
[23] Marwan, W. and Oesterhelt, D. (1987) J. Mol. Biol. 195, 333–342.
[24] McCain, D.A., Amici, L.A. and Spudich, J.L. (1987) J. Bacteriol. 169, 4750–4758.
[25] Spudich, J.L. and Stoeckenius, W. (1979) Photobiochem. Photobiophys. 1, 43–53.
[26] Hildebrand, E. (1977) Biophys. Struct. Mechanism 3, 69–77.
[27] Stoeckenius, W., Wolff, E.K. and Hess, B. (1988) J. Bacteriol. 170, 2790–2795.
[28] Hildebrand, E. and Dencher, N. (1975) Nature 257, 46–48.
[29] Wolff, E.K., Bogomolni, R.A., Scherrer, P., Hess, B. and Stoeckenius, W. (1986) Proc. Natl. Acad. Sci. U.S.A. 83, 7272.
[30] Wagner, G., Rothärmel, T. and Traulich, B. (1991) In: General and Applied Aspects of Halophilic Microorganisms (Rodriguez-Valera, F., Ed.) Plenum press.
[31] Sperling, W. and Schimz, A. (1980) Biophys. Struct. Mech. 6, 165–169.
[32] Traulich, B., Hildebrand, E., Schimz, A., Wagner, G. and Lanyi, J. (1983) Photochem. Photobiol. 37, 577–579.
[33] Takahashi, T., Mochizuki, Y., Kamo, N. and Kobatake, Y. (1985) Biochem. Biophys. Res. Commun. 127, 99–105.
[34] Takahashi, T., Tomioka, H., Kamo, N., and Kobatake, Y. (1985) FEMS Microbiol. Lett. 28, 161–164.
[35] Otomo, J., Marwan, W., Oesterhelt, D., Desel, H. and Uhl, R. (1989) J. Bacteriol. 171, 2155–2159.
[36] Ehrlich, B.E., Schen, C.R. and Spudich, J.L. (1984) J. Membrane Biol. 82, 89–94.
[37] Oesterhelt, D. and Marwan, W. (1987) J. Bacteriol. 169, 3515–3520.
[38] Wagner, G. and Marwan, W. (1992) Progress in Botany, Vol. 53 (Ziegler, H., series Ed.), pp. 126–149, Springer, Heidelberg.
[39] Marwan, W., Hegemann, P. and Oesterhelt, D. (1988) J. Mol. Biol. 199, 663–664.
[40] Hildebrand, E. and Schimz, A. (1986) J. Bacteriol. 167, 305–311.
[41] Marwan, W., Schäfer, W. and Oesterhelt, D. (1990) EMBO J. 9, 355–362.
[42] Marwan, W. and Oesterhelt, D. (1991) Naturwiss. 78, 127–129.
[43] Barak, R. and Eisenbach, M. (1992) J. Bacteriol. 174, 643–645.
[44] Schimz, A. (1981) FEBS Lett. 125, 205–207.
[45] Bibikov, S.I., Baryshev, V.A. and Glagolev, A.N. (1982) FEBS Lett. 146, 255–258.
[46] Spudich, E.N., Hasselbacher, C.A. and Spudich, J.L. (1988) J. Bacteriol. 170, 4280–4285.
[47] Alam, M., Lebert, M., Oesterhelt, D. and Hazelbauer, G.L. (1989) EMBO J. 8, 631–639.
[48] Kehry, M.R., Doak, T.G. and Dahlquist, F.W. (1984) J. Biol. Chem. 259, 11828–11835.
[49] Thoelke, M.S., Bedale, W.A., Nettleton, D.O. and Ordal, G.W. (1987) J. Biol. Chem. 262, 2811–2816.
[50] Sundberg, S.A., Bogomolni, R.A. and Spudich, J.L. (1985) J. Bacteriol. 164, 282–287.
[51] Stewart, R.C. and Dahlquist, F.W. (1987) Chem. Rev. 87, 997–1025.
[52] Schimz, A. and Hildebrand, E. (1987) Biochim. Biophys. Acta 923, 222–232.
[53] Hazelbauer, G.L., Yaghmai, R., Burrows, G.G., Baumgartner, J.W., Dutton, D.P. and Morgan, D.G. (1990) In: Symposium of Soc. Gen. Microbiol. (Armitage, J.P. and Lackie, J.M., Eds.) Vol. 46, pp. 219–239, Cambridge University Press.
[54] Alam, M. and Hazelbauer, G.L. (1991) J. Bacteriol. 173, 5837–5842.
[55] Schimz, A., Hinsch, K.-D. and Hildebrand, E. (1989) FEBS Lett. 249, 59–61.

[56] Bogomolni, R.A. and Spudich, J.L. (1982) Proc. Natl. Acad. Sci. U.S.A. 79, 6250–6254.
[57] Spudich, E.N. and Spudich, J.L. (1982) Proc. Natl. Acad. Sci. U.S.A. 79, 4308–4312.
[58] Scherrer, P., McGinnis, K. and Bogomolni, R.A. (1987) Proc. Natl. Acad. Sci. U.S.A. 84, 402–406.
[59] Schegk, E.S. and Oesterhelt, D. (1988) EMBO J. 7, 2925–2933.
[60] Blanck, A., Oesterhelt, D., Ferrando, E., Schegk, E.S. and Lottspeich, F. (1989) EMBO J. 8, 3963–3971.
[61] Spudich, J.L. and Bogomolni, R.A. (1984) Nature 312, 509–513.
[62] Marwan, W. and Oesterhelt, D. (1990) J. Mol. Biol. 215, 277–285.
[63] Wagner, G., Rothärmel, T. and Traulich, B. (1991) In: General and Applied Aspects of Halophilic Microorganisms (Rodriguez-Valera, F., Ed.), pp. 139–147, Plenum Press, New York.
[64] Wagner, G., Traulich, B. and Hartmann, K.M. (1987) Photochem. Photobiol. 45, 299–303.
[65] Wagner, G., Hesse, V. and Traulich, B. (1991) Photochem. Photobiol. 53, 701–706.
[66] Hildebrand, E. and Schimz, A. (1986) Trends Biochem. Sci. 11, 402.
[67] Yan, B., Takahashi, T., Johnson, R., Derguini, F., Nakanishi, K. and Spudich, J.L. (1990) Biophys. J. 57, 807–814.
[68] Yan, B., Nakanishi, K. and Spudich, J.L. (1991) Proc. Natl. Acad. Sci. U.S.A. 88, 9412–9416.
[69] Hegemann, P., Gärtner, W. and Uhl, R. (1991) Biophys. J. 60, 1477–1489.
[70] Lawson, M.A., Zacks, D.N., Derguini, F., Nakanishi, K. and Spudich, J.L. (1991) Biophys. J. 60, 1490–1498.
[71] Ganter, U.M., Schmidt, E.D., Perez-Salo, D., Rando, R.R. and Siebert, F. (1989) Biochemistry 28, 5954–5962.

CHAPTER 6

Ion transport rhodopsins (bacteriorhodopsin and halorhodopsin): Structure and function

Janos K. LANYI

Department of Physiology and Biophysics, University of California, Irvine, CA 92717, USA

1. Introduction

The bacteriorhodopsins, which are proton pumps, and the halorhodopsins, which are chloride ion pumps, constitute a class of unique light-energy-transducing systems. They serve bioenergetic functions driven more commonly by redox reactions or chlorophyll-based photosynthesis in other kinds of membranes. As far as known, these ion-motive bacterial rhodopsins are found only in the extremely halophilic archaea, but interest in them in the last two decades has far exceeded what would be normally accorded any single protein, even in an evolutionarily distinct class of organisms. The general interest and the intense and voluminous research in this field originated instead from the expectation that the study of these simple active transport systems, based on the light-induced isomerization of retinal in small integral membrane proteins, would reveal general principles of translocation mechanisms and energy coupling in ionic pumps.

To some extent this has indeed happened. Results so far suggest that bacteriorhodopsin functions in an unexpectedly simple way while incorporating all of what one would expect from an effective energy-transducing system: proton transfer across the internal hydrophobic barrier generates a proton-motive force independent of external proton concentration; it is essentially irreversible and resists proton back-pressure yet proceeds with minimal thermal losses. The mechanism of chloride transport in halorhodopsin is not yet as clear, but there is reason to believe there is a far-reaching analogy between this system and bacteriorhodopsin.

Recent results have provided a much improved understanding of bacteriorhodopsin. As discussed in this review, they include: (1) the high-resolution diffraction structure of the protein, 2) the thorough description of the photoreaction intermediates by vibrational spectroscopy, which have revealed bond rotations and distortions in the retinal, as well as changes in the ionization, hydrogen bonding, and possibly conformational states of critical aspartate and other protein residues, (3) better appreciation of the interactions of several charged residues which are located on the extracellular side of the retinal Schiff base and together may function as its counterion and proton acceptor, (4) fitting of kinetic data for the photocycle to a self-consistent reaction sequence and a thermodynamic description, and (5) the use of in vivo and in vitro produced mutations to examine the role of individual residues in forming the characteristic purple color of the chromophore and in the step-wise transfer of the proton during its transit from one membrane surface to the other. Another line of investigation, related to (5), which has shown promise is the comparison of bacteriorhodopsins and halorhodopsins in various halophilic archaea, since these species and proteins appear to be sufficiently diverse to allow functional and evolutionary comparisons.

This review will describe recent advances, as well as current views and controversies, rather than give an exhaustive account of the vast literature of the bacterial rhodopsins. For the latter, and for aspects not covered here, the reader should consult the numerous earlier and more specialized reviews [1–7].

2. Structure

Aligned protein sequences for three bacteriorhodopsins and two halorhodopsins are shown in Fig. 1; the secondary structure of bacteriorhodopsin is in Fig. 2. It is evident that these are very hydrophobic proteins with long runs of residues lacking any charged groups, in keeping with the fact that they are tightly membrane bound and solubilized only by strong detergents. The proteins contain a large proportion of α-helix [8,9]. Predictions of secondary structure by a variety of algorithms have uniformly produced seven transmembrane helices connected by short loops [10,11], a model consistent with structures from electron diffraction, first at 7 Å [12,13] and more recently at 3.5 Å [14–16] resolution. Residue identities among the proteins are pronounced inside the helical regions, but infrequent in the loops. Two categories of conserved residues are worth noting in Fig. 1: identity in all five proteins, and identity in the three proton pumps and in the two chloride pumps. Both are most numerous in helices C and G. Presumably, the residues in the first category have to do with the overall architecture of the proteins, and the structural requirements for locating the retinal chromophore and accommodating its shape changes during isomerization. Residues which seem to denote helical ends on various helices, the site of retinal attachment (by a Schiff base formed with lys216 in the middle of helix G), the three conserved trp residues (in helices C and F) which surround the retinal ring, and probably the three conserved buried pro residues, are examples. The residues in the second category must relate to the differences inherent in the fact that proton transfer requires different functional groups than chloride transfer in the protein. Some of these differences are obvious: for example, asp85 and asp96, which play essential roles in proton transport (cf. below) are replaced by thr and ala, respectively, in the halorhodopsins. Other consistent differences are more puzzling: in nearly half of the residues in this category the replacement is conservative (ala or ser for gly, val for ile, ile for leu, etc.). Given the fact that the gene sequences are much more divergent than protein sequences [17] and therefore any conservation at the protein level must originate from selection for function, the pattern reveals the importance of the exact size of side-chains which are packed densely inside these proteins. However, in spite of the fact that evolution conserved these residues, replacement of many of them

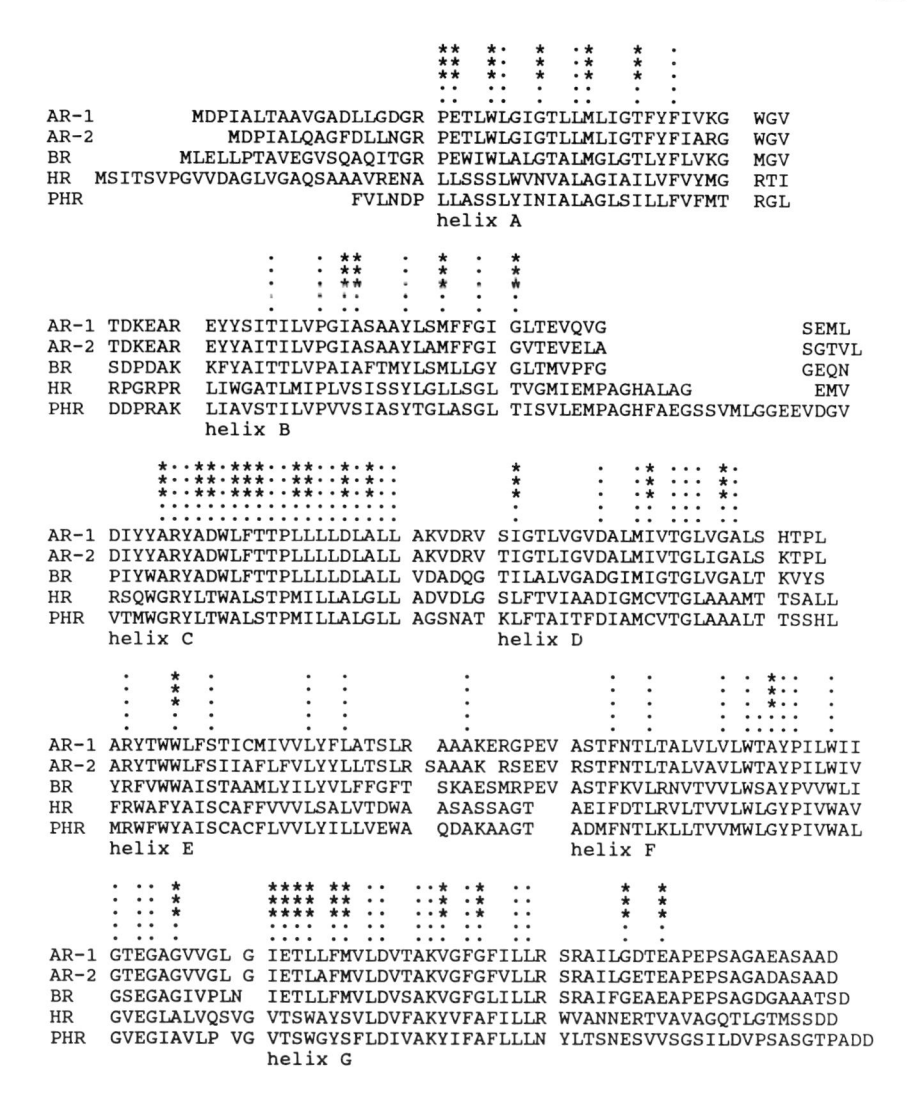

```
                              **  *·   *   ·*     *     ·
                              **  *·   *   ·*     *     ·
                              **  *·   *   ·*     *     ·
                              ··  ··   ·   ··     ·     ·
AR-1             MDPIALTAAVGADLLGDGR  PETLWLGIGTLLMLIGTFYFIVKG   WGV
AR-2             MDPIALQAGFDLLNGR     PETLWLGIGTLLMLIGTFYFIARG   WGV
BR               MLELLPTAVEGVSQAQITGR PEWIWLALGTALMGLGTLYFLVKG   MGV
HR     MSITSVPGVVDAGLVGAQSAAAVRENA   LLSSSLWVNVALAGIAILVFVYMG   RTI
PHR                          FVLNDP  LLASSLYINIALAGLSILLFVFMT   RGL
                                     helix A


                        ·    ·  **    ·     *    ·    *
                        ·    ·  **    ·     *    ·    *
                        ·    ·  **    ·     *    ·    *
                        ·    ·  ··    ·     ·    ·    ·
AR-1   TDKEAR    EYYSITILVPGIASAAYLSMFFGI  GLTEVQVG                SEML
AR-2   TDKEAR    EYYAITILVPGIASAAYLAMFFGI  GVTEVELA                SGTVL
BR     SDPDAK    KFYAITTLVPAIAFTMYLSMLLGY  GLTMVPFG                GEQN
HR     RPGRPR    LIWGATLMIPLVSISSYLGLLSGL  TVGMIEMPAGHALAG         EMV
PHR    DDPRAK    LIAVSTILVPVVSIASYTGLASGL  TISVLEMPAGHFAEGSSVMLGGEEVDGV
                 helix B


       *··*  ***·***··**···*··        *               ·  *···  *·
       *··**·***·***··**···*··        *               ·  *···  *·
       *··**·***·***··**···*··        *               ·  *···  *·
       ···············              · · · · · · ·       · · · · ·
AR-1   DIYYARYADWLFTTPLLLLDLALL   AKVDRV   SIGTLVGVDALMIVTGLVGALS  HTPL
AR-2   DIYYARYADWLFTTPLLLLDLALL   AKVDRV   TIGTLIGVDALMIVTGLIGALS  KTPL
BR     PIYWARYADWLFTTPLLLLDLALL   VDADQG   TILALVGADGIMIGTGLVGALT  KVYS
HR     RSQWGRYLTWALSTPMILLALGLL   ADVDLG   SLFTVIAADIGMCVTGLAAAMT  TSALL
PHR    VTMWGRYLTWALSTPMILLALGLL   AGSNAT   KLFTAITFDIAMCVTGLAAALT  TSSHL
       helix C                             helix D


              ·    *   ·                                     · ·*··   ·
              ·    *   ·                                     · ·*··   ·
              ·    *   ·                                     · ·*··   ·
              ·    ·   ·                                     · ····   ·
AR-1   ARYTWWLFSTICMIVVLYFLATSLR   AAAKERGPEV  ASTFNTLTALVLVLWTAYPILWII
AR-2   ARYTWWLFSIIAFLFVLYYLLTSLR   SAAAK RSEEV RSTFNTLTALVAVLWTAYPILWIV
BR     YRFVWWAISTAAMLYILYVLFFGFT   SKAESMRPEV  ASTFKVLRNVTVVLWSAYPVVWLI
HR     FRWAFYAISCAFFVVVLSALVTDWA   ASASSSAGT   AEIFDTLRVLTVVLWLGYPIVWAV
PHR    MRWFWYAISCACFLVVLYILLVEWA   QDAKAAGT    ADMFNTLKLLTVVMWLGYPIVWAL
       helix E                                 helix F


       · ··  *      ****  **  ··   ··* ·*   ··       * *
       ·  ·  *      ****  **  ··   ··* ·*   ··       * *
       ·  ·  *      ****  **  ··   ··* ·*   ··       * *
       ·  ·  ·      ····  ··  ··   ··· ··   ··       · ·
AR-1   GTEGAGVVGL G  IETLLFMVLDVTAKVGFGFILLR  SRAILGDTEAPEPSAGAEASAAD
AR-2   GTEGAGVVGL G  IETLAFMVLDVTAKVGFGFVLLR  SRAILGETEAPEPSAGADASAAD
BR     GSEGAGIVPLN   IETLLFMVLDVSAKVGFGLILLR  SRAIFGEAEAPEPSAGDGAAATSD
HR     GVEGLALVQSVG  VTSWAYSVLDVFAKYVFAFILLR  WVANNERTVAVAGQTLGTMSSDD
PHR    GVEGIAVLP VG  VTSWGYSFLDIVAKYIFAFLLLN  YLTSNESVVSGSILDVPSASGTPADD
                     helix G
```

Fig. 1. Sequence alignment for three bacteriorhodopsins and two halorhodopsins. For brevity the single-letter amino acid code is used. Alignment and helical assignments are as in ref. [17]. Designations: AR-1, a bacteriorhodopsin from *Halobacterium* sp. aus-1 [44]; AR-2, a bacteriorhodopsin from *Halobacterium* sp. aus-2 [45]; BR, bacteriorhodopsin from *H. halobium* [5]; HR, halorhodopsin from *H. halobium* [43]; PHR, a halorhodopsin from *Natronobacterium pharaonis* [17]. The patterns of dots and asterisks indicate either identity in all five sequences (dots only) or identity among the bacteriorhodopsins and halorhodopsins only (dots and asterisks).

by site-specific mutations, e.g. serines or threonines, does not diminish proton transport in bacteriorhodopsin [18].

The arrangement of bacteriorhodopsin into trimers and the two-dimensional extended hexagonal lattice formed by the trimers has made it possible to describe the structure

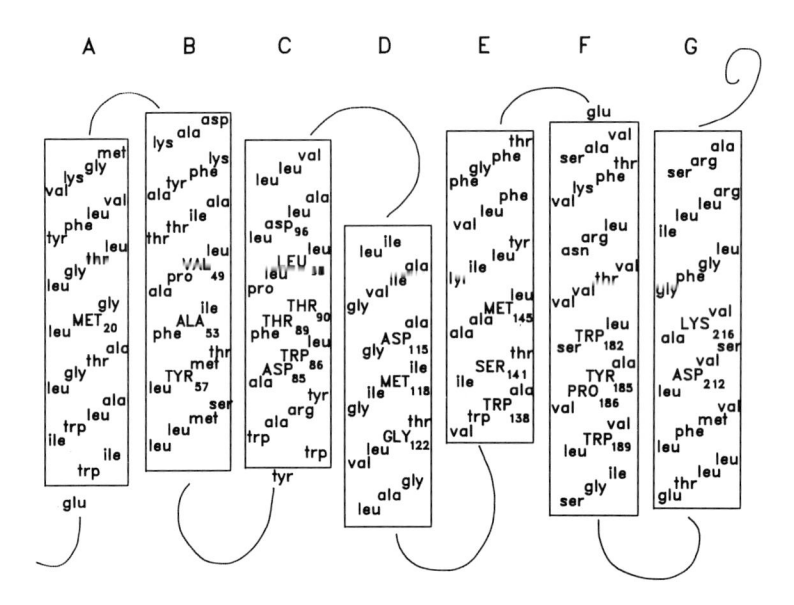

Fig. 2. Suggested secondary structure of bacteriorhodopsin, based on cryo-electron microscopy [16]. The four blocks represent transmembrane helices A through G. Residues in the retinal binding pocket are shown in capitals; these and asp96 are numbered for orientation. Residues in the N and C terminal segments and in the interhelical loops are not shown.

of the protein with a unique electron diffraction method [19]. The locations of bulky residues in the recent 3.5 Å resolution density map were used to orient the helices and thereby produce the first structural model at atomic resolution [16]. The seven helices surround the retinal. The functionally important polar residues face inward while a rather hydrophobic outer surface faces the lipid bilayer. The model identifies the residues which form the retinal binding site extending from lys216 on helix G and near the middle of the membrane, at a 23° angle to the plane of the membrane, toward helices A and B. It identifies also the residues which form the proton channel, in a roughly perpendicular direction to the plane of the membrane, above and below the Schiff base. The positioning of these channels (Fig. 3) immediately suggests their function in the proton pump. One is a wide hydrophilic opening from the Schiff base toward the cell exterior and the other a narrow hydrophobic channel connecting the Schiff base with the cytoplasmic side. Inside the former channel the protonatable residues asp212, asp85, arg82, and the Schiff base form a roughly tetrahedral cage, whose functional significance as a possible complex counterion and proton acceptor is made likely by various observations [20–28]. These residues, together with asp96 [20,21,27–34] and possibly tyr57 [35,36] and tyr185 [35,37–42], appear to be important in proton transport.

The sequence similarities between bacteriorhodopsins and halorhodopsins from various species [17,43–45] argue for extensive structural similarities between the two kinds of proteins. It seems likely that the tertiary structure of the halorhodopsins is a generalized bacteriorhodopsin-like arrangement, but two- or three-dimensional crystalline halorhodopsin has not been produced as yet. A crude model for halorhodopsin, based on

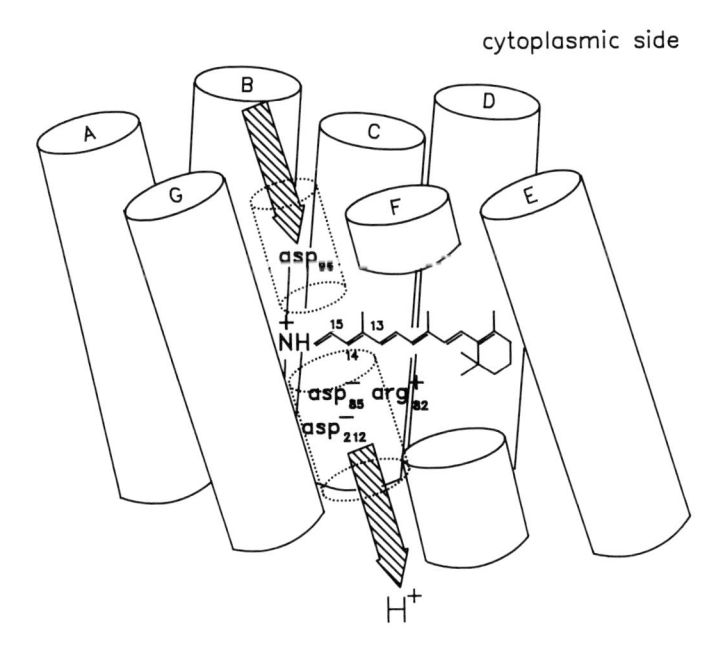

Fig. 3. Approximate structure of bacteriorhodopsin based on ref. [16]. Helices A through G span the lipid bilayer, and surround the retinal which is attached via Schiff-base linkage to lys216 which is located near the middle of helix G. The retinal is shown as all-*trans*. The two proton channels connecting the Schiff base to the extracellular and cytoplasmic surfaces are indicated as inclined dotted cylinders. In the former region the residues of interest (cf. the text) are arg82, asp85, and asp212; in the latter region it is asp96.

the bacteriorhodopsin structure, is given in Fig. 4, together with the residues of probable importance in transport.

3. Chromophore

The ion-motive bacterial rhodopsins contain one retinal per molecule, in a protonated Schiff base linkage to a lysine residue near the middle of helix G. The absorption maxima are 140–150 nm red-shifted from model retinal compounds in solution; this "opsin shift" is attributed mostly to the fact that the intrinsic counterion (in bacteriorhodopsin most likely a complex hydrogen-bonded structure which contains asp85, asp212, and arg82) is several Å removed from the Schiff base nitrogen. This leaves the positive charge of the Schiff base partly uncompensated and confers some double-bond character to the alternating single bonds in the conjugated chain of the retinal, thereby creating a diffuse delocalized electron orbital over the length of the ethylenic chain [46–48]. Other influences which have been proposed to contribute to the opsin shift include decrease of the angle between the β-ionone ring and the plane of the methyl groups [49], and Coulombic effects from polar residues or dipoles near the ring [50].

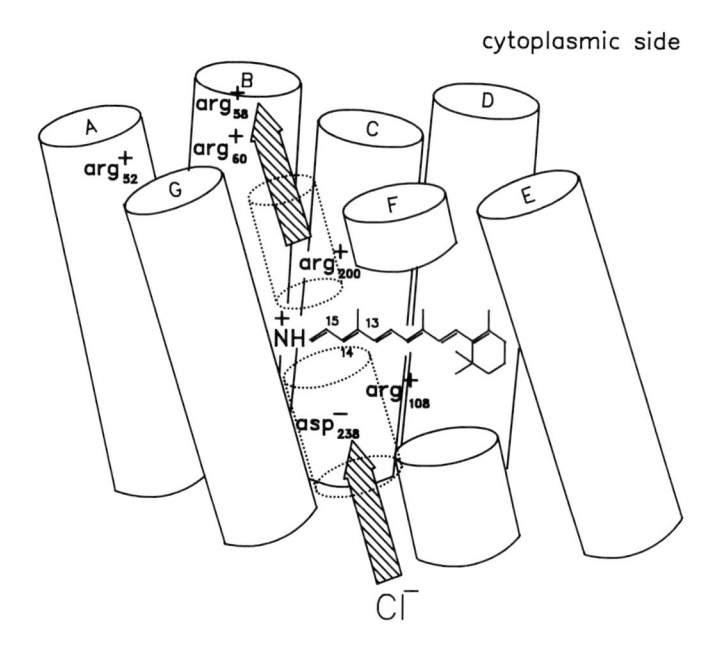

Fig. 4. Approximate structure of halorhodopsin, based on analogy with bacteriorhodopsin. Putative chloride channels are as indicated with inclined dotted cylinders. arg108 probably participates in chloride uptake; arg200 probably participates in chloride release [50].

In halorhodopsin the absorption maximum of the chromophore is influenced by very similar factors, since replacement of the retinal with various dihydro analogues results in similar wavelength shifts as in bacteriorhodopsin [50]. In halorhodopsin the chromophore is affected also by the presence or absence of a transported anion. The absorption maximum in the presence of chloride is at 578 nm, while in nitrate it is at 565 nm [51]. Competition between chloride and nitrate for binding is indicated by the fact that absorption shifts reflecting this difference in the maxima can be produced in both directions upon adding one of the anions to a sample containing the other anion [52]. Another measurement which shows such effects is the C=N stretch vibrational frequency [53]. Determined by resonance Raman spectroscopy, this frequency is influenced by coupling to N–H rocking, i.e., by the strength of the counterion near the Schiff base. The C=N stretch in halorhodopsin is distributed into two discrete values, at 1633 cm^{-1} and at 1645 cm^{-1}. In the presence of chloride only the low-frequency band is seen, while with nitrate both bands are obtained. At saturating nitrate concentrations the two bands have approximately equal amplitude. These results also suggest competition between the anions. In the simplest model [52] there are two sites with influence on the Schiff base; one will bind either chloride or nitrate, the other only nitrate. Importantly, only one of these sites can be occupied at a time. The two sites bring to mind asp85 and asp96 in bacteriorhodopsin, i.e., they suggest the idea of an uptake and a release site for the transported ion. Consistent with this idea: (1) nitrate is transported at a maximal rate about half that of chloride [54], and 2) the photocycle in the presence of nitrate appears to be a mixture of two parallel cycles, one of which resembles the cycle in the presence of chloride [55].

Although bacteriorhodopsin contains all of the buried arginine residues which could possibly play a role in binding anions in halorhodopsin, its absorption does not show anion-dependent effects except at very low pH where protonation of asp85 (with a pK of 2.5) causes a shift from 568 nm to about 605 nm [22,56–63]. Addition of chloride to this blue chromophore shifts the maximum back to 565 nm [56,61,64–66]. A sustained photocurrent was not seen at the low pH, but after addition of chloride the photocurrent reappeared. It is tempting to compare these chloride-dependent effects to the behavior of halorhodopsin; the possibility of chloride transport by bacteriorhodopsin at low pH was mentioned [67,68]. However, there are discrepancies. Chloride in halorhodopsin causes a red-shift rather than a blue-shift, and the photocycle of bacteriorhodopsin at low pH with bound chloride is quite different from the photocycle of halorhodopsin with bound chloride [61].

The extended chain of the all-*trans* retinal is accommodated well by the residues which form the binding pocket. The conserved prolines P50, P91 [69] and P186 [37,69] apparently determine the size and shape of the cavity since their replacement with bulky residues caused slower reconstitution of the chromophore with added all-*trans* retinal. The structural model of bacteriorhodopsin shows that conserved tryptophanes W12, W86, W182 and W189 surround the retinal in such a way that the ring π-electron orbitals effectively overlap those of the retinal chain [16], sandwiching the retinal in the way suggested by earlier data [70–72].

After a few minutes illumination ("light-adaptation") bacteriorhodopsin immobilized in the purple membrane lattice contains 100% all-*trans* retinal [73–78]. Light-adapted solubilized bacteriorhodopsin [79,80] and halorhodopsin [81–83], on the other hand, contain a mixture of about $\frac{2}{3}$ all-*trans* and $\frac{1}{3}$ 13-*cis* retinal. Dark-adaptation which takes minutes or hours, depending on conditions, results in thermally stable mixtures of $\frac{1}{3}$ all-*trans* and $\frac{2}{3}$ 13-*cis* chromophores in all cases. The dark-adapted 13-*cis* chromophores are stable because the overall shape of retinal is not very different from that of the all-*trans* chain, the C=N bond having assumed the *syn* rather than the *anti* configuration [84,85].

Most of the information on the photoreactions of the chromophores (cf. below) is for the all-*trans* form. The 13-*cis* chromophores produced in the photoreaction are C=N *anti* [85] and not as well accommodated by the binding pocket; this results in the thermal relaxation to all-*trans* and thus the cyclic reaction. Because the position of the β-ionone ring is fixed, the displacement caused by the bond isomerization around the C13–C14 bond is confined to the chain near the Schiff base. As will be discussed below, it is the movements near the Schiff-base region of the retinal which store excess free energy to drive the reactions of the photocycle and the accompanying proton transport.

4. Properties of the Schiff base

The pK_a of the protonated Schiff base in bacteriorhodopsin is shifted from around 7 in model compounds in solution to well above 10 [86,87]. In contrast, the halorhodopsin Schiff-base pK_a is not raised and a pH-dependent equilibrium at 7.4 between the protonated chromophore (absorbing near 580 nm) and the deprotonated chromophore (absorbing near 410 nm) is readily detected [51,88,89]. Since in halorhodopsin the most conspicuous sequence difference in the vicinity of the Schiff base is the lack of asp85, and replacement of asp85 in bacteriorhodopsin lowers the pK_a to near 7 [22,23], the reasonable suggestion was made that the pK_a shift originates from interaction of the Schiff base and asp85. Replacement of asp212, which is also nearby, with asparagine did not affect the Schiff-base pK_a [22,26]. It is very likely that stabilization of the Schiff-base proton is by the unfavorable free energy of an uncompensated charge at asp85. According to this argument asp85 is the primary counterion for the protonated Schiff base.

The Schiff base counterion does not appear to be as simple, however, as the idea of an ion pair would suggest. NMR investigations of the Schiff base nitrogen resonance indicated that the counterion should be described as a weak and diffuse entity rather than a single charged residue [24,25]. For this reason it was suggested that the Schiff base participates in a complex structure which contains also a centrally located water, two acidic residues (asp85 and asp212 are the obvious candidates), and an additional basic residue (arg82 is the obvious candidate). Transfer of the Schiff-base proton would be to this three-dimensional hydrogen-bonded structure, where the excess positive charge could be effectively delocalized. The observed 16 nm red-shift in the absorption maximum of the asp212asn bacteriorhodopsin [26] suggests indeed that asp212 is part of such a complex, although asp212 is not quite equivalent to asp85, whose replacement caused a red-shift of twice this magnitude [20,22,23]. As expected, proton transfer from the Schiff base of the asp212asn protein during the photocycle (cf. below) was either totally or partially inhibited, depending on pH [26]. Analysis of the proton transfer reaction indicated that this inhibition originated from a changed proton affinity (pK_a) of either the Schiff base or asp85. Thus, asp212 affects, but not necessarily participates in, the proton transfer.

In halorhodopsin the Schiff-base pK_a is raised by as much as 1.5 units in the presence of chloride, and less with other anions, suggesting specific binding in the vicinity of the Schiff base [51,89], consistent with other information (cf. above).

5. Photoreactions and photocycles

In free retinal, illumination will produce a mixture of several single- and double-bond isomeric configurations. Protein-bound retinal is more restricted in its motional possibilities; in the rhodopsins the binding pocket is tailored to fit only a few specific retinal configurations. Indeed, the light-initiated reaction cycles of the bacterial rhodopsins are based exclusively on all-*trans* to 13-*cis* isomerization of their retinal chromophores. When under some conditions other isomers are produced upon illumination, such as 9-*cis*, they accumulate in limited quantities [90–92].

Absorption of a photon by the all-*trans* chromophore produces an excited state which relaxes within 500 femtoseconds to the J intermediate (in the case of bacteriorhodopsin at least), in which rotation around the C_{13}–C_{14} double bond had occurred [93,94]. It has been suggested that there is isomerization also around the C_{14}–C_{15} single bond [95], but others dispute this [96]. All further intermediates are thermal (dark) products of J. Some reports distinguish, on the basis of their absorption maxima, between an early K

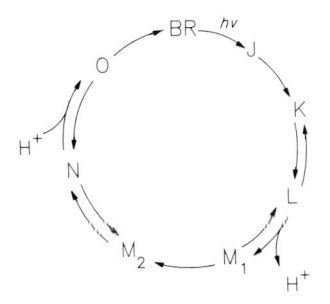

Fig. 5. Suggested scheme for the photocycle of bacteriorhodopsin [114–116]. The photoreaction produces state J, the other reactions represent the pathway of thermal relaxation back to the initial state. Where H$^+$ release and uptake are shown, they refer to exchange of the proton with the aqueous phase.

(lifetime in the picosecond range) and a late K, or KL (lifetime in the nanosecond range). These red-shifted intermediates have "strained" 13-*cis* configurations [85,97] and exhibit strong HOOP modes [98–100]. The late-K state relaxes in about 1 μs to a blue-shifted state called L. Vibrational spectra indicate that the retinal in L is clearly in the 13-*cis* configuration [101,102]. From this point the reactions of the bacteriorhodopsins and the halorhodopsins diverge. In bacteriorhodopsin, L is followed by a strongly blue-shifted state M [103], in which the Schiff base is deprotonated. Following M but partly overlapping it in time are the slightly blue-shifted N and the red-shifted O states, whose decay finally regenerates the bacteriorhodopsin. N still contains 13-*cis* retinal [104,105], but in O the retinal will have reisomerized to all-*trans* [106].

Many reports show this photocycle as a linear unidirectional scheme,

$$BR \xrightarrow{h\nu} J \to K \to L \to M \to N \to O \to BR,$$

but this represents an oversimplification as it does not account for the multiexponential rise and decay kinetics for several of the intermediates. In particular, the rise and decay of M, which are easiest to follow since the absorption maximum of M is far from the maximum of bacteriorhodopsin, are both described by 2 or 3 time constants. The explanations proposed for this are numerous and have included multiple initial bacteriorhodopsin forms [107,108], branching reactions prior to M [109–112], and a two-photon photocycle [113]. Recently, however, a consensus has begun to develop in favor of a simple linear scheme which contains several reverse reactions of significant rates [114–119].

Fig. 5 shows a kinetic model of the bacteriorhodopsin photocycle. It accounts quantitatively for time-resolved difference spectra measured between 100 ns and 100 ms after flash excitation at temperatures between 5 and 30°C. The kinetic connection between M, L, and K through reversible reactions introduces two additional time constants into the rise kinetics of M, and the reverse reactions which connect M, N, and O likewise affect the decay of M. An important part of this scheme is the existence of two consecutive M substates, M_1 and M_2, connected by an irreversible reaction [114].

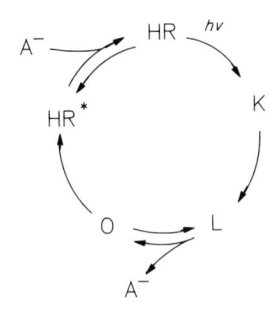

Fig. 6. Suggested scheme for the photocycle of halorhodopsin [120,121,123] (cf. comments on the bacteriorhodopsin photocycle in Fig. 4). A⁻ refers to chloride, bromide, or nitrate, but with nitrate an equilibrium mixture of halorhodopsins develops and an additional photocycle appears in which L is absent [3,52,55].

This is the simplest explanation for the observation that when L and M have come to an equilibrium which contains these species in comparable amounts, the concentration of L decreases to near zero even while M remains at its maximal accumulation. Recent measurements of the quasi-equilibrium which develops in asp96asn bacteriorhodopsin before the delayed reprotonation of the Schiff base confirm this kinetic paradox [115]. Two M states have been suggested also on the basis that the rise of N did not correlate with the decay of M [117]. In monomeric bacteriorhodopsin the two proposed M states in series have been distinguished spectroscopically as well [115]. It is well known, however, that kinetic data of the complexity exhibited by this system do not necessarily have a single mathematical solution. Thus, assurance that a numerically correct model represents the true behavior of the reaction must come from testing it for consistencies with physical principles. It is encouraging therefore that the model in Fig. 5 predicts spectra for the intermediates much as expected from other, independent measurements, and the rate constants produce linear Arrhenius plots and a self-consistent thermodynamic description [116].

The proposed halorhodopsin photocycle is shown in Fig. 6. In this protein the Schiff base of L does not deprotonate (except as a side reaction, cf. below), but a chloride-dependent equilibrium is established, within a few milliseconds, between L and a red-shifted O-like state [120–123]. From the kinetics it is inferred that the O → L back-reaction is accompanied by uptake of a chloride ion [120,121,123]. Decay of O by a non-reversible reaction regenerates halorhodopsin in tens of milliseconds, but to balance the loss of the chloride it is assumed that first the chloride-free form, HR*, is formed, which produces the original state after uptake of chloride on a time scale too rapid to measure. In the absence of chloride the L state is not observed [55,121]; it appears that O is directly produced in a truncated photocycle. In the presence of nitrate, an anion poorly transported by halorhodopsin from *H. halobium*, both photocycles are seen simultaneously. Halorhodopsin from *Natronobacterium pharaonis* transports nitrate as well as chloride, and in this system O cannot be detected in the photocycle [54].

Deprotonation of the halorhodopsin Schiff base does, in fact, occur during sustained illumination, as indicated by the slow pH-dependent accumulation of a species which absorbs near 410 nm [124,125]. This state contains 13-*cis* retinal [81–83] and its subsequent reprotonation in the dark occurs over hours. The chromophore is thus trapped in a fairly stable 410 nm-absorbing state. The stability of this species must originate with the 13-*cis* isomeric configuration of the retinal, since in all-*trans* halorhodopsin the dark protonation equilibrium is very rapid [51,124] and its illumination with blue light causes rapid protonation [126]. It seems clear that in both dark and light it is the reisomerization to all-*trans* which allows the reprotonation of the Schiff base. Because the rate at which the photostationary state develops is dependent on the light intensity but not its composition, the deprotonation is a branch reaction rather than part of the photocycle [125]. Some of these results suggest that it is O in which the Schiff base deprotonates reversibly in this way. Both de- and reprotonation are accelerated by added weak anionic bases, such as azide or cyanate, which appear to penetrate into the protein near the Schiff base from the cytoplasmic side and exchange protons with it [125,127]. The transient production of an M-like state superficially resembles the photocycle of bacteriorhodopsin, but the proton transfer under these conditions is not vectorial and there is no net transport. Thus, azide fulfills the function of asp96 in bacteriorhodopsin, a residue not present in halorhodopsin. Indeed, in bacteriorhodopsins where asp96 is replaced with a non-protonable residue, azide restores both rapid M decay and transport activity [32].

The proton transfer reaction between azide and other weak bases with various pK_as and the Schiff base allowed estimation of the pK_a of the halorhodopsin Schiff base at that time in the photocycle where the proton is exchanged. The ratio of forward and reverse proton transfer was dependent on the pK_a of the weak base and reached a value of unity at a pK_a of about 4.3 [127]. This is then the pK_a of the Schiff base at the time of proton transfer, representing a drop of about 3 pH units from its dark value. In bacteriorhodopsin it must be a similar pK_a shift, but of much greater magnitude, which allows the Schiff-base deprotonation in the L → M step.

6. Transport mechanism

The initial photochemistry of bacteriorhodopsin and halorhodopsin is essentially the same. As the excited state of the retinal chromophore relaxes, rotation around the C_{13}–C_{14} bond changes the shape of the extended conjugated chain from slightly bent to acutely bent at carbon 13. Since the β-ionone ring is spatially fixed it is the protonated Schiff base which is displaced by this motion, and removed from its counterion(s). This charge separation and decreased electronegativity of the Schiff base, owing to the rearranged chain, cause a decrease in affinity for the Schiff-base proton. In bacteriorhodopsin the proton will thus be transferred to a nearby acceptor group, thereby initiating a series of proton transfer reactions which result in the net translocation of a proton from the cytoplasmic side to the exterior side during the photocycle. In halorhodopsin, the lack of a suitable proton acceptor prevents deprotonation of the Schiff base, but it has been proposed [4,82] that its movement will cause the approach of a chloride ion, initiating in turn the subsequent chloride translocation.

A minimal description of the two transport mechanisms would identify the donor and acceptor groups in the ion transfer reactions, and establish the connection between these reactions and the steps of the photocycles. For bacteriorhodopsin there is enough evidence to pose at least the outline of such a proton transport mechanism; for halorhodopsin the relevant data are sparse and allow only some speculations.

Low-temperature and time-resolved FTIR spectra identified several aspartate residues in bacteriorhodopsin which undergo reversible protonation or deprotonation during the photocycle [30]. Examination of site-specific and in vivo mutated proteins identified these as asp85, asp96, and probably asp212 [20–34]. In addition, asp115 appears to change hydrogen-bonding state. Glutamates do not participate in proton transfer; the protonation of a tyrosinate residue [35,37–42] during the photocycle is currently a controversial question. Replacement of asp212 affects the pK_a of the Schiff base and/or asp85 in such a way as to make proton transfer less likely [26]. Replacement of asp85 and asp96 with non-protonable residues greatly reduces transport [20–23,27–29,32–33,36].

Lack of a negative charge at residue 85 shifts the absorption maximum to the blue, and the photocycle ends at L, i.e. before the Schiff base deprotonates [23,28]. This is much like the situation with blue membrane produced either by lowering the pH or by deionization. The results thus suggest that asp85 (1) plays an important role in neutralization of the charge on the Schiff-base nitrogen, and (2) is likely to be the acceptor in the initial proton transfer. The FTIR spectra indicate that asp85 becomes protonated indeed during the L → M reaction [31,128]. However, since these measurements also show that asp85 remains protonated until a later time, and according to a large number of reports the released proton appears in the medium already during the rise of M [129–133], at least one other residue must participate in the proton transfer. It is possible that arg82 is this residue. The location of this arginine would allow ion-pair formation between the guanidinium group and asp85 and the required proton exchange, but its pK_a would have to decrease from an aqueous value of 10.5 to at least 4 in the protein in order to allow it to release its proton.

It is generally agreed that in the M → N reaction the deprotonated Schiff base regains the proton from asp96; asp96 is in the protonated state until this step of the photocycle, and FTIR spectra indicate that it deprotonates during the M → N reaction [30,31]. Replacement of this residue with a non-protonable residue, as in the asp96asn and asp96ala mutants, slows down the decay of the M state and confers a pH dependency to it; in these mutated proteins the decay rate is proportional to [H^+] at the surface [29,32,33]. In the absence of a protonable residue in position 96 therefore the reprotonation of the Schiff base occurs not by an internal reaction but directly from protons in the medium. Azide, an artificial proton donor/acceptor which allows for example the reversible deprotonation of the halorhodopsin Schiff base upon illumination and thus inhibits chloride transport [125,127], restores near-normal M decay and proton transport to the mutated bacteriorhodopsins [32,33]. Thus, functionally azide substitutes for asp96 as the internal proton donor; it accelerates proton transfer by shuttling HN_3 from the medium directly to the Schiff base.

The proton conduction between asp96 and the Schiff base takes place in a narrow channel lined with mostly hydrophobic residues [16]. The distance between the donor and acceptor groups is about 10 Å; the means by which proton movement is achieved are uncertain. Much has been speculated on the existence of hydrogen-bonded chains in proteins through which protons can move by a mechanism much like that occurring in ice [134,135]. In bacteriorhodopsin there are not enough residues in this region to form a continuous proton-transfer chain. A string of trapped water molecules could fit into the gaps and serve this function, however. A difference-density map for D_2O in the protein revealed four tightly bound water molecules coincident with the projected position of this region [136]. Recent results (Yi, Váró, Chang, Ni, Needleman and Lanyi, unpublished experiments) indicate that a large part of the barrier to the transfer of a proton from asp96 to the Schiff base consists of the enthalpy of the initial proton–aspartate ion pair formed in the reaction. The effects of osmotically active solutes on the proton transfer indicate that this transition state is stabilized by at least 15 moles of loosely bound water.

During the following step, the decay of the N state, the retinal reisomerizes to all-*trans*. At the same time asp96 is reprotonated from the medium on the cytoplasmic side [33,113, 132]. The reprotonation of asp96 is facilitated by the negative charge of this residue: in the asp96asn protein a large negative activation entropy considerably slows the proton capture at the opening of the cytoplasmic channel. The protonation and isomerization processes appear to be coupled to one another as no substates of N have been observed so far. The protonation of asp96 must be the initial reaction as the dependence on pH indicates that it is the rate-limiting step. In the asp96asn mutant when the decay of M is accelerated with azide a normal N will accumulate (as cited in ref. [132]). As expected however, in this protein which lacks asp96 the decay of N is not very pH dependent.

The N decay step completes the transport cycle even though the recovery of bacteriorhodopsin has not yet taken place. Until this point in the photocycle, absorption of a second photon (e.g. by M in the "blue-light effect") defeats the transport process as it recovers bacteriorhodopsin without net proton translocation [137,138]. Absorption of a second photon by N, however, results in completion of the transport cycle and this takes place more rapidly than it does by the thermal route [139]. It appears then that the reactions after N serve simply to recover the chromophore.

7. Energetics and coupling

Only about 10% of the energy of the absorbed photon is conserved in the form of proton-motive force, but most of the losses in the photocycle occur before the K state. Once the all-*trans* to 13-*cis* retinal configurational change is accomplished, the return to the initial state, and the translocation of the proton which accompanies it, are highly efficient. The enthalpy conserved in K was determined to be 49 kJ/ml (Birge, personal communication); the excess free energy in this state would be greater if it contained also a significant entropy term. This seems unlikely. Large entropy changes are not expected in the reactions preceding K as they are so rapid as to preclude conformational consequences in the protein which would occur slower, and the retinal itself does not possess the required degrees of freedom. A proton-motive force of -200 mV is equivalent to about 17 kJ/mol, giving a thermodynamic efficiency of about 35% for this transport system. Calorimetric measurements of enthalpy changes, combined with enthalpies and entropies of activation calculated from the temperature dependencies of the elementary rate constants, provide a thermodynamic description of the photocycle which explains the high efficiency [116]. It appears from this description that in bacteriorhodopsin, as in other enzyme reaction cycles, most of the interconversions take place without a change in free energy, i.e., near equilibrium. This ensures minimal thermal losses.

Operating near equilibrium also means that in the proton transfer steps the pK_as of donor and acceptor groups are closely matched. Since titration of color changes in the chromophore suggest that the initial pKa of the Schiff base may be above 10 [86] and the pK_a of asp85 is about 3 [128,140], at least 40 kJ/mol of the excess free energy in K is used to keep the Schiff-base pK_a low (or that of asp85 high) during the $L \to M_1$ step. This would be largely so even if these pK_as depended on the ionization state of other residues, so that the proton affinities in the highly specific proton transfer reaction between asp85 and the Schiff base were somewhat closer together than given by the titration results.

Large decreases of free energy occur only in the $M_1 \to M_2$ reaction, and in the recovery of the initial chromophore. These ensure irreversibility of the proton transport and full recovery of the chromophore, respectively. The consequence of the large free-energy drop at the $M1 \to M_2$ step is that, although the model requires that the free-energy drop over the entire K–N segment drive the proton translocation, proton-motive force will not reverse the entire sequence but piles up M_2 against the large reverse free-energy barrier. This is indeed observed: proton-motive force causes a slowing of M decay [141].

Since bacteriorhodopsin forms a crystalline lattice in the purple membrane the question naturally arises as to the effects of motional constraints [142,143] on the photocycle reactions. A model in which conformational changes in the protein begin with M_1 and continue until the recovery of BR would predict that in monomers the entropies of K, L, and M_1 would be as in the purple-membrane lattice but after M_1 they would change. This is in fact what was observed for the entropy levels, and for enthalpy levels as well [144]. Another effect of solubilization was that the transition states of all forward reactions had lowered entropies. Thus, the crystalline lattice confers an entropic advantage to the photocycle reactions: the conformation of the protein is fixed to provide optimal access to all transition states. The price paid for this is higher thermal barriers. The latter indicate that in the lattice larger molecular motions in the protein are necessary to complete the reactions. Time-resolved diffraction has recently indicated that conformational changes do occur during the decay of M [145].

8. Summary

Proton transport by bacteriorhodopsin is now understood to be linked mainly to transient pK_a changes of the Schiff base, which occur in response to the light-driven isomerization of the retinal and its relaxation. Proton-motive force is created as the Schiff-base proton is transferred to asp85 and a proton is released into the extracellular medium. The rest of the reactions, which reset the system, consist of reprotonation of the Schiff base from asp96, reisomerization of the retinal, as well as reprotonation of asp96 from the cytoplasmic side. The mechanism of chloride transport by halorhodopsin is less certain, but a similar scheme is likely, in which binding of the transported anion might be by arginine residues which play the roles the aspartate residues have in the proton pump.

References

[1] Trissl, H.W. (1990) Photochem. Photobiol. 51, 793–818.
[2] Kouyama, T., Kinosita, K. and Ikegami, A. (1988) Adv. Biophys. 24, 123–175.
[3] Lanyi, J.K. (1990) Physiol. Rev. 70, 319–330.
[4] Oesterhelt, D. and Tittor, J. (1989) TIBS 14, 57–61.
[5] Khorana, H.G. (1988) J. Biol. Chem. 263, 7439–7442.
[6] Hegemann, P., Tittor, J., Blanck, A. and Oesterhelt, D. (1987) Progress in halorhodopsin. In: Retinal Proteins (Ovchinnikov, Yu.A., Ed.), p. 333–352, VNU Science Press, Utrecht.
[7] Smith, S.O., Lugtenburg, J. and Mathies, R.A. (1985) J. Membr. Biol. 85, 95–109.
[8] Jap, B.K., Maestre, M.F., Hayward, S.B. and Glaeser, R.M. (1983) Biophys. J. 43, 81–89.
[9] Vogel, H. and Gärtner, W. (1987) J. Biol. Chem. 262, 11464–11469.
[10] Engelman, D.M., Steitz, T.A. and Goldman, A. (1986) Annu. Rev. Biophys. Biophys. Chem. 15, 321–353.
[11] White, S.H. and Jacobs, R.E. (1990) J. Membr. Biol. 115, 145–158.
[12] Henderson, R. and Unwin, P.N. (1975) Nature 257, 28–32.
[13] Engelman, D.M., Henderson, R., McLachlan, A.D. and Wallace, B.A. (1980) Proc. Natl. Acad. Sci. U.S.A. 77, 2023–2027.
[14] Henderson, R., Baldwin, J.M., Downing, K.H., Lepault, J. and Zemlin, F. (1986) Ultramicroscopy 19, 147–178.
[15] Ceska, T.A. and Henderson, R. (1990) J. Mol. Biol. 213, 539–560.
[16] Henderson, R., Baldwin, J.M., Ceska, T.A., Zemlin, F., Beckmann, E. and Downing, K.H. (1990) J. Mol. Biol. 213, 899–929.
[17] Lanyi, J.K., Duschl, A., Hatfield, G.W., May, K.M. and Oesterhelt, D. (1990) J. Biol. Chem. 265, 1253–1260.
[18] Mart, T., Otto, H., Mogi, T., Rösselet, S.J., Heyn, M.P. and Khorana, H.G. (1991) J. Biol. Chem. 266, 6919–6927.
[19] Unwin, P.N. and Henderson, R. (1975) J. Mol. Biol. 94, 425–440.
[20] Mogi, T., Stern, L.J., Marti, T., Chao, B.H. and Khorana, H.G. (1988) Proc. Natl. Acad. Sci. U.S.A. 85, 4148–4152.
[21] Marinetti, T., Subramaniam, S., Mogi, T., Marti, T. and Khorana, H.G. (1989) Proc. Natl. Acad. Sci. U.S.A. 86, 529–533.
[22] Subramaniam, S., Marti, T. and Khorana, H.G. (1990) Proc. Natl. Acad. Sci. U.S.A. 87, 1013–1017.
[23] Otto, H., Marti, T., Holz, M., Mogi, T., Stern, L.J., Engel, F., Khorana, H.G. and Heyn, M.P. (1990) Proc. Natl. Acad. Sci. U.S.A. 87, 1018–1022.
[24] De Groot, H.J.M., Harbison, G.S., Herzfeld, J. and Griffin, R.G. (1989) Biochemistry 28, 3346–3353.
[25] De Groot, H.J.M., Smith, S.O., Courtin, J., Van den Berg, E., Winkel, C., Lugtenburg, J., Griffin, R.G. and Herzfeld, J. (1990) Biochemistry 29, 6873–6883.
[26] Needleman, R., Chang, M., Ni, B., Váró, G., Fornes, J., White, S.H. and Lanyi, J.K. (1991) J. Biol. Chem. 266, 11478–11484.
[27] Butt, H.-J., Fendler, K., Bamberg, E., Tittor, J. and Oesterhelt, D. (1989) EMBO J. 8, 1657–1663.
[28] Miercke, L.J.W., Betlach, M.C., Mitra, A.K., Shand, R.F., Fong, S.K. and Stroud, R.M. (1991) Biochemistry 30, 3088–3098.
[29] Holz, M., Drachev, L.A., Mogi, T., Otto, H., Kaulen, A.D., Heyn, M.P., Skulachev, V.P. and Khorana, H.G. (1989) Proc. Natl. Acad. Sci. U.S.A. 86, 2167–2171.
[30] Engelhard, M., Gerwert, K., Hess, B., Kreutz, W. and Siebert, F. (1985) Biochemistry 24, 400–407.

[31] Gerwert, K., Hess, B., Soppa, J. and Oesterhelt, D. (1989) Proc. Natl. Acad. Sci. U.S.A. 86, 4943–4947.
[32] Tittor, J., Soell, C., Oesterhelt, D., Butt, H.-J. and Bamberg, E. (1989) EMBO J. 8, 3477–3482.
[33] Otto, H., Marti, T., Holz, M., Mogi, T., Lindau, M., Khorana, H.G. and Heyn, M.P. (1989) Proc. Natl. Acad. Sci. U.S.A. 86, 9228–9232.
[34] Stern, L.J., Ahl, P.L., Marti, T., Mogi, T., Duñach, M., Berkovitz, S., Rothschild, K.J. and Khorana, H.G. (1989) Biochemistry 28, 10035–10042.
[35] Mogi, T., Stern, L.J., Hackett, N.R. and Khorana, H.G. (1987) Proc. Natl. Acad. Sci. U.S.A. 84, 5595–5599.
[36] Soppa, J., Otomo, J., Straub, J., Tittor, J., Meessen, S. and Oesterhelt, D. (1989) J. Biol. Chem. 264, 13049–13056.
[37] Ahl, P.L., Stern, L.J., During, D., Mogi, T., Khorana, H.G. and Rothschild, K.J. (1988) J. Biol. Chem. 263, 13594–13601.
[38] Hackett, N.R., Stern, L.J., Chao, B.H., Kronis, K.A. and Khorana, H.G. (1987) J. Biol. Chem. 262, 9277–9284.
[39] Ahl, P.L., Stern, L.J., Mogi, T., Khorana, H.G. and Rothschild, K.J. (1989) Biochemistry 28, 10028–10034.
[40] Duñach, M., Berkowitz, S., Marti, T., He, Y.-W., Subramaniam, S., Khorana, H.G. and Rothschild, K.J. (1990) J. Biol. Chem. 265, 16978–16984.
[41] Rothschild, K.J., Braiman, M.S., He, Y.-W., Marti, T. and Khorana, H.G. (1990) J. Biol. Chem. 265, 16985–16991.
[42] Jang, D.-J., El-Sayed, M.A., Stern, L.J., Mogi, T. and Khorana, H.G. (1990) Proc. Natl. Acad. Sci. U.S.A. 87, 4103–4107.
[43] Blanck, A. and Oesterhelt, D. (1987) EMBO J. 6, 265–273.
[44] Sugiyama, Y., Maeda, M., Futai, M. and Mukohata, Y. (1989) J. Biol. Chem. 264, 20859–20862.
[45] Uegaki, K., Sugiyama, Y. and Mukohata, Y. (1991) Arch. Biochem. Biophys. 286, 107–110.
[46] Kakitani, H., Kakitani, T., Rodman, H. and Honig, B. (1985) Photochem. Photobiol. 41, 471–479.
[47] Warshel, A. (1978) Proc. Natl. Acad. Sci. U.S.A. 75, 2558–2562.
[48] Nakanishi, K., Balogh-Nair, V., Arnaboldi, M., Tsujimoto, K. and Honig, B. (1980) J. Am. Chem. Soc. 102, 7945–7947.
[49] Harbison, G.S., Smith, S.O., Pardoen, J.A., Courtin, J., Lugtenburg, J., Herzfeld, J., Mathies, R.A. and Griffin, R.G. (1985) Biochemistry 24, 6955–6962.
[50] Lanyi, J.K., Zimányi, L., Nakanishi, K., Derguini, F., Okabe, M. and Honig, B. (1988) Biophys. J. 53, 185–191.
[51] Schobert, B., Lanyi, J.K. and Oesterhelt, D. (1986) J. Biol. Chem. 261, 2690–2696.
[52] Lanyi, J.K., Duschl, A., Váró, G. and Zimányi, L. (1990) FEBS Lett. 265, 1–6.
[53] Pande, C., Lanyi, J.K. and Callender, R.H. (1989) Biophys. J. 55, 425–431.
[54] Duschl, A., Lanyi, J.K. and Zimányi, L. (1990) J. Biol. Chem. 265, 1261–1267.
[55] Zimányi, L. and Lanyi, J.K. (1989) Biochemistry 28, 5172–5178.
[56] Fischer, U. and Oesterhelt, D. (1979) Biophys. J. 28, 211–230.
[57] Edgerton, M.E., Moore, T.A. and Greenwood, C. (1980) Biochem. J. 189, 413–420.
[58] Mowery, P.C., Lozier, R.H., Chae, Q., Tseng, Y.W., Taylor, M. and Stoeckenius, W. (1979) Biochemistry 18, 4100–4107.
[59] Moore, T.A., Edgerton, M.E., Parr, G., Greenwood, C. and Perham, R.N. (1978) Biochem. J. 171, 469–476.
[60] Duñach, M., Padros, E., Seigneuret, M. and Rigaud, J.-L. (1988) J. Biol. Chem. 263, 7555–7559.
[61] Váró, G. and Lanyi, J.K. (1989) Biophys. J. 56, 1143–1151.
[62] Heyn, M.P., Dudda, C., Otto, H., Seiff, F. and Wallat, I. (1989) Biochemistry 28, 9166–9172.
[63] Albeck, A., Friedman, N., Sheves, M. and Ottolenghi, M. (1989) Biophys. J. 56, 1259–1265.

[64] Szundi, I. and Stoeckenius, W. (1988) Biophys. J. 54, 227–232.
[65] Renthal, R., Shuler, K. and Regalado, R. (1989) Biochim. Biophys. Acta 1016, 378, 384.
[66] Drachev, A.L., Drachev, L.A., Kaulen, A.D., Khitrina, L.V., Skulachev, V.P., Lepnev, G.P. and Chekulaeva, L.N. (1989) Biochim. Biophys. Acta 976, 190–195.
[67] Dér, A., Tóth-Boconádi, R. and Keszthelyi, L. (1989) FEBS Lett. 259, 24–26.
[68] Keszthelyi, L., Száraz, S., Dér, A. and Stoeckenius, W. (1990) Biochim. Biophys. Acta Bio-Energetics 1018, 260–262.
[69] Mogi, T., Stern, L.J., Chao, B.H. and Khorana, H.G. (1989) J. Biol. Chem. 264, 14192–14196.
[70] Rothschild, K.J., Braiman, M.S., Mogi, T., Stern, L.J. and Khorana, H.G. (1989) FEBS Lett. 250, 448–452.
[71] Lin, S.W. and Mathies, R.A. (1989) Biophys. J. 56, 653–660.
[72] Rothschild, K.J., Gray, D., Mogi, T., Marti, T., Braiman, M.S., Stern, L.J. and Khorana, H.G. (1989) Biochemistry 28, 7052–7059.
[73] Pettei, M.J., Yudd, A.P., Nakanishi, K., Henselman, R. and Stoeckenius, W. (1977) Biochemistry. 16, 1955–1959.
[74] Kamo, N., Hazemoto, N., Kobatake, Y. and Mukohata, Y. (1985) Arch. Biochem. Biophys. 238, 90–96.
[75] Dencher, N.A., Kohl, K.D. and Heyn, M.P. (1983) Biochemistry 22, 1323–1334.
[76] Ohno, K., Takeuchi, Y. and Yoshida, M. (1977) Biochim. Biophys. Acta 462, 575–582.
[77] Sperling, W., Carl, P., Rafferty, C.N. and Dencher, N.A. (1977) Biophys. Struct. Mech. 3, 79–94.
[78] Scherrer, P., Mathew, M.K., Sperling, W. and Stoeckenius, W. (1989) Biochemistry 28, 829–834.
[79] Casadio, R. and Stoeckenius, W. (1980) Biochemistry 19, 3374–3381.
[80] Casadio, R., Gutowitz, H., Mowery, P.C., Taylor, M. and Stoeckenius, W. (1980) Biochim. Biophys. Acta 590, 13–23.
[81] Lanyi, J.K. (1986) J. Biol. Chem. 261, 14025–14030.
[82] Oesterhelt, D., Hegemann, P., Tavan, P. and Schulten, K. (1986) Eur. Biophys. J. 14, 123–129.
[83] Maeda, A., Ogurusu, T., Yoshizawa, T. and Kitagawa, T. (1985) Biochemistry 24, 2517–2521.
[84] Harbison, G.S., Smith, S.O., Pardoen, J.A., Winkel, C., Lugtenburg, J., Herzfeld, J., Mathies, R.A. and Griffin, R.G. (1984) Proc. Natl. Acad. Sci. U.S.A. 81, 1706–1709.
[85] Smith, S.O., Myers, A.B., Pardoen, J.A., Winkel, C., Mulder, P.P.J., Lugtenburg, J. and Mathies, R.A. (1984) Proc. Natl. Acad. Sci. U.S.A. 81, 2055–2059.
[86] Druckmann, S., Ottolenghi, M., Pande, A., Pande, J. and Callender, R.H. (1982) Biochemistry 21, 4953–4959.
[87] Ehrenberg, B. and Lewis, A. (1978) Biochem. Biophys. Res. Commun. 82, 1154–1159.
[88] Steiner, M., Oesterhelt, D., Ariki, M. and Lanyi, J.K. (1984) J. Biol. Chem. 259, 2179–2184.
[89] Schobert, B. and Lanyi, J.K. (1986) Biochemistry 25, 4163–4167.
[90] Fischer, U., Towner, P. and Oesterhelt, D. (1981) Photochem. Photobiol. 33, 529–537.
[91] Chang, C.-H., Liu, S.-Y., Jonas, R. and Govindjee, R. (1987) Biophys. J. 52, 617–623.
[92] Zimányi, L. and Lanyi, J.K. (1987) Biophys. J. 52, 1007–1013.
[93] Van den Berg, R., Du-Jeon-Jang, Bitting, H.C. and El-Sayed, M.A. (1990) Biophys. J. 58, 135–141.
[94] Polland, H.J., Franz, M.A., Zinth, W., Kaiser, W., Kölling, E. and Oesterhelt, D. (1984) Biochim. Biophys. Acta 767, 635–639.
[95] Gerwert, K. and Siebert, F. (1986) EMBO J. 5, 805–811.
[96] Fodor, S.P., Pollard, W.T., Gebhard, R., van den Berg, E.M., Lugtenburg, J. and Mathies, R.A. (1988) Proc. Natl. Acad. Sci. U.S.A. 85, 2156–2160.
[97] Braiman, M.S. and Mathies, R.A. (1982) Proc. Natl. Acad. Sci. U.S.A. 79, 403–407.
[98] Smith, S.O., Hornung, I., van der Steen, R., Pardoen, J.A., Braiman, M.S., Lugtenburg, J. and Mathies, R.A. (1986) Proc. Natl. Acad. Sci. U.S.A. 83, 967–971.

[99] Rothschild, K.J., Marrero, H., Braiman, M.S. and Mathies, R.A. (1984) Photochem. Photobiol. 40, 675–679.
[100] Siebert, F. and Mantele, W. (1983) Eur. J. Biochem. 130, 565–573.
[101] Alshuth, T. and Stockburger, M. (1986) Photochem. Photobiol. 43, 55–66.
[102] Aton, B., Doukas, A.G., Callender, R.H., Becher, B. and Ebrey, T.G. (1977) Biochemistry 16, 2995–2999.
[103] Lozier, R.H., Bogomolni, R.A. and Stoeckenius, W. (1975) Biophys. J. 15, 955–963.
[104] Fodor, S.P., Ames, J.B., Gebhard, R., van der Berg, E.M., Stoeckenius, W., Lugtenburg, J. and Mathies, R.A. (1988) Biochemistry 27, 7097 7101.
[105] Braiman, M.S., Bousché, O. and Rothschild, K.J. (1991) Proc. Natl. Acad. Sci. U.S.A. 88, 2388–2392.
[106] Smith, S.O., Pardoen, J.A., Mulder, P.P.J., Curry, B., Lugtenburg, J. and Mathies, R.A. (1983) Biochemistry 22, 6141–6148.
[107] Hanamoto, J.H., Dupuis, P. and El-Sayed, M.A. (1984) Proc. Natl. Acad. Sci. U.S.A. 81, 7083–7087.
[108] Dancsházy, Z., Govindjee, R. and Ebrey, T.G. (1988) Proc. Natl. Acad. Sci. U.S.A. 85, 6358–6361.
[109] Sherman, W.V., Slifkin, M.A. and Caplan, S.R. (1976) Biochim. Biophys. Acta 423, 238–248.
[110] Sherman, W.V., Eicke, R.R., Stafford, S.R. and Wasacz, F.M. (1979) Photochem. Photobiol. 30, 727–729.
[111] Lozier, R.H., Niederberger, W., Ottolenghi, M., Sivorinovsky, G. and Stoeckenius, W. (1978) In: Energetics and Structure of Halophilic Microorganisms (Caplan, S.R. and Ginzburg, M., Eds.), pp. 123–139, Elsevier, Amsterdam.
[112] Beach, J.M. and Fager, R.S. (1985) Photochem. Photobiol. 41, 557–562.
[113] Kouyama, T., Nasuda-Kouyama, A., Ikegami, A., Mathew, M.K. and Stoeckenius, W. (1988) Biochemistry 27, 5855–5863.
[114] Váró, G. and Lanyi, J.K. (1990) Biochemistry 29, 2241–2250.
[115] Váró, G. and Lanyi, J.K. (1991) Biochemistry 30, 5008–5015.
[116] Váró, G. and Lanyi, J.K. (1991) Biochemistry 30, 5016–5022.
[117] Gerwert, K., Souvignier, G. and Hess, B. (1990) Proc. Natl. Acad. Sci. U.S.A. 87, 9774–9778.
[118] Ames, J.B. and Mathies, R.A. (1990) Biochemistry 29, 7181–7190.
[119] Milder, S.J., Thorgeirsson, T.E., Miercke, L.J.W., Stroud, R.M. and Kliger, D.S. (1991) Biochemistry 30, 1751–1761.
[120] Oesterhelt, D., Hegemann, P. and Tittor, J. (1985) EMBO J. 4, 2351–2356.
[121] Lanyi, J.K. and Vodyanoy, V. (1986) Biochemistry 25, 1465–1470.
[122] Tittor, J., Oesterhelt, D., Maurer, R., Desel, H. and Uhl, R. (1987) Biophys. J. 52, 999–1006.
[123] Zimányi, L., Keszthelyi, L. and Lanyi, J.K. (1989) Biochemistry 28, 5165–5172.
[124] Lanyi, J.K. and Schobert, B. (1983) Biochemistry 22, 2763–2769.
[125] Hegemann, P., Oesterhelt, D. and Steiner, M. (1985) EMBO J. 4, 2347–2350.
[126] Hegemann, P., Oesterhelt, D. and Bamberg, E. (1985) Biochim. Biophys. Acta 819, 195–205.
[127] Lanyi, J.K. (1986) Biochemistry 25, 6706–6711.
[128] Braiman, M.S., Mogi, T., Marti, T., Stern, L.J., Khorana, H.G. and Rothschild, K.J. (1988) Biochemistry 27, 8516–8520.
[129] Ort, D.R. and Parson, W.W. (1979) Biophys. J. 25, 341–353.
[130] Grzesiek, S. and Dencher, N.A. (1986) FEBS Lett. 208, 337–342.
[131] Drachev, L.A., Kaulen, A.D. and Skulachev, V.P. (1984) FEBS Lett. 178, 331–335.
[132] Váró, G. and Lanyi, J.K. (1990) Biochemistry 29, 6858–6865.
[133] Heberle, J. and Dencher, N.A. (1990) FEBS Lett. 277, 277–280.
[134] Nagle, J.F. and Tristram-Nagle, S. (1983) J. Membr. Biol. 74, 1–14.
[135] Schulten, Z. and Schulten, K. (1986) Methods Enzymol. 127, 419–438.

[136] Papadopoulos, G., Dencher, N.A., Zaccai, G. and Büldt, G. (1990) J. Mol. Biol. 214, 15–19.
[137] Dancsházy, Z., Drachev, L.A., Ormos, P., Nagy, K. and Skulachev, V.P. (1978) FEBS Lett. 96, 59–63.
[138] Ormos, P., Dancsházy, Z. and Keszthelyi, L. (1980) Biophys. J. 31, 207–213.
[139] Kouyama, T. and Nasuda-Kouyama, A. (1989) Biochemistry 28, 5963–5970.
[140] Rothschild, K.J., Zagaeski, M. and Cantore, W.A. (1981) Biochem. Biophys. Res. Commun. 103, 483–489.
[141] Groma, G.I., Helgerson, S.L., Wolber, P.K., Beece, D., Dancsházy, Z., Keszthelyi, L. and Stoeckenius, W. (1984) Biophys. J. 45, 985–992.
[142] Cherry, R.J., Heyn, M.P. and Oesterhelt, D. (1977) FEBS Lett. 78, 25–30.
[143] Czégé, J., Dér, A., Zimányi, L. and Keszthelyi, L. (1982) Proc. Natl. Acad. Sci. U.S.A. 79, 7273–7277.
[144] Váró, G. and Lanyi, J.K. (1991) Biochemistry 30, 7165–7171.
[145] Koch, M.H.J., Dencher, N.A., Oesterhelt, D., Plöhn, H.-J., Rapp, G. and Büldt, G. (1991) EMBO J. 10, 521–526.

M. Kates et al. (Eds.), *The Biochemistry of Archaea (Archaebacteria)*
© 1993 Elsevier Science Publishers B.V. All rights reserved

CHAPTER 7

Proteins of extreme thermophiles

R. HENSEL

FB 9 Mikrobiologie, Universität GHS Essen, Universitätsstr. 5, 45117 Essen 1, Germany

Abbreviations

ATPase	adenosine triphosphate hydrolase	Asp	aspartic acid
GAPDH	glyceraldehyde-3-phosphate dehydrogenase	Gln	glutamine
		Glu	glutamic acid
PGK	3-phosphoglycerate kinase	Ile	isoleucine
cDPG	cyclic 2,3-diphosphoglycerate	Val	valine
Arg	arginine	Tyr	tyrosine
Asn	asparagine	Phe	phenylalanine

1. Introduction

The discovery of archaea with growth temperatures near to or above 100°C over the last decade [1–6] brings up evident questions about the biochemical background of the adaptation to these extreme temperatures.

Like the other macromolecular constituents of the cell (membranes, polynucleotides), the proteins of these extreme thermophiles must also be stable enough to resist heat-induced destruction of conformation and covalent structure. Since virtually all proteins (functional as well as structural proteins) exhibit dynamic properties to fulfill the demands of the living cell, their structure must provide a compromise between rigidity and flexibility, allowing not only stability but also conformational freedom for their biological function at the respective temperature. This means they are not only thermoresistant but require the higher temperature for optimal function.

Here I will deal only with proteins from archaea growing at higher temperatures (temperature optima significantly above 60°C) than those normally found for thermophiles from bacteria or eucarya. The study of proteins from these organisms provides the opportunity to test the validity of the adaptation rules already deduced from proteins of organisms with lower thermophilic potential (mostly members of bacteria) and to detect new features of thermoadaptation specific for the high-temperature range. Thus, the following questions arise: How are proteins from archaea, growing at or above 100°C (so-called hyperthermophiles [6]), protected from covalent damage at temperatures which induce chemical modifications in mesophilic proteins? Do post-translational modifications play a significant role in stabilizing these proteins? Are extrinsic factors – within the

TABLE 1

Enzymes isolated from extremely thermophilic archaea and characterized with respect to their thermophilic properties[a]

Enzyme species [b]	Source	Thermostability	T_{opt}	Dependence on extrinsic factors	Structure information	Ref.
1. Extracellular enzymes						
Hydrolases						
Amylase	P. furiosus	$t_{0.5}(120°C) = 2$ h	100°C	–	–	[77]
	P. woesei	$t_{0.5}(100°C) = 6$ h	100°C	–	monomer; aa comp.	[58]
Protease	D strain $Tok_{12}S_1$	$t_{0.5}(95°C) \approx 1.5$ h	95°C	–	–	[63]
	S. acidocaldarius	100% after 1.5 h at 90°C	90°C	–	monomer, gene seq.	[13]
	P. furiosus	$t_{0.5}(100°C) = 4$ h	115°C	stab. by Ca^{2+} (1 species)	mixture of diff. species	[65-67]
2. Membrane-bound or -associated enzymes						
Oxidoreductases						
NADH dehydrogenase	S. acidocaldarius	70% after 0.25 h at 100°C	–	–	homomeric dimer; aa comp.	[78]
Succinate dehydrogenase	S. acidocaldarius	–	81°C	–	tetramer of 4 diff. subs	[79]
Hydrolases						
ATPase	S. acidocaldarius	100% after 0.3 h at 89°C	>75°C	Mg^{2+}-, Mn^{2+}-dep.	multimer of 4 diff. subs	[80,81]
3. Intracellular enzymes						
Oxidoreductases						
Alcohol dehydrogenase	S. solfataricus	$t_{0.5}(70°C) = 5$ h	>90°C	Zn^{2+}-dep.	homomeric dimer	[72]
Malate dehydrogenase	S. acidocaldarius	$t_{0.5}(90°C) = 0.25$ h	–	–	homomeric tetramer; aa comp.	[82]
	M. fervidus	$t_{0.5}(90°C) = 9$ min	>85°C	stab. by K^+ and cDPG	homomeric dimer; gene seq.	[83]
Malic enzyme	S. solfataricus	$t_{0.5}(85°C) = 5$ h	85°C	Me^{2+}-dep.	dimer; aa comp.	[84]
Glucose dehydrogenase	S. solfataricus	$t_{0.5}(70°C) = 45$ h	77°C	–	homomeric tetramer; aa comp.	[70]
Glyceraldehyde-3-phosphate dehydrogenase	M. fervidus	$t_{0.5}(85°C) = 0.25$ h	>75°C	stab. by K^+ and cDPG	homomeric tetramer; gene seq.	[85]
	P. woesei	$t_{0.5}(100°C) = 0.75$ h	90°C	stab. by K^+-salts	homomeric tetramer; gene seq.	[10]
$NADP^+$-spec. enzyme	T. tenax	$t_{0.5}(100°C) = 35$ min	>85°C	–	homomeric tetramer; aa comp.	[86]
NAD^+-spec. enzyme	T. tenax	$t_{0.5}(100°C) = 20$ min	>85°C	–	homomeric tetramer; aa comp.	[86]
Glutamate dehydrogenase	S. solfataricus	–	60–70°C	–	homomeric hexamer; prot. seq.	[8,87]
	P. furiosus	$t_{0.5}(100°C) = 12$ h	>80°C	–	homomeric hexamer	[88]
Aldehyde ferredoxin oxidoreductase	P. furiosus	–	>95°C	–	monomer; aa comp.	[89]
Hydrogenase	P. furiosus	$t_{0.5}(100°C) = 2$ h	>95°C	–	homomeric trimer; aa comp.	[90]
N^5,N^{10}-Methylene-H_4MPT dehydrogenase	M. kandleri	–	>90°C	salt-dep.	homomeric tetramer; N-terminal seq.	[33]

Enzyme	Organism	Thermostability	T_opt	Cofactors/Stab.	Structure	Ref.
N^5,N^{10}-Methylene-H$_4$MPT reductase	M. kandleri	$t_{0.5}(90°C) \approx 1-2$ min	90°C	salt-dep.	homomeric multimer; N-terminal seq.	[34]
Sulfur oxygenase reductase	D. ambivalens	–	85°C	–	homomeric multimer	[91]
Transferases						
5′-Methyl-thioadenosine phosphorylase	S. solfataricus	100% after 1 h at 100°C	93°C	–	–	[92]
Propylamine transferase	S. solfataricus	100% after 1 h at 100°C	90°C	–	homomeric trimer	[93]
Aspartate amino transferase	S. solfataricus	$t_{0.5}(100°C) = 2$ h	95°C	–	homomeric dimer; gene seq	[94,95]
DNA-dep. RNA polymerase	S. acidocaldarius	$t_{0.5}(83.5°C) = 10$ min	55–70°C	Mg^{2+}-dep.	multimer of 12 subs; gene seq.	[96,97]
	S. solfataricus	"stable" at 60–75°C	60–80°C	Mg^{2+}-dep.	heteromeric multimer	[98]
	T. tenax	$t_{0.5}(100°C) = 2.25$ h	85°C	Mg^{2+}-dep.	heteromeric multimer	[97]
	D. mucosus	$t_{0.5}(95°C) = 23$ min	75–85°C	Mg^{2+}-dep.	heteromeric multimer	[97]
	P. woesei	100% after 10 min at 100°C	–	–	heteromeric multimer	[5]
DNA polymerase	S. acidocaldarius	$t_{0.5}(87°C) = 0.25$ h	75°C	Mg^{2+}-dep.	monomer	[99,100]
	S. solfataricus	$t_{0.5}(85°C) = 35$ min	~75°C	Mg^{2+}-dep.	monomer?	[101]
	T. litoralis	$t_{0.5}(100°C) = 2$ h	–	Mg^{2+}-dep.	–	[74]
Hydrolases						
Endonuclease	S. acidocaldarius	$t_{0.5}(80°C) = 0.5$ h	60–70°C	Mg^{2+}-dep.	–	[37]
Reverse gyrase	D. amylolyticus	–	100°C	Mg^{2+}-dep.	monomer*	[102]
Topoisomerase III	D. amylolyticus	–	60–99°C	Mg^{2+}-dep.	monomer	[103]
ATPase complex	P. occultum	$t_{0.5}(110°C) = 35$ min	100°C	Mg^{2+}-dep.	16mer of 2 diff. subs	[53]
Carboxyl-esterase	S. acidocaldarius	$t_{0.5}(100°C) = 1$ h	–	–	homomeric tetramer	[71]
α-D-Glucosidase	P. furiosus	$t_{0.5}(98°C) = 48$ h	105–115°C	–	–	[60]
β-D-Galactosidase	S. solfataricus	$t_{0.5}(95°C) = 3$ h	90°C	–	monomer or tetramer; gene seq.	[61,104]
Aminopeptidase	S. solfataricus	$t_{0.5}(87°C) = 0.25$ h	75°C	Co^{2+}-, Mn^{2+}-dep.	homomeric tetramer	[36]
Lyases						
Citrate synthase	S. acidocaldarius	$t_{0.5}(90°C) = 10$ min	–	–	dimer	[105]
N^5,N^{10}-Methenyl-H$_4$MPT cyclohydrolase	M. kandleri	100% after 1 h at 90°C	95°C	salt-dep.	–	[35]

[a] Abbreviations and symbols used: –, not given; aa comp., amino acid composition; dep., dependent; diff., different; gene seq., sequence deduced from the nucleotide sequence; prot. seq., sequence determined on protein level; stab, stabilized; subs, subunits; $t_{0.5}$ (°C), half-life of inactivation at the given temperature; percentage values in the 'Thermostability' column indicate residual activity after incubation at the given temperature; T_{opt}, temperature optimum.
[b] Boldface italic headings specify the localization, boldface roman headings the class of the enzyme.

cell – involved in thermoadaptation? Does life at these extreme conditions require specific proteins for protecting the cell machinery from heat injury or for repairing heat-denatured molecules?

Although these questions are mainly motivated by curiosity concerning the molecular adaptation mechanism and the endurance of the living cell to high temperatures, considerable interest is also stimulated by the direct or indirect biotechnological application of proteins from these thermophiles. For instance, some – mostly extracellular – enzyme proteins from hyperthermophilic archaea show properties suitable for direct use in biotechnological processes. As a prospect for the future, with a closer insight into the structure details of proteins from these thermophiles, it would also be possible to use this information for improving the stability of proteins with biotechnological interest.

2. Features of protein thermoadaptation in archaea

Most thermophilic archaeal proteins have been isolated from the Sulfolobales (*Sulfolobus solfataricus, S. acidocaldarius*) and Thermococcales (mainly *Pyrococcus furiosus* and *P. woesei*) probably because of the rather simple cultivation procedures for these organisms. Other frequent sources are the thermophilic methanogens (*Methanothermus fervidus* and *Methanopyrus kandleri*) and the Thermoproteales (*Thermoproteus tenax*, different species of the genus *Desulfurococcus*).

Enzymes prove to be especially suitable for analyzing protein thermoadaptation because their catalytic activity provides a suitable measure not only for temperature-induced changes of function but also for the stability of the native state.

Although a considerable number of extracellular and intracellular proteins have been isolated and described from thermophilic archaea, few detailed studies concerning the structure and thermophilic properties of the respective proteins are available. Of the approximately 40 different enzymes isolated from the extremely thermophilic archaea and characterized with respect to basic thermophilic properties (Table 1), only eight have been analyzed with respect to their primary structure, mostly using the nucleotide sequence of the coding genes, and in no case could the three-dimensional structure of the proteins be resolved.

2.1. How structurally different are proteins from the extreme thermophiles as compared to their mesophilic counterparts?

Considering the astonishing intrinsic stabilities towards heat damage and resistance against chemicals of some proteins from hyperthermophiles (e.g. α-glucosidase from *P. furiosus*, proteinases and α-amylase from *P. woesei*, Table 1) one would expect that these proteins might be equipped with unique structural features. However, from our rather preliminary knowledge about the structure of mainly intracellular thermophilic archaeal proteins, these proteins – even those from the hyperthermophilic archaea – do not differ in general structure from "normal" mesophilic proteins. Thus, the respective amino-acid sequences deduced mainly from the nucleotide sequences of the coding genes do not show striking peculiarities. Also, no stringent indications are available that the proteins are adapted to high temperatures by post-translational modification. Nevertheless, methylation of lysines has been described for the ferredoxin from *Thermoplasma acidophilum* [7] and glutamate dehydrogenase from *S. solfataricus* [8]. But it is unclear whether these modifications play any role in the adaptation of these proteins to higher temperature. Considering that intracellular enzymes from extreme thermophiles [9,10] can be expressed in *E. coli* with the same phenotypic properties as shown by the respective proteins from the original organism, it is unlikely that post-translational modifications play a significant role in thermoadaptation of these intracellular proteins. On the other hand, post-translational modification (e.g., glycosylation, crosslinking via cystine bridges) is rather a common feature of extracellular proteins. It is not surprising then that extracellular proteins from extremely thermophilic archaea are also modified. This is true in the case of the S-layer protein from the hyperthermophile *M. fervidus*, where glycosylation of the protein was shown by Bröckl et al. [12]. Indications for glycosylation of the extracellular protease of *S. acidocaldarius* "thermopsin" have also been reported [13]. To what extent these modifications are related to thermoadaptation, and differ from modifications of "normal" mesophilic proteins, remains to be determined.

To analyze the subtle specific structural features of thermophilic archaeal proteins, comparisons with closely related mesophilic proteins are necessary. Since the Crenarchaeota comprise exclusively thermophilic strains, respective comparisons are only possible with proteins from Euryarchaeota. Structural comparisons between thermophilic archaeal proteins and mesophilic bacterial or eucaryal proteins are unlikely to be useful because of the large evolutionary distance between different domains, which blur the thermoadaptive features.

Comparative sequence and functional studies were performed with glyceraldehyde-3-phosphate dehydrogenases (GAPDH) and 3-phosphoglycerate kinases (PGK) (refs. [10, 14], and Krüger, K., Hess, D., Knappik, J. and Hensel, R., manuscript in preparation) from the mesophile *Methanobacterium bryantii* and the thermophiles *Methanothermus fervidus* (growth optimum: 83°C [15]) and *Pyrococcus woesei* (growth optimum: 100°C [15]), one of the most thermophilic strains so far isolated. Sequence information from meso- and thermophilic enzyme homologues is also available for component C of methyl–coenzyme M reductases (sources of mesophilic enzyme: *Methanosarcina barkeri* and *Methanococcus vannielii*; sources of thermophilic enzyme: *Methanobacterium thermoautotrophicum* and *Methanothermus fervidus* [16]).

Analysis of the sequence data obtained from various GAPDH and PGK molecules shows that adaptation to the higher temperature is accompanied by an increase of average hydrophobicity and a decrease of chain flexibility. Similar trends have also been deduced from comparisons of bacterial proteins adapted to different temperatures [17,18]. Although the difference in these parameters cannot be quantitatively correlated with changes in thermophilicity, nevertheless these trends are indicative of the importance of hydrophobic interactions and chain rigidity for the structure of thermophilic proteins.

The limited number of comparable sequences does not permit a statistical analysis for identifying "thermophilic" residues in archaeal proteins. Considering the data obtained for the GAPDH and PGK sequences, similar trends in the archaeal and bacterial thermoadaptation can be recognized: The hydrophobic residues Ile, Val, Tyr, Phe and the charged residue Glu are preferred in the thermophilic sequence, whereas the residues Asp, Asn and Cys are discriminated against [19,20]. Differences observed in other residues between the PGK and GAPDH systems[1] or between the archaeal and eubacterial or eucaryal systems may be explained – at least partially – on the basis of different functions of the respective enzymes.

Strikingly, in the *P. woesei* enzymes the proportion of aromatic residues increases, especially that of Phe (ref.[10], and Krüger, K., Hess, D., Knappik, J. and Hensel, R., manuscript in preparation). Thus, one might speculate that these residues play a special role in stabilizing the proteins from thermophiles growing near or above the boiling point of water, possibly by aromatic–aromatic interactions [21] or perhaps by strengthening the hydrophobic interactions.

2.2. How are the proteins from thermophiles growing at or above the boiling point of water stabilized towards heat-induced covalent modifications of the peptide chain?

The high growth temperatures of archaeal thermophiles raise questions not only about the stability of the protein conformation but also about the protection of the peptide chain from covalent damage which occurs in mesophilic proteins. As shown by several authors [22–28] these chemical modifications mainly comprise, (a) deamidation of Asn and, to a minor extent, Gln; (b) hydrolysis of Asp-containing peptide bonds (limited to the acidic pH range) and Asn–X bonds; (c) destruction of cysteine and cystine residues.

Experiments with glyceraldehyde-3-phosphate dehydrogenases from *M. fervidus* and *P. woesei* as well as with different mutant enzymes [29] indicate that the peptide chain is largely protected from covalent modification in the native protein conformation. It would appear that the native conformation protects the weak links in the chain against attacking water molecules by steric hindrance and by restricting the torsional freedom of the chain.

The labilized conformation of mutated thermophilic enzymes results in a significantly higher susceptibility towards chemical modification. Increased susceptibility to thermogenic destruction is also observed after the native conformation is disrupted by preincubation with denaturing reagents [29].

Asn residues prove to be especially weak links within the peptide chain. These residues tend to deamidate, and the Asn–X bonds are labile to hydrolysis even at neutral pH. The observed decrease in the content of Asn, as well as Cys residues in proteins adapted to high temperatures may be interpreted as protection against such chemical modification. Presumably, these weak links of the chain are substituted especially in regions which tend to unfold (e.g., in surface loops).

On the other hand, the observation that unfolded proteins undergo irreversible chemical destruction at these extremely high temperatures would mean that a real equilibrium between native and unfolded state cannot exist in these hyperthermophiles. Once unfolded, the proteins become irreversibly denatured. One might speculate that the strategy of protein stabilization at these temperatures aims mainly at an increase of the activation energy of unfolding, i.e., at a retardation of the unfolding process.

---

[1] The most striking difference between the archaeal systems of GAPDH and PGK concerns the Arg residues, which follow the classic thermophilic trend [19,20] in PGK but not in GAPDH.

2.3. Extrinsic factors stabilizing the native state of proteins at high temperatures

As is generally known, the conformational stability of proteins depends on intramolecular interactions and on environmental factors (the surrounding solvent; different low- and high-molecular weight compounds). Often, the thermostability of isolated proteins measured in vitro is unexpectedly low, suggesting that the in vitro conditions are lacking stabilizing factors present in vivo.

Hints of the importance of the intracellular milieu in stabilizing proteins against heat denaturation were obtained for *Methunothermus fervidus* and *Methanopyrus kandleri*. Both organisms contain high concentrations of K^+ and cyclic 2,3-diphosphoglycerate (cDPG) (*M. fervidus*, 1000 mM K^+, 300 mM cDPG [30]; *M. kandleri*, 2300 mM K^+, 600 mM cDPG [31]). As demonstrated with the *M. fervidus* enzymes (glyceraldehyde-3-phosphate dehydrogenase, malate dehydrogenase, 3-phosphoglycerate kinase) the tripotassium salt of cDPG increases the protein thermostability in vitro as well. Also, the methyl–Coenzyme M reductase [32], N^5,N^{10}-methylenetetrahydromethanopterin dehydrogenase [33], N^5,N^{10}-methylenetetrahydromethanopterin reductase [34] and the N^5,N^{10}-methyltetrahydromethanopterin cyclohydrolase [35] from *M. kandleri* showed a striking salt dependence in their thermostability and activity (Table 1). Contrary to the rather uniform salt response of the *M. fervidus* enzymes, the *M. kandleri* enzymes differ from one another in their reaction to different inorganic salts.

Since in *M. fervidus* an increase of growth temperature was followed by an increase in K^+ and cDPG concentration, and since a clear trend of increasing K^+ and cDPG concentration with increasing optimal growth temperature can be observed in extremely thermophilic methanogens, it has been suggested that these ions play an important role in thermoadaptation [30]. Whether cDPG fulfills functions other than its stabilizing action on proteins (as energy or metabolite reservoir) remains to be determined.

Interestingly, high K^+ concentration could also be found in *P. woesei* (600mM [10]) counterbalanced by a new, not yet described phosphoderivative of inositol. Like the potassium salt of cDPG this potassium salt also increases the thermostability of proteins [107].

In organisms without strikingly high intracellular ion concentrations (e.g. members of the Crenarchaeota), some of their proteins show a low intrinsic stability in vitro (e.g., aminopeptidase from *S. solfataricus* [36] or endonuclease SuaI from *S. acidocaldarius* [37]), suggesting that more or less specific interactions with other cell constituents are likely important to stabilize the native state of the intracellular enzymes at high temperatures.

3. Proteins with suggested thermoadaptive functions

Although very attractive, the suggestion that hyperthermophilic archaea possess special protein equipment for heat protection or repair of heat-induced damage at extreme temperatures is based only on a few findings.

3.1. Proteins which presumably protect DNA

An enzyme which may specifically prevent heat denaturation of DNA is thought to be the reverse gyrase [38–41]. Although not specific for archaea, this enzyme seems to be specific for thermophilic prokaryotes (archaea as well as bacteria). The reverse gyrase introduces positive supercoils into double-stranded, covalently closed DNA. But as mentioned by Bouthier de la Tour et al. [41], "there is no direct evidence for the involvement of this type of supercoiling in the stability of DNA at high temperature".

Another type of DNA-stabilizing proteins are represented by histones or histone-like proteins. Although most reports on these proteins in archaea come from investigations on thermophilic species [42–46], DNA-binding proteins were also observed in mesophiles [47]. Considering the low GC content of DNA of some thermophiles (e.g., *M. fervidus* with a temperature optimum of 83°C exhibits only a GC content of 33% [15]) the presence of DNA-binding proteins for stabilizing the DNA duplex seems to be essential. Although their stabilizing effect on double-stranded DNA could be demonstrated, no information is available about the specific protein–DNA interaction in archaea with high temperature optima. Sandman et al. [48] speculate that the DNA–protein complexes in *M. fervidus* form a nucleosome-like structure. Possibly, these complexes are specific for hyperthermophiles with growth temperatures near or above 100°C.

3.2. Proteins which presumably protect proteins

The universality of the heat shock response has been proved by comparative studies not only within the domains of bacteria and eucarya but also, more recently, within the domain of archaea. Thus the phenomenon of acquired thermotolerance associated with the synthesis of specific proteins could also be found in mesophilic and thermophilic archaea [49–52].

The high level of a protein complex with ATPase activity in *P. occultum* at temperatures near the upper growth limit (11% of the soluble protein at the growth optimum of 100°C; 73% at 108°C; [53]) led to the suggestion that this protein may be essential for growth at the upper temperature limits of life. Unfortunately, no experimental data are yet available giving information about its physiological role. From its quaternary structure it was concluded that this abundant protein may function as a chaperone.

A chaperone-like function for a similar ATPase complex from *S. shibatae* [54], which showed a preferred binding to newly unfolded proteins, seems probable. To understand the real function of this ATPase complex, however, more detailed studies about its binding specificity are necessary.

4. Proteins with biotechnological potential

The usefulness of thermophilic enzymes in biotechnical applications has been reviewed by several authors [55–57]. The main advantages result from their general high thermostability often combined with a remarkable chemoresistance and a longer shelf life as well as from their ability to catalyze reactions at higher temperature.

Generally, for a broader application of archaeal thermophilic proteins in biotechnology, more detailed knowledge about the proteins of these organisms would be necessary. The culturing difficulties and the low cell yields of some thermophilic archaea may be due to the rather limited knowledge of their proteins. The demonstration that no general restraint limits the expression of proteins even from hyperthermophilic archaea in mesophilic hosts like *E. coli* (e.g., expression of the GAPDH from *P. woesei* in *E. coli* [10]) will make the proteins of these organisms more attractive for industrial application. The heterologous expression of these proteins also opens the possibility to tailor them for special purposes. For rational engineering, however, a detailed structural analysis of the enzyme in question would be necessary.

On the worldwide market a high demand for thermophilic polymer-degrading enzymes exists. This is especially true in the starch-processing industry where larger amounts of carbohydrate-hydrolyzing enzymes are used. There are still only a few examples of the industrial use of archaeal thermophilic enzymes, such as the α-amylase from *P. woesei* [58] (cf. Table 1) which is characterized by an exceedingly high thermostability ($t_{0.5}(120°C) = 6$ h) and a broad range of activity with respect to pH and temperature (over 60% maximal activity found between 70 and 125°C and between pH 3.5 and pH 6.5).

Of potential interest are also the cellulolytic, xylanolytic and pollulan-hydrolyzing activities detected in *Thermophilum* strains and in AN1, a member of the Thermococcales [59]. An additional candidate for biotechnical application is also represented by the α-glucosidase from *P. furiosus*, which exhibits the highest thermostability ($t_{0.5}(98°C) = 46-48$ h) and temperature optimum for activity (105–115°C) compared to all enzymes investigated so far [60] (Table 1).

Of special interest may be the β-glycosidase activity found in *S. sulfolobus*, which hydrolyzes both β-galactosides and β-glucosides and possesses a striking resistance against organic solutes and denaturants like urea [61,62]. Because of its low substrate specificity this enzyme could be applied not only for hydrolyzing lactose from the dairy industry but also for hydrolyzing cellobiose, the main product of cellulose degradation by cellulases.

The proteolytic enzymes constitute another class of hydrolyzing enzymes that ranks high in the industrial enzyme market. Various proteolytic activities of industrial interest have been found in several thermophilic archaeal strains. For utilization in the detergent industry a rather broad substrate spectrum combined with resistance against high pH values and surfactants would be desirable. Candidates for this application are the proteases from *Desulfurococcus* strain Tok$_{12}$S$_1$ [63] or various *Thermococcus* strains [64] with a rather alkaline pH optimum. Also interesting in this regard are the proteases of *P. furiosus* [65–67], some of which show not only an extreme thermostability but also a remarkable resistance against SDS. Further examples for industrial application may be the proteolytic activities (aminopeptidase, carboxypeptidases, endopeptidases) of *Sulfolobus* strains [36,68,69].

Other thermophilic enzymes from archaea may be used for analytical and preparative purposes. For instance, the glucose dehydrogenase from *S. solfataricus* [70] may serve as a suitable tool for glucose determination, whereas the relatively broad substrate spectrum of the esterase from *S. acidocaldarius* (catalyzes the acyl transfer to various alcohols and amines; [71]) or of the alcohol dehydrogenase from *S. solfataricus* (oxidizes various aliphatic and aromatic alcohols; [72]) makes these enzymes rather attractive for preparative purposes. The applicability of the thermoresistant and chemoresistant alcohol dehydrogenase for the synthesis of compounds of commercial value was recently shown by experiments performed in continuous-flow membrane reactors using macromolecular derivatives of NAD^+ or $NADP^+$ [73].

Examples of enzymes suitable for use in molecular biology are the DNA polymerases from *Thermococcus litoralis* [74] and *P. furiosus* [75]. With respect to fidelity of the polymerase reaction both enzymes surpass the DNA polymerase from *Thermus aquaticus* ("Taq polymerase") currently used for DNA sequencing and DNA amplification (via PCR reaction).

Not only functional but also structural proteins from archaeal thermophiles are of economic interest. Thus, the S-layer proteins of thermophilic archaea possess some technical potential. As outlined by Sleytr and Sara [76], the S-layers of different members of bacteria are suitable as molecular sieves. The presumed robust S-layers of the hyperthermophilic archaea will certainly enlarge their applicability.

5. Conclusions

The isolation of extremely thermophilic archaea over the last decade stimulated interest in the proteins of these remarkable organisms. Up to now a considerable number of different proteins from these organisms has been purified and characterized (Table 1). But an understanding of the structural and environmental factors which enable them to be stable and biologically active at temperatures around 100°C is still sparse. Information about the three-dimensional structure of these proteins will be necessary to understand the structure–function relationship of proteins constructed for the upper temperature limits of life.

Examples of thermophilic archaeal proteins successfully used in biotechnology already indicate their inherent economical potential (see section 4). However, our knowledge about the proteins of these extreme thermophiles is still too limited to estimate their real potential. Thus the proteins of thermophilic archaea represent an open field for fundamental and applied research over the next few decades.

Acknowledgement

I thank Mrs. B. Siebers and Mr. D. Hess for critically reading the manuscript.

References

[1] Stetter, K.O., König, H. and Stackebrandt, E. (1983) Syst. Appl. Microbiol. 4, 535–551.
[2] Huber, R., Kristijansson, J.K. and Stetter, K.O. (1987) Arch. Microbiol. 149, 95–101.
[3] Huber, R.M., Kurr, M., Jannasch, H.W., and Stetter, K.O. (1989) Nature, 342, 833–834.
[4] Fiala, G. and Stetter, K.O. (1986) Arch. Microbiol. 145, 56–61.
[5] Zillig, W., Holz, I., Klenk, H.-P., Trent, J., Wunderl, S., Janekovic, D., Imsel, E. and Haas, B. (1987) Syst. Appl. Microbiol. 9, 62–70.

[6] Stetter, K.O., Fiala, G., Huber, G., Huber, R. and Segerer, A. (1990) FEMS Microbiology Reviews 75, 117–124.
[7] Minami, Y., Wakabayashi, S., Wada, K., Matsubara, H., Kerscher, L. and Oesterhelt, D. (1985) J. Biochem. 97, 745–753.
[8] Maras, B., Consalvi, V., Chiaraluce, R., Politi, L., De Rosa, M., Bossa, F., Scandurra, R. and Barra, D. (1992) Eur. J. Biochem. 203, 81–87.
[9] Fabry, S., Lehmacher, A., Bode, W. and Hensel, R. (1988) FEBS Lett. 237, 213–217.
[10] Zwickl, P., Fabry, S., Bogedain, Ch. Haas, A. and Hensel, R. (1990) J. Bacteriol. 172, 4329–4338.
[11] Rossi, M., Rella, R., Pisani, Cubellis, M.V., Moracci, M., Nucci,R. and Vaccaro, C. (1991) In: Life under Extreme Conditions (Di Prisco, G., Ed.), pp. 115–123, Springer, Berlin.
[12] Bröckl, G., Behr, M., Fabry, S., Hensel, R., Kaudewitz, H., Biendl, E. and König, H. (1991) Eur. J. Biochem. 199, 147–152.
[13] Lin, X. and Tang, J. (1990) J. Biol. Chem. 265, 1490–1495.
[14] Fabry, S., Heppner, P., Dietmaier, W. and Hensel, R. (1990) Gene 91, 19–25.
[15] Stetter, K.O., Thomm, M., Winter, J., Wildgruber, G., Huber, H., Zillig, W., Janecovic, D., König, H., Palm, P. and Wunderl, S. (1981) Zbl. Bakteriol. Mikrobiol. Hyg. I. Abt. Orig. C 2, 166–178.
[16] Weil, C.F., Cram, D.S., Sherf, B.A. and Reeve, J. (1988) J. Bacteriol. 170, 4718–4726.
[17] Merkler, D.J., Farrington, G.K. and Wedler, F.C. (1981) Int. J. Peptide Protein Res. 18, 430–442.
[18] Vihinen, M. (1987) Protein Eng. 1, 477–480.
[19] Zuber, H. (1988) Biophys. Chem. 29, 171–179.
[20] Menendez-Arias, L. and Argos, P. (1989) J. Mol. Biol. 206, 397–406.
[21] Burley, S.K. and Petsko, A. (1985) Science 229, 23–29.
[22] Inglis, A.S. (1983) Methods Enzymol. 91, 324–332.
[23] Clarke, S. (1985) Annu. Rev. Biochem. 54, 479–506.
[24] Ahern, T.J. and Klibanov, A.M. (1985) Science, 228, 1280–1284.
[25] Zale, S.E. and Klibanov, A.M. (1986) Biochemistry, 25, 5432–5444.
[26] Geiger, T. and Clarke, S. (1987) J. Biol. Chem. 262, 785–794.
[27] Vooter, Ch.E.M., de Haard-Hoekman, W.A., van den Oetalaar, P.J.M., Bloemendal, H. and de Jong, W.W. (1988) J. Biol. Chem. 263, 19020–19023.
[28] Wriht, T. (1991) Crit. Rev. Biochem. Mol. Biol. 26, 1–52.
[29] Hensel, R., Jakob, I., Scheer, H. and Lottspeich, F. (1992) Biochem. Soc. Symp. 58, 127–133
[30] Hensel, R. and König, H. (1988) FEMS Microbiol. Lett. 49, 75–79.
[31] Lehmacher, A. and Hensel, R. (1990) In: DECHEMA Biotechnology Conferences, Vol. 4, pp. 415–418, VCH Verlagsgesellschaft, Heidelberg.
[32] Rospert, S., Breitung, J., Ma, K., Schwörer, B., Zirngibl, C., Thauer, R.K., Linder, D., Huber, R. and Stetter, K.O. (1991) Arch. Microbiol. 156, 49–55.
[33] Ma, K., Zirngibl, C., Linder, D., Stetter, K.O. and Thauer, R.K. (1991) Arch. Microbiol. 156, 43–48.
[34] Ma, K., Linder, D., Stetter, K.O. and Thauer, R.K. (1991) Arch. Microbiol. 155, 593–600.
[35] Breitung, J., Schmitz, R.A. Stetter, K.O. and Thauer, R.K. (1991) Arch. Microbiol. 156, 517–524.
[36] Hanner, M., Redl, B. and Stöffler, G. (1990) Biochim. Biophys. Acta 1033, 148–153.
[37] Prangishvili, D.A., Vashakidze, R.P., Chelidze, M.G. and Gabridze, I.Yu. (1985) FEBS Lett. 192, 57–60.
[38] Kikuchi, A. and Asai, K. (1984) Nature 309, 677–681.
[39] Forterre, P., Mirambeau, G. Jaxel, C., Nadal, M. and Duguet, M. (1985) EMBO J. 4, 2123–2128.
[40] Bouthier de la Tour, C., Portemer, C., Nadal, M., Stetter, K.O., Forterre, P. and Duguet, M. (1990) J. Bacteriol. 172, 6803–6808.

[41] Bouthier de la Tour, C., Portemer, C., Huber, R., Forterre, P. and Duguet, M. (1991) J. Bacteriol. 173, 3921–3923.
[42] Searcy, D.G. (1975) Biochim. Biophys. Acta 395, 535–547.
[43] Stein, D.B. and Searcy, D.G. (1978) Science 202, 219–221.
[44] Searcy, D.G and Delange, R.J. (1980) Biochim. Biophys. Acta 609, 197–200.
[45] Thomm, M., Stetter, K.O. and Zillig, W. (1982) Zbl. Bakt. Hyg. C 3 128–139.
[46] Reddy, T.R. and Suryanarayana, T. (1989) J. Biol. Chem. 264, 17298–17308.
[47] Imbert, M., Laine B., Prensier, G., Touzel, J.-P. and Sautiere, P. (1988) Can. J. Microbiol. 34, 931–937.
[48] Sandman, K. Krzycki, J.A. Dobrinski, B. Lurz, R. and Reeve, J.N. (1990) Proc. Natl. Acad. Sci. U.S.A. 87, 5788–5791.
[49] Daniels, C.J., McKee, A.H.Z. and Doolittle, W.F. (1984) EMBO J. 3, 745–749.
[50] Hebert, A.M., Kropinski, A.M. and Jarrell, K.F. (1991) J. Bacteriol. 173, 3224–3227.
[51] Jerez, C.A. (1988) FEMS Microbiol. Lett. 56, 289–294.
[52] Trent, J.D., Osipiuk, J. and Pinkau, T. (1990) J. Bacteriol. 172, 1478–1484.
[53] Phipps, B.M., Hoffmann, A., Stetter, K.O. and Baumeister, W. (1991) EMBO J. 10, 1711–1722.
[54] Trent, J.D., Nimmersgern, E., Wall, J.S. Hartl, F.U. and Horwich, A.L. (1991) Nature 354, 490–493.
[55] Ng, T.K. and Kenealy, W.F. (1986) In: Thermophiles: General, Molecular, and Applied Microbiology (Brock, T.D., Ed.), pp. 197–217, Wiley, New York.
[56] Kristjansson, J.K. (1989) TIBTECH 7, 349–353.
[57] Arbige, M.V. and Pitcher, W.H. (1989) TIBTECH 7, 330–335.
[58] Koch, R., Speinart, A., Lemke, K. and Antranikian, G. (1991) Arch. Microbiol. 155, 572–578.
[59] Bragger, J.M., Daniel, R.M., Coolbear, T. and Morgan, H.W. (1989) Appl. Microbiol. Biotechnol. 31, 556–561.
[60] Costantino, H.R., Brown, S.H. and Kelly, R.M. (1990) J. Bacteriol. 172, 3654–3660.
[61] Pisani, F.M., Rella, R., Raia, C.A., Rozzo, C., Nucci, R., Gambacorta, A., De Rosa, M. and Rossi, M. (1990) Eur. J. Biochem. 187, 321–328.
[62] Grogan, D.W. (1991) Appl. Environ. Microbiol. 57, 1644–1649.
[63] Cowan, D.A., Smolensky, K.A., Daniel, R.M. and Morgan, H.W. (1987) Biochem. J. 247, 121–133.
[64] Klingeberg, M., Hashwa, F. and Antranikian, G. (1991) Appl. Microbiol. Biotechnol. 34, 715–719.
[65] Blumentals, I.I., Robinson, A.S. and Kelly, R.M. (1990) Appl. Environ. Microbiol. 56, 1992–1998.
[66] Eggen, R., Geering, A., Watts, J. and de Vos, W.M. (1990) FEMS Microbiol. Lett. 71, 17–20.
[67] Connaris, H., Cowan, D.A. and Sharp, R.J. (1991) J. Gen. Microbiol. 137, 1193–1199.
[68] Fusek, M., Lin, X. and Tang, J. (1990) J. Biol. Chem. 265, 1496–1501.
[69] Fusi, P., Villa, M., Burlini, N., Tortora, P. and Guerritore, A. (1991) Experientia 47, 1057–1063.
[70] Giardina, P., De Biasi, M.-G., De Rosa, M., Gambacorta, A. and Buonocore, V. (1986) Biochem. J. 239, 517–522.
[71] Sobek, H. and Görisch, H. (1989) Biochem. J. 261, 993–998.
[72] Rella, R., Raia, C.A. Pensa, M., Pisani, F.M., Gambacorta, A., De Rosa, M. and Rossi, M. (1987) Eur. J. Biochem. 167, 475–479.
[73] Guagliardi, A., Raia, C., Rella, R., Bückmann, A.F., D'Auria, S., Rossi, M. and Bartolucci, S. (1991) Biotech. Appl. Biochem. 13, 25–35.
[74] Mattila, P. Korpela, J., Tenkanen, T. and Pitkänen, K. (1991) Nucleic Acids Res. 19, 4967–4973.
[75] Lundberg, K.S., Shoemaker, D.D., Adams, M.W.W., Short, J.M., Sorge, F.A. and Mathur, E.J. (1991) Gene 108, 1–6.
[76] Sleytr, V.B. and Sara, M. (1986) Appl. Microbiol. Biotech. 25, 83–90.

[77] Koch, R., Zablowski, P., Speirat, A. and Antranikian, G. (1990) FEMS Microbiol. Lett. 71, 21–26.
[78] Wakao, H., Wagaki, T. and Oshima, T. (1987) J. Biochem. 102, 255–262.
[79] Moll, R. and Schäfer, G. (1991) Eur. J. Biochem. 201, 593–600.
[80] Wagaki, T. and Oshima, T. (1985) Biochim. Biophys. Acta 817, 33–41.
[81] Lübben, M., Lünsdorf, H. and Schäfer, G. (1988) Biol. Chem. Hoppe–Seyler 369, 1259–1266.
[82] Hartl, T., Grossebrüter, W., Görisch, H. and Stezowski, J.J. (1987) Biol. Chem. Hoppe–Seyler 368, 259–267.
[83] Honka, E., Fabry, S., Niermann, T., Palm, P. and Hensel, R. (1990) Eur. J. Biochem. 188, 623–632.
[84] Bartolucci, S., Rella, R., Guarliardi, A., Raia, C., Gambacorta, A., De Rosa, M. and Rossi, M. (1987) J. Biol. Chem. 262, 7725–7731.
[85] Fabry, S. and Hensel, R. (1988) Gene 64, 189–197.
[86] Hensel, R., Laumann, S., Lang, J., Heumann, H. and Lottspeich, F. (1987) Eur. J. Biochem. 170, 325–333.
[87] Consalvi, V., Chiaraluce, R., Politi, L., Gambacorta, A., De Rosa, M. and Scandurra, R. (1991) Eur. J. Biochem. 196, 459–467.
[88] Consalvi, V., Chiaraluce, R., Politi, L., Vaccaro, R., De Rosa, M. and Scandurra, R. (1991) Eur. J. Biochem. 202, 1189–1196.
[89] Mukund, S. and Adams, M.W.W. (1991) J. Biol. Chem. 266, 14208–14216.
[90] Bryant, F.O. and Adams, M.W.W. (1988) J. Biol. Chem. 264, 5070–5079.
[91] Kletzin, A. (1989) J. Bacteriol. 171, 1638–1643.
[92] Carteni-Farina, M., Oliva, A., Romeo, G., Napolitano, G., De Rosa, M., Gambacorta, A. and Zappia, V. (1979) Eur. J. Biochem. 317–342.
[93] Cacciapuoti, G., Porcelli, M., Carteni-Farina, M., Gambacorta, A. and Zappia, V. (1986) Eur. J. Biochem. 161, 263–271.
[94] Marino, G., Nitti, G., Arnone, M.I. Sannia, G., Gambacorta, A. and De Rosa, M. (1988) J. Biol. Chem. 263, 12305–12309.
[95] Cubellis, M.V., Rozzo, C., Marino, G., Nitti, G., Arnone, M.I. and Sannia, G. (1989) Eur. J. Biochem. 186, 375–381.
[96] Pühler, G., Leffers, H., Gropp, F., Palm, P., Klenk, H.-P., Lottspeich, F., Garrett, R. and Zillig, W. (1989) Proc. Natl. Acad. Sci. U.S.A. 86, 4569–4573.
[97] Prangishvili, D.A., Zillig, W., Gierl, A., Biesert, L. and Holz, I. (1982) Eur. J. Biochem. 122, 471–477.
[98] Zillig, W., Stetter, K.O., Wunderl, S., Schulz, W., Priess, H. and Scholz, I. (1980) Arch. Microbiol. 125, 259–269.
[99] Prangishvili, D.A. (1986) Mol. Biol. USSR 20, 477–488.
[100] Elie, Ch., De Rocondo, M. and Forterre, P. (1989) Eur. J. Biochem. 178, 619–626.
[101] Rossi, M., Pensa, M. Bartolucci, S., De Rosa, M., Gambacorta, A., Raia, C.A. and Orabona, N.D'A. (1986) System. Appl. Microbiol. 7, 337–341.
[102] Slesarev, A.I. (1988) Eur. J. Biochem. 173, 395–399.
[103] Slesarev, A.I. Zaitzev, D.A., Kopylov, V.M., Stetter, K.O. and Kozyavkin, S.A. (1991) J. Biol. Chem. 266, 12321–12328.
[104] Cubellis, M.V., Rozzo, C., Montecucchi, P. and Rossi, M. (1990) Gene 94, 89–94.
[105] Grossebüter, W. and Görisch, H. (1985) System. Appl. Microbiol. 6, 119–124.
[106] Breitung, J., Schmitz, R.A. Stetter, K.O. and Thauer, R.K. (1991) Arch. Microbiol. 156, 517–524.
[107] Scholz, S., Sonnenbichler, J., Schäfer, W. and Hensel, R. (1992) FEBS Lett. 306, 239–242.

M. Kates et al. (Eds.), *The Biochemistry of Archaea (Archaebacteria)*
© 1993 Elsevier Science Publishers B.V. All rights reserved

CHAPTER 8

Cell envelopes of archaea: Structure and chemistry

Otto KANDLER[1] and Helmut KÖNIG[2]

[1]*Botanisches Institut der Ludwig-Maximillians-Universität München, 80638 München, Germany,* [2]*Abteilung für Angewandte Mikrobiologie und Mykologie, Universität Ulm, 89069 Ulm, Germany*

1. Introduction

Cell envelopes[1] of archaea differ distinctly from those of bacteria and show remarkable structural and chemical diversity. Murein, the typical sacculus-forming polymer of bacteria, and lipopolysaccharide-containing outer membranes, characteristic of gram-negative bacteria, are not found in archaea. Crystalline surface layers (S-layers) are common in both prokaryotic domains and they consist of protein or glycoprotein subunits (Table 1). However, S-layers in archaea have a form-stabilizing function especially when they are the only envelope layer outside the cytoplasmic membrane, while in bacteria S-layers have no distinct form-stabilizing function.

Several of the methanogenic and the extremely halophilic genera of the kingdom Euryarchaeota exhibit cell-wall sacculi resembling morphologically those of bacteria consisting of non-murein polymers which exhibit different chemical structures in each of the respective groups (Table 1). Two methanogenic genera are characterized by the formation of filaments held together by proteinaceous sheaths of unique fine structure.

2. Structure and chemistry of cell walls of gram-positive archaea

2.1. Methanobacteriales and Methanopyrus

2.1.1. Morphology
The gram-positive rods or cocci of the order Methanobacteriales [3] and of the genus *Methanopyrus* [4] possess an electron-dense cell wall sacculus, mostly 15–20 nm in width,

---

[1] Irrespective of the chemical and physical nature of the layers surrounding the cells, their sum will be referred to as the "cell envelope". The term "cell wall" will be restricted to mean a rigid layer forming a sacculus enclosing the plasma membrane of an individual cell, which, after isolation, still exhibits the shape of the cell and is insensitive to detergents. Arrays of protein or glycoprotein subunits covering the surface of the cells will be referred to as "surface layers" (S-layers) [1,2]. Layers of distinct mechanical stability extending over several cells will be termed 'sheats'. Mucous substances surrounding single cells or groups of cells will be termed 'capsulus'.

TABLE 1
Survey on the distribution of cell envelope components among archaea

Species	Rigid sacculus	S-layer (sheath)	Polymer
I. Euryarchaeota			
Methanobacteriaceae	+	–	Pseudomurein
Methanothermaceae	+	+	Pseudomurein/glycoprotein[a]
Methanopyrus kandleri	+	+	Pseudomurein[a]
Methanosarcina	±	±	Methanochondroitin or/and S-layer
Methanococcales	–	+	Protein
Methanomicrobium mobile	–	+	Protein
Methanogenium cariacii	–	+	Protein
Methanoculleus marisnigri	–	+	Glycoprotein[b]
Methanoplanus limicola	–	+	Glycoprotein[b]
Methanolobus tindarius	–	+	Glycoprotein[c]
Methanothrix soehngenii	–	Sheath	Glycoprotein[a]
Methanospirillum hungatei	–	Sheath	Glycoprotein[d]
Archaeoglobus fulgidus	–	+	Glycoprotein[b]
Halococcus morrhuae	+	–	Sulfated heteropolysaccharide
Halobacterium halobium	–	+	Glycoprotein[a]
Haloferax volcanii	–	+	Glycoprotein[a]
Natronococcus occultus	+	–	"Glycosaminoglycan"
Thermoplasma acidophilum	–	(Glycocalyx)	Glycoprotein[a]
II. Crenarchaeota			
Sulfolobus acidocaldarius	–	+	Glycoprotein[c]
Thermoproteus tenax	–	+	Glycoprotein[d]
Pyrodictium occultum	–	+	Glycoprotein[b]
Staphylothermus marinus	–	+	Glycoprotein[b]

[a] Binding between protein and carbohydrate demonstrated.
[b] Periodate-Schiff reaction (PAS) positive bands in PAGE of solubilized S-layers.
[c] Presence of carbohydrates demonstrated in hydrolysates of purified proteins of S-layers.
[d] Presence of carbohydrates demonstrated in isolated S-layers or sheaths.

which participates in septum formation during cell division as it is well known in the case of the murein sacculus of bacteria. The sacculi of *Methanothermus fervidus* often exhibit channels at the two cell poles (Fig. 1), whose diameters correspond with the thickness of the flagella [5]. Thus, they may be considered as passage holes for the flagella. Whereas the members of the genera *Methanobacterium*, *Methanobrevibacter* and *Methanosphaera* exhibit no additional surface layer, the members of the genera *Methanothermus* [6,7] and *Methanopyrus* [4] have, in addition to the cell wall sacculus, a proteinaceous S-layer, which consists of hexagonally arranged subunits in the case of *Methanothermus fervidus*.

2.1.2. Chemical structure and modifications of pseudomurein

The pseudomurein sacculi could be isolated using the same methods as those usually applied for the isolation of the murein sacculi of gram-positive bacteria [8]. The harvested

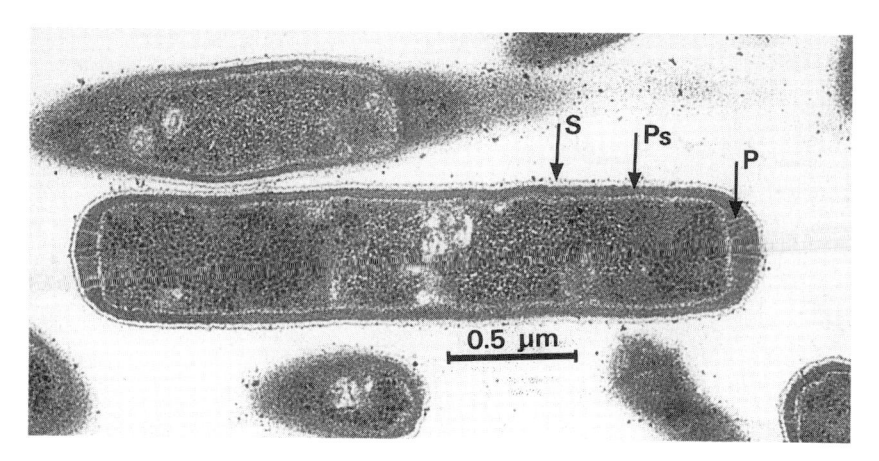

Fig. 1. Thin section of *Methanothermus fervidus* (symbols: S, S-layer; Ps, pseudomurein sacculus; P, pores).

cells were disrupted with glass beads by ultrasonication and further purified by digestion with trypsin and/or by treatment with sodium dodecyl sulfate. Isolated cell walls exhibit the shape of the original cells.

Chemical analysis of the isolated cell wall sacculi from all species of Methanobacteriales investigated so far has revealed that none of them contains the typical murein constituents muramic acid, diaminopimelic acid, D-glutamic acid or D-alanine. Instead, they contain a set of three-L-amino acids (Lys, Glu, Ala), the N-acetylated amino sugars glucosamine and/or galactosamine and N-acetyl-L-talosaminuronic acid [8–10]. Detailed studies to elucidate the chemical structure of the cell-wall polymer [10] have shown that it is made up of glycan strands consisting of alternating $\beta(1 \rightarrow 3)$-linked N-acetyl-D-glucosamine and $\beta(1 \rightarrow 3)$-linked N-acetyl-L-talosaminuronic acid residues, which are cross-linked via short peptides attached to the carboxylic group of N-acetyl-L-talosaminuronic acid (Fig. 2) [9–11]. Thus, the overall chemical structure of the cell-wall polymer of Methanobacteriales resembles that of murein and has to be classified as a peptidoglycan [12]. It is, however, a basically different type of peptidoglycan, not merely one of the many modifications of murein found in bacteria [13,14]. It has been named pseudomurein in order to indicate the similarities as well as the dissimilarities compared with the bacterial murein [15].

The main differences between pseudomurein and murein are the occurrence of talosaminuronic acid instead of muramic acid; the presence of the $\beta(1 \rightarrow 3)$- instead of the $\beta(1 \rightarrow 4)$-linkage of the glycan components; the partial or total replacement of glucosamine by galactosamine; the lack of D-amino acids; and the accumulation of the unusual ε- and γ-peptide bonds. As a consequence of its different composition, pseudomurein is resistant to cell-wall antibiotics, such as penicillin, D-cycloserine or vancomycin [16–18], which interact with various steps of murein biosynthesis. Pseudomurein is also insensitive to lysozymes and common proteases, because of the different glycosidic bonds of its glycan strands and the occurrence of various ε- and

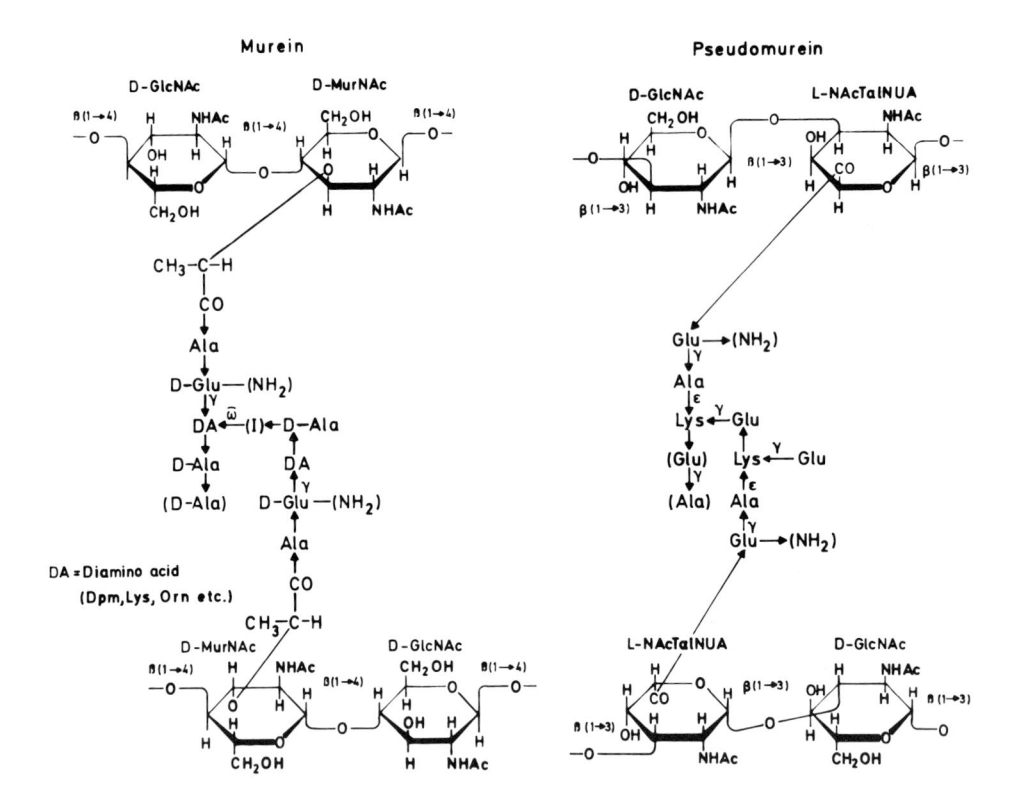

Fig. 2. Primary structure of dimers of murein and pseudomurein (compounds in parentheses may be missing in some cases). From ref. [46].

γ-bonds in its peptide moiety [10]. As in murein [13,14,19], chemical modifications are also found in the glycan as well as the peptide moiety of pseudomurein (Table 2) [10,20].

Regarding the glycan moiety, glucosamine may be partly or completely replaced by galactosamine (e.g., *Methanobrevibacter arboriphilus* strain AT; Table 2). The interchange of the two amino sugars without altering the glycan structure is possible, since carbon 3 of both glucosamine and galactosamine possess the same anomeric configuration. In some species the peptide moiety also exhibits distinct variations (Fig. 3):

– Alanine may be partially or completely replaced by threonine (e.g., *Methanobrevibacter ruminantium* strain M1) or by serine (e.g., *Methanosphaera stadtmanae* strain MCB-3).

– Ornithine is found in addition to lysine (e.g., *Methanobrevibacter smithii* strain PS and *Methanopyrus kandleri*). In *Methanobrevibacter smithii*, ornithine was found to be attached to the α-carboxylic group of a glutamic acid residue (Fig. 3).

– One of the two glutamic acid residues of pseudomurein is replaced by aspartic acid (e.g., *Methanobacterium alcaliphilum*; O. Kandler, F. Fiedler and J. Winter, paper in preparation).

TABLE 2
Molar ratios of the components of the cell walls of Methanobacteriales and *Methanopyrus*

Species	Amino acids[a]								N-Acetylamino sugars[a]				Neutral sugars[a]				%Ph[b]	Ref.
	Ala	Thr	Ser	Glu	Lys	Orn	NH_3		Gal-NH_2	Glc-NH_2	Tal-NUA[c]	Gal	Glc	Man	Rha			
Methanobrevibacter																		
Mb. arboriphilus DH1	1.52	–	–	2.13	1.00	–	1.36		0.67	–	(1.00)	–	–	0.02	0.02		4.54	10
Mb. arboriphilus AZ	1.17	–	–	2.32	1.00	–	3.92		0.91	–	(1.00)	0	0	0	0		0	10
Mb. smithii PS	1.42	–	–	2.04	1.00	0.40	0.81		0.44	0.56	(1.00)	0.11	0.14	–	0.09		1.00	10
Mb. ruminantium M1[d]	–	0.90	–	1.85	1.00	–	0.85		0.80	0.60	(1.00)	0.29	0.02	0.02	0.21		4.65	10
Mb. ruminantium M1[d]	0.70	0.40	–	1.80	1.00	–	1.00		0.80	0.50	(1.00)	0	0	0	0		0	10
Methanobacterium																		
M. bryantii M.o.H.	1.39	–	–	2.39	1.00	–	1.12		0.45	1.25	(1.00)	0.84	0.09	1.18	+		0.20	10
M. bryantii M.o.H.G.	1.32	–	–	2.21	1.00	–	0.84		–	0.84	(1.00)	0.53	0.26	0.26	+		0	10
M. formicicum MF	1.51	–	–	2.11	1.00	–	1.53		1.00	1.16	(1.00)	0.76	0.27	0.37	0.62		0.64	10
M. thermoautotrophicum ΔH	1.20	–	–	2.27	1.00	–	0.61		0.16	1.18	(1.00)	0.02	0.13	0	+		0.28	10
M. thermoautotrophicum Marburg	1.32	–	–	2.37	1.00	–	0.77		0.87	0.23	(1.00)	0	0	0	0		0	10
M. thermoautotrophicum JW 500	1.43	–	–	2.36	1.00	–	1.64		1.29	1.14	(1.00)	0	0	0	0		0	10
M. thermoautotrophicum JW 501	1.29	–	–	2.57	1.00	–	1.07		0.86	1.07	(1.00)	0	0	0	0		0	10
M. thermoautotrophicum JW 510	1.25	–	–	2.25	1.00	–	1.25		1.42	1.00	(1.00)	0	0	0	0		0	10
M. alcaliphilum DSM 2387	1.06	–	0.91[e]	0.90	1.00	–	1.27		0.05	1.02	(1.00)	0	0	0	0		0	10
Methanothermus																		
M. fervidus V24S	1.47	–	–	2.23	1.00	–	1.64		0.35	0.49	(1.00)	–	0.01	–	0		0	6
M. sociabilis KF1-Fl	1.14	–	–	2.51	1.00	–	0.96		0.36	1.30	(1.00)	0	0	0	0		0	7
Methanosphaera																		
M. stadtmanae MCB-3	0.26	–	0.77	1.94	1.00	–	1.56		0.32	0.59	(1.00)	0.04	0.24	–	–		0	20
M. cuniculi 1R-7	–	–	1.13	2.47	1.00	–	0.29		0.52	0.97	(1.00)	0	0	0	0		0	–[g]
Methanopyrus																		
M. kandleri AV 19	1.51	–	–	2.33	1.00	1.06	0		0.91	–	(1.00)	0	0	0	0		0	4

[a] Symbols: –, not present; +, trace; 0, not determined. [b] Phosphate (% d.W.). [c] The molar ratio of N-acetyltalosaminuronic acid was only estimated from the analysis of partial acid hydrolyzates, since it is completely destroyed in total hydrolyzates. [d] Two different batches of cells, both obtained from R. Wolfe and W. Balch (Urbana, IL, USA). [e] Aspartic acid. [f] Kandler, O., Winter, J., Fiedler, F., in preparation. [g] König, unpublished.

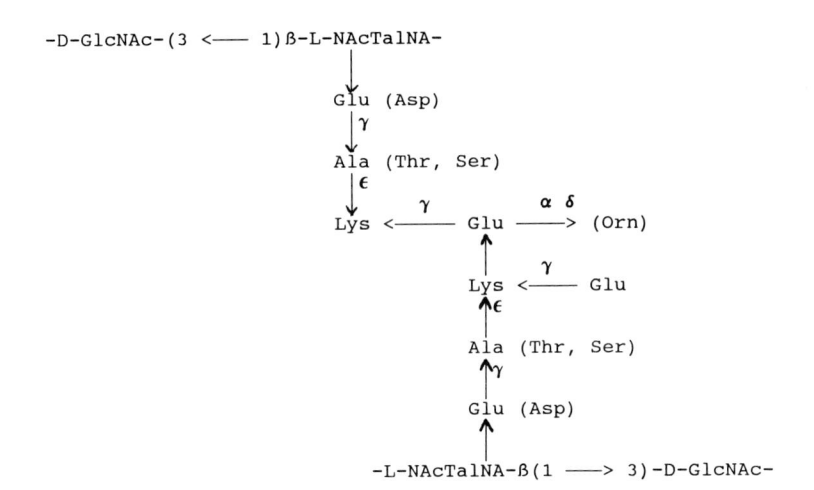

```
-D-GlcNAc-(3 <—— 1)ß-L-NAcTalNA-
                      │
                      ↓
                    Glu (Asp)
                    │γ
                    ↓
                    Ala (Thr, Ser)
                    │ε
                    ↓      γ              α  δ
                    Lys <——————— Glu ————————> (Orn)
                    ↑                 ↑
                    │                 │γ
                    │                 Lys <——————— Glu
                    │                 ↑ε
                    │                 │
                    │                 Ala (Thr, Ser)
                    │                 ↑γ
                    │                 │
                    │                 Glu (Asp)
                    │                 ↑
                    │                 │
                    -L-NAcTalNA-ß(1 ————> 3)-D-GlcNAc-
```

Fig. 3. Proposed amino acid sequences of cross-linked subunits of pseudomurein and its modifications. Compounds in parentheses indicate alternative sequences observed in some species. Modified from ref. [10].

Modifications of the amino-acid composition of pseudomurein could be induced by addition of glycine, threonine, ornithine, or aspartic acid in elevated concentrations to the culture medium [21]. The isolated sacculi from several species also contain varying amounts of monosaccharides (Table 2). They could be extracted with hot formamide, indicating that they belong to polysaccharides associated with but not covalently bound to pseudomurein. Teichoic acid or teichuronic acid-like polymers are, however, not present in gram-positive methanogens.

2.1.3. Secondary and tertiary structure of pseudomurein
X-ray diffraction measurements and structural calculations on murein [22–25] and pseudomurein [26–28] have revealed several common structural features in both polymers. Murein and pseudomurein sacculi possess a density of $\rho = 1.39-1.46\,\text{g/cm}^3$, which is characteristic of highly ordered material. A much lower density, in the range of $\rho = 1.24-1.32\,\text{g/cm}^3$, is to be expected for amorphous polymers [26]. X-ray diffraction showed diffuse Debye–Scherer rings with Bragg periodicities of about 0.45 nm and 0.94 nm in the planes and of 4.3–4.5 nm vertically to the planes of both types of cell walls. These data have been interpreted in two different ways:

(i) The peptide moieties of either murein or pseudomurein are radially oriented around the glycan strands, which form a screw rather than a ribbon. Three to four individual layers are needed to form a 4.5 nm periodicity vertical to the planes of the cell walls [24,26].

(ii) The glycan strands perform twofold screw axes (ribbons). Consequently, all peptide moieties point in the same direction. Two superimposed layers, facing each other, either the glycan or the peptide phase, give rise to a periodicity of 4.5 nm vertical to the planes of the cell walls. In this case the X-ray data may be interpreted as dimensions of an oblique-angled elementary cell with a height of 2.23 nm and an area of $0.45 \times 1\,\text{nm}^2$, which includes one disaccharide subunit [22,27]. The twofold screw axis of the glycan strands of murein has been assumed to be similar to that of chitin [22,25,29]. In the pseudomurein, a twofold screw axis is also possible, if the β-N-acetyl-D-glucosamine residues are in $^4C_1$ conformation and the β-N-acetyl-L-talosaminuronic acid residues in $^1C_4$ conformation.

2.1.4. Lysis of pseudomurein

Information on lytic enzymes is still somewhat anecdotal. *Methanobacterium bryantii* undergoes spontaneous lysis when the NH^{4+} or Ni^{2+} ions are exhausted in the culture medium [30].

In the presence of D-sorbitol, *Methanobacterium thermoautotrophicum* forms protoplasts after addition of lysozyme to growing cells [31]. Growing cells of *Methanobacterium thermoalcaliphilum* were also found to lyse in the absence of substrate [32]. The mechanism responsible for the lysis has not been elucidated in any of these cases. An extracellular enzyme, produced by a streptomycete, was found to lyse *Methanobacterium formicicum* [33].

An autolytic enzyme, which hydrolyzes the pseudomurein sacculi after depletion of the energy source (H_2), could be induced in *Methanobacterium wolfei* [34]. The cell lysate or the isolated enzyme from *Methanobacterium wolfei* also lyses cells of other pseudomurein containing species such as *Methanobacterium thermoautotrophicum* strain ΔH or strain Marburg or freshly grown cells of *Methanothermus fervidus*. The isolated lytic enzyme was found to be a peptidase hydrolysing the ε-Ala–Lys bond of the peptide subunits [35]. The enzyme has proved to be useful for the isolation of high yields of chromosomal and plasmid DNA [35] and the preparation of protoplasts for membrane studies [36]. Recently, the enzyme has been partially purified and further characterized with respect to temperature and oxygen tolerance [37].

2.1.5. Biosynthesis of the pseudomurein

The first steps of murein biosynthesis take place in the cytoplasm, whereas the two final steps occur at the inner and outer face of the cytoplasmic membrane, respectively [38]. This may also be true for the pseudomurein, according to a tentative scheme of the biosynthesis of pseudomurein (Fig. 4) [39–43], proposed on the basis of the structure of putative precursors isolated from cell extracts.

UDP-N-acetylgalactosamine was found in relatively high amounts in cell extracts of *Methanobacterium thermoautotrophicum* strain ΔH even though the content of N-acetyl-galactosamine in the intact glycan of this species was very low. However, monomeric talosaminuronic acid or its UDP-activated derivative was missing in the cell extracts while UDP-GlcNAc-(3 ← 1)β-NAc-TalNA was present. It is therefore assumed that N-acetyl-talosaminuronic acid is formed during the synthesis of the disaccharide by epimerization and oxidation of UDP-N-acetylgalactosamine.

The synthesis of the pentapeptide moiety is supposed to start with a UDP-activated glutamic acid residue followed by the stepwise formation of UDP-activated peptides up to a pentapeptide. UDP was directly linked to the N^α-amino group of the glutamic acid residue, a type of activation of amino acids previously not found in nature. Finally, the UDP-activated pentapeptide is linked to the disaccharide to give a UDP-activated disaccharide pentapeptide.

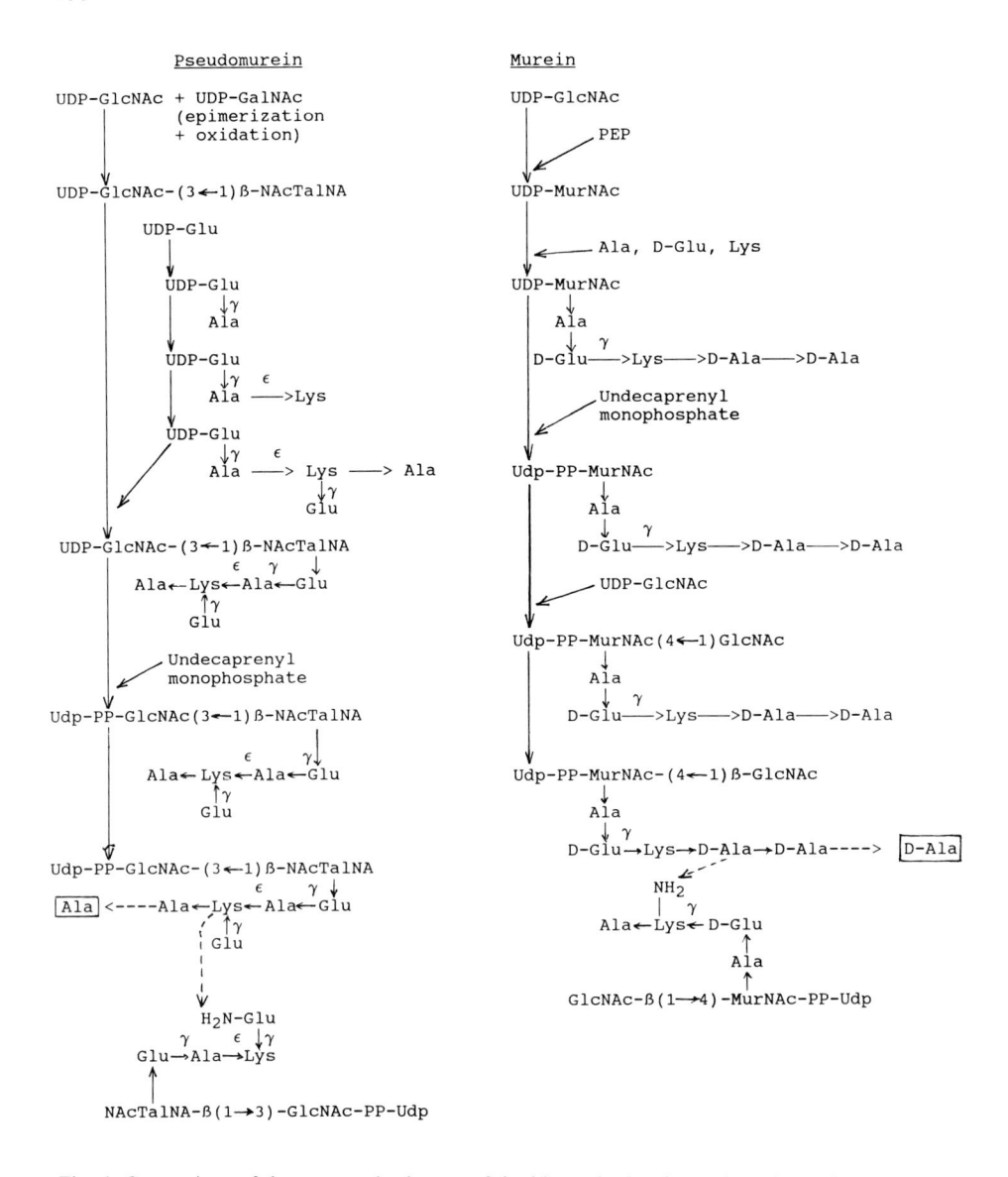

```
       Pseudomurein                          Murein
UDP-GlcNAc + UDP-GalNAc                  UDP-GlcNAc
          (epimerization                     │
          + oxidation)                       │    ╱PEP
          │                                  ▼  ╱
          ▼                                UDP-MurNAc
UDP-GlcNAc-(3←1)β-NAcTalNA                   │
          │                                  │
          │  UDP-Glu                         │    ╱─ Ala, D-Glu, Lys
          │   │                              ▼  ╱
          │   ▼                            UDP-MurNAc
          │  UDP-Glu                          │
          │   │↓γ                             ▼
          │   ▼ Ala                          Ala
          │  UDP-Glu                          │  γ
          │   │↓γ  ε                          ▼
          │   ▼ Ala ──>Lys                  D-Glu──>Lys──>D-Ala──>D-Ala
          │  UDP-Glu                          ╱Undecaprenyl
          │   │↓γ   ε                        ╱ monophosphate
          │   ▼ Ala ──> Lys ──> Ala         ╱
          │            │γ                   ▼
          │            ▼                  Udp-PP-MurNAc
          │           Glu                    │
          │                                  ▼
UDP-GlcNAc-(3←1)β-NAcTalNA                  Ala
          ε  γ  │                            │  γ
          Ala←Lys←Ala←Glu                    ▼
               ↑γ                          D-Glu──>Lys──>D-Ala──>D-Ala
               Glu                            ╱
                                             ╱─ UDP-GlcNAc
              ╱Undecaprenyl                 ╱
             ╱ monophosphate                ▼
            ╱                             Udp-PP-MurNAc(4←1)GlcNAc
Udp-PP-GlcNAc(3←1)β-NAcTalNA                 │
          │                                  ▼
          │  ε      γ↓                      Ala
          │  Ala←Lys←Ala←Glu                 │  γ
          │       ↑γ                         ▼
          │       Glu                      D-Glu──>Lys──>D-Ala──>D-Ala
          │                                  │
          ▼                                  ▼
Udp-PP-GlcNAc-(3←1)β-NAcTalNA             Udp-PP-MurNAc-(4←1)β-GlcNAc
          ε      γ │                          │
 ┌───┐                                        ▼
 │Ala│<────Ala←Lys←Ala←Glu                   Ala
 └───┘         ↑γ                             │ γ
               │Glu                           ▼                    ┌─────┐
               │                            D-Glu→Lys→D-Ala→D-Ala──>│D-Ala│
               ▼                                  ╱                 └─────┘
             H₂N-Glu                           NH₂
           γ    ε ↓γ                            │  γ
           Glu→Ala→Lys                        Ala←Lys← D-Glu
                ↑                                      ↑
                │                                     Ala
                │                                      ↑
   NAcTalNA-β(1→3)-GlcNAc-PP-Udp             GlcNAc-β(1→4)-MurNAc-PP-Udp
```

Fig. 4. Comparison of the proposed schemes of the biosynthesis of murein and pseudomurein [40].

In a further step, the nucleotide-activated disaccharide pentapeptide is most likely transferred to undecaprenyl monophosphate resulting in an undecaprenyl pyrophosphate-activated disaccharide pentapeptide, which penetrates the cytoplasmic membrane and is transferred to the glycan strand. The last step of pseudomurein biosynthesis, the cross-linking of adjacent glycan strands, is probably brought about by transpeptidation. Thereby, the terminal alanyl residue is split off and a glutamic acid residue is linked with a lysine residue of an adjacent peptide, resembling the mechanisms observed in the case of murein biosynthesis [38,44]. The excretion of alanine by growing cultures of *Methanobacterium thermoautotrophicum* into the medium has been observed [45].

The participation of a nucleotide-activated disaccharide and UDP-activated peptide intermediates are unique features of pseudomurein biosynthesis. Usually, oligosaccharide precursors of bacterial cell-wall polymers are formed at the lipid stage [38,44] and amino acids carrying a nucleotide residue at the N^α-amino group have not been found in nature so far. The distinct differences between the two biosynthetic routes support the hypothesis that murein and pseudomurein represent independent inventions made after the domains bacteria and archaea had been separated from each other during evolution [40,46].

2.1.6. Biological activity of pseudomurein

Murein components are known to exhibit a series of biological activities in humans and higher animals [47]; so far much less information on the biological activities of pseudomurein components is available.

The antigenic determinants of the bacterial murein have been well investigated [48–52]. It has been shown that pseudomurein is also antigenic in animals [53]. The immunochemical characterization with monoclonal antibodies revealed four distinctive determinants in the pseudomurein, namely three components of the glycan strand (*N*-acetylglucosamine, *N*-acetylgalactosamine, *N*-acetyltalosaminuronic acid) and the C-terminal sequence γ-glutamyl-alanine of the peptide moiety.

Bacterial infections elicit a series of acute-phase responses which include central nervous system effects such as changes in body temperature and increased slow-wave sleep. Dead bacteria [54] and murein preparations [55] as well as other bacterial cell-wall products induce similar responses. Intravenous injections of rabbits with suspensions of pseudomurein from *Methanobacterium thermoautotrophicum* also alter sleep and brain temperature. The mechanisms responsible for these somnogenic and pyrogenic effects are unknown [56]. It has been demonstrated that in a rat arthritis model, intra-articular injection of high doses of pseudomurein-polysaccharide fragments from *Methanobacterium formicicum* caused an acute inflammation [57].

2.1.7. Chemical structure and biosynthesis of the S-layer glycoprotein of Methanothermus fervidus

Methanothermus fervidus has a double-layered cell envelope. The inner pseudomurein sacculus is covered by an S-layer composed of glycoprotein subunits, which are hexagonally arranged. The S-layer glycoprotein could be isolated by extracting whole cells with trichloroacetic acid (5–10%) and subsequent reversed-phase chromatography using aqueous formic acid (10%) as eluant [5].

The primary sequence of the peptide was derived from the nucleotide sequence of the corresponding gene [58]. The mature glycoprotein consists of 593 amino acids. Compared to mesophilic S-layer glycoproteins this hyperthermophilic glycoprotein possesses significant higher amounts of, e.g., isoleucine, asparagine and cysteine. It also has 14% more β-sheet structures. A typical leader peptide and 20 sequon structures (Asn–Xaa–Ser/Thr) as potential *N*-glycosylation sites are present. The isolated heterosaccharide consist of mannose, 3-*O*-methylmannose/3-*O*-methylglucose and *N*-acetylgalactosamine, while in the glycoprotein also small amounts of galactose and *N*-acetylglucosamine were found. The heterosaccharides, which are linked to asparagine via *N*-acetylgalactosamine, consist of alternating mannose and 3-*O*-methylmannose/3-*O*-methylglucose residues [Kärcher and König, in preparation].

Based on isolated precursors [59] the biosynthesis of the glycan chains is suggested to start with the formation of C-1-phosphate derivatives of the corresponding sugars. They are then converted to the corresponding nucleotide activated derivatives, which form UDP activated oligosaccharides. Glucose and 3-*O*-methylmannose/3-*O*-methylglucose are not found as monomers, but they are constituents of the oligosaccharides. The oligosaccharides are most probably transferred to C_{55}-dolichyl-phosphate, forming dolichyl-pyrophosphate-activated oligosaccharides. The lipid-activated oligosaccharides contain additional Glc residues, which are removed during further processing.

2.2. Methanosarcina

2.2.1. Morphology

Cells of *Methanosarcina* species are usually surrounded by a rigid, sometimes laminated, cell wall consisting of methanochondroitin [60] and exhibiting a variable width of 20–200 nm. In many cases the cells form globoidal packets which share a common wall (Fig. 5). The formation of macro- and microcysts surrounded by a very thick wall was observed in a strain of the so-called *Methanosarcina* biotype 3. The cyst skin may not merely be a thickened regular cell wall. However, no chemical analysis of such cyst skins exists as yet [61].

Rigid cell walls are absent in some species, for instance in *Methanosarcina frisia* [62] and in *Methanosarcina acetivorans* [63], which exhibit only a proteinaceous S-layer outside the cytoplasmic membrane, although both species are closely related to other members of *Methanosarcina* on the basis of several metabolic and molecular characteristics.

An S-layer was also found between the cytoplasmic membrane and the rigid cell wall of *Methanosarcina mazei* [64]. Consequently, the authors do not consider the thick rigid cell wall layer to be homologous to the murein sacculus of bacteria and thus call it a "matrix". In fact, the very variable thickness of the methanochondroitin "cell wall" suggests a more or less distinct swelling capacity typical of slimy bacterial capsules. Thus, the methanochondroitin layer may be understood as an intermediate between a classical cell wall and a capsule with a very restricted swelling capacity, which may or may not be separated from the cytoplasmic membrane by an S-layer.

2.2.2. Chemical structure of methanochondroitin

Protein-free cell walls of several strains of *Methanosarcina* species have been isolated using the techniques applied to the isolation of cell walls of gram-positive bacteria [60,65]. The isolated rigid sacculi were found to be resistant to cellulolytic enzymes and chondroitin hydrolase. They could be solubilized by incubation in 10% sodium borohydride at 60°C for 12 h. Acid hydrolysates of the dialyzed soluble fraction (molecular mass $\sim$10 000 Da) exhibited the same composition as the original, not solubilized cell-wall material, namely D-galactosamine, D-glucuronic acid or D-galacturonic acid, D-glucose and acetic acid. Mannose was present in trace amounts [60]. Muramic acid, talosaminuronic acid, glucosamine or significant amounts of amino acids were not found. The phosphate and sulfate contents were negligible. The chemical

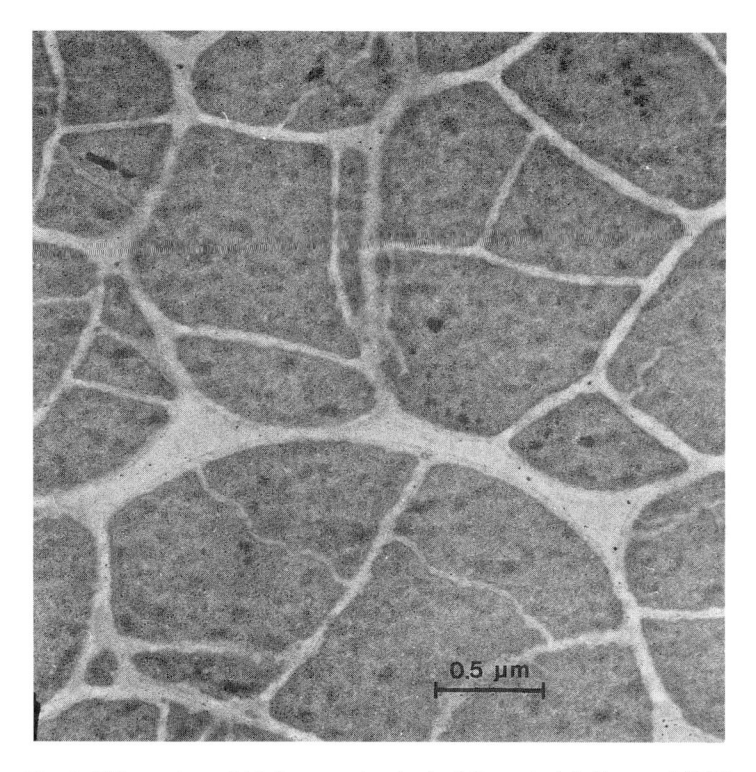

Fig. 5. Thin section of *Methanosarcina barkeri* (type strain). From ref. [105].

data for several strains are listed in Table 3. In all strains, including *Methanosarcina vacuolata*, whose cell wall was described as appearing triple-layered [66], the molar ratio of N-acetyl-D-galactosamine to uronic acid was about 2:1, but it was 3:1 in *Methanosarcina mazei*. In two strains, galacturonic acid is replaced by glucuronic acid [60].

As a structural element, chondrosin has been isolated from partial acid hydrolysates. In addition, a trimer was obtained, which could be identified as being the building block of the glycan [60]. The proposed structure of the polymer is shown in Fig. 6.

Thus, the cell-wall polymer of *Methanosarcina barkeri* is reminiscent of chondroitin sulfate, a common component of the connective tissue in humans and animals. In contrast with chondroitin, it is not sulfated and contains two instead of one N-acetylgalactosamine residue in the repeating unit. Because of its similarities to chondroitin sulfate (Fig. 6), the shape-maintaining cell-wall polymer of *Methanosarcina* has been named methanochondroitin [60].

2.2.3. Autolysis

In substrate-depleted cultures of *Methanosarcina barkeri*, methanochondroitin is lysed autolytically, yielding spontaneous protoplasts [67,68]. Also during later stages of the life

TABLE 3
Chemical composition of isolated cell walls of *Methanosarcina* [60]

Strain[a]	Components[b,c]								
	D-Glc	D-GalN	D-GlcA	D-GalA	N-acetyl groups	Ash	H_2O	PO_4^{3-}	SO_4^{2-}
DSM 800	3.75	27.5	16.0	–	8.3	20	14	1.6	0.35
	0.20	1.54	0.82	–	1.9			0.17	0.036
DSM 804	5.0	35	14.8	–	8.7	12.2	8.4	1.4	0.18
	0.27	1.95	0.76	–	2.0			0.15	0.018
DSM 3338	3.8	35	3.5	14	8.6	10	14	0.53	0.19
	0.21	1.95	0.18	0.72	2.0			0.056	0.019
DSM 1232	2.2	30.8	18.0	–	8.7	2.5	16	0.67	0.18
	0.12	1.72	0.92	–	2.0			0.07	0.018
DSM 1825	0.5	41.45	18.8	–	10.5	8.35	14.6	0.14	0.20
	0.027	2.3	0.97	–	2.4			0.015	0.020
DSM 2053	0.42	40	6.9	6.9	9.6	10	16.8	0.112	0.18
	0.023	2.23	0.35	0.35	2.23			0.112	0.018
DSM 1538	3.5	20.6	11.2	–	nd	nd	nd	nd	nd
	0.19	1.15	0.58	–	nd	nd	nd	nd	nd

[a] DSM 800, 804, 1538: *Methanosarcina barkeri*; DSM 2053: *Methanosarcina mazei*; DSM 1232: *Methanosarcina vacuolata*; DSM 1825: *Methanosarcina thermophila*; DSM 3338: *Methanosarcina* sp. strain G1.
[b] For each strain values are given in %d.w. (top row) and mmol/g d.w. (bottom row)
[c] Abbreviations: nd, not determined; –, not present.

A) [——>)-ß-D-GlcA-(1——>3)-ß-D-GalNAc-(1——>4)-ß-D-GalNAc-(1——>]$_n$

B) [D-GlcA-(1——>3)-GalNAc-(1——>4)-]$_{23-25}$

C) [ß-D-GlcA-(1——>3)-ß-D-GalNAc-(1——>4)-ß-D-GlcA-1——>]$_n$
(4 or 6 sulfate)

Fig. 6. Comparison of building blocks of (A) methanochondroitin of *Methanosarcina* sp.; (B) teichuronic acid of *Bacillus licheniformis*; (C) chondroitin sulfate of animal connective tissue.

cycle of *Methanosarcina mazei*, methanochondroitin becomes degraded and single cells are released from the aggregates [69–71]. For complete disaggregation at the end of the growth phase, elevated concentrations of substrate (100 mM acetate or methanol) and divalent cations (Ca^{++}) were required [72].

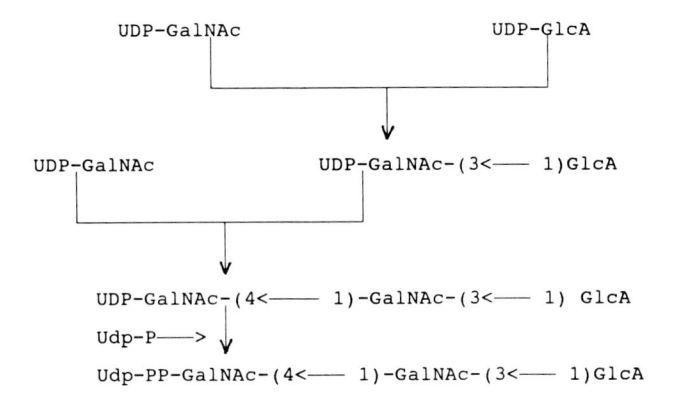

```
        UDP-GalNAc                    UDP-GlcA
            |_____|
                          |
                          v
    UDP-GalNAc        UDP-GalNAc-(3<——— 1)GlcA
        |_____|
                |
                v
        UDP-GalNAc-(4<——— 1)-GalNAc-(3<——— 1) GlcA
        Udp-P——>  |
        Udp-PP-GalNAc-(4<——— 1)-GalNAc-(3<——— 1)GlcA
```

Fig. 7. Proposed scheme for the biosynthesis of methanochondroitin [73].

2.2.4. Biosynthesis of methanochondroitin

Based on 4 UDP-activated and 1 undecaprenyl pyrophosphate-activated putative precursors of methanochondroitin isolated from extracts of *Methanosarcina barkeri*, a tentative pathway for methanochondroitin biosynthesis has been proposed (Fig. 7) [73]. According to this proposal, UDP-N-acetylchondrosin is formed from UDP-N-acetylgalactosamine and UDP-glucuronic acid, followed by the transfer of N-acetylchondrosin to UDP-N-acetylgalactosamine, thus yielding a UDP-activated trisaccharide [73], which possesses the same structure as the suggested repeating unit of methanochondroitin [60]. In the next step, the UDP residue of the trisaccharide is most probably replaced by undecaprenyl pyrophosphate, and the trisaccharide may then be polymerized onto methanochondroitin at the outer face of the cytoplasmic membrane.

This proposed pathway of methanochondroitin biosynthesis differs significantly from known biosynthetic pathways of polysaccharides in bacteria and eucarya, where no nucleotide-activated oligosaccharides are involved. Also, the biosynthesis of the structurally very similar eukaryotic chondroitin sulfate, which is attached to protein, follows a quite different mode. Here, the oligosaccharides are directly formed from their nucleotide-activated monomers [74] at distinct serine or threonine residues of the growing polypeptide chain. Also, teichuronic acid of *Bacillus licheniformis* (Fig. 6), exhibiting a structure very similar to methanochondroitin [75], is built from monomers at the lipid stage with undecaprenyl pyrophosphate as lipid carrier like other cell-wall polymers of eubacteria. Thus, it appears that nucleotide-activated oligosaccharides, also involved in the biosynthesis of the pseudomurein [40], and the S-layer glycoprotein of *Methanothermus fervidus* [59] are a typical common feature of the biosynthesis of glycan components in methanogens. Furthermore, nucleotide oligosaccharide conjugates are also assumed to be involved in methanogenesis [76,77]. It is noteworthy that nucleotide-activated oligosaccharides have also been isolated from hen oviduct and from human and goat colostrum and milk (cf. ref. [78]) as well as from bacteria (Hartmann and König, unpublished). While in bacteria nucleotide-activated oligosaccharides are involved

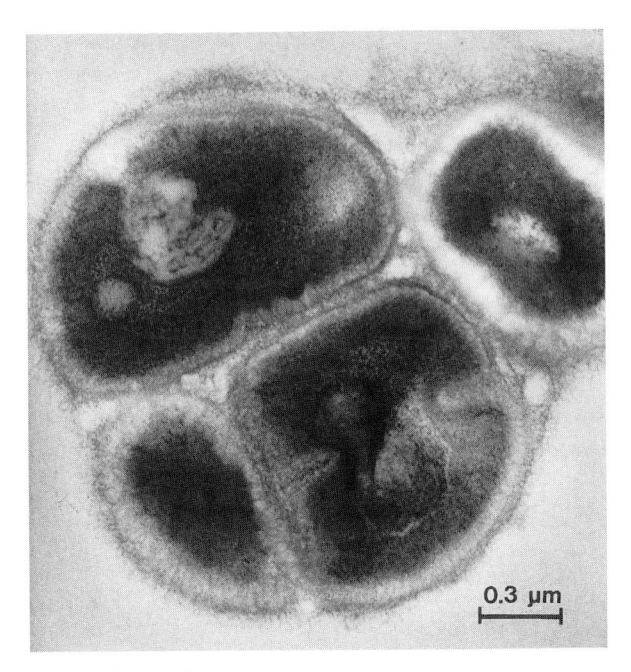

Fig. 8. Thin section of *Halococcus morrhuae*. From ref. [105].

in glycoprotein synthesis (Hartmann, Messner and König, unpublished) their function in eucarya has not been demonstrated.

2.3. Halococcus

2.3.1. Morphology

Halococcus morrhuae forms single cells or cuboidal packets. The cells are surrounded by one electron-dense layer outside the cytoplasmic membrane, 50–60 nm in width (Fig. 8). This layer forms a rigid cell wall sacculus [79] and tends to become laminated as was also observed for *Methanosarcina barkeri* [61,64]. The rigid cell-wall sacculi prevent lysis of the cocci in media with low ionic strength and have been isolated by the usual techniques applied for gram-positive bacteria [80].

2.3.2. Chemical structure of halococcal heteropolysaccharide

The isolated cell-wall sacculi are rather stable and could not be lysed by 13 different endo- and exoglycosidases, but alkaline treatment (0.5 N NaOH, 60°C, 12 h) led to partially solubilized material.

The cell walls consist of highly sulfated heteropolysaccharide which consists of a complex mixture of neutral and amino sugars, uronic acids, glycine and an aminuronic acid (gulosaminuronic acid) (Table 4) [80–84]. The typical components of murein or pseudomurein are missing.

TABLE 4
Chemical composition of purified (trypsin-treated) cell walls of three strains of *Halococcus morrhuae*[a]

Component	μmol/mg cell wall		
	CCM 859	CCM 537	CCM 889
Glucose	0.44	0.39	0.30
Mannose	0.35	0.21	0.30
Galactose	0.27	0.18	0.17
Total neutral sugars	1.06	0.78	0.77
Glucosamine	0.38	0.21	0.35
Galactosamine	0.20	0.08	0.24
Gulosaminuronic acid	0.11	0.09	0.10
Total amino sugars	0.69	0.38	0.69
Uronic acids[b]	0.67	0.24	0.63
Acetate	0.62	nd	nd
Glycine	0.10	0.13	0.10
Sulfate	1.47	0.80	0.18

[a] From Schleifer et al. [19]
[b] Total uronic acids determined as glucuronic acid. Glucuronic and galacturonic acids were present in cell walls of CCM 859 in a molar ratio of about 2.3:1.

The proposed structure of the complex heterosaccharide is shown in Fig. 9. Fragmentation studies suggest that the carbohydrate monomers are supposed to be arranged in three domains. Glycyl bridges may link N-acetylglucosamine and uronosyl residues of the different domains [85]. No data on the biosynthesis of the *Halococcus* cell-wall heteropolysaccharide are available.

2.4. Natronococcus

2.4.1. Morphology
Cells of *Natronococcus occultus* form aggregates of 2 to 10 coccoid cells. Single cells possess a homogeneous electron-dense cell-wall layer up to 120 nm in width, which participates in septum formation (Fig. 10). The cells are embedded in a capsule [86,87].

2.4.2. Chemical composition of the natronococcal "glycosaminoglycan"
The cell-wall sacculi of *Natronococcus occultus* could be isolated as described for gram-positive eubacteria [14]. The protein-free isolated cell-wall polymer contains D-glucosamine, an uronic acid and L-glutamic acid in nearly equimolar amounts. D-galactose, D-glucose and D-mannose may be present in smaller amounts (Table 5; Kandler, Tindall, Winter, Fiedler, and König, paper in preparation). The overall chemical composition indicates that the cell wall of *Natronococcus occultus* consists of a glycosaminoglycan-like polymer, which is substituted with glutamic acid. More detailed structural studies have to be done. In addition to *Methanosarcina barkeri*, *Natronococcus*

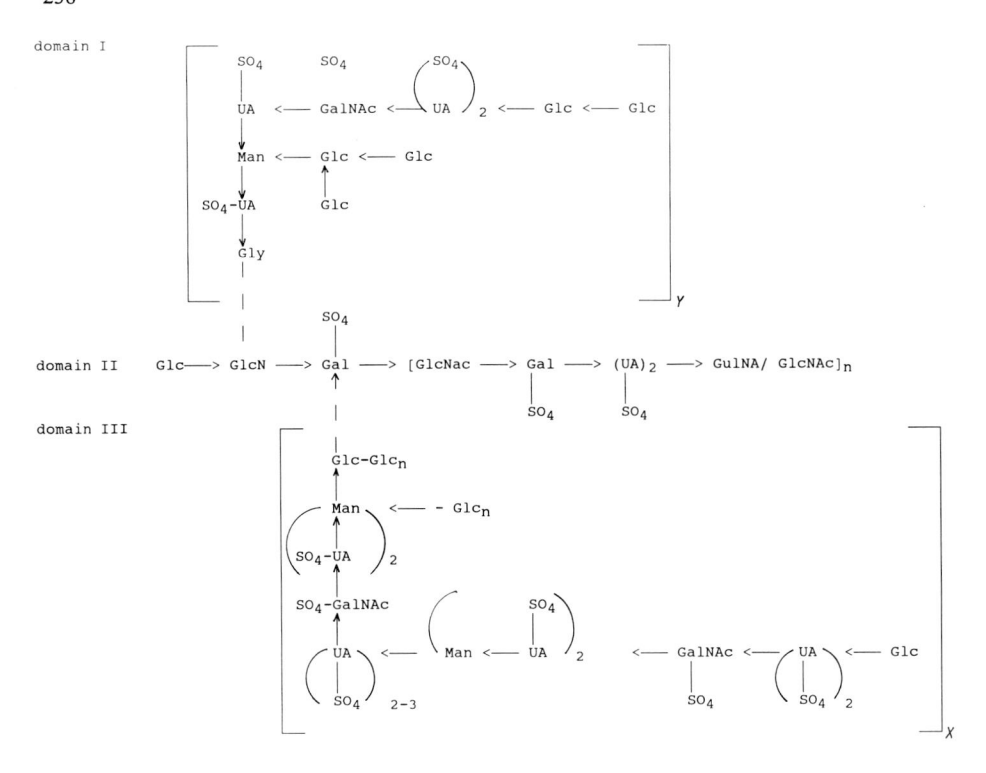

Fig. 9. Proposed linkage of the sugar residues within the cell wall of *H. morrhuae* CCM 859. Abbreviations: Gal, galactose; GalNAc, *N*-acetylgalactosamine; Glc, glucose; GlcNAc, *N*-acetylglucosamine; Gly, glycine; GulNA, *N*-acetylgulosaminuronic acid; Man, mannose; UA, uronic acid. Modified from ref. [84].

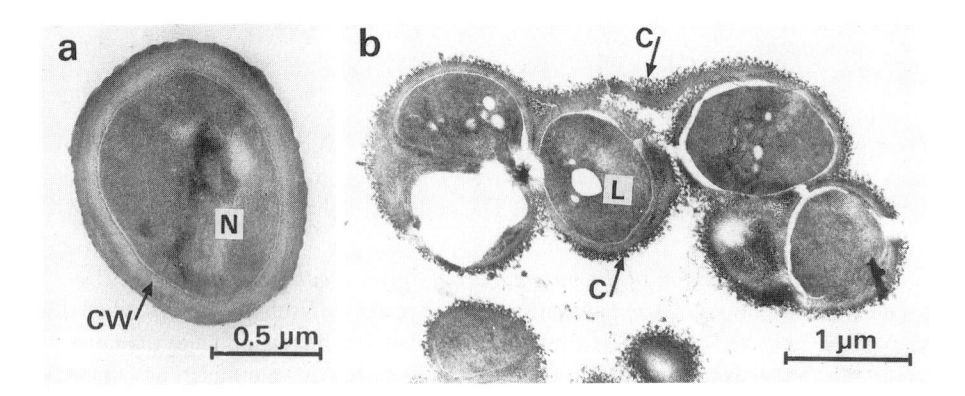

Fig. 10. Thin sections of *Natronococcus occultus*. Stationary phase. (a) single coccoid cell; (b) package of several cells in a common capsule. Abbreviations: CW, cell wall; N, nucleoid; C, capsule; L, lipid vacuoles. From ref. [87].

TABLE 5
Chemical composition of the cell walls of *Natronococcus occultus* SP4 [a]

Compound	Molar ratio	
	Batch 1	Batch 2
GlcN	1.00	1.00
UA[b]	1.13[c]	1.35[c]
Glu	0.82	0.75
Gal	0.06	–
Glc	–	0.01
Man	–	0.01

[a] From Kandler, O., Tindall, B., Winter, J. and König, H., in preparation.
[b] UA, uronic acids.
[c] Colour development in the phenylphenol test.

occultus may be another archaeon containing a cell-wall polymer reminiscent of polymers occurring in human and animal connective tissue.

3. Structure and chemistry of cell envelopes of gram-negative archaea

3.1. Proteinaceous sheaths

3.1.1. Methanospirillum hungatei
The typical long "spirilla" observed under the light microscope consist of several (mostly 5–10) cells (Figs. 11, 12) held together by a proteinaceous sheath [15,88]. Each single cell is surrounded by an electron-dense layer – the "inner cell wall" in the terminology of Zeikus and Bowen [88] – presumably proteinaceous in nature. Cell division occurs by septation in a similar manner as in gram-positive bacteria: the plasma membrane and the "inner cell wall" grow inward to partition the cell in two [89]. Subsequently, up to three plates (structural elements, Fig. 12b) are formed at the new cell poles. They consist of electron-dense layers exhibiting an 18.0 nm periodicity [89]. The plates join the sheath and thus form plugs between the cells which are gradually separated from each other by increasing amounts of amorphous spacer material. When the spacer exceeds about 0.5 µm in length, small lesions may become apparent in the sheath adjacent to the amorphous zone of the spacer, and the filament finally fractures in two. Thus, the spacer plugs become terminal plugs of the filaments.

Sheath material showing the original morphology of the filaments can easily be prepared by the usual techniques for the isolation of gram-positive cell walls [8]. The analysis of total hydrolysates of isolated sheath material by two authors [8,90], using different isolation procedures, has shown a complex spectrum of 18 amino acids (65–72%) and 5 different neutral sugars (6.03–8.2%) and ash. It is not known whether the carbohydrates are directly bound to the protein as in the case of glycoproteins. The isolated sheath material is resistent to detergents, and all proteases and glycosidases

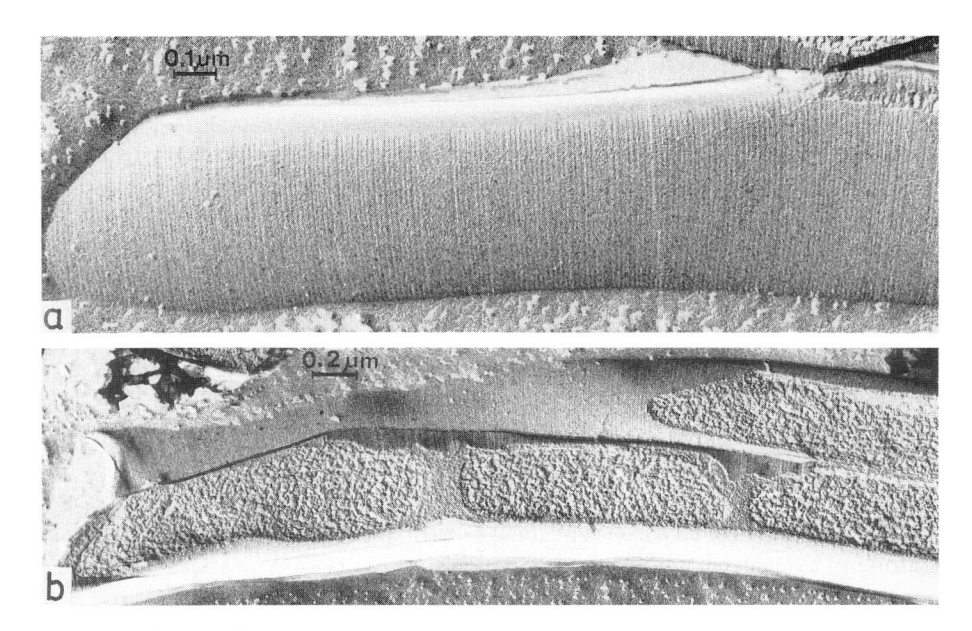

Fig. 11. *Methanospirillum hungatei.* Freeze-etching of (a) the sheath surface and (b) a longitudinal break through a filament. From ref. [105].

commonly used to solubilize or break down cell-wall material. It also resists 6 M urea and other powerful denaturants [91].

Computer image processing of tilted-view electron micrographs of negative stains of such isolated sheaths have revealed a two-dimensional S-layer-like paracrystalline structure with subunits exhibiting P1 symmetry and unit cells of $a = 12.0$ nm, $b = 2.9$ nm, and $\gamma = 93.7°$ [92], while Stewart et al. [93] found subunits of P2 symmetry and unit cells of $a = 5.66$ nm, $b = 2.81$ nm, and $\gamma = 85.6°$. Treatment of isolated sheaths at 90°C with a combination of β-mercaptoethanol and sodium dodecyl sulfate under alkaline conditions results in the solubilization of "glue peptides" and the liberation of single hoops, the essential structural component of the cylindrical sheaths [94]. Imaging the inner and outer surfaces of isolated sheaths with the bimorph scanning tunneling electron microscope confirmed that the sheaths form a paracrystalline structure, and that they consist of a series of stacked hoops of ~ 2.5 nm in width [95]. This study also showed that the sheath possesses minute pores and therefore is impervious to solutes of hydrated radius of > 0.3 nm.

3.1.2. Methanothrix concilii (recently renamed Methanosaeta concilii) [96]

The gram-negative organism forms very long flexible proteinaceous sheaths containing up to several hundred single rods of 1.04–4.0 μm in length and 0.5–0.8 μm in width. The single cells are surrounded by a "veil" which resembles a cell wall but is unable to maintain the rod shape when the cell is extruded from the sheath. As in the case of *Methanospirillum,* the single cells are separated from each other by spacer plugs

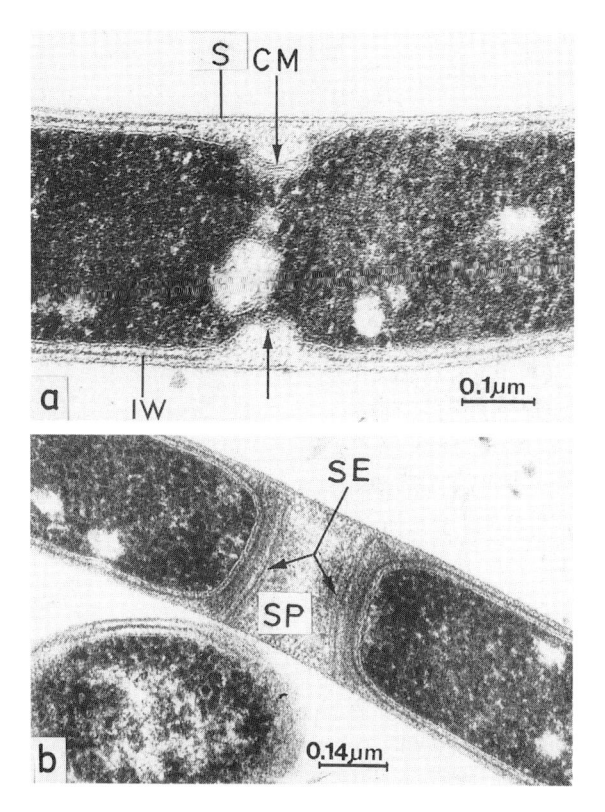

Fig. 12. Thin section of *Methanospirillum hungatei* (a) during and (b) after cell division. Abbreviations: IW, inner wall; S, sheath; SE, structural elements (plugs); SP, spacer. Modified from ref. [88].

which are formed during cell division [97]. Here, the plugs are two-layered structures; one layer shows concentric rings (Fig. 13) and is the first to be laid down during the division ingrowth. The second layer consists of larger, raised, concentric ribbons which progressively follow the advance of the first layer during division. Thus, the plug structure of *Methanothrix* differs significantly from that of *Methanospirillum* whose plugs are composed of two and more plates each exhibiting hexagonal subunit patterns. On the other hand, the sheath of *Methanothrix* consists of stacked hoops (Fig. 13) which possess a 2.8 nm lattice of subunits on their surface [98]. The analysis of hydrolysates of isolated sheaths of three strains of *Methanothrix soehngenii* has shown the presence of 18 amino acids and of neutral sugars [99,100], which suggests the presence of glycoprotein. This suggestion is strongly supported by recent work of Pellerin et al. [101]. These authors obtained asparagine-rich glycoprotein fractions by hydrozinolysis of sheath preparations from *Methanothrix soehngenii* strain FE and proposed the presence of asparaginyl-rhamnose linkages on Asn–X–Ser glycosylation sites of the sheath glycoprotein.

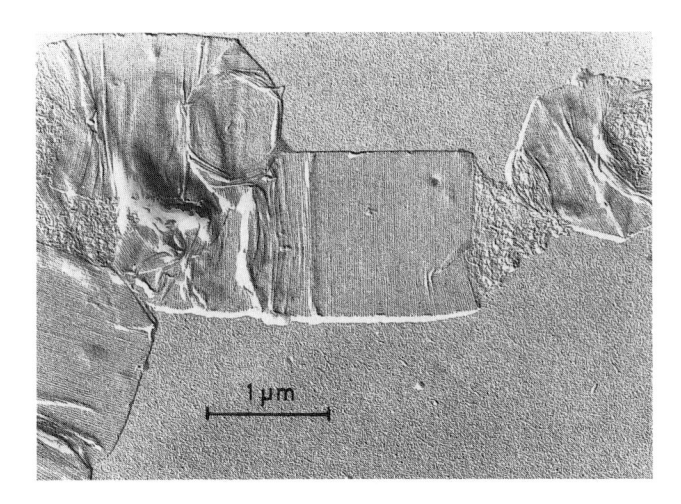

Fig. 13. Platinum-shadowed sheath of *Methanothrix soehngenii*. From ref. [105].

3.2. S-layers of gram-negative methanogenic rods and cocci

The cell envelopes of the coccoid or rod-shaped members of 6 genera of methanogens (*Methanococcus, Methanomicrobium, Methanogenium, Methanoculleus, Methanoplanus, Methanolobus*) consist of S-layers covering the cytoplasmic membrane. The S-layers show hexagonal arrays of protein or glycoprotein subunits (Fig. 14) and disintegrate when treated with 2% SDS for 30 min at 100°C [8]. Among the various taxa, S-layers are heterogeneous in the molecular weights of their proteins [102] and in their antigenicity [103]. The S-layers of the various species compared to date have similar chemical and physical stabilities and overall amino-acid compositions. The acidic proteins are predominantly composed of hydrophobic amino acids. The average hydrophobicity of the S-layer protein is lower than in many cellular proteins [102] although hydrophobic interactions contribute to the stability of the S-layer [104].

The chemical nature of S-layers of the above mentioned genera of methanogens has only undergone preliminary study, either by analysis of their amino-acid and carbohydrate content or by gel electrophoretic separation and periodate-Schiff reaction of the separated protein bands of the solubilized S-layers. The results of this largely unpublished work by König and Stetter are listed in Table 1 and are discussed in more detail in a previous review [105]. These data show that S-layers of the species *Methanococcus* consist of non-glycosylated protein, while those of the other genera are most likely composed of glycoproteins.

A recent three-dimensional reconstruction from a tilted series of negatively stained and of metal-shadowed preparations of the S-layer of *Methanoplanus limicola* [106] has shown P6 symmetry, a lattice constant of 14.7 nm, and a thickness of ~4.5 nm. Chemical analysis of these S-layers revealed an apparent molecular weight of the protein of 135 kDa, and the neutral sugar content of 240 mg/g polypeptide corroborates earlier suggestions

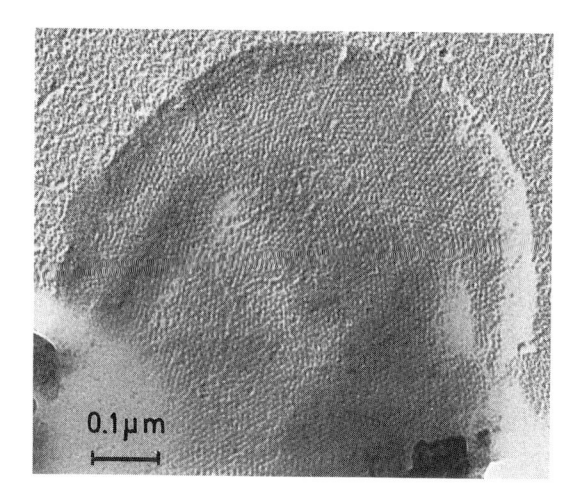

Fig. 14. Platinum-shadowed isolated S-layer of *Methanococcus vannielii*. From ref. [105].

based on a positive periodate-Schiff reaction [107] that the S-layer of *Methanoplanus limicola* consists of glycoprotein.

3.3. S-layers of gram-negative halobacteria

Gram-negative halobacteria comprise extremely halophilic rods and irregular plates [108]. Their cell envelopes consist solely of S-layers made up of glycoproteins.

3.3.1. Chemical structure

The surface glycoprotein of *Halobacterium salinarium* was the first glycosylated protein detected in prokaryotes [109]. More recent work on the chemical structure of glycopeptides [110–113] and the primary structure of the polypeptide moiety of the surface glycoprotein of *Halobacterium halobium* – an organism related to *Hb. salinarium* at the species level – has led to a detailed picture (Fig. 15A) of this molecule. Its main features are:

(i) The polypeptide chain of the mature glycoprotein shows a single hydrophobic stretch of 21 amino acids, which is only 3 amino acids away from the C-terminus (Fig. 15A). This hydrophobic peptide may serve as a membrane anchor [114].

(ii) Three different types of glycosyl linkages, each involving a different carbohydrate, are found: (a) A glycosaminoglycan chain, consisting of a repeating sulfated pentasaccharide block (Fig. 15B, left), is linked to position 2 of the polypeptide via the novel N-glycosyl linkage unit asparaginyl–GalNAc [111]. (b) Ten sulfated oligosaccharides (Fig. 15 B, centre) that contain glucose, glucuronic acid, and iduronic acid are bound to the protein via the hitherto unknown N-glycosyl linkage unit asparaginylglucose [110]. (c) About 15 glucosylgalactose disaccharides (Fig. 15B, right) are O-glycosidically linked to a cluster of threonine residues close to the postulated

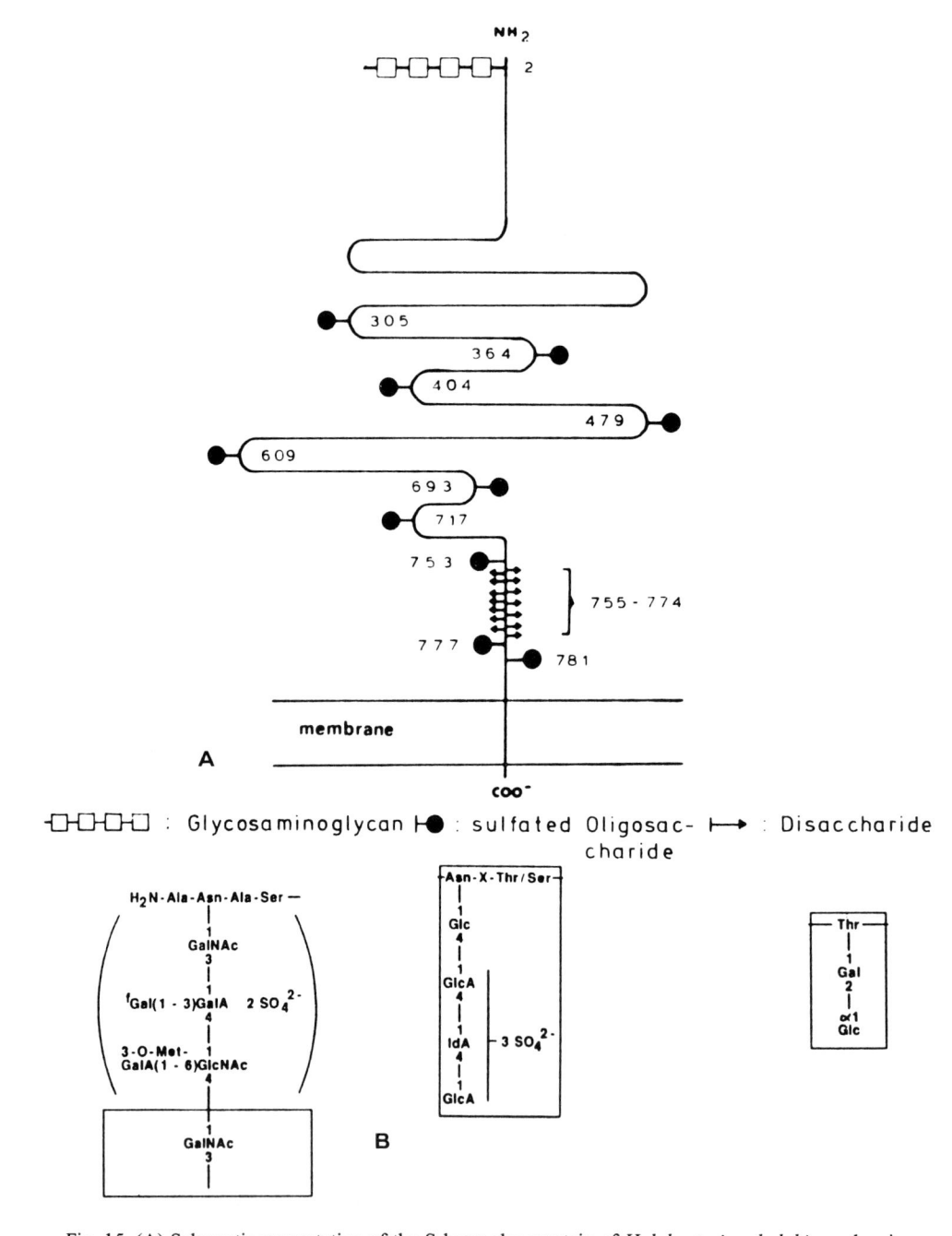

Fig. 15. (A) Schematic presentation of the S-layer glycoprotein of *Halobacterium halobium*, showing the different glycosylation sites and the membrane binding domain at the C-terminal end. Numbering indicates amino acid positions. (B) Schematic presentation of the three different glycoconjugates (B1–B3). From ref. [117].

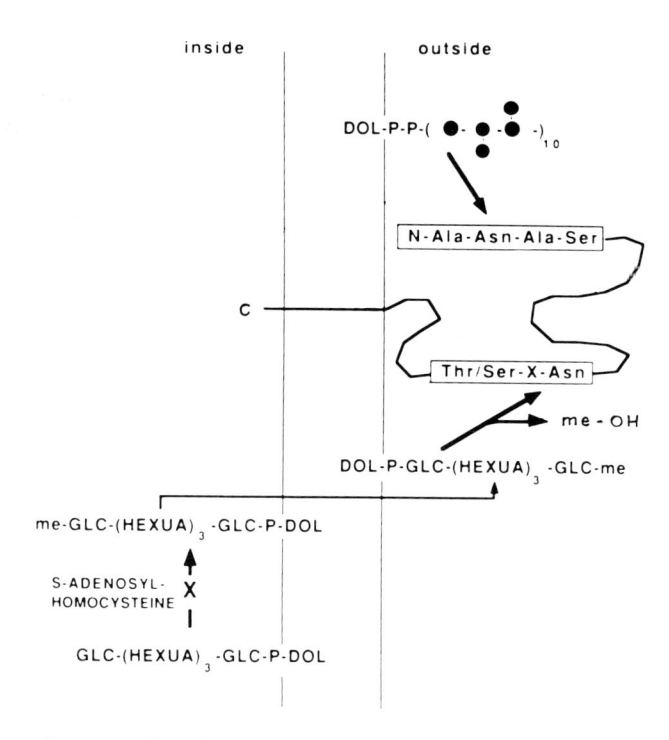

Fig. 16. Schematic presentation of the present hypotheses on the biosynthesis of glycoprotein of halobacteria (for details see the text). From ref. [117].

transmembrane domain. In structure and composition, these disaccharides resemble a type of oligosaccharide linked to hydroxylysine in animal collagens [109,115].

(iii) The surface glycoprotein of *Halobacterium salinarium* (120 kd; core protein 87 kd) is extraordinarily acidic. Several factors contribute to this acidity: (a) more than 20% of its amino acids are aspartic and glutamic acids [109]; (b) one mole of glycoprotein contains about 40–50 moles of uronic acids within its glycoconjugates [111]; and (c) about 40–50 moles of sulfate residues per one mole of glycoprotein are esterified [116] to its carbohydrates.

3.3.2. Biosynthesis

Studies on the biosynthesis of the surface glycoprotein [117] revealed that, for the glycosaminoglycan, dolicholpyrophosphate serves as lipid carrier and Asn–Ala–Ser serves as acceptor peptide (Fig. 16), while dolicholmonophosphate is the lipid carrier for the second sulfated oligosaccharide. A transient methylation of a glucose residue is required for the transfer of the dolychyl oligosaccharides onto the protein outside the cytoplasmic membrane (Fig. 16) [117].

3.3.3. Halobacterial versus eukaryotic glycoproteins

The comparison of the chemical structure and biosynthesis of the halobacterial and the eukaryotic glycoproteins shows similarities as well as distinct differences. As pointed out by Lechner and Wieland [117], the most distinctive differences are:

(i) The presence of two additional, hitherto unknown types of N-glycosyl linkages (Asn–Glc; Asn–GalNAc) in halobacteria, and the absence of the N-glycosidically linked mannose-containing branched glycans common in eukaryotic glycoproteins.

(ii) Eukaryotic glycoproteins are covalently modified at the protein-linked level in the Golgi apparatus, while in halobacteria, lacking a Golgi apparatus, the oligosaccharides are completed and sulfated while still attached to dolichol on the cytosolic side of the cell membrane. Thereafter, they are translocated to the cell surface, possibly by mechanisms that involve transient methylation.

(iii) Glycosylation of the protein occurs outside the cell membrane. Thus, the halobacterial cell surface is functionally equivalent to the luminal side of the endoplasmatic reticulum membrane in eukaryotic cells.

3.3.4. Three-dimensional structure

For a long time, morphological study of halobacterial S-layers by electron microscopy has been hampered by the high salt concentrations required to maintain the integrity of the regular surface arrays. Only the finding that S-layers of the moderate halophile *Haloferax volcanii* [118], formerly named *Halobacterium volcanii* [119] from the Dead Sea could be maintained intact in concentrations of divalent cations as low as 10 mM $CaCl_2$ alone [120], allowed a three-dimensional reconstruction from electron micrographs of negatively stained S-layers from this halophile. As shown by Kessel et al. [121], the glycoprotein is arranged on a P6 lattice with a center-to-center distance of 16.8 nm. It forms 4.5 nm high dome-shaped complexes like an inverted funnel (Fig. 17) with a pore at the apex. The authors suggest that the narrow end of the funnel points to the outside of the cell and, consequently, the wide opening of the funnel is directed towards the cytoplasmic membrane.

The model shown in Fig. 17 is based on the integration of morphological data obtained with S-layers from *Haloferax volcanii* [121] and of chemical data obtained with S-layers from *Halobacterium halobium* [117]. The integration of experimental data from different organisms in one and the same model is unsatisfactory. However, a recent study of the chemical structure of the glycoprotein of *Haloferax volcanii* by Sumper et al. [122] has shown that the most important domains for model building found in *Halobacterium halobium* are also present in the glycoprotein of *Haloferax volcanii*:

(i) the hydrophobic stretch at the C-terminal end, the presumed membrane anchor; and

(ii) the cluster of threonine-rich residues near the C-terminal end, the presumed spacer area of the model, which may be regarded as analogous to the periplasmic space of gram-negative bacteria.

The most distinctive differences between the two glycoproteins are found in the glycosylation pattern of the N-terminal portion of the molecules, which forms the bulky dome. Here, variations of the chemical structure may have little or no morphological impact.

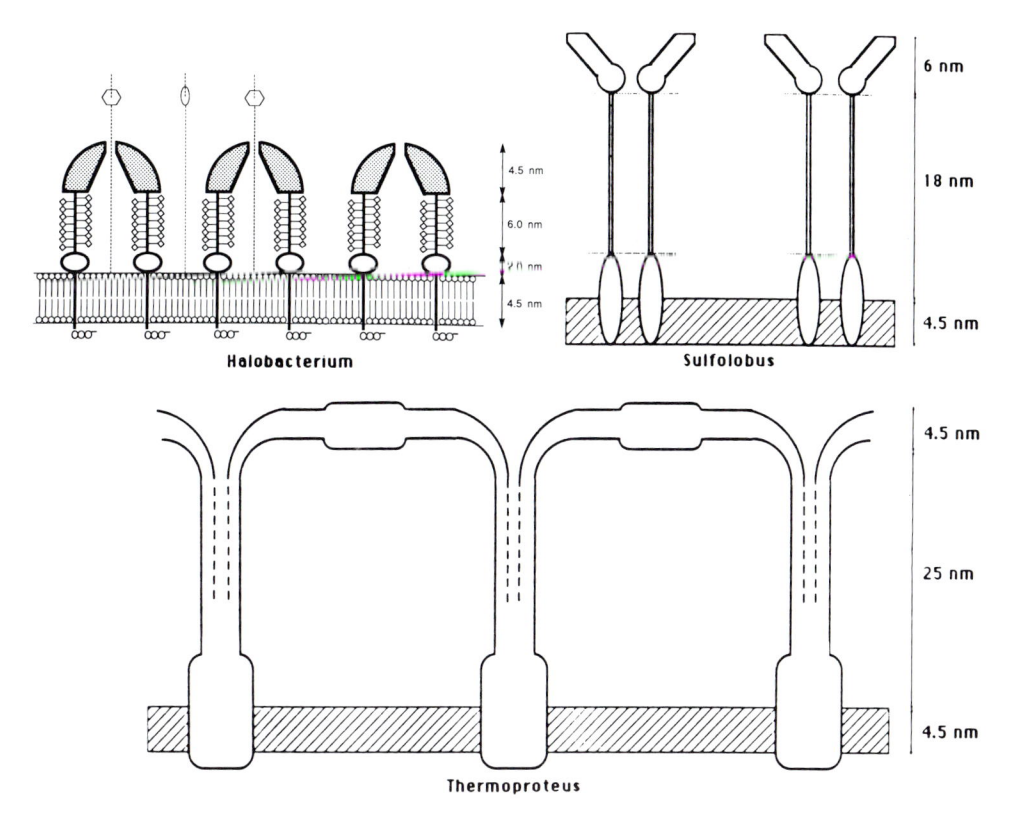

Fig. 17. Models of the pseudo-periplasmatic space in archaea. The pseudoperiplasmatic space is a region between the membrane and the porous outer canopy of the S-layer, which is maintained by regularly disposed spacer elements. The S-layer protein is either anchored directly in the membrane, for instance in *Thermoproteus* and *Halobacterium,* or interacts with a distinct membrane-embedded protein, as presumably in *Sulfolobus*. Modified from ref. [140].

3.4. S-layers of Archaeoglobus

The genus *Archaeoglobus* [123] comprises strains of hyperthermophilic (optimal growth in the range of 80–110°C) sulfate-reducing archaea. All strains possess S-layers of hexagonally arrayed morphological units as the sole envelope component outside the cytoplasmic membrane. Only the S-layer of one strain, *Archaeoglobus fulgidus* strain Z, has been studied in more detail, by Zellner et al. [124]. The center-to-center spacings of the morphological units were 17.5 nm. A periodate-Schiff positive band of an apparent molecular weight of 132 000 representing the presumed S-layer glycoprotein was observed on 10% slab gels.

The three-dimensional reconstruction from a tilt series of a negatively stained cell envelope revealed a dome-shaped morphological unit with a wide opening toward the cytoplasmic membrane and a narrow opening to the exterior at the apex [125]. This design resembles that of *Haloferax volcanii* [121] and, to a lesser extent, that of *Sulfolobus*

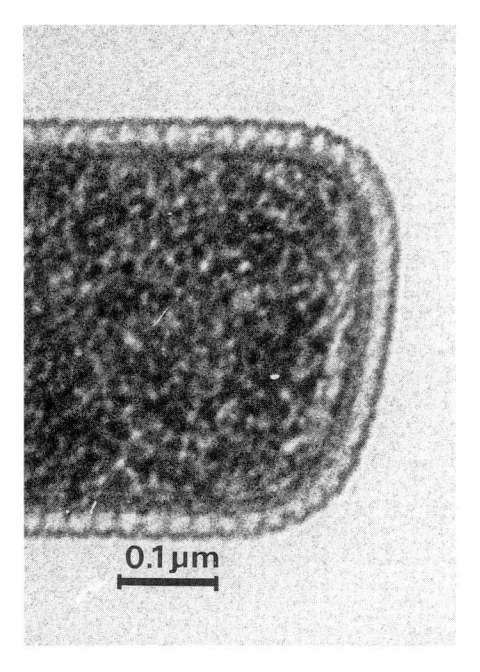

Fig. 18. Thin section of *Thermoproteus tenax* showing an extremely wide S-layer. Courtesy of W. Zillig.

solfataricus [126]. However, while in *Sulfolobus* and in *Halobacterium* the morphological complexes are tightly interconnected via the two-fold or three-fold axis, respectively, there is no evidence for any extensive interconnections via direct protein–protein interaction in *Archaeoglobus*. This is in agreement with the observation that S-layers of *Archaeoglobus* resist detergent treatment only when they are preabsorbed onto a carbon support. This is in sharp contrast to the behaviour of S-layers of most crenarchaeota [127], which resist detergent treatment without support even at high temperatures [125].

3.5. S-layers of sulfur-metabolizing hyperthermophilic archaea

All known representatives of sulfur-metabolizing archaea (Crenarchaeotes and Thermococcales) are hyperthermophilic and possess S-layers of hexagonally arrayed morphological units as the sole cell-envelope component outside the cytoplasmic membrane. The center-to-center spacing varies among the known species between 12.5 nm in *Pyrococcus furiosus* [128] and ~30 nm in *Thermoproteus tenax* [129]. The molecular masses of the surface proteins range from 40 to 325 kDa [2]. As indicated by their carbohydrate content and the presence of periodate-Schiff-positive bands in gel electropherograms (Table 1), the S-layers of the majority of the investigated species consist of glycoprotein. However, the glycosidic linkages are not as rigorously established as in the S-layer of *Halobacterium* [117].

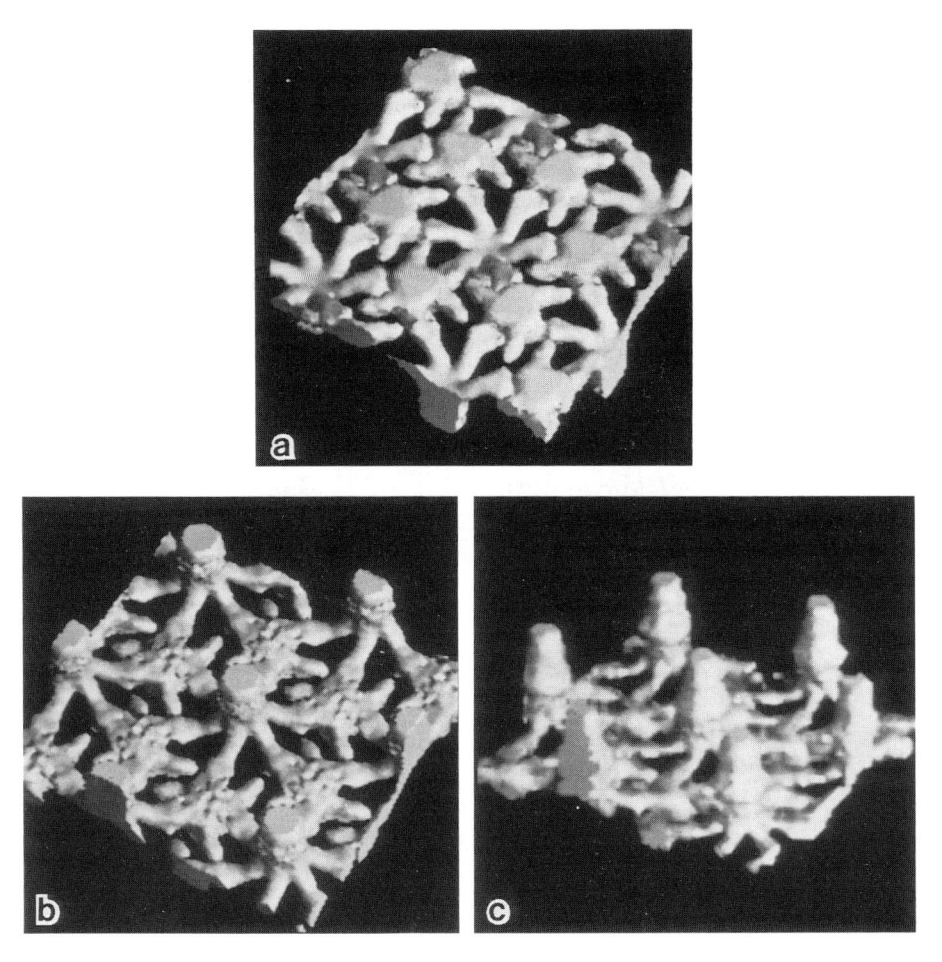

Fig. 19. Three-dimensional reconstruction of the (a) outer and (b,c) inner surface of the S-layer of *Thermoproteus tenax*. The 'pillars' have been truncated into (a) and (b) near their bases. They are shown in their full height in (c), the view at a glancing angle of the inner surface. From ref. [132].

Much attention was attracted by the three-dimensional structure of S-layers, which has been studied in the following species: *Sulfolobus acidocaldarius* [130,131], *Sulfolobus solfataricus* [126], *Thermoproteus tenax* [129,132], *Desulfurococcus mobilis* [133], *Pyrobaculum organotrophum* [134], *Pyrobaculum islandicum* [135], *Pyrodictium brockii* [136], *Pyrodictium occultum* [137], *Acidianus brierleyi* [138], and *Hyperthermus butylicus* [139]. Although these studies revealed structural similarities between closely related strains and species, for instance between strains of *Sulfolobus* [126] or between the two species of *Thermoproteus* [129], no conserved feature analogous to the funnel-shaped core structure in bacterial S-layers [140] was found in hyperthermophilic sulfur-metabolizing archaea. Even the interspace between the cytoplasmic membrane and the outer surface of the S-layer, seen in cross-sections of all strains studied, has no common conserved structural

Fig. 20. Computer-generated view of the outer surface of *Hyperthermus butylicus* surface protein. From ref. [139].

basis. In *Thermoproteus tenax*, the extremely wide interspace (Fig. 18) is due to about 25 nm long protrusions extending from the relatively thin (3–4 nm) filamentous network of the outer surface of the S-layer (Fig. 19) toward the cytoplasmic membrane. The distal ends of these pillar-like protrusions appear to penetrate the membrane, thus serving as membrane anchors reminiscent of the structure found in *Halobacterium* (Fig. 17). Similar protrusions or spikes are seen in S-layers of *Pyrobaculum islandicum* [135], an organism closely related to *Thermoproteus tenax* [141], and that of the more distantly related *Desulfurococcus mobilis* [133]. However, in other species, the S-layer protein may be attached to the membrane via an interaction with a second, membrane-integral protein, for instance in *Sulfolobus*, schematically depicted in Fig. 17. No connection between the S-layer and cytoplasmic membrane was seen in *Hyperthermus butylicus* [139], although electron micrographs of cross-sections of cells show an envelope profile similar to that of other hyperthermophiles [142]. The S-layer of *Hyperthermus butylicus*, obtained by detergent extraction, shows hexagonal arrays with a lattice constant of approximately 25.8 nm, and unusual compact cores (Fig. 20). The protein–protein interactions within the surface array must be strong to maintain the integrity of the lattice in the absence of an underlying plasma membrane. The distinct variation of S-layer structures, not only within the hyperthermophiles but in archaea in general, becomes obvious when comparing the projection structures of various S-layers of archaea (Fig. 21).

Considering the multitude of patterns, it is difficult to think of any common function for S-layers. In some cases, the S-layer contributes to the stability of the cell, similar to murein or pseudomurein cell-wall sacculi, for instance in *Thermoproteus tenax* which possesses an extremely stable S-layer. In many cases, the interspace between the

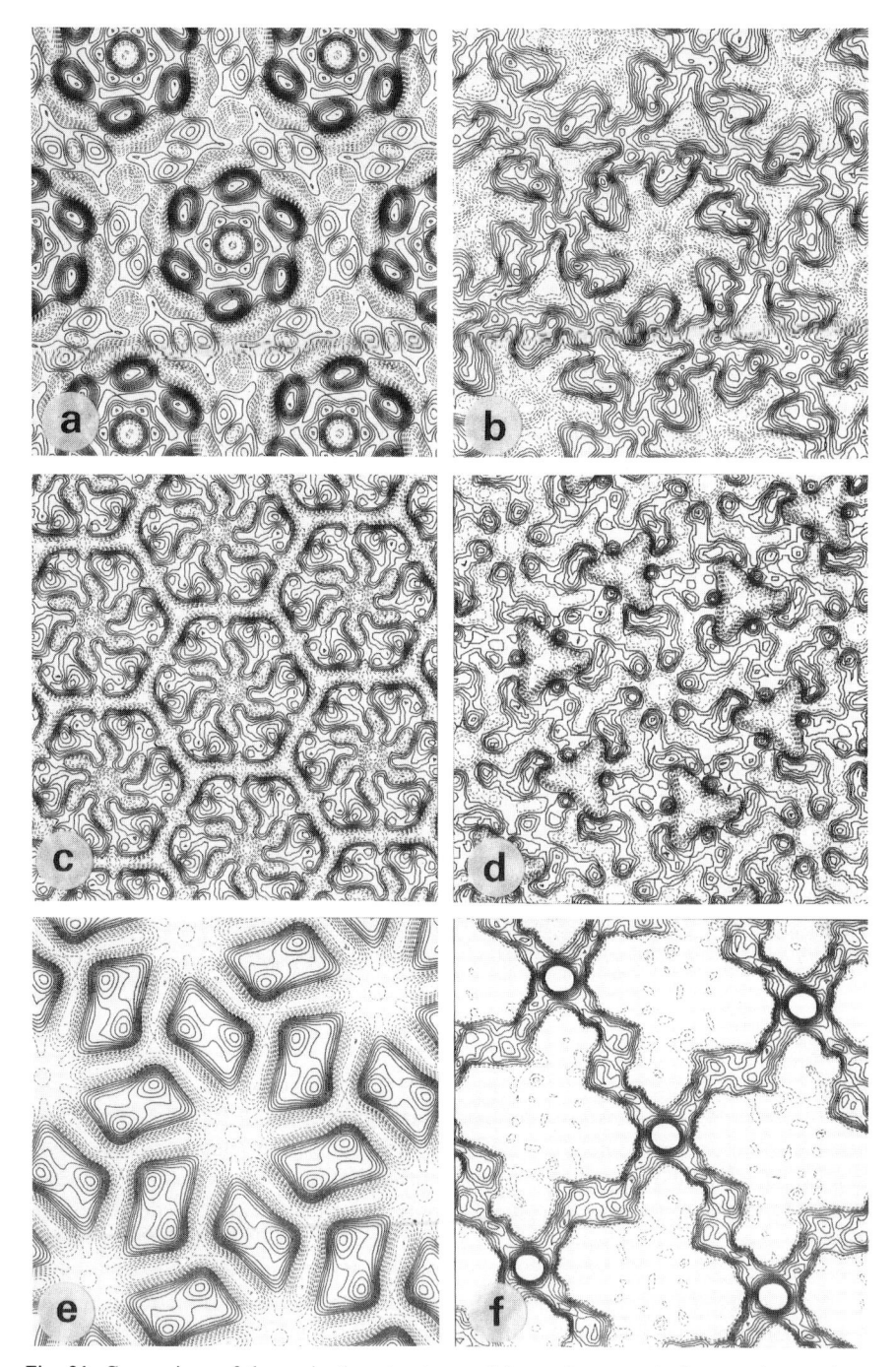

Fig. 21. Comparison of the projection structures of the surface protein from various archaea. From ref. [139]. (A) *Hyperthermus butylicus* (lattice constant 25.8 nm); (B) *Pyrodictium occultum* (lattice constant 21.8 nm); (C) *Archaeoglobus fulgidus* (lattice constant 17.5 nm); (D) *Sulfolobus* sp. (strain B6/2) (lattice constant 21.0 nm); (E) *Thermococcus celer* (lattice constant 18.0 nm); (F) *Desulfurococcus mobilis* (lattice constant 18.0 nm).

cytoplasmic membrane and the outer surface of the S-layer may at least partly fulfill the role of a periplasmic space reminiscent of gram-negative bacteria, or the S-layer may act as a protective molecular sieve. However, in most cases the meshwork of the S-layer is much too loose to fulfill such functions. The possible physiological functions and presumed technical applications of S-layers have recently been discussed in more detail by Messner and Sleytr [2].

4. Surface structure of archaea without cell envelopes

The members of the genus *Thermoplasma* are the only archaea not possessing any cell envelope layers outside their cytoplasmic membrane [2]. Like their bacterial counterpart, the mycoplasms, *Thermoplasma* forms pleomorphic coccoid to filamentous cells. They were first discovered by Darland et al. [143] at the surface of burning coal refuse piles. Recently, they have been found world-wide in solfatara fields [144]. They grow microaerobically or, as recently found, anaerobically by sulfur respiration at pH 0.5–4.0 in a temperature range of 33°C to 67°C [144]. The stability of the cell membrane required for growth under these harsh environmental conditions is supported not only by diphytanyl ethers common to all archaea [145] but also by a combination of unique mannose-rich glycoproteins [146] and lipoglycans [145,147] (Fig. 22) found in isolated cell membranes, which consisted of 60% protein, 25% lipid, and 10% carbohydrate [146]. The major protein band obtained by SDS polyacrylamide gel electrophoresis exhibited a molecular mass of 152 kDa and was periodate-Schiff-positive. The carbohydrate component consisted of mannose, glucose, and galactose (molar ratio 31:9:4) and was found to be linked to asparagine via a disaccharide composed of *N*-acetylglucosamine (Fig. 22). The glycan chains of the lipoglycan and the glycoprotein are directed to the outside of the cell, as evidenced by employing peroxidase-labeled concanavalin A [148]. Thus, the surface structure of *Thermoplasma* is reminiscent of the glycocalyx of eukaryotic cells.

5. Concluding remarks

The lack of a murein cell-wall sacculus and the discovery of different cell-envelope polymers and structures in some physiologically unusual prokaryotes, was one of the first biochemical and cytological evidences in favour of Carl Woese's archaebacteria concept [46,149,150]. Since then, increasingly more unique cell-envelope polymers and new types of biosynthetic pathways have become known. These findings corroborate the proposal that the archaea represent a third lineage of organisms [150] in addition to bacteria and eucarya, and that the common ancestor or ancestral population of the archaea did not evolve any cell-wall polymer before it radiated into the various sublineages known today [46,151].

---

[2] A penicillin-insensitive, cell wall-less mycoplasma-like methanogen has been isolated from a sewage sludge fermenter and named *Methanoplasma elizabethii* [153]. However, no pure culture has been deposited in a culture collection, and no further data on this organism are so far available.

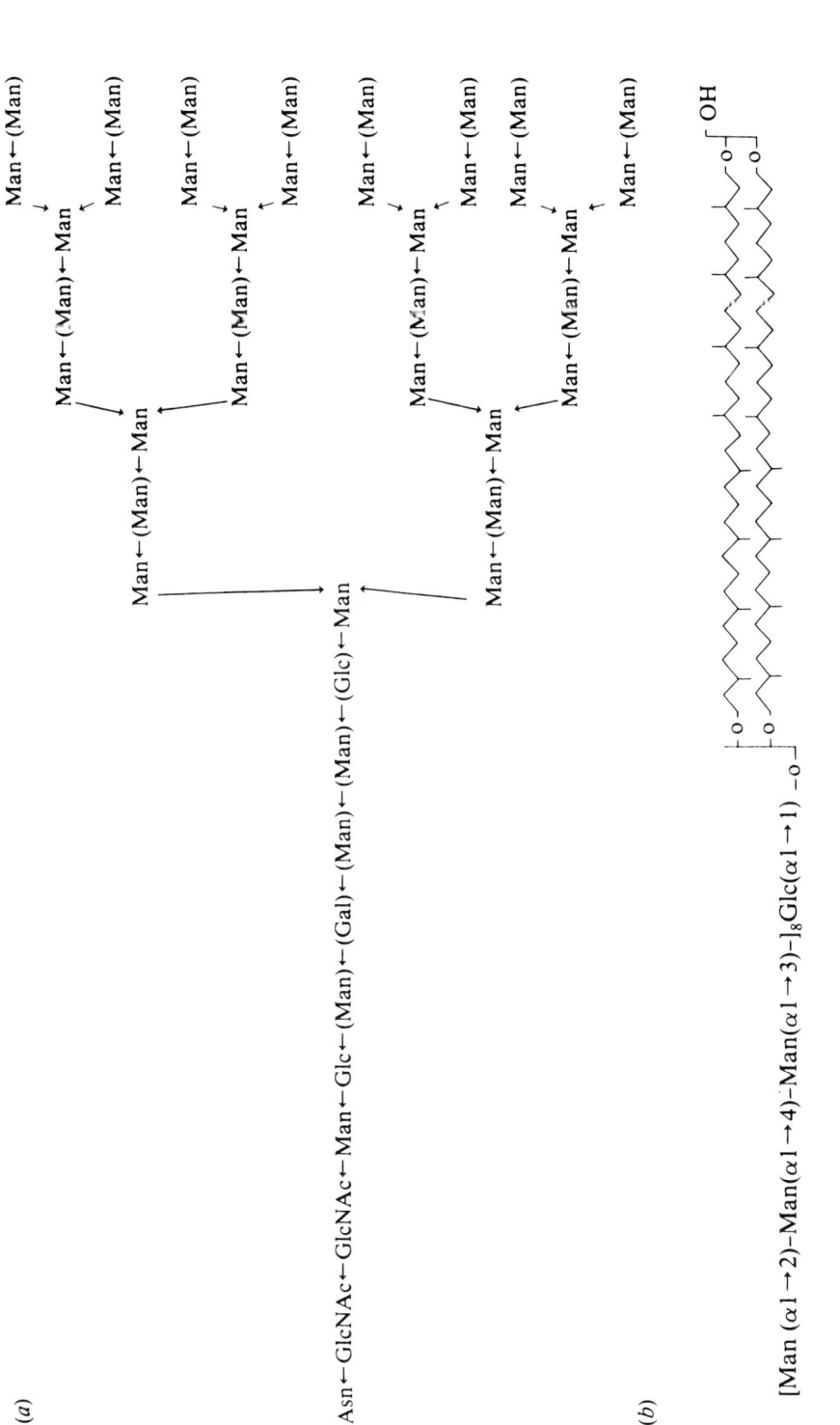

(a)

Asn←GlcNAc←GlcNAc←Man←Glc←(Man)←(Gal)←(Man)←(Man)←(Glc)←Man

(b)

[Man(α1→2)-Man(α1→4)-Man(α1→3)-]₈Glc(α1→1)

Fig. 22. Glycan structures of cytoplasmic membrane components of *Thermoplasma acidophilum*: (a) glycoprotein (parentheses indicate variable numbers of mannose residues); (b) lipoglycan. Modified from ref. [145].

ARCHAEA

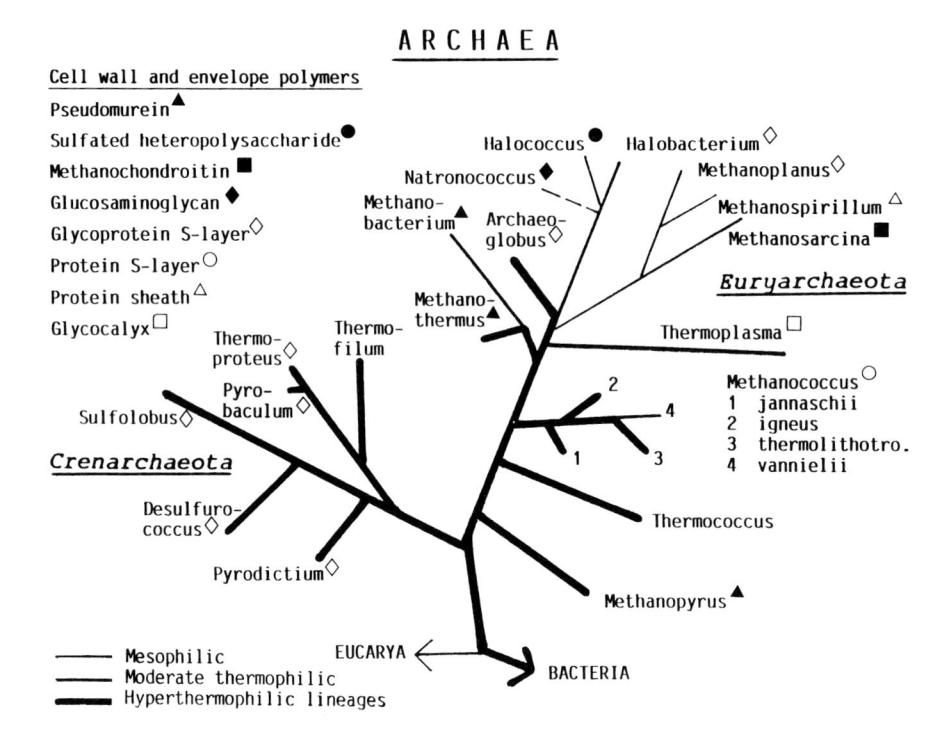

Fig. 23. Distribution of cell-wall and cell-envelope polymers among archaea. The depicted portion of the phylogenetic tree is taken from Kandler[151]. Branching order and branch lengths are based upon 16S rRNA sequence comparison[127].

The distribution of the cell-wall and envelope polymers within the phylogenetic tree of archaea (Fig. 23) reflects the physiological homogeneity of the kingdom Crenarchaeota, which comprises exclusively hyperthermophilic sulfur-metabolizing phenotypes, possessing (glyco-)proteinaceous S-layers as sole envelope component, and the physiological inhomogeneity of the kingdom Euryarchaeota, which comprises biochemically and ecologically diverse phenotypes, such as the anaerobic methanogens and the aerobic extreme halophiles, possessing a variety of cell wall polymers and S-layer types. Even within the clusters of the particular phenotypes, fundamentally different cell-wall polymers and S-layer proteins were developed. Thus, cell-wall formation in euryarchaeotes is considered to be a good example of parallelophyly[151] which, according to Mayr and Ashlock[152], is characterized by "parallel but independent, evolutionary changes in several sublineages", due to the common propensities and genetic potentials of the members of the ancestral population (founder population sensu E. Mayr[152]).

The diversity of envelope chemistry and structure in archaea resembles that in eucarya which also have evolved chemically different rigid cell walls in plants, fungi and algae, as well as a variety of flexible, so-called "pellicles" in various lineages of protists.

References

[1] Sleytr, U.B., Messner, P., Pum, D. and Sára, M., Eds. (1988) Crystalline bacterial cell surface layers, Springer, Heidelberg.
[2] Messner, P. and Sleytr, U.B. (1992) Adv. Microbial Physiol. 33, 213–274.
[3] Balch, W.E., Fox, G.E., Magrum, L.J., Woese, C.R. and Wolfe, R.S. (1979) Microbiol. Rev. 43, 260–296.
[4] Kurr, M., Huber, R., König, H., Jannasch, H.W., Fricke, H., Trincone, A., Kristjansson, J.K. and Stetter, K.O. (1991) Arch. Microbiol. 156, 239–247.
[5] Nußer, E., Hartmann, E., Allmeier, H., König, H., Paul, G. and Stetter, K.O. (1988) In: Crystalline Bacterial Cell Surface Layers (Sleytr, U.B., Messner, P., Pum, D. and Sàra, M., Eds.), pp. 21–25, Springer, Berlin.
[6] Stetter, K.O., Thomm, M., Winter, J., Wildgruber, G., Huber, H., Zillig, W., Janecovic, D., König, H., Palm, P. and Wunderl, S. (1981) Zentralbl. Bakteriol., Mikrobiol., Hyg., Abt. 1, Orig. C 2, 166–178.
[7] Lauerer, G., Kristjansson, J.K., Langworthy, T.A., König, H. and Stetter, K.O. (1986) System. Appl. Microbiol. 8, 100–105.
[8] Kandler, O. and König, H. (1978) Arch. Microbiol. 118, 141–152.
[9] König, H. and Kandler, O. (1979) Arch. Microbiol. 123, 295–299.
[10] König, H., Kralik, R. and Kandler, O. (1982) Zentralbl. Bakteriol., Mikrobiol., Hyg. Abt. 1, Orig. C 3, 179–191.
[11] König, H., Kandler, O., Jensen, M. and Rietschel, E.T. (1983) Hoppe-Seyler's Z. Physiol. Chem. 364, 627–636.
[12] Sharon, N. (1986) Eur. J. Biochem. 159, 1–6.
[13] Schleifer, K.H. (1985) In: Methods in Microbiology (Gottschalk, G., Ed.), pp. 123–156, Academic Press, New York.
[14] Schleifer, K.H. and Kandler, O. (1972) Bacteriol. Rev. 36, 407–477.
[15] Kandler, O. (1979) Naturwissenschaften 66, 95–105.
[16] Hammes, W.P., Winter, J. and Kandler, O. (1979) Arch. Microbiol. 123, 275–279.
[17] Hilpert, R., Winter, J., Hammes, W. and Kandler, O. (1981) Zentralbl. Bakteriol., Mikrobiol., Hyg., Abt. 1, Orig. C 2, 11–20.
[18] Böck, A. and Kandler, O. (1985) In: The Bacteria, Vol. 8 (Woese, C.R. and Wolfe, R.S., Eds.), pp. 525–544, Academic Press, New York.
[19] Schleifer, K.H. and Seidl, P.H. (1984) In: Chemical Methods in Bacterial Systematics (Goodfellow, M. and Minnikin, D.E., Eds.), pp. 201–219, Academic Press, New York.
[20] König, H. (1986) System. Appl. Microbiol. 8, 159–162.
[21] König, H. (1985) J. Gen. Microbiol. 131, 3271–3275.
[22] Formanek, H. (1982) Z. Naturforsch., C: Biosci. 37C, 226–235.
[23] Burge, R.E., Fowler, A.G. and Reaveley, D.A. (1977) J. Mol. Biol. 117, 927–953.
[24] Labischinski, H., Barnickel, G., Bradaczek, H. and Giesbrecht, P. (1979) Eur. J. Biochem. 95, 147–155.
[25] Formanek, H., Formanek, S. and Wawra, H. (1974) Eur. J. Biochem. 46, 279–294.
[26] Labischinski, H., Barnickel, G., Leps, B., Bradaczek, H. and Giesbrecht, P. (1980) Arch. Microbiol. 127, 195–201.
[27] Formanek, H. (1985) Z. Naturforsch. 40c, 555–561.
[28] Leps, B., Labischinski, H., Barnickel, G., Bradaczek, H. and Giesbrecht, P. (1984) Eur. J. Biochem. 144, 279–286.
[29] Kelemen, M.V. and Rogers, H.J. (1971) Proc. Natl. Acad. Sci. U.S.A. 68, 992–996.
[30] Jarrel, K.F., Colvin, J.R. and Sprott, G.D. (1982) J. Bacteriol. 149, 346–353.
[31] Sauer, F.D., Mahadevan, S. and Erfle, J.D. (1984) Biochem. J. 221, 61–69.

[32] Blotevogel, K.-H., Fischer, U., Mocha, M. and Jannsen, S. (1985) Arch. Microbiol. 142, 211–217.
[33] Bush, J.W. (1985) J. Bacteriol. 163, 27–36.
[34] König, H., Semmler, R., Lerp, C. and Winter, J. (1985) Arch. Microbiol. 141, 177–180.
[35] Kiener, A., König, H., Winter, J. and Leisinger, Th. (1987) J. Bacteriol. 169, 1010–1016.
[36] Mountford, D.O., Mörschel, E., Beimborn, D.B. and Schönheit, P. (1986) J. Bacteriol. 168, 892–900.
[37] Morii, H. and Koga, Y. (1992) J. Ferment. Bioeng. 73, 6–10.
[38] Tipper, K.J. and Wright, A. (1979) In: The Bacteria: A Treatise on Structure and Function, Vol. 7 (Gunsalus, I.C., Sokatch, J.R. and Ornston, L.N., Eds.), pp. 291–426, Academic Press, New York.
[39] König, H., Kandler, O. and Hammes, W. (1989) Can. J. Microbiol. 35, 176–181.
[40] Hartmann, E. and König, H. (1990) Naturwissenschaften 77, 472–475.
[41] Hartmann, E. and König, H. (1990) Arch. Microbiol. 153, 444–447.
[42] Hartmann. E., König. H., Kandler, O. and Hammes, W. (1989) FEMS Microbiol. Lett. 61, 323–328.
[43] Hartmann, E., König, H., Kandler, O. and Hammes, W. (1990) FEMS Microbiol Lett 69, 271–276.
[44] Wright, A. and Tipper, K.J. (1979) In: The Bacteria. A Treatise on Structure and Function, Vol. 7 (Gunsalus, J.C., Sokatch, J.R. and Ornston, L.N., Eds.), pp. 427–485, Academic Press, New York.
[45] Schönheit, P. and Thauer, R.K. (1980) FEMS Microbiol. Lett. 9, 77–80.
[46] Kandler, O. (1982) Zentralbl. Bakteriol., Mikrobiol., Hyg., Abt. 1, Orig. C 3, 149–160.
[47] Seidl, P.H. and Schleifer, K.H., Eds. (1986), Biological Properties of Peptidoglycan, Walter de Gruyter, Berlin.
[48] Karakawa, W.W., Lackland, H. and Krause, R.M. (1967) J. Immunol. 99, 1179–1182.
[49] Schleifer, K.H. and Krause, R.M. (1971a) J. Biol. Chem. 246, 986–993.
[50] Schleifer, K.H. and Krause, R.M. (1971b) Eur. J. Biochem. 19, 471–478.
[51] Seidl, H.P. and Schleifer, K.H. (1977) Eur. J. Biochem. 74, 353–363.
[52] Seidl, H.P. and Schleifer, K.H. (1978) Arch. Microbiol. 118, 185–192.
[53] Conway de Macario, E., Macario, A.J.L, Magarinos, M.C., König, H. and Kandler, O. (1983) Proc. Natl. Acad. Sci. U.S.A. 80, 6346–6350.
[54] Toth, L.A., Krueger, J.M. (1988) Infect. Immun. 56, 1785–1791.
[55] Johannsen, L., Toth, L.A., Rosenthal, R.S., Opp, M.R., Obál Jr., F., Cady, A.B. and Krueger, J.M. (1990) Am. J. Physiol. 259, R182–R186.
[56] Johannsen, L., Labischinski, H. and Krueger, J.M. (1991) Infect. Immun. 59, 2502–2504.
[57] Stimpson, S.A., Brown, R.R., Anderle, S.K., Klapper, D.G., Clark, R.L., Cromartie, W.J. and Schwab, J.H. (1986) Infect. Immun. 51, 240–249.
[58] Bröckl, G., Behr, M., Fabry, S., Hensel, R., Kaudewitz, H., Biendl, E. and König, H. (1991) Eur. J. Biochem. 199, 147–152.
[59] Hartmann, E. and König, H. (1989) Arch. Microbiol. 151, 274–281.
[60] Kreisl, P. and Kandler, O. (1986) Syst. Appl. Microbiol. 7, 293–299.
[61] Zhilina, T.N. and Zavarzin, G.A. (1973) Mikrobiologiya 42, 235–241.
[62] Blotevogel, K.-H. and Fischer, U. (1989) Int. J. System. Bacteriol. 39, 91–92.
[63] Sowers, K.R., Baron, S.F. and Ferry, J.G. (1984) Appl. Environ. Microbiol. 47, 971–978.
[64] Aldrich, H.C., Robinson, R.W. and Williams, D.S. (1986) System. Appl. Microbiol. 7, 314–319.
[65] Kandler, O. and Hippe, H. (1977) Arch. Microbiol. 113, 57–60.
[66] Zhilina, T. (1971) Mikrobiologiya 40, 674–680.
[67] Archer, D.B. and King, N.R. (1984) J. Gen. Microbiol. 130, 167–172.
[68] Davis, R.P. and Harris, J.E. (1985) J. Gen. Microbiol. 131, 1481–1486.

[69] Robinson, R.W. (1986) Appl. Environ. Microbiol. 52, 17–27.
[70] Robinson, R.W., Aldrich, H.C., Hurst, S.F. and Bleiweis, A.S. (1985) Appl. Environ. Microbiol. 49, 321–327.
[71] Liu, Y., Boone, D.R., Sleat, R., and Mah, R.A. (1985) Appl. Environ. Microbiol. 49, 608–613.
[72] Boone, D.R. and Mah, R.A. (1987) Appl. Environ. Microbiol. 53, 1699–1700.
[73] Hartmann, E. and König, H. (1991) Biol. Chem. Hoppe-Seyler 372, 971–974.
[74] Róden, L. (1970) In: Metabolic Conjugation and Metabolic Hydrolysis, Vol. 2 (Fishman, W.H., Ed.), pp. 345–442, Academic Press, New York.
[75] Rogers, H.J., Perkins, H.R. and Ward, J.B. (1980) Microbial Cell Walls and Membranes. Chapman & Hall, New York.
[76] Marsden, B.J. Sauer, F.D., Blackwell, B.A. and Kramer, J.K.G. (1989) Biochem. Biophys. Res. Commun. 159, 1404–1410.
[77] Keltjens, J.T., Kraft, H.J., Damen, W.G., van der Drift, C. and Vogels, G.D. (1989) Eur. J. Biochem. 184, 395–403.
[78] Nakanishi, Y., Shimizu, S., Takahashi, N., Sugiyama, M. and Suzuki, S. 1967. J. Biol. Chem. 242, 967–976.
[79] Kocur, M., Smid, B. and Martinee, T. (1972) Microbios 5, 101–107.
[80] Steber, J. and Schleifer, K.H. (1975) Arch. Microbiol. 105, 173–177.
[81] Brown, A.D. and Cho, K.J. (1970) J. Gen. Microbiol. 62, 267–270.
[82] Reistadt, R. (1972) Arch. Microbiol. 82, 24–30.
[83] Reistadt, R. (1974) Carbohydr. Res. 36, 420–423.
[84] Schleifer, K.H., Steber, J. and Mayer, H. (1982) Zentralbl. Bakteriol., Mikrobiol., Hyg., Abt. 1, Orig. C 3, 171–178.
[85] Steber, J. and Schleifer, K.H. (1979) Arch. Microbiol. 123, 209–212.
[86] Kostrikina, N.A., Zvyagintseva, I.S. and Duda, V.I. (1991) Mikrobiologiya 59, 1019–1023.
[87] Kostrikina, N.A., Zvyagintseva, J.S. and Duda, V.J. (1991) Arch. Microbiol. 156, 344–349.
[88] Zeikus, J.G. and Bowen, V.G. (1975) J. Bacteriol. 121, 373–380.
[89] Beveridge, T.J., Harris, B.J. and Sprott, G.D. (1987) Can. J. Microbiol. 33, 725–732.
[90] Sprott, G.D. and McKellar, R.C. (1980) Can. J. Microbiol. 26, 115–120.
[91] Beveridge, T.J., Stewart, M., Doyle, R.J. and Sprott, D.G. (1985) J. Bacteriol. 162, 728–737.
[92] Shaw, P.J., Hills, G.J., Henwood, J.A., Harris, J.E. and Archer, D.B. (1985) J. Bacteriol. 161, 750–757.
[93] Stewart, M., Beveridge, T.J. and Sprott, G.D. (1985) J. Mol. Biol. 183, 509–515.
[94] Sprott, G.D., Beveridge, T.J., Patel, B.G. and Ferrante, G. (1986) Can. J. Microbiol. 32, 847–854.
[95] Beveridge, T.J., Southam, G., Jericho, M.H. and Blackford, B.L. (1990) J. Bacteriol. 172, 6589–6595.
[96] Patel, G.B. and Sprott, G.D. (1990) Int. J. Bacteriol. 40, 79–82.
[97] Beveridge, T.J., Harris, B.J., Patel, G.B. and Sprott, G.D. (1986) Can. J. Microbiol. 32, 779–786.
[98] Beveridge, T.J., Patel, G.B., Harris, B.J. and Sprott, G.D. (1986) Can. J. Microbiol. 32, 703–710.
[99] König, H. and Stetter, K.O. (1986) System. Appl. Microbiol. 7, 300–309.
[100] Patel, G.B., Sprott, G.D., Humphrey, R.W. and Beveridge, T.J. (1986) Can. J. Microbiol. 32, 623–631.
[101] Pellerin, P., Fournet, B. and Debeire, P. (1990) Can. J. Microbiol. 36, 631–636.
[102] Nußer, E. and König, H. (1987) Can. J. Microbiol. 33, 256–261.
[103] Conway de Macario, E., König, H., Macario, A.J.L. and Kandler, O. (1984) J. Immunol. 132, 883–887.
[104] Jaenicke, R., Welsch, R., Sara, M. and Sleytr, U.B. (1985) Biol. Chem. Hoppe-Seyler 366, 663–670.

[105] Kandler, O. and König, H. (1985) In: The Bacteria. A Treatise on Structure and Function, Vol. 8, Archaebacteria (Woese, C.R. and Wolfe, R.S., Eds.), pp. 413–457, Academic Press, New York.
[106] Cheong, G.W., Cejka, Z., Peters, J., Stetter, K.O. and Baumeister, W. (1991) System. Appl. Microbiol. 14, 209–217.
[107] Wildgruber, G., Thomm, M., König, H., Ober, K., Ricchiuto, T. and Stetter, K.O. (1982) Arch. Microbiol. 132, 31–36.
[108] Grant, W.D. and Ross, H.N.M. (1986) FEMS Microbiol. Rev. 39, 9–15.
[109] Mescher, M.F. and Strominger, J.L. (1976) J. Biol. Chem. 251, 2005–2014.
[110] Wieland, F., Heitzer, R. and Schaefer, W. (1983) Proc. Natl. Acad. Sci U.S.A. 80, 5470–5474.
[111] Paul, G., Lottspeich, F. and Wieland, F. (1986) J. Biol. Chem. 261, 1020–1024.
[112] Wieland, F., Lechner, J. and Sumper, M. (1986) FEBS Lett. 195, 77–81.
[113] Paul, G. and Wieland, F. (1987) J. Biol. Chem. 262, 9587–9593.
[114] Lechner, J. and Sumper, M. (1987) J. Biol. Chem. 262, 9724–9729.
[115] Wieland, F., Lechner, J. and Sumper, M. (1982) Zentralbl. Bakteriol. Mikrobiol. Abt. Orig. C 3, 161–170.
[116] Wieland, F., Dompert, W., Bernhardt, G. and Sumper, M. (1980) FEBS Lett. 120, 120–124.
[117] Lechner, J. and Wieland, F. (1989) Annu. Rev. Biochem. 58, 173–194.
[118] Torreblanca, M., Rodriguez-Valera, F., Juez, G., Ventosa, A., Kamekura, M. and Kates, M. (1986) System. Appl. Microbiol. 8, 89–99.
[119] Mullakhanbhai, M.F. and Larson, H. (1975) Arch. Microbiol. 104, 207–214.
[120] Cohen, S. (1987) Ph.D. Thesis, Hebrew University, Jerusalem.
[121] Kessel, M., Wildhaber, I., Cohen, S. and Baumeister, W. (1988) EMBO J. 7, 1549–1554.
[122] Sumper, M., Berg, E., Mengele, R. and Strobel, I. (1990) J. Bacteriol. 172, 7111–7118.
[123] Stetter, K.O. (1988) System. Appl. Microbiol. 10, 172–173.
[124] Zellner, G., Stackebrandt, E., Kneifel, H., Messner, P., Sleytr, U.B., Conway de Macario, E., Zabel, H.P., Stetter, K.O. and Winter, J. (1989) System. Appl. Microbiol. 11, 151–160.
[125] Kessel, M., Volker, S., Santarius, U., Huber, R. and Baumeister, W. (1990) System. Appl. Microbiol. 13, 207–213.
[126] Prüschenk, R. and Baumeister, W. (1987) Eur. J. Cell Biol. 45, 185–191.
[127] Woese, C.R., Kandler, O. and Wheelis, M.L. (1990) Proc. Natl. Acad. Sci. U.S.A. 87, 4576–4579.
[128] Fiala, G. and Stetter, K.O. (1986) Arch. Microbiol. 145, 56–61.
[129] Messner, P., Pum, D., Sara, M., Stetter, K.O. and Sleytr, U.B. (1986) J. Bacteriol. 166, 1046–1054.
[130] Taylor, K.A., Deatherage, J.F. and Amos, L.A. (1982) Nature 299, 840–842.
[131] Lembcke, G., Dürr, R., Hegerl, R. and Baumeister, W. (1990) J. Microscopy 161, 263–278.
[132] Wildhaber, I. and Baumeister, W. (1987) EMBO J. 6, 1475–1480.
[133] Wildhaber, I., Santarius, U. and Baumeister, W. (1987) J. Bacteriol. 169, 5563–5568.
[134] Phipps, B.M. Huber, R. and Baumeister, W. (1991) Mol. Microbiol. 5, 253–265.
[135] Phipps, B.M., Engelhardt, H., Huber, R. and Baumeister, W. (1990) J. Struct. Biol. 103, 152–163.
[136] Dürr, R., Hegerl, R., Volker, S., Santarius, U. and Baumeister, W. (1991) J. Struct. Biol. 106, 181–190.
[137] Hegerl, R. and Baumeister, W. (1988) Electron. Microsc. Tech. 9, 413–419.
[138] Baumeister, W., Volker, S. and Santarius, U. (1991) System. Appl. Microbiol. 14, 103–110.
[139] Baumeister, W., Santarius, U., Volker, S., Dürr, R., Lemcke, G. and Engelhardt, H. (1990) System. Appl. Microbiol. 13, 105–111.
[140] Baumeister, W., Wildhaber, I. and Phipps, B.M. (1989) Can. J. Microbiol. 35, 215–227.

[141] Kjems, J., Larsen, N., Dalgaard, J.Z., Garrett, R.A. and Stetter, K.O. (1992) System. Appl. Microbiol. 15, 203–208.
[142] Zillig, W., Holz, I., Janekovic, D., Klenk, H.P., Imsel, E., Trent, J., Wunderl, S., Forjaz, V.H., Coutinho, R. and Ferreira, T. (1990) J. Bacteriol. 172, 3959–3965.
[143] Darland, G., Brock, T.D., Samsonoff, W. and Conti, S.F. (1970) Science 170, 1416–1418.
[144] Segerer, A., Langworthy, T.A. and Stetter, K.O. (1988) System. Appl. Microbiol. 10, 161–171.
[145] Langworthy, T.A., Tornabene, T.G. and Holzer, G. (1982) Zentralbl. Bakteriol., Mikrobiol. Hyg., Abt. I. Orig. C 3, 228–244.
[146] Yang, L.L. and Haug, A. (1979) Biochem. Biophys. Acta 556, 277–365.
[147] Smith, P.F. (1984) CRC Crit. Rev. Microbiol. 11, 157–186.
[148] Mayberry-Carson, K.J., Jewell, M.J. and Smith, P.F. (1978) J. Bacteriol. 133, 1510–1513.
[149] Woese, C.R., Magnum, L.J. and Fox, G.E. (1978) J. Mol. Evol. 11, 245–252.
[150] Woese, C.R. (1982) Zentralbl. Bakteriol., Mikrobiol. Hyg., Abt. I., Orig. C 3, 1–17.
[151] Kandler, O. (1993) In: Early Life on Earth – Proc. Nobel Symposium, Vol. 84 (Bengtson, S., Ed.), Columbia University Press, New York, in press.
[152] Mayr, E. and Ashlock, P.D. (1991) Principles of Systematic Zoology, 475 pp., McGraw-Hill, New York.
[153] Rose, C.S. and Pirt, S.J. (1981) J. Bacteriol. 147, 248–254.

Membrane lipids of archaea

Morris KATES

Department of Biochemistry, University of Ottawa, Ottawa, Ont. K1N 6N5, Canada

1. Introduction

According to the accepted Singer–Nicolson "Fluid Mosaic Model" of the cell membrane [1], membrane polar lipids form a closed liquid-crystalline bilayer structure, into which the membrane proteins are inserted in a mosaic pattern. Such a structure effectively controls the passage of nutrients, metabolites, etc., into and out of the cell and also serves as a hydrophobic milieu for optimal conformation and functioning of the membrane proteins/enzymes under the environmental growth conditions of the organism. Since the hydrophobic properties of the lipid bilayer are determined by the kind of hydrocarbon chains present, and the permeability properties are determined by the hydrocarbon chains and the polar head groups, it has long been postulated that the structure and composition of the membrane lipids might be generally characteristic of related organisms under their particular growth conditions [2]. This generalization has been found to apply well to bacteria, on the level of the family and even the genus, so that lipid structure and composition, or the lipid "profile", can serve as a bacterial taxonomic marker (see reviews [3–6]).

For the archaebacteria [7] (now designated archaea [8]), membrane-lipid patterns [4,5, 9–12], together with their 16S ribosomal RNA sequences (see the Introduction, and Chapter 14 of this volume) and their cell wall structures (see Chapter 8 of this volume), are particularly useful taxonomically to distinguish between the three groups of archaebacteria (extreme halophiles, methanogens and thermoacidophiles) and to clearly delineate the archaea from all other organisms. Archaeal membrane lipids are unique in consisting of derivatives of a C_{20},C_{20}-isopranyl glycerol diether (diphytanylglycerol diether [11]) and its dimer (dibiphytanyldiglycerol tetraether [12])(see sections 2.1 and 2.2, and Figs. 1 and 2). In contrast, eubacteria and plants contain largely diacylglycerol-derived membrane lipids and eukarya contain mostly diacylglycerol lipids and some monoacyl-monoalkyl- or monoacyl-monoalk-1-enyl glycerol-derived lipids. The diether-derived archaeal complex lipids are comprised of phospholipids, glycolipids, sulfoglycolipids and sulfophospholipids; the tetraether-derived complex lipids include phospholipids, phosphoglycolipids, glycolipids and sulfoglycolipids (see sections 3.1–3.3).

The existence of such a large variety of unusual lipid structures in archaea raises questions concerning the biosynthetic pathways for these lipids and their function in archaeal membranes, and also concerning the evolutionary relationships within the archaea and between archaea, eubacteria and eukarya. This chapter will first review the chemical structures of the polar lipids in extreme halophiles, methanogens and thermoacidophiles, then deal with the biosynthetic pathways for these lipids and their probable role in membrane structure and function, and finally discuss the evolutionary implications of their unique structures. Archaebacterial lipids have been the subject of several previous reviews [4,5,9–21].

2. "Core" archaeal lipids

"Core" lipids are defined as the hydrophobic portion of complex polar membrane lipids, and are usually obtained from archaeal lipids by strong-acid methanolysis or acetolysis to remove polar groups such as phosphate esters or sugars, without cleavage of the alkyl ether linkage [21,22]. Alkaline methanolysis or hydrolysis with cold 48% hydrofluoric acid can also be used for phospholipids [21,22]. For core lipids containing 3-hydroxyisopranyl groups, milder degradative conditions [23], such as weak methanolic-HCl or hydrogen fluoride [24,25], or enzymatic hydrolysis [24], are necessary to avoid the formation of degradative artifacts and subsequent misassignment of ether lipid cores [21]. Archaeal core lipids include the monomeric diphytanylglycerol diether and its variants (see Fig. 1) and the dimeric dibiphytanyldiglycerol tetraether and its variants (see Fig. 2).

Present nomenclature of these core lipids and their variants or derivatives is cumbersome and trivial names appear to be required. Nishihara et al. [26] have suggested that diphytanylglycerol diether (structure **1**, Fig. 1) and its variants (**1A–1E**, Fig. 1) be called "archaeol" (modified by adding the appropriate alkyl group designations, such as C_{20}, C_{25}, 3-hydroxy, etc.), and dibiphytanyldiglycerol tetraether (**2**) and its variant (**2A**) be called "caldarchaeol" and "nonitolcaldarchaeol", respectively. This forms the basis of a useful and practical system of nomenclature for archaebacterial lipids, which will be followed here.

It should be noted that the extreme halophiles contain only archaeol-derived lipids, the methanogens contain both archaeol-and caldarchaeol-derived lipids, in about 2:1 to 1:1 proportion, respectively, and the thermoacidophiles contain largely caldarchaeol- and nonitolcaldarchaeol-derived lipids, with a maximum of 1–10%a of archaeol-derived lipids [4,10,21,27,28].

2.1. Diphytanylglycerol diether (archaeol) and variants

The structure and configuration of archaeol (diphytanylglycerol) derived from the complex lipids of the extreme halophile, *Halobacterium cutirubrum,* was established as *sn*-2,3-di-*O*-3R,7R,11R,15-tetramethylhexadecyl-glycerol (structure **1**, Fig. 1) in 1963 [11], long before the archaea were recognized as a separate domain [7,8]. The structure and configuration of archaeol (**1**) has been confirmed by chemical synthesis [11]. Archaeol occurs in all three groups of archaea, extreme halophiles, methanogens and extreme thermophiles, but only in the extreme halophiles is it unaccompanied by caldarchaeol [4,10,21,27,28]. The archaeol and caldarchaeol derived-lipids may be considered as reliable markers for distinguishing the archaea from all

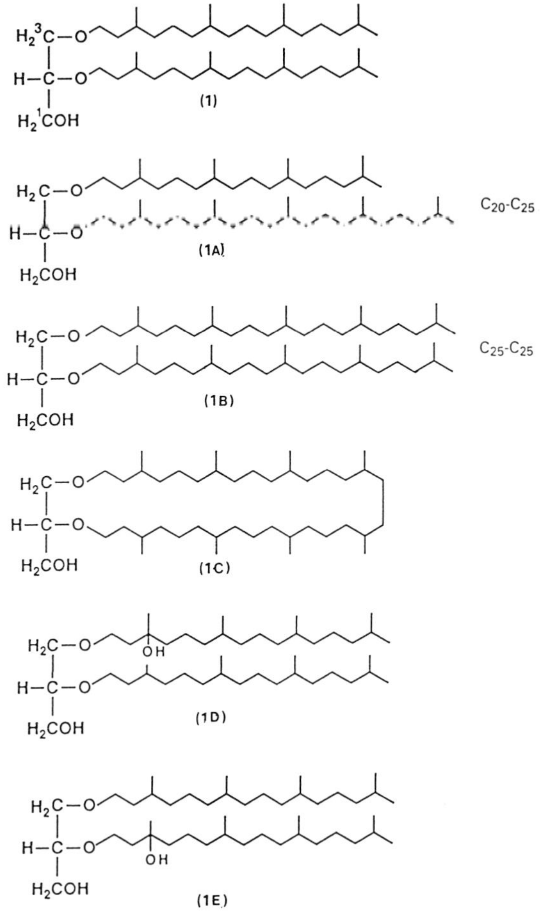

Fig. 1. Diphytanylglycerol ether (**1**, archaeol) and variants (**1A–1E**, see the text): core lipids of extreme halophiles and some methanogens.

other organisms [4,5,9,10]. The rare presence of plasmalogen (monoacyl-monoether) and dialkylether glycerolipids in some eubacterial obligate anaerobes [4], as well as the presence of a geranylgeranyl glycerol monoether in a brown alga [29] should not detract from the reliability of the above-mentioned criterion, as long as lipids derived from archaeol and caldarchaeol or their variants have not been detected in eubacteria or eukarya.

Variants of the diphytanylglyceroldiether (**1**) found in some archaea are as follows: (a) both the C_{20}–C_{25} diether (structure **1A**, Fig. 1) and C_{25}–C_{25} diether (structure **1B**) core lipids occur in alkaliphilic species of extreme halophiles (*Natronobacterium* and *Natronococcus*) [30,31], and the C_{20}–C_{25} diether (**1A**) also occurs in *Halococci* [32] and in *Halobacterium, Haloferax* and *Natronobacterium* strains from hypersaline environments in India [33]; (b) a macrocyclic C_{40}-diether (structure **1C**, Fig. 1) [34] is

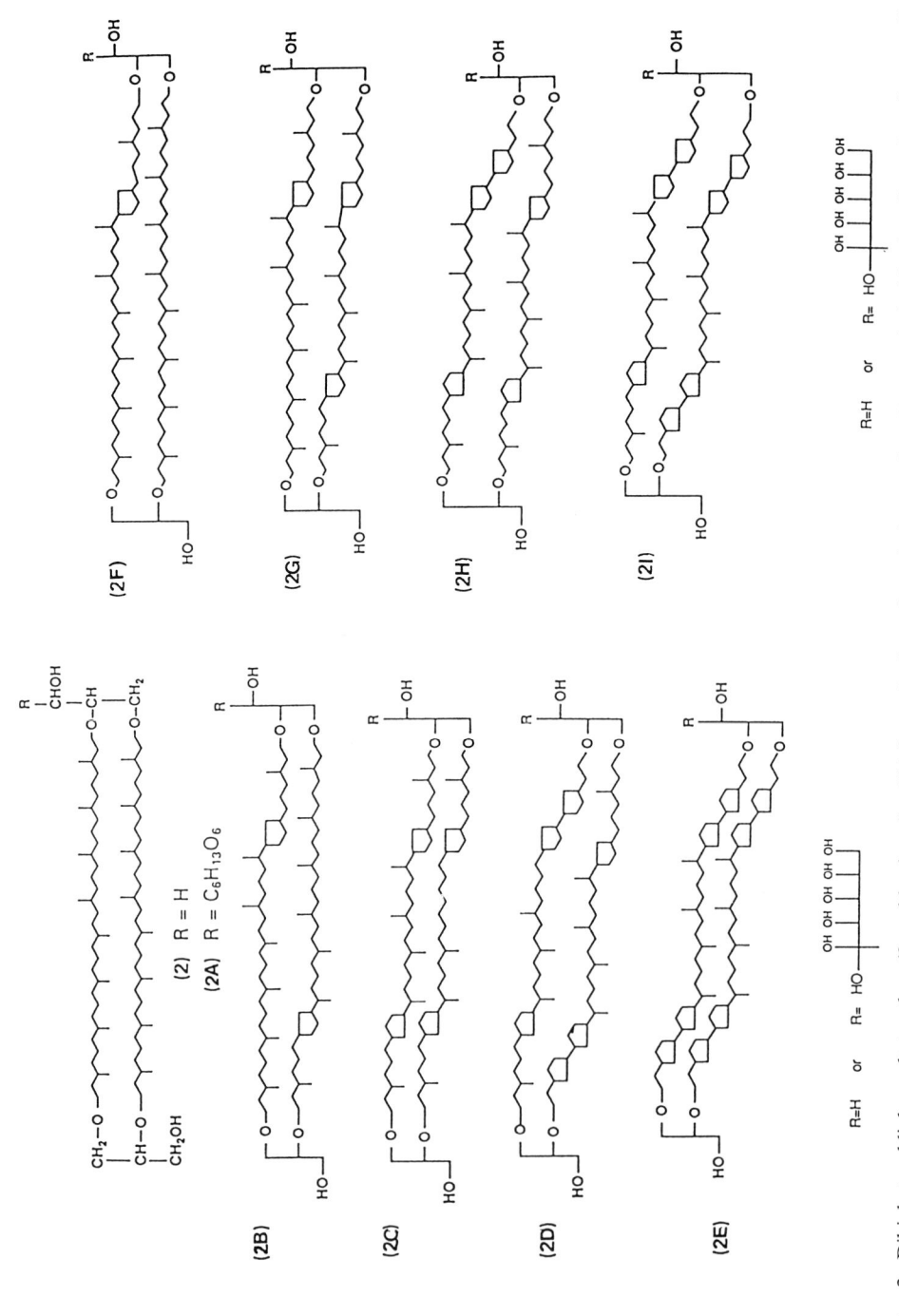

Fig. 2. Dibiphytanyldiglyceroltetraether (**2**, caldarchaeol), dibiphytanyl glycerol nonitol tetraether (**2A**, nonitolcaldarchaeol), and various cyclized derivatives (**2B–2I**, cyclized caldarchaeols and nonitolcaldarchaeols): core lipids in thermoacidophiles and some methanogens [10,12,20].

the major core lipid in the thermophilic methanogen *Methanococcus jannaschii* [21]; and (c) *Methanosaeta* (*Methanothrix*) and *Methanosarcina* species contain novel 3-hydroxydiether core lipids (structures **1D** and **1E**, respectively) [21,25,35,36], which have recently been identified in a haloalkaliphile (*Natronobacterium*) from India [33].

2.2. Dibiphytanyldiglycerol tetraether (caldarchaeol) and variants

The structure and configuration of caldarchaeol (structure **2**, Fig. 2), derived from the complex lipids of *Thermoplasma acidophilum,* in which two C_{20}–C_{20} diether moieties are linked head-to-head, was established by Langworthy [37] in 1977, although the presence of ether lipids in *Thermoplasma* and *Sulfolobus* was recognized as early as 1972 and 1974, respectively (see ref. [37]). The configuration of the two glycerols in the tetraether [9,38] is the same as in the diether (**1**) [11]. Also, the biphytanyl groups in the glycerol tetraether have the configuration 3R,7R,11R,15R,15'R,11'R,7'R,3'R [39], analogous to the 3R,7R,11R configuration of the phytanyl group [11].

Variants of the dibiphytanyldiglyceroltetraether (**2**) core-lipid structures are also found in some archaea; for example, species of the thermoacidophilic *Sulfolobus* genus contain lipids derived from a dibiphytanylglycerol nonitoltetraether (nonitolcaldarchaeol, structure **2A**, Fig. 2) as well as from dibiphytanyldiglyceroltetraether (caldarchaeol, **2**) [10,12,17,18,20]. Both caldarchaeol and nonitolcaldarchaeol may contain one to four cyclopentane rings in each of the C_{40} biphytanyl groups (structures **2B–2I**, Fig. 2) [10,12].

3. Polar lipids

3.1. Extreme halophiles

3.1.1. Phospholipids
The polar lipids of the extreme halophiles consist of one major and two or more minor phospholipids (all are acidic lipids, amino lipids being absent), and one major and several minor glycolipids. The phospholipid structures have been shown [11,15,32] to be archaeol analogs of: phosphatidylglycerol (PG, structure **3**, Fig. 3), phosphatidylglycerophosphate (PGP, structure **4**, Fig. 3), phosphatidylglycerosulfate (PGS, structure **5**, Fig. 3) and phosphatidic acid (PA, structure **6**, Fig. 3) [32,40]. However, the major phospholipid in *Halobacterium halobium* [41] and *H. cutirubrum* [42,43] has now been identified by FAB-mass spectrometry (FAB-MS) and nuclear magnetic resonance (NMR) spectroscopy as the monomethylated derivative of PGP (PGP-Me, structure **4A**, Fig. 3). PGP-Me and PGP can be distinguished by thin-layer chromatography (TLC) in alkaline solvent systems, which greatly retard the mobility of PGP but not that of PGP-Me [43]. PGP-Me (**4A**) is most likely the major phospholipid in all extreme halophiles and extreme haloalkaliphiles, having been identified by FAB-MS and TLC in several genera of extreme halophiles, including *Halobacteria, Haloarcula, Haloferax, Halococci,* and *Natronobacteria* and *Natronococci* [32,33,43], but a more extensive survey is needed to establish this generalization unambiguously.

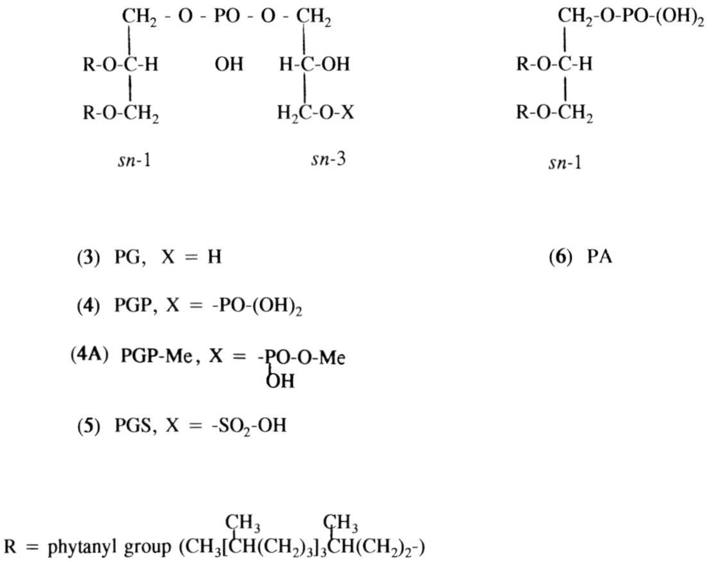

```
        CH₂ - O - PO - O - CH₂                    CH₂-O-PO-(OH)₂
             |                                         |
        R-O-C-H    OH    H-C-OH                   R-O-C-H
             |         |                               |
        R-O-CH₂        H₂C-O-X                    R-O-CH₂

          sn-1           sn-3                        sn-1
```

(3) PG,   X = H   (6) PA

(4) PGP, X = -PO-(OH)₂

(4A) PGP-Me, X = -PO-O-Me
                  |
                  OH

(5) PGS, X = -SO₂-OH

```
                       CH₃        CH₃
                       |          |
R = phytanyl group (CH₃[CH(CH₂)₃]₃CH(CH₂)₂-)
```

Fig. 3. Structures of archaeol phospholipids in extreme halophiles.

It should be noted that PGP as well as PA are minor components, probably present as biosynthetic intermediates [44,45] (see section 4.1). Another minor phospholipid, that has been identified in *Natronococcus occultis*, is the cyclic form of PGP [46].

Stereochemically, the structures of the archaeol analogs of PG, PGP, PGP-Me and PGS are unusual in that both glycerol moieties have the opposite configuration to those in the corresponding diacylglycerol forms of PG and PGP found in eubacteria and eukaryotes [11] (see Fig. 3).

3.1.2. Glycolipids

The glycolipids of *Halobacteria* include a major sulfated triglycosyl archaeol, galactosyl-3-sulfate-mannosyl-glucosyl-archaeol (S-TGA-1, structure **7**, Fig. 4) [11], and several minor glycolipids, such as: the desulfated TGA-1 (structure **7A**, Fig. 4); a sulfated tetra-glycosyl archaeol (S-TeGA, structure **8**) and its desulfated product TeGA (structure **8A**, Fig. 4) [47].

Haloarcula species contain a major triglycosyl archaeol, glucosyl-mannosyl-glucosyl-archaeol (TGA-2, structure **9**, Fig. 4), distinguishable from TGA-1 by TLC, and a minor diglycosyl archaeol (mannosyl-glucosyl archaeol, DGA-2) of unidentified structure [48]. *Haloferax* species contain a sulfated diglycosylarchaeol, mannosyl-6-sulfate-glucosyl-archaeol (S-DGA-1, structure **10**, Fig. 4) and its desulfated product, DGA-1 (structure **10A**, Fig. 4) [49]. It is of interest that *Halobacterium saccharovorum* [50] and a *Halococcus* strain, *Hc. saccharolyticus* [32], both contain S-DGA-1, but not DGA-1.

The structures of the glycolipids (**7–10A**) appear to be derived from a basic diglycosyl archaeol, mannosyl-glucosyl-archaeol (DGA-1, **10A**) by substitution of sugar or sulfate

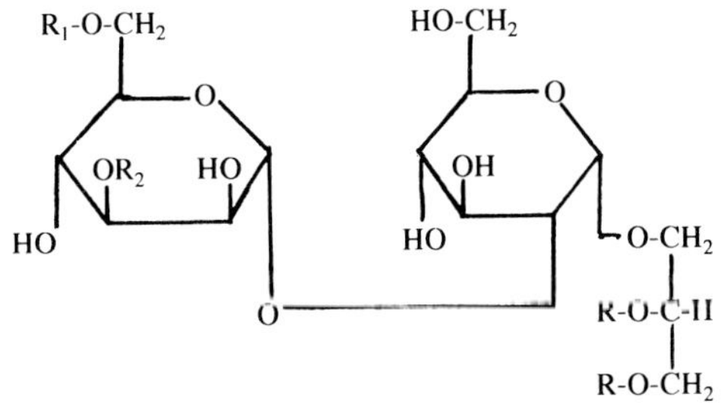

Genus		Lipid	R1	R2
Halobacterium	(7)	S-TGA-1	3-SO$_3$-β-Gal$p$	H
Halobacterium	(7A)	TGA-1	β-Gal$p$	H
Halobacterium	(8)	S-TeGA	3-SO$_3$-β-Gal$p$	α-Gal$f$
Halobacterium	(8A)	TeGA	β-Gal$p$	α-Gal$f$
Haloarcula	(9)	TGA-2	β-Glc$p$	H
Haloferax	(10)	S-DGA-1	–SO$_2$OH	H
Haloferax	(10A)	DGA-1	H	H

R = phytanyl group: CH$_3$[CH(CH$_2$)$_3$]$_3$CH(CH$_2$)$_2$–

Fig. 4. Structures of archaeol glycolipids in extreme halophiles.

groups at the 3 or 6 positions of the mannose residue (see Fig. 4) [51,52]. However, such a structural relationship between the known halophilic glycolipids does not necessarily hold for some newly discovered glycolipids; for example: (a) the "S-DGA-2" of an extreme halophile from Japan (strain 172) [15], which is a mannosyl-2,6-disulfate-glucosyl archaeol (structure **11**, Fig. 5) [53]; (b) the mannosyl-2-sulfate-glucosyl-archaeol (S-DGA-3, structure **12**, Fig. 5) of *Halobacterium sodomense* [54]; and (c) an unsulfated glucopyranosyl-1,6-glucopyranosyl-archaeol (DGA-4, structure **13**, Fig. 5) isolated from *Natronobacterium* strain SSL1 from salt locales in India [33]; this glycolipid appears to be present in traces also in other *Natronobacterium* strains (SSL2–SSL6) from India, and in *Nb. gregoryi* [33]. It appears that some of these strains may require reclassification [50,52,54], or the criteria for relating lipid composition to taxonomic classification may have to be modified (see section 3.1.3).

3.1.3. Taxonomic relations

The polar lipid composition of extreme halophiles appears to be correlated with their taxonomic classification on the level of the genera so far distinguished [4,5,10,32, 50–52,54]: *Halobacterium*, *Haloarcula*, *Haloferax*, *Halococcus*, *Natronobacterium* and *Natronococcus* (Table 1). All of these genera were found to contain C$_{20}$–C$_{20}$-archaeol PGP-Me (**4A**) as the major phospholipid, C$_{20}$–C$_{20}$-archaeol PG (**3**) and PGS (**5**) as minor phospholipids, and small to trace amounts of archaeol PA (**6**) [5,10,11,32,40] (see Table 1), with the following exceptions: (a) PGS is characteristically absent

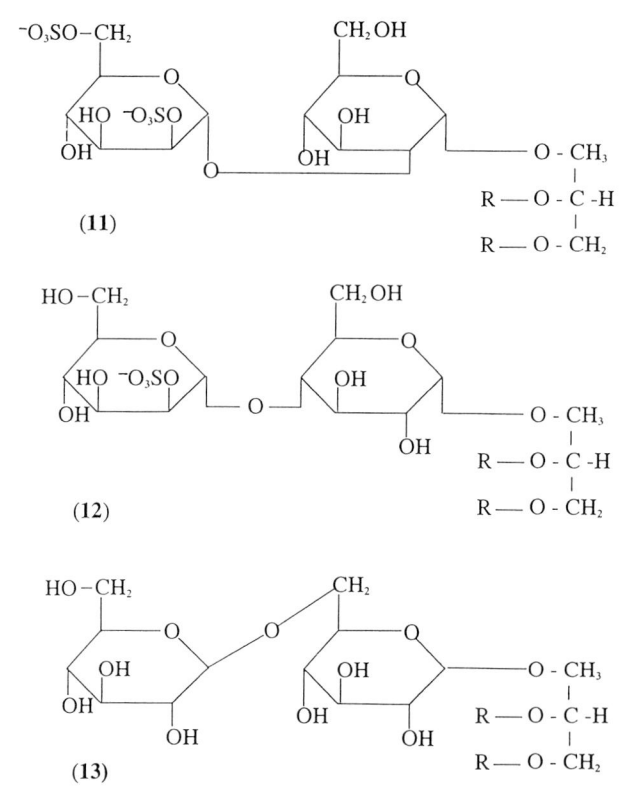

Fig. 5. Structures of novel glycolipids in extreme halophiles: **11**, Man$p$-2,6-disulfate-α1-2-Glc$p$α1-1-archaeol (S-DGA-2) in *Halobacterium* strain 172; **12**, Man$p$-2-sulfate-α1-4-Glc$p$α1-1-archaeol (S-DGA-3) in *H. sodomense*; and **13**, Glc$p$-1-6-Glc$p$-archaeol (DGA-4) in *Natronobacterium* strain SSL1. R represents a phytanyl group.

in *Haloferax*, *Halococci*, *Natronobacteria* and *Natronococci* species, and it is unexpectedly absent in a *Haloarcula* strain (A4-1) from Granada, Spain [Monteoliva-Sanchez, M., Ruiz Rodriguez, C. and Kates, M., unpublished data]; (b) both C_{20}–C_{20} and C_{20}–C_{25} species of PGP-Me and PG occur in the haloalkaliphiles [46,57], including *Natronobacterium* species from salt locales in India [33], and in *Halococcus saccharolyticus* [32]; (c) a cyclic form of PGP was found in *Natronococcus occultis* but not in any other halophilic archaebacterium [46] (Table 1).

The glycolipid composition is further discriminating (see Table 1 and Fig. 4), and the following generalizations may be made: (a) *Halobacteria* contain S-TGA-1 (**7**) as major glycolipid and TGA-1 (**7A**), S-TeGA (**8**), and TeGA (**8A**) as minor glycolipids; (b) *Haloarcula* contain major amounts of TGA-2 (**9**) and minor amounts of an unidentified diglycosylarchaeol, DGA-2 (containing glucose and mannose); (c) *Haloferax* species contain S-DGA-1 (**10**) as major glycolipid and DGA-1 (**10A**) as minor glycolipid; (d) some *Halococci* [55,56] are reported to contain S-TGA-1 (**7**), TGA-1 (**7A**), S-TeGA (**8**), but only S-DGA-1 (**10**) and an unidentified phosphoglycolipid were detected in *Hc. saccharolyticus* [32]; and (e) neither the *Natronobacteria* nor the

TABLE 1

Distribution of polar lipids in known genera of extreme halophiles

Genus	PGP-Me (4A)	PG (3)	PGS (5)	PA[a] (6)	S-TGA-1 (7)	TGA-1 (7A)	S-TeGA (8)	TGA-2 (9)	S-DGA-1 (10)	DGA-1 (10a)
Halobacterium [b]	+++	+	+	tr	++	+	+	–	–	–
Haloarcula [c]	+++	+	++	tr	–	–	–	++	–	–
Haloferax [d]	++	+++	–	tr	–	–	–	–	++	+
Halococcus [e,f]	+++[i]	+[i]		+[f]	+[e]	tr[e]	–	–	+[f]	–
Natronobacterium [g,h]	+++[i]	+[i]	–	tr	–	–	–	–	–	–
Natronococcus [g]	+++[i]	+[i]	–	tr	–	–	–	–	–	–

[a] First reported in a halophilic archaebacterium from a salt mine [40].
[b-h] Type species: [b] *Hb. cutirubrum*, *Hb. halobium*, *Hb. salinarium*, *Hb. saccharovorum* [50-52,54]; [c] *Ha. marismortui*, *Ha. vallismortis*, *Ha. hispanica*, *Ha. californiae*, *Ha. sinaiiensis* [48,51,52]; [d] *Hf. mediterranei*, *Hf. volcanii*, *Hf. gibbonsii* [51,52]; [e] *Hc. morrhuae* [55], *Sarcina literalis* and a *Sarcina* sp. [56]; [f] *Hc. saccharolyticus* [32] contains archaeol PG, PGP-Me, PA, S-DGA-1 and several unidentified glycolipids including a phosphoglycolipid; [g] *Nb. pharaonis*, *Nb. magadii*, *Nb. gregoryi* and *Nc. occultis* [46,57], which also contains a cyclic form of PGP; [h] *Natronobacterium* strains SSL1-SSL6 have, in addition, a novel glycolipid DGA-4 (13) also detected in *Nb. gregoryi* [33].
[i] These lipids occur as both C_{20}–C_{20} and C_{20}–C_{25} species. For structures of lipids see Figs. 3 and 4.

Natronococci examined [10,46,57] were reported to contain any glycolipid components, but DGA-4 (13) can be detected on TLC plates when overloaded [33].

The glycolipid composition together with the phospholipid composition appear to be sufficient to delineate each of the known genera of extremely halophilic bacteria, as indicated in Table 1, but not necessarily new or uncertain genera. Examples of the latter are *Halobacterium sodomense* which contains a new glycolipid, S-DGA-3 (12, Fig. 5) [54], which is probably also present in unclassified "halobacterial" species (3.5 rp 4 and 3.1 palp 4) [51,52]. The latter species and *H. sodomense* should be further studied to determine whether they should be reassigned to another genus [54]. *Halobacterium saccharovorum* [50] and *Halococcus saccharolyticus* [32] contain S-DGA-1 (10), the *Haloferax* marker glycolipid, suggesting that these organisms should be reexamined for possible reclassification into new genera [4,50,52,54].

3.2. Methanogens

Of the ten genera of anaerobic methanogens known so far, the following have been examined for their lipid composition: *Methanobacterium*, *Methanococcus*, *Methanospirillum*, *Methanosarcina*, and *Methanosaeta* (*Methanothrix*) [4,9,10,21]. The complex lipids of methanogenic bacteria are derived both from archaeol (1) and from caldarchaeol (2). The archaeol derivatives (Fig. 6) consist of: (a) phospholipids (e.g., archaeol analogs of phosphatidylinositol (API, 14, Fig. 6), phosphatidylethanolamine (APE, 15, Fig. 6), and phosphatidylserine (APS, 16, Fig. 6) and (b) glycolipids (mono- and diglycosylarchaeols). The caldarchaeol lipids (Fig. 7) consist of: (a) phospholipids with phosphoethanolamine attached to one hydroxyl; (b) glycolipids with sugar groups attached to one of the hydroxyl groups; and (c) phosphoglycolipids with sugar groups

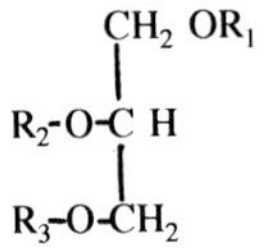

$$\begin{array}{c} CH_2\ OR_1 \\ | \\ R_2\text{-O-C H} \\ | \\ R_3\text{-O-CH}_2 \end{array}$$

	Lipid	R_1	R_2		R_3
(14)	API	P-myo-inositol	C_{20}		C_{20}
(14A)	A_{OH}PI	P-myo-inositol	3-OH-C_{20}		C_{20}
(15)	APE	P-ethanolamine	C_{20}		C_{20}
(15A)	A_{cyc}PE	P-ethanolamine	C_{20}	– cyclic –	C_{20}
(15B)	A_{OH}PE	P-ethanolamine	C_{20}		3-OH-C_{20}
(16)	APS	P-serine	C_{20}		C_{20}
(16A)	A_{OH}PS	P-serine	3-OH-C_{20}		C_{20}
(17)	P-MGA-1	P-1-βGlcN.Ac	C_{20}		C_{20}
(17A)	P-MGA-2	3-P-Glc	C_{20}		C_{20}
(18)	PPDA	P-pentanetetrol-N(CH$_3$)$_2$	C_{20}		C_{20}
(19)	PPTA	P-pentanetetrol-N(CH$_3$)$_3$	C_{20}		C_{20}
(20)	MGA	Glc$p$β-1-1	C_{20}		C_{20}
(20A)	MGA$_{cyc}$	Glc$p$β-1-1	C_{20}	– cyclic –	C_{20}
(20B)	MGA$_{cyc}$-PE	6-P-ethanolamine-Glc$p$β-1-1	C_{20}	– cyclic –	C_{20}
(21)	DGA-5	Glc$p$ α-1-2-Gal$f$β-1-1	C_{20}		C_{20}
(22)	DGA-6	Gal$f$β-1-6-Gal$f$β-1-1	C_{20}		C_{20}
(23)	DGA-7	Glc$p$β-1-6-Glc$p$β-1-1	C_{20}		C_{20}
(23A)	DGA$_{cyc}$-7	Glc$p$β-1-6-Glc$p$β-1-1	C_{20}	– cyclic –	C_{20}
(24)	DGA-8	Man$p$α-1-3-Gal$p$β-1-1	C_{20}		C_{20}
(25)	DGA$_{OH}$-5	Gal$p$β-1-6-Gal$p$β-1-1	C_{20}		3-OH-C_{20}
(26)	A-P	P	C_{20}		C_{20}

Fig. 6. Complex archaeol lipids in methanogens.

attached to one hydroxyl and a phosphoglycerol or phosphoethanolamine group linked to the other hydroxyl.

Studies of the lipids of four families of methanogens [58–60] have identified the archaeol analogs of phosphatidylinositol (API, **14**, Fig. 6) [61], phosphatidylethanolamine (APE, **15**, Fig. 6) and phosphatidylserine (APS, **16**, Fig. 6), as major components in *Methanobacteriaceae* but not in *Methanomicrobiaceae*, and *Methanosarcinaceae*, and only APS (**16**) was found in *Methanococcaceae*. Another phospholipid, caldarchaeolphosphoethanolamine (CPE, **34**, Fig. 7) was found only in the genera *Methanobacterium* and *Methanosarcina*.

3.2.1. Methanobacteriaceae
In a complete analysis of the lipids of *Methanobacterium thermoautotrophicum* [62], the major polar lipids could be grouped into three "pentads" based on the

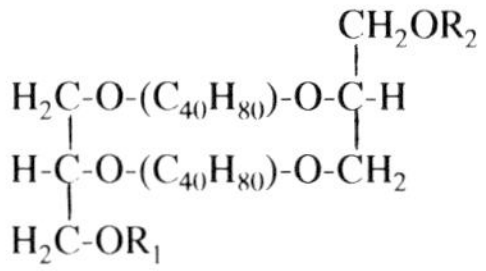

$$\begin{array}{l}
\phantom{H_2C-O-(C_{40}H_{80})-O-}CH_2OR_2\\
H_2C\text{-}O\text{-}(C_{40}H_{80})\text{-}O\text{-}\overset{|}{C}\text{-}H\\
H\text{-}\overset{|}{C}\text{-}O\text{-}(C_{40}H_{80})\text{-}O\text{-}CH_2\\
H_2C\text{-}OR_1
\end{array}$$

	Lipid	R_1	R_2
(27)	C-P	P	H
(28)	DGC-1	Glc$p\alpha$-1-2-Gal$f\beta$-1-1	H
(28A)	PGC-1	Glc$p\alpha$-1-2-Gal$f\beta$-1-1	P-sn-3-glycerol
(29)	DGC-2	Gal$f\beta$-1-6-Gal$f\beta$-1-1	H
(29A)	PGC-2	Gal$f\beta$-1-6-Gal$f\beta$-1-1	P-sn-3-glycerol
(30)	PGC-3	Glc$p\alpha$-1-2-Gal$f\beta$-1-1	PPTA
(30A)	PGC-4	Gal$f\beta$-1-6-Gal$f\beta$-1-1	PPTA
(31)	PGC-5	Glc$p\alpha$-1-2-Gal$f\beta$-1-1	PPDA
(31A)	PGC-6	Gal$f\beta$-1-6-Gal$f\beta$-1-1	PPDA
(32)	DGC-3	Glc$p\beta$-1-6-Glc$p\beta$-1-1	H
(32A)	PGC-7	Glc$p\beta$-1-6-Glc$p\beta$-1-1	P-ethanolamine
(33)	CPI	H	P-myo-inositol
(33A)	PGC-8	Glc$p\beta$-1-6-Glc$p\beta$-1-1	P-myo-inositol
(34)	CPE	H	P-ethanolamine
(35)	CPS	H	P-serine
(35A)	PGC-9	Glc$p\beta$-1-6-Glc$p\beta$-1-1	P-serine

Fig. 7. Complex caldarchaeol lipids in methanogens.

polar head groups P-ethanolamine, P-serine and P-inositol. Each pentad consists of two archaeol lipids (an archaeol phospholipid and an archaeol glycolipid) and three caldarchaeol lipids (a caldarchaeol phospholipid, a caldarchaeol glycolipid, and a caldarchaeol phosphoglycolipid). Thus, the phospholipids were API (**14**), APE (**15**) and APS (**16**) (see Fig. 6) and the corresponding caldarchaeol phospholipids CPI, CPE and CPS (**33,34,35**, respectively, Fig. 7); the glycolipids were a diglycosyl(gentiobiosyl)archaeol, Glc$p\beta$1-6Glc$p\beta$-1-1 archaeol (DGA-7, **23**, Fig. 6) and the corresponding gentiobiosylcaldarchaeol (DGC-3, **32**, Fig. 7); and the phosphoglycolipids (PGCs, the major lipids) were gentiobiosylcaldarchaeol-P-ethanolamine (PGC-7, **32A**, Fig. 7), gentiobiosylcaldarchaeol-P-inositol (PGC-8, **33A**, Fig. 7) and gentiobiosylcaldarchaeol-P-serine (PGC-9, **35A**, Fig. 7). A gentiobiosyl/P-inositol "pentad" (API, DGA-3, DGC-3, CPI and PGC-8, structures **14,23,32,33,33A**, respectively, Figs. 6 and 7), identical to that in *M. thermoautotrophicum* [62] has been identified in *Methanobrevibacter arboriphilicus* [59]. The phosphoglycolipid (**32A**) has been proposed as a "signature" lipid for the family *Methanobacteriaceae* [59,62]. Two minor lipids, probably biosynthetic intermediates [63], have been identified in *M. thermoautotrophicum* as the archaeol and caldarchaeol analogs of phosphatidic acid (structures **26** and **27**, respectively, Figs. 6 and 7). The biosynthetic mechanisms [62] underlying the formation of these "pentads" of polar lipids will be discussed in sections 4.2.2. and 4.4.

3.2.2. Methanomicrobiaceae

The lipids of *Methanospirillum hungatei* [64,65] are derived from both archaeol and caldarchaeol and consist largely of two sets of triads of: diglycosylarchaeols (DGAs), diglycosylcaldarchaeols (DGCs) and phosphodiglycosylcaldarchaeols (PGCs), each set containing one of the following diglycosyl groups: Group 1, Glc$p\alpha$1-2Gal$f\beta$, or Group 2, Gal$f\beta$-6Gal$f\beta$ (Figs. 6, 7). The glycosylarchaeols (DGA-5 and DGA-6, **21** and **22**, Fig. 6) have the diglycosyl-1 or -2 attached to the OH group of archaeol [64]. The glycosylcaldarchaeols (DGC-1 and DGC-2, **28** and **29**, Fig. 7) have the diglycosyl-1 or -2 attached to one hydroxyl of caldarchaeol. The phosphoglycolipids (PGC-1 and PGC-2, **28A** and **29A**, Fig. 7) have the diglycosyl-1 or -2 attached to one hydroxyl of caldarchaeol, and a *sn*-3-phosphoglycerol attached to the second hydroxyl. In addition, minor amounts of C_{20}–C_{20}-archaeol-derived phospholipids are also present: phosphatidyl-*N,N*-dimethylamino- (PPDA, **18**, Fig. 6) and *N,N,N*-trimethylaminopentanetetrol (PPTA, **19**, Fig. 6) [65]. The presence of minor amounts of PGCs containing P-trimethylaminopentanetetrol attached to one hydroxyl of caldarchaeol and either diglycosyl-1 or -2 attached to the second hydroxyl (PGC-3 and PGC-4, **30** and **30A**, Fig. 7) and P-dimethylaminopentanetetrol attached to one caldarchaeol OH and either diglycosyl-1 or -2 attached to the other OH (PGC-5, PGC-6, **31** and **31A**, Fig. 7) has been reported [21] (Ferrante, G., Ekiel, I. and Sprott, G.D., unpublished data). The reported presence of the archaeol analog of phosphatidyl-*sn*-1-glycerol (APG) [64] was not confirmed [65], but recent FAB-MS and NMR studies (see ref. [21]) provide tentative evidence for the existence of APG and conclusive evidence for the presence of A_{OH}PG in *Methanosarcina* spp.

3.2.3. Methanococcaceae

The lipids of one species of *Methanococcus* examined, *Mco. voltae*, are derived only from archaeol [66]. The major glycolipid was identified as gentiobiosylarchaeol (DGA-7, **23**, Fig. 6) found also in *M. thermoautotrophicum* [62]; minor glycolipids were Glc$p\beta$-1-1 archaeol (MGA, **20**, Fig. 6), and a novel phosphoglycosylarchaeol, GlcNAc-1-P-archaeol (P-MGA-1, **17**, Fig. 6).

Lipids of the thermophilic deep-sea methanogen, *Methanococcus jannaschii* [67] are based largely on the macrocyclic diether (cyc-archaeol, **1C**) and consist of mono- and di-glucosyl-cyc-archaeol (**20A, 23A**, respectively, Fig. 6), P-ethanolamine-glucosyl-cyc-archaeol (**20B**) and cyc-archaeol-PE (**15A**); a small amount of archaeol-PE (**15**) is also present. *Mco. jannaschii* is also capable of forming caldarchaeol-derived polar lipids [68], and the proportions of diether, macrocyclic diether and tetraether lipids can vary as a function of growth temperature (see section 5.2).

3.2.4. Methanosarcinaceae

Ferrante et al. [25,69] have shown that the lipids of *Methanothrix* (now *Methanosaeta*) *concilii* are derived from both archaeol and hydroxyarchaeol which has a hydroxyl group at C-3 of the phytanyl chain on the *sn*-3-position (structure **1D**, Fig. 1) lipid cores. The major components are a Man$p$-Gal-p-archaeol (DGA-8, **24**, Fig. 6), a Gal$p$-Gal$p$-hydroxyarchaeol (DGA$_{OH}$-9, **25**, Fig. 6), and archaeol-P-inositol (**14**, Fig. 6). Some minor components have now been identified [70] as archaeol-P-ethanolamine (**15**,

Fig. 6), hydroxyarchaeol-P-ethanolamine (**15B**, Fig. 6), β-galactopyranosyl-archaeol or -hydroxyarchaeol, and a triglycosyl-hydroxyarchaeol.

Sprott et al. [35] have shown that the lipids of *Methanosarcina barkeri* and *M. mazei* are derived from two lipid cores: archaeol and an isomeric hydroxyarchaeol (**1E**, Fig. 1) in which the hydroxyl group is located at C-3 in the phytanyl chain on the *sn*-2-position of glycerol. The presence of this isomeric hydroxyarchaeol lipid core in *M. barkeri* has been confirmed by Nishihara and Koga [36] who also identified the major hydroxyarchaeol lipids as hydroxyarchaeol P-inositol (**14A**, Fig. 6) and hydroxyarchaeol P-serine (**16A**, Fig. 6). Recently, a novel archaeol phosphoglycolipid was identified in *M. barkeri* as archaeol-P-inositol-6-glucosamine [71], the polar headgroup of which is identical with that of the analogous eukaryal glycosyl phosphatidylinositol [72].

The large variety of core lipids reported previously to be present in *M. barkeri* by De Rosa et al. [73], have not been confirmed by the studies of Sprott et al. [35] and Nishihara and Koga [36], who did not detect any C_{20}–C_{20}-tetritol diether, C_{20}–C_{25} archaeol or caldarchaeol with 1-3 cyclopentane rings. The failure to detect hydroxyarchaeol core lipids [73,74] in *M. barkeri* is probably due to the drastic hydrolysis conditions used [21,23].

3.2.5. Taxonomic relations
The data available so far indicate that several methanogenic genera may have distinctive polar lipid patterns [21,60,62–71] which might be taxonomically useful. For example, the polar lipid pattern of *Methanospirillum hungatei* [21] is based on two diglycosyl headgroups (archaeol structures **21, 22**, and caldarchaeol structures **28, 29**) and two non-eukaryal phosphorylated headgroups (**18,19**) (see Figs. 6 and 7) and is quite distinct from that of any other methanogenic genus or family. This pattern may be characteristic of the *Methanospirillum* genus or the *Methanomicrobiaceae* family.

The polar lipids of several *Methanobacteria* and *Methanobrevibacteria* are characterized by a preponderance of eukaryal-type phosphorylated amino- or polyol headgroups (structures **14, 15, 16**) and a gentiobiosyl glycosylated headgroup (structures **23, 32, 32A**) [58–60,62], suggesting that the major phosphoglycolipid (**32A**) can be considered as a "signature" lipid for the *Methanobacteriaceae* family [59,62]. However, it should be noted that the gentiobiosyl headgroup also appears as DGA-7 (**23**) in *Methanococcus voltae,* a member of the *Methanococcaceae* family, and also in DGC-3 (with cyclopentane rings, structure **41**) in *Thermoproteus* (see Fig. 8, and section 3.3.3). Another member of the *Methanococcaceae, Mco. jannaschii,* is clearly distinguished from *Mco. voltae* by the presence of macrocyclic-archaeol- as well as archaeol- and caldarchaeol-derived lipids [21,34,67]. Whether these seemingly taxonomically related differences in polar lipid structures are generally valid will depend on more detailed lipid-structural studies of a wider range of methanogenic genera.

3.3. Extreme thermophiles

The taxonomical classification of thermoacidophiles is not as complex as that of the methanogens: eight genera are known so far, distributed among four orders: Thermoplasmatales, Sulfolobales, Thermoproteales and Thermococcales. Only a relatively small number of thermoacidophiles and thermococcales have been examined

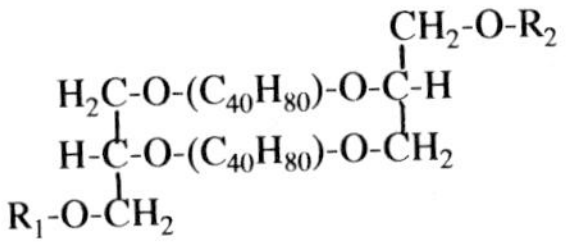

$$\text{H}_2\text{C-O-(C}_{40}\text{H}_{80})\text{-O-}\overset{\overset{\text{CH}_2\text{-O-R}_2}{|}}{\underset{|}{\text{C}}}\text{-H}$$
$$\text{H-}\overset{|}{\underset{|}{\text{C}}}\text{-O-(C}_{40}\text{H}_{80})\text{-O-CH}_2$$
$$\text{R}_1\text{-O-CH}_2$$

	Lipid	R_1	R_2	No. of cyclopentane rings
(36)	MGC-1	monoglycosyl	H	0–2
(36A)	PGC-10	monoglycosyl	P-sn-3-glycerol	0–2
(37)	MGC-2	Glcpβ	H	0–4
(37A)	PGC-11	Glcpβ	P-inositol	0–2
(38)	DGC-4	Glcpβ-Galpβ	H	0
(38A)	PGC-12	Glcpβ-Galpβ	P-inositol	0
(39)	DGC-5	Glcpβ-1-4-Galpβ	H	0
(39A)	PGC-13	Glcpβ-1-4-Galpβ	P-inositol	0
(40)	PGC-14	Galpβ	P-inositol	0
(41)	DGC-3	Glcpβ-1-6-Glcpβ	H	0–4
(41′)	DGC-3′	Glcpβ-1-2-Glcpβ	H	
(42)	TGC	(Glcpβ1-6)$_2$Glcpβ	H	0–4
(42A)	PGC-15	(Glcpβ-1-6)$_2$Glcpβ	P-inositol	0–4
(42′)	TGC′	(Glcpβ-1-2)$_2$Glcpβ	H	
(42A′)	PGC-15′	(Glcpβ-1-2)$_2$Glcpβ	P-inositol	
(43)	DGC-6	Glcpα-1-4-Galpβ	H	0
(43A)	PGC-16	Glcpα-1-4-Galpβ	P-inositol	0

$$\text{H}_2\text{C-O-(C}_{40}\text{H}_{80})\text{-O-}\overset{\overset{\text{CH}_2\text{OR}_2}{|}}{\underset{|}{\text{C}}}\text{-H}$$
$$\text{H-}\overset{|}{\underset{|}{\text{C}}}\text{-O-(C}_{40}\text{H}_{80})\text{-O-CH}_2$$
$$\text{HO-}\overset{|}{\underset{|}{\text{C}}}\text{-H}$$
$$\text{HO--CH}_2\text{-CH--CH--CH--}\overset{|}{\underset{|}{\text{C}}}\text{--CH}_2\text{OH}$$
$$\quad\quad\quad\overset{|}{\text{OH}}\ \ \overset{|}{\text{OR}_1}\ \ \overset{|}{\text{OH}}\ \ \overset{|}{\text{OH}}$$

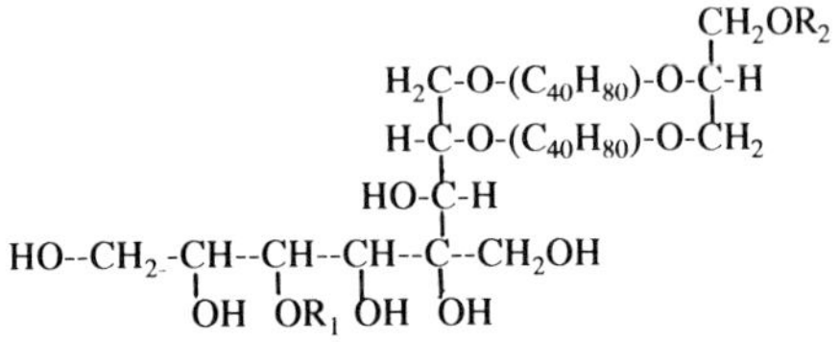

(44)	GNC	Glcpβ	H	0–4
(44A)	PGNC	Glcpβ	P-inositol	0–4
(44B)	SGNC	HSO$_3$–Glcpβ	H	0–4

Fig. 8. Complex caldarchaeol and nonitolcaldarchaeol lipids of thermoacidophiles.

for their lipids, but their core lipids are more complex and varied than any of the other archaebacteria [9,10,12] (see Fig. 8), consisting mostly of caldarchaeol (**2**) or nonitolcaldarchaeol (**2A**) with zero to four cyclopentane rings per C_{40} biphytanyl chain (structures **2B–2I**, Fig. 2).

3.3.1. Thermoplasmatales

Thermoplasma acidophilum, a wall-less mycoplasma-like aerobe, is the one species of *Thermoplasma* that has been examined for lipids [9]. Its polar lipids contain at least six different glycolipids and at least seven phosphorus-containing lipids, all based on caldarchaeol with different degrees of cyclization [9]. The major component (80% of the polar lipids), is a phosphoglycolipid (PGC-10, **36A**, Fig. 8) based on caldarchaeol, having a single unidentified sugar attached to one hydroxyl of the tetraether and a *sn*-3-glycerophosphate group attached to the other [9,16]. The corresponding monoglycosyl caldarchaeol containing the unidentified sugar (**36**, Fig. 8) as well as monoglucosyl and diglucosylcaldarchaeol derivatives of uncharacterized structure (**37** and **38**, respectively) are also present. The rest of the polar lipids are mostly derived from caldarchaeol but a few are derived from archaeol [9]. Detailed structural studies on the glycolipids of *Thermoplasma* have still to be done.

3.3.2. Sulfolobales

Species of the sulfur-dependent *Sulfolobus* genus live in sulfataric hot springs at temperatures between 60°C and 100°C and pH values between 1 and 5, and are sulfur- and iron-oxidizing. The polar lipids of only three species, *Sulfolobus acidocaldarius*, *Sulfolobus solfataricus* (formerly *Caldariella acidophila*) and *Sulfolobus brierleyi* (also called "ferrolobus") have been investigated in detail [12,16,17,31]. These lipids are derived almost exclusively from either caldarchaeol (**2**) or nonitolcaldarchaeol (**2A**). Both types of tetraether may also contain one to four cyclopentane rings in their biphytanyl chains (Fig. 8), the extent of cyclization increasing with increasing growth temperature [76].

The polar lipids of *S. acidocaldarius*, grown either heterotrophically or autotrophically, and "ferrolobus", a strict autotroph, are predominantly glycolipids (35%) and phosphoglycolipids (45%) [77,78]. The major glycolipids in *S. acidocaldarius* and "ferrolobus" were tentatively identified as glycolipid DGC-4 (**38**, Fig. 8), a glucosylgalactosyl caldarchaeol, and glycolipid GNC (**44**, Fig. 8), a glucosyl nonitolcaldarchaeol, the glucosyl group being linked to one of the OH groups of the nonitol group [78].

The phosphoglycolipids in both *S. acidocaldarius* and "ferrolobus" [78] consisted predominantly (72%) of the phosphoinositol derivatives of glycolipids DGC-4 (**38**) and GNC (**44**): Glc-Gal-caldarchaeol-P-inositol (PGC-11, **38A**, Fig. 8), and Glc-nonitolcaldarchaeol-P-inositol (PGNC, **44A**, Fig. 8), in which the glucosyl group is linked to one OH in the nonitol group and the phosphoinositol is linked to the free OH in the glycerol residue of caldarchaeol. A minor acidic glycolipid detected in *S. acidocaldarius* cells, but not in "ferrolobus", was partially characterized as a monosulfate derivative of glycolipid GNC (**44B**, Fig. 8); the sulfate group is presumably attached to the glucosyl group but its precise location is not known [17]. The main glycolipids of *S. acidocaldarius* and "ferrolobus" are probably similar if not identical to those in *S. solfataricus* (see below), but further structural studies are necessary to establish this unambiguously [9,12].

The glycolipids of *S. solfataricus* [9,10,12,75,79] have been characterized in more detail than those of *S. acidocaldarius*, but their structures are still not unambiguously defined. The major glycolipids are the caldarchaeol glycolipid DGC-4 (6%) (structure **38**, Fig. 8) and the nonitolcaldarchaeol glycolipid GNC (14%) (structure **44**, Fig. 8); the linkage positions of the disaccharide in structure (**38**) and the position of attachment of the glucosyl group to the nonitol group in structure (**44**) have still to be determined. The phosphoglycolipids consisted mainly of the phosphoinositol derivatives of glycolipids DGC-4 (**38**) and GNC (**44**): Glc$p\beta$-Gal$p\beta$ caldarchaeol-P-inositol (8%; PGC-11, **38A**,

Fig. 8), and Glcpβ-nonitolcardarchaeol-P-inositol (55%; PGNC, **44A**, Fig. 8) respectively; the sulfated derivative of glycolipid GNC (**44**), HSO_3-Glcpβ-nonitolcaldarchaeol (10%; SGNC, **44B**, Fig. 8) was also identified, as well as the phospholipid caldarchaeol-P-inositol (CPI, **33**, Fig. 7)[10].

The same complex lipids (structures **33, 38, 38A, 44, 44A, 44B**) were identified in *Desulfurolobus ambivalens*, grown both aerobically and anaerobically, except for differences in the number of cyclopentane rings in the biphytanyl chains[20,79a].

Metallosphaera sedula[79b], which represents a new genus of aerobic metal-mobilizing thermophilic archaea in the order Sulfolobales, was found to contain a similar pattern of caldarchaeol and nonitolcaldarchaeol-derived glycolipids and the corresponding P-inositol phosphoglycolipids to those of *Sulfolobus solfataricus* (e.g., see structures **38, 38A, 44** and **44B**, Fig. 8[75,79]). However, the relative proportions of the major glycolipids (**38** and **44**) and the minor complex lipids of *M. sedula* differ somewhat from those of *S. solfataricus*.

3.3.3. Thermoproteales

One species of the anaerobic, sulfur-dependent *Thermoproteus* genus, *T. tenax*, was found to contain 5% neutral lipids, 20% glycolipids and 75% phospholipids and phosphoglycolipids[80]. The ether core lipids consist mainly (95%) of caldarchaeols, with minor proportions (5%) of archaeols. The caldarchaeols are acyclic (**2**), or contain one (**2B**) or two (**2C**) cyclopentane rings in each chain, as well as mixed numbers of rings (structures **2F, 2G, 2H**, Fig. 2), similar to those found in *Sulfolobus* [12]. The polar lipid composition resembles that of *Sulfolobus* but is characteristically different with respect to the glycosyl groups of the glycolipids. The glycolipids of *T. tenax* consist of mono-, di-, and triglucosyl-caldarchaeols, the sugar groups being β1-6-linked (MGC-2, DGC-3 and TGC; **37, 41** and **42**, respectively, Fig. 8), and the caldarchaeol core lipids having 0–4 rings (Fig. 8).

It is of interest that, except for the presence of rings in the caldarchaeol core, the diglucosyl(gentiobiosyl)caldarchaeol (DGC-3, **41**) of *T. tenax* is identical to that found in the methanogen *M. thermoautotrophicum* (DGC-3, **29**, Fig. 7), and the monoglucosyl-caldarchaeol (MGC-2, **37**, Fig. 8) of *T. tenax* is similar to MGC-1 (**36**, Fig. 8) in *Thermoplasma*. However, the triglucosylcaldarchaeol (TGC, **42**, Fig. 8) has not been reported in any other archaebacterium[80]. The major phosphoglycolipids are the monoglucosylcaldarchaeol with a phosphoinositol attached to the free OH group (PGC-11, **37A**, Fig. 8), which is a variation of PGC-10 (**36A**) in *Thermoplasma*, and the phosphoinositol derivative (PGC-15, **42A**, Fig. 8) of the triglucosylcaldarchaeol (**42**). No nonitolcaldarchaeol glyco- or phosphoglycolipids appear to be present in *T. tenax* [80].

Similar lipid patterns to those in *T. tenax* (except for the sugar linkages) have been found in two *Pyrobaculum* species[81]: a monoglucosyl-caldarchaeol (MGC-2, **37**, Fig.8); a diglucosylcaldarchaeol and a triglucosylcaldarchaeol with the sugars being β1-2-linked (DGC-3', **41'** and TGC', **42'**, Fig. 8); and two phosphoglycolipids, βGlcp-caldarchaeol-PI (PGC-11, **37A**, Fig. 8) and (Glcpβ1-2)$_2$Glcpβ-PI (PGC-15', **42A'**, Fig. 8).

The lipids of one species of *Desulfurococcus* (anaerobic sulfur-reducing thermo-acidophiles), *Desulfurococcus mobilis*[82], are derived from caldarchaeol and resemble those in *Sulfolobus*, except that the lipid cores do not contain any nonitolcaldarchaeols, nor are there any cyclopentane rings in the biphytanyl chains. The major glycolipid is

Glcpα1-4Galpβ-caldarchaeol (DGC-6, **43**, Fig. 8), and the phosphoglycolipids are Galpβ-caldarchaeol-P-inositol (PGC-14, **40**, Fig. 8) and Glcpα1-4Galpβ-caldarchaeol-P-inositol (PGC-16, **43A**, Fig. 8) [82].

3.3.4. Thermococcales

Thermococcus and *Pyrococcus* species and an AN1 isolate are unique among the thermophilic archaea in that their lipids are derived only from archaeol [10,83,84]. Until recently they were classified with the methanogens and extreme halophiles, but they are now included in the sulfur-dependent branch of extreme thermophiles [7,10]. In *Thermococcus celer*, the major polar lipid (85% of total) is the archaeol analogue of phosphatidylinositol (API, **14**, Fig. 6) [83], a phospholipid also found in methanogens (e.g., *M. thermoautotrophicum* [62]). This same phospholipid (API, **14**) is also found in *Pyrococcus* species and an AN1 isolate as a major component, together with a novel phosphoglycolipid, 3-P-glucosylarchaeol (P-MGA-2, **17A**, Fig. 6) [84], which is analogous to the P-1-GlcN.Ac-archaeol (P-MGA-1, **17**, Fig. 6) in *Mco. voltae* [66].

3.3.5. Taxonomic relations

Despite the limited number of extreme thermophile species that have been examined for lipids, it appears that a few genera, *Thermoplasma, Sulfolobus, Thermoproteus, Desulfurococcus, Thermococcus* and *Pyrococcus* may be distinguished by their glycolipid and phosphoglycolipid compositions. However, it should be noted that genera in the Order Sulfolobales examined so far, such as *Sulfolobus, Desulfurolobus* and *Metallosphaera* have very similar lipid patterns, and it would be difficult to differentiate between them on the basis only of their lipids.

Thermoproteus, Pyrobaculum and *Desulfurococcus,* on the other hand, can be clearly distinguished from *Sulfolobus* by the absence of any nonitolcaldarchaeol-derived lipids, as well as by the structure of the glycosyl groups in the glycolipids and phosphoglycolipids; the latter feature may also be useful in distinguishing between the above three Thermoproteales genera. As was mentioned previously, the Thermococcales, *Thermococcus* and *Pyrococcus* are clearly distinguished from other sulfur-dependent archaea by the absence of any caldarchaeol- or nonitolcaldarchaeol-derived lipids, and from methanogens having only archaeol-derived lipids by the presence of an unusual P-glucosyl-archaeol (**17A**, Fig. 6).

In general, the lipid analyses of archaea available so far (see Figs. 3–8) allow a clear distinction between the extreme halophiles, the methanogens and the extreme thermophiles (see Fig. 9), based on the following criteria: (a) the extreme halophiles contain only archaeol-based lipids, the polar headgroups being phosphoglycerol-derived, or sulfated/unsulfated glycosyl (glucose, mannose, galactose) groups; (b) the methanogens have lipids based both on archaeol and on caldarchaeol (without rings), with phosphoamino polar headgroups (mostly serine and ethanolamine) and unsulfated sugar groups (glucose, galactose); and (c) the extreme thermophiles have mostly lipids based on caldarchaeol and nonitolcaldarchaeol, with or without rings, with phosphoinositol polar headgroup as well as sugars (glucose, galactose, usually unsulfated).

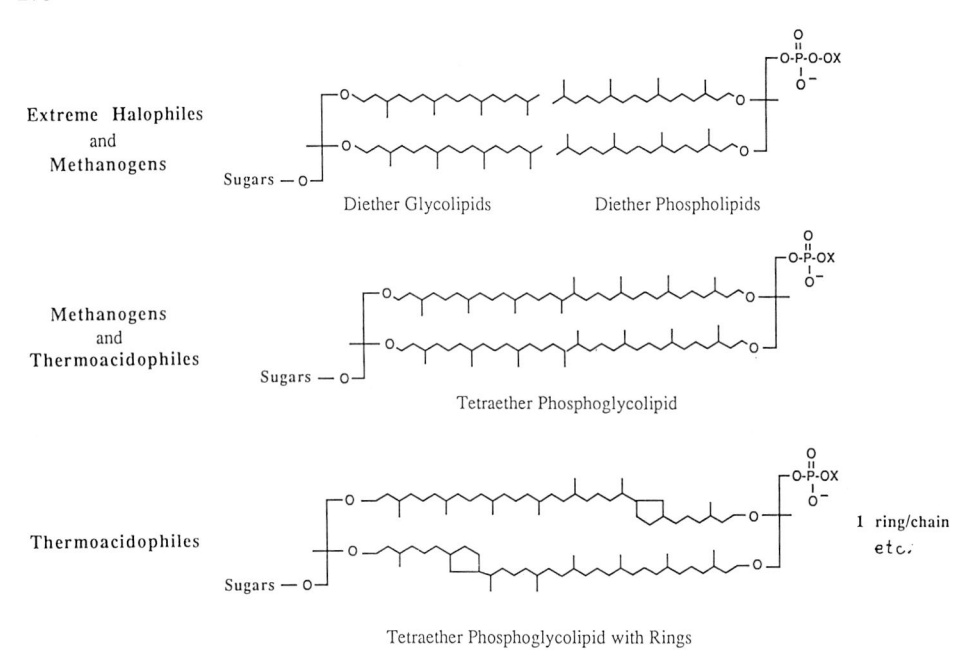

Fig. 9. Lipid structural relations among archaea. X represents H, glycerol, glycerol-P-OMe, inositol, ethanolamine, or serine.

4. Biosynthetic pathways

From the foregoing review of archaeal membrane-lipid structure, it is clear that the features distinguishing archaeal lipids from those of all other organisms are the presence only of derivatives of archaeol, caldarchaeol and/or their variants (Fig. 9). Archaeol or diether phospholipids and glycolipids are present in all archaebacteria; caldarchaeol or tetraether phosphoglycolipids are present in both methanogens and thermoacidophiles; and tetraether and nonitoltetraether phosphoglycolipids, with or without cyclopentane rings, are present only in the thermoacidophiles and extreme thermophiles (Fig. 9).

The question then arises: by what pathways are archaeal ether lipids biosynthesized and how were these pathways selected rather than those used by all other organisms for acyl ester lipid synthesis? The available information on lipid biosynthesis in archaea is based largely, with a few exceptions, on labelling studies with whole cells (see previous reviews [4,5,9,10,13,15,85]). Biosynthetic pathways for archaeol, caldarchaeol and their complex lipid derivatives will now be discussed.

4.1. Archaeol lipid cores

In extreme halophiles, biosynthesis of the archaeol analogues of phospholipids and glycolipids proceeds by complex pathways in a multienzyme, membrane-bound system that is absolutely dependent on 4 M salt concentration [4,13,15]. Synthesis of the isoprenoid/isopranoid chains in the archaeol lipid core takes place via the mevalonate pathway for isoprenoids, also absolutely dependent on 4 M salt concentration [13].

When cells of *H. cutirubrum* were grown in the presence of [^{14}C]acetate, over 98% of the label was incorporated into the isopranyl (phytanyl) groups of the polar lipids, and less than 0.5% was found in long-chain fatty acids [86]. Cell-free studies subsequently demonstrated the presence of a fatty acid synthetase (FAS) which is largely ($\sim$80%) inhibited by 4 M NaCl or KCl [87]. However, sufficient FAS activity remains for the formation of saturated fatty acids (14:0, 16:0, 18:0) recently found to be esterified to proteins of the red membrane but not to bacteriorhodopsin in the purple membrane of extreme halophiles nor to the polar lipids (Pugh and Kates, unpublished data).

Thus, the lipid biosynthetic enzyme system evolved in extreme halophiles to utilize the (halophilic) mevalonate pathway for synthesis of virtually all of its hydrocarbon (isoprenoid/isopranoid) chains, rather than the (non-halophilic) fatty-acid synthetase system which was retained only for synthesis of normal fatty acid chains required for incorporation into proteins of the red membrane (Pugh and Kates, unpublished data). Starting from acetate and involving lysine, which provides the branch-methyl and methine carbons [88]), the mevalonate pathway proceeds to geranylgeranyl-PP (GG-PP) [13,15,89] as follows:

$$\text{acetate} \xrightarrow{\text{lysine}} \to \text{mevalonate} \to\to\to \text{isopentenyl-PP} \quad (1)$$
$$\rightleftarrows \text{dimethylallyl-PP} \to\to \text{geranyl-PP} \to\to \text{farnesyl-PP} \to\to \text{geranylgeranyl-PP}.$$

Final reduction of geranylgeranyl-PP to the phytanyl group takes place only after linkage of the prenyl groups to the glycerol backbone to form a di-geranylgeranyl glycerol ether derivative ("pre-diether", the precursor of both phospholipids and glycolipids), and is the final step in the synthesis of the individual phospholipids and glycolipids [44,45] (Figs. 10, 11). Evidence supporting the involvement of an isoprenyl-PP as donor of the C_{20}-isoprenyl group is provided by the demonstration that bacitracin, which complexes with polyprenylpyrophosphates, is a powerful inhibitor of the biosynthesis of the "pre-diether" and of phospholipids and glycolipids in whole cells of *H. cutirubrum* [45].

Another pertinent question at this point is how the halophiles form lipids having the *sn*-1-glycerol configuration rather than the ubiquitous *sn*-3-configuration? *H. cutirubrum* has been shown to contain glycerokinase and glycerophosphate dehydrogenase activities, associated with the cell envelope and cytoplasm, that form only the *sn*-3-glycerophosphate (GP) stereoisomer [90,91]. So *sn*-3-GP would not be a suitable acceptor for alkylation by geranylgeranyl-PP unless it undergoes an oxidation (dehydrogenation) step followed by a stereospecific reduction step to form the *sn*-1-glycerol derivative [91]. Consistent with this mechanism, it was found in whole-cell labelling studies with 1(3)- and 2-tritium-glycerol precursors that the glycerol moiety

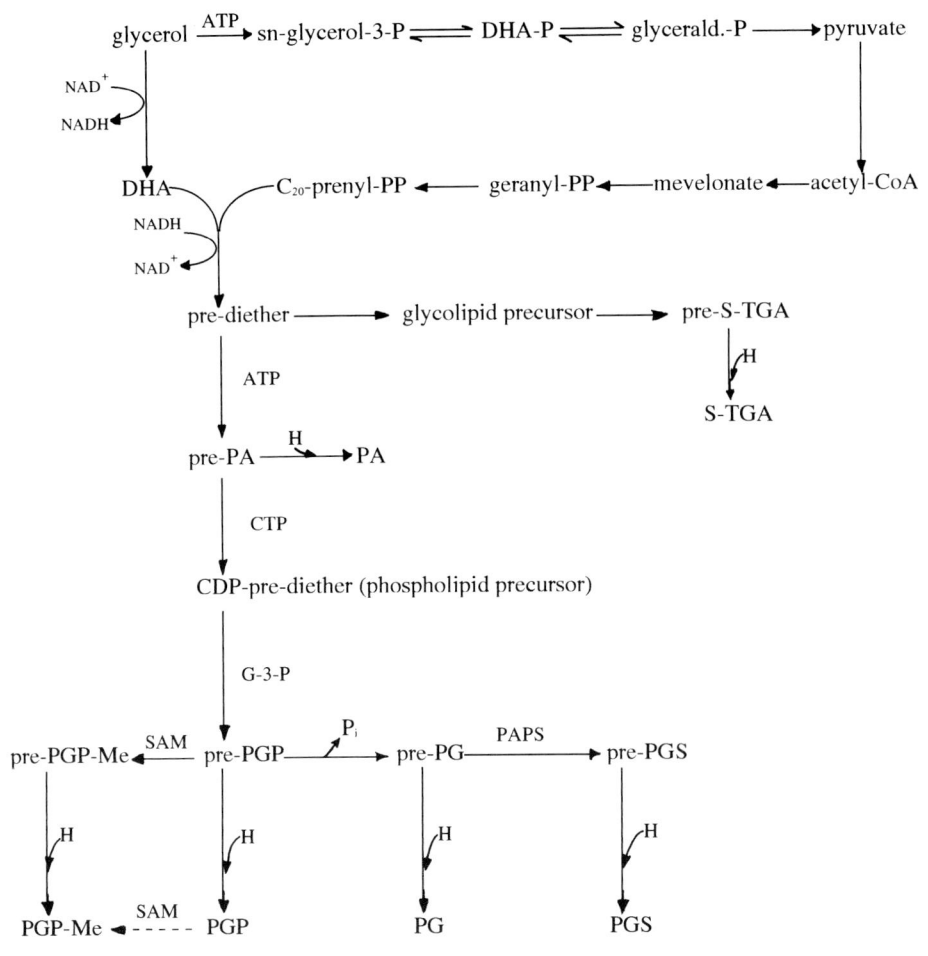

Fig. 10. Proposed pathways for biosynthesis of archaeol phospholipids and glycolipids in extreme halophiles.

of diphytanylglycerol undergoes dehydrogenation at C-2 but not at C-1 (or C-3) [15,92]. These restraints would appear to eliminate the participation of triose phosphates (DHA-P and glyceraldehyde-P) and glycerol-P as possible acceptors in the synthesis of the "pre-diether", since they would exchange hydrogen at C-1 by aldo–keto or keto–enol isomerizations (see Fig. 10). However, these phosphate esters could act as precursors of the glycerol moiety provided they were kept in a pool physically separate from the cytoplasm, such as in the cell membrane [92] (see Fig. 12).

With these considerations in mind, it was hypothesized [4,5,15,44,45] that the "pre-diether" may be formed by alkylation of a glycerol derivative, tentatively dihydroxyacetone (DHA) or dihydroxyacetone phosphate (DHAP), with geranylgeranyl-PP; concomitant stereoselective reduction of the ketogroup would form the sn-2,3-geranylgeranylglycerol derivative (Figs. 10–12). sn-3-Glycerophosphate (GP) or

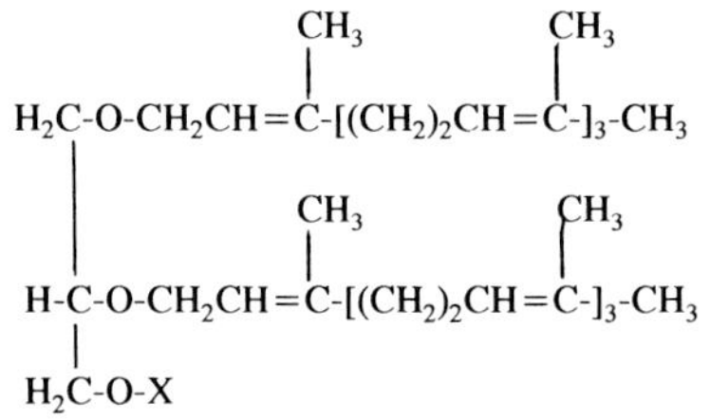

$$H_2C\text{-}O\text{-}CH_2CH=\overset{\overset{CH_3}{|}}{C}\text{-}[(CH_2)_2CH=\overset{\overset{CH_3}{|}}{C}\text{-}]_3\text{-}CH_3$$

$$H\text{-}\overset{|}{C}\text{-}O\text{-}CH_2CH=\overset{\overset{CH_3}{|}}{C}\text{-}[(CH_2)_2CH=\overset{\overset{CH_3}{|}}{C}\text{-}]_3\text{-}CH_3$$

$$H_2C\text{-}O\text{-}X$$

pre-diether	X = unidentified
pre-PA	X = -PO-(OH)$_2$
pre-PG	X = -PO(OH)-O-CH$_2$CH(OH)CH$_2$-OH
pre-PGP	X = -PO(OH)-O-CH$_2$CH(OH)CH$_2$-O-PO(OH)$_2$
pre-PGP-Me	X = -PO(OH)-O-CH$_2$CH(OH)CH$_2$-O-PO(OH)-OMe
pre-PGS	X = -PO(OH)-O-CH$_2$CH(OH)CH$_2$-O-SO$_3$-OH
Phospholipid precursor	X = -P(O)(O-)-O-P(O)(O-)-O-cytidine
Glycolipid precurosr	X = unidentified
pre-S-TGA	X = -Glc-Man-Gal-3-SO$_3$-

Fig. 11. Structures of isoprenyl precursors of archaeol phospholipids and glycolipids of extreme halophiles.

glycerol [10,12] might also serve as an acceptor, provided they underwent oxidation (dehydrogenation) and stereospecific reduction at C-2 during the alkylation.

Recent whole-cell studies with *H. halobium* by Kakinuma et al. [93,94] have confirmed the loss of hydrogen from C-2 of the glycerol in archaeol, probably through oxidation by NAD$^+$, and stereospecific replacement of this hydrogen by reduction with NADH. These authors have suggested that synthesis of archaeol occurs by stepwise alkylation of *sn*-3-GP, as follows (with GG standing for geranylgeranyl):

$$sn\text{-}3\text{-}GP \xrightarrow{NAD^+ / NADH} DHAP \xrightarrow{GG\text{-}PP / PP} GG\text{-}DHAP$$

$$\xrightarrow{NADH / NAD^+} sn\text{-}3\text{-}GG\text{-}1\text{-}GP \xrightarrow{GG\text{-}PP / PP} sn\text{-}2,3\text{-}di\text{-}GG\text{-}1\text{-}GP. \quad (2)$$

Archaeol variants, such as C_{20}–C_{25}- or C_{25}–C_{25}-archaeol could presumably be formed by substituting the C_{25}-isoprenyl-PP for geranylgeranyl-PP as prenyl donor at the appropriate steps in the alkylation process (Reaction 2).

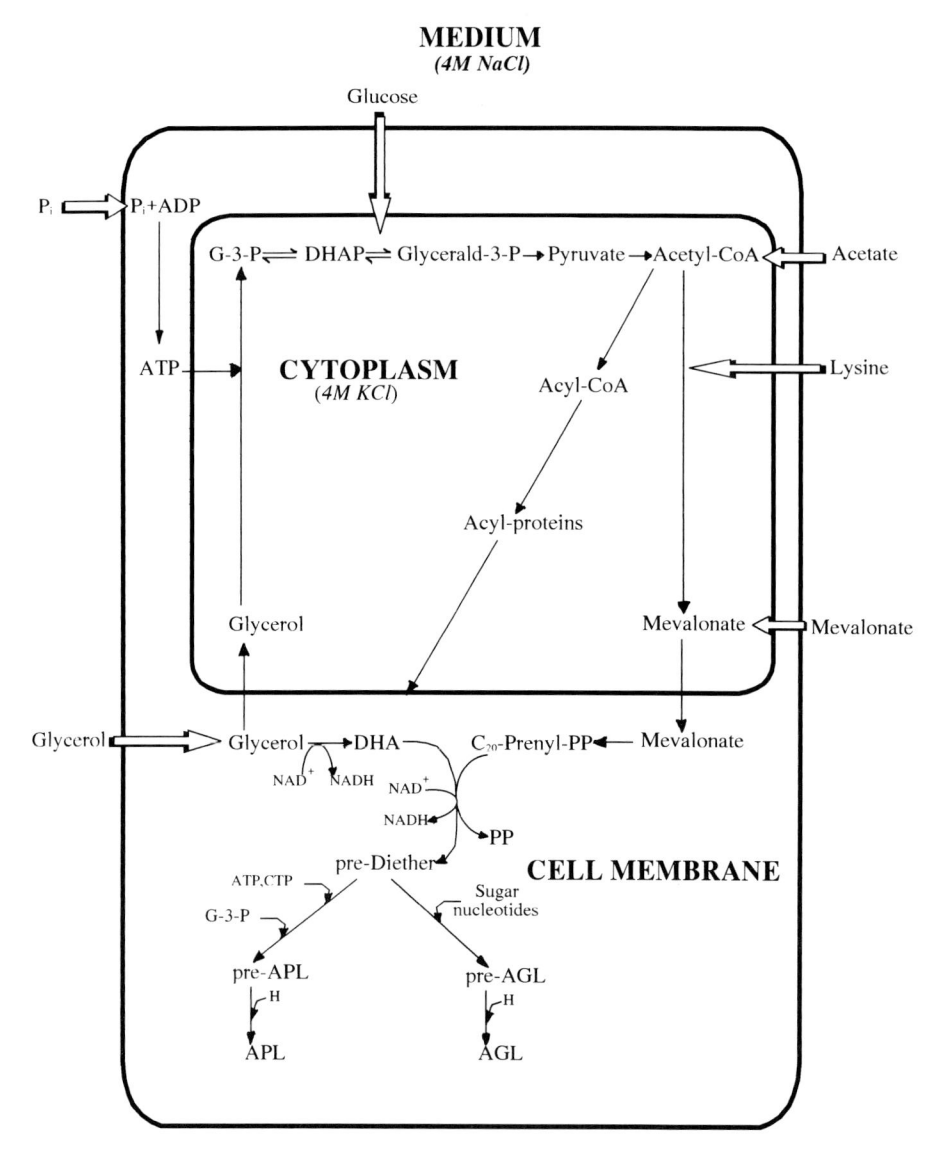

Fig. 12. Summary of lipid biosynthetic pathways in an extremely halophilic cell. APL, archaeol phospholipids; AGL, archaeol glycolipids.

In contrast to the extreme halophiles, methanogens are apparently capable of synthesizing sn-1-GP and utilizing it as an acceptor for alkylation by geranylgeranyl-PP. Recent studies with cell-free extracts of *Methanobacterium thermoautotrophicum* [95] have demonstrated the synthesis of sn-2,3-di-geranylgeranylglycerol-1-P by two stepwise prenyltransferase (PT) reactions with sn-1-GP as acceptor, as follows:

$$sn\text{-}1\text{-}GP \xrightarrow[PT\text{-}I]{GG\text{-}PP \quad PP} sn\text{-}3\text{-}GG\text{-}1\text{-}GP \xrightarrow[PT\text{-}II]{GG\text{-}PP \quad PP} sn\text{-}2,3\text{-}di\text{-}GG\text{-}1\text{-}GP. \qquad (3)$$

Neither DHAP nor glycerol could act as substrates for the prenyltransferases, and phytanyl-PP did not act as a prenyl donor; however, low but significant activities (3%) were obtained with sn-3-GP as prenyl acceptor, and moderate (18%) activities were obtained with phytyl-PP as prenyl donor [95]. The final product, sn-2,3-di-geranylgeranyl-1-GP, actually is the "pre-PA" (Figs. 10, 11), which may be converted to archaeol phospholipids as described in section 4.2.1. for extreme halophiles. There still remains the question as to how the unusual sn-1-GP stereoisomer is formed in methanogens. It is conceivable that the glycerol kinase and/or the GP-dehydrogenase in methanogens give rise to the sn-1-GP rather than the sn-3-GP formed in extreme halophiles, but this still requires further study.

In thermoacidophiles, glycerol alone has been suggested [10,12] to serve as an acceptor for prenyl ether formation, without having to undergo dehydrogenation and reduction at C-2 to form the S-glycerol configuration. Presumably, the prenyl transferase is capable of stereospecifically selecting the sn-C-3 of glycerol for alkylation [10].

4.2. Archaeol phospholipids

4.2.1. In extreme halophiles
Pulse-labelling studies with whole cells of *H. cutirubrum* using [^{32}P]phosphate, [^{14}C]glycerol or [^{14}C]mevalonate as precursors showed that the phospholipids were labelled in the following order [44]:

$$\text{pre-PA} > \text{phospholipid precursor} > \text{pre-PGP-Me} > \text{pre-PG} > \text{pre-PGS}. \qquad (4)$$

These results suggest the following pathway for biosynthesis of phospholipids in extreme halophiles (Figs. 10, 11), based on the analogous pathway for diacyl PGP and PG in eubacteria [96]:

(a) the pre-diether is phosphorylated with ATP, and the pre-PA formed is then converted to the "phospholipid precursor", most likely cytidine diphosphate archaeol (CDPA)[44];
(b) CDPA is then reacted with glycerophosphate (or DHAP [15]) to form pre-PGP;
(c) pre-PGP is methylated by S-adenosylmethionine (SAM) to form pre-PGP-Me;
(d) pre-PGP is dephosphorylated by a specific phosphatase to give pre-PG;
(e) pre-PG is converted to pre-PGS by sulfation with PAPS;
(f) all of the "pre-" phospholipids are subsequently hydrogenated to give the final, saturated archaeol analogues of PGP-Me, PGP, PG and PGS.
(g) PGP-Me may also be formed by methylation of PGP with SAM.

Further studies are needed to establish these pathways unambiguously.

4.2.2. In methanogens
In regard to the pathways for biosynthesis of individual archaeol phospholipids in methanogens, a study on the incorporation and turnover of ^{32}P in archaeol and

caldarchaeol phospholipids and in caldarchaeol phosphoglycolipids of growing cells of *Methanobacterium thermoautotrophicum* has been carried out [62]. The archaeol phospholipids APS, API and APE were rapidly labelled with ^{32}P with little or no lag, whereas the caldarchaeol phospholipids and phosphoglycolipids, after a lag of 15–90 min, were labelled more slowly. Also, rapid turnover of APS, API and APA, relative to the caldarchaeol lipids, was observed. The relative rates of APS and APE labelling suggested that APE was being formed from APS [62], presumably by the decarboxylation reaction well known in eubacteria and eukariots [96]:

$$\text{APS} \xrightarrow{\text{CO}_2} \text{APE.} \tag{5}$$

The results of this labelling study also suggest that the archaeol phospholipids are the precursors of the caldarchaeol phospholipids and phosphoglycolipids [62], and this will discussed further in section 4.4.

4.3. Archaeol glycolipids

In extreme halophiles, the "pre-diether" (see section 4.2) can also be converted to archaeol glycolipids [44], presumably by stepwise glycosylation with glucose, mannose and galactose followed by sulfation with PAPS to give the pre-S-TGA, which is finally reduced to the saturated S-TGA (**7**) (see Figs. 10, 11). Evidence in support of a stepwise glycosylation and sulfation pathway has been provided by pulse-labelling studies with whole cells of *H. cutirubrum* grown in the presence of [^{35}S]sulfate or [^{14}C]glycerol, which established the following product–precursor relationship between the glycolipids of this bacterium (Deroo and Kates, unpublished data) (see Fig. 4):

$$\text{glycolipid precursor} \xrightarrow{\text{UDP-Glc}} \text{Glc-archaeol(MGA-1)} \xrightarrow{\text{UDP-Man}} \text{DGA-1} \xrightarrow{\text{UDP-Gal}} \text{TGA-1} \xrightarrow{\text{PAPS}} \text{S-TGA-1.} \tag{6}$$

The minor glycolipid S-TeGA (**8**) could be biosynthesized by galactofuranosylation of S-TGA-1 (**7**) or by galactofuranosylation of TGA-1 (**7A**) followed by sulfation with PAPS. The minor non-sulfated glycolipids TGA-1 (**7A**) and TeGA (**8A**) could be formed by deletion of the appropriate sulfation steps or by the action of sulfatases on S-TGA-1 and S-TeGA, respectively. Such pathways would account for the presence of the glycolipids found in the *Halobacteria* (Fig. 4). In *Haloferax* species, the major glycolipid S-DGA-1 (**10**) (Fig. 4) might be formed by deletion of the galactosylation step of DGA-1 and insertion of a specific DGA sulfation step. In *Haloarcula*, the major glycolipid TGA-2 (**9**) (Fig. 4) might be formed by replacement of the galactosylation step of DGA-1 with a glucosylation reaction using UDP-Glc.

Thus the characteristic glycolipid composition of the three genera of *Halobacteria* and of *Halococcus* genera (Table 1), as well as of newly discovered glycolipids of uncertain genera (Fig. 5), could be achieved by deletion and/or insertion of the genes for the appropriate glycosylating enzymes and sulfating enzymes, as suggested above. Studies on the isolation of the enzymes and the corresponding genes involved in glycolipid and

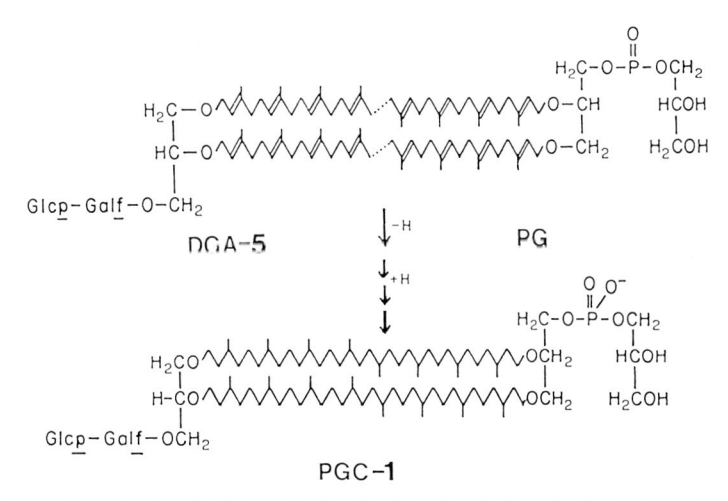

Fig. 13. Proposed mechanism for biosynthesis of caldarchaeol phosphoglycolipids in methanogens, e.g., *M. hungatei* [64].

phospholipid biosynthesis would help to establish the biosynthetic pathways of these lipids with greater certainty and to elucidate the taxonomic relationships between the genera of extreme halophiles.

4.4. Caldarchaeol phospholipids, glycolipids and phosphoglycolipids

4.4.1. In methanogens
Examination of the known structures of methanogen and extreme thermophile lipids (Figs. 6–8) suggests possible biosynthetic relationships between archaeol and caldarchaeol complex lipids that could be tested experimentally. It may be assumed first that the archaeol molecule would be synthesized by the same or similar routes as in extreme halophiles (see Reactions 2–4 and Figs. 10, 11). On the basis of the presence in *M. hungatei* of analogous archaeol and caldarchaeol glyco- and phosphoglycolipids, Kushwaha et al. [64] proposed that the biosynthesis of the phosphoglycosylcaldarchaeols (PGC-1, **28A** and PGC-2, **29A**) might occur by head-to-head condensation of a molecule of archaeol-PG (**3**) with DGA-5 (**21**) or DGA-6 (**22**), respectively. More likely, condensation would occur between the corresponding isoprenyl ether precursors followed by hydrogenation (see Fig. 13). However, it should be noted that the required archaeol-*sn*-3-glycerophosphate (**3**) (or its isoprenyl ether derivative) has not been identified in *M. hungatei* [64,65], although recent evidence for its existence in methanogens is forthcoming [21].

The pronounced lag observed between labelling of the diether and tetraether polar lipids, in recent [^{32}P]phosphate pulse-chase studies with *M. thermoautotrophicum* cells [62], is consistent with the generalized mechanism proposed by the authors (Fig. 4 of ref. [62]; see also Fig. 13). This mechanism would account for the biosynthesis in *M. thermoautotrophicum* of the three families of "heptads" of archaeol and caldarchaeol polar lipids containing P-ethanolamine, P-serine and P-inositol, since all of the corresponding archaeol phospholipids and glycolipids, as well as archaeol itself, are present [62] (see Figs. 6, 7).

No direct experimental evidence is available concerning the mechanism of this unique head-to-head condensation (Fig. 13), other than that the two carbons involved in the coupling are derived from C-2 of mevalonate [10,12]. It is plausible that the head-to-head condensation would involve coupling of terminal carbons on isoprenyl intermediates, presumably isoprenyl ether archaeol derivatives ("pre-phospholipids" and "pre-glycolipids") similar to those detected in extreme halophiles (see Figs. 10, 11). Recent labelling studies with *M. hungatei* have indeed eliminated the possibility of coupling of saturated archaeol termini, and favor the coupling of termini in the isoprenyl analogues [97].

4.4.2. In thermoacidophiles
Biosynthesis of caldarchaeol and nonitolcaldarchaeol lipids in thermoacidophiles has been studied in whole cells of *S. solfataricus* using U-^{14}C, 1(3)-^{3}H glycerol and U-^{14}C, 2-^{3}H glycerol as precursors [12,98]. It was found that the biphytanyl moieties of both caldarchaeol and nonitolcaldarchaeol undergo complete loss of ^{3}H from 2-^{3}H glycerol and 50% loss from 1(3)-^{3}H glycerol, as expected for biosynthesis of acetate from glycerol via the glycolytic pathway, and consistent with the findings for phytanyl groups in *H. cutirubrum* [15]. The glycerol moieties of caldarchaeol and nonitolcaldarchaeol, however, do not undergo loss of hydrogen either from C-1(3) or from C-2, nor does the free glycerol of glycerophosphate pools undergo any dehydrogenation [12]. This is in contrast to the situation in extreme halophiles in which the archaeol glycerol moiety loses hydrogen at C-2 but not at C-1(3). The glycerol precursors of caldarchaeol or nonitolcaldarchaeol therefore cannot undergo dehydrogenation at C-2, nor aldo–keto or keto–enol isomerizations, thus eliminating the involvement of DHA, DHAP, and glyceraldehyde-P. It was concluded [10,12,98] that the ether-forming step could occur in thermoacidophiles by a direct stereospecific alkylation of glycerol by geranylgeranyl-PP or similar allylic pyrophosphates. Presumably, the products of this alkylation reaction would be isoprenyl ether intermediates similar to those detected in the biosynthesis of archaeol lipids in *H. cutirubrum* (Figs. 10, 11) [44] and in *M. hungatei* [97].

However, in thermoacidophiles these prenyl ether intermediates would probably not be reduced directly, but would first participate in head-to-head condensations to form the caldarchaeol polar lipids [97] (see Fig. 13). The introduction of cyclopentane rings would also occur before final reduction of the chains [10,99]. The symmetrical arrangement of cyclopentane rings in the biphytanyl chains suggests that the rings were closed in a concerted way on both alkyl chains starting from the middle of the chains toward the ether linkages [10]. The mechanism of cyclopentane ring formation has been studied in *S. solfataricus* using strategically labelled [^{13}C, ^{2}H]mevalonolactones as precursors of tetraether lipids [99]. The results obtained were compatible with a mechanism based on initial isomerization of an isoprenyl double bond and concerted cyclization and hydride reduction; other mechanisms based on hydroxylation of the isoprenyl chains could not be eliminated [10,99].

The nonitol group in nonitolcaldarchaeol is found only in the Sulfolobales, including the genera *Desulfurolobus*, *Acidianus* and *Metallosphaera* [10,20]. Whole-cell studies showed that the biosynthesis of nonitol in *S. solfataricus* [20,98] could occur by an aldol- or acetoin-type condensation between a triose and a hexose precursor, followed by reduction. Evidence favouring this mechanism is the complete loss of ^{3}H from [2-^{3}H]glycerol, 70% retention of ^{3}H from [1(3)-^{3}H]glycerol at carbons 1 to 3, and the incorporation of labelled glucose or fructose into carbons 4 to 9 in the nonitol skeleton [10,12,98].

It is clear that detailed studies with whole-cell and cell-free systems are now required to elucidate the novel mechanisms underlying the complex biosynthetic pathways of archaebacterial lipids, which have been outlined in the foregoing sections. Perhaps the time is now ripe for a concerted effort to be made to clone the genes involved in archaebacterial lipid biosynthesis, thereby aiding greatly in identifying the enzymes involved.

5. Membrane function of archaeal lipids

The fact that all archaebacterial membranes contain lipids derived from diphytanylglycerol diether (**1**) or its dimer dibiphytanyldiglycerol tetraether (**2**) suggests that these "peculiar" lipids have functions that are common to all archaebacteria [4,9,10]. Thus, in general, the alkyl-ether structure, in contrast to the usual acyl-ester structure, should impart stability to the lipids over the wide range of pH encountered by these bacteria; the saturated alkyl chains would impart stability towards oxidative degradation, particularly in the extreme halophiles that are exposed to air and sunlight; the branched isopranyl structure of archaeol and caldarchaeol would ensure that the membrane lipids are in the liquid-crystalline state at ambient temperatures; and the covalent linking of the ends of the chains together with the introduction of cyclopentane rings would keep the fluidity fairly constant as the temperature increased [76,100].

Furthermore, the "unnatural" *sn*-1 configuration of the backbone glycerol would impart resistance to attack by phospholipases released by other organisms and would thus have a survival value for the archaebacteria. More specific functions of archaebacterial lipids will now be considered.

5.1. Archaeol-derived lipids in extreme halophiles

In extreme halophiles the sulfated triglycosylarchaeol (S-TGA-1, **7**) is associated exclusively with the purple membrane of *Halobacteria* [101] in which it is located entirely in the exterior surface layer of the membrane [102]. S-TGA-1, together with the major phospholipid PGP-Me (**4A**) (in mole ratio 1:1 [4]) are capable of participating in proton conductance pathways [103]. Such pathways, involving the polar headgroup sulfate of S-TGA-1 and the phosphate groups of PGP-Me, would serve to transport the protons transduced by light-activated bacteriorhodopsin across the outer surface of the purple membrane (PM) to the red membrane (RM), where the PGP-Me headgroup phosphates would conduct the protons to the sites of the H$^+$-ATPases in the red membrane, to drive ATP synthesis [104]. The participation of the free OH of the headgroup glycerol of PGP-Me in the H-bonding network has been demonstrated by FT-IR studies of PGP-Me [105].

Another point that should be noted is the maintenance of a high negative charge surface density by the high concentration of acidic lipids in the membranes of *Halobacteria*, *Haloarcula*, *Haloferax* and *Halococci* (Table 1). Such a high negative charge surface density would be shielded by the high Na$^+$-ion concentration (4 M), thus preventing disruption of the lipid bilayers due to charge–charge repulsion [105a,b]. A highly negatively charged membrane surface would thus appear to be required for the survival of extreme halophiles in media of high salt concentration.

The sulfated triglycosylarchaeol may have another function in the purple membrane, related to proton pumping action of bacteriorhodopsin, since it has been demonstrated that reconstitution of BR in PGP-Me vesicles containing S-TGA-1 results in increased rates of proton pumping [106]. Similar functions might be envisaged for the sulfated diglycosylarchaeol (S-DGA-1, **10**) in species of *Haloferax,* which also contain the purple membrane, but no experimental evidence is available concerning the function of this glycolipid.

5.2. Caldarchaeol-derived lipids in methanogens and thermoacidophiles

As was discussed above, the membranes of caldarchaeol-containing methanogens and thermoacidophiles consist of essentially bipolar monolayer structures. These structures would impart stability to the membrane at the high growth temperatures of thermophilic methanogens and of thermoacidophiles [10]. In the methanogens, possession of both archaeol and caldarchaeol-based membrane lipids would also be an advantage under conditions of high methane concentration that must be present in these cells and might lead to membrane disruption. However, it should be noted that some species of methanogens and thermophiles, e.g. *Mco. voltae,* and *Thermococcus,* manage to survive with only the archaeol-type lipids in their membranes.

Another methanogen, *Methanococcus jannaschii,* isolated from a deep-sea hydrothermal vent at temperatures of 85°C or higher, has membrane lipids based on *cyc*-archaeol (**1C**), which was not found in any other species of methanogens surveyed and may be unique to methanogens from deep-sea hydrothermal vents [34,67]. It is possible that the presence of *cyc*-archaeol-based lipids may be related to the high pressures under which these deep-sea methanogens live.

The high proportions of glycosylated caldarchaeols present in membranes of both methanogens and thermoacidophiles may further stabilize the membrane structure by inter-glycosyl headgroup hydrogen bonding. In addition, as was discussed above, the presence of cyclopentane rings in the biphytanyl chains of caldarchaeol may serve to fine-tune the rigidity of the membrane monolayer in direct response to the growth temperature of the thermoacidophile [10,12,76,100]. In contrast, there does not appear to be any specific structural feature of the polar lipids of thermoacidophiles that would be related to the low pH of their growth environment; for example, no amino or other basic groups are present that might serve to protect the membrane against the high H^+ concentration in the external environment. Another puzzling point concerns the asymmetric orientation at the exterior membrane surface of the glycosyl groups in the caldarchaeol lipids of thermoacidophiles and some methanogens, and hence the orientation of the anionic groups (phosphates in phosphoglycocaldarchaeols) into the inner membrane layer, thus placing a highly negatively charged surface density on one side of the membrane. It is difficult to understand how the membrane would remain stable under such an arrangement, unless the negative charges on the interior side were neutralized or shielded by protonated amino groups in the membrane proteins.

6. Evolutionary considerations and conclusions

The observation that the fatty acid synthetase (FAS) in *H. cutirubrum* is strongly inhibited by high salt concentration while the mevalonate enzyme system for isoprenoid biosynthesis has an absolute requirement for high salt concentration (see section 4.1) may offer a clue to the mechanism of evolution of extreme halophiles from non-halophilic, or moderately halophilic precursors, and possibly also for the evolution of the methanogens and extreme thermophiles. The following hypothetical scenarios are offered for discussion (see ref. [107] and Figs. 9, 14):

(a) In scenario 1, an anaerobic precursor synthesizing acyl ester phospholipids (and perhaps also monoacyl-monoalkyl ether phospholipids), as well as isoprenoid compounds, was exposed to gradually increasing salt concentrations such as in evaporating salt lakes or ponds. As the intracellular salt concentration increased, progressive inhibition of the FAS, the complex acyl lipid-synthesizing enzymes (particularly the acyl–CoA:*sn*-3-GP acyl transferase), and the mevalonate enzyme system occurred, resulting eventually in a near-lethal deficiency of membrane lipids. However, before this state was actually reached, a mutant arose, designated "pre-archaebacterium" (or "pre-extreme halophile"), in which the mevalonate enzyme system and the alkyl ether lipid synthetases, but not the FAS, were modified for effective functioning in increasing salty environments. The driving force here was the increased survival value afforded by the more stable isopranyl ether phospholipids and glycolipids.

(b) As the salt concentration of the environment approached saturation, the mevalonate enzyme system and the alkyl ether lipid synthetases of the "pre-archaebacterium" were further modified for optimal synthesis of isoprenyl/isopranyl chains and diphytanyl glycerol diether (archaeol) lipids, respectively, in nearly saturated (4 M) NaCl, to form the "pro-extreme halophile" (Fig. 14). The FAS, which was only slightly modified, was now largely inhibited, but still able to produce sufficient fatty acids for membrane protein acylation (see Fig. 12). Replacement of the membrane acyl lipids by the phytanyl glycerol diether lipids rectified the deficiency in acyl polar lipids and at the same time provided a membrane lipid bilayer that was more suited to a high-salt environment. Replacement of the typical eubacterial cell wall by the archaebacterial glycosylated protein cell wall may also have happened at this stage ("pro-extreme halophile").

(c) Further mutation of the "pro-extreme halophile" occurred to produce bacteriorhodopsin and bacterioruberin, enabling the mutant to function anaerobically in the light. The aerobic system of fermentation evolved later to form the extreme halophiles known today.

(d) The methanogens and perhaps the extreme thermophiles may have evolved from either the pre-archaebacterium or the pro-extreme halophile to form a "pro-methanogen" or "pro-thermoacidophile" (Fig. 14), a process that would have involved re-adaptation of the mevalonate enzyme system and the alkyl ether synthetases to function at lower salt concentrations. In this connection it is of interest that a halophilic methanogen growing in 2 M salt has recently been described [108]; a halophilic thermophile has not yet been isolated [7]. At this stage, development of the novel enzyme system for head-to-head coupling of the ends of the isoprenyl chains to form tetraether lipids (see Fig. 13) may have occurred in the "pro-methanogen" and "pro-thermoacidophile", as an adaptation to high temperatures. Finally, the cyclopentane ring-forming enzyme system developed in

Scenario 1.

```
                        ANAEROBIC PRECURSOR
                                 |
                          increasing salt
                                 |
                                 ▼
                   Pre-ARCHAEBACTERIUM (halophile)
       saturated salt     /     |
                         /      |
                        ▼       ▼
  Pro-EXTREME HALOPHILE – – – ➤ Pro-METHANOGEN – – – ➤ Pro-EXTREME THERMOPHILE
           |                    |                              |
           ▼                    ▼                              ▼
  EXTREME HALOPHILE       METHANOGEN                EXTREME THERMOPHILE
```

Scenario 2.

```
                        ANAEROBIC PRECURSOR
                                 |
                    low pH, increasing temperature
                                 |
                                 ▼
                  Pre-ARCHAEBACTERIUM (thermophile)
                         /       |
                        /        |
                       ▼         ▼
  Pro-EXTREME THERMOPHILE ◄ – – ➤ Pro-METHANOGEN – – ➤ Pro-EXTREME HALOPHILE
           |                     |                              |
           ▼                     ▼                              ▼
  EXTREME THERMOPHILE       METHANOGEN                 EXTREME HALOPHILE
```

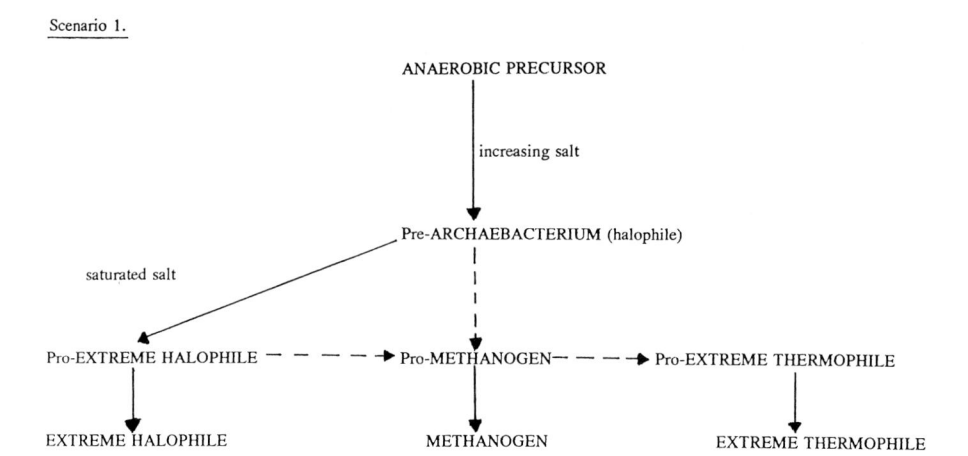

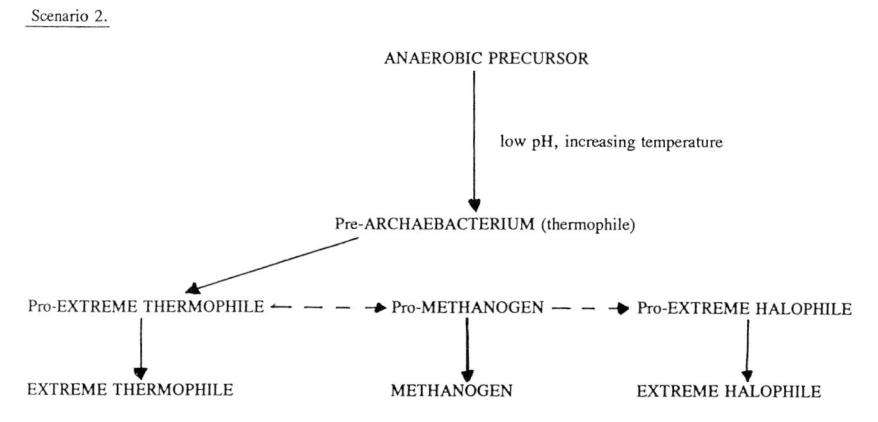

Fig. 14. Proposed evolutionary relationships between extreme halophiles, methanogens and extreme thermophiles based on lipid composition.

the "pro-thermoacidophile", thus leading to the extreme thermophiles being able to grow at temperatures near 100°C.

(e) It may be argued (scenario 2) that the thermophiles could have developed independently of the halophiles as a result of inhibition of the FAS in a "pre-archaebacterium" by low pH and/or high temperature. This possibility implies that the mevalonate enzyme system and ether lipid synthetases, but not the FAS, would have been modified appropriately to function optimally at low pH and high temperature. Furthermore, the enzyme systems for synthesis of archaeol-, caldarchaeol- and nonitolcaldarchaeol-derived lipids would have developed first in a pro-thermophile and have then been passed on to a pro-methanogen, and then to a pro-extreme halophile, with deletion of the caldarchaeol and nonitolcaldarchaeol synthetases (Fig. 14).

The first scenario (a–d) postulates a pre-archaebacterium precursor which was halophilic ("pre-extreme halophile") and would imply an evolutionary relationship:

pre-archaebacterium (halophile) → extreme halophile
→ methanogen → extreme thermophile.

The second scenario (e) implies a thermophilic pre-archaebacterium and the resulting evolutionary relationship (Fig. 14):

thermophile precursor → extreme thermophile → methanogen → extreme halophile.

It would be of interest to test these hypotheses by examining the methanogens and thermophiles for the presence of FAS and acyl transferases and studying the effect of salt concentration, low pH and high temperature on their activities and on those of the mevalonate enzyme system. It may be noted that preliminary studies with *M. thermoautotrophicum* have revealed the presence of a functional FAS producing fatty acids for acylation only of membrane proteins (Pugh and Kates, unpublished data).

The foregoing review of membrane lipids in archaebacteria has revealed a remarkable variety of polar lipids classes, including phospholipids, glycolipids, phosphoglycolipids and sulfolipids, all derived from the one basic core structure, diphytanylglycerol (**1**, Fig. 1), and an equally remarkable set of novel pathways for their biosynthesis. Even with the relatively limited knowledge that we have of the physical properties of these lipids, it is clear that they are well-adapted as membrane components to the particular environmental conditions of the three groups of archaebacteria: extreme halophiles, methanogens, and extreme thermophiles. However, much remains to be learned concerning the precise asymmetric arrangement of the lipids in the membrane bilayers or monolayers, the interaction of the lipids with the membrane proteins, and the function of this membrane-lipid asymmetry with respect to ion transport, permeability to nutrients, proton transport and conductance, and energy transduction. Perhaps then these unusual lipids will not appear so strange and our knowledge of them will help us understand the function of the more familiar lipids in the eubacteria and eukaryotes.

Acknowledgement

Support by the Natural Sciences and Engineering Research Council of Canada and the Medical Research Council of Canada is acknowledged.

Note added in proof

A novel glycolipid has been identified in *Hb. trapanicum* as a 2-sulfate-Man$p$-α1-2-Glc$p$-α1-1-archaeol [109], suggesting that this extreme halophile, along with others mentioned in section 3.1.3, may require reclassification.

References

[1] Singer, S.J. and Nicolson, G.L. (1972) Science 175, 720–731.
[2] Kates, M. (1964) Adv. Lipid Res. 2, 17–90.
[3] Asselineau, C. and Asselineau, J. (1990) Biochem. Cell Biol. 68, 319–386.
[4] Kates, M. (1990) In: Glycolipids, Phosphoglycolipids and Sulfoglycolipids (Kates, M., Ed.), pp. 1–122, Plenum Press, New York.
[5] Kates, M. (1992) Biochem. Soc. Symp. 58, 51–72.
[6] Lechevalier, H. and Lechevalier, M.P. (1988) In: Microbial Lipids, Vol. 1 (Ratledge, C. and Wilkinson, S.G., Eds.), pp. 869–902, Academic Press, New York.
[7] Woese, C.R. & Wolfe, R.S., Eds. (1985) The Bacteria, Vol. 8: Archaebacteria, Academic Press, New York.
[8] Woese, C.R., Kandler, O. and Wheelis, M.L. (1990) Proc. Natl. Acad. Sci. U.S.A. 87, 4576–4579.
[9] Langworthy, T.A. (1985) In: The Bacteria, Vol. 8 (Woese, C.R. and Wolfe, R.S., Eds.), pp. 459–497, Academic Press, New York.
[10] De Rosa, M., Tricone, A., Nicolaus, B. and Gambacorta, A. (1991) In: Life Under Extreme Conditions (di Prisco, G., Ed.), pp. 61–87, Springer, Berlin.
[11] Kates, M. (1978) Prog. Chem. Fats Other Lipids 15, 301–342.
[12] De Rosa, M., Gambacorta, A. and Gliozzi, A. (1986) Microbiol. Rev. 50, 70–80.
[13] Kates, M. and Kushwaha, S.C. (1978) In: Energetics and Structure of Halophilic Microorganisms (Caplan S.R., and Ginzburg, M., Eds.), pp. 461–480, Elsevier, Amsterdam.
[14] Kates, M. (1988) In: Biological Membranes: Aberrations in Membrane Structure and Function (Karnovsky, M.L., Leaf, A. and Bolis, L.C., Eds.), pp. 357–384, Alan Liss, New York.
[15] Kamekura, M. and Kates, M. (1988) In: Halophilic Bacteria, Vol. II (Rodriguez-Valera, F., Ed.), pp. 25–54, CRC Press, Boca Raton, FL.
[16] Langworthy, T.A., Tornabene, T.G. and Holzer, G. (1982) Zentralbl. Bakteriol. Hyg. Abt. 1 Orig. C 3, 228–244.
[17] Langworthy, T.A. (1982) Curr. Top. Membr. Transp. 17, 45–77.
[18] Langworthy, T.A. and Pond, J.L. (1986) In: Thermophiles: General, Molecular and Applied Microbiology (Brock, T.D., Ed.), pp. 107–134, Wiley-Interscience, New York.
[19] Ross, H.N.M., Grant, W.D. and Harris, J.E. (1985) In: Chemical Methods in Bacterial Systematics (Goodfellow, M. and Minnikin, D.E., Eds.), pp. 289–300, Society for Applied Microbiology, Florida.
[20] De Rosa, M. and Gambacorta, A. (1988) Prog. Lipid Res. 27, 153–175.
[21] Sprott, G.D. (1992) J. Bioenerg. Biomembr. 24, 555–566.
[22] Kates, M. (1986) Techniques of Lipidology, 2nd revised edition, pp. 163–164, Elsevier, Amsterdam.
[23] Ekiel, I. and Sprott, G.D. (1992) Can. J. Microbiol. 38, 764–768.
[24] Morii, H., Nishihara, M., Ohga, M. and Koga, Y. (1986) J. Lipid Research 27, 724–730.
[25] Ferrante, G., Ekiel, I., Patel, G.B. and Sprott, G.D. (1988) Biochim. Biophys. Acta. 963, 173–182.
[26] Nishihara, M., Morii, H. and Koga, Y. (1987) Biochem. 101, 1007–1115.
[27] Makula, R.A. and Singer, M.E. (1978) Biochem. Biophys. Res. Commun. 82, 716–722.
[28] Tornabene, T.G. and Langworthy, T.A. (1979) Science 203, 51–53.
[29] Amico, V., Oriente, G., Piatelli, M., Tringali, C., Fattorusso, E., Mango, S. and Mayol, L. (1977) Experientia 33, 989–990.
[30] De Rosa, M., Gambacorta, A., Nicolaus, B., Ross, H.N.M., Grant, W.D. and Bu'Lock, J.D. (1982) J. Gen. Microbiol. 128, 343–348.

[31] De Rosa, M., Gambacorta, A., Nicolaus, B. and Grant, W.D. (1983) J. Gen. Microbiol. 129, 2333–2337.
[32] Moldoveanu, N., Kates, M., Montero, C.G. and Ventosa, A. (1990) Biochim. Biophys. Acta 1046, 127–135.
[33] Upasani, V.N., Desai, S.G., Moldoveanu, N. and Kates, M. (1993) J. Gen. Microbiol., in press.
[34] Comita, P.B. and Gagosian, R.B. (1983) Science 222, 1329–1331.
[35] Sprott, G.D., Ekiel, I. and Dicaire, C. (1990) J. Biol. Chem. 265, 13735–13740.
[36] Nishihara, M. and Koga, Y. (1991) Biochim. Biophys. Acta 1082, 211–217.
[37] Langworthy, T.A. (1977) Biochim. Biophys. Acta 487, 37–50.
[38] Kushwaha, S.C., Kates, M., Sprott, G.D. and Smith, I.C.P. (1981) Science 211, 1163–1164.
[39] Heathcock, C.H., Finkelstein, B.L., Aoki, T. and Poulter, C.D. (1985) Science 229, 862–864.
[40] Lanzotti, V., Nicolaus, B., Trincone, A., De Rosa, M., Grant, W.D. and Gambacorta, A. (1989) Biochim. Biophys. Acta 1002, 398–400.
[41] Tsujimoto, K., Yorimitsu, S., Takahashi, T. and Ohashi, M. (1989) J. Chem. Soc. Chem. Commun. 668–670.
[42] Fredrickson, H.L., de Leeuw, J.W., Tas, A.C., van der Greef, J., LaVos, G.F. and Boon, J.J. (1989) Biomed. Environ. Mass Spectrom. 18, 96–105.
[43] Kates, M., Moldoveanu, N. and Stewart, L.C. (1993) Biochim. Biophys. Acta 1169, 46–53.
[44] Moldoveanu, N. and Kates, M. (1988) Biochim. Biophys. Acta 960, 164–182.
[45] Moldoveanu, N. and Kates, M. (1989) J. Gen. Microbiol. 135, 2503–2508.
[46] Lanzotti, V., Trincone, A., De Rosa, M., Grant, W.D. and Gambacorta, A. (1989) Biochim. Biophys. Acta 1001, 31–34.
[47] Smallbone, B.W. and Kates, M. (1981) Biochim. Biophys. Acta 665, 551–558.
[48] Evans, R.W., Kushwaha, S.C. and Kates, M. (1980) Biohim. Biophys. Acta 19, 533–544.
[49] Kushwaha, S.C., Kates, M., Juez, G., Rodriguez-Valera, E. and Kushner, D.J. (1982) Biochim. Biophys. Acta 711, 19–25.
[50] Lanzotti, V., Nicolaus, B., Trincone, A. and Grant, W.D. (1988) FEMS Microbiol. Lett. 55, 223–228.
[51] Kushwaha, S.C., Juez-Perez, G., Rodriguez-Valera, E., Kates, M. and Kushner, D.J. (1982) Can. J. Microbiol. 28, 1365–1372.
[52] Torreblanca, M., Rodriguez-Valera, F., Juez, G., Ventossa, A., Kamekura, M. and Kates, M. (1986) System. Appl. Microbiol. 8, 89–99.
[53] Matsubara, T., Tanaka, N., Kamekura, M., Moldoveanu, N., Ishizuka, I., Onishi, H., Kushner, D.J., Hayashi, A. and Kates, M. (1993) Biochim. Biophys. Acta, in press.
[54] Tricone, A., Nicolaus, B., Lama, L., De Rosa, M., Gambocorta, A. and Grant, W.D. (1990) J. Gen. Microbiol. 136, 2327–2331.
[55] Kocur, M. and Hodgkiss, W. (1973) Int. J. Syst. Bacteriol. 23, 151–156.
[56] Kates, M., Palameta, B., Joo, C.N., Kushner, D.J. and Gibbons, N.E. (1966) Biochemistry 5, 4092–4099.
[57] De Rosa, T., Gambacorta, A., Grant, W.D., Lanzotti, V.D. and Nicolaus, B. (1988) J. Gen. Microbiol. 134, 205–211.
[58] Koga, Y., Ohga, M., Nishihara, M. and Morii, H. (1987) System. Appl. Microbiol. 9, 176–182.
[59] Morii, H., Nishihara, M. and Koga, Y. (1988) Agric. Biol. Chem. 52, 3149–3156.
[60] Grant, W.D., Pinch, G., Harris, J.E., De Rosa, M. and Gambacorta, A. (1985) J. Gen. Microbiol. 131, 3277–3286.
[61] Kramer, J.K.G., Sauer, F.D. and Blackwell, B.A. (1987) Biochem. J. 245, 139–143.
[62] Nishihara, M., Morii, H. and Koga, Y. (1989) Biochemistry 28, 95–102.
[63] Nishihara, M., and Koga, Y. (1990) Biochem. Cell Biol. 68, 91–95.
[64] Kushwaha, S.C., Kates, M., Sprott, G.D. and Smith, I.C.P. (1981) Biochim. Biophys. Acta 664, 156–173.

[65] Ferrante, G., Ekiel, I. and Sprott, G.D. (1987) Biochim. Biophys. Acta 921, 281–291.
[66] Ferrante, G., Ekiel, I. and Sprott, G.D. (1986) J. Biol. Chem. 261, 17062–17066.
[67] Ferrante, G., Richards, J.C. and Sprott, G.D. (1990) Biochem. Cell Biol. 68, 274–283.
[68] Sprott, G.D., Meloche, M. and Richards, J.C. (1991) J. Bacteriol. 173, 3907–3910.
[69] Ferrante, G., Ekiel, I., Patel, G.B. and Sprott, G.D. (1988) Biochim. Biophys. Acta 963, 162–172.
[70] Ferrante, G., Brisson, J.-R., Patel, G.B., Ekiel, I. and Sprott, G.D. (1989) J. Lipid Res. 30, 1601–1609.
[71] Nishihara, M., Utagawa, M., Akutsu, H. and Koga, Y. (1992) J. Biol. Chem. 267, 12432–12435.
[72] Ferguson, M.A.J. and Williams, A.F. (1988) Annu. Rev. Biochem. 57, 285–320.
[73] De Rosa, M., Gambacorta, A., Lanzotti, V., Trincone, A., Harris, E. and Grant, W.D. (1986) Biochim. Biophys. Acta 875, 487–492.
[74] Hedrick, D.B., Guckert, J.B. and White, D.C. (1991) J. Lipid Res. 32, 659–666.
[75] De Rosa, M., Gambacorta, A. and Nicolaus, B., (1983) J. Membr. Sci. 16, 287–294.
[76] De Rosa, M., Esposito, E., Gambacorta, A., Nicolaus, G. and Bu'Lock, J.D. (1980) Phytochemistry 19, 827–831.
[77] Langworthy, T.A., Mayberry, W.R. and Smith, P.F. (1974) J. Bacteriol. 119, 106–116.
[78] Langworthy, T.A. (1977) J. Bacteriol. 130, 1326–1332.
[79] De Rosa, M., Gambacorta, A., Nicolaus, B. and Bu'Lock, J.D. (1980) Phytochemistry 19, 821–826.
[79a] Trincone, A., Lamzotti, V., Nicolaus, B., Zillig, W., De Rosa, M., and Gambacorta, A. (1989) J. Gen. Microbiol. 135, 2751–2757.
[79b] Huber, G., Spinler, C., Gambacorta, A. and Stetter, K.O. (1989) System. Appl. Microbiol. 12, 38–47.
[80] Thurl, S. and Schäfer, W. (1988) Biochem. Biophys. Acta 961, 233–238.
[81] Trincone, A., Nicolaus, B., Palmieri, G., De Rosa, M., Huber, R., Stetter, K.O. and Gambacorta, A. (1992) System. Appl. Microbiol. 15, 11–17.
[82] Lanzotti, V., De Rosa, M., Trincone, A., Basso, A., Gambacorta, A. and Zillig, W. (1987) Biochim. Biophys. Acta 922, 95–102.
[83] De Rosa, M., Gambacorta, A., Trincone, A., Basso, A., Zillig, W., and Holz, I. (1987) System. Appl. Microbiol. 9, 1–5.
[84] Lanzotti, V., Trincone, A., Nicolaus, B., Zillig, W., De Rosa, M., Grant, W. and Gambacorta, A. (1989) Biochim. Biophys. Acta 1004, 44–48.
[85] Bu'Lock, J.D., De Rosa, M. and Gambacorta, A. (1983) In: Biosynthesis of Isoprenoid Compounds (Porter, J.W., Ed.), pp. 159–189, Wiley, New York.
[86] Kates, M., Wassef, M.K. and Kushner, D.J. (1968) Can. J. Biochem. 46, 971–977.
[87] Pugh, E.L., Wassef, M.K. and Kates, M. (1971) Can. J. Biochem. 49, 953–958.
[88] Ekiel, I., Sprott, G.D. and Smith, I.C.P. (1986) J. Bacteriol. 166, 559–564.
[89] Kushwaha, S.C. and Kates, M. (1978) Phytochemistry 17, 2029–2030.
[90] Wassef, M.K., Kates, M. and Kushner, D.J. (1970) Can. J. Biochem. 48, 63–67.
[91] Wassef, M.K., Sarner, J. and Kates, M. (1970) Can. J. Biochem. 48, 69–73.
[92] Kates, M., Wassef, M.K. and Pugh, E.L. (1970), Biochim. Biophys Acta 202, 206–208.
[93] Kakinuma, K., Yamagishi, M., Fujimoto, Y., Ikekawa, N. and Oshima, T. (1988) J. Am. Chem. Soc. 110, 4861–4863.
[94] Kakinuma, K., Yamagishi, M., Fujimoto, Y. Ikekawa, N. and Oshima, T. (1990) J. Am. Chem. Soc. 112, 2740–2745.
[95] Zhang, D.-L., Daniels, L. and Poulter, C.D. (1990) J. Am. Chem. Soc. 112, 1264–1265.
[96] Kennedy, E.P. (1961) Fed. Proc. 20, 934–940.
[97] Poulter, C.D., Aoki, T. and Daniels, L. (1988) J. Am. Chem. Soc. 110, 2620–2624.

[98] Nicolaus, B., Trincone, A., Esposito, E., Vaccaro, M.R., Gambacorta, A. and De Rosa, M. (1990) Biochem. J. 226, 785–791.
[99] Trincone, A., Gambacorta, A., De Rosa, M., Scolastico, C., Sydimov, A., and Potenza, D. (1989) In: Microbiology of Extreme Environments and the Potential for Biotechnology (Da Costa, M.S., Duarte, J.C. and Williams, R.A.D., Eds.), pp. 180–186, Applied Science Publishers, Barking, UK.
[100] Gliozzi, A., Paoli, G., Rolandi, R., De Rosa, M. and Gambacorta, A. (1983) Biochim. Biophys. Acta 735, 234–242.
[101] Kushawaha, S.C., Kates, M. and Martin, W.G. (1975) Can. J. Biochem. 53, 284–292.
[102] Henderson, R., Jubb, J.S. and Whytock, S. (1978) J. Mol. Biol. 123, 259–274.
[103] Teissie, J., Prats, M., Lemassu, A., Stewart, L.C. and Kates, M. (1990) Biochemistry 29, 59–65.
[104] Falk, K.-E., Karlsson, K.-A. and Samuelsson, B.E. (1990) Chem. Phys. Lipids 27, 9–21.
[105] Stewart, L.C., Yang, P.W., Mantsch, H.H. and Kates, M. (1990) Biochem. Cell Biol. 68, 266–273.
[105a] Chen, J.S., Barton, P.G., Brown, D. and Kates, M. (1974) Biochim. Biophys. Acta 352, 202–217.
[105b] Quinn, P.J., Brain, A.P.R., Stewart, L.C. and Kates, M. (1986) Biochim. Biophys. Acta 863, 213–223.
[106] Lind, C., Hojeberg, B. and Khorana, H.G. (1981) J. Biol. Chem. 256, 8298–8306.
[107] Kates, M. (1986) FEMS Microbiol. Rev. 39, 95–101.
[108] Paterek, J.R. and Smith, P.H. (1985) Appl. Environ. Microbiol. 50, 877–881.
[109] Trincone, A., Trivellone, E., Nicolaus, B., Lama, L., Pagnotta, E., Grant, W.D. and Gambacorta, A. (1993) Biochim. Biophys. Acta, in press.

CHAPTER 10

The membrane-bound enzymes of the archaea

Lawrence I. HOCHSTEIN

Ames Research Center, Moffett Field, CA 94035-1000, USA

1. Introduction

Studying membrane-bound enzymes in isolation of their membrane milieu is not without problems. There are numerous examples of how function changes when enzymes are separated from membranes; a phenomenon referred to as allotopy. Removing membrane-bound enzymes with detergents may result in a preparation where it is not possible (or desirable) to remove the detergent, thus affecting subsequent behavior. In addition, it is not clear to what extent the membrane environment moderates the effects that are potentially destructive. The archaea, which are found in ecosystems characterized by extremes of pH, temperature, and ionic strength, contain membranes whose lipids are unlike those found in the Bacteria and Eucarya [1]. This situation raises the possibility of studying how this lipid environment moderates the effects of these extremes on the enzymes associated with the membranes. In addition, these enzymes can serve as models for understanding the molecular changes that account for their stability and function in such unusual environments.

The most thoroughly studied of the archaeal membrane-bound enzymes are the proton-translocating ATPases. Interest in these enzymes comes from the observation that they bear a close relationship to the vacuolar ATPases of the Eucarya [2], and that archaeal ATPases may be more comparable to the "primitive" version of the progenitor proton-translocating ATPase than are the others [3,4]. Less information exists about the electron transport chain in the archaea. Many of the early studies were carried out with *Halobacterium salinarium* [5,6] which probably reflect the relative ease of growing this organism (the names *H. halobium* and *H. cutirubrum* are currently considered to be synonyms of *H. salinarium* [7] and to avoid confusion the original appellations will be used parenthetically). Since that time, extreme halophiles of various physiological types have been described and the nature of their electron systems remains to be characterized. The recent studies of electron transport in *Sulfolobus acidocaldarius* suggest that it is a relatively simple system [8]. It is not known whether similar simplicity occurs in other sulfur-dependent acidophilic thermophiles. The discovery of cytochromes in the methanogens has not gone beyond describing their properties. There remains the question as to their function.

It is becoming increasingly evident that there are a number of enzymes that carry out similar functions yet behave as if they are different entities (i.e., the ATPases and NADH dehydrogenases). To what extent these differences reflect the methods used to isolate the enzymes is not known.

2. Methods

2.1. Preparation of membranes

Any of the methods used to disrupt bacteria are successful when applied to archaeal cells. The only qualifier is that conditions used during breakage take into account the peculiar properties of archaeal proteins (e.g., sensitivity to oxygen; requirement for high salt concentrations).

Sonication of extremely halophilic bacteria produces inverted [9], "inside-in" vesicles, as well as randomly oriented vesicles [10] from *H. salinarium* (*halobium*) [10,11]. In addition, sonication may produce membranes that sediment as an oily, poorly packed, pellet. Sonication is also successful for disrupting *S. acidocaldarius* [12]. Freezing and thawing disrupts halobacterial cells [13–15], and cells of *S. acidocaldarius* [16], and when used in conjunction with passage through a French Pressure Cell is effective for disrupting *Pyrodictium brockii* [17]. A particularly useful method for disrupting archaeal cells is to pass them through a French pressure cell [2,18,19]. Disruption of extreme halophiles at low pressure (e.g., 5 psi) produces vesicles that are predominantly oriented "inside-in" [20]. A common difficulty experienced when disrupting cells at low pressures is the unmanageable viscosity due to the presence of DNA. Incubating extracts with traces of DNAase at room temperature effectively deals with this problem. This is true even in the case of extracts of extreme halophiles where the disrupted cells are suspended in solutions that are 4 M with respect to NaCl.

2.2. Isolation of membrane components

Detergents are commonly used to solubilize membrane proteins (see refs. [21,22] for the properties of detergents and their use for solubilizing membrane proteins). There is no particular detergent or a set of conditions that is optimal. Membrane-bound enzymes can be selectively extracted by sequentially treating membranes with detergents. An interesting example is provided by the solubilization of the ATPase from *H. saccharovorum* using octyl-β-D-glucopyranoside followed by sodium desoxycholate [23]. The cytochrome oxidase from *S. acidocaldarius* can also be selectively extracted from membranes when they are first treated with sodium cholate followed by decanoyl-*N*-methyl glucamide (MEGA-10) [24].

The ATPase from *H. halobium* is more firmly membrane-bound than is the enzyme from *H. saccharovorum*, and solubilization of the former requires treating membranes with EDTA before sonication [25]. EDTA by itself is effective for solubilizing the ATPases from *Methanosarcina barkeri* [26], *Methanolobus tindarius* [27], and *S. acidocaldarius* [28]. On the other hand, the membrane-bound ATPase from *Methanobacterium thermoautotrophicum* requires extraction with EDTA followed by

2 M LiCl [29]. Several of the membrane-bound enzymes in *Sulfolobus* are loosely bound and are solubilized following high pressure decompression [2,30].

Successful solubilization of the membrane components of the extreme halophiles can also be achieved by manipulating the ionic strength of the solutions in which the membranes are suspended. Sequential solubilization of constituents from membranes of *H. cutirubrum* is observed at various salt concentrations [31]. Flavoproteins are removed between 1 and 2.2 M NaCl; cytochrome b between 0 and 1.2 M NaCl; and cytochrome oxidase between 0 and 0.5 M NaCl. The membrane-bound NADH dehydrogenase from an unidentified extreme halophile, strain AR-1, is solubilized when membranes are suspended in buffer that is 2 M with respect to NaCl [32]. The residual membrane-fraction still oxidizes α-glycerophosphate and TMPD-ascorbate, indicating that the respiratory chain upstream to the NADH dehydrogenase is relatively unaffected [Hochstein, L.I., unpublished data]. The NADH dehydrogenase from *S. acidocaldarius* is solubilized after membranes are suspended in 0.5 M KCl [33] whereas the membrane-bound ATPase requires EDTA [28]. Solubilization can effect the structure and function of membrane-bound enzymes. Differences between the membrane-bound and soluble ATPase from *H. saccharovorum* [34] and the hydrogenase from *P. brockii* [35] indicate that this may not be a trivial issue.

Once soluble, the membrane-bound enzymes are much like cytoplasmic enzymes and can be purified using similar methods. Excellent descriptions of such methods can be found in ref. [36]. These methods are only limited by circumstances such as the salt dependence of the proteins from extreme halophiles and the "cold-sensitivity" exhibited by some membrane-bound enzymes when solubilized.

3. The ATPases

Three types of ATP-driven cation pumps can be distinguished on the basis of their structure and their sensitivity to inhibitors. They are the E_1E_2-ATPases, the F_1F_0-proton-translocating ATP synthase, and the vacuolar ATPases which are designated as P-, F-, and V-type ATPases, respectively [37]. Several of the distinguishing characteristics of these enzymes are summarized in Table 1. The F- and V-ATPases can be differentiated by the sensitivity of the former to azide and the latter to nitrate and N-ethylmaleimide (NEM) [38]. In addition, the V-ATPases are exquisitely sensitive to the antibiotic bafilomycin A_1 [39,40].

The membrane-bound archaeal ATPases have attracted considerable interest since patterns of inhibitor sensitivity, subunit structure, and amino-acid sequence homologies suggest a close relationship to the V-type ATPases [3,41]. In addition, the membrane-bound ATPases from *S. acidocaldarius*, *H. salinarium (halobium)*, *Methanosarcina barkeri,* and the V-type ATPase from *Saccharomyces cerevisiae* are immunologically related [42].

V-type and P-type ATPases are present in the archaea. Although the V-like archaeal ATPases have been asserted to be ATP synthases [4,43] the evidence for this is at best inferential [44]. The ATPase from *H. saccharovorum* is proposed to be an F-type ATPase on the basis of the mechanism of ATP hydrolysis [44]. However, there is increasing evidence that the archaea possess a variety of ATPases, and with the possible exception

TABLE 1
Comparison of proton-translocating ATPases[a]

Property	Type of ATPase		
	P	F	V
Molecular mass (kDa)			
Native enzyme	100	500	500
Subunits		α (55)	A (70)
		β (50)	B (60)
		γ (31)	C (41)
		δ (20)	D (34)
		ε (15)	E (33)
		a (30)	a (20)
		b (17)	
		c (8)	c (16)
Stoichiometry		$(\alpha\beta)_3\gamma\delta\varepsilon$	$(AB)_3CDE$
Catalytic subunit		β	A
Inhibitors[b]	vanadate	azide, NBD-Cl, DCCD	nitrate, NEM, NBD-Cl, DCCD, bafilomycin

[a] Adapted from ref. [41].
[b] Abbreviations: NBD-Cl, 7-chloro-4-nitrobenzo-2-oxa-1,3-diazole; NEM, N-ethylmaleimide.

of the enzyme from *H. salinarium* (*halobium*) [45] their functions are unknown [2,46]. There is presumptive evidence for an F-type ATPase in *H. saccharovorum* that is unlike the membrane-bound enzyme previously characterized in that organism [47].

3.1. The ATPases of the methanogens

Methanococcus voltae contains a membrane-bound vanadate-sensitive ATPase [48] that is inhibited by diethylstilbestrol, an inhibitor of eukaryotic P-type ATPases. The purified enzyme is composed of a single subunit (M_r 74 000), forms a covalent acyl-phosphate enzyme intermediate, and is not inhibited by nitrate or bafilomycin [49]. No such ATPase activity has been reported in other archaea. The presence of a second ATPase in *M. voltae* has been inferred since membranes react with antiserum prepared against the β subunit from the V-type ATPase of *S. acidocaldarius* [50]. Two peptides are detected whose M_r values (51 000 and 65 000) correspond to the masses for the two largest subunits of the *S. acidocaldarius* ATPase [51]. There is evidence that ATP synthesis in the *M. voltae* enzyme is due to the operation of a sodium-translocating ATPase [50]. The relationship of the putative V-like ATPase to the sodium-translocating ATPase has not been established.

The best evidence for the existence of an F-type ATPase in the archaea comes from experiments with a methylotropic methanogen [52]. Membranes contain particles that react with polyclonal antiserum against the β subunit of the *Escherichia coli* F-ATPase. In addition, negative staining reveals particles attached to the membranes with a stalk whose appearance is suggestive of an F-type ATPase [53]. Unfortunately, no further studies have been carried out with this ATPase to determine its sensitivity to F-type and V-type ATPase inhibitors.

M. thermoautotrophicum cells synthesize ATP in the presence of an artificially imposed pH gradient [18]. Proton uptake is not detected following acidification. The addition of valinomycin results in the synthesis of ATP and is accompanied by the extrusion of K^+ but not protons. ATP synthesis is unaffected by DCCD and is stimulated by uncouplers such as 2,4-dinitrophenol and *m*-chlorophenyl hydrazone. Membrane vesicles from *M. thermoautotrophicum* synthesize ATP when conditions are anaerobic in response to the membrane potential since the addition of K^+ suppresses synthesis. ATP synthesis is inhibited by 100 µM CCCP and partially inhibited by DCCD (53% at 100 µM). ATP synthesis also takes place in response to a ΔpH produced by the oxidation of hydrogen. In this case, ATP synthesis is inhibited by 10 µM DCCD and CCCP. Unlike cells, vesicles do not synthesize ATP in response to an artificially imposed ΔpH or in the presence of valinomycin [54]. *M. thermoautotrophicum* membranes have an ATPase activity that hydrolyzes ATP, GTP, and UTP at approximately the same rate. The enzyme loses activity at $-90°C$ which is due to aggregation, and activity is restored following sonication. ATPase activity is partially inhibited by DCCD (40% at 100 µM) when membranes are incubated at pH 8 for 10 min [18]. A similar ATPase is found in a different strain of *M. thermoautotrophicum* [29]. The enzyme is most active at an alkaline pH and it is not significantly inhibited by ADP, 5 mM NEM, or 150 µM DCCD. The absence of NEM inhibition suggests that the enzyme may not be a V-type ATPase.

M. barkeri contains a membrane-bound sulfite-activated ATPase that is most active at pH 5.2 and is inhibited some 87% by 10 µM DCCD [26]. As with the ATPases from other methanogens, the enzyme from *M. barkeri* is not sensitive to oxygen. ATP hydrolysis is not significantly inhibited by 7-chloro-4-nitrobenzo-2-oxa-1,3-diazole (NBD-Cl), azide, or NEM. The native enzyme (M_r 420 000) is composed of two subunits, designated as α (M_r 62 000) and β (M_r 49 000). The native enzyme dissociates into its subunits when incubated at $-50°C$ and the subunits can subsequently be separated by high-performance liquid chromatography [55]. Trypsin digestion of the α subunit from *M. barkeri* results in fragments that are not detected in a 14% acrylamide gel. When digestion is carried out in the presence of either ADP or ATP, the major product is an M_r 54 000 fragment. More of this material accumulates when trypsin digestion is carried out in the presence of Mg^{2+} and nucleotides. Trypsin digestion of the β subunit results in the production of several smaller peptides (between M_r 30 000 and 20 000), but in this case nucleotides (with or without Mg^{2+}) have no effect on the extent of digestion. These observations imply that the α subunit is the catalytic subunit (as is the case with V-ATPases).

The membrane-bound ATPase from *M. barkeri* is inhibited by DCCD and the inhibitor is bound to a small hydrophobic peptide (M_r 6000). A DCCD-sensitive ATPase, solubilized by octylglucoside, has been purified in the presence of the detergent. This form of the ATPase contains six subunits (M_r 62 000, 49 000, 40 000, 27 000, 23 000, and 6 000). When incubated with ^{14}C-DCCD, all the radioactivity is associated with the M_r 6000 subunit. [56]. This subunit is smaller than the DCCD-binding peptides from *S. acidocaldarius* [57] and coated vesicles [58]. The function of the M_r 6000 peptide has yet to be established, although its properties suggest that it may be equivalent to the subunit c, the DCCD-binding peptide of F-type ATPases.

The membrane-bound ATPase from *M. tindarius* has a pH optimum of 5, hydrolyzes a variety of nucleotides (ATP ≫ GTP, ITP > CTP, UTP) and is not activated by sulfite [27]. The activity in the presence of Mg^{2+} is greater than with Mn^{2+}, while Zn^{2+} and Ca^{2+} support a low level of activity. The response to divalent cations and the absence of sulfite activation distinguishes this enzyme from the halobacterial [19], the *S. acidocaldarius* [12,28], and the *M. barkeri* [26] ATPases, respectively. The enzyme from *M. tindarius* (M_r 445 000) is inhibited by nitrate, but not azide or vanadate, and is composed of four subunits designated α–δ in order of decreasing relative molecular masses (M_r 67 000, 52 000, 20 000, and <10 000). The β subunit reacts with an antiserum directed against the β subunit from the *S. acidocaldarius* ATPase. The membrane-bound ATPase is most sensitive to DCCD when incubated at pH 6.5. A ratio of 260 nmol DCCD per mg membrane protein is required to effect 50% inhibition. The enzyme is also inhibited at pH 5.0 and 8.5 but requires higher DCCD/membrane-protein ratios. Incubating the membranes at pH 6.5 with ^{14}C-DCCD results in the labeling of an M_r 5 5000 proteolipid. Incubating membranes at either pH 5.0 or 8.9 resulted in no incorporation of radioactivity in the proteolipid (even though ATPase activity is inhibited).

3.2. The ATPases of Sulfolobus

Several membrane-bound ATPases occur in the genus *Sulfolobus*. There are two ATP-hydrolyzing activities in *S. acidocaldarius* strain 7. One has a pH optimum at 6.5 in the absence of sulfate, and the presence of that anion activates the enzyme and shifts the pH optimum to 5.0. ATP hydrolysis is unaffected by DCCD, azide, NEM, *p*-hydroxymercuribenzoate, or vanadate [59]. The other ATPase is most active at pH 2.5, is inhibited by sulfate, and appears to be a pyrophosphatase [16]. The purified sulfate-activated ATPase (M_r 360 000) is composed of three subunits (M_r 69 000, 54 000, and 28 000). It is most active at 85°C, stimulated some three-fold by sulfate, sulfite, and bicarbonate, but is unaffected by chloride. There are two pH optima. One is located at pH 5 and the other at pH 8.5 and neither is affected by sulfite. ATPase activity is inhibited by nitrate (63% at 20 mM) and NBD-Cl (90% at 1 mM) but is not significantly affected by azide (5 mM), vanadate (100 µM), and NEM (100 µM) [28].

Another membrane-bound ATPase is found in a different strain of *S. acidocaldarius* (DSM 639) [12]. The enzyme also has two pH optima (pH 5.5 and 8) when assayed without sulfite. However, the presence of sulfite enhances enzyme activity and produces a single pH optimum which is located at about pH 6.3. This activity represents the enhancement of activity at the pH minimum when the enzyme is assayed in the absence of sulfite. There is no activation at pH 8, and about a 1.4-fold stimulation occurs at pH 5.5 in the presence of sulfite. The enzyme is unaffected by azide or vanadate but is inhibited by nitrate and *p*-hydroxymercuribenzoate [51]. The enzyme (M_r 430 000) is composed of two subunits (M_r 65 000 and 51 000) that are designated the α and β subunits. Antiserum prepared against the β subunit reacts with membrane extracts from *S. solfataricus*, *Acidianus brierlyi*, *A. infernus*, *P. occultum*, *Staphylothermus marinus*, *M. thermoautotrophicum*, *M. tindarius* and *M. barkeri* [51]. The reaction is with a subunit with an M_r between 51 000 and 56 000. An ATPase with a different subunit composition (but similar with respect to other properties) is obtained when pyrophosphate is used to remove the enzyme from membranes [60]. The M_r is slightly lower (380 000) and the enzyme is composed of four subunits (M_r 65 000, 51 000, 20 000, and 12 000). Electron microscopy of the *S. acidocaldarius* ATPase reveals particles that have a hexagonal arrangement with a central structure that is reminiscent of the F_1 moiety of F-type ATPases. The antiserum against the β-*Sulfolobus* subunit gives equally strong reactions with the β subunits from the F-type ATPases of *E. coli, Propionigenium modestum,* spinach chloroplasts, beef-heart submitochondrial particles, and subunit II from the ATPase from *H. saccharovorum*. There is a weak reaction with subunit B of V-ATPases from chromaffin granules, coated vesicles, *Saccharomyces cerevisiae* vacuoles, but not the plasma membrane P-type ATPase from *Schizosaccharomyces pombe* [60].

Amino-acid sequences for the α and β subunits from *S. acidocaldarius* (strain 7) have been deduced from the nucleotide sequences [61,62]. Neither of the sequences is closely homologous to the α and β subunits of F-type ATPases, but they are 50% identical to sequences of subunits A and B from the V-type ATPases [3]. These homologies suggest that archaeal and V-type ATPases share a common ancestral V-type ATPase [41]. Implicit in this notion is that archaeal ATPases synthesize ATP [4]. A more complete picture of the *Sulfolobus* ATPase is provided by the structure of the ATPase operon in *S. acidocaldarius* (strain 7). At least five genes have been detected. Besides those coding for the α and β subunits, three additional genes that code for subunits designated as γ, δ, and ε (M_r 25 000, 13 000, and 7 000), are present. It is interesting that the order of the genes differs from the order in F-type operon from bacteria [63]. The molecular masses of the subunits as deduced from the amino-acid compositions are in good agreement with the values obtained by SDS-PAGE [28,60]. The M_r 7000 subunit is distinct from the membrane-associated proteolipid (subunit c) that reacts with DCCD [64].

A tenuously membrane-bound V-type ATPase is found in *S. solfataricus* [2]. The enzyme occurs in the membrane and cytoplasmic fractions following high-pressure decompression and probably is related to the force used to disrupt the cells. The ATPase (M_r 370 000) is composed of three subunits (M_r 63 000, 48 000, and 24 000). These values are in relatively good agreement with the values for *S. acidocaldarius* ATPase. The *S. solfataricus* enzyme is inhibited by NEM but the presence of ATP does not affect the inhibition. NEM is bound to the largest subunit although traces are associated with the M_r 45 000 subunit. Nitrate, NBD-Cl, and *p*-chloromercuriphenyl sulfonate are also inhibitory. The inhibition by the latter is reversed by cysteine, which also stimulated ATPase activity, and suggests that thiols are located at or near the substrate binding site that is associated with the largest subunit. The properties of the three *Sulfolobus* enzymes suggest that while the enzymes are structurally similar, their collective properties are sufficiently different (Table 2) to warrant considering them to be different ATPases. Implicit in this suggestion is that they are ATPases of unknown function.

S. acidocaldarius (DSM 639) extrudes protons and synthesizes ATP while respiring endogenous substrates [65]. DCCD inhibits this ATP synthesis and the inhibitor binds to a membrane proteolipid [57]. This proteolipid (M_r 34 000) can be purified using procedures for isolating DCCD-binding proteolipids [66]. Two subunits are observed after SDS-electrophoresis (M_r 17 000 and 19 000) but a single peptide (M_r 6000) is detected when SDS-electrophoresis is carried out in the presence of 8 M urea. The amino-acid sequence from the N-terminal end of an M_r 2000 CNBr-peptide exhibits considerable homology to sequences of subunit c from the F_0 of *E. coli* and the thermophilic bacterium, PS3 [64].

TABLE 2
Comparison of the *Sulfolobus* ATPases

Property	Hochstein [2]	Konishi [28]	Lübben [12]
pH optimum			
Without sulfite	6.5–7.2	5.0; 8.5	5.3; 8.0
With sulfite		6.5–7.2	5.5–7.0
Effect of			
Sulfate	inhibits	stimulates	inhibits
Chloride	inhibits	no effect	stimulates
Bicarbonate	no effect	stimulates	not reported
NEM	inhibits	no effect	not reported
Substrate specificity	ATP=GTP	GTP ≫ ATP	ATP > GTP
Cation specificity	Mg=Mn	not reported	Mn > Mg

The nucleotide sequence derived from the gene for the DCCD-reactive proteolipid from *S. acidocaldarius* (strain 7) indicates that the peptide has an M_r of about 10 000 [67]. A partial amino-acid sequence derived from the nucleotide sequence shows considerable homology with subunit c of F-ATPases but not the M_r 16 000 proteolipid of V-type ATPases. It is proposed that the proteolipid associated with V-ATPases is a product of a gene that duplicated after the V- and archaeal ATPases had differentiated [67].

3.3. *The ATPases of Thermoplasma*

Thermoplasma acidophilum membranes hydrolyze a variety of phosphate esters most actively at pH 4.6. ATPase activity is stimulated by sulfate and is unaffected by DCCD, azide, or nitrate. Cells extrude sulfate when the intracellular compartment is acidified, and it is suggested that the enzyme functions as a sulfate-translocating ATPase [46].

3.4. *The ATPases of the extreme halophiles*

The membrane fractions of various extreme halophiles contain a cryptic ATPase activity that is activated by detergents including Triton X-100 [68]. The activity, even in the presence of detergent, is relatively low, with values of 3–24 nmol of P_i produced per minute and per mg protein. Since such membranes are predominantly in the "inside-in" orientation [20,69], detergents most likely dissipate the permeability barrier and make ATP accessible to the enzyme. However, there is an additional effect since once solubilized the ATPase is still activated by detergents [19].

The membrane-bound *H. saccharovorum* ATPase can be purified to a specific activity of 3.6 μmol P_i/min/mg protein [46]. The native enzyme (M_r 350 000) is composed of four subunits. Molecular-mass determinations in the presence of sodium dodecylsulfate (SDS) provide M_r estimates of 87 000, 60 000, 29 000, and 20 000 for the subunits, designated I, II, III, and IV, respectively [23]. SDS gel electrophoresis overestimates the molecular masses of acidic proteins [70] and more accurate values can be obtained by electrophoresis in the presence of cetyltrimethylammonium bromide [71]. When this is done, the resulting values for M_r agree with the those obtained using SDS [46].

The membrane-bound and purified enzymes lose little activity over a 2–3 month period when stored at room temperature in the presence of 4 M NaCl. The purified ATPase is rapidly inactivated at 4°C and this is due to the dissociation of the enzyme [23]. The *H. saccharovorum* ATPase is more active in NaCl than in KCl, and most active when assayed between pH 9 and 10 in the presence of at least 3.5 M NaCl [19]. The membrane-bound enzyme rapidly loses activity when stored in 400 mM NaCl but is considerably more stable when either spermine (30 mM) or $MgCl_2$ (100 mM) is present [34]. The purified ATPase is rapidly inactivated when stored in 1 M NaCl but is as stable in 500 mM NaCl/1 M $(NH_4)_2SO_4$ as in 4 M NaCl. The halobacterial ATPase is not inhibited by vanadate, azide, or Dio-9 [19] but is inhibited by DCCD when the enzyme is preincubated at pH 6 or less and at relatively high quantities of DCCD (125 nmol/mg protein). When membranes are incubated with ^{14}C-DCCD, virtually all the radioactivity is associated with subunit II [19]. The conditions of inhibition are similar to those that inhibit F_1 activity [72–74]. This suggests that subunit II may be similar to the β subunit of F-type ATPases, an assumption consistent with the immunological reactivity of the subunit. Antiserum against the β-subunit from the ATPase from *S. acidocaldarius* which reacts with the β subunit from F-type ATPases [51] also reacts with subunit II from *H. saccharovorum* [60].

However, the following evidence indicates that *H. saccharovorum* ATPase is not an F-type ATPase. The amino-acid compositions of subunits I and II differ considerably from those of the α and β subunits from the *E. coli* F-type ATPase, and a comparison of chymotrypsin peptide maps of subunits I and II and the α and β subunits from *E. coli* confirm that the sequences differ [46]. These differences are significant since the amino-acid composition and sequences of the β subunit from various F-ATPases are highly conserved [75,76]. Antiserum prepared against the largest of the V-type ATPase subunits from *Neurospora crassa* (subunit A) reacts with subunit I from the *H. saccharovorum* ATPase. The *H. saccharovorum* enzyme is inhibited by nitrate and NEM, and inhibition by the latter is attenuated when either ADP or ATP is present [77]. NEM binds to subunit I, and less is associated with the subunit when ATP or ADP are present [78]. In the case of *N. crassa*, NEM inhibits the V-ATPase where it binds primarily to subunit A. Binding is also diminished when inhibition is carried out in the presence of ATP or ADP, and this is taken to suggest that the ATP binds with subunit A [79]. The ATPase from *H. saccharovorum* is inhibited by NBD-Cl [19], and adenine nucleotides prevent binding [77]. The inhibitor binds to subunits I and II although more of the inhibitor is associated with subunit II. The apparent discrepancy between the labeling patterns observed with NEM and NBD-Cl has not been resolved.

ATP hydrolysis is non-linear with respect to time. This is not due to enzyme inactivation since progress curves, where the product of variable enzyme concentration and time is plotted against the amount of product, yield points that fall on a single line. This is a response expected if a competitive inhibition occurs during catalysis [80] and suggests an explanation for non-linearity since ADP is a competitive inhibitor with respect to ATP [19]. However, the presence of an ATP-generating system, which increases the rate of phosphate production by a factor of three, does not result in a linear rate [81].

The membrane-bound ATPase from *H. saccharovorum* exhibits different properties when solubilized with Triton X-100 and subsequently purified in the presence of 4 M NaCl [82]. The specific activity following purification is only 0.86 μmol P_i/min/mg protein). The enzyme is activated by various detergents, with Triton X-165 being the most effective. The enzyme is most active when assayed in the presence of 3.5 M KCl, and significant ATPase activity is detected when the enzyme is assayed in the presence of 1.5 M Na_2SO_4. The enzyme (M_r 450 000) is composed of five subunits (M_r 110 000, 71 000, 31 000, 22 000, and 14 000) although subsequent values of 98 000 and 71 000 were reported for the two largest subunits [83]. The M_r values for the two largest subunits differ significantly from those reported previously [23] which were obtained using the Laemmli method [84]. The former values were obtained using the procedure described by Neville [85]. The method is extremely hyperbolic in the very range of M_r where maximum disagreement with the earlier values occurs. It is reasonable to assume that small changes in relative mobility could result in large changes in M_r.

ATP hydrolysis by the *H. saccharovorum* ATPase prepared by modified procedure is also non-linear with respect to time. The non-linearity is attributed to the presence of two rates that, in turn, reflect the existence of two forms of the enzyme. One (E_1) is highly active but unstable and is rapidly and irreversibly converted during catalysis to a less active but more stable form (E_2). This model explains the action of nitrate as being due to two effects: it increases the rate at which E_1 is converted to E_2 and inhibits the activity of the latter [82]. This is a different explanation than what is proposed for the V-ATPase from *N. crassa*. In this case, nitrate acts by causing the release of subunits peripheral to the membrane in a process that occurs only when nucleotides are present [86]. An analysis of ATP hydrolysis suggests that the *H. saccharovorum* ATPase is an F-ATPase [44]. Azide (an inhibitor of F-type but not V-ATPases) inhibits ATPase activity where it acts at the level of E_2 activity. However, the effect requires relatively high concentrations of azide (30% inhibition at 300 mM azide). Sulfite affects the *H. saccharovorum* ATPase by decreasing the rate at which E_1 is converted to E_2, as well as enhancing the activity of the E_2 form of the enzyme. In addition, sulfite reverses the inhibition observed in the presence of nitrate and relieves ADP inhibition. Subsequent studies on the nucleotide binding sites indicate that a high-affinity site is located on subunit I [83] and confirm results obtained with NEM [78] that nucleotide binding sites are associated with that subunit. A non-catalytic nucleotide-binding site is also present in subunit II that probably corresponds to the site that binds the small quantity of NEM [78] as well as the NBD-Cl that binds to that site [77]. The site on subunit II is proposed to be physiologically insignificant [83]. This is contrary to the observation that the DCCD inhibition of ATPase activity is associated with the covalent binding of the inhibitor to subunit II [68] with approximately 1.2 mol DCCD bound per mol of subunit [Stan-Lotter, H. and Hochstein, L.I., unpublished data].

An ATPase isolated from *H. salinarium (halobium)* superficially resembles the enzyme from *H. saccharovorum*. The purified enzyme (M_r 320 000) is composed of two subunits (M_r 86 000 and 64 000) but there is no evidence that other subunits are associated with the enzyme [25]. The purified enzyme is most active when assayed in the presence of 2 M $(NH_4)_2SO_4$ and is about twice as active when assayed in the presence of 1.5 M Na_2SO_3. Little if any activity is detected when the enzyme is assayed in the presence of 4 M NaCl or KCl. ATPase activity is not affected by azide (1 mM) but is inhibited by NEM (1 mM), DCCD (300 μM), and NBD-Cl (500 μM). The inhibition by NBD-Cl is antagonized by ADP. The enzyme from *H. salinarium (halobium)* rapidly loses activity when incubated at room temperature in absence of NaCl and slowly

TABLE 3

Comparison of the ATPases from *H. saccharovorum* and *H. salinarium*

Property	ATPase from	
	H. saccharovorum [19,23]	*H. salinarium* [25]
No. of subunits	4	2
Substrate specificity	ATP=GTP > ITP > CTP	ITP=ATP > GTP > CTP
pH optimum	9–10	5.5 6
Activation by		
Detergent	+	–
Chloride	+	–
Sulfate	–	+

when incubated in the presence of 4 M NaCl. The ATPase is more stable in a mixture of 1 M $(NH_4)_2SO_4$ and 1 M Na_2SO_4. The enzyme loses activity when incubated at 4°C in 4 M NaCl as a result of dissociating into its subunits. The ATPases from *H. saccharovorum* and *H. salinarium* (*halobium*) appear similar with respect to the M_r of the native enzymes and the two largest subunits [46] and the stability of both enzymes in 1 M $(NH_4)_2SO_4$ [34]. However, the enzymes differ in a number of significant properties that are summarized in Table 3.

The properties of the *H. salinarium* (*halobium*) enzyme are those expected of a V-type ATPase. Polyclonal antibodies prepared against the *H. salinarium* (*halobium*) ATPase react with the two largest subunits from the ATPases from *S. acidocaldarius* and beet root tonoplast, but fail to react with the subunits from chloroplast F-ATPase [87]. This result contradicts the observation that polyclonal antiserum against the β subunit from *S. acidocaldarius*, which reacts with various β subunits from F-type ATPases, reacts with subunit II from *H. saccharovorum* and a M_r 67 000 subunit from the *H. salinarium* (*halobium*) ATPase [60].

The amino-acid sequences of the two subunits from the *H. salinarium* (*halobium*) ATPase have been determined [88]. The largest subunit exhibits 63% and 49% sequence homology with the largest subunit from the *M. barkeri* and *S. acidocaldarius* ATPases, respectively. The sequence of the smaller subunit is 66% and 55% homologous with the "β" subunits of these organisms. The homology between the two subunits from the *H. salinarium* (*halobium*) ATPase and the α and β subunits from F-type ATPases is only 30%, whereas the sequences are better than 50% homologous when compared to the sequences of subunits A and B from V-type ATPases. The M_r values of the two subunits (64 104 and 51 956), as deduced from their amino-acid compositions, are considerably less than those obtained by SDS-PAGE [25]. These differences reflect the unreliability of determining the size of acidic proteins in the presence of SDS [71]. When the subunits are amidated, they migrate at rates consistent with smaller relative molecular masses and agree well with values inferred from the amino composition of the *S. acidocaldarius* ATPase subunits [61,62].

4. The electron transport system

As is true of the ATPases, the early and most intensive studies on the respiratory systems of the archaea were carried out in the extreme halophiles. These studies focused on pathways associated with oxygen-linked respiration. It is surprising that the extremely halophilic bacteria were first viewed as being obligately aerobic organisms even though the ecological niches they inhabit are oxygen-limited [88]. Subsequent observations that retinal-containing strains grow phototropically under anaerobic conditions [89]; that arginine fermentation supports anaerobic growth [90]; that representatives of the genera *Halobacterium, Haloferax,* and *Haloarcula* grow anaerobically in the presence of dimethylsulfoxide, trimethylamine N-oxide [91] and fumarate [92]; that fermentative growth is widespread among the extreme halophiles [93]; and that denitrification is a common mode of respiration in *Haloferax* and *Haloarcula* [94] dispels that view. The enzymology associated with growth under these conditions is yet be elucidated. The electron transport system has been studied in the various strains of *H. salinarium (halobium)* [5,6,95–97]. Membranes contain dehydrogenases for NADH, succinate and α-glycerophosphate (α-GP), a complex of b-type cytochromes and oxidases. The participation of quinones during electron transport has not been demonstrated, but since menaquinones with eight isoprene units are the predominate quinone in the extreme halophiles [98], presumably they are involved during electron transport.

Members of the Sulfolobaceae (*Sulfolobus* and *Acidianus*) are facultative heterotrophs that can oxidize hydrogen sulfide to elemental sulfur and the latter to sulfuric acid. Nothing is known about the enzymology of these processes and to what extent membrane-bound enzymes are involved. What information there is relates to the electron transport system of *Sulfolobus* which is relatively simple consisting as it does of dehydrogenases for succinate and NADH, a quinone pool, a complex of b-cytochromes, and several oxidases.

4.1. NADH oxidases

4.1.1. The NADH oxidase of Sulfolobus

Membranes from *S. acidocaldarius* (strain 7) catalyze a slow oxidation of NADH. The activity is not significantly affected by antimycin A, azide, or cyanide whereas the oxidation of reduced horse-heart cytochrome c is inhibited by cyanide and azide [16]. Since NADH oxidase activity was assayed by the disappearance of NADH rather than oxygen consumption, this discrepancy probably may reflect an assay that measured the NADH dehydrogenase rather than the oxidase.

The addition of NADH to membranes from *S. acidocaldarius* (DSM 639) results in the reduction of cytochromes. Antimycin A does not affect cytochrome reduction, while the absence of complete reduction in the presence of cyanide suggests the presence of a branched electron transport chain [65].

4.1.2. The NADH oxidase of the extreme halophiles

Membranes prepared by high-pressure disruption of *Halobacterium* have an NADH oxidase as judged by the presence of the cyanide-sensitive oxidation of

NADH [6,99]. The NADH oxidase from an extremely halophilic bacterium, strain AR-1, is most active at high solute concentrations. This salt dependency, in part, reflects the dissociation of the NADH dehydrogenase from the membrane [100]. The cation requirement for oxidase activity is relatively non-specific when the cations are added as the chloride salts. NADH oxidase activity is profoundly affected by the nature of the anion. The order of decreasing effectiveness (as the Na^+ salt) mimics the lyotropic series [101]:

citrate > sulfate > acetate > chloride > nitrate > bromide > iodide,

the significance of which relates to the enhancement of hydrophobic interactions [102].

In *H. salinarium* (*cutirubrum*), NADH is oxidized through c-type cytochromes and an cytochrome a oxidase. 2-heptyl-4-hydroxyquinoline-N-oxide (HOQNO) inhibits electron transport chain at the level of the NADH dehydrogenase. The salt dependence of the NADH oxidase system indicates that the salt dependence of the NADH oxidase system is a reflection of the sensitivity of the NAD dehydrogenase [95].

4.2. NADH dehydrogenases

4.2.1. NADH dehydrogenase from Sulfolobus

The NADH dehydrogenase activity of *S. acidocaldarius* (strain 7) is distributed between the soluble and membrane fractions when cells are disrupted by repeated freezing and thawing [59]. The electrophoretically homogeneous enzyme (M_r 95 000) is composed of two subunits (M_r 50 000). The pH optimum with DCIP as the acceptor is 7.5, whereas the optimum with ferricyanide (the most active acceptor) is pH 4.5. The increase in dehydrogenase activity at pH 4.5 is not due to an effect on V_{max} but to a decrease in the apparent K_m for ferricyanide (from 1.5 mM to 140 μM). The enzyme contains 1.4 mol of flavin per mol and is assumed to be an FAD-containing flavoprotein since the addition of FAD enhances enzyme activity when DCIP is the oxidant. Caldariella quinone is reduced at about 2% of the rate observed in the presence of ferricyanide [33]. Since NADH oxidation is a major source of the reducing equivalents that flow through the caldariella quinone pool in *S. acidocaldarius* [8], it would be interesting to determine whether the in-vivo rate of quinone turnover during NADH oxidation is consistent with the operation of the electron transport system.

4.2.2. NADH dehydrogenase of extreme halophiles

Membranes of extremely halophilic bacteria oxidize NADH in the presence of a variety of redox dyes, including menadione [103] and DCIP [100]. The enzyme, which is located on the inner aspect of the cytoplasmic membrane [104], has been used as a marker of vesicle orientation [20]. However, a second marker should be used in conjunction with menadione reductase as there are indications that the location of the dehydrogenase may be randomized during membrane preparation [105]. A similar situation occurs with respect to the ATPase activities in *Micrococcus lysodeikticus* [106] and *E. coli* [75,107] as well as the NADH and succinate dehydrogenases from *E. coli* [75,108]. The orientation of the NADH dehydrogenase in the case of *H. saccharovorum* does not correspond to the "in-side in" orientation of the membrane-bound ATPase as determined by the susceptibility of these activities to proteolytic digestion with trypsin [69].

The NADH dehydrogenase from the extreme halophile strain AR-1 is unstable at low salt concentrations. The rate of inactivation is considerably reduced by mM concentrations of $MgCl_2$ and μM concentrations of spermine [101]. Inactivation is also retarded by 100 μM NADH. Significant reactivation takes place when the partially inactivated (but not completely inactivated) enzyme is incubated at room temperature in the presence of 3 M NaCl. This suggests that the enzymatically active form of the enzyme is in equilibrium with an inactive form (or forms) and it is the latter form that is irreversibly denatured. An analogous model is proposed to account for the inactivation of the menadione reductase from *H. salinarium* (*cutirubrum*) [103]. The NADH dehydrogenase [32] contains FMN as the major flavin although 15% of the total flavin is FAD. An M_r of 60 000 is obtained from gel filtration compared to a value of 50 700 obtained on the basis of the FMN content. Maximum activity occurs in the presence of 2.5 M NaCl, and the extrapolated rate is 43% of the maximum in the absence of NaCl. The enzyme reduces DCIP, menadione, and 5-hydroxy-1,4-naphthoquinone but not several other naphthoquinone, ubiquinone analogs, horse-heart cytochrome c, or ferricyanide. A different NADH dehydrogenase is present in *H. salinarium* (*cutirubrum*) [103]. This dehydrogenase, which is light sensitive, is most active in the presence of 1.5 M NaCl when menadione is the acceptor. Cations do not affect the activity of the enzyme whereas anions enhance the activity and stability. Their effectiveness follows a lyotropic series and suggests that high salt concentrations act by stabilizing hydrophobic interactions. The ability of polyvalent cations such as spermine to stabilize the enzyme is due to counter-ion shielding while the enzyme is in a partially unfolded conformation [109]. The NADH dehydrogenases from *H. salinarium* (*cutirubrum*) and strain AR-1 are most likely different enzymes since the latter is most active in 500 mM NaCl when menadione is the acceptor, is not sensitive to light, and has a lower pH optimum.

The salt-dependent properties of the NADH dehydrogenase from strain AR-1 provide a different view of the effect of salts on halophilic enzymes. Kinetic analysis [110] reveals that the effect of decreasing the NaCl concentration on the observed velocity is to increase the apparent K_m for DCIP but is without significant effect on V_{max} or the apparent K_m for NADH. Replacing NaCl with the chloride salts of the other alkali metals is without effect on any of the parameters. Enzyme activity is affected by the nature of the anion. The effect is at the level of the apparent K_m for NADH, the value of which at any concentration is smallest in the presence of sulfate and largest in the presence of nitrate. These results suggest that salts do no activate the enzyme in the sense of affecting V_{max} but affect the enzyme at the level of the substrate binding sites. Cations neutralize the charges at or near the DCIP binding site allowing the approach of the anionic form of the dye whereas anions affect the conformation of the enzyme by altering the accessibility of the NADH binding site to the bulk solvent.

4.3. Succinic dehydrogenases

4.3.1. Succinic dehydrogenases of Sulfolobus
The succinate dehydrogenase of *S. acidocaldarius* (DSM 639) is located in the cytoplasmic and membrane fractions when cells are disrupted either by sonication or decompressive disruption. About 10–30% of the activity is associated with the membrane fraction [30]. The purified membrane-bound succinate dehydrogenase activity (M_r 141 000) consists of four subunits (M_r 66 000, 31 000, 28 000, and 12,800). The enzyme contains a covalently-bound flavin as well as iron and acid-labile sulfide but no cytochrome [111]. The dehydrogenase reduces the following acceptors (listed in order of decreasing V_{max}):

TMPD > phenazine methosulfate/DCIP=DCIP, ferricyanide ≫ caldariella quinone.

The rate with the quinone is less than 1% of the rate with TMPD so that it would be interesting to determine whether the in-vivo rate of quinone turnover during succinate oxidation is consistent with the operation of the electron transport system. Although the *Sulfolobus* enzyme is composed of four subunits, there is no evidence that all are associated with the enzyme other than that they copurify with enzyme activity. The succinate dehydrogenase from eucarya and bacteria have two subunits whose M_r are approximately 70 000 and 28 000 [112]. They appear to be analogous to the M_r 66 000 and 31 000 subunits from the *Sulfolobus* succinic dehydrogenase. The M_r 66 000 subunit from *Sulfolobus* is catalytically active by itself, and like the M_r 70 000 subunit contains covalently-bound FAD. Antisera against the M_r 66 000 *Sulfolobus* subunit cross-reacts with an M_r 67 000 constituent in membranes from *T. acidophilum, S. solfataricus*, beef-heart submitochondrial particles, and *Bacillus subtilis*.

4.3.2. Succinate dehydrogenase from the extreme halophiles
Crude extracts from *H. salinarium* prepared by sonication contain succinate dehydrogenase activity with about 70% associated with the membrane fraction [113]. From 80 to 95% of the total succinate dehydrogenase activity is associated with the membrane fraction following disruption by freezing and thawing. The specific activity of the membrane-bound enzyme is from 12.5 to 50 times greater than the activity of the cytoplasmic enzyme [114]. The membrane-bound and cytoplasmic forms of the enzymes differ with respect to the apparent K_m for succinate and the ability of the membrane-bound form of the enzyme to reduce DCIP in the absence of phenazine methosulfate. The M_r (90 000) is the same when determined by gel exclusion chromatography or ultracentrifugation using malate and isocitrate dehydrogenases from *H. marismortui* and *H. cutirubrum* as internal standards. This value is in reasonable agreement with the M_r for the enzyme from beef-heart mitochondria [115]. The relationship of the minor quantity of succinic dehydrogenase found in the cytoplasmic fraction to the membrane-bound enzyme is not established.

4.4. The cytochromes

4.4.1. The cytochromes of the methanogens

Cytochromes have been identified in methanogenic bacteria that use methanol, methylamines, or acetate but not in those that grow only with H_2 and CO_2 or formate. The cytochromes are associated with the membrane fraction and spectral characteristics suggest they are b- and c-type cytochromes [116]. A CO-binding b-type cytochrome, characterized by an α band located at 559 nm and the protoheme, is present in *M. barkeri* [117]. Low-temperature and derivative spectroscopy of this and other strains of *Methanosarcina* show a diverse cytochrome composition. Cells grown with methanol contain two b-type and a c-type cytochrome. The b-type cytochromes have a very low mid-point potential (-325 mV and -183 mV). A covalently-bound heme is present in membranes after they are extracted with HCl-acetone. The pyridine hemochromogen has a dithionite reduced minus oxidized spectrum with an absorption maximum at 551 nm. Low-temperature dithionite-reduced minus oxidized spectra have α bands at 553, 556, and 565 nm, which indicate the presence of three b-type cytochromes. Derivative spectra also indicate the presence of three absorption bands and suggest that the absorption band of cytochrome c is buried within the 553 nm absorption band [118]. The function of these cytochromes is not clear. Since only those methanogens that use methanol, various methylamines, or acetate contain cytochromes, the cytochromes are proposed to be involved in oxidation of the methyl group [116,119].

4.4.2. The cytochromes of Sulfolobus

Membranes from *S. acidocaldarius* (DSM 639) contain a malonate-sensitive succinate oxidase that is inhibited by cyanide [30]. The difference spectrum of membranes is indicative of a b-type cytochrome and o- and a-type cytochrome oxidases but provides no indication of a c-type cytochrome. NADH partially reduces the cytochromes when cyanide is present, indicating the presence of a cyanide-resistant pathway [65].

A cytochrome b (M_r 30 000) from *S. acidocaldarius* (DSM 639) contains a single copper atom and has an α band located at 558 nm when determined at room temperature and a split α band (553 and 562 nm) when measured at liquid-nitrogen temperatures. This cytochrome reacts with CO and may function as an o-type cytochrome oxidase. Solubilization results in the loss of catalytic activity with either TMPD-ascorbate or caldariella quinone as a consequence of either the loss of lipids or polypeptides that constituted a integral part of the oxidase system.

There is another oxidase present in *S. acidocaldarius* (DSM 639) [120], a cytochrome aa_3 oxidase (α peak at 605 nm). The cytochrome oxidizes horse-heart cytochrome c and TMPD-ascorbate. The ability to oxidize cytochrome c is lost during purification, whereas the oxidation of TMPD-ascorbate increases. The purified cytochrome (M_r 120 000) is composed of a single subunit (M_r 38 000) and contains two heme molecules as well as two copper atoms per subunit. TMPD-ascorbate oxidation is inhibited by cyanide, sulfide (1–3 mM) and azide (20 mM). The purified enzyme also oxidizes reduced cytochrome c but at too slow a rate to be physiologically significant. The absence of cytochrome c in *S. acidocaldarius* and the loss of the ability to oxidize cytochrome c upon purification indicates this cytochrome is not a cytochrome c oxidase. The cytochrome oxidizes the indigenous quinone found in the membranes, caldariella quinone. Oxidation is inhibited by cyanide which suggests that the physiological role of the enzyme is that of a quinol oxidase [121]. An overview of the electron transport chain in *S. acidocaldarius* (DSM 639) is that it consists of NADH and succinate dehydrogenases which funnel electrons into a caldariella quinone pool after which electron flow is partitioned between a pathway involving the quinol cytochrome aa$_3$ oxidase and a second pathway where electrons pass through b- and o-type cytochromes [8]. At this time, it is not clear what controls the flow of electrons through the branched chain.

S. acidocaldarius (strain 7) contains a cyanide-sensitive cytochrome oxidase [24]. The purified cytochrome (M_r 150000) is composed of three subunits (M_r 37000, 23000, and 14000). Difference spectra following reduction with dithionite show a Soret band at 441 nm and a maximum at 603 nm characteristic of aa$_3$-type cytochromes. In addition, there is a band at 558 nm whose connection to the oxidase is not clear. This oxidase is stimulated by cholate, but unlike the oxidase from the DSM 639 strain it is inhibited by low concentrations of cyanide (μM as opposed to mM) and oxidizes horse-heart cytochrome c, TMPD-ascorbate, and caldariella quinol. The rates of oxidation (μmol/min/mg protein) for cytochrome c, TMPD-ascorbate, and quinol are 63, 6.1, and 0.2, respectively. Another cytochrome oxidase that has an absorption maximum at 602 nm, oxidizes caldariella quinol, but does not oxidize cytochrome c, is also present in strain 7 so that the terminal portion of the electron transport system in *S. acidocaldarius* consists of at least three oxidases. It is suggested [8] that the presence of three oxidases in *S. acidocaldarius* is unlikely and that the cyanide-sensitive oxidase was isolated from a different species, namely *S. solfataricus*. There is little taxonomic information in this assertion to judge whether strain 7 and DSM 639 are indeed different species. However, based on growth conditions reported by the investigators [12,28], which are unique for *S. acidocaldarius* and *S. solfataricus* [122], there is no reason to suspect that these organisms are different species.

4.4.3. The cytochromes of Thermoplasma

T. acidophilum contains a menaquinone [123], and spectral data suggests the presence of c-type and o-type cytochromes [124]. The respiratory chain of *T. acidophilum* has been studied in early stationary cells in aerobic and oxygen-limited cultures. Aerobic cells contain α bands located at 554 nm and 619 nm, and shoulders at 559 and 549 nm. The 549 nm and 554 nm bands are associated with c-type cytochromes while the 559 nm and 619 nm bands are ascribed to b and d(a$_2$) cytochromes, respectively. In oxygen-limited cultures, the spectral features associated with the c-type cytochrome and the cytochrome oxidase decrease and those related to the b-type cytochromes increase [125].

Membranes prepared from *T. acidophilum* oxidize succinate, malate, lactate, and NADH, consuming oxygen in the process, but only contain a single cytochrome. It appears to be a b-type cytochrome based on its absorption spectrum and the solubility of the heme in acid-acetone [126]. Since aeration during growth affects the cytochrome composition in *T. acidophilum*, failure to detect a more complex cytochrome pattern may reflect the level of aerobiosis during growth.

Membranes from *T. acidophilum* [124] oxidize ubiquinol Q_{10}. Menaquinones are oxidized at about a 10-fold greater rate which is in keeping with the observation that a menaquinone derivative (thermoplasmaquinone) occurs in *Thermoplasma* membranes [127]. Quinol oxidase can be extracted using detergents and activity is enhanced by detergents, but the enzyme is less stable when detergents are present. Although HOQNO is inhibitory, antimycin A and myxothiazol are not, which, taken with the absence of detectable cytochrome c, suggests the absence of a cytochrome bc_1 complex. Difference spectra of dithionite reduced minus oxidized membranes show maxima at 559 and 625 nm which correspond to the α peaks of b- and d-type cytochromes. In addition, a small absorption maximum present at 595 nm is suggestive of an a_1-cytochrome. There is no evidence for the presence of a c-type cytochrome. At low concentrations of CO, a spectrum typical of an o-type cytochrome is observed while at higher concentrations, additional CO-binding pigments are detected. Redox titrations suggest the existence of four b-type cytochromes with midpoint potentials of 40, −90, −215, and −250 mV. The cytochrome with the highest midpoint potential has an α band at 558 nm; other maxima are located at 560, 562, and 553 nm. A b-type cytochrome (M_r 18 000) contains copper as well as low- and high-potential hemes. The α peak of the high-potential heme has a maximum at 558 nm, whereas the low-potential heme has a split α peak with maxima located at 562 and 553 nm. One of the conclusions from these results is that the cytochrome composition in *T. acidophilum* is more closely related to *E. coli* than *S. acidocaldarius*.

4.4.4. The cytochromes of the extreme halophiles
Membranes from *H. salinarium* (*cutirubrum*) oxidize NADH, α-GP, and succinate. Spectral studies resolve the electron transport system into two pathways: one where NADH and α-GP reduce cytochromes b_{563}, c_{550}, and a CO-binding cytochrome a_{592}; the other where succinate reduces cytochromes b_{559}, c_{555}, and the oxidase, cytochrome a_{592} [6]. Succinate oxidation goes through a (HOQNO)-insensitive pathway whereas NADH is oxidized through a fast HOQNO-sensitive (where the inhibitor acts at the level of the NADH dehydrogenase) and a slow HOQNO-insensitive pathway. The latter involves the same constituents reduced by succinate. Cytochrome c $_{555}$is reduced by TMPD-ascorbate and in turn is oxidized by the oxidase, cytochrome a_{592}. The kinetics of NADH oxidation is consistent with the notion that cytochrome b_{563} is off the main path of electron flow [95]. The striking features of this pathway are the presence of a single terminal oxidase and the location of the site of HOQNO inhibition.

A more complex electron transport system occurs in *H. salinarium* (*cutirubrum*) [96] and a colorless mutant of *H. salinarium* [5]. The pathways are characterized by a complex of b-type cytochromes and a c-type cytochrome. Cytochrome o occurs in the electron transport chain of both organisms. In *H. salinarium* (*cutirubrum*), cytochrome aa_3 serves as an additional oxidase; in *H. salinarium*, cytochromes a, a_1, and a_3 terminate the pathway. In the former case cytochrome c and cytochrome a_3 are reduced by TMPD-ascorbate and ascorbate, respectively. In *H. salinarium*, cytochrome b_{555} is reduced by TMPD-ascorbate whereas cytochrome b_{557} reduces cytochrome o. Reducing equivalents from the cytochrome b complex also reduce cytochrome c which is oxidized either by cytochrome a or by cytochrome a_3. In *H. salinarium* (*halobium*) [97], cytochromes o and a_1 are the terminal oxidases. Ascorbate and TMPD-ascorbate reduce cytochrome b_{561} which either reduces cytochromes c or o, and the former is oxidized by cytochrome a_1. HOQNO and antimycin A are inhibitory and act between cytochromes b and c. It is not clear why such ostensibly similar organisms should have such a different suite of cytochromes and possess such different HOQNO and antimycin A sensitive sites.

Two of the b-type cytochromes from *H. salinarium* (*halobium*) were purified [15] and one (b_{559}) is partially reduced by TMPD-ascorbate, suggesting it may be similar to the TMPD-ascorbate reducible cytochrome b $_{(561)}$ from *H. salinarium* (*halobium*) [97]. Potentiometric titration indicates the presence of low- and high-potential cytochromes [128]. An additional insight into the complexity of the cytochrome b complex is provided by potentiometric titration of *H. salinarium* (*halobium*) membranes that indicates the presence of four b-type cytochromes [129]. The two with the highest mid-point potentials (+261 and +161 mV) are reduced by succinate as well as NADH, which also reduces a b-type cytochrome with a mid-point potential of +30 mV. The b-type cytochrome with the lowest potential (−153 mV) is reduced by dithionite but not NADH. Antimycin A inhibits electron transport by affecting the reoxidation of the cytochrome (+180 mV) that is part of the succinate pathway but not either of the cytochromes associated with the oxidation of NADH. This is consistent with the earlier observation that this agent inhibits between cytochrome b and c [5]. The electron transport system in *H. salinarium* (*halobium*) is terminated by two CO-binding cytochromes: a cytochrome aa_3 and the b-type cytochrome with the highest mid-point potential.

An unusual aa_3-type cytochrome oxidase occurs in *H. salinarium* (*halobium*). The enzyme is thought to occur in early exponential phase cultures and virtually to disappear after the mid-exponential phase of growth [14]. However, replotting the growth curve in the semi-log rather than the linear form shows that what is taken to be the early exponential phase is the late exponential phase and the putative mid-exponential phase is the early stationary phase. Unlike the cytochrome oxidase from *H. salinarium* (*cutirubrum*) which is more active and stable as the NaCl concentration is increased [130], this cytochrome aa_3 oxidase does not require salt for activity. In fact, the oxidation of reduced cytochrome c decreases with increasing salt concentration. This recalls an earlier observation that the cytochrome c reductase from *Paracoccus halodenitrificans* becomes less active with increasing salt concentration [131]. In this case, the effect reflects an increase in the apparent K_m for cytochrome c and is without effect on V_{max}. Whether a similar effect is responsible for the inhibitory action of NaCl on the cytochrome aa_3 oxidase from *H. salinarium* (*halobium*) is not known. The aa_3-cytochrome oxidase can be solubilized by a variety of detergents, but only sodium deoxycholate provides a preparation whose spectral properties are similar to those of the membrane-bound cytochrome [132]. The soluble cytochrome is composed of a single subunit (M_r 40 000) and does not exhibit any oxidase activity. No copper is detected following purification but two a-type hemes are present. Both are reduced by dithionite but only one is reduced by TMPD-ascorbate. The heme reduced by TMPD-ascorbate is not oxidized by oxygen whereas one of the two dithionite-reduced hemes is so oxidized. The amino-acid sequence derived from the nucleotide sequence of the cloned gene [133] gives a higher M_r (65 000) than the value obtained from SDS-gel electrophoresis; this is assumed to be due to the hydrophobicity of the peptide. The amino-acid sequence of the *H. salinarium* (*halobium*) aa_3-type cytochrome appears more homologous to the aa_3-cytochrome oxidase of the eucarya than the oxidase from bacteria. The identities are as follows: subunit I of maize mitochondria, human and yeast aa_3-cytochrome oxidases (44%, 42% and 41%, respectively); thermophilic bacterium PS3, *P. denitrificans*, *E. coli* (41%, 36% and 36%). The cytochrome aa_3 oxidase from *H. salinarium* (*halobium*) also possesses the consensus sequences associated with metal-binding residues even though no copper is associated with the purified oxidase so it is thought that the copper is lost during purification.

5. Hydrogenases

Hydrogenases catalyze the reversible oxidation of molecular hydrogen. As such, the enzyme functions to supply reducing power or provide means for disposing excess electrons. Hydrogenases are widely distributed and vary in a number of characteristics such as sensitivity to oxygen, catalytic activity, and relative molecular mass (for a review see ref. [134]). Hydrogenase activity is present in several hyperthermophilic anaerobic archaea but to date hydrogenases have not been found in the sulfur-dependent acidophilic thermophiles or the extreme halophiles.

5.1. Hydrogenases in methanogens

A variety of hydrogenases are found in the methanogens where they occur as soluble and membrane-bound enzymes [135]. What function hydrogenases play in the physiology of the methanogens is not clear, particularly since the hydrogen atoms in methane originate from water and not hydrogen [136]. The membrane-bound hydrogenase from *Methanobacterium,* strain G2R (M_r 900 000) is composed of three subunits (M_r, $\sim$80 000, 51 000, and 39 000). The enzyme reduces a variety of acceptors, including methylviologen, benzylviologen, methylene blue and DCIP. The hydrogenase is inactivated by oxygen but can be reactivated when incubated in a hydrogen or nitrogen atmosphere either in the presence of dithionite or a combination of glucose and glucose oxidase. Whereas the electrophoretic mobilities of the cytoplasmic and membrane-bound hydrogenases of strain G2R are similar, the electrophoretic mobilities of the cytoplasmic hydrogenases from other methanogens differ [137]. *M. thermoautotrophicum* contains three hydrogenases. Two are soluble [138]; the third is membrane-bound and is present in vesicles that oxidize hydrogen with the concomitant production of ATP [54]. The relationship of the other hydrogenases to ATP synthesis remains to be clarified.

5.2. Hydrogenase in Pyrodictium

P. brockii is an anaerobic organism that grows at 105°C by reducing S^0 in the presence of CO_2 and H_2 [139]. The reduction of S^0 to S^{2-} and the path of electron transport from H_2 to S^0 is not known. The oxidation of H_2 is brought about by a membrane-bound hydrogenase that is not irreversibly inhibited by O_2. The enzyme couples the oxidation of H_2 to a variety of electron acceptors with those having the most positive potential (i.e., methylene blue) being more active than the low-potential acceptors (i.e., methylviologen). No H_2 evolution is observed in the presence of either reduced methylviologen or benzylviologen, suggesting the enzyme is an H_2-uptake hydrogenase. Antiserum prepared against the M_r 65 000 subunit from the *Bradyrhizobium japonicum* hydrogenase reacts with crude extracts from *P. brockii*. However, antibodies against the M_r 33 k subunit from *B. japonicum* do not recognize the analogous subunit in such extracts. *P. brockii* genomic digests hybridize to a probe that is specific for the large subunit of the *B. japonicum* hydrogenase. This indicates that at least at the level of the large subunits the two hydrogenases are somewhat related [17].

The membrane-bound enzyme can be purified without taking any special precautions with respect to O_2. Some of the properties of the membrane-bound and purified enzymes are sufficiently different to suggest the importance of the membrane environment. The purified hydrogenase (M_r 118 000) is composed of two subunits (M_r 66 000 and 45 000) and is most active with methylene blue as the acceptor; however, benzylviologen is as good an acceptor as any of the others and in some cases better. In addition, the hydrogenase evolves H_2 from reduced methylviologen but not benzylviologen. The membrane-bound enzyme is thermally more stable than the purified enzyme over the short term. However, the enzyme, whether membrane-bound or purified, is unstable at the optimal growth temperature of the organism when incubated for one or more hours. Since the oxidation of H_2 is essential for the growth of *P. brockii*, this raises the interesting question as to the stability of the enzyme during growth.

6. The enzymes of denitrification

Denitrification, coupled to oxidative phosphorylation, is used by a variety of organisms. To date, the process is restricted to the extreme halophiles among the archaea. Nitrate reduction is widespread and occurs in members of the genera *Haloarcula*, *Haloferax*, and *Halobacterium* where the reduction of nitrate leads to the accumulation of nitrite or dinitrogen [140]. It is suggested that denitrification by the sulfur-dependent acidophilic thermophiles, leading to the production of nitrite which is a powerful mutagen, would not have any evolutionary value and would not persist [141]. However, this presumes that nitrate reduction necessarily leads to nitrite accumulation. A search for denitrification among the sulfur-dependent acidophilic thermophiles might be warranted.

6.1. Nitrate reductase

Nitrate is reduced to nitrite, where it accumulates, or to the other products of denitrification. Nothing is known about the nitrate reductases that occur in organisms that reduce nitrate only to the level of nitrite. There is no evidence that such organisms grow anaerobically at the expense of this process, although this possibility has not been explored.

The nitrate reductase from *Haloferax denitrificans* was purified to electrophoretic homogeneity aided by the ease with which the enzyme is solubilized from membranes and the stability of the reductase in solutions of low ionic strength [142]. The enzyme is composed of two subunits that resemble the α and β subunits of the dissimilatory nitrate reductases found in bacteria [143]. Dissimilatory nitrate reductases have a third subunit that contains a b-type cytochrome. No such subunit is detected in the nitrate reductase from *H. denitrificans*. However, this observation has no significance since this subunit is often lost during purification. The most striking property of the enzyme is its response to salt concentration, both when membrane-bound [144] and following purification [142]. Nitrate reduction is most active in the absence of added salt and the enzyme is stable for weeks on end in the absence of salt. Similar nitrate reductase activities occur in

Haloarcula marismortui [145], *Haloferax mediterranei*, and another extreme halophile, strain Baja-12, chemotaxonomically identified as a member of the genus *Haloferax* [146]. The enzyme activities from these organisms are stable as well as most active in the absence of salt and the enzymatic properties are indistinguishable from those of *H. denitrificans* [144].

There are other proteins from extremely halophilic bacteria that do not require high concentrations of salt for stability or activity [147–149]. This raises the possibility that there are additional proteins in the extreme halophiles that do not require conditions of high ionic strength for stability or activity. The failure to detect such proteins could be due to the conditions used during their isolation and characterization (i.e., high concentrations of salt) that are inimical to such enzymes.

6.2. Nitrite reductase activity

H. denitrificans reduces nitrite to nitric oxide in the presence of phenazine methosulfate-NADPH. When dithionite is used as the reductant, nitrous oxide is the sole product. However, this should be taken with a great deal of caution since dithionite chemically reduces nitric oxide to nitrous oxide [150]. The production of nitric oxide in the presence of phenazine methosulfate-NADPH is inhibited by a variety of metal-chelating agents including diethylthiocarbamate, which inhibits the dissimilatory copper nitrite reductases in other organisms [151]. Absolute spectra of membranes have an absorption band located at 570 nm. This maximum disappears upon reduction with phenazine methosulfate-NADPH and reappears upon the addition of nitrite, suggesting that this absorption band is associated with nitrite reduction [152]. The production of nitric oxide is more vigorous in the presence of salt; activity is rapidly lost in the absence of salt but is stable when the NaCl is 500 mM or greater [Hochstein, L.I., unpublished data]. Polyclonal antibodies prepared against the copper-containing nitrite reductase from *Achromobacter cycloclastes* that react with other copper-containing nitrite reductases do not recognize the nitrite reductase from *H. denitrificans* though [153].

A nitrite reductase in *H. marismortui* [145] causes the disappearance of nitrite in the presence of dithionite with either methylviologen or the organism's ferredoxin as the electron mediators. The product of reduction is presumed to be nitric oxide.

7. Summary

With the exception of the extreme halophiles, the membrane-bound enzymes from the archaea do not possess properties that suggest they are unique when compared to analogous enzymes from the Bacteria and Eucarya. In the latter case, the differences reflect the requirement for high ionic strength for activity and stability. Although denitrification is a widespread property among the extreme halophiles, only the nitrate reductase has been studied in any detail. The other enzymes associated with denitrification remain to be characterized. It is interesting that the membrane-bound archaeal ATPases, while probably the most thoroughly studied of the archaeal membrane-bound enzymes, are the least understood. There still remain a number of questions to be addressed: what are the functions of these enzymes that structurally resemble and appear to be related to eucaryal V-type ATPases; is the putative Na-dependent ATPase of *M. voltae* a V-type ATPase, or does an F-type ATPase couple ATP synthesis to sodium ions rather than protons; do the extreme halophiles synthesize ATP using any one of the various V-type ATPases or an F-type ATPase? These questions are the objects of intense study at this time and presumably answers will be forthcoming.

References

[1] Kamekura, M. and Kates, M. (1988) In: Halophilic Bacteria, Vol. II (Rodriguez-Valera, F., Ed.), pp. 25–54, CRC Press, Boca Raton, FL.
[2] Hochstein, L.I. and Stan-Lotter, H. (1992) Arch. Biochem. Biophys. 295, 153–160.
[3] Gogarten, J.P., Kibak, H., Dittrich, P., Taiz, L., Bowman, E.J., Bowman, B.J., Manolson, M.F., Poole, R.J., Date, T., Oshima, T., Konishi, J., Denda, K. and Yoshida, M. (1989) Proc. Natl. Acad. Sci. U.S.A. 86, 6661–6665.
[4] Nelson, N. (1992) Biochim. Biophys. Acta 1100, 109–124.
[5] Cheah, K.S. (1970) Biochim. Biophys. Acta 216, 43–53.
[6] Lanyi, J.K. (1968) Arch. Biochem. Biophys. 716–724.
[7] Larsen, H. (1984) In: Bergey's Manual of Systematic Bacteriology, Vol. 1 (Krieg, N.R., Ed.), pp. 261–266, Williams & Wilkins, Baltimore, MD.
[8] Schäfer, G., Anemüller, S., Moll, R., Meyer, W. and Lübben, M. (1990) FEMS Microbiol. Rev. 75, 335–348.
[9] Garty, H., Danon, A. and Caplan, S.R. (1980) Eur. J. Biochem. 111, 411–418.
[10] MacDonald, R.E. and Lanyi, J.K. (1975) Biochemistry 2882–2889.
[11] Mukohata, Y., Isoyama, M. and Fuke, A. (1986) J. Biochem. 99, 1–8.
[12] Lübben, M. and Schäfer, G. (1987) Eur. J. Biochem. 164, 533–540.
[13] Esser, A.F. and Lanyi, J.K. (1973) Biochemistry 12, 1933–1939.
[14] Fujiwara, T., Fukumori, Y. and Yamanaka, T. (1987) Plant Cell Physiol. 29–36.
[15] Hallberg, C. and Baltscheffsky, H. (1981) FEBS Lett. 125, 201–204.
[16] Wakagi, T. and Oshima, T. (1986) System. Appl. Microbiol. 7, 342–345.
[17] Pihl, T.D., Schicho, R.N., Kelly, R.M. and Maier, R.J. (1989) Proc. Natl. Acad. Sci. U.S.A. 86, 138–141.
[18] Doddema, H.J., Hutten, T.J., van der Drift, C. and Vogels, G.D. (1978) J. Bacteriol. 136, 19–23.
[19] Kristjansson, H. and Hochstein, L.I. (1985) Arch. Biochem. Biophys. 241, 590–595.
[20] Lanyi, J.K. and MacDonald, R.E. (1979) Methods Enzym. 56 (Fleischer, S. and Packer, L., Eds.), pp. 398–407, Academic Press, New York.
[21] Hjelmeland, L.M. (1990) Methods Enzym. 182 (Deutscher, M.P., Ed.), pp. 277–282, Academic Press, San Diego, CA.
[22] Neugebauer, J. (1990) Methods Enzym. (Deutscher, M.P., Ed.) 182, 239–253, Academic Press, San Diego, CA.
[23] Hochstein, L.I., Kristjansson, H. and Altekar, W. (1987) Biochem. Biophys. Res. Commun. 147, 295–300.
[24] Wakagi, T., Yamauchi, T., Oshima, T., Müller, M., Azzi, A. and Sone, N. (1989) Biochem. Biophys. Res. Commun. 165, 1110–1114.
[25] Nanba, T. and Mukohata, Y. (1987) J. Biochem. 102, 591–598.
[26] Inatomi, K.-I. (1986) J. Bacteriol. 167, 837–841.
[27] Scheel, E. and Schäfer, G. (1990) Eur. J. Biochem. 187, 727–735.
[28] Konishi, J., Wakagi, T., Oshima, T. and Yoshida, M. (1987) J. Biochem. 102, 1379–1387.
[29] Roth, R., Duft, R., Binder, A. and Bachofen, R. (1986) System. Appl. Microbiol. 7, 346–348.

[30] Moll, R. and Schäfer, G. (1989) Biol. Chem. Hoppe-Seyler 370, 936–937.
[31] Lanyi, J.K. (1971) J. Biol. Chem. 246, 4552–4559.
[32] Hochstein, L.I. and Dalton, B.P. (1973) Biochim. Biophys. Acta 302, 216–228.
[33] Wakao, H., Wakagi, T. and Oshima, T. (1987) J. Biochem. 102, 255–262.
[34] Altekar, W., Kristjansson, H., Ponnamperuma, C. and Hochstein, L.I. (1984) Origins of Life 14, 733–738.
[35] Pihl, T.D. and Maier, R.J. (1991) J. Bacteriol. 1839–1844.
[36] Deutscher, M.P., Ed. (1990) Protein Purification, Methods Enzym., Vol. 182, 894pp, Academic Press, San Diego, CA.
[37] Pedersen, P.L. and Carafoli, E. (1987) Trends Biochem. Sci. 12, 146–150.
[38] Forgac, M. (1989) Physiol. Rev. 69, 765–603.
[39] Hanada, H., Moriyama, Y., Maeda, M. and Futai, M. (1990) Biochem. Biophys. Res. Commun. 170, 873–878.
[40] Bowman, E.J., Siebers, A. and Altendorf, K. (1988) Proc. Natl. Acad. Sci. U.S.A. 85, 7972–7976.
[41] Nelson, N. and Taiz, L. (1989) Trends Biochem. Sci. 14, 113–116.
[42] Konishi, J., Denda, K., Oshima, T., Wakagi, T., Uchida, E., Ohsumi, Y., Anraku, Y., Matsumoto, T., Wakabayashi, Y., Mukohata, Y., Ihara, K., Inatomi, K.-I., Kato, K., Ohta, T., Allison, W.S. and Yoshida, M. (1990) J. Biochem. 108, 554–559.
[43] Schäfer, G. and Meyering-Vos, M. (1992) Biochim. Biophys. Acta 1101, 232–235.
[44] Schobert, B. (1991) J. Biol. Chem. 266, 8008–8014.
[45] Mukohata, Y. and Yoshida, M. (1987) J. Biochem. 102, 797–802.
[46] Stan-Lotter, H. and Hochstein, L.I. (1989) Eur. J. Biochem. 179, 155–160.
[47] Hochstein, L.I. (1992) FEMS Microbiol. Lett. 97, 155–160.
[48] Dharmavaram, R.M. and Konisky, J. (1987) J. Bacteriol. 169, 3921–3925.
[49] Dharmavaram, R.M. and Konisky, J. (1989) J. Biol. Chem. 264, 14085–14089.
[50] Dybas, M. and Konisky, J. (1992) J. Bacteriol. 174, 5575–5583.
[51] Lübben, M., Lünsdorf, H. and Schäfer, G. (1987) Eur. J. Biochem. 167, 211–219.
[52] Jussofie, A., Mayer, F. and Gottschalk (1986) Arch. Microbiol. 146, 245–249.
[53] Mayer, F., Jussofie, A., Salzmann, M., Lübben, M., Rohde, M. and Gottschalk, G. (1987) J. Bacteriol. 169, 2307–2309.
[54] Doddema, H.J., van der Drift, C., Vogels, G.D. and Veenhuis, M. (1979) J. Bacteriol. 140, 1081–1089.
[55] Inatomi, K.-I. and Maeda, M. (1988) J. Bacteriol. 170, 5960–5962.
[56] Inatomi, K.-L., Maeda, M. and Futai, M. (1989) Biochem. Biophys. Res. Commun. 162, 1585–1590.
[57] Lübben, M., Anemüller, S. and Schäfer, G. (1987) Biol. Chem. Hoppe-Seyler 368, 555.
[58] Zhang, J., Myers, M. and Forgac, M. (1992) J. Biol. Chem. 267, 9773–9778.
[59] Wakagi, T. and Oshima, T. (1985) Biochim. Biophys. Acta 817, 33–41.
[60] Lübben, M., Lünsdorf, H. and Schäfer, G. (1988) Biol. Chem. Hoppe-Seyler 369, 1259–1266.
[61] Denda, K., Konishi, J., Oshima, T., Date, T. and Yoshida, M. (1988) J. Biol. Chem. 263, 6012–6015.
[62] Denda, K., Konishi, J., Oshima, T., Date, T. and Yoshida, M. (1988) J. Biol. Chem. 263, 17251–17254.
[63] Denda, K., Konishi, J., Hajiro, K., Oshima, T., Date, T. and Yoshida, M. (1990) J. Biol. Chem. 265, 21509–21513.
[64] Lübben, M. and Schäfer, G. (1989) J. Bacteriol. 171, 6106–6116.
[65] Anemüller, S., Lübben, M. and Schäfer, G. (1985) FEBS Lett. 193, 83–87.
[66] Sebald, W. and Hoppe, J. (1981) Curr. Top. Bioenerg. 12, 1–64.

[67] Denda, K., Konishi, J., Oshima, T., Date, T. and Yoshida, M. (1989) J. Biol. Chem. 264, 7119–7121.
[68] Kristjansson, H., Sadler, M.H. and Hochstein, L.I. (1986) FEMS Microbiol. Rev. 39, 151–157.
[69] Kristjansson, H. and Hochstein, L.I. (1986) FEMS Microbiol. Lett. 35, 171–175.
[70] Dunker, A.K. and Rueckert, R.R. (1969) J. Biol. Chem. 244, 5074–5080.
[71] Eley, M.H., Burns, P.C., Kannapell, C.C. and Campbell, P.S. (1979) Anal. Biochem. 92, 411–419.
[72] Satre, M., Lunardi, J., Pougeois, R. and Vignais, P.V. (1979) Biochemistry 18, 3134–3140.
[73] Pougeois, R., Satre, M. and Vignais, P.V. (1979) Biochemistry 18, 1408–1413.
[74] Pougeois, R., Satre, M. and Vignais, P.V. (1980) FEBS Lett. 117, 344–348.
[75] Futai, M. (1974) J. Membr. Biol. 15, 15–28.
[76] Walker, J.E., Fearnley, I.M., Gay, N.J., Gibson, B.W., Northrop, F.D., Powell, S.J., Runswick, M.J., Saraste, M. and Tybulewicz, V.L.J. (1985) J. Mol. Biol. 184, 677–701.
[77] Stan-Lotter, H., Bowman, E.J. and Hochstein, L.I. (1991) Arch. Biochem. Biophys. 284, 116–119.
[78] Sulzner, M., Stan-Lotter, H. and Hochstein, L.I. (1992) Arch. Biochem. Biophys. 296, 347–349.
[79] Bowman, E.J., Mandala, S., Taiz, L. and Bowman, B.J. (1986) Proc. Natl. Acad. Sci. U.S.A. 83, 48–52.
[80] Selwyn, M.J. (1965) Biochim. Biophys. Acta 105, 193–195.
[81] Kristjansson, H. (1983) Partial purification and characterization of an ATPase in the archaebacterium *Halobacterium saccharovorum*, Ph.D. Thesis, University of Maryland.
[82] Schobert, B. and Lanyi, J.K. (1989) J. Biol. Chem. 264, 12805–12812.
[83] Bonet, M.L. and Schobert, B. (1992) Eur. J. Biochem. 207, 369–376.
[84] Laemmli, U.K. (1970) Nature 227, 680–685.
[85] Neville, D.M. (1971) J. Biol. Chem. 246, 6328–6334.
[86] Bowman, B.J., Dschida, W.J., Harris, T. and Bowman, E.J. (1989) J. Biol. Chem. 264, 15606–15612.
[87] Mukohata, Y., Ihara, K., Yoshida, M., Konishi, J., Sugiyama, Y. and Yoshida, M. (1987) Arch. Biochem. Biophys. 259, 650–653.
[88] Rodriguez-Valera, F. (1988) In: Halophilic Bacteria, Vol. II (Rodriguez-Valera, F., Ed.), pp. 3–30, CRC Press, Boca Raton, FL.
[89] Oesterhelt, D. and Krippahl, G. (1983) Ann. Microbiol. (Inst. Pasteur) 134B, 137–150.
[90] Hartmann, R., Sickinger, H.-D. and Oesterhelt, D. (1980) Proc. Natl. Acad. Sci. U.S.A. 77, 3821–3825.
[91] Oren, A. and Trüper, H.G. (1990) FEMS Microbiol. Lett. 70, 33–36.
[92] Oren, A. (1991) J. Gen. Microbiol. 137, 1387–1390.
[93] Javor, B.J. (1984) Appl. Environ. Microbiol. 48, 352–360.
[94] Hochstein, L.I. and Tomlinson, G.A. (1985) FEMS Microbiol. Lett. 27, 329–331.
[95] Lanyi, J.K. (1969) J. Biol. Chem. 244, 2864–2869.
[96] Cheah, K.S. (1969) Biochim. Biophys. Acta 180, 320–333.
[97] Cheah, K.S. (1970) Biochim. Biophys. Acta 205, 148–160.
[98] Collins, M.D., Ross, H.N.M., Tindall, B.J. and Grant, W.D. (1981) J. Appl. Bacteriol. 50, 559–565.
[99] Hochstein, L.I. and Dalton, B.P. (1968) J. Bacteriol. 95, 37–42.
[100] Hochstein, L.I. (1976) In: Extreme Environments. Mechanisms of Microbial Adaptation (Heinrich, M.R., Ed.), pp. 213–225, Academic Press, New York.
[101] Hochstein, L.I. and Dalton, B.P. (1968) Biochim. Biophys. Acta 167, 638–640.
[102] Lanyi, J.K. (1974) Bacteriol. Rev. 38, 272–290.
[103] Lanyi, J.K. (1969) J. Biol. Chem. 244, 4168–4173.
[104] Lanyi, J.K. (1972) J. Biol. Chem. 247, 3001–3007.

[105] Clark, R.D. and MacDonald, R.E. (1980) Biochem. Biophys. Res. Commun. 97, 1467–1473.
[106] Oppenheim, J.D. and Salton, M.R.J. (1973) Biochim. Biophys. Acta 297–322.
[107] van Thienen, G. and Postma, P.W. (1973) Biochim. Biophys. Acta 323, 429–440.
[108] Weiner, J.H. (1974) J. Membr. Biol. 15, 1–14.
[109] Lanyi, J.K. and Stevenson, J. (1970) J. Biol. Chem. 245, 4047–4080.
[110] Hochstein, L.I. (1975) Biochim. Biophys. Acta 403, 58–66.
[111] Moll, R. and Schäfer, G. (1991) Eur. J. Biochem. 201, 593–600.
[112] Hatefi, Y. (1985) Annu. Rev. Biochem. 1015–1069.
[113] Aitken, D.M. and Brown, A.D. (1969) Biochim. Biophys. Acta 177, 351–354.
[114] Gradin, C., Hederstedt, L. and Baltscheffsky, H. (1985) Arch. Biochem. Biophys. 239, 200–205.
[115] Davis, K.A. and Hatefi, Y. (1971) Biochemistry 10, 2509–2516.
[116] Kühn, W., Fiebig, K., Hippe, H., Mah, R.A., Huser, B.A. and Gottschalk, G. (1983) FEMS Microbiol. Lett. 20, 407–410.
[117] Kühn, W., Fiebig, K., Walther, R. and Gottschalk, G. (1979) FEBS Lett. 105, 271–274.
[118] Kühn, W. and Gottschalk, G. (1983) Eur. J. Biochem. 135, 89–94.
[119] Blaut, M. and Gottschalk, G. (1985) Trends Biochem. Sci. 10, 486–489.
[120] Anemüller, S. and Schäfer, G. (1989) FEBS Lett. 244, 451–455.
[121] Anemüller, S. and Schäfer, G. (1990) Eur. J. Biochem. 191, 297–305.
[122] Segerer, A. and Stetter, K.O. (1991) In: The Prokaryotes, Vol. I (Balows, A., Trüper, H.G., Dworkin, M., Harder, W. and Schleifer, K-H., Eds.), pp. 684–701, Springer, New York.
[123] Holländer, R., Wolf, G. and Mannheim, W. (1977) Antonie v. Leeuwenhoek 43, 177–185.
[124] Belly, R.T., Bohlool, B.B. and Brock, T.D. (1973) Ann. N.Y. Acad. Sci. 225, 94–107.
[125] Holländer, R. (1978) J. Gen. Microbiol. 165–167.
[126] Searcy, D.G. and Whatley, F.R. (1982) Zbl. Bakt. Hyg. I. Abt. Orig. C 3, 245–257.
[127] Collins, M.D. (1985) FEMS Microbiol. Lett. 28, 21–23.
[128] Gradin, C.H. and Colmsjö, A. (1987) Arch. Biochem. Biophys. 256, 515–522.
[129] Gradin, C.H. and Colmsjö, A. (1989) Arch. Biochem. Biophys. 272, 130–136.
[130] Lieberman, M.M. and Lanyi, J.K. (1971) Biochim. Biophys. Acta 245, 21–33.
[131] Miller, J.E. and Hochstein, L.I. (1972) J. Bacteriol. 112, 656–659.
[132] Fujiwara, T., Fukimori, Y. and Yamanaka, T. (1989) J. Biochem. 105, 287–292.
[133] Denda, K., Fujiwara, T., Seki, M., Yoshida, M., Fukumori, Y. and Yamanaka, T. (1991) Biochem. Biophys. Res. Commun. 181, 316–322.
[134] Adams, M.W.W., Mortenson, L.E. and Chen, J.-S. (1981) Biochim. Biophys. Acta 594, 105–176.
[135] Whitman, W.B. (1985) In: The Bacteria, Vol. VIII (Woese, C.R. and Wolfe, R.S., Eds.), pp. 3–84, Academic Press, San Diego, CA.
[136] Daniels, L., Fulton, G., Spencer, R.W. and Orme-Johnson, W.H. (1980) J. Bacteriol. 141, 694–698.
[137] McKellar, R.C. and Sprott, G.D. (1979) J. Bacteriol. 139, 231–238.
[138] Fox, J.A., Livingston, D.J., Orme-Johnson, W.H. and Walsh, C.T. (1987) Biochemistry 26, 4219–4227.
[139] Stetter, K.O. (1982) Nature (London) 300, 258–260.
[140] Tindall, B.J. (1991) In: The Prokaryotes, Vol. I (Balows, A., Trüper, H.G., Dworkin, M., Harder, W. and Schleifer, K-H., Eds.), pp. 768–808, Springer, New York.
[141] Betlach, M.R. (1982) Antonie van Leeuwenhoek 48, 585–607.
[142] Hochstein, L.I. and Lang, F. (1991) Arch. Biochem. Biophys. 288, 380–385.
[143] Hochstein, L.I. and Tomlinson, G.A. (1988) Annu. Rev. Microbiol. 42, 231–261.
[144] Hochstein, L.I. (1991) In: General and Applied Aspects of Halophilic Microorganisms (Rodriguez-Valera, F., Ed.), pp. 129–137, Plenum Press, New York.
[145] Werber, M.M. and Mevarech, M. (1978) Arch. Biochem. Biophys. 186, 60–65.

[146] Tindall, B.J., Tomlinson, G.A. and Hochstein, L.I. (1987) System. Appl. Microbiol. 9, 6–8.
[147] Danson, M.J., McQuattie, A. and Stevenson, K.J. (1986) Biochemistry 25, 3880–3884.
[148] May, B.P. and Dennis, P.P. (1987) J. Bacteriol. 169, 1417–1422.
[149] Torreblanca, M., Meseguer, I. and Rodriguez-Valera, F. (1989) J. Gen. Microbiol. 135, 2655–2661.
[150] Kakutani, T., Watanabe, H., Arima, K. and Beppu, T. (1981) J. Biochem. 89, 453–461.
[151] Shapleigh, W.J. and Payne, W.J. (1985) FEMS Microbiol. Lett. 26, 275–279.
[152] Hochstein, L.I. (1988) In: Halophilic Bacteria, Vol. II (Rodriguez-Valera, F., Ed.), pp. 67–83, CRC Press, Boca Raton, FL.
[153] Coyne, M.S., Arunakumari, A., Averill, B.A. and Tiedje, J.M. (1989) Appl. Environ. Microbiol. 55, 2924–2931.

CHAPTER 11

Chromosome structure, DNA topoisomerases, and DNA polymerases in archaebacteria (archaea)

Patrick FORTERRE and Christiane ELIE

Institut de Génétique et Microbiologie, CNRS, URA 1354, Université Paris-Sud, 91405 Orsay Cedex, France

1. Introduction

Our knowledge about chromosome structure and enzymes involved in DNA replication, recombination and repair in archaebacteria lags behind that on enzymes and proteins involved in RNA and protein syntheses. This can be explained both on the basis of historical factors, such as the early involvement of several laboratories in the study of transcription and translation in archaebacteria, and by more subjective factors. For example, deciphering the mechanism of gene expression in archaebacteria is considered a prerequisite to the exploitation of their putative biotechnological potential, whereas analysis of their chromosome structure and replication seems, a priori, more academic. Still another reason may be that scientists working in the DNA field are in general less interested in the problem of the origin of life than those dealing with RNA and the genetic code. However, the study of DNA metabolism in archaebacteria should give us important new insights into their relationship with the other two domains. Indeed, since DNA has been most probably a "recent" acquisition in the history of biomolecules, the main features of the DNA world were probably still evolving at the time of the last common ancestor of archaebacteria, eubacteria and eukaryotes. This could account for the major differences observed today between eubacterial and eukaryotic chromosomes and enzymes involved in DNA metabolism.

In this chapter we will review the archaebacterial DNA areas which have been more thoroughly investigated during the last ten years: chromosome structure and putative histone-like proteins, DNA topoisomerases and DNA polymerases, with some emphasis on the problem of DNA stability in thermophiles. Since only a small number of laboratories are involved in these studies, many results are still preliminary or dispersed, and some are awaiting confirmation; however, significant and sometimes surprising findings have already emerged, indicating that this field of research is worthy of active investigation.

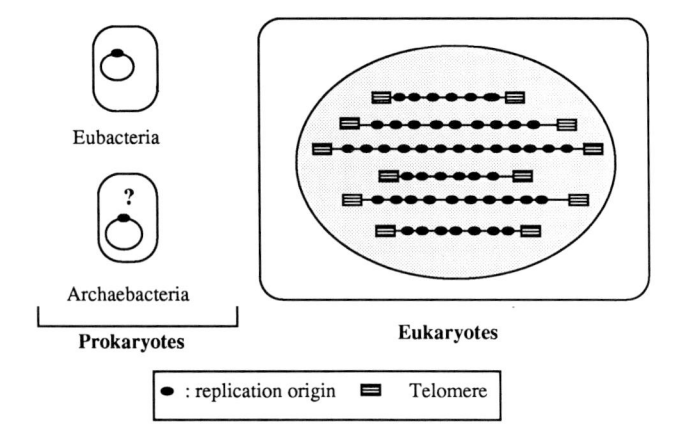

Fig. 1. Comparative chromosomal organization of prokaryotes and eukaryotes. The question mark indicates that we do not know whether archaebacteria have a single origin or multiple replication origins. The diagrams correspond to the average situation: a few eubacteria have linear chromosomes but without telomeres whereas some eubacteria and *Haloferax* species have several circular chromosomes and/or giant plasmids. The cell size is also only indicative: a few eubacteria are bigger than typical eukaryotic cells.

2. Chromosome structure

2.1. Genome size and organization

The genomes of six archaebacteria have been physically mapped. They are representative of the different archaebacterial phenotypes, including two methanogens, *Methanococcus voltae* [1] and *Methanobacterium thermoautotrophicum* strain Marburg [Leisinger, personal communication], two halophiles, *Haloferax volcanii* [2] and *Hf. mediterranei* [3], and two non-methanogenic extreme thermophiles (sulfothermophiles), *Thermococcus celer* [4] and *Sulfolobus acidocaldarius* [5]. Their sizes (1.9, 1.6, 4.1, 3.8, 1.9, and 3×10^6 bp, respectively) are in the same range as those of eubacterial chromosomes. A high-resolution physical and genetic map was obtained for *Haloferax volcanii* with a cosmid library covering nearly completely the whole genome [2]. In the two methanogens and the sulfothermophiles, the genomes are composed of a single circular chromosome, whereas in both halophiles, two circular "giant plasmids" or "small chromosomes" with sizes ranging from 690 to 320 kbp are present, in addition to a circular chromosome of 2.9 Mbp. This genomic organization into one or a few circular chromosomes resembles strikingly those of most eubacteria. This genomic structure could therefore define a "positive" prokaryotic feature, in striking contrast to that of eukaryotes which is characterized by the association of several different linear chromosomes with telomeric ends (Fig. 1). However, it is not yet known if replication of archaebacterial chromosomes starts at a single origin, as in eubacteria, or at multiple origins, as in eukaryotes. The similar range in generation times and genome size of archaebacteria and eubacteria nevertheless suggests a similar mechanism to control the timing of DNA replication.

2.2. Putative histone-like proteins and nucleosomes

A major difference between eubacteria and eukaryotes is the mode of DNA compaction in the chromosome, i.e. the existence of histones and nucleosomes in most eukaryotes and their absence in eubacteria. Histone-like proteins have been isolated from eubacteria, but they do not form stable and regular nucleosomes and their role in DNA packaging has not yet been demonstrated. Several putative "histone-like" proteins with molecular masses in the range of 7 to 15 kDa have also been isolated from chromosomal fractions of different archaebacteria, including *Thermoplasma acidophilum* [6–13], several Methanosarcinaceae [14–22], *Methanothermus fervidus* [23–25], *S. acidocaldarius* and *Sulfolobus solfataricus* [27–34]. Relationships between these various DNA-binding proteins have not always been established; in particular, it is not clear whether *Sulfolobus* histone-like proteins isolated by different laboratories are similar or not (a situation reminiscent of the early work on eubacterial histone-like proteins). Furthermore, the criteria used to define histone-like proteins in prokaryotes may be misleading [35]: any small, abundant and basic double-stranded DNA-binding protein is sometimes considered as a putative histone. Obviously, many proteins unrelated to chromosome structure also fit these criteria; for example, two putative *E. coli* histones, proteins H and HLP-1, turned out recently to be a ribosomal protein [36] and an outer-membrane protein [37], respectively. Another criterion to define histone-like proteins is their isolation from "chromosomal fractions"; however such fractions are often contaminated by other cellular components. For example, several *S. acidocaldarius* histone-like proteins called HSNP (Helix-Stabilizing Nucleoid Protein) have been isolated from the "nucleoid" of this archaebacterium [33,34], but further studies have shown that *Sulfolobus* "nucleoids" contain the cellular glycoprotein envelope (S-layer), in addition to the chromosome [38,39]. Accordingly, in this chapter, we will review only the archaebacterial DNA-binding proteins which have been either investigated in detail, or which are clearly homologous to eubacterial histone-like proteins or eukaryotic histones.

2.2.1. The protein HTa
The first archaebacterial DNA-binding protein to be described and extensively studied has been the protein HTa (Histone *Thermoplasma acidophilum*) isolated by Searcy from *Thermoplasma acidophilum* [6–8] (for a review see ref. [9]). HTa can be isolated as an oligomer of four subunits of 9.9 kDa, either in association with DNA or free in solution. HTa is an archaebacterial homologue of the eubacterial histone-like HU proteins since the amino-acid sequences of HTa and HU proteins exhibit a high degree of similarity and can be aligned without any gap (Fig. 2A). In eubacteria, HU is involved in the formation of various nucleoprotein complexes (for example between initiator proteins and DNA replication origins) probably via its capacity to facilitate DNA bending. However, mutants lacking HU are viable, suggesting that other DNA-binding proteins can take over its function.

A general role for HU/HTa proteins in DNA packaging is not clear: these proteins are not related to eukaryotic histones in terms of amino-acid sequence similarity (see Fig. 2) and they form pseudo-nucleosomes in vitro whose structure is still controversial [40–42]. HTa tetramers condense the DNA into small particles of 5.5 nm, which is only half

A) - Alignment of eubacterial HU proteins sequences with *Thermoplasma acidophilum* protein HTa

```
HU  BSu   MNKTELINAVAEASELSKKDATKAVDSVFDTILDALKNGDKIQLIGFGNFEVRERSARKGRNPQTGEEIEIPASKVPAFKPGKALKDAVAGK
HU  Cpe   MNKAELITSMAEKSKLTKKDAELALKALIESVEEALEKGEKVQLVGFGTFETRERAAREGRNPRTKEVINIPATTVPVFKAGKEFKDKVNK
HU  Rme   MNKNELVAAVADKAGLSKADASSAVDAVFETIQGELKNGGDIRLVGFGNFSVSRREASKGRNPSTGAEVDIPARNVPKFTAGKGLKDAVN
HU  Ecol  MNKSQLIDKIAAGADISKAAAGRALDAIIASVTESLKEGDDVALVGFGTFAVKERAARTGRNPQTGKEITIAAAKVPSFRAGKALKDAVN
HTA       MVGISELSKEVAKKANTTQKVARTVIKSFLDEIVSEADGGQKINLAGFGIFERRTQGPRKARNPQTKKVIEVPSKKKFVFRASSKIKYQQ
          ***  **  **   *    *        *       * ***  ***   **  ***         ** ***********   *
           **  **  **                  *                  * ***  ***  **  *****         *  *  *
                *   *                                          ***    *   *  *
                                                                              *           *
```

Bu *Bacillus subtilis*, Cpe *Clostridium pasteurianum*; Rme *Rhizobium meliloti*; Ecol *Escherichia coli*(HU)
HTA *Thermoplasma acidophilum*.

B) - Putative alignment of eukaryotic histones with *Methanothermus fervidus* protein HMf

```
H2A    NH2--22aa--LQFPVGVRHRLLRKGNYAERVGAG---APVYLAAVLEYLTAEILELA---GNAARDNKKTRIIPRHLQLAIRNDE----38aa--COOH
H3     NH2--59aa--LLIRKLPFQRLVREIAQDFKTDLRFQ-SSAVMA-LQEASEAYLVGLFEDTNLCAIHAKRVTIMPKDIQLARRIRG----3aa--COOH
H2B    NH2--33aa--KEYSI-YIYKVLKQ--VHPDTGISSK-AMGIMNSFVNDIKERIAGEA--SRLAHYNKRSTITSREIQTAVRLLL----23aa--COOH
H4     NH2--26aa--QGITKPAIRRLARGGVKRISGLIYEETRGVLKVFLENVIRDSVTYTE----HAKRKTVTAMDVVYALKRQG----8aa--COOH
HMF-2                MELPIAPIGRIIKDAG-AERVS---DDARITLAKILEEMGRDIASEA---IKLARHAGRKTIKAEDIELAVRRFKK
                  *  **   *  ****         **   **     ****  **    *  *************
                            *                  **    **        ** ***  ***  *****   ** **
                            *                                   * **  **  ***                 *
                                                                                  *
```

Concensus sequences of eukaryotic histones, from ref.22, HMf, *Methanothermus fervidus*

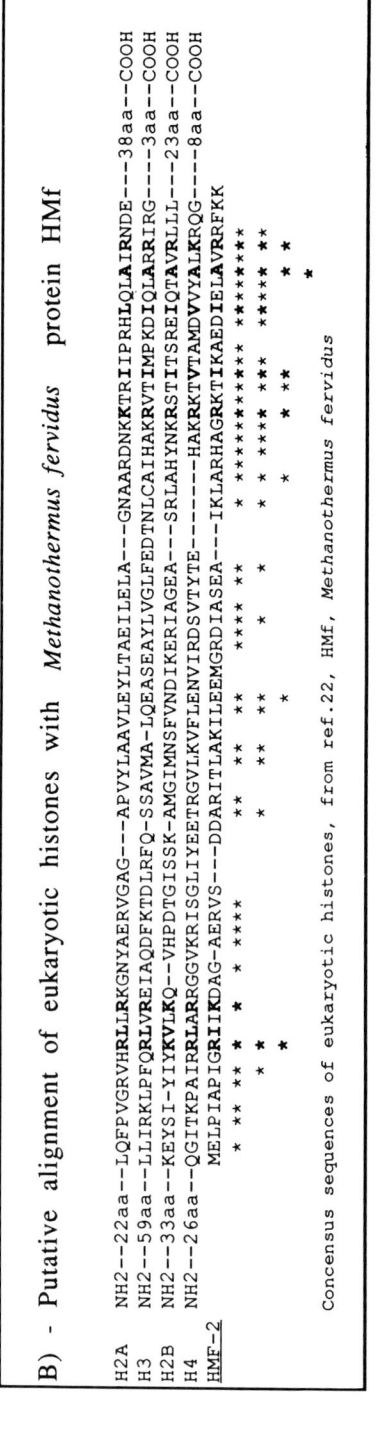

Fig. 2. Alignments of histones and histone-like proteins from archaebacteria, eubacteria and eukaryotes. Conserved amino acids in all sequences are in bold (R=K and I=L=V). Adapted from refs. [23,41]. The numbers of asterisks correspond to the number of amino acids identical with the archaebacterial sequence for each position.

the diameter of eukaryotic nucleosomes [10]. Searcy and Stein [10] suggested that each tetramer is associated in these particles with 40 base pairs of DNA (in contrast to about 150 base pairs in eukaryotic nucleosomes). On the other hand, the HU protein of *Bacillus stearothermophilus* has been crystallized and a different model for protein/protein, and protein/DNA interactions has been proposed [41]. The authors proposed that HU is a dimeric molecule which binds to the phosphate backbone of DNA through two symmetry-related arms. Bending of the DNA should be induced by the cooperative binding of several adjacent dimeric molecules. Conclusive evidence is lacking to make a choice between these models.

2.2.2. The protein MC1
In methanogens, a chromosomal DNA-binding protein of 11 kDa called MC1 (Methanogen Chromosomal protein 1) has been isolated from different species of Methanosarcinaceae [14–22]. MC1 compacts the DNA [19] and induces DNA bending and torsional constraints [21], but it does not form nucleosome-like structures in vitro [16] and has no amino-acid sequence similarities with any known proteins. The DNA-binding domain of MC1 has been determined by cross-linking experiments and peptide sequencing [22]. Depending on the protein/DNA ratio, MC1 stimulates or inhibits DNA transcription in vitro by *E. coli* RNA polymerase; it could therefore participate in the regulation of gene expression via structural alteration of the DNA. A gene encoding a protein homologous to MC1 has just been discovered close to the unique ribosomal RNA operon of *H. salinarium* (Mankin, A., personal communication). This gene is cotranscribed with the rRNA operon, suggesting that MC1 has an important cellular regulatory function.

2.2.3. The protein HMf
Reeve and coworkers [23] have isolated a putative histone from the hyperthermophilic methanogen *Methanothermus fervidus*, using an agarose gel mobility shift assay. This protein, called HMf (Histone *M. fervidus*), increases the mobility of DNA in agarose gels. It is a dimer composed of two closely related isoforms, HMf-1 and HMf-2, of 7.2 kDa. The gene encoding HMf-2 has been cloned and sequenced [23]. HMf is a member of a protein family which includes eukaryotic histones H2A, H2B, H3 and H4. However, the alignments in Fig. 2B show that several gaps are required to align eukaryotic histones between each other and with HMf; accordingly, HMf is more distantly related to eukaryotic histones than is HTa from eubacterial HU proteins. One can speculate that the diversification of the HMf/histone family occurs before the divergence between eukaryotes and archaebacteria.

Reeve and coworkers [24] noticed that additional N-terminal and C-terminal amino-acid residues which are present in the yeast histones H4, but not in HMf, are precisely those which are not essential for the formation of nucleosomes in yeast. This suggests that HMf has conserved the minimal structure required to built a nucleosome. Indeed, DNA restriction fragments bound to HMf migrate as discrete bands in agarose gel, indicating that HMf forms uniform and stable structures (in contrast, similar fragments bound to HTa migrate as diffuse bands) [24]. Furthermore, electron microscopy of HMf–DNA complexes shows quasi-spherical structures, separated by protein-free regions of DNA (Fig. 3A, C). Omission of glutaraldehyde fixation results in a partial loss of HMf from

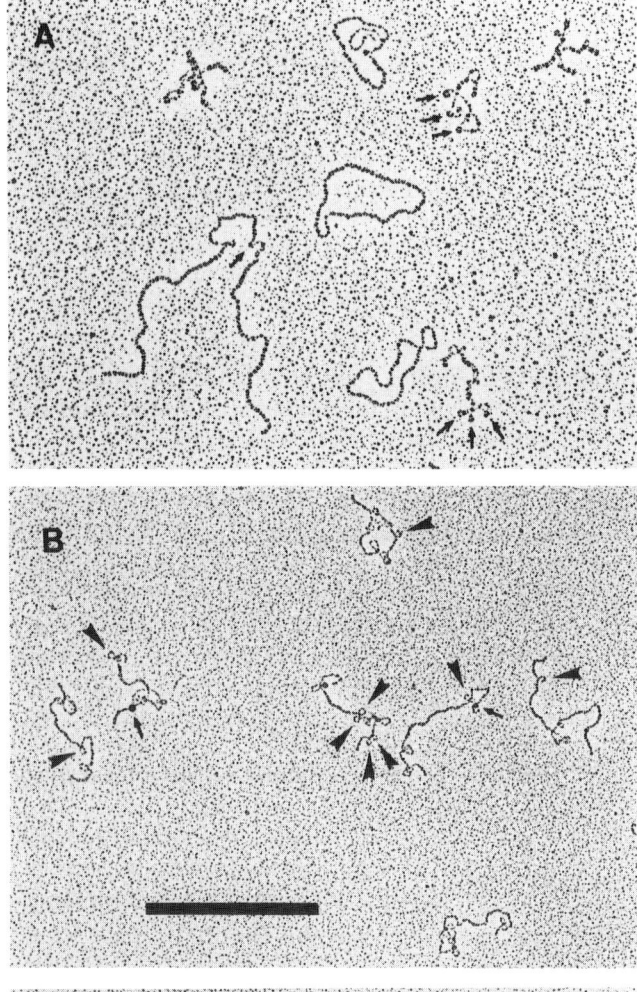

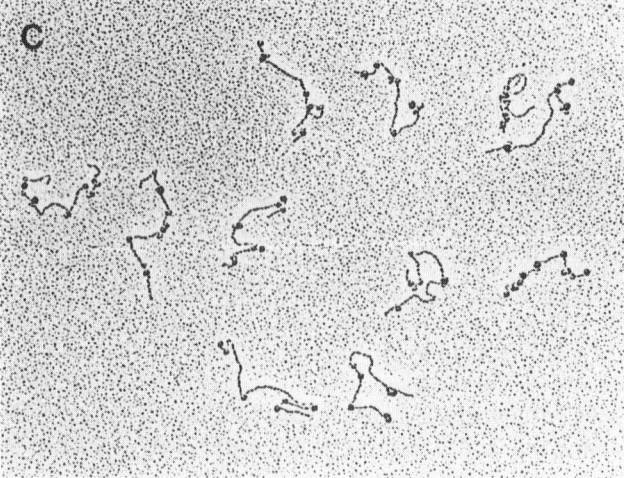

Fig. 3. Binding of HMf protein from *Methanothermus fervidus* to pUC18–DNA. (Adapted from ref. [23].) Panel A shows the stronger binding of HMf to supercoiled DNA compared to open circular or linear DNA. Panel B shows the small loops found with unfixed material. Panel C shows the formation of nucleosome-like structures on linearized fragments with fixation of glutaraldehyde. The bar in panel B represents 0.5 μm. Courtesy R. Lurz.

the preparation, revealing that DNA in these structures was constrained into small loops (Fig. 3B). Finally, HMf binds preferentially to supercoiled DNA (Fig. 3C). All these data strongly suggested that DNA was wrapped in a negative supercoil around the HMf core, as in eukaryotic nucleosomes. However, recent experiments by Musgrave et al. [25] indicate that this may not be the case. These authors show that HMf introduces negative superturns into a relaxed DNA only at low protein/DNA ratio (<0.2) and that it introduces positive superturns at the protein/DNA ratios required for "nucleosome" formation (> 0.5). Since HMf binding increases the helical periodicity of DNA molecules from 10.5 bp to about 11 bp, they suggest that negative supercoiling at low HMf/DNA ratio is due to underwinding of the double helix by HMf monomers, whereas positive supercoiling at higher HMf/DNA ratio is due to wrapping of the DNA in a positive superturn around a core of HMf tetramers [25]. Considering that HMf belongs to the same family as eukaryotic histones, it is remarkable that these proteins have such different effects on DNA structure. Proteins similar to HMf have been recently isolated by Reeve and coworkers from two other Methanobacteriales, the thermophile *Methanobacterium thermoautotrophicum* and the mesophile *Methanobacterium formicicum* [Reeve, personal communication].

2.2.4. Putative nucleosomal organization
Shioda et al. [43,44] visualized by electron microscopy both regions of naked DNA and of DNA covered with particles in the chromosome of *Halobacterium salinarium* isolated from gently lysed cells. In a control experiment, they did not detect such particles in *E. coli*. They also reported the existence of nucleosome-like structures in *S. acidocaldarius* and methanogens (unpublished results cited in ref. [43]). The size of the particles detected in *H. salinarium* (9.5 nm) is similar to that of eukaryotic nucleosomes (10.3 nm); however, this putative archaebacterial "chromatin" is not as regular as eukaryotic chromatin, since not all of the DNA is covered with nucleosomes and since the length of the DNA spacer between the particles is not uniform. In contrast to these results, Bohrmann and coworkers [45] did not visualize nucleosome-like structures in isolated chromosome fibers of *Thermoplasma acidophilum*. These authors also reported that in situ the nucleoid of *T. acidophilum* appears to be highly dispersed in the cytoplasm.

The detection of "nucleosomes" in archaebacteria should be taken with caution since particles have been also visualized on the *E. coli* chromosome in vitro by some investigators [46]. Such structures could be artefacts of specimen preparation (compactosomes) produced by DNA supercoiling [47]. In fact, it is likely that the archaebacterial chromatin resemble more the dynamic chromatin of eubacteria, in which all genes can be activated at any time, than the stable chromatin of eukaryotes. Reeve and coworkers [24] have shown for example that the binding of HMf to the DNA is dynamic and readily reversible. If archaebacterial nucleosomes exist, they should be therefore less stable than eukaryotic ones. This could explain contradictory results about their existence in archaebacteria.

2.3. DNA stability in hyperthermophiles

Linear DNA molecules are denatured in vitro when they are exposed to current growth temperatures of hyperthermophilic archaebacteria. Accordingly, it is often argued that specific devices should exist to prevent separation of DNA strands in these microorganisms, such as specific DNA-binding proteins, polyamines, high salt concentrations (in the case of methanogens), or positive supercoiling (see next section). Indeed, HTa and HMf, for example, increase the T_m of linear DNA in vitro by as much as 40°C and 25°C, respectively [7,23]. However, this does not prove that these proteins are specifically required for thermal adaptation, since histone-like proteins isolated from mesophilic eubacteria also protect DNA against thermal denaturation in vitro [48]. Moreover, the thermostability of intracellular DNA cannot be compared to that of a linear DNA in vitro since the DNA is divided in vivo into topologically closed domains by transcribing RNA polymerases and other DNA-binding proteins which prevent the free rotation of the two DNA strands around each other. Such DNA is intrinsically resistant to denaturation because the two strands cannot be unlinked without breaking at least one of them [49]. We have recently observed in our laboratory that covalently closed circular plasmids remain double-stranded up to 110°C in buffer conditions preventing DNA cleavage [Marguet and Forterre, unpublished data]. The stability of the double helical structure in hyperthermophiles may therefore simply be due to the universal topological properties of the DNA. This hypothesis would explain why there is no correlation between the GC content and the growth temperatures of archaebacteria (Table 1, below) whereas stable RNAs are more GC-rich in hyperthermophilic archaebacteria than in mesophilic ones [50,51]. One can suppose that the secondary structure of stable RNA molecules needs to be protected against thermal denaturation by a high GC content precisely because they do not belong to a topologically closed structure.

A problem which has not yet been addressed in extreme and hyperthermophiles is how these microorganisms avoid heat-induced covalent modifications of the DNA. Exposure of DNA molecules to very high temperatures in vitro induces depurination [52], deamination of cytosines [53], and cleavage of the phosphodiester bonds [54]. Several observations suggest that these covalent modifications probably also occur in vivo; for example, cytosine deamination at high temperature is mutagenic [55], and C-5-methyl-cytosine, which is thermolabile, is replaced by the thermoresistant base N4-methyl-cytosine both in thermophilic archaebacteria and eubacteria [56]. A priori, heat-induced cleavage of phosphodiester bonds may be very harmful since it removes the topological barriers to denaturation previously discussed [39]. Recent studies in our laboratory indicate that cleavage and depurination of covalently closed circular DNA at high temperature occurs at the same rate as depurination. As a consequence, such DNA is quite resistant to thermally induced hydrolysis in buffers which maintain a neutral pH at high temperature, such as phosphate buffer. At low pH (Tris buffer) DNA is partly protected at high temperature by the divalent salts $MgCl_2$ and $ZnCl_2$, and completely protected by very high concentrations of monovalent KCl salt (ref. [39], and Marguet and Forterre, unpublished result). The latter result suggests that the high intracellular concentration of K^+ in some hyperthermophilic methanogens could play a role in the protection of DNA against thermodegradation. The protective effect of Mg^{2+} and Zn^{2+} on DNA at high temperature is in striking contrast to the catalysis by the same salts of heat-induced RNA hydrolysis [57,58]. This suggests that RNA stability could be more of a problem than DNA stability in hyperthermophiles.

3. DNA topoisomerases and DNA topology

DNA supercoiling is an essential parameter of chromosome structure, which has a profound influence on the expression and replication of the genetic material [59,60]. In one superturn, the two strands of a DNA duplex are coiled around each other. This superturn is said to be positive if the two strands are coiled in the same sense as in a classical turn (B-DNA) and negative if they are coiled in the opposite direction. The additional linking between the two DNA strands, due to supercoiling, intuitively accounts for the famous relationship

$$Lk = Tw + Wr, \tag{1}$$

in which Lk is the total number of links between the two DNA strands (Linking number), while Tw (Twist number) and Wr (writhing number) are approximations of the numbers of duplex turns and superturns, respectively. The linking number is a topological invariant which cannot be changed as far as the DNA is topologically closed (as in a circular duplex), whereas the twist can change with temperature, salt concentration and interaction of the double helix with DNA-intercalating agents and DNA-binding proteins. Accordingly, any difference between the linking number of a covalently closed DNA duplex and the twist number will induce supercoiling.

Supercoiling helps DNA bending required for the formation of nucleoprotein complexes. Furthermore, the change in the twist, which is concomitant to any change in supercoiling, according to

$$\Delta Lk = \Delta Tw + \Delta Wr,$$

influences specific DNA/protein interactions. Finally, negative supercoiling (which corresponds to a deficit in the linking number) facilitates all processes which require untwisting or unwinding of DNA.

In all cells, the DNA is supercoiled in the course of transcription and replication: waves of positive and negative superturns are produced in front of and behind the ongoing polymerases, respectively (Fig. 4) [61–63]. These superturns are processed by DNA topoisomerases that prevent their accumulation and regulate the level of supercoiling. Supercoiling can be also produced by wrapping the DNA around proteins (for example in the nucleosomes), or by specific DNA topoisomerases (Fig. 4).

DNA topoisomerases are nicking-closing enzymes which change the linking number of a DNA molecule [64,65]. They also catalyze intramolecular knotting–unknotting of DNA and catenation–decatenation of DNA duplexes. DNA topoisomerases are involved in the resolution of the topological problems which occur during DNA replication, transcription, and recombination, and in the course of chromosome condensation and segregation. They are also essential for chromosome stability by reducing the level of recombination [66].

DNA topoisomerases have been classified as two types according to their mechanistic properties: type I catalyzes the crossing of two DNA strands through each other via a transient single-stranded break, whereas type II catalyzes the crossing of two DNA duplexes via a transient double-stranded break (Fig. 5). Type I DNA topoisomerases are monomeric enzymes, usually ATP-independent, whereas type II DNA topoisomerases

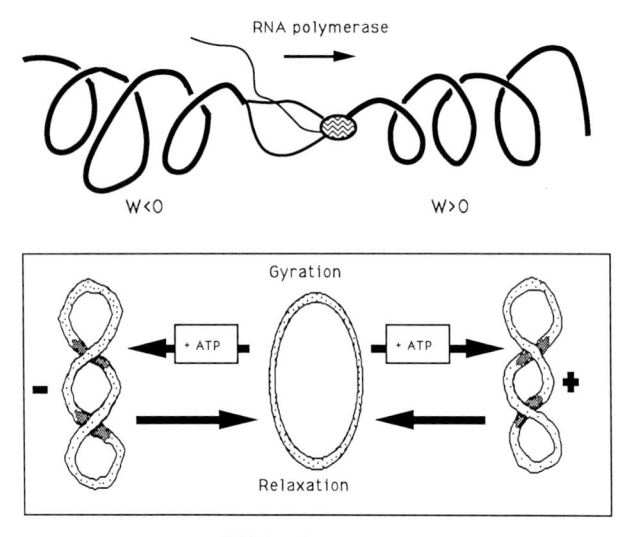

Fig. 4. The level of cellular DNA supercoiling is dependent on both DNA-tracking processes, such as transcription, and the activity of DNA topoisomerases [61–63]. W is the writhing number; $W > 0$ corresponds to positive supercoiling and $W < 0$ corresponds to negative supercoiling.

are ATP-dependent and exhibit a multi-subunit structure with a dyad symmetry. In the course of the reaction of topoisomerization, DNA topoisomerases are linked by a covalent tyrosine-phosphate bond to either the $3'$ end or the $5'$ end of the transient break.

It was recognized early that DNA topoisomerases are interesting enzymes from an evolutionary point of view, since they catalyze different reactions in eukaryotes and in eubacteria, with probably important consequences on chromosome structure and on relationships between DNA topology and gene expression [67]. In eukaryotes, the major type I DNA topoisomerase relaxes either positive or negative superturns, whereas the two eubacterial type I DNA topoisomerases that have been described (protein ω and *E. coli* DNA topoisomerase III) relax only negative superturns (Fig. 6). In addition, the eukaryotic type I DNA topoisomerase is transiently linked to the $3'$ end of the DNA break, whereas the two eubacterial type I DNA topoisomerases are covalently linked to the $5'$ end. These ($3'$) and ($5'$) type I DNA topoisomerases are not only mechanistically but also phylogenetically unrelated, as indicated by the complete absence of similarity in their amino-acid sequences.

The eubacterial and eukaryotic type II DNA topoisomerases are also very different from each other mechanistically, since only the eubacterial enzyme exhibits a DNA gyrase activity (i.e., the introduction of negative superturns into the DNA in the presence of ATP) (Fig. 6). However, they exhibit extensive similarities in their amino-acid sequences [68]. As a consequence of gyrase activity, the level of supercoiling can be adjusted rapidly in eubacteria by changing the intracellular ATP pool [69]. This could play a major role in the regulation of eubacterial gene expression. In contrast, negative

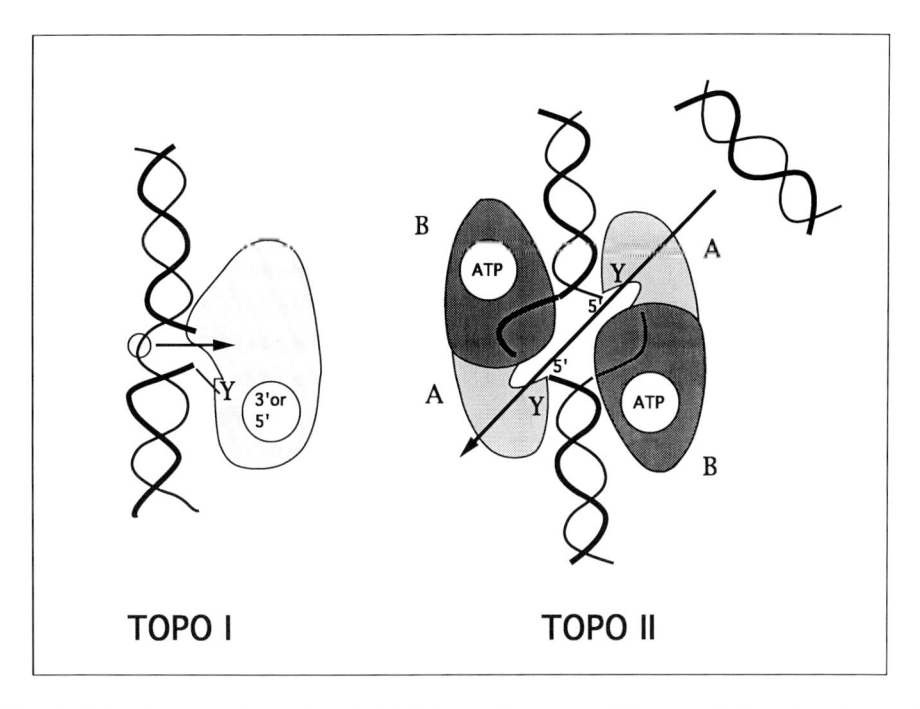

Fig. 5. Major features of type I and II DNA topoisomerases. The A and B regions in type II DNA topoisomerases correspond either to two polypeptides (DNA gyrase) or to two domains of a single polypeptide (see also Fig. 10). In the course of the reaction of topoisomerization, the enzymes are transiently linked by a tyrosine (Y)–phosphate bond either to the 3′ end or to the 5′ end of the DNA break. The circle labelled ATP corresponds to the ATP binding site.

supercoiling of the chromosome in eukaryotes is due to the wrapping of the DNA around nucleosomes (Fig. 6): there is no detectable free torsional tension and the involvement of DNA supercoiling in gene expression has not yet been clearly demonstrated in vivo.

Recently, the above pattern of DNA topoisomerase activities in eubacteria and eukaryotes has become more complex with the identification of several new genes encoding DNA topoisomerases. Surprisingly, a yeast gene, TOP3, encodes a protein homologous to the eubacterial type I (5′) DNA topoisomerases [66]. Also unexpectedly, two *E. coli* genes, parC and parE, encode proteins highly similar to the DNA-gyrase subunits gyrA and gyrB, respectively [70]. This new type II DNA topoisomerase (named *E. coli* DNA topoisomerase IV) has no gyrase activity. The yeast DNA topoisomerase encoded by TOP3 is involved in suppression of recombination, whereas *E. coli* Topo IV is required for chromosome segregation. Otherwise, it is not yet known if these new DNA topoisomerases play a role in the overall regulation pattern of DNA topology.

Type I and type II DNA topoisomerases have also been identified in archaebacteria (for a previous review see ref. [71]). They exhibit both classical and novel features. Up to now, biochemical studies have been performed only with enzymes isolated from thermophiles, whereas physiological and genetic studies have been performed mainly with halophiles.

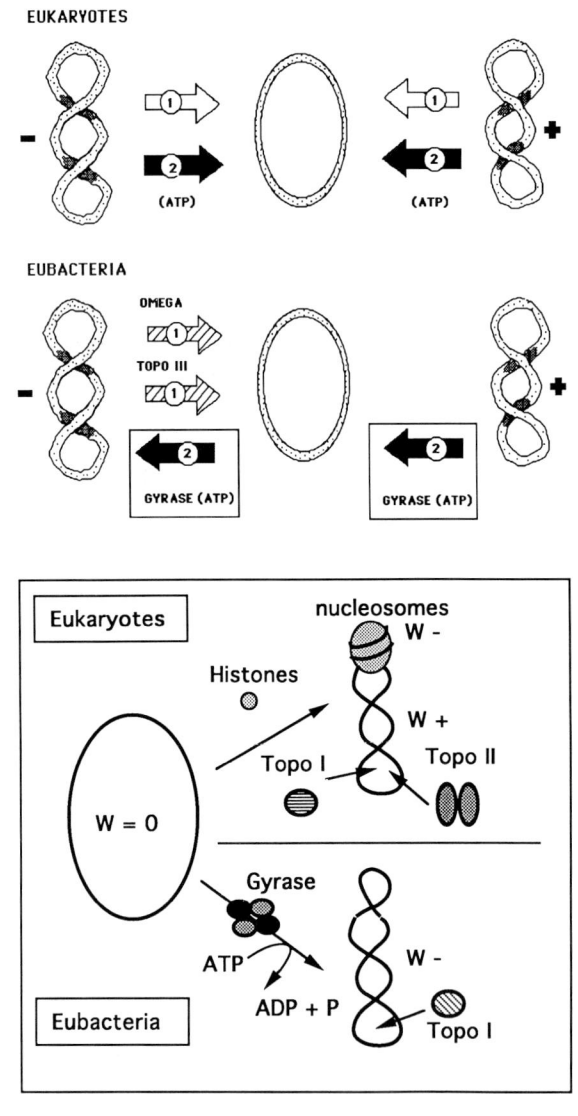

Fig. 6. DNA topoisomerases and DNA topology in eubacteria and eukaryotes. Upper panel: DNA topoisomerases of the two domains; the arrows with the same shading correspond to homologous DNA topoisomerases, 1 and 2 are for type I and type II DNA topoisomerases, respectively. Lower panel: origin of negative supercoiling in eukaryotes and eubacteria (see the text). $W > 0$, positive supercoiling; $W < 0$, negative supercoiling.

3.1. Reverse gyrase

3.1.1. Discovery

In 1984, Kikuchi and Asai[72] discovered that partially purified protein fractions of *S. acidocaldarius* transform negatively supercoiled DNA into a positively supercoiled form at high temperature in the presence of ATP [72]. They attributed this activity to a new DNA topoisomerase, which they called reverse gyrase, in reference to the canonical eubacterial DNA gyrase which catalyzes the opposite reaction. At the same time, Forterre, Duguet and coworkers [73,74] independently undertook the search for DNA topoisomerases in *Sulfolobus* and also detected this positive supercoiling activity. They found that reverse gyrase can supercoil relaxed DNA [74]. These early studies demonstrated that ATP-dependent positive supercoiling occurred in a catalytic manner [72,74]. Accordingly, reverse gyration could not be explained by the combination of a DNA-binding protein, wrapping the DNA into positive superturns, and a conventional DNA topoisomerase, relaxing negative superturns. Since positively supercoiled DNA is overlinked, it was readily suggested that reverse gyrase could be specifically required to prevent DNA denaturation in extreme thermophiles (see below).

3.1.2. Biochemical characterization
Kikuchi and Asai initially described reverse gyrase as a type II DNA topoisomerase with four subunits [72], in agreement with its ATP-dependence and supercoiling capacity (both reminiscent of DNA gyrase). However, Mirambeau et al. [73] noticed that reverse gyrase is resistant to novobiocin, an ubiquitous inhibitor of type II DNA topoisomerases, and can function with ATP concentrations as low as $10\,\mu M$, whereas type II DNA topoisomerases require ATP concentrations in the millimolar range. Subsequently, Forterre et al. [74] and Nakasu and Kikuchi [75] demonstrated that reverse gyrase is not a type II but a type I DNA topoisomerase, since it changes the linking number of DNA by steps of 1. They also found that reverse gyrase activity was not associated with a multimeric structure, but with a polypeptide of 120–130 kDa, in agreement with the classical monomeric structure of type I DNA topoisomerases (the four polypeptides detected by Kikuchi and Asai were probably the major RNA polymerase subunits, according to their size and abundance). Reverse gyrase from *S. acidocaldarius* was purified to near homogeneity by Nakasu and Kikuchi [75] and Nadal et al. [76]. Purified fractions indeed only contained the 120–130 kDa polypeptide.

A DNA-dependent ATPase activity was found associated with the purified enzyme [77]. In contrast, an ATP-independent relaxation activity detected in partially purified fractions of reverse gyrase [74] was absent from completely purified fractions and probably corresponds to a distinct DNA topoisomerase (see below). A reverse gyrase with similar structure and properties was purified later by Slezarev [78] from another extremely thermophilic archaebacterium, *Desulfurococcus amylolyticus* [78].

3.1.3. Mechanistic studies
Incubation of a stoichiometric amount of reverse gyrase with DNA, in the absence of either magnesium or ATP, induced single-stranded DNA cleavage [75,79]. The induction of single-stranded DNA breaks was expected for a type I DNA topoisomerase, but the initiation of DNA cleavage by omitting ATP once more outlined the specificity of reverse gyrase. After cleavage, reverse gyrase was covalently linked by a tyrosine phosphate bond [80] to the 5′ end of the break [79,81], indicating that reverse gyrase could belong to the same family as other (5′) type I DNA topoisomerases. Indeed, reverse gyrase shares other features with protein ω, the prototype eubacterial member of this family: (i) it relaxes only negative superturns, whereas the eukaryotic (3′) DNA topoisomerase I

relaxes both positive and negative superturns; (ii) it cleaves the DNA between the third and the fourth bases after a cytosine residue [81]; and (iii) it is inhibited by excess of single-stranded DNA [77,82]. The latter data suggest that protein ω and reverse gyrase bind preferentially onto transiently melted regions in a DNA duplex, in contrast to the eukaryotic (3′) type I DNA topoisomerases that bind preferentially to perfectly matched regions of DNA duplex. Since transiently melted regions occur in negatively supercoiled DNA but not in positively supercoiled DNA, this could explain why protein ω normally only relaxes negatively supercoiled DNA. Indeed, protein ω can relax a positively supercoiled DNA harboring an inserted heteroduplex region, producing a single-stranded bubble [83]. Similarly, Slezarev et al. [82] have shown that reverse gyrase can introduce additional positive superturns in a DNA already positively supercoiled if the latter harbors such single-stranded bubble. In that case, the coupling of ATP hydrolysis to the topoisomerase activity reverses the trends of the reaction, compared to those performed by the ω protein.

The binding of a stoichiometric amount of reverse gyrase onto the DNA in the absence of ATP at high temperature introduces negative superturns [71,79]. In a similar experiment, eubacterial DNA gyrase introduces positive superturns because of the right-handed wrapping of the DNA around gyrase molecules [84]. In the case of reverse gyrase, it has not been determined whether the structural change of DNA induced by the binding of the enzyme corresponds to left-hand wrapping of the DNA around reverse gyrase, to the stabilization of single-stranded regions, or to the partial unwinding or untwisting of the double helix.

3.1.4. Primary structure

The gene encoding reverse gyrase was recently cloned using antibodies prepared against the purified enzyme [85]. This gene encodes a protein of 143 kDa. Examination of the amino-acid sequence indicates that reverse gyrase is composed of two distinct domains: a C-terminal domain which exhibits clear-cut similarities with type I (5′) DNA topoisomerases, in agreement with the mechanistic properties of reverse gyrase, and an N-terminal domain which exhibits no sequence similarities with other DNA topoisomerases, but which contains a "type A nucleotide binding site" reading AxxGxGKT. This result definitively demonstrates that reverse gyrase is a unique DNA topoisomerase, very different both from other type I DNA topoisomerases which are ATP independent, and from type II DNA topoisomerases which have a different nucleotide binding site.

Comparison of the amino-acid sequence of the topoisomerase domain of reverse gyrase with those of the three other (5′) type I DNA topoisomerases sequenced to date indicates that reverse gyrase is slightly more related to protein ω than to *E. coli* DNA topoisomerase III and yeast TOP3 protein [85]. This observation, and the fact that several gaps are required to correctly align these four proteins, suggest that their genes diverged from each other before the split between archaebacteria, eubacteria and eukaryotes.

3.1.5. Mechanism of reverse gyration

How does a type I DNA topoisomerase, such as reverse gyrase, catalyze DNA supercoiling? Slezarev and Kozyavkin [82] suggested that reverse gyrase binds onto melted (single-stranded) regions of DNA duplex and catalyzes the unidirectional crossing of one strand through another to reform the duplex, continuously increasing the linking number. This hypothesis was suggested by experiments showing that reverse gyrase requires single-stranded DNA to initiate positive supercoiling (see above). In this model, ATP is required for the conformational change of the enzyme associated with the unidirectional strand passage.

In a different model, Zhang et al.[86] suggested that reverse gyrase activity results from the combined action of an ATP dependent helicase and a "classical" type I DNA topoisomerase: the helicase would create two waves of positive and negative superturns, respectively, whereas the DNA topoisomerase I would relax only the negative ones, leading to positive supercoiling. Effectively, analysis of the amino-acid sequence of the N-terminal domain of reverse gyrase strongly suggests that this enzyme exhibits a helicase activity[85]. In addition to a putative ATP-binding site, this domain harbors several amino-acid motifs which are found at similar locations and in the same arrangement in different RNA and DNA helicases. Accordingly, reverse gyrase activity could be explained by the "helicase plus topoisomerase" model, but with the two activities on the same polypeptide.

Two versions of the "helicase plus topoisomerase" model can be proposed. In the dynamic scenario, the enzymes slide onto the DNA to produce the waves of positive and negative superturns, as in the case of the RNA polymerase or SV40 T antigen helicase[88]. Another possibility is that the helicase activity helps to create the single-stranded sites required to initiate the reaction[89]. In that case, reverse gyrase would supercoil the DNA without moving along the molecule. In both the dynamic and the static models, the single-stranded region required to initiate the reaction could produce the entry site for the helicase domain of reverse gyrase.

3.1.6. Distribution of reverse gyrase in the living world
Reverse gyrase could not be detected in crude extracts of *Sulfolobus acidocaldarius* because of a strong nuclease activity present in this strain[73]. In contrast, Collin et al.[89] have shown that ATP-dependent positive supercoiling (probably reverse gyrase activity) can be easily detected in crude extracts of various other extreme and hyperthermophilic sulfothermophilic archaebacteria which lack such nuclease activity. They reported the presence of reverse gyrase in all the sulfothermophiles tested, with the exception of the moderate thermophile *T. acidophilum,* and its absence in moderate thermophilic and mesophilic methanogens[89]. Forterre et al.[90] also reported the absence of reverse gyrase in partially purified fractions of *T. acidophilum*. Later on, Bouthier de la Tour et al[91] extended these results by showing that reverse gyrase activity was present not only in extremely thermophilic sulfothermophiles but also in the hyperthermophilic methanogens *Mt. fervidus* (Fig. 7) and *Methanopyrus kandleri*. These authors first considered reverse gyrase to be a hallmark of hyperthermophilic archaebacteria because they failed to detect this activity in the hyperthermophilic eubacterium *Thermotoga maritima*[89,91]. However, Bouthier de la Tour et al.[92] later showed that reverse gyrase was in fact present in *T. maritima* and in three other Thermotogales. In their first experiments, reverse gyrase was hidden in crude extracts of *T. maritima* by a very active ATP-independent relaxation activity which was stimulated by the high salt concentrations used to suppress nuclease activities. A reverse gyrase with

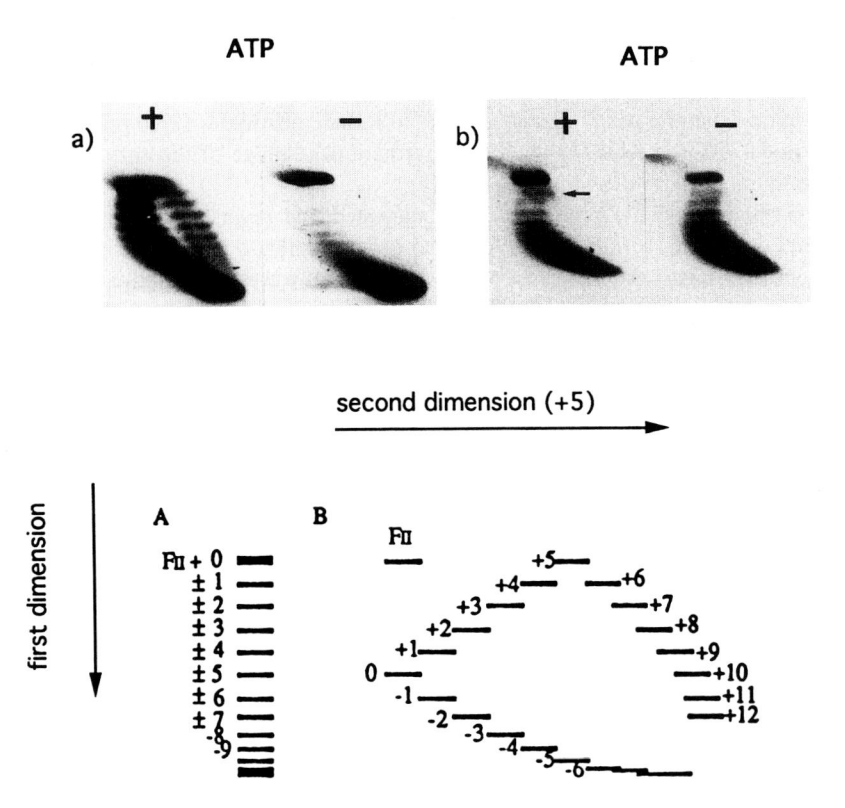

Fig. 7. Detection of reverse gyrase activity in crude extract of extremely thermophilic archaebacteria by two-dimensional gel electrophoresis. (Adapted from ref. [91].) Plasmid DNA (negatively supercoiled) was incubated with cellular crude extracts. Upper panel: (a) reverse gyrase activity in *Methanothermus fervidus*; the right-hand branch of the +ATP-arch corresponds to positively supercoiled topoisomers, the gel without ATP shows the position of the control plasmid. (b) ATP-independent relaxation activity in *Methanococcus thermolithotrophicus*. Lower panel: schematic representation of (A) one- and (B) two-dimensional gel electrophoresis of a mixture of positively and negatively supercoiled topoisomers. (From ref. [76]). The second dimension is performed in the presence of a DNA-intercalating drug (such as ethidium bromide) which has removed five double helical turns (in this example), thereby increasing the writhing number by +5. Accordingly, a DNA which was relaxed in the absence of drug has now 5 positive superturns.

structure and properties very similar to that in archaebacteria has been recently purified from *Calderobacterium hydrogenophilum* [Mikulik, personal communication] which belongs to a group of extreme thermophilic eubacteria distinct from the Thermococcales. Table 1 summarizes the distribution of reverse gyrase in prokaryotes.

3.1.7. Putative roles of reverse gyrase

The excellent correlation between the presence of reverse gyrase and the extreme thermophilic phenotype, both in archaebacteria and in eubacteria (Table 1) strongly suggests that reverse gyrase helps to maintain the DNA in a suitable conformation at

TABLE 1
Distribution of reverse gyrase activity in prokaryotes[a]

Prokaryote[b]	e/c [c]	T_{opt} [c]	G+C (%)	RvG act. [c]
Extremely thermophilic archaebacteria				
Pyrodictium occultum	c	105	62	+
Pyrococcus furiosus	c	100	38	+
Pyrobaculum islandicum	c	100	46	+
Methanopyrus kandleri	e	100	60	+
Staphylothermus marinus	c	90	35	+
Thermococcus celer (3)	e	90	56	+
Acidianus infernus	c	90	31	+
Thermoproteus tenax (2)	c	88	56	+
Methanothermus fervidus	e	83	33	+
Desulfurococcus mobilis (4)	c	85	51	+[d]
Archaeoglobus fulgidus	e	83	46	+
Sulfolobus acidocaldarius (4)	c	75	37	+[d]
Extremely thermophilic eubacteria				
Thermotoga maritima		80	40	+
Thermotoga thermarum		80	40	+
Fervidobacterium islandicum		75	40	+
Thermosipho africanus		75	30	+
Calderobacterium hydrogenophilum		75	?	+[d]
Moderately thermophilic archaebacteria				
Mc. thermolitotrophicus	e	65	32	ND
M. thermoautotrophicum	e	65	49	ND
Thermoplasma acidophilum (2)	e	60	46	ND
Halobacterium salinarium	e	44	67	ND
Mesophilic archaebacteria				
Methanobacterium ivanovii	e	37	37	ND
Methanolobus siciliae	e	37	40	ND
Methanosarcina barkeri	e	37	37	ND

[a] Data are from references [89,91,92] and from Mikulik (personal communication) for *Calderobacterium hydrogenophilum*. Reverse gyrase activity has not been detected in moderately thermophilic and mesophilic eubacteria from the genera *Bacillus* and *Thermus* [89]. In all cells lacking reverse gyrase activity, except *Halobacterium halobium*, one can detect an ATP-independent relaxation activity.

[b] Number of species of the same genus checked is given in parentheses.

[c] Abbreviations: RvG act., reverse-gyrase activity; +, detection of ATP-dependent positive supercoiling in crude extract; ND, not detected (see also Fig. 7); e, euryarchaeota; c, crenarchaeota; T_{opt}, optimal growth temperature.

[d] Reverse gyrase has been purified to homogeneity.

extremely high temperatures. Slezarev and Kozyavkin [82] suggest that reverse gyrase senses the topological state of the DNA in extreme thermophiles via its preferential binding to partially melted DNA at high temperature, and introduces positive superturns in order to maintain the DNA in a double-stranded state. However, we did not detect melted DNA regions in covalently closed DNA up to 110°C [Marguet and Forterre, unpublished data]. Furthermore, we have already noticed that any topologically closed DNA (negatively or positively supercoiled) is highly resistant to thermal denaturation. Furthermore, we observed that a DNA positively supercoiled by reverse gyrase is not significantly more resistant to depurination and cleavage than a negatively supercoiled DNA [Marguet and Forterre, unpublished data].

It remains that single-stranded DNA regions are produced in the course of DNA replication, recombination and repair, or could be associated with transient waves of negative supercoiling. Reverse gyrase could be required to promote rapid renaturation of such melted DNA in order to prevent abnormal protein/DNA interactions at regulatory sequences leading to uncontrolled gene expression, and/or to prevent covalent modifications of single-stranded DNA at high temperatures, since single-stranded DNA is more susceptible to degradation than double-stranded DNA [51–53].

Besides its putative role in DNA protection, Kikuchi and Asai suggested [72] that reverse gyrase may be used to eliminate DNA structures, such as cruciforms and Z-DNA, otherwise stabilized by negative supercoiling. These authors also suggested that reverse gyrase helps to dissociate nucleosome-like structures in which the DNA is negatively supercoiled. Indeed, it has been shown that the wave of positive supercoiling produced by transcription partly destabilizes eukaryotic nucleosomes [93]. Nevertheless, eukaryotic nucleosomes can also be formed on positively supercoiled DNA (although less efficiently than on negatively supercoiled DNA) [94].

Finally, the composite helicase–topoisomerase structure of reverse gyrase suggests that this unusual enzyme could also be involved in any one of the mechanisms which requires the concerted action of such activities [Duguet, M., personal communication].

3.2. Other DNA topoisomerases in thermophilic archaebacteria

In their first paper, Kikuchi and Asai [72] reported that *S. acidocaldarius* contained, besides a reverse gyrase, one ATP-independent and two ATP-dependent thermophilic DNA topoisomerases (including one DNA gyrase). However, their purification procedure lacked a step to remove DNA, so that at least one of the ATP-dependent topoisomerases probably corresponded to the reverse gyrase copurifying with DNA (discussed in ref. [74]). Later work did not confirm the presence of a classical gyrase but demonstrated the presence of at least one type II DNA topoisomerase, and probably one ATP-independent type I DNA topoisomerase in *Sulfolobus* and in other thermophilic archaebacteria.

3.2.1. Sulfolobus type II DNA topoisomerase
A type II DNA topoisomerase has been detected by Kikuchi and coworkers in *S. acidocaldarius* [95]. This enzyme seems to be much less abundant and active than reverse gyrase. It exhibits the usual ATP requirement of type II DNA topoisomerases (in the millimolar range) and can relax either negatively or positively supercoiled DNA.

The unknotting reaction (typical of type II DNA topoisomerases) is ten times more efficient than relaxation. The most purified fraction contained two major polypeptides of 40 and 60 kDa. The *S. acidocaldarius* type II DNA topoisomerase has no DNA gyrase activity. This suggested that it could specifically resemble the eukaryotic type II DNA topoisomerase; however, taking into account its putative dimeric structure, it could also resemble the new type II DNA topoisomerase (topo IV) recently discovered in *E. coli* [70]. We have recently detected in our laboratory a type II DNA topoisomerase in *Sulfolobus shibatae*. This enzyme catalyzes the same reactions as the enzyme from *Sulfolobus solfataricus* and exhibits a pattern of drug sensitivity very similar to that of the eukaryotic enzyme [Bergerat, A., this laboratory].

3.2.2. ATP-independent DNA topoisomerases

ATP-independent DNA relaxation was detected by Forterre et al. [74] in partially purified fraction of *Sulfolobus* reverse gyrase. At that time, it was not clear whether it corresponded to a residual activity of reverse gyrase in the absence of ATP, or to another DNA topoisomerase. In later experiments, significant ATP-independent relaxation activity was not observed in purified fraction of reverse gyrase [75,79]. Recently, Slezarev et al. [96] reported that an ATP-independent relaxing activity which contaminated their first preparation of reverse gyrase from *D. amylolyticus* can be separated from the latter by phenyl-sepharose chromatography. This activity is specific for negative superturns and thus resembles classical eubacterial type I DNA topoisomerases. Slezarev et al. described this activity as a new type I DNA topoisomerase, associated with a monomer of 108 kDa, which they call DNA topoisomerase III. However, the association of this monomer with the topoisomerase activity has not been demonstrated, and it is not known whether the putative enzyme is phylogenetically more related to eubacterial DNA topoisomerase III than to eubacterial protein ω.

An ATP-independent DNA topoisomerase activity has also been described in crude extracts of archaebacteria lacking reverse gyrase activity [91] (Fig. 7) and has been partially purified from *T. acidophilum* [90]. This enzyme specifically relaxes negatively supercoiled DNA and could be phylogenetically related either to reverse gyrase and/or to the DNA topoisomerase III of *D. amylolyticus*.

Fig. 8 summarizes the different topoisomerase activities discovered up to now in extremely thermophilic archaebacteria.

3.3. Topological state of the DNA in extremely thermophilic archaebacteria

Since reverse gyrase is the major DNA topoisomerase in extremely thermophilic archaebacteria, it was tempting to speculate that DNA in these unusual microorganisms is positively supercoiled. For a long time, the only topologically closed circular DNA from extremely thermophilic archaebacteria available for topological analysis has been the genome of the UV-inducible virus SSV1 from *Sulfolobus shibatae* (previously named *Sulfolobus* sp. B12, see ref. [97]). SSV1 viruses and cells of *S. shibatae* harbor a covalently closed circular double-stranded DNA molecule of 15 kb. Nadal et al. [98] made the exciting discovery that this DNA is indeed positively supercoiled. The viral particles contained only highly positively supercoiled DNA, whereas intracellular SSV1 DNA exhibited a broad distribution of topoisomers, with still a majority of

EXTREME THERMOPHILIC ARCHAEBACTERIA

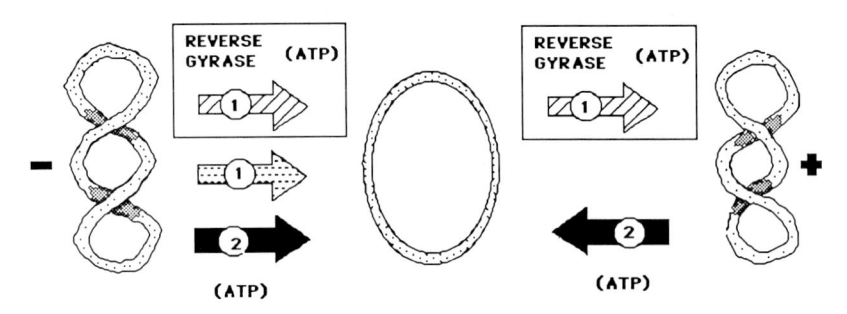

Fig. 8. DNA topoisomerases in extremely thermophilic archaebacteria. The arrows with the same shading correspond to homologous DNA topoisomerases, 1 and 2 are for type I and type II DNA topoisomerases, respectively. $W > 0$, positive supercoiling; $W < 0$, negative supercoiling.

positively supercoiled forms. The production of viral SSV1 particles and intracellular SSV1 plasmids is amplified by irradiation of the cells with UV [99]. The first studies were performed with DNA purified from such UV-irradiated cells. We have recently shown that the few SSV1 DNA molecules which can be isolated from cells of S. shibatae without UV-irradiation are much less positively supercoiled than those isolated from UV-treated cells [39] (Fig. 9). This indicates that exposure to UV triggers positive supercoiling.

A small plasmid of 3.4 kb, called pGT5, which could be a better indicator of DNA topology in hyperthermophiles, was recently discovered in a new strain of sulfur-dependent archaebacterium (GE5) [100]. Although crude extracts of strain GE5 exhibit a very strong reverse gyrase activity, the plasmid pGT5 appears to be negatively supercoiled at room temperature. However, the superhelical density of pGT5 is only half of those from typical eubacterial plasmids [101]. Furthermore, if one takes into account the effect of temperature on the DNA twisting angle in the range from 25°C to 95°C ($-0.01°/°C$) [Charbonnier, F., this laboratory and Duguet, M., personal communication], pGT5 should be relaxed at the physiological temperature of 95°C [101]. The topology of pGT5 resembles therefore those of SSV1 plasmids isolated from cells which have not been UV irradiated. This indicates that DNA topology is indeed different in extremely thermophilic archaebacteria compared to all other organisms (including halophilic archaebacteria whose plasmids are highly negatively supercoiled [Charbonnier, F., this laboratory]). This unusual topology is probably due to reverse gyrase. However, the relaxed state of pGT5 also demonstrates the existence in extreme thermophiles of a mechanism which should partly counteract the activity of reverse gyrase in vivo. A simple possibility is that the DNA is partitioned between protein-free regions of DNA that are positively supercoiled by reverse gyrase, and regions in which the DNA is negatively supercoiled by interaction with DNA-binding proteins [39].

Attempts have been made recently to analyze directly the topological state of chromosomal DNA in extremely thermophilic archaebacteria. A nucleoid-like structure has been isolated from S. acidocaldarius using procedures similar to those previously used to analyze DNA topology of E. coli chromosome [32,33,38,39]. Unfortunately,

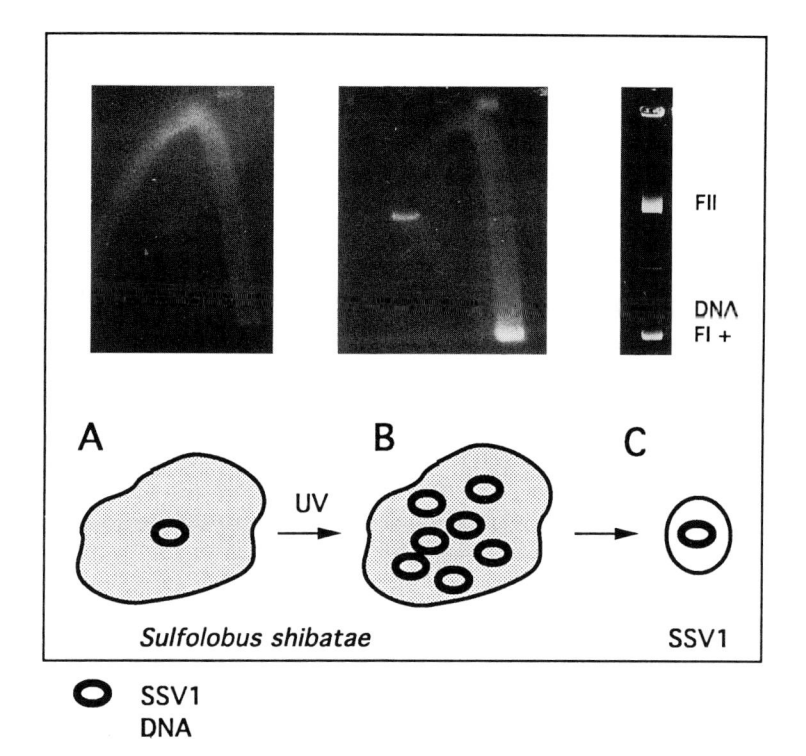

Fig. 9. Positive supercoiling of SSV1 DNA. Panels A and B show two-dimensional agarose gel electrophoresis of SSV1 DNA isolated from cells of *Sulfolobus shibatae,* (A) before and (B) after UV induction (for methods see refs. [39,98]). The left-hand branch of the arch visible in A corresponds to negatively supercoiled DNA, the top of the arch corresponds to relaxed DNA and the right-hand branch corresponds to positively supercoiled DNA. The upper bands in A and B correspond to form II (open circular) and the middle band in B corresponds to form III (linear SSV1) (pictures courtesy of G. Mirambeau). Panel C shows a one-dimensional agarose gel electrophoresis of SSV1 DNA isolated from viral particles.

these pseudo-nucleoids are very heavy, since they contain the S-layer in addition to chromosomal DNA [38,39]; as a consequence, their sedimentation coefficient does not change in the presence of ethidium bromide or netropsin, preventing the determination of DNA superhelicity. A different and powerful method to analyze chromosomal DNA supercoiling in vivo is to measure the rate of trimethyl-psoralen (TMP) photobinding onto the DNA (UV-induced cross-linking), before and after relaxation of chromosomal DNA by γ-ray irradiation [102]. In eubacteria, relaxation of the chromosome reduces TMP photobinding because TMP binds preferentially to negatively supercoiled DNA. On the contrary, cleavage of eukaryotic DNA by γ-rays does not reduce TMP photobinding since the DNA in the internucleosomal regions is relaxed, and the DNA in nucleosomes is not accessible to TMP. In the case of *S. shibatae,* preliminary experiments indicate that DNA cleavage did not reduce the rate of TMP photobinding [39]. Taking into account that TMP binding is performed at low temperature, this suggests that the DNA accessible to TMP in vivo is slightly positively supercoiled at physiological temperatures.

In the case of *Mt. fervidus,* an interesting question is how positive supercoiling of the DNA by reverse gyrase interacts with the formation of "reverse nucleosomes" by HMf? Reeve and coworkers [25] suggest that formation of HMf nucleosomes antagonizes the activity of reverse gyrase by introducing negative superturns into free DNA. Another possibility is that HMf nucleosomes stabilize the positive superturns introduced by reverse gyrase. However, it remains to be known whether reverse nucleosomes are formed in vivo.

3.4. DNA topology in halophilic archaebacteria

3.4.1. Sensitivity of halobacteria to DNA topoisomerase II inhibitors

It has not yet been possible to detect any DNA topoisomerase activity in halobacteria in vitro, probably because of the difficulty to study protein/DNA interactions in the presence of high intracellular salt concentrations. Nevertheless, the presence of a type II DNA topoisomerase in halobacteria was suspected from their sensitivity to several classical inhibitors of these enzymes. Interestingly, halobacteria are inhibited both by drugs which are otherwise specific for eubacterial DNA gyrase, such as coumarin (novobiocin and coumermycin) and fluoroquinolones [103,104] and by antitumoral compounds which are otherwise known as typical inhibitors of eukaryotic type II DNA topoisomerases, such as epipodophyllotoxins (VP16 and VM26) [103,105] and several DNA intercalators: adryamicin, ellipticin and actinomycin D [106,107].

Coumarins compete with ATP for its binding site on type II DNA topoisomerases. These drugs are specific for eubacterial DNA gyrase at low concentrations (0.1–10 µg/ml) but also inhibit eukaryotic type II DNA topoisomerase at higher concentrations. Halobacteria are sensitive to very low doses of novobiocin (below 0.1 µg/ml), suggesting that the target of this drug in vivo is a type II DNA topoisomerase of the gyrase family. This has been confirmed by genetic analysis (see below).

Several data indicate that fluoroquinolones and epipodophyllotoxins probably also inhibit the growth of halobacteria by interacting with their type II DNA topoisomerase. In eubacteria and eukaryotes, these drugs inhibit type II DNA topoisomerases by preventing the relegation step of the topoisomerization reaction. They stabilize a cleavable complex with the enzyme covalently linked to the 5' ends of the DNA break. The cleaved molecules pull apart during DNA isolation, as a consequence of protein denaturation. In halobacteria, fluoroquinolones and epipodophyllotoxins also induce DNA cleavage in vivo [105,108, 109]. In the case of the epipodophyllotoxins, it has been demonstrated that a protein is linked at the 5' end of the breaks [108]. Interestingly, fluoroquinolones (otherwise inhibitors of the eubacterial enzyme) and epipodophyllotoxins (otherwise inhibitors of the eukaryotic enzyme) induce DNA cleavage at the same sites in vivo on the plasmid pGRB-1 of *Halobacterium* GRB [109]. Another indication that the target of epipodophyllotoxins in halobacteria is a type II DNA topoisomerase is that some *Haloferax volcanii* mutants resistant to these drugs exhibit novobiocin cross-resistance, whereas a novobiocin-resistant mutant of *Haloferax* A2.2 is also slightly more resistant to epipodophyllotoxins [110].

3.4.2. Gene structure and primary sequence of a halobacterial type II DNA topoisomerase

Holmes and Dyall-Smith [111] have cloned the gene conferring novobiocin resistance in the halophilic archaebacterium *Haloferax* strain Aa2 by complementation of a wild-type strain with DNA from a novobiocin-resistant mutant. This gene encodes a protein highly homologous to the B subunit of eubacterial DNA gyrase (gyrB) which is the target of novobiocin in eubacteria [112]. They isolated the wild-type gene and showed that it differs from the novobiocin-resistant one by changes in three amino acids which are precisely localized in a region corresponding to the putative ATP-binding site in eubacterial DNA gyrase. A gene encoding a protein homologous to the A subunit of eubacterial DNA gyrase (gyrA) was found at the 3' end of the *Haloferax* gyrB [112]. The archaebacterial gyrA and gyrB are cotranscribed, forming the first type II DNA topoisomerase operon yet described. In eubacteria, gyrB and gyrA are unlinked in *E. coli* but linked in *B. subtilis* [113]. In eukaryotes, a single gene encodes the type II DNA topoisomerase, and the N- and C-terminal parts of this protein are homologous to the gyrB and the gyrA subunit of DNA gyrases, respectively [68] (Fig. 10, upper panel).

Alignment of the amino-acid sequence of the *Haloferax* type II DNA topoisomerase with those of eukaryotic and eubacterial type II DNA topoisomerases and the construction of phylogenetic trees indicates that the archaebacterial enzyme clusters with eubacterial type II DNA topoisomerases, in agreement with the extreme sensitivity of halobacteria to novobiocin (Fig. 10B). It is striking that *E. coli* DNA gyrase aligns better with *Haloferax* type II DNA topoisomerase than with the recently discovered *E. coli* type II DNA topoisomerase (topo IV) which lacks gyrase activity. This strongly suggests that the *Haloferax* type II DNA topoisomerase is a DNA gyrase.

The grouping of *Haloferax* type II DNA topoisomerase with eubacterial type II DNA topoisomerases contradicts both the rRNA phylogenetic tree, in which the three domains are clearly separated, and protein trees in which archaebacteria and eukaryotes are grouped together (for example RNA polymerases or elongation factors). Several hypotheses can explain this data. Firstly, one could imagine the loss of either the eubacterial or archaebacterial type II DNA topoisomerase gene early in evolution, and its replacement by the homologous gene from the other prokaryotic domain via lateral gene transfer. Secondly, one could suppose the accelerated evolution of the type II DNA topoisomerase gene produced by a modification of its function (gain or loss of gyrase activity) either in the eukaryotic lineage or in a common branch to eubacteria and archaebacteria. However, one should note that *E. coli* DNA topoisomerase IV, which lacks gyrase activity, clearly belongs to the prokaryotic family [70]. Finally, one or several duplications of the ancestral type II DNA topoisomerase gene could have occurred before the divergence of the three domains. The latter hypothesis is consistent with the great differences observed in the primary structure of the eukaryotic and prokaryotic type II DNA topoisomerases (numerous gaps are required to align them correctly) whereas all prokaryotic sequences (eubacteria and archaebacteria) can be aligned with very few gaps. The duplication hypothesis would also explain the existence of a type II DNA topoisomerase without gyrase activity in *Sulfolobus*.

Interestingly, the type II DNA topoisomerase of bacteriophages T4 is located between the eukaryotic and the eubacterial-archaebacterial groups in the phylogenetic tree of type II DNA topoisomerases (Fig. 10, lower panel). This could suggest that viruses

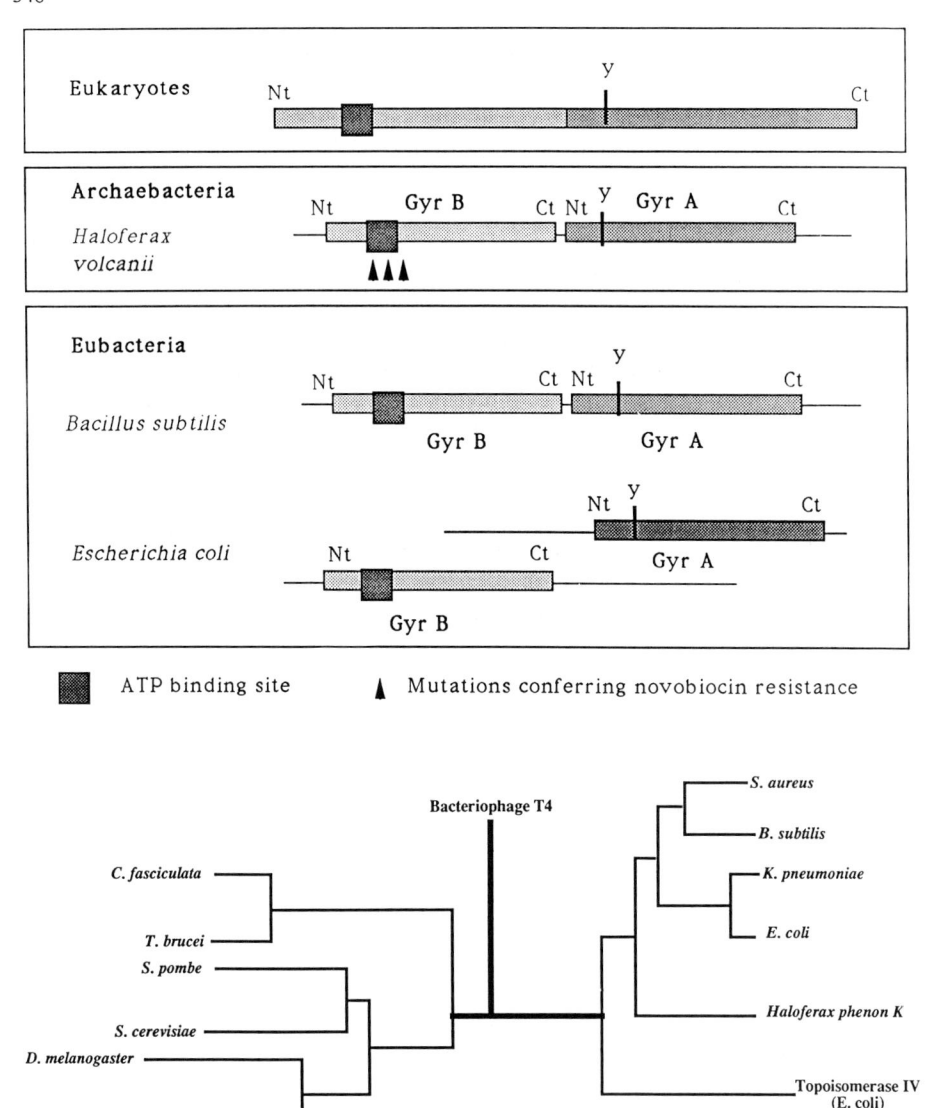

Fig. 10. Evolution of type II DNA topoisomerases. Upper panel: comparison of DNA topoisomerase II genes arrangement in eubacteria, archaebacteria and eukaryotes. (Adapted from refs. [68,111,112].) Lower panel: Phylogenetic tree grouping the gyrA proteins from *Haloferax* phenon K and eubacteria with the similar gyrA-like regions (C-terminal) of eukaryotic type II DNA topoisomerases, and the gene 52 of bacteriophage T4 [Holmes, Dyall-Smith, Labedan and Forterre, unpublished data]. The tree was constructed from the alignment of conserved similar regions in the different proteins, using a distance method (Fish–Margoliash). Other distance or parsimony methods give a similar tree topology for all proteins except *Haloferax* gyrA which is grouped in some cases with gram-positive eubacteria. Similar trees are obtained with gyrB subunits and gyrB-like regions of type II DNA topoisomerases. Abbreviations: C, *Crithidia*; T, *Trypanosoma*; S, *Saccharomyces*; D, *Drosophila*; H, *Homo*; S, *Staphylococcus*; B, *Bacillus*; K, *Klebsiella*; E, *Escherichia*.

became independent living forms before the divergence between prokaryotes and eukaryotes [114].

3.4.3. Biological roles of type II DNA topoisomerase in halophilic archaebacteria
The inhibition of the halobacterial type II DNA topoisomerase by novobiocin has several effects in vivo: as in eubacteria, novobiocin rapidly inhibits chromosomal DNA replication [104], indicating that the type II DNA topoisomerase is absolutely required to relax positive superturns which otherwise would accumulate at the replication forks. In contrast, both type I and type II DNA topoisomerases can function as swivels for DNA replication in eukaryotes [115]. Novobiocin also interferes with plasmids maintenance and replication: in the presence of this drug the high-molecular-weight plasmids of halobacteria disappeared [104] whereas a single-stranded form of the small multicopy plasmid pGRB-1 accumulated [116]. Finally, novobiocin induces positive supercoiling of small halobacterial plasmids [117]. Again, this is reminiscent of the situation in eubacteria. In *E. coli*, positive supercoiling induced by novobiocin is due to the accumulation of positive superturns in front of moving RNA polymerases when DNA gyrase is inhibited [62]. In halophilic archaebacteria, positive supercoiling induced by novobiocin could also be related to transcription since it is inhibited by doses of actinomycin D which otherwise specifically inhibit transcription in *E. coli* [106]. In contrast to the situation in eubacteria and halobacteria, novobiocin does not induce positive supercoiling in eukaryotes since positive superturns can be relaxed by eukaryotic type I DNA topoisomerases in the absence of DNA topoisomerase II activity [63]. Positive supercoiling in eukaryotes was only observed in yeast mutants lacking eukaryotic type I DNA topoisomerase and expressing cloned eubacterial protein ω [63]. Both inhibition of DNA replication by novobiocin and positive supercoiling induced by this drug therefore strongly suggest that halobacteria do not have a type I DNA topoisomerase resembling the eukaryotic enzyme. This is in agreement with the resistance of halobacteria to camptothecin [106], a specific inhibitor of the eukaryotic type I DNA topoisomerase.

In eubacteria, DNA gyrase could play a crucial role in gene regulation by controlling the level of chromosomal superhelicity, which in turn determines the recognition of promoters and regulatory sequences by RNA polymerase and transcription factors [69]. The type II DNA topoisomerase of halophilic archaebacteria could play a similar role: Yang and DasSarma [118] reported that novobiocin inhibits the induction of purple membrane and gas vesicle synthesis in *H. salinarium* but does not affect the transcription of a non-inducible gene used as control [118]. Furthermore, novobiocin increases the amount of DNA gyrase mRNA in *Haloferax* cells [119], suggesting that DNA gyrase expression in halobacteria is subjected to homeostatic regulation, as in eubacteria [120].

3.5. An overview of DNA topology in archaebacteria

DNA topoisomerases and the mechanisms which control DNA topology appear to be quite different in thermophilic and halophilic archaebacteria. The situation in halophilic archaebacteria strikingly resembles that in mesophilic eubacteria, with a putative DNA gyrase and highly negatively supercoiled plasmids, whereas DNA topology in extremely thermophilic archaebacteria is unusual, with a reverse gyrase and relaxed DNA. The presence of reverse gyrase activity in both extremely thermophilic eubacteria and archaebacteria

TABLE 2
Eubacterial and eukaryotic features in DNA topology and DNA topoisomerases detected in archaebacteria

Eubacterial features	
Extremely thermophilic archaebacteria	(1) Reverse gyrase activity detected in extreme thermophiles
Mesophilic and moderately thermophilic archaebacteria	(2) The major DNA topoisomerase activity is the specific ATP-independent relaxation of negative superturns
Halophilic archaebacteria	(3) Similarity between *Haloferax* DNA topoisomerase II and eubacterial DNA gyrases (correlate with extreme novobiocin sensitivity)
	(4) Positive supercoiling induced by novobiocin
	(5) Absolute requirement of DNA topoisomerase II for DNA replication
	(6) Resistance to camptothecin
	Points (4)–(6) suggest the absence of an eukaryotic-like type I DNA topoisomerase relaxing positive superturns.
Eukaryotic features	
DNA topoisomerase II lacking gyrase activity in extreme thermophiles	

suggests that this enzyme was present in the common ancestor of all prokaryotes. It remains to be determined whether reverse gyrase is homologous in both domains, and whether its role is indeed to protect the DNA at very high temperature. If the answer is yes for both questions, the case for a common extremely thermophilic progenitor to all prokaryotes will be very strong.

The composite structure of reverse gyrase recalls those of some viral proteins involved in DNA replication, such as helicase-primase which bear two different activities on the same polypeptide. This raises the exciting evolutionary problems of how and when the formation of reverse gyrase gene occur: did reverse gyrase arise as a fusion between a helicase and a type I DNA topoisomerase, or is reverse gyrase an ancient protein from which classical (5′) type I DNA topoisomerases emerged after losing the helicase domain?

Preliminary data also suggest a correlation between the presence of reverse gyrase activity and the level of intracellular DNA supercoiling. It is important now to check the DNA topology of a wider range of archaebacterial plasmids and chromosomes, and to determine whether the above statement also holds for thermophilic eubacteria containing reverse gyrase. How the interactions between transcription, reverse gyrase, other DNA topoisomerase activities and histone-like proteins cooperate to establish a precise level of DNA supercoiling remains to be determined. Finally, one would like to know how different levels of supercoiling affect the critical intracellular DNA/protein interactions which are dependent on the level of intracellular supercoiling. In particular, it should be important to analyze the effect of supercoiling on transcription in hyperthermophiles harboring reverse gyrase.

At the moment, the overall pattern of DNA topology and DNA topoisomerases in archaebacteria is clearly more similar to the eubacterial one than to the eukaryotic one (Table 2). One can speculate that the similarities between DNA topology in eubacteria and archaebacteria are related to the presence of the same type of "prokaryotic" chromosomal organization in these two domains. An exception to this pattern is the presence of a type II DNA topoisomerase without gyrase activity in *Sulfolobus*. It remains to be determined whether the latter enzyme is indeed specifically related to the eukaryotic type II DNA topoisomerases or to *E. coli* DNA topoisomerase IV, and which kinds of type II DNA topoisomerases exist in other sulfothermophiles and methanogens.

4. DNA polymerases

DNA polymerases are ubiquitous enzymes required for replication, repair and recombination of DNA genomes [121]. Most DNA polymerases are multi-functional enzymes which exhibit, in addition to the polymerase activity, a 3′ to 5′ exonuclease activity involved in the fidelity of DNA synthesis (proof-reading).

In eubacteria and eukaryotes, several types of DNA polymerases have been characterized: three in eubacteria (DNA polymerases I, II and III), and five in eukaryotes (DNA polymerases α, β, δ, ε and γ). Some of these enzymes, named "DNA replicases", are specifically involved in DNA-chain elongation at the replication fork. They have a multi-subunit structure and can prime and perform DNA replication in a processive way when they are associated with the other replicative proteins. In eubacteria, only one DNA replicase has been isolated (DNA polymerase III), whereas several DNA replicases co-exist in eukaryotes: DNA polymerases α, δ and ε, which are essential for the replication of nuclear DNA, and DNA polymerase γ, which is responsible for the replication of the mitochondrial genome. The other eubacterial and eukaryotic DNA polymerases are monomeric and are preferentially involved in mechanisms which require replication of short DNA fragments, in the course of either DNA repair (DNA polymerases I and II from *E. coli*, eukaryotic DNA polymerase β), or DNA replication (maturation of Okazaki fragments by *E. coli* DNA polymerase I).

Comparative analyses of the primary sequences of DNA polymerases allow one to define four DNA polymerase families [122]: family A groups eubacterial DNA polymerases I; family B groups all DNA polymerases sensitive to the drug aphidicolin on the basis of short conserved motifs (the three eukaryotic DNA replicases but also *E. coli* DNA polymerase II, a typical repair enzyme); family C groups eubacterial DNA replicases, and family X groups eukaryotic DNA polymerases β. All DNA polymerases that exhibit a 3′ to 5′ exonuclease activity belong to families A, B or C and have in common three short consensus sequences in the same arrangement, the exo boxes [123]. These DNA polymerases could therefore belong to a superfamily of DNA polymerases. However, they cannot be aligned with confidence between the exo boxes and in other regions from one family to another, indicating that DNA polymerases are extremely divergent enzymes.

In archaebacteria, the first studies on DNA polymerases and DNA replication have shown that aphidicolin, a specific inhibitor of eukaryotic DNA replication, also

TABLE 3
Pattern of sensitivity of archaebacteria to aphidicolin

Species[a]	Residual growth in vivo (%)		Res. DNA synth.[b] (%)	Ref.
	5 µg/ml	20 µg/ml		
Halobacterium halobium	30	0	60[c]	[124,125,128]
Other halobacteriales (6)	30	0	ND	[124,125,128]
Methanococcus vannielii	5	0	30	[126]
Other Methanococcales (5)	ND	ND	30–100[d]	[127,128]
Archaeoglobales (2)	ND	ND	50	[128]
M. thermoautotrophicum	ND	ND	100	[126,128]
Other Methanobacteriales (3)	100	0–100	100	[126,128]
Methanomicrobiales (5)	40–100	0–100	75–100	[126,128]
Thermoplasma acidophilum	ND	ND	100	[128]
Sulfolobus acidocaldarius	100	100	100	[124]

[a] The numbers in parentheses correspond to the number of species tested of the same order.
[b] Residual DNA synthesis, crude extracts, 10 mg/ml.
[c] The DNA polymerase activity is very low in crude extract of halophilic archaebacteria compared to other strains in ref. [128].
[d] The DNA polymerase activity is inhibited in all strains at 100 µg/ml of aphidicolin [125], except in *Methanococcus voltae* [126], however in the latter, aphidicolin induces morphological changes.

inhibits DNA synthesis and/or cell growth in halobacteria and some methanogens [124–127]. These data suggested that replicases of these archaebacteria could be related to eukaryotic replicases. Indeed, an aphidicolin-sensitive DNA polymerase has been isolated from these archaebacteria. The genes encoding three of them have been recently cloned and sequenced, demonstrating that these enzymes belong to the B family of DNA polymerases. However, their function has not yet been determined, and in other archaebacteria only one aphidicolin-resistant DNA polymerase has been isolated.

4.1. Sensitivity of archaebacteria to aphidicolin

Archaebacteria exhibit different sensitivities to aphidicolin (Table 3): growth of halobacteria and of some methanogens is completely or partially inhibited with 20 µg/ml of aphidicolin, whereas that of other methanogens and of *S. acidocaldarius*, the only thermophile tested, is resistant [124–128]. In the case of *H. halobium*, aphidicolin specifically inhibits in vivo DNA replication and induces cell filamentation [124,125]. This result indicates that the target of aphidicolin in halobacteria is implicated in the replication of the genome.

DNA synthesis detected in crude extracts of archaebacteria also exhibits different sensitivity to aphidicolin: it is partially inhibited by 10 µg/ml of aphidicolin in some methanogens [126,128] but completely resistant in other methanogens and in the thermoacidophiles tested [127,128]. There is some correlation between the in vivo and in vitro results with aphidicolin (Table 3), suggesting that the activity detected in crude extracts of archaebacteria could correspond to a DNA replicase. However, further work is required to ascertain this correlation since several DNA polymerases sensitive to aphidicolin could be present in the same cell, and one cannot exclude the existence of a repair enzyme sensitive to aphidicolin in archaebacteria. For example, whereas aphidicolin has no effect on *E. coli* DNA replication, it inhibits *E. coli* DNA polymerase II [129].

4.2. DNA polymerases from sulfothermophiles and methanogens

4.2.1. Aphidicolin-sensitive DNA polymerases

An aphidicolin sensitive DNA polymerase, with a catalytic subunit of about 100 kDa and most probably being a monomeric enzyme, has been isolated from the methanogen *Mc. vannielii* [126], the thermoacidophile *Sulfolobus solfataricus* [130] and the Thermococcales *Thermococcus littoralis* and *Pyrococcus furiosus* (unpublished results [New England Biolabs and Stratagene, respectively]). The DNA polymerase of *Mc. vannielii* is sensitive to low doses of aphidicolin, and Zabel et al. [131] have shown that the inhibition is competitive with dCTP, as is the case for eukaryotic replicases α. Higher concentrations of aphidicolin are required to inhibit the DNA polymerases of the three sulfothermophiles (ref. [130], and Gadelle, D., this laboratory, unpublished observations) and the mechanism of inhibition has not yet been studied.

The DNA polymerases of *T. littoralis* and *P. furiosus* have been marketed for use in DNA amplification by the polymerase chain reaction (PCR) method as the Vent and pfu DNA polymerases, respectively. These enzymes are more accurate in vitro than the *Thermus aquaticus* (Taq) DNA polymerase, both in classical fidelity tests [132] and in PCR [133,134]. Indeed, they have an associated 3' to 5' exonuclease activity involved in proof-reading, whereas the Taq polymerase is devoid of such activity.

The genes encoding the DNA polymerases of *S. solfataricus*, *T. littoralis* and *P. furiosus* have recently been cloned and sequenced [135–137]. The *S. solfataricus* and *T. littoralis* genes encode polypeptides of 100 and 93 kDa, respectively. The complete nucleotide sequence from the gene encoding *S. solfataricus* DNA polymerase has been published [135], but for the two other cases only partial amino-acid sequence data have been reported. All these DNA polymerases exhibit some of the amino-acid motifs typical for DNA polymerases of the B family. These motifs are located in the same spatial arrangement as their counterparts in eubacterial and eukaryotic family B DNA polymerases. In the case of *S. solfataricus* and *T. littoralis*, the three exonuclease boxes have been identified. This suggests that the DNA polymerase of *S. solfataricus* also harbors a 3' to 5' exonuclease activity.

The two short stretches of amino-acid sequences of the DNA polymerase from *P. furiosus* which have been published were aligned with eukaryotic DNA polymerases of the B family to claim another "eukaryotic feature" of archaebacteria [137]. However, Fig. 11 shows that these amino-acid stretches of *P. furiosus* DNA polymerase and of other archaebacterial DNA polymerases align as well with *E. coli* DNA polymerase II, a prokaryotic member of the B family [138]. The presence of family B DNA polymerases in the three domains strongly suggests that the divergence between the DNA polymerases of families A, B and C occurred before the divergence between archaebacteria, eubacteria

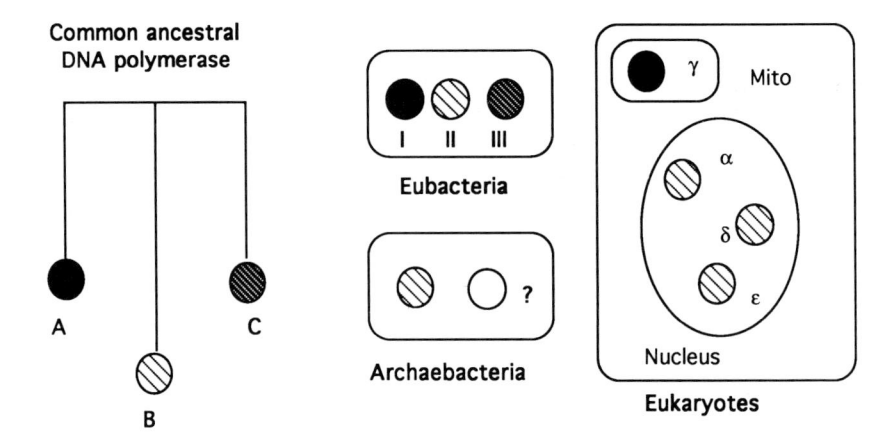

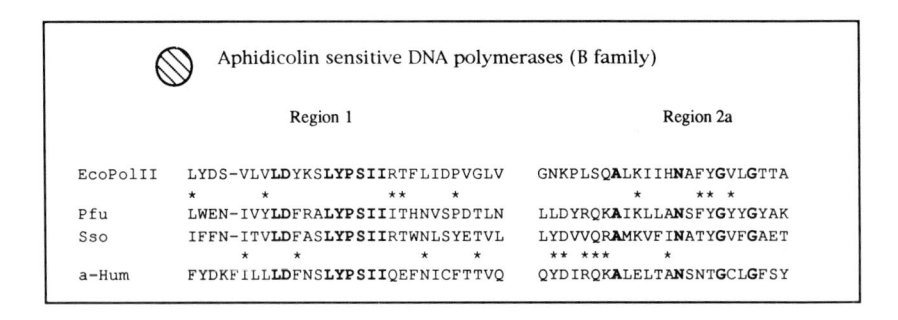

Fig. 11. Evolution of DNA polymerases. Upper panel: distribution of DNA polymerases from A, B and C subfamilies between eubacteria, archaebacteria and eukaryotes. Lower panel: alignments of regions 1 and 2a of *E. coli* DNA polymerase II (EcoPol II), *Pyrococcus furiosus* (Pfu), *Sulfolobus solfataricus* (Sso), and human DNA polymerase-α (α-Hum). (Adapted from refs. [114,135-138].)

Conserved amino-acids in all sequences are in bold (R = K and I = L = V). Asterisks indicate the amino acids identical between the archaebacterial sequences and only the eubacterial or the eukaryotic sequences.

and eukaryotes. This means that the last common ancestor of the three domains contained already several DNA polymerases with proofreading activity. Since DNA polymerases of families A and C have been discovered up to now only in eubacteria, this also suggests that some genes encoding DNA polymerases have been lost during the divergence of the three domains.

Most interestingly, the gene encoding the DNA polymerase of *T. littoralis* is interrupted by two intervening sequences IVS1 and IVS2, of 539 and 391 bp, respectively [136]. This is the first example of putative introns in genes coding for proteins in prokaryotes. Surprisingly, the sequences of IVS1 and IVS2 form one continuous open reading frame with the three DNA polymerase exons. Furthermore, these intervening sequences are inserted in two of the motifs characteristic of the B family. IVS2 encodes an endonuclease, I-Tli1, that has features in common with endonucleases of group I introns.

In eukaryotes, those enzymes are supposed to promote intron mobility by cleavage at the "homing site" in genes that lack introns. Indeed, deletion of IVS2 creates an I-Tli1 site at the exon2–exon3 junction. However, besides the I-Tli1 endonuclease, IVS2 lacks other features characteristic of group I intron, and IVS1 and IVS2 are also different from group II or archaebacterial pre-tRNA introns. Actually, it is not known whether production of the mature polymerase is achieved via mRNA or via protein splicing. However, two results support the latter hypothesis. Firstly, IVS2 self-splices in *E. coli* to yield active polymerase, but processing is abolished if the IVS2 reading frame is disrupted. Secondly, silent changes in the DNA sequence at the exon2–IVS2 junction that maintain the original protein sequence do not inhibit splicing.

4.2.2. Aphidicolin-resistant DNA polymerases

Klimczak et al.[139] have purified to homogeneity a monomeric DNA polymerase which is resistant to aphidicolin from the thermoacidophile *S. acidocaldarius*[139] and the methanogen *M. thermoautotrophicum*[140]. The DNA polymerase from *S. acidocaldarius*, and more recently an aphidicolin-resistant DNA polymerase from *T. acidophilum*, were purified to homogeneity in our laboratory[141,142]. Depending on the strain, the molecular mass of these monomeric DNA polymerases lies in the range of 70–100 kDa, which is close to the molecular mass of the aphidicolin-sensitive DNA polymerases isolated from other archaebacterial species. These enzymes are associated with a 3' to 5' exonuclease activity that could be involved in a proofreading mechanism (refs. [139,140,142], and our laboratory, unpublished results). In addition, the DNA polymerase from *M. thermoautotrophicum* is associated with a 5' to 3' exonuclease activity[140], as is the case for eubacterial DNA polymerase I. In contrast, this activity has not been detected with DNA polymerases of thermoacidophiles. It has been shown that the 100 kDa DNA polymerase from *S. acidocaldarius* can be used in PCR[143,144].

As for aphidicolin-sensitive DNA polymerases, the functions of aphidicolin-resistant DNA polymerases have not been determined. However, DNA synthesis performed by the 100 kDa DNA polymerase of *S. acidocaldarius* has been characterized in some detail using as substrate a natural single-stranded DNA (derived from M13) singly primed by a short oligonucleotide[143,145]. At 70°C, the optimal temperature for in vitro DNA synthesis, this enzyme can efficiently replicate long stretches of single-stranded DNA in the absence of any accessory protein. Analysis of the polymerization products shows that all primers are extended at the same rate in a wide range of polymerase/template ratio, indicating that the DNA polymerase is randomly recycled on the template molecules. These properties are similar to those of eubacterial DNA polymerases I and not to those of eubacterial or eukaryotic DNA replicases.

An additional aphidicolin-resistant DNA polymerase composed of several 35–40 kDa polypeptides has been reported in *S. acidocaldarius*[146] and in *T. acidophilum*[147]. The authors have proposed that this DNA polymerase could be homologous to the eukaryotic DNA polymerase β, which is a monomeric enzyme composed of a 40 kDa polypeptide. However, we have shown that in each case, the 35–40 kDa polypeptides are devoid of DNA polymerase activity when they are totally separated from the high-molecular-mass DNA polymerase already described[141,142].

4.3. DNA polymerases from halophiles

In the halophile *H. halobium,* Nakayama et al. [148] have described an aphidicolin-sensitive DNA polymerase that they named "DNA polymerase α". The enzyme has a high sedimentation coefficient value, is able to synthesize RNA on a synthetic DNA template, and is associated with a 3′ to 5′ exonuclease activity. Two major polypeptides of 60 and 70 kDa were detected in a purified fraction of *H. halobium* DNA polymerase α [149]. The authors concluded from these results that this enzyme corresponds to a multi-subunit DNA polymerase able to perform DNA priming like eukaryotic DNA polymerase α. However, the association of the 60 and 70 kDa polypeptides with the polymerase activity has not been demonstrated and the specific activity of the enzyme is very low compared to any other purified DNA polymerases.

Kohiyama and coworkers [150] also reported the existence, in *H. halobium,* of a so-called "DNA polymerase β" which is resistant to aphidicolin. Purified fractions of this enzyme contain a major polypeptide of 65 kDa and are associated with 3′ to 5′ and 5′ to 3′ exonuclease activities [150].

The report of two DNA polymerases with different sensitivity to aphidicolin in the same species is strikingly different from the situation observed in methanogens and sulfothermophiles. Furthermore, the structure and size of the aphidicolin-sensitive DNA polymerase from *H. halobium* depart from those of the monomeric aphidicolin-sensitive DNA polymerases isolated in other archaebacteria. The reason for such discrepancies is not clear at the moment.

4.4. Conclusions

Table 4 summarizes the properties of the different DNA polymerases which have been purified from archaebacteria. It is clear from studies performed in vivo that halophiles and some methanogens have at least one aphidicolin-sensitive DNA polymerase involved in DNA replication (most probably a replicase); on the other hand, only one DNA polymerase, either sensitive or resistant to aphidicolin, has been detected in various archaebacteria (with the exception of *H. halobium*). What are the phylogenetic relationships between these aphidicolin-resistant and -sensitive enzymes? In collaboration with F. Lottspeich, we obtained the amino-acid sequences of several peptides from *S. acidocaldarius* DNA polymerase (resistant) and we observed that these sequences are present in the primary structure of *S. solfataricus* DNA polymerase (sensitive). This suggests that aphidicolin-resistant and aphidicolin-sensitive DNA polymerases detected in archaebacteria are homologous. It remains to determine whether aphidicolin-resistant enzymes contain the amino-acid motifs typical for DNA polymerases of the B family.

Other important questions remain to be solved: are the single monomeric DNA polymerases detected in different species unique or do they correspond to a predominant activity in vitro which hides the activities of one or several other DNA polymerases (as DNA polymerase I hides DNA polymerases II and III in eubacteria); does this monomeric DNA polymerase correspond to a replicase (or to the catalytic subunit of a replicase complex), or to a repair enzyme? The in vitro study of the replication of an extrachromosomal DNA would be useful for such investigations, but a definitive answer would require the isolation of conditionally lethal mutants.

TABLE 4
Archaebacterial DNA polymerases

Species	Aph. sens. [a,b]	m_{sub}[a] (kDa)	#$_{sub}$[a]	Spec. act.[a,c] (units/mg)	exo 3'–5'	exo 5'–3'	Ref.
M. thermoautotrophicum	R	70	1	40 000	+	+	[140]
Thermoplasma acidophilum	R	88	1	175 000	+	+	[142]
Sulfolobus acidocaldarius	R	100	1	715 000	+	+	[139,141]
Halobacterium halobium β	R	65[d]	1[d]	1300	+	+	[150]
Sulfolobus solfataricus	S	100	1[d]	?	ND[e]	ND[e]	[130,135]
Pyrococcus furiosus	S	93	1[d]	?	+	ND	[133]
Thermococcus littoralis	S	88[d]	1[d]	?	+	–	[132,136]
Methanococcus vannielii	S	100	1	?	ND	ND	[126]
Halobacterium halobium α	S	60+70[d]	2[d]	350	+	–	[148,149]

[a] Abbreviations: Aph. sens., aphidicolin sensitivity; m_{sub}, subunit molecular masses; #$_{sub}$, number of subunit; Spec. act., specific activity.
[b] Effect of 50 μg/ml of aphidicolin: R, resistant, S, sensitive.
[c] Definition of unit: nmoles of dNTPs incorporated on DNAse I treated DNA in 30 minutes.
[d] Results not demonstrated rigorously.
[e] ND, not determined.

One has therefore to hope that future work on archaebacterial DNA polymerases will not be exclusively concerned with the interest of these enzymes for PCR and other biotechnological applications. Further work is also required to determine the molecular mechanisms and the nature of the other proteins involved in replication and repair of archaebacterial genomes such as DNA ligases, helicases and so on.

5. General discussion

5.1. Future prospects

The exploration of the archaebacterial DNA world will require additional studies. Fascinating questions still remain to be answered: how do extreme thermophiles cope with extensive depurination which should affect DNA at temperatures near the boiling point of water? What are the enzymatic mechanisms of DNA recombination and repair under extremophilic conditions? What is the nature and the architecture of the archaebacterial replicases? What is the meaning of the presence of intervening sequences splitting conserved motifs of some DNA polymerase genes? etc. Even those questions already under investigation are far from being correctly understood. What is the exact role of reverse gyrase? Which of the histone-like proteins are really significant for chromosome structure? HMf is a good candidate, but is it ubiquitous in archaebacteria? Do, for example, HMf and HTa coexist in some archaebacteria? What are their respective roles? These are difficult questions to approach since the nature and the role of histone-like proteins in eubacteria are not yet clear, despite the existence of mutants and genetic tools not so far available for archaebacteria. And yet, histone-like proteins, very abundant by definition, are easy to work with, compared to other proteins involved in DNA manipulation. The situation with DNA polymerases and topoisomerases is more difficult since, unlike histone-like proteins, RNA polymerases, or ribosomal proteins, they are present in small amounts in the cell. This explains the first mistakes in characterizing the structure of *S. acidocaldarius* reverse gyrase [72] or DNA polymerases [103,146,147]. In both cases, partially purified fractions contained abundant contaminating polypeptides which were confused with the enzyme under investigation: RNA polymerase subunits in the case of reverse gyrase, and a 35–40 kDa polypeptide in the case of DNA polymerases from thermoacidophiles. Accordingly, different approaches should be combined to confirm the structural identity of such enzymes, such as activity gels, co-migration of polypeptide and activity on sedimentation gradients, neutralization by antibodies against electroeluted polypeptides, etc. In eubacteria, over-expression of cloned genes has often been a prerequisite for the identification and structural studies of DNA replication proteins. The recent design of genetic tools in different archaebacterial systems should help to introduce similar strategies with these microorganisms.

5.2. Evolutionary considerations

5.2.1. Eukaryotic versus eubacterial features of archaebacteria
Although the data on the archaebacterial DNA world are far from exhaustive, it is nevertheless already possible to draw tentative phylogenetic considerations. In their transcription and translation machineries, archaebacteria exhibit several eukaryotic-like features, compared to those in eubacteria, such as an RNA polymerase and elongation factors of the eukaryotic type (see ref. [151], and other chapters of this volume). Several features of the DNA world have been frequently considered to be "eukaryotic": the presence of histone-like proteins, first HTa [152], more recently HMf [23], the sensitivity of halobacteria to drugs otherwise specific of the eukaryotic DNA topoisomerase II [103], and the existence of an aphidicolin-sensitive DNA polymerase [124]. However, it appears that HTa is a close relative of eubacterial HU proteins (see Fig. 2), that *Haloferax* Type II DNA topoisomerase is very similar to eubacterial DNA gyrases (see Fig. 9) and that aphidicolin also inhibits an eubacterial DNA polymerase. HMf is clearly homologous to eukaryotic histones, but "HMf nucleosomes" are drastically different from eukaryotic ones; furthermore, one cannot exclude the existence of HMf-related proteins in eubacteria.

The wish to discover eukaryotic features in prokaryotes and to emphasize the uniqueness of archaebacteria has often been misleading. For example, introns in prokaryotic tRNA genes, first detected in archaebacteria and considered to be an eukaryotic feature, have later been found in eubacteria as well [153–154]; reverse gyrase, once considered to be a hallmark of archaebacteria, was later found in extremely thermophilic eubacteria. In fact, if one considers the present data on chromosome structure (see Fig. 1) and DNA topology (Table 2), archaebacteria resemble more closely the eubacteria than the eukaryotes. Additional similarities between the two prokaryote groups, which have not been reviewed here, are the presence of similar types of plasmids [2,100,155,156], modification–restriction systems [157–158], head–tailed bacteriophages [159–161] (for a review see ref. [162]), numerous metabolic pathways [163], operons, Shine–Dalgarno sequences, etc. (see reviews in other chapters of this volume and in ref. [164]). This is one of the reasons why we decided not to use the term "archaea" in the present review, since that term has been specifically designed to emphasize differences between eu- and archaebacteria [165]. Another reason is that the term "archaea" suggests that archaebacteria represent the most ancient phenotype present on Earth, which remains speculative. One cannot decide yet whether the phenotypic resemblance between archaebacteria and eubacteria is analogy from convergent evolution, homology testifying to the existence of a common specific prokaryotic ancestor for both groups, or stems from the fact that today's prokaryotes resemble the common ancestor of the three domains of life.

5.2.2. The root of the tree of life and other phylogenetic problems
It has been claimed recently by Iwabe, Gogarten and coworkers [166,167] that the root of the universal tree of life should be located in the eubacterial branch. This would favor the hypothesis of a prokaryotic-like universal ancestor. This claim was based on the construction of composite phylogenetic trees of putative paralogous proteins (i.e., proteins which arose by duplication before the divergence between the three domains). The two couples of protein models used were the two elongation factors (EF-1α Tu) and EF-2 (G) and the α and β subunits of membrane-bound ATPases. However, two eubacterial ATPases have been recently discovered which are much more similar to the archaebacterial and eukaryotic V-type ATPases than to the eubacterial F_1/F_0 ATPases used in the tree of Iwabe, Gogarten and coworkers [168,169]. This suggests that the V-like and the F_1/F_0 ATPases are themselves paralogous. Accordingly, the eubacterial rooting of the composite ATPase subunit tree of Iwabe, Gogarten and coworkers do not reflect the rooting of the tree of life but a fortuitous combination of orthologous (sister) and paralogous (cousin) proteins (V-type and F_1/F_0 type) in the same trees [170]. In the case of elongation factors, a striking observation is that EF-1α (Tu) and EF-2 (G) are very divergent proteins (EF-2 is twice the size of EF-1) which can be aligned only in a short N-terminal region (about 150 amino-acids) corresponding to the GTP binding site [171]. Furthermore, a cladistic analysis of this alignment shows that one elongation factor cannot be used to root the universal tree of the other [171]. The eubacterial rooting obtained by Iwabe et al. by a distance method of tree construction only reflects the fact that the archaebacterial elongation factors resemble their eukaryotic counterparts more closely. In our laboratory, we have tried to root the tree relating glutamate dehydrogenases from the three domains using a paralogous family of glutamate dehydrogenases as an outgroup [172]. We obtained different rootings according to the method of tree construction used: in one case the root was in the eukaryotic branch, whereas in another case, it was inside the archaebacterial domain [172]. This could be due to the fact that GDHs from the three domains are equidistant, resembling the situation with rRNA. This seems to confirm the prediction of Meyer et al. [173] that protein phylogeny is difficult to use for the determination of phylogenetic relationships between very distantly related organisms. Accordingly, the actual rooting of the tree of life is still an open question.

Another hot topic among evolutionary biologists is the monophyletic versus polyphyletic status of archaebacteria. At the moment, there are no data in the archaebacterial DNA world which support the division of archaebacteria into two kingdoms (euryarchaeota and crenarchaeota) [165] or in two clades (eocytes and "true" archaebacteria) [174]. As for the rooting of the tree of life, the question of the monophyletic versus polyphyletic status of archaebacteria is a difficult one. Lake [175] recently noticed that a specific insertion in the EIF-1α elongation factor argues in favor of the eocyte tree. However, a specific insertion in one heat-shock protein suggests the grouping together of archaebacteria and gram-positive eubacteria! [176]. Once more, such contradictions could be explained by confusions between orthologous and paralogous proteins.

5.2.3. The last common universal ancestor
The finding of several families of paralogous proteins such as DNA polymerases A, B and C, V-like and F_1/F_0 ATPases, GDH families I and II, (5′) type I DNA topoisomerases I, type II DNA topoisomerases, histones, etc. suggest that numerous protein duplications occurred before the divergence between the three domains. Some of these proteins seem to have been lost during the subsequent diversification of the domains (for example, DNA gyrase and HU-like proteins may have been lost in eukaryotes, and the type I DNA topoisomerase relaxing positive superturns may have been lost in prokaryotes). It is therefore likely that the last universal ancestor of the three domains was not a progenote but a bona fide member of the DNA world, with a faithfully replicated DNA genome containing several histones, DNA polymerases with proofreading capacity and type I and type II DNA topoisomerases [171]. The complex nature of the last universal ancestor also suggests that the DNA world evolved considerably, from the emergence of the first DNA-cell, up to the appearance of the last universal ancestor (first age of the DNA world, see ref. [114]). Accordingly, one can hope that further analyses of DNA metabolic enzymes in archaebacteria will not only help to determine the relationships between the three domains but also will give us new insights about the nature of the last universal ancestor and the first age of the DNA world.

Acknowledgements

We are grateful to J. Reeve, M. Dyall-Smith, S. Takayanagi, K. Mikulik and A. Mankin for personal communications and delivery of manuscripts before publication, to G. Mirambeau and C. Boutier de la Tour for pictures and to M. Nadal for critical reading of the manuscript. We are especially grateful to R. Lurtz for electron microscopic pictures. We thank all our present and past collaborators in studying DNA topoisomerases and polymerases in archaebacteria, especially M. Duguet, A.M. De Recondo, their students and coworkers. The work in our laboratory has been supported by grants from the Association de Recherche contre le Cancer (ARC), the Ligue Française de Recherche contre le Cancer (LRC), the Ministère de la Recherche et de la Technology (MRT) and the Institut Français de Recherche et d'Exploitation de la Mer (IFREMER).

References

[1] Sitzmann, J. and Klein, A. (1991) Mol. Microbiol. 5, 505–513.
[2] Charlebois, R., Schalkwyk, L.C., Hofman, J.D. and Doolittle, W. (1991) J. Mol. Biol. 222, 509–524.
[3] Lopez-Garcia, P., Abad, J.P., Smith, C. and Amils, R. (1992) Nucl. Acids Res. 20, 2459–2464.
[4] Noll, K. (1989) J. Bacteriol. 171, 6720–6725.
[5] Yamagishi, A. and Oshima, T. (1990) Nucl. Acids Res. 18, 1133–1135.
[6] Searcy, D.G. (1975) Biochem. Biophys. Acta 395, 535–547.
[7] Stein, D.B. and Searcy, D.G. (1978) Science 202, 219–221.
[8] DeLange, R. (1981) J. Biol. Chem. 256, 900–904.
[9] Searcy, D.G. (1986) In: Bacterial Chromatin (Gualerzi, C.O. and Pon, C.L., Eds.), pp. 175–184, Springer, Berlin.
[10] Searcy, D.G. and Stein, D.B. (1980) Biochem. Biophys. Acta 609, 180–195.
[11] DeLange, R. and Williams, L. (1981) J. Biol. Chem. 256, 905–911.
[12] Searcy, D.G., Montenay-Garestier, T., Laston, D. and Hélène, C. (1988) Biochem. Biophys. Acta 953, 321–333.
[13] Searcy, D.G., Montenay-Garestier, T. and Hélène, C. (1989) Biochemistry 28, 9058–9065.
[14] Chartier, F., Laine, B., Sautière, P., Touzel, J.P. and Albagnac, G. (1985) FEBS Lett. 183, 119–123.
[15] Chartier, F., Laine, B. and Sautière, P. (1988) Biochem. Biophys. Acta 951, 149–156.
[16] Imbert, M., Laine, B., Prensier, G., Touzel, J.P. and Sautière, P. (1988) Can. J. Microbiol. 34, 931–937.
[17] Chartier, F., Laine, B., Belaïche, D., Touzel, J.P. and Sautière, P. (1989) Biochem. Biophys. Acta 1008, 309–314.
[18] Chartier, F., Laine, B., Belaïche, D. and Sautière, P. (1989) J. Biol. Chem. 264, 17006–17015.
[19] Imbert, M., Laine, B., Helbecque, N., Mornon, J.P., Henichart, J.P. and Sautière, P. (1990) Biochem. Biophys. Acta 1038, 346–354.
[20] Laine, B., Chartier, F., Imbert, B. and Sautière, P. (1990) FEMS Symposium (Belaich, J.P., Bruschi, M. and Garcia, J.L., Eds.), Plenum Press, New York.
[21] Laine, B., Culard, F., Maurizot, J.C. and Sautière, P. (1991) Nucl. Acids Res. 19, 3041–3045.
[22] Katouzian-Safadi, M., Laine, B., Chartier, F., Cremet, J.-Y., Belaiche, D., Sautière, P. and Charlier, M. (1991) Nucl. Acids Res. 19, 4937–4941.
[23] Sandman, K., Krzycki, J.A., Dobrinski, B., Lurz, R. and Reeve, J.N. (1990) Proc. Natl. Acad. Sci. U.S.A. 87, 5788–5791.
[24] Krzycki, J.A., Sandman, K. and Reeve, J.N. (1990) In: Proc. 6th Int. Symp. on Genetics of Industrial Microorganisms, Vol. 2 (Heslot, H., Davies, J., Florent, J., Bobichon, L., Durand, G. and Penasse, L., Eds.), pp. 603–610, Société Française de Microbiologie, Strasbourg, France.
[25] Musgrave, D.R., Sandman, K.M. and Reeve J.N. (1991) Proc. Natl. Acad. Sci. U.S.A. 88, 10397–10401.
[26] Thomm, M., Stetter, K.O. and Zillig, W. (1982) In: Archaebacteria (Kandler, O., Ed.), pp. 128–139, Gustav Fisher Verlag, Stuttgart.
[27] Green, G.R., Searcy, D.G. and Delange, R.J. (1983) Biochem. Biophys. Acta 741, 251–257.
[28] Kimura, M., Kimura, J., Davie, P., Reinhardt, R. and Dijk, V. (1984) FEBS Lett. 176, 176–178.
[29] Grote, M., Dijk, J. and Reinhardt, R. (1986) Biochem. Biophys. Acta 873, 405–413.
[30] Lurz, R., Grote, M., Dijk, J., Reinhardt, R. and Dobrinski, B. (1986) EMBO J. 5, 3715–3721.
[31] Choli, T., Henning, P., Wittman-Liebold, B. and Reinhardt, R. (1988) Biochem. Biophys. Acta 950, 193–203.
[32] Choli, T., Wittman-Liebold, B. and Reinhardt, R. (1989) J. Biol. Chem. 263, 7087–7093.
[33] Reddy, T.R. and Suryanarayana, T. (1988) Biochem. Biophys. Acta 949, 87–96.

[34] Reddy, T.R. and Suryanarayana, T. (1989) J. Biol. Chem. 264, 17298–17308.
[35] Schmid, M.S. (1990) Cell 63, 451–453.
[36] Hirvas, L., Coleman, J., Koski, P. and Vaara, M. (1990) FEBS Lett. 262, 123–126.
[37] Bruckner, R.C. and Cox, M.M. (1989) Nucl. Acids Res. 17, 3145–3161.
[38] Collin R.G. (1991) Ph.D. Thesis, University of Waikato.
[39] Forterre, P., Charbonnier, F., Marguet, E., Harper, F. and Henckes, G. (1991) Biochem. Soc. Symp., Vol. 58, pp. 99–112, Portland Press.
[40] Rouvière-Yaniv, J. and Yaniv, M. (1979) Cell 17, 265–274.
[41] Tanaka, I., Appelt, K., Djik, J., White, W. and Wilson, K.S. (1984) Nature 310, 376–381.
[42] Broyles, S.S. and Petitjohn, D.E. (1988) J. Mol. Biol. 187, 47–60.
[43] Shioda, M., Sugimori, K., Shiroya, T. and Takanagi, S. (1989) J. Bacteriol. 171, 4515–4517.
[44] Shioda, M., Shiroya, T. and Takayanagi, S. (1989) Mol. Gen. 8, 163–166.
[45] Bohrmann, B., Arnold-Schulz-Gahmen, B. and Kellenberger, E. (1990) J. Struct. Biol. 104, 112–119.
[46] Griffith, J.D. (1976) Proc. Natl. Acad. Sci. U.S.A. 73, 563–567.
[47] Kellenberger, E. (1991) Res. Microbiol. 142, 229–238.
[48] Rouvière-Yaniv, J., Gros, F., Haselkorn, R. and Reiss, C. (1977) In: The Organization and Expression of the Eukaryotic Genome (Javaherian, E.M.B.K., Ed.), pp. 211–231, Academic Press, New York.
[49] Vinograd, J., Lebowitz, J. and Watson, R. (1968) J. Mol. Biol. 33, 173–197.
[50] Kaine, B., Schurke, C. and Stetter, K. (1989) Syst. Appl. Microbiol. 12, 8–14.
[51] Weisburg, W.D., Giovannoni, S.J. and Woese, C.R. (1989) Syst. Appl. Microbiol. 11, 128–134.
[52] Lindahl, T. and Nyberg, B. (1972) Biochem. 11, 3610–3618.
[53] Lindahl, T. and Nyberg, B. (1974) Biochem. 13, 3405–3410.
[54] Eigner, J., Boedtker, H. and Michaels, G. (1961) Biochem. Biophys. Acta 51, 165–168.
[55] Andreev, O. and Kaboev, O. (1982) Biochim. Biophys. Acta 698, 100–101.
[56] Ehrlich, M., Gama-Sosa, M.A., Carreira, L.H., Ljungdahl, L.G., Kuo, K.C. and Gehrke, C.W. (1985) Nucl. Acids Res. 13, 1399–1412.
[57] Lindahl, T. (1967) J. Biol. Chem. 242, 1970–1973.
[58] Butzow, J.J. and Eichhorn, G. (1975) Nature 254, 358–359.
[59] Wang, J.C. (1987) Biochem. Biophys. Acta 909, 1–9.
[60] Cozzarelli, N.R. and Wang, J.C., Eds. (1990) DNA Topology and its Biological Effects, Cold Spring Harbour Laboratory Press, Cold Spring Harbor, NY.
[61] Liu, L.F. and Wang, J.C. (1987) Proc. Natl. Acad. Sci. U.S.A. 84, 7024–7027.
[62] Wu, H.Y., Shy, S., Wang, J.C. and Liu, L.F. (1988) Cell 53, 433–440.
[63] Giaever, G. and Wang, J.C. (1988) Cell 55, 849–856.
[64] Wang, J.C. (1985) Annu. Rev. Biochem. 54, 665–697.
[65] Maxwell, A. and Gellert, M. (1986) Adv. Protein Chem. 38, 68–107.
[66] Wang, J.C., Caron, P.R. and Kim, R.A. (1990) Cell 10, 403–406.
[67] Forterre, P., Assairi, L., and Duguet, M. (1983) In: New Approaches in Eukaryotic DNA Replication (De Recondo, A.M., Ed.), pp. 123–178, Plenum Press, New York.
[68] Wyckoff, E., Natalie, D., Nolan, J.M., Lee, M. and Hsieh, T. (1989) J. Mol. Biol. 205, 1–13.
[69] Drlica K. (1990) TIG 6, 433–437.
[70] Kato, J., Nishimura, Y., Imura, R., Niki, H., Hiraga, S. and Suzuki, H. (1990) Cell 63, 394–404.
[71] Kikuchi, A. (1990) In: DNA Topology and its Biological Effects (Cozzarelli, N.R. and Wang, J., Eds.), pp. 285–298, Cold Spring Harbor Laboratory Press, Cold Spring Harbor, NY.
[72] Kikuchi, A. and Asai, K. (1984) Nature 309, 677–681.
[73] Mirambeau, G., Duguet, M. and Forterre, P. (1984) Mol. Biol. 179, 559–563.
[74] Forterre, P., Mirambeau, G., Jaxel, C., Nadal, M. and Duguet, M. (1985) EMBO J. 4, 2123–2128.

[75] Nakasu, S. and Kikuchi, A. (1985) EMBO J. 4, 2705–2710.
[76] Nadal, M., Jaxel, C., Portemer, C., Forterre, P., Mirambeau, G. and Duguet, M. (1988) Biochemistry 27, 9102–9108.
[77] Shibata, T., Nakasu, S., Yasui, K. and Kikuchi, A. (1987) J. Biol. Chem. 262, 10419–10421.
[78] Slezarev, A.I. (1988) Eur. J. Biochem. 173, 395–399.
[79] Jaxel, C., Nadal, M., Mirambeau, G., Forterre, P., Takahashi, M. and Duguet, M. (1989) EMBO J. 8, 3135–3139.
[80] Nadal, M. (1990) Ph.D. Thesis, University Paris VI
[81] Kovalsky, O.I., Kozyavkin, S.A. and Slesarev, A.I. (1990) Nucl. Acids Res. 18, 2801–2805.
[82] Slezarev, A.I. and Kozyavkin, S.A. (1989) J. Biomol. Struct. & Dynam. 7, 935–942.
[83] Kirkegaard, K. and Wang, J.C. (1985) J. Mol. Biol. 185, 625–637.
[84] Liu, L.F. and Wang, J.C. (1978) Cell 15, 979–984.
[85] Confalonieri, F., Elie, C., Nadal, M., Bouthier de la Tour, C., Forterre, P. and Duguet, M. (1993) Proc. Natl. Acad. Sci. U.S.A. 90, 4753–4757.
[86] Zhang, H., Hesse, C.B. and Liu, L.F. (1990) Proc. Natl. Acad. Sci. U.S.A. 87, 9078–9082.
[87] Yang, L., Jessee, C.B., Lau, K., Zhang, H. and Liu, L.F. (1989) Proc. Natl. Akad. Sci. U.S.A. 86, 6121–6125.
[88] Forterre, P. (1985) Ph.D. Thesis, University Paris VII.
[89] Collin, R.G., Morgan, H.W., Musgrave, D.R. and Daniel, R.M. (1988) FEMS Lett. 55, 235–239.
[90] Forterre, P., Elie, C., Sioud, M. and Hamal, A. (1989) Can. J. Microbiol. 35, 228–233.
[91] Boutier de la Tour, C., Portemer, C., Nadal, M., Stetter, K.O., Forterre, P. and Duguet, M. (1990) J. Bacteriol. 172, 6803–6808.
[92] Boutier de la Tour, C., Portemer, C., Huber, R., Forterre, P. and Duguet, M. (1991) J. Bacteriol. 173, 3921–3923.
[93] Lee, M-S. and Garrard, W.T. (1991) Proc. Natl. Acad. Sci. U.S.A. 88, 9675–9679.
[94] Clark, D.J. and Felsenfeld, G. (1991) EMBO J. 10, 387–395.
[95] Kikuchi, A., Shibata, T. and Nakasu, S. (1986) Syst. Appl. Microbiol. 7, 72–77.
[96] Slezarev, A., Zaitzev, D., Kopylov, V., Stetter, K. and Kozyavkin, S. (1991) J. Biol. Chem. 266, 12321–12328.
[97] Grogan, D., Palm, P. and Zillig, W. (1990) Arch. Microbiol. 154, 594–599.
[98] Nadal, M., Mirambeau, G., Forterre, P., Reiter, W.D. and Duguet, M. (1986) Nature 321, 256–258.
[99] Martin, A., Yeats, S., Janekovic, D., Reiter, W.-D., Aicher, W. and Zillig, W. (1984) EMBO J. 3, 2165–2168.
[100] Erauso, G., Charbonnier, F., Barbeyron, T., Forterre, P. and Prieur, D. (1992) C.R. Acad. Sci. 314, 387–393.
[101] Charbonnier, F., Erauso, G., Barbeyron, T., Prieur, D. and Forterre, P. (1992) J. Bacteriol. 174, 6103–6108.
[102] Sinden, R., Carlson, J.O. and Pettijohn, D. (1980) Cell 21, 773–783.
[103] Forterre, P., Nadal, M., Elie, C., Jaxel, C., Mirambeau, G. and Duguet, M. (1986) Syst. Appl. Microbiol. 7, 67–71.
[104] Sioud, M., Possot, O., Elie, C., Sibold, L. and Forterre, P. (1988) J. Bacteriol. 170, 946–953.
[105] Sioud, M., Baldacci, G., Forterre, P. and De Recondo, A.M. (1987) Eur. J. Biochem. 169, 231–236.
[106] Forterre, P., Gadelle, D., Charbonnier, F. and Sioud, M. (1991) In: General and Applied Aspects of Halophilic Microorganisms (Rodriguez-Valera, F., Ed.), pp. 333–338, Plenum Press, New York.
[107] Gadelle, D. and Forterre, P. (1990) In: Proc. 5th European Congress on Biotechnology, Vol. 1 (Christiansen, C., Monck, L. and Villatsen, J., Eds.), pp. 229–232, Munksgaard International Publisher, Copenhagen.

[108] Sioud, M., Baldacci, J., Forterre, P. and De Recondo, A.M. (1987) Nucl. Acids Res. 15, 8217–8234.
[109] Sioud, M. and Forterre, P. (1989) Biochemistry 28, 3638–3641.
[110] Labedan, B. and Forterre, P. (1991) Biochem. Life Sci. Adv. 10, 17–24.
[111] Holmes, M.L. and Dyall-Smith, M.L. (1990) J. Bacteriol. 172, 756–761.
[112] Holmes, M.L. and Dyall-Smith, M.L. (1991) J. Bacteriol. 173, 642–648.
[113] Reece, R.J. and Maxwell, A. (1991) CRC in Biochem. & Mol. Biol. 26, 335–375.
[114] Forterre, P. (1992) In: Frontiers of Life (Trân Thanh Vân, J. and K., Mounolou, J.C., Schneider, J. and McKay, C., Eds.), Editions Frontières, Gif sur Yvette, France, in press.
[115] Kim, R.A. and Wang, J.C. (1989) J. Mol. Biol. 208, 257–267.
[116] Sioud, M., Baldacci, G., Forterre, P. and De Recondo, A.M. (1988) Nucl. Acids Res. 16, 7833–7842.
[117] Sioud, M., Baldacci, G., De Recondo, A.M. and Forterre, P. (1988) Nucl. Acids Res. 16, 1379–1392.
[118] Yang, C.F. and DasSarma, S. (1990) J. Bacteriol. 172, 4118–4121.
[119] Holmes, M.L., Nuttall, S. and Dyall-Smith, M.L. (1991) J. Bacteriol. 173, 3807–3813.
[120] Menzel, R. and Gellert, M. (1983) Cell 34, 105–113.
[121] Kornberg, A. and Baker, T. (1991) DNA Replication, Freeman and Co, New York.
[122] Ito, J. and Braithwaite, D. (1991) Nucl. Acids Res. 19, 4045–4057.
[123] Morrison, A., Bell, J.B., Kunkel, T.A. and Sugino, A. (1991) Proc. Natl. Acad. Sci. U.S.A. 88, 9473–9477.
[124] Forterre, P., Elie, C. and Kohiyama, M. (1984) J. Bacteriol. 159, 800–802.
[125] Schinzel, R. and Burger, K.J. (1984) FEMS Microbiol. Lett. 25, 187–190.
[126] Zabel, H.P., Fischer, H., Holler, E. and Winter, J. (1985) Syst. Appl. Microbiol. 6, 111–118.
[127] Possot, O., Gernhardt, P., Klein, A. and Sibold, L. (1988) Appl. Env. Microbiol. 54, 734–740.
[128] Zellner, G., Stackebrandt, E., Kneifel, H., Messner, P., Sleytr, U.B., Conway de Macario, E., Zabel, H.P., Stetter, K.O. and Winter, J. (1989) Syst. Appl. Microbiol. 11, 151–160.
[129] Chen, H., Lawrence, C.B., Bryan, S.K. and Moses, R.E. (1990) Nucl. Acids Res. 18, 7185–7186.
[130] Rossi, M., Rella, R., Pensa, M., Bartolucci, S., De Rosa, M., Gambacorta, A., Raia, C.A. and Dell'aversano, O. (1986) Syst. Appl. Microbiol. 7, 337–341.
[131] Zabel, H.P., Holler, E. and Winter, J. (1987) Eur. J. Biochem. 165, 171–175.
[132] Mattila, P., Korpela, J., Tenkanen, T. and Pitkänen, K. (1991) Nucl. Acids Res. 19, 4967–4973.
[133] Lundberg, K.S., Shoemaker, D.D., Adams, M.W., Short, J.M., Sorge, J.A. and Mathur, E.J. (1991) Gene 108, 1–6.
[134] Cariello, N.F., Swenberg, J.A. and Skopek, T.R. (1991) Nucl. Acids Res. 19, 4193–4198.
[135] Pisani, F.M., De Martino, C. and Rossi, M. (1992) Nucl. Acids Res. 20, 2711–2716.
[136] Perler, F., Comb, D., Jack, W., Moran, L., Qiang, B., Kucera, R., Benner, J., Slatko, B., Nwankwo, D., Hempstead, S.K., Carlow, C. and Jannasch, H. (1992) Proc. Natl. Acad. Sci. U.S.A. 89, 5577–5581.
[137] Mathur, E.J., Adams, M.W.W., Callen, W.N. and Cline, J.M. (1991) Nucl. Acids Res. 19, 24.
[138] Forterre, P. (1992) Nucl. Acids Res. 20, 1811.
[139] Klimczak L.T., Grummt, F. and Burger, K.J. (1985) Nucl. Acids Res. 13, 5269–5282.
[140] Klimczak, L.J., Grummt, F. and Burger, K.J. (1986) Biochemistry 25, 4850–4855.
[141] Elie, C., De Recondo, A.M. and Forterre, P. (1989) Eur. J. Biochem. 178, 619–626.
[142] Hamal, A., Forterre, P. and Elie, C. (1990) Eur. J. Biochem. 190, 517–521.
[143] Elie, C., Salhi, S., Rossignol, J.M., Forterre, P. and De Recondo, A.M. (1988) Biochem. Biophys. Acta 951, 280–289.
[144] Salhi, S., Elie, C., Jean-Jean, O., Meunier-Rotival, M., Forterre, P., Rossignol, J.M. and De Recondo, A.M. (1990) Biochem. Biophys. Res. Commun. 167, 1341–1347.

[145] Salhi, S., Elie, C., Forterre, P., De Recondo, A.M. and Rossignol, J.M. (1989) J. Mol. Biol. 209, 635–644.
[146] Prangishvili, D.A. (1986) Mol. Biol. 20, 380–389.
[147] Chinchaladze, D.Z., Prangishvili, D.A., Kachabaeva, L.A. and Zaalishvili, M.M. (1986) Mol. Biol. 19, 1193–1201.
[148] Nakayama, M. and Kohiyama, M. (1985) Eur. J. Biochem. 152, 293–297.
[149] Nakayama, M., Ben Mahrez, K. and Kohiyama, M. (1988) Eur. J. Biochem. 175, 265–270.
[150] Sorokine, I., Ben-Mahrez, K., Nakayama, M. and Kohiyama, M. (1991) Eur. J. Biochem. 197, 781–784.
[151] Zillig, W., Palm, P., Reiter, W.-D., Gropp, F., Pühler, G. and Klenk, H-P. (1988) Eur. J. Biochem. 173, 473–482.
[152] Searcy, D.G. (1986) Syst. Appl. Microbiol. 7, 198–201.
[153] Kuhsel, M., Strickland, R. and Palmer, J.D. (1990) Science 250, 1570–1572.
[154] Xu, M.Q., Kathe, S.D., Heidi, G.B., Nierzwicki-Bauer, S.A. and Shub, D.A. (1990) Science 250, 1566–1570.
[155] Meile, L., Kiener, A. and Leisinger, T. (1983) Mol. Gen. Genet. 191, 480–484.
[156] Nölling, J., Frijlink, M. and De Vos, W.M. (1991) J. Gen. Microbiol. 137, 1981–1986.
[157] Prangishvili, D.A., Vashakidze, R.P., Chelidze, M.G. and Gabriadze, I.U. (1983) FEBS Lett. 192, 57–60.
[158] Schmid, K., Thomm, M., Laminet, A., Lane, F.G., Kessler, C., Stetter, K.O. and Schmitt, R. (1984) Nucl. Acids Res. 12, 2619–2628.
[159] Lunnen, K., Morgan, R., Timan, C. Krzycki, J., Reeve, J. and Wilson, G. (1989) Gene 77, 11–19.
[160] Jordan, M., Meile, L., Leisinger, T. (1989) Mol. Gen. Genet. 220, 161–164.
[161] Vogelsang-Wenke, H. and Oesterhelt, D. (1988) Mol. Gen. Genet. 211, 407–414.
[162] Zillig, W., Reiter, W., Palm, P., Gropp, F., Neumann, H. and Rettenberger, M. (1988) In: The Bacteriophages (Calendar, R., Ed.), pp. 517–558, Plenum Press, New York.
[163] Zillig, W. (1991) Curr. Opinion Genet. & Dev. Biol. 4, 544–551.
[164] Forterre, P. (1993) In: Archaebacteria Biotechnology Handbook, series D (Cowan, D., Ed.), Plenum Press, New York, in press.
[165] Woese, C.R., Kandler, O. and Wheelis, M.L. (1990) Proc. Natl. Acad. Sci. U.S.A. 87, 4576–4579.
[166] Iwabe, N., Kuma, K.I., Hasegawa, M., Osawa, S.E. and Miyata, T. (1989) Proc. Natl. Acad. Sci. U.S.A. 86, 9355–9359.
[167] Gogarten, J.P., Kibak, H., Dittrich, P., Taiz, L., Bowman, E., Manolson, M.F., Poole, R.J., Date, T., Oshima, T., Konishi, J., Denda, K. and Yoshida, M. (1989) Proc. Natl. Acad. Sci. U.S.A. 86, 6661–6665.
[168] Kakinuma, Y., Igarashi, K., Konishi, K. and Yamato, I. (1991) FEBS Lett. 292, 64–68.
[169] Tsutsumi, S., Denda, K., Yokoyama, K., Oshima, T., Date, T. and Yoshida, M. (1991) Biochem. Biophys. Acta 1098, 13–20.
[170] Forterre, P. (1992) Nature 335, 305.
[171] Forterre, P., Benachenhou, N., Confalonieri, F., Duguet, M., Elie, C. and Labedan, B. (1993) Biosystem 28, 15–32.
[172] Benachenhou, N., Forterre, P. and Labedan, B. (1993) J. Mol. Evol. 36, 335–346.
[173] Meyer, T.E., Cusanovich, M.A. and Kamen, M.D. (1986) Proc. Natl. Acad. Sci. U.S.A. 83, 217–220.
[174] Lake, J. (1991) TIBS 16, 46–50.
[175] Rivera, M.C. and Lake, J. (1992) Science 257, 74–76.
[176] Gupta, R. and Singh, B. (1992) J. Bacteriol. 174, 4594–4605.

CHAPTER 12

Transcription in archaea

Wolfram ZILLIG, Peter PALM, Hans-Peter KLENK,
Doris LANGER, Uwe HÜDEPOHL, Johannes HAIN,
Martin LANZENDÖRFER and Ingelore HOLZ

Max-Planck-Institut für Biochemie, D-82152 Martinsried, Germany

Abbreviations

GAPDH	glyceraldehydephosphate dehydrogenase	PGK	phosphoglycerate kinase
		RNAP	DNA-dependent RNA polymerase
MDH	malate dehydrogenase		

1. Introduction

On the basis of sequence comparison of 16S rRNA sequences Carl Woese and coworkers had divided "the prokaryotic domain" into two urkingdoms of life, the eubacteria and the archaebacteria, each a coherent entity distinct from the other as well as from the third urkingdom, the eukaryotes [1,2]. Recently Woese et al. [3] have proposed to call the highest taxon level a domain instead of an Urkingdom, and to rename the three domains of life Bacteria, Archaea and Eucarya in order to emphasize their distinction as equally separate major entities.

Phylogenetic dendrograms of other ubiquitous and sufficiently conserved macromolecules, e.g. 23S rRNAs [4], DNA-dependent RNA polymerases (RNAPs) [5,6], translation factors [7,8], ribosomal proteins [7] and ATPases [9,10] confirmed the coherence of the Archaea and the distinction of the three domains. The comparison of the specific shaping of an increasing number of molecular and metabolic features, exemplarily listed in Table 1, yielded further evidence. The first 7 of the 23 listed features contrast the Archaea, including *Halobacterium, Methanococcus* and *Sulfolobus,* both with the Bacteria and the Eucarya, thus supporting the validity of the concept. The Archaea share a further 14 of the listed features. They resemble the Eucarya in 12 and the Bacteria in 2 of these, which therefore do not yield independent trees but support the concept according to the rules of numeric taxonomy. An opposing view separating the Archaea into three phyla and deriving the Bacteria from one and the Eucarya from another of them [33–35] has been refuted [35] and is in complete discrepancy with these data.

Many of the known distinctions between Archaea and Bacteria concern the genetic apparatus where an archaeal shaping of a feature can often be confronted with a

TABLE 1

Distribution of discrete characteristics among Bacteria, Eucarya and three phyla of Archaea[a]

Characteristic	Bacteria (E. coli)	Archaea Hb.	Archaea Mc.	Archaea S.	Eucarya (Eukaryotes)
Acylester lipids [12,13,14]	+	−	−	−	+
Flagellins (b = bacterial, a = archaeal, e = eucaryal) [15]	b	a	a	?	e
Isopranyl ether lipids [12,13,16]	−	+	+	+	−
A′+A″ split of RNAP component A(β′) [17,18]	−	+	+	+	−
RNAP gene order rpoHBA1A2-X-rpsLG [17,18]	−	+	+	+	−
Unique modified nucleotides in tRNA [19]	−	+	+	+	−
Unique sequences flanking rRNA genes [20]	−	+	+	+	−
7S RNA [21]	−	+	+	+	+
RNAP comp. H corresponds to yeast ABC27 kDa subunit [18]	−	+	+	+	+
EFII ADP ribosylatable [22]	−	+	+	+	+
DNAPα aphidicolin sensitive [23,24]	−	+	+	+	+
Promoter type (e = boxA, b = bacterial) [22,26,27]	b	e	e	e	e[b]
RNAP type (e = complex, b = bacterial) [17]	b	e	e	e	e
Protein splicing of tRNAs [29,30,31]	−	(+)	?	+	(+)
Translation starts with N-formyl-methionyl-tRNA [32]	+	−	−	−	−
RNAP gene order rplKAJLrpoBC [32]	+	−	−	−	−
Mureine	+[c]	−	−	−	−
RNAP rifampicin sensitive [32]	+	−	−	−	−
Ribosome streptomycin sensitive [32]	+	−	−	−	−
Transcription units [32]	+	+	+	+	−
Shine Dalgarno sequences [32]	+	+	+	+	−
B″+B′ split of RNAP component B(β) [17]	−	+	+	−	−
Gene order in rRNA operon, b = rrs,ala,rrl,rrf; e = rrs,rrl (rrf unlinked) [17]	b	b	b	e	e

[a] Abbreviations: Hb., Halobium.
[b] TATA box in pol 2 and some pol 3 promoters.
[c] Except mycoplasms.

bacterial design. One of the most revealing manifestations of this character is in the transcription machinery including evidence for distinct genome organizations (see below). The lipid structure [12,13,37,38] (see also Chapter 9 of this volume) is a further characteristic distinguishing Archaea and Bacteria. With the exception of their lipids, Eucarya resemble Archaea in most of these characters. On the other hand, there is considerable overlap in the complex mosaic of feature designs in Archaea and Bacteria, especially in central metabolism [39] (see also Chapter 1 of this volume). Some enzymes of the central metabolism of eucarya, like GAPDH[1] [40], MDH[2] [41] and PGK[3] [42], resemble their bacterial homologs much more than their archaeal correspondents.

Resolving these paradoxes could lead to a better understanding of early steps in biotic evolution, including the nature of the common ancestor, the separation of the domains, speciation and the origin of the Eucarya. A thorough investigation of the archaeal transcriptional apparatus and its comparison with those of Bacteria and Eucarya should help in this endeavour: first, because transcription reveals one of the clearest distinctions between the latter domains but striking similarities between Eucarya and Archaea, and second, because RNAP is a large, ubiquitous and highly conserved macromolecule allowing sequence comparison between patterns of 20 amino-acid residues rather than of only 4 nucleotides. This chapter will review and discuss available information on transcription in Archaea as compared to Bacteria and Eucarya.

2. DNA-dependent RNA polymerase

2.1. Composition

RNAPs have been isolated from many species from all established orders of Archaea either by variations of a standard procedure [43] or, in case of enzymes of low stability, by methods using phase partitioning and reverse phase chromatography respectively for the separation of the enzyme from the DNA [44]. As first revealed by immunochemical techniques [45] and later confirmed by sequencing [46–48], the enzymes were of two types (Fig. 1): Those of the extremely thermophilic, often sulfur-dependent crenarchaeota [3], and of the Thermococcales [49] and *Thermoplasma* [50] both belonging to the euryarchaeota, consisted of about 12 stably linked components in stoichiometric proportion which have been named B, A', A'', D through I, K, L and N, in order of decreasing apparent molecular weight in SDS PAGE. The enzymes of the other phyla of the euryarchaeota, including three orders of methanogens, the *Halobacteriae* and *Archaeoglobus,* however, contain instead of a single B component two fragments, B' and B'', roughly corresponding to the second and the first half of the homologous B subunit. Whereas all other archaeal enzymes isolated so far contained an A' and an A'' subunit, in *Archaeoglobus* these subunits appeared linked to a coherent A component according to immunochemical evidence [51].

Immunochemical analysis [52] and later sequencing of cloned genes revealed similarities between archaeal, bacterial and eucaryal nuclear RNAP components. The archaeal A' and A'' components (previously termed A and C) thus correspond to the largest components of the diversified eucaryal RNAPs and to bacterial β'. The archaeal components B, or B'' and B', in this order, correspond to the second-largest components of the three eucaryal enzymes and to bacterial β, respectively (Fig. 1b). The archaeal small component H is homologous to about the last third of the sequence of subunit ABC27 of *Saccharomyces cerevisiae* [18,53,54] which is one of five components shared by all three eucaryal nuclear enzymes (Fig. 2). The others are ABC23, 14 and 10. The archaeal small component K is homologous to about the second half of subunit ABC23 [Langer, D., unpublished data from our laboratory] present in yeast pol 1, pol 2 and pol 3 [53,54]. The homologies between components of different archaeal enzymes are indicated by designation with the same capital Roman letters.

Some RNAPs of methanogens and most prominently the enzyme from *H. halobium* appear to lack some of the smaller components and to contain substoichiometric amounts of others, most probably as a result of lower stability.

370

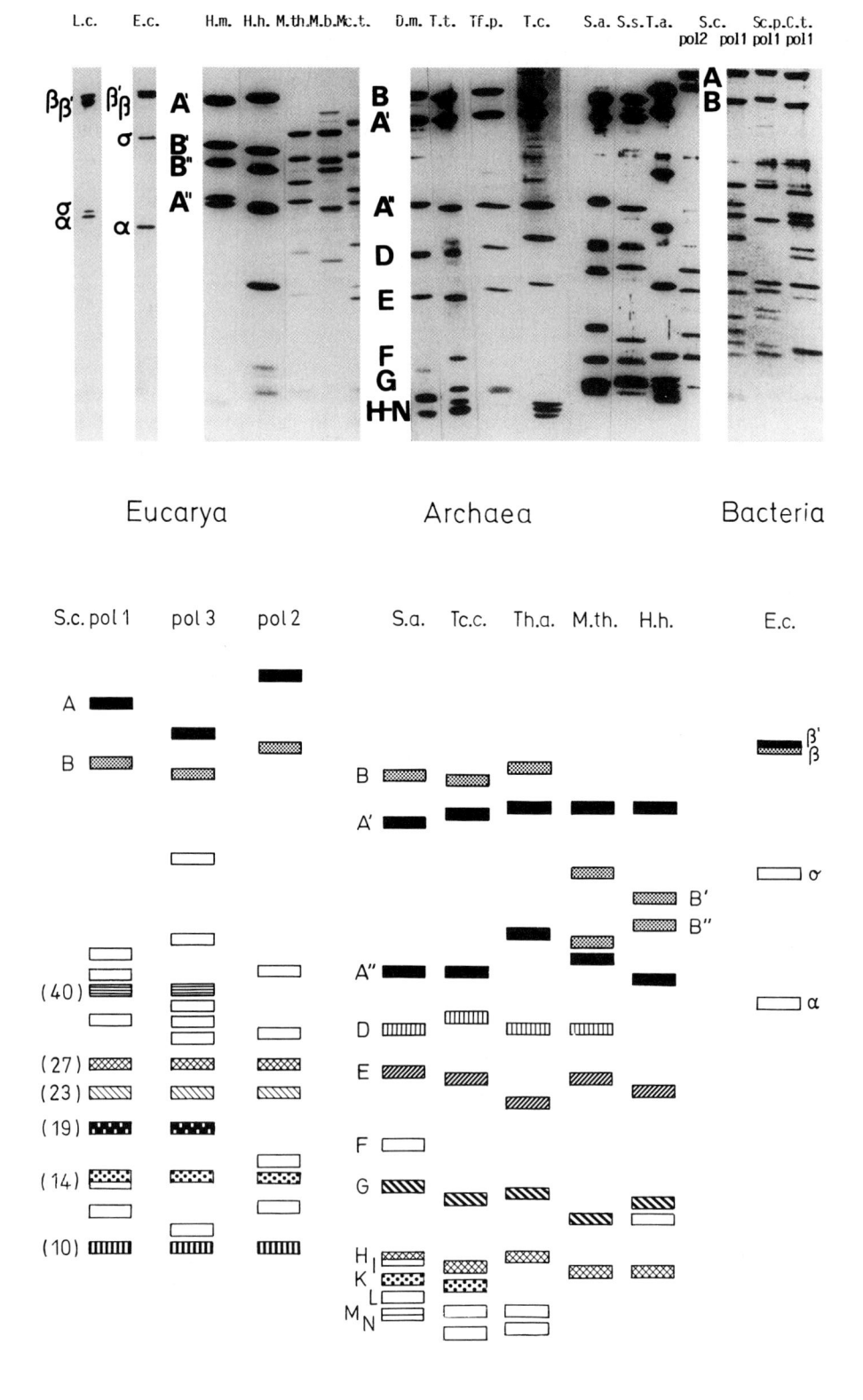

Eucarya   Archaea   Bacteria

S.c. pol 1   pol 3   pol 2   S.a.   Tc.c.   Th.a.   M.th.   H.h.   E.c.

Fig. 1. SDS PAGE component patterns of RNAPs from various archaea in comparison to patterns from representatives of bacteria and of eucaryal (yeast) nuclear RNAPs.

(a) Actual patterns: L.c., *Lactobacillus curvatus*; E.c., *Escherichia coli*; H.m., *Halococcus morrhuae*; H.h., *Halobacterium halobium*; M.th., *Methanobacterium thermoautotrophicum*; M.b., *Methanosarcina barkeri*; Mc.t., *Methanococcus thermolithoautotrophicus*; D.m., *Desulfurococcus mucosus*; T.t., *Thermoproteus tenax*; Tf.p., *Thermofilum pendens*; T.c., *Thermococcus celer*; S.a., *Sulfolobus acidocaldarius*; S.s., *Sulfolobus solfataricus*; T.a., *Thermoplasma acidophilum*; S.c., *Saccharomyces cerevisiae*; Sc.p., *Schizosaccharomyces pombe*; C.t., *Candida tropicalis*.

(b) Schematic patterns showing homologies between components (by equal marking). The numbers in parentheses are apparent molecular weights in kDa.

2.2. Organization of RNAP component genes

Clusters of contiguous genes containing one small component gene, *rpoH* followed by the genes of the large components, *rpoB* or *rpoB2* and *rpoB1* (these are the genes for subunits B'' and B', respectively), *rpoA1* and *rpoA2* (these are the genes for subunits A' and A'', respectively) have been found in *Sulfolobus acidocaldarius* [47], *Thermococcus celer,* [150] and *Thermoplasma acidophilum* [151], *Methanococcus vannielii* [Palm, P. and Arnold, I., unpublished results from this laboratory], and *Halobacterium halobium* (Fig. 2). Similar data are also available for *Methanobacterium thermoautotrophicum* [48] (the published sequence does, however, not include *rpoH* and the region downstream of *rpoA2*), and part of *Halococcus morrhuae* [46].

In *E. coli,* an operon comprising the genes for the ribosomal proteins L11, L1, L10 and L12 is found immediately upstream of *rpoB*. An ORF which in archaea is usually found immediately downstream of *rpoA2* (exceptions are *Halobacterium* and *Thermoplasma*), encodes a putative protein homologous to the ribosomal protein L30 of yeast but not to any known bacterial protein. This gene is immediately followed by an ORF encoding a putative protein X highly conserved within archaea but so far without correspondence in bacteria and eucarya. The genes normally following downstream encode the ribosomal proteins S12 and S7, and the translation factors EF-2 and EF-1α. These are in the same order as in the streptomycin operon in *E. coli* where they, however, occur in opposite orientation relative to the RNAP large component operon and far upstream rather than directly downstream of the latter [7]. In some archaeal gene clusters, the genes for one or the other or even both translation factors are absent [7,55–8] (see Fig. 2).

Promoters were found immediately upstream of *rpoH* and transcripts of the whole region were detected either directly or indirectly [46,47,150]. In the case of *Sulfolobus,* an additional promoter was mapped immediately upstream of *rpoA2*.

The archaeal *rpoB* (or *B2B1,* respectively) *A1A2* transcription unit thus corresponds closely to the bacterial *rpoBC* operon but its neighborhood is entirely different. A similar statement can be made for the streptomycin genes, though the context in this gene cluster appears less rigid. At least one of the two genes preceding this cluster encodes a ribosomal protein (L30) apparently missing in bacteria. If the similarity of context of such gene clusters indicates their existence in the common ancestor, it appears possible that the genes for the L30 homolog and for protein X originally belonged to this cluster, as they still do in extant archaea, and were lost by "streamlining" in the bacterial lineage, like most of the components of the ancestral RNAP. Similarly, three of

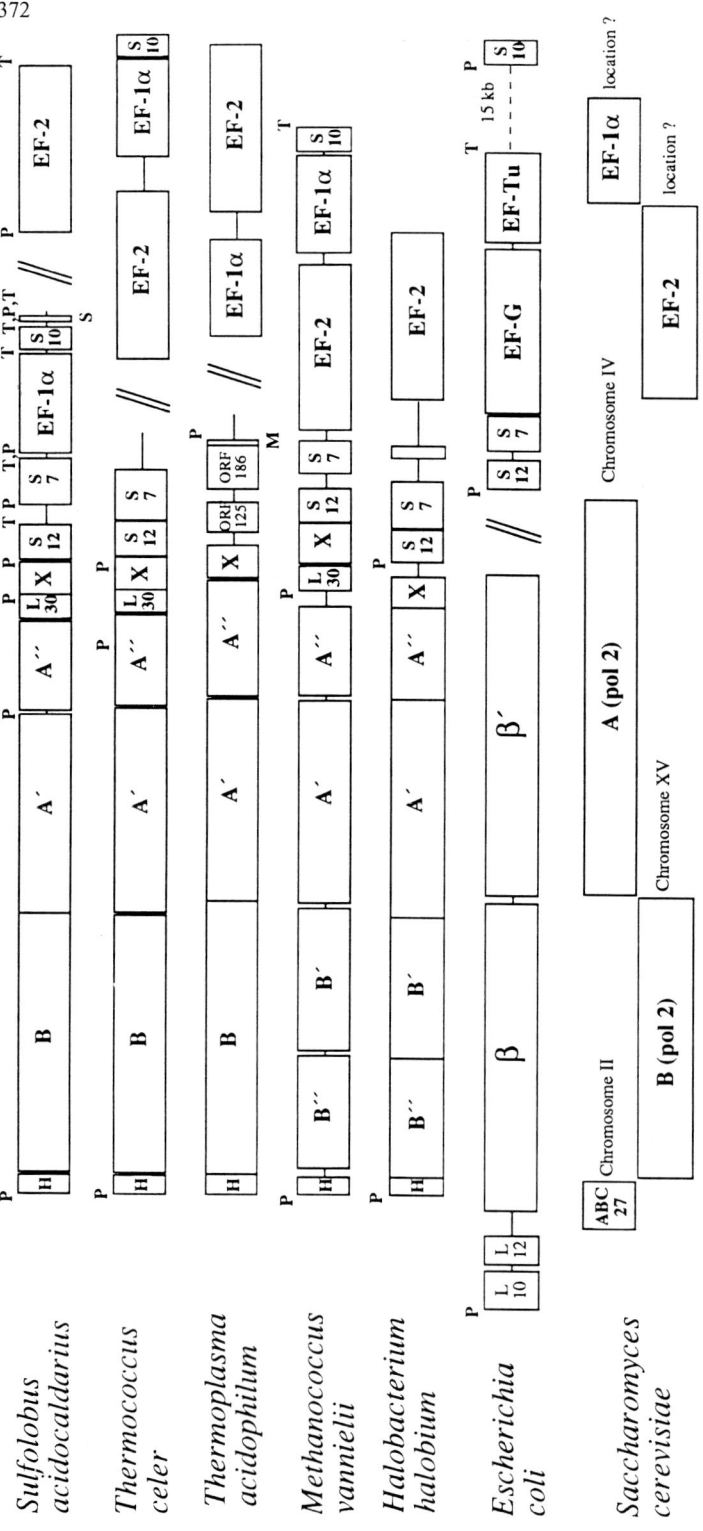

Fig. 2. Organization of gene clusters containing the genes of the large components of DNA-dependent RNA polymerases of various Archaea, including their immediate neighborhood, in comparison with the organisation of the corresponding genes in *E. coli* and *Saccharomyces cerevisiae* (pol2). Component H is homologous to the last third of S.c.ABC27 [18,53]. Symbols: P, mapped promoter; T, mapped terminator. Homologies as indicated by letter designations are described in the text.

the genes of the spectinomycin operon of archaea encode ribosomal proteins shared with eucarya but absent in bacteria [59,60], and an archaeal gene cluster corresponding to the *E. coli* S10 operon [59,61] also contains one gene encoding a protein corresponding to a eucaryal but not to any bacterial ribosomal protein. The gene encoding protein S10, which is the first in the S10 operon of *E. coli,* has been found in a different neighborhood in archaea, namely directly downstream of the gene encoding EF-1α within the streptomycin gene cluster [59].

It should be emphasized that, except for the loss of the genes for the L30 homolog and for EF-1α, the context [58] and the neighborhood of the RNAP transcription unit in Halobacteriales correspond to those in other archaea. The A'A" gene split is a general characteristic of Archaea, whereas the B"B' gene split is a distinctive character of *Archaeoglobus* [51] and of methanogenic and extremely halophilic archaea. Two phyla in the major branch of the Archaea termed Euryarchaeota, namely the Thermococcales [62] and the genus *Thermoplasma* [Klenk, H.-P., unpublished data from this laboratory], however, resemble the Crenarchaeota [3] in possessing unsplit B genes indicating that the B gene split occurred within the euryarchaeotal lineage after the separation of *Thermococcus* and *Thermoplasma* but before the branching off of *Methanococcus* and *Methanobacterium.*

2.3. Similarities between sequences of RNAP components

Several sequences of genes of large components of bacterial RNAPs and of each of the three eucaryal RNAPs are available for sequence comparison with the archaeal homologs. The alignment of protein sequences both of β' and eucaryal A (=largest) components (similarities listed in Table 2) and of β and eucaryal B (=second-largest) components with those of archaeal A'+A" subunits and B or B"+B' subunits, respectively, is simplified by many highly conserved sequence blocks and signatures. The sequences of the archaeal components are highly similar and colinear to eucaryal pol 2 and pol 3 subunits and can thus be aligned to these with few gaps and inserts. The similarity between the archaeal and these eucaryal sequences is indeed often higher than that between components of the two different eucaryal RNAPs themselves. In contrast, both the similarity and the colinearity between archaeal and bacterial sequences are low though clearly significant. Sequences of pol 1 components appear equally distant from those of their archaeal homologs and from those of their eucaryal paralogs (same origin, different function). They show distinct local similarities, in sequence as well as in colinearity, to either correspondents.

2.4. The phylogeny of RNAP components

Because a complete collection of sequences of all domains, and specifically of the three different eucaryal nuclear RNAPs, was first available for the β', the A and the A'+A" components, respectively, a thorough phylogenetic analysis was originally confined to the latter. Unrooted phylogenetic dendrograms have been constructed by various algorithms, including the distance-matrix method according to Fitch and Margoliash [78], different parsimony programs [79], and two procedures designed to correct for errors resulting from differences in the rate of evolution: the maximum-likelihood procedure of Felsenstein [80,81] and the evolutionary parsimony algorithm of

TABLE 2

Identities and similarities between sequences of RNAP large components from the three domains of life[a,b]

	Mv	Mt	Hh	Tc	Sa	Mm2	Sc2	Ce2	Pf2	Tb2	Tb3	Gl3	Sc3	Scl	Tbl	Spl	Ec	MpCh	SoCh	Pp
Mv	100.0	59.3	49.9	54.6	50.6	35.8	37.4	35.6	35.5	33.3	33.9	33.6	37.0	28.9	25.1	29.5	30.4	26.1	25.8	29.2
Mt	64.6	100.0	55.0	58.6	50.9	39.2	39.4	37.7	36.9	33.6	37.5	34.4	38.3	30.2	26.9	29.5	30.1	26.1	26.2	30.1
Hh	54.7	63.4	100.0	52.1	46.8	35.5	37.2	34.4	33.2	31.2	33.8	32.8	34.7	28.5	25.2	29.9	29.6	26.0	25.3	28.3
Tc	60.4	64.1	58.8	100.0	53.6	38.5	39.6	38.1	36.1	34.7	36.2	36.0	38.9	31.4	27.2	26.8	30.2	26.0	27.1	28.5
Sa	54.1	56.0	52.7	58.3	100.0	39.1	39.8	38.3	36.8	34.8	36.6	37.1	39.8	32.0	26.2	30.7	29.3	25.6	26.7	28.1
Mm2	36.3	42.2	37.0	39.1	41.1	100.0	53.6	71.2	50.5	40.8	33.7	30.5	35.0	28.8	27.5	27.8	28.1	25.7	25.6	26.9
Sc2	37.4	41.4	39.2	41.4	40.6	60.0	100.0	52.3	46.6	40.5	33.0	32.2	35.2	28.6	27.4	28.2	28.8	26.1	26.7	27.1
Ce2	35.8	40.5	35.7	39.6	39.9	79.1	57.8	100.0	49.0	40.8	32.5	30.6	35.4	28.3	25.8	28.2	25.9	23.7	23.7	25.5
Pf2	34.5	37.0	32.0	35.1	37.1	53.2	48.3	51.7	100.0	39.7	30.5	29.5	32.4	28.3	23.1	27.6	24.4	22.2	21.7	23.3
Tb2	34.0	35.9	32.6	36.1	37.1	42.4	42.5	41.2	39.9	100.0	32.9	28.4	33.4	27.6	24.9	26.6	25.7	23.4	24.3	25.4
Tb3	33.4	37.9	34.2	36.9	37.1	30.2	30.9	29.4	27.3	30.2	100.0	36.4	40.4	29.5	27.0	27.8	26.0	24.8	25.5	25.3
Gl3	32.2	34.3	32.6	34.1	36.2	26.1	27.0	26.4	23.8	25.6	34.3	100.0	39.9	25.7	24.4	24.8	25.3	23.3	23.6	23.3
Sc3	38.0	40.4	37.0	39.9	40.7	35.2	35.6	34.5	31.9	32.2	40.3	38.7	100.0	28.6	26.3	26.7	26.7	25.1	25.2	25.9
Scl	26.1	27.5	23.0	28.9	29.9	25.3	25.5	26.1	24.1	23.1	23.9	21.0	25.8	100.0	31.4	53.4	25.4	22.2	22.0	23.8
Tbl	19.4	22.7	18.5	22.2	22.1	20.1	20.1	19.1	15.0	17.8	18.0	17.6	20.3	27.7	100.0	32.4	21.4	19.7	20.8	22.3
Spl	24.9	24.5	21.4	26.7	27.5	22.7	23.4	23.6	21.4	20.6	21.0	19.4	23.3	60.9	27.4	100.0	24.5	21.8	21.8	22.7
Ec	27.9	27.7	28.0	26.2	25.4	26.4	25.8	25.1	18.6	22.4	18.4	17.5	21.1	21.1	12.8	18.3	100.0	41.5	38.6	73.5
MpCh	19.3	22.6	19.3	20.2	21.1	18.9	18.5	18.3	14.5	15.8	16.2	16.2	17.0	15.6	11.9	14.3	41.6	100.0	68.8	39.5
SoCh	20.3	21.7	19.3	20.8	21.7	19.0	19.6	18.2	13.8	16.6	15.3	15.1	16.5	14.9	10.5	13.9	38.3	74.7	100.0	37.6
Pp	25.9	25.6	25.4	23.8	23.8	24.9	23.2	23.1	17.1	20.5	16.9	14.7	18.8	18.8	11.3	16.1	79.3	38.9	36.5	100.0

[a] Percent similarity between sequences of archaeal A' plus A'', bacterial β' and eucaryal nuclear largest (=A) components of DNA-dependent RNA polymerases calculated from identities (upper right) and similarities according to Dayhoff et al. [63] (lower left).
[b] Abbreviations used: Mv, *Methanococcus vannielii* (Palm, P., Auer, J. and Arnold-Ammer, I., unpublished data from this laboratory); Mt, *Methanobacterium thermoautotrophicum* [47]; Hh, *Halobacterium halobium* [45]; Tc, *Thermococcus celer* [150]; Sa, *Sulfolobus acidocaldarius* [46]; Mm2, *Mus musculus* pol 2 [63]; Sc2, *Saccharomyces cerevisiae* pol 2 [64]; Ce2, *Caenorhabditis elegans* pol 2 [65]; Pf2, *Plasmodium falsiparum* pol 2 [66]; Tb2, *Trypanosoma brucei* pol 2 [67]; Tb3, *Trypanosoma brucei* pol 3 [68]; Gl3, *Giardia lamblia* pol 3 [28]; Sc3, *Saccharomyces cerevisiae* pol 3 [64]; Scl, *Saccharomyces cerevisiae* pol 1 [69]; Tbl, *Trypanosoma brucei* pol 1 [70,71]; Spl, *Schizosaccharomyces pombe* pol 1 [72]; Ec, *Escherichia coli* [73]; MpCh, *Marchantia polymorpha* chloroplast [74]; SoCh, *Spinacia oleracea* chloroplast [75]; Pp, *Pseudomonas putida* [76].

Lake [82,83]. Significance analysis of the resulting dendrograms was performed with the DNA bootstrap program of Felsenstein [79,84] and by investigating the influence of the degree of conservation of fractions of the aligned sequences on the tree topology.

As expected, the resulting dendrogram (Fig. 3) showed the Archaea and the Bacteria each on its own branch thus confirming the coherence of the Archaea. Unexpectedly, however, the three paralogous eucaryal nuclear RNAPs each formed a distinct branch, of which those of pol 2 and/or pol 3 appeared as direct neighbors of the Archaea, whereas

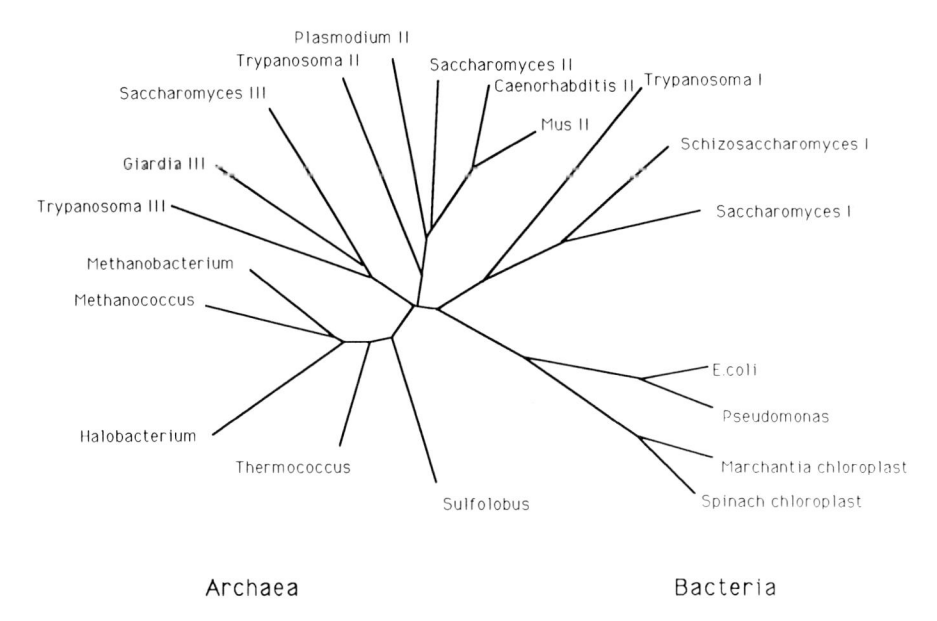

Fig. 3. Phylogenetic dendrogram of DNA-dependent RNA polymerase A (Eucarya), A' plus A" (Archaea) and β' (Bacteria) components constructed by the distance-matrix method of Fitch and Margoliash from the data of Table 1 using a program by Felsenstein [79]. Features are discussed in the text.

pol 1 shared a bifurcation with the Bacteria. Significance analysis using the program DNA boot of Felsenstein [79,84] corroborated the separate ramification of the pol 1 branch and its bifurcation with the Bacteria (193 of 200 trees) but could not decide whether pol 2 or pol 3 or a composite branch carrying both pol 2 and pol 3 on a short shared stem is the next neighbor of the archaeal branch. The similarity of the consensus of pol 2 and some pol 3 promoters [85,86] corroborates the possibility that both polymerases originated from an early duplication diversification in the ancestral eukaryote.

Two opposing explanations are in line with this unexpected though significant branching topology: First, the Archaea and/or the Bacteria could have arisen from the Eucarya by one or two reduction events after the diversification of pol 2 and pol 3 from pol 1. The archaeal enzymes could then have been derived from pol 2/3 and the bacterial polymerases from pol 1. This reduction would include recovery of general transcription potential by one of the diversified polymerases concomitant with the loss of the other(s). This appears as unlikely as the origin of the archaeal and/or bacterial domains by reduction.

Alternatively, the Eucarya could have resulted from some sort of fusion event of unknown nature between early archaeal and possibly bacterial genomes, pol 2/3 derived from an archaeal, pol 1 from a bacterial or very early archaeal ancestor (Fig. 4). Only the analysis of complex and composite features of apparently diverse origin, in this case

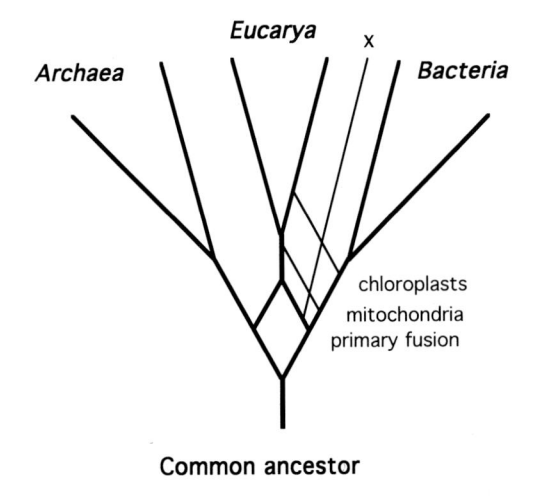

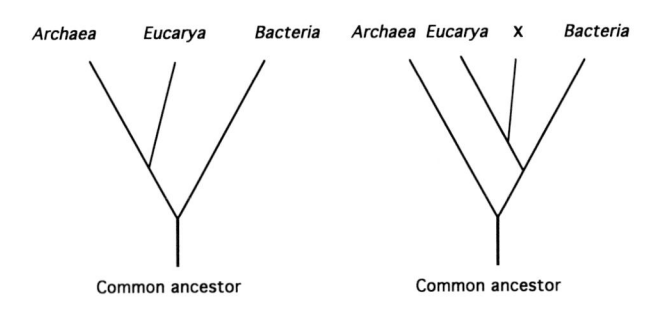

Fig. 4. Sketch of the fusion hypothesis explaining the generation of the eucaryal chimera by fusion between genomes of different ancestral archaea and/or bacteria. Upper part: Phylogenetic tree of the three domains of life showing the acquisition of eucaryal genes both from the early archaeal and the early bacterial lineage. X represents an extant branch from the bacterial lineage which participated in the fusion. Lower part: The two types of phylogenetic trees of genes, the superposition of which yields the tree in the upper part. On the left is the typical dendrogram of genes acquired from an ancestral *Archaeon,* e.g. ATPase [9,10], elongation factors [8], and pol 2 and pol 3 larger component genes [17]. On the right is the typical tree for a gene acquired from an ancestral *Bacterium,* e.g. GAPDH, MDH, PGK and pol 1 large component genes.

the existence of diversified polymerases, can yield indications for such relationships. Some enzymes of central metabolism, GAPDH [40], MDH [41] and PGK [42] of eucarya resemble their bacterial much more than their archaeal homologs. Even the GAPDH of *Giardia lamblia,* a putatively primitive eukaryote which is thought to never have possessed mitochondria, shows this similarity to bacterial GAPDH sequences [40]. This is further evidence supporting the fusion hypothesis and specifically the assumption that the events creating the chimeric nuclear genome occurred before the acquisition of the mitochondria. The chimeric nature of eucarya as a result of their generation by fusion was already postulated in 1910, on the ground of their cellular organization, by Mereschkowsky [87]. The existence of eucarya devoid of organelles but containing archaeal and bacterial endosymbionts, e.g. *Pelomyxa palustris* [88], shows that associations like those assumed in the fusion hypothesis do actually occur.

The construction of composite dendrograms of two different pairs of paralogous sequences (same origin, different function) existing in all three domains, namely ATPase α and β subunits [9,10] and EF-Tu (EF-1α) and EF-G (EF-2) [8] has opened the possibility to root phylogenetic trees. In both cases, the Bacteria branch off before the Eucarya separate from the Archaea. These dendrograms have been suggested to contradict the fusion hypothesis [89]. It should, however, be kept in mind that they describe the phylogeny of genes which in our interpretation have been donated to the Eucarya by ancestral Archaea, and that the phylogeny of eucaryal organisms should result from the superposition of dendrograms of many genes of archaeal and bacterial origin.

The L-MDH and L-LDH sequences of extant bacteria and eucarya have resulted from a duplication diversification event which probably occurred prior to the separation of these domains. Archaea seem to be devoid of L-LDH but possess two types of L-MDH [40,90]. In a phylogenetic tree, one of these appears on the basis of the bacterial and eucaryal L-MDH branch, but the other branches off below the separation of bacterial and eucaryal L-LDHs. Assuming that the diversification of the L-MDH types was the event from which the L-MDH and L-LDH lineages originated, this tree can be rooted between the two archaeal L-MDHs. The resulting rooted tree shows the Archaea branching off prior to the separation of Bacteria and Eucarya, opposite to the trees of ATPase subunits [9,10] and translation factors [8]. It is indeed a prediction of the fusion hypothesis that rooted phylogenetic trees of genes introduced by the early archaeal fusion partner show the Archaea branching off first. On the contrary, trees of genes imported by the early bacterial fusion partner should show the Archaea as the deepest branch as is the case in the L-MDH/L-LDH tree.

The striking deviation of sequences of many archaeal and bacterial genes seems inconsistent with the correspondence between Archaea and Bacteria in genome size and gene stock and in the composition and order of genes in gene clusters. The strongly contrasting apparent disorganization of eucaryal genomes could have resulted from a different mode of evolution initiated by the fusion, namely by inventive addition rather than economized adaptation to a changing environment. The fusion of two genomes would create an abundance of genes allowing excessive diversification which should soon lead to practical irreversibility. One way to cope with this situation could have been the emergence of further playgrounds for this mode of evolution, e.g., the separation of domains by introns and the recombination of domains.

2.5. The structure of the RNAP of S. acidocaldarius

As mentioned above, highly purified RNAP of *S. acidocaldarius* consists of 12 different components, B, A', A", D, E, F, G, H, I, K, L and N in near stoichiometric ratio which have been separated by SDS PAGE [91] (Fig. 5). Different N-terminal sequences have been obtained for all of these components, except for F and I which both appear N-terminally blocked [Lottspeich, F., unpublished results]. Internal peptides from F have been partially sequenced [Lottspeich, F. and Meitinger, C., unpublished]. Using

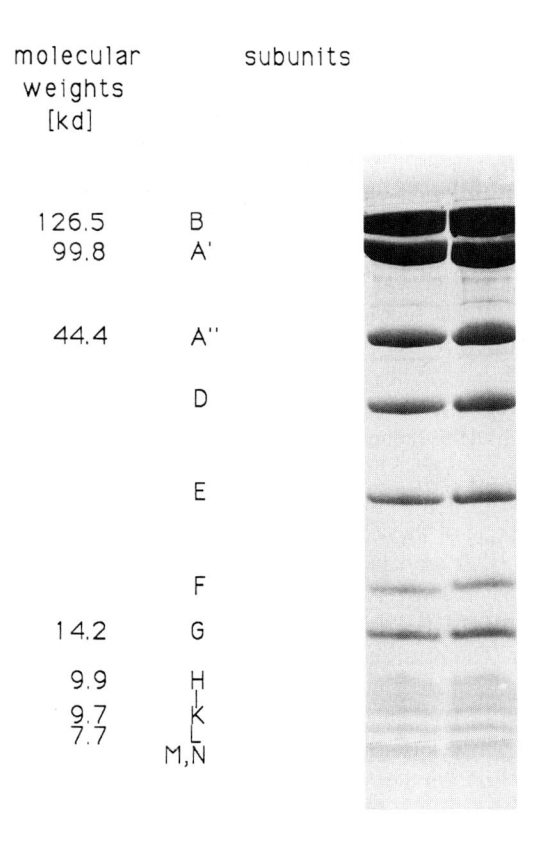

molecular weights [kd]	subunits
126.5	B
99.8	A'
44.4	A"
	D
	E
	F
14.2	G
9.9	H
9.7	K
7.7	L
	M,N

Fig. 5. SDS PAGE (according to Schägger[91]) of DNA-dependent RNA polymerase of *S. acidocaldarius*.

oligonucleotides corresponding to the amino-acid sequences, the genes for components G, K and L were cloned and sequenced (ref.[92], and Langer, D., unpublished results from this laboratory]. Subunits H and K turned out to closely resemble two of the four components shared by the three eucaryal nuclear RNAPs, ABC27 (nomenclature of Sentenac, ref.[53]) and ABC14, respectively (ref.[18], and Langer, D., unpublished data from this laboratory].

In buffers containing 6M urea and 25% formamide, the RNAP is dissociated into single subunits and two specific complexes, D–L and E–I, which have been separated by cellogel electrophoresis according to charge density [93] (Fig. 6). Component F is a mixture of rather acidic charge density variants. A polymerase fraction devoid of F and five fractions each containing a different F variant have been separated by chromatography using a Mono Q column (Pharmacia). The enzyme lacking F is unimpaired in its capacity to correctly initiate transcription from archaeal promoters in vitro (see below), indicating that F is neither required for specific initiation nor for basic functions in elongation. Subunit G is phosphorylated but does not contain a nucleotide residue. Subunits K and more so H and N are positively charged even at pH 9.2. The polymerase carries a

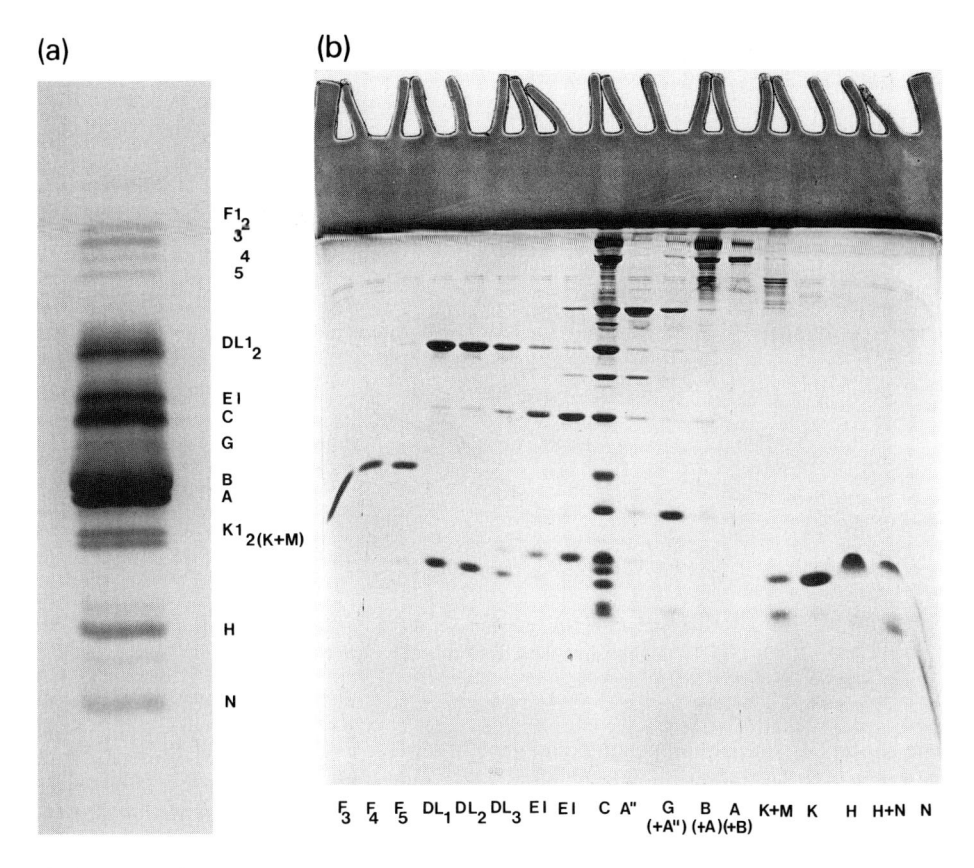

Fig. 6. Pattern of components of DNA-dependent RNA polymerase of *S. acidocaldarius* (a) after separation by cellogel electrophoresis [93] and (b) identification of the bands by SDS PAGE.

yellow chromophore showing an absorption maximum at 400 nm, which after dissociation remains linked to the D–L complex. No metal ion could be found in this complex by inductive coupled plasma mass spectroscopy. The first successful attempts to renature dissociated enzyme [Holz, I. and Lanzendörfer, M., unpublished data from this laboratory] are the basis for systematic reconstitution attempts.

Oblong crystals with hexagonal cross section could be obtained from highly purified enzyme (Fig. 7). Improvement of their quality is required to obtain an X-ray diffraction pattern of sufficient resolution.

3. Transcription signals: Promoters and terminators

3.1. Promoters

Both in Archaea and in Bacteria, correlated genes, e.g. cooperating in the same pathway or encoding components of the same complex structure, have often been found to constitute

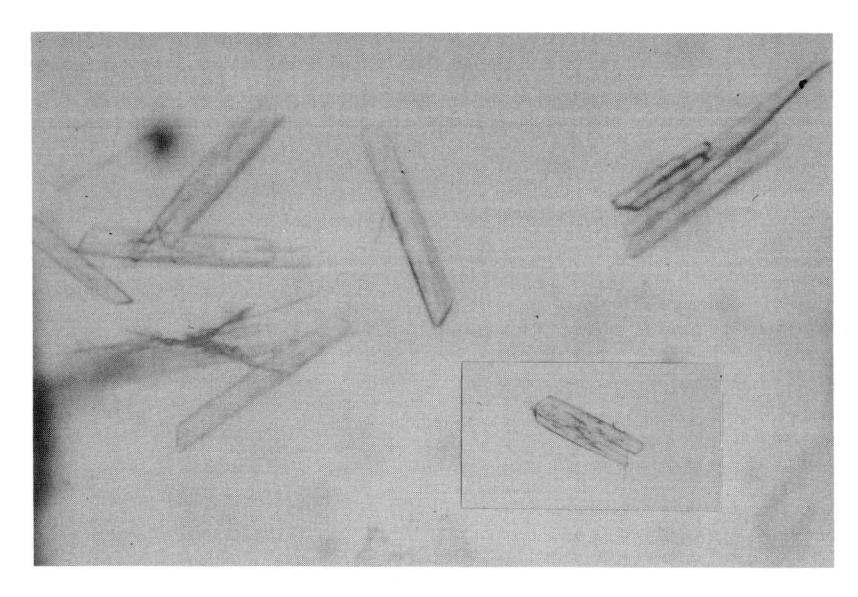

Fig. 7. Crystals of DNA-dependent RNA polymerase of *S. acidocaldarius* [Lanzendörfer, M., unpublished data from this laboratory].

gene clusters of similar composition and gene order (for a review see ref. [32]). Some of them proved to be transcription units allowing to map transcript starts and termini.

A consensus sequence for the promoters of rRNA and tRNA genes in *Methanococcus vannielii* was first derived by Wich et al. [94]. By the addition of mapped promoter sequences from the *Sulfolobus* virus SSV1 [94] and other sources it was expanded to a general archaeal promoter consensus [25] (Table 3). In vitro function was first shown as protection of two archaeal promoters by bound purified RNAP from *M. vannielii* [26]. The striking resemblance of the archaeal to the eucaryal TATA box containing pol 2 promoter consensus [121] was independently observed by Reiter et al. [25] and by Thomm and Wich [26]. In a thorough mutational analysis of a *Sulfolobus* rRNA promoter, originally using deletions, insertions and linkers [27], and later single nucleotide and sequence block exchanges [122], promoter efficiency was measured employing an in vitro transcription system. A distal promoter element containing a strongly conserved box A, the center of which is situated at a distance of optimally 27($\pm$4) nucleotides from the actual start on a purine following a pyrimidine. The expanded promoter list (Table 3) does, however, not support a general importance of the previously postulated box B which sometimes does even not include the start site. In the strong promoters of rRNA genes of *Sulfolobus shibatae*, promoter strength appeared to depend on a third sequence motif termed proximal promoter element. Whether weaker promoters contain weaker versions of the latter or lack it entirely remains to be shown.

Promoter efficiency depends, among other prerequisites, on the conservation of the sequence TA in the center of the box A consensus TTTA$^T/_A$A. Replacement of the standard box A by a eucaryal pol 2 consensus TATA box led to a reduction of transcription

TABLE 3
List of mapped archaeal promoter sequences[a,b]

Organism and gene	Sequence	Ref.[c]
D.a. DNA-Ligase	CATCATCAAATATATTATATCGT AATTCAAACT TTTATTTATG GAG	152
D.a. SOR (aer)	AAAGAAAGAATATAAAAGTAGAC AGAAATGTAT ATTTGACCAA AAA	153
D.a. sor-ORF2 (const.)	AATTCACTTATTTATACATTTGT TAAGGATGGG TTATATTATG CAT	153
D.a. sor-ORF3 (aer)	GAAGAATACATTTATTCCTGAA TAAAGTACAT TAGATTATGG TCG	153
D.a. sor-ORF4 (anaer)	CTCAACTATTTTTAAATACTAGT GCAAGAAAGA ATAGCATATG AGT	153
D.m. 5S rRNA	CCTAACACACTATACAATATATT GATGCTCGCA ATAGTGGTAG CCC	80
D.m. rRNA-P1	TATATAAACCCTTATCGCATAGA GTAAGATTCC AGACGCTTAC AGC	80
D.m. rRNA-P2	ACCCGTCATGATTAATACCCTTG GAGCAAATAG ATTCATCAAG CGG	80
H.c. rRNA P1	TCGACGGTGTTTTATGTACCCCA CCACTCGGAT GAGATGCGAA CGA	81
H.c. rRNA P2	GTCCGATGCCCTTAAGTACAACA GGGTACTTCG GTGGAATGCG AAC	81
H.c. rRNA P3	ATTCGATGCCCTTAAGTAATAAC GGGTGTTCCG ATGAGATGCG AAC	81
H.c. L11e	AAAGACAAGGGTTAAACCCGCGG CGGCGGTTTC TCGGAGTATG GCT	82
H.c. L1e	CTTCGACGCTTTTAAGCCCGGGA TCACCGTCTG TAGAACCGAG ACA	82
H.c. NAB	GTTTCGACACGTTAATACGCCGA GTGAAGCCAT CGCATAGTGA TGG	82
H.c. SOD	CGGAAACCACCATAAGCAGCGCC GACGTACGAC ACACTGTATG TCC	83
H.h. RNAP	GTTGACAAGGCTTAAATGCTGTG GGCAGCAACT GGCCTGTAGT ATC	45
H.h. S12	AAAGGTCGGGCTTAAGTGCCTCC GCGGGATACT CGGCTGTATG ACG	45
H.h. HOP	GTTGGGGAGGTTATTTAATGGC GTGCCGTGTC CTTCCGAACA CAT	84
H.h. BOP	GGGTCGTAGAGTTACACACATAT CCTCGTTAGG TACTGTTGCA TGT	85
H.h. Glycoprotein	GCAGAAAGCATTTACCAGTGGCC GGGTATAGTC TGGAGCACCC CTA	86
H.h. flgA	GGGACAAACGTTTATTAGTCAGC TGTCACACTC ACGTCAAACA CTC	87
H.h. flgB	CCTACACTTTTGTATCGATGGCC GATCTGTATG GGTAAGCCCC AGC	87
H.h. Mn-SOD	CTCACGAACATTTAACATGACGC CGCGTGATCA CTGATCCGGT GGA	88
H.h. p-vac	CATACACATCCTTATGTGATGCC CGAGTATAGT TAGAGATGGG TTA	89
H.h. p-gvpD	GAGTCGTACTTCTAAGTACGGAG AGTGTAAAGC TTCTTAGACA GTC	90
H.h. φH T1,2,3	GAGGGTCAATTTTATTATACTGG GGTTCCACGG ACATGACAGA GCA	91
H.h. φH T4	TAGGAATAGATATAAGTTAGACC CCTCGTAAAG TCCAGACTGA CGA	91
H.h. φH T6	ATGGGAACACGTTATGATGGGCC AAAAACCTCT TTTAGGTCAT GCA	91
H.h. φH T7	CTGACGGGATTTATCTTTGCAC GCATGGAAGT CCACTCGTTC CAT	91
H.h. φH T8	ATCAAACAGATTTAATAGGTAGG GGCCTCTAAT GTTTGACAAG GTA	154
H.h. φH T9/10	TGGCACGAAATTTACTTACCGAT GCGTCTATTC TATTCTGACA TGA	155
H.mm. 16S/23S rRNA P1	CTTCGACGGCGTTAAGTGTGGCT CACCCATCGG AATGAAATGC GAA	92
H.mm. 16S/23S rRNA P2	TTCCGACGCCCTTAAGTGTAACA GGGCGTTCGG AATGAACGCA AAG	92
H.mm. 16S/23S rRNA P3	ATCCGACGCCCTTAAGTGTAACA GGGTGCTCGA ATGAACGCGA ACG	92
M.t. mvhDGAB	ATAAGTAACCTTTATACTTACCA CTCCTAAACC ATATACCAAG TAC	93
M.t. mcr	AGAAAGAAATCTTAATTAATTAT GATCAAGTTA AATATGCACC GTA	94
M.t. purE	GGTGCCATCTATTAAATATGACC TCCGATAGAT TCTCTATATA TAA	95
M.v. 16S/23S rRNA	TACCTAAAACAATACATATTACA ACACGTTTTC ATATTATGCA AAT	78
M.v. tRNA & 5S rRNA	TACCGAAAACTTTATATATTATA ACACTAGTAT TCAGTATGCG AAC	78
M.v. tRNA[Arg]	AACCGAAATATTTATATACTAGA ATACCCTTCC TATACTATGC TCT	78
M.v. ORFa	TAAGCATAATTTTAAATACCATT AACCAGTATG TCTATCGTAT AGA	96
M.v. S17	CTATATTAAACTTATATACCGT AGTTCGTATT AAATCTGGAA CTT	96
M.v. hisA	TAGGTACCAATATATATGTTAAA ACCTAATTTA ACATAGTTTT AAT	97
M.v. mcr	TGAAAACTTGAATATATCTTCCT TTAATAATGT TATGATGTAT ATA	98

continued on next page

TABLE 3, continued

Organism and gene	Sequence	Ref.[c]
M.v. ORF1	TATGCAAAAG<u>TATAAAA</u>ATAACA TCTAACTATT AACTTACTC**G** TAT	99
S.a. tRNA[Ser]	TTTGATTATT<u>TTTATA</u>AGCATCT CGTATAATCT TATTA**A**TGGC CGG	54
S.a. S12	CTAAGTTAAT<u>ATTAAAA</u>AGGTAT TTGGATATTG ATTCTGTT**GT** TGT	46
S.a. S7	CACAAACTCT<u>ATTAAA</u>CCCTTCA TAATATATTT AAGGT**A**GCAG TAA	54
S.a. ORF88 (rpoH)	CTAACATAAT<u>TTTATT</u>AAGCTTA AATGAGTAGG ATGATATT**A**C TGT	46
S.a. rpoC	CTGTGAGAAG<u>TTTATA</u>TGGAGAT ATTGTTCAAG TAGTATA**TG**G TGA	46
S.a. ORF-X	CGGTAAACC<u>TTTTATG</u>GTATCTG TTATTAGGTA TAGTT**G**ATGA AGG	46
S.a. Ef-1α	TAGCTAAACT<u>TTTAAT</u>GCTAGTC TAGTTATAAA AATGTA**A**GAG GTA	54
S.a. Ef-2	TCCAACACAC<u>TTTAAT</u>ACTTAAA AGTTAACTTA CTATTC**G**AAA CTG	55
S.a. SOD	AATCAAAATA<u>TTTAAT</u>CCACCAT TGGCTATTTG TTTGT**A**TGAC CCA	156
S.s. 16S/23S rRNA	AGAAGTTAGA<u>TTTATA</u>TGGGATT TCAGAACAAT ATGTATAAT**G** CGG	24
S.s. 5S rRNA	TAGTTAATT<u>TTTTATA</u>TGTGTTA TGAGTACTTA ATTTT**G**CCCA CCC	24
S.s. tRLNLA[Arg]	TGTCTAACAG<u>TTTAAA</u>ATATCCG ATGAGGCATA TTTATGTTGG ACC	149
S.s. SSV1 T1,2	ACTGGAGGG<u>TTTAAAA</u>ACGTAA GCGGGAAGCC GATATT**G**ACC AAG	24
S.s. SSV1 T3	TTAGGCTCTT<u>TTTAAAG</u>TCTACC TTCTTTTTCG CTTAC**A**ATGA GGA	24
S.s. SSV1 T4	GATAGCCCTT<u>TTTAAAG</u>CCATAA ATTTTTTATC GCTT**A**ATGAA GTG	24
S.s. SSV1 T9	AAGTAGGCCC<u>TTTATA</u>AAGTCAT ATTCTTTTTC TTTCCCT**G**AT GAG	24
S.s. SSV1 T5	TAGAGTAAGAC<u>TTAAAT</u>ACTAAT TTATACATAG AGTATAGAT**A** GAG	24
S.s. SSV1 T6	TAGAGTAAAG<u>TTTAAAT</u>ACTTAT ATAGATAGAG TATAGAT**A**GA GGG	24
T.a. 16S rRNA	GCCTTCGAAA<u>GTTATA</u>TATACTG ATTTGCTATT CTTTACTTT**G** CAC	100
T.a. 23S rRNA	GATCAAAATG<u>CTTATAT</u>CCCTCT TAATGATATA GTCCAT**A**CAC GCT	100
T.a. 5S rRNA	TCACGAAAAT<u>CTTATA</u>TAGATGT GTTCTATATA GTGT**T**CGGCA ACG	100
T.a. tRNA[Met]	CCCTACAGCC<u>ATTATT</u>CGTCCAC TATGCTGGAT TGAAGCGGGG TGG	151
T.c. rpoH	ACTCGAAAAG<u>CTTATA</u>TACCGAA ACCCGCAGGG GGAGGT**A**GTG ATA	18
T.c. rpl30	ACCCGAAAAG<u>GTTAAAA</u>AGCACC GCCTTAAGGG TCTTTC**G**CGC GTC	157
T.c. rps12	CTAGGAACAT<u>TTTATA</u>AGAGCCT CGGGGTTAAG GTTAAA**A**GGG TCC	157
T.p. 16S/23S rRNA	GCATAATATT<u>CATATA</u>ACCCCCC GTTTACTAAC TAGATT**G**CCG CCA	101
T.p. tRNA[Met]	ATGCTAAAG<u>GTTTATT</u>ACCCAGG AAGTATTCCG GTCAT**G**GGGG GTT	101
T.p. tRNA[Gly]	GGGAAAAGC<u>TTTTAAG</u>CATGCCT TTTTACTTCC TTCTAGAGGC TC**A**	101
T.p. ORF1	GCATTCGACA<u>AATATTT</u>TAACCT CGTTCGAGGA GTAT**A**GAGGT GGT	101
T.p. ORF2	TAAATAGCT<u>TTTTAAC</u>GTAGGAG GTGCTACGTC CAC**G**GGATCG TAA	101
T.t. 16S/23S rRNA	GCGAAAATT<u>TTTTAATT</u>TAGGGT GTTTTAGGAT GGTC**G**CGCCT TAA	102
T.t. tRNA[Ala]	AGCGAAAAAA<u>TTTAAAT</u>CGGTGA GTAAGTACGC TC**G**GGCCGG TAG	102
T.t. tRNA[Met]	ACAAAAGCTT<u>TTTAAA</u>TTCGCGC AAAGCTTAGA CCT**A**GCGGGG TAG	102
H.c. ORF	GGAAAGCGCTTTTCGGCGCTTGCTGTCTACGGGCACGT**G**ATG	82
H.h. BRP	GGTCTTTTTTGATGCTCGGTAGTGACGTGTGTATTCAT**A**TGAGCA	103
H.m. mc-gvpA	CACGAATGATTTTGTTACTTGCCAACACGTTTTCAGAT**G**GGTA	104
S.s. SSV1 T[ind]	GTCGACTCTGTGTATCTTATGTATCTTATACAAAAAAT**A**TGGGA	24

[a] The list shows the transcription start sites of mapped archaeal genes and the promoter sequences (box A sequences are underlined; the transcription start sites are bold and underlined). The upper sequences are aligned on the box A element. The last four sequences are aligned on the transcription start sites hence no consensus box A could be found.

continued on next page

TABLE 3, continued

[b] Abbreviations used for organisms: H.c., *Halobactearium cutirubrum*; H.h., *Halobacterium halobium*; H.mm., *Halobacterium marismortui*; H.me., *Halobacterium mediterranei*; M.t., *Methanobacterium thermoautotrophicum*; M.v., *Methanococcus vannielii*; T.a., *Thermoplasma acidophilum*; D.a., *Desulfurolobus ambivalens*; D.M., *Desulfurococcus mobilis*; S.a., *Sulfolobus acidocaldarius*; S.s., *Sulfolobus shibatae*; T.c., *Thermococcus celer*; T.P., *Thermofilum pendens*; T.t., *Thermoproteus tenax*.

efficiency to 13% [122]. A distance of about 27($\pm$4) nucleotides appears to be measured between the center of box A and the transcription start site on a distinguished purine R following a pyrimidine y. Often only one such yR sequence is available within this distance window. In cases where one of several yR sites situated within the window is selected, a weak consensus $^A/_T$ $^T/_C$ $^G/_A$ or a weakly conserved alternating yR sequence starting at position -7 and including the start site could exert a further directing influence. In the absence of such directing motifs, e.g., in insertion or deletion mutants moving the normal start site out of the window, start-site selection becomes ambiguous, although starts remain on purines following pyrimidines within the distance window situated 27($\pm$4) nucleotides downstream of the center of box A [122]. The distance between box A and the transcription start site appears to be reduced when several T or A residues precede the box A consensus sequence, possibly due to bending. In the case of the 16S rRNA promoter of *S. shibatae*, the proximal promoter element stretching from position -11 to position -6 upstream of the start site reads ATATGT. Replacement of the sequence between positions -11 and -3 by the complementary sequence or by runs of T or A of the same length resulted in a reduction of the efficiency to about 5% each, indicating that the AT content and thus the ease of strand separation was insufficient to determine the quality of this element.

The promoters of some strongly regulated genes, e.g. the *brp* gene and the chromosome-encoded gene for the major gas vesicle protein, both of *H. halobium,* and the promoter for the UV-inducible transcript of the *Sulfolobus* virus SSV1, deviate from the consensus, the latter to the extent that no box A is recognizable any more (see Table 3).

The existence of a general standard promoter characteristic for Archaea is not only suggested by correspondence to the consensus but also by the capacity of the RNAP from *Sulfolobus* to specifically initiate transcription in vitro from either of two heterologous constitutive promoters operating in *Halobacterium* [123], namely that of the plasmid-encoded p-*vac* gene for the major gas vesicle protein [106] and that of the major early transcript T4 of the *Halobacterium* phage ΦH [100]. The promoter of the regulated chromosome-encoded c-*vac* gene for the major gas vesicle protein of *Halobacterium* [106] is not accepted by the *Sulfolobus* system, indicating different recognition or the requirement of additional components.

3.2. Terminators

With the transcripts of SSV1, the sequence TTTTTTT or TTTTTCT was found immediately upstream of multiple transcription termination sites [124]. It is not clear yet, whether additional elements are required for function. Similar sequences were identified at the 3' termini of other transcription units (e.g., ref. [47]) and also at eucaryal pol 2 and pol 3 transcription termini [125,126]. In some cases no motifs of this type could be detected in the vicinity of termination sites (see, e.g., ref. [55]). The involvement of a similar element for in vitro transcription termination in a *M. vannielii* system speaks for its actual participation in termination rather than post-transcriptional processing [Thomm, M., personal communication]. Post-transcriptional processing could, however, be responsible for the lack of terminator sequences downstream of the 5' ends of some RNAs found in vivo.

In several instances other or more complicated types of elements, especially containing stem and loop structures, were proposed as termination signals [102,127,128]. Many different sequences in the vicinity of termination sites have been compiled by Brown et al. [129].

4. In vitro transcription systems

The first step in the development of an in vitro transcription system was the demonstration of specific binding of the *M. vannielii* RNAP to corresponding promoters in the absence of additional factors [26]. Such factors were, however, required for transcription of cloned genes in an in vitro system in which specific initiation and termination resulted in the synthesis of defined transcripts [130]. In a cell-free system using the RNAP of *S. shibatae* and cloned promoters, e.g., of the corresponding 16S rRNA gene [131] or of the transcription unit encoding transcript T3 of the *Sulfolobus* virus SSV1 [25], transcription efficiency was estimated by quantitative S1 analysis [132]. Specific initiation from the weaker T3 promoter did not require factors [158]. On the strong rRNA promoter, specific initiation by RNAP only amounted to about 25% of total transcription, the rest constituting semispecific initiation at several yR sequences closely upstream of the start of specific transcription. In a partially purified cell-free transcription system from *M. thermoautotrophicum,* initiation occurred exclusively at one site 18 nucleotides upstream of the in vivo transcript start [133]. In contrast to specific initiation, semispecific transcription in the *Sulfolobus* system did not require the promoter box A [132]. A soluble factor prohibited semispecific but not truly specific initiation. Different from the case of eucaryal pol 2, the factor was thus not involved in promoter recognition. Instead, it competed with RNAP for the weaker box-A-independent binding preceding semispecific initiation but not for the strong box-A-dependent binding allowing specific initiation. Because of the similarity of promoters and especially of box A and the TATA box, a component of the archaeal RNAP should therefore correspond to the eucaryal TATA box binding factor TFIID [134,135].

Specific termination of transcription has been observed in the in vitro system from *M. vannielii* [Thomm, M., personal communication]. It requires a terminator sequence identified in vivo, although it does not occur exactly at the same site as in vivo.

5. Control of transcription

Controls of gene expression have been observed though not elucidated in several archaeal virus host systems [136,137].

Halobacterium halobium phage ΦH is a temperate virus capable of establishing lysogeny. As for *Salmonella* phage P1, the prophage in lysogens is the free circular phage genome. The product of the transcript T6 encoded in the L region of the phage genome appears to be a repressor binding to an operator immediately upstream of the promoter of the major early gene which triggers lytic multiplication [138]. This is the clearest example of negative control so far known for an archaeal gene. A phage mutant in which a 1 kbp insert has separated operator and promoter is able to break the immunity suppression observed in defective lysogens carrying the circularized L region [100]. Two further genes in the L region are exclusively transcribed in the lysogenic state [100]. A second level of immunity not broken by the insertion upstream of the T4 promoter is established by the complete prophage [100]. Several transcripts from the L region are constitutive [154,155]. The early/late switch in the transcription of lytic genes involves protein synthesis and thus possibly positive control [100].

Production of the virus SSV1 in its lysogenic host *Sulfolobus shibatae* is induced by UV irradiation [139]. The gene encoding transcript T_{ind} which was not expressed in late log cells before induction was transcribed about three hours after induction [140]. The promoter of this gene does not contain a recognizable box A [25]. Induction might thus involve positive control. This reading unit harbors a short ORF of unknown significance. Its expression somehow leads to virus DNA replication which might be the immediate cause of virus production [141]. The promoters of the other viral transcription units are of three types [25,95]. Preliminary evidence indicates different transcription controls. The transcription unit for the transcripts T1 and T2 encoding three structural proteins appears to be constitutively transcribed [140]. Its promoter conforms to the standard consensus. The promoters of the units transcribed into transcripts T3, T4, T7, T8 and T9, respectively, resemble each other strongly in the sequence of the inter-box region. These transcripts were jointly switched up after UV induction [95]. Transcription from the promoters for T5 and T6, which exhibit characteristic highly conserved repeats flanking their boxes A [25], was induced shortly after T_{ind} [95]. None of these controls has been understood.

More than 50 genes are switched on and off respectively in major metabolic switches from the aerobic (= sulfur oxidation) to the anaerobic (= sulfur reduction) mode of autotrophic existence of *Desulfurolobus ambivalens* [142,143] and vice versa [144]. These include a sulfur oxidoreductase (aerobic mode) [145] and the amplification of a 7 kbp plasmid (anaerobic mode) [142]. Synthesis of the chromosome-encoded major gas vesicle protein of *H. halobium* is switched on in the approach to stationary growth phase [106]. In *Haloferax mediterranei*, the corresponding gene is only expressed at high salt but not at low salt [117]. Transcription of the *bop* and *brp* genes involved in purple-membrane synthesis of *H. halobium* is induced at low oxygen partial pressure [146]. The β-galactosidases of *Sulfolobus* species are induced by substrate [Schleper, C., unpublished data from this laboratory]. One of the two superoxide dismutase genes each in *H. cutirubrum* and *Haloferax volcanii* is induced by paraquat, an intracellular generator of superoxide [147]. Two different methyl–coenzyme M reductases were found in *Methanobacterium thermoautotrophicum* in a ratio depending on the growth conditions [148].

Although it appears probable that the mechanisms of these controls resemble those operating in Bacteria, experimental verification is required.

6. Summary

This chapter describes structure, phylogeny and function of DNA-dependent RNA polymerases from Archaea. The consideration of structure concerns mainly the component composition. The discussion of RNAP phylogeny is based on the description of the organization of the large component genes. Phylogenetic trees of the sequences of the large components and their genes, and the implications of the branching topology of these trees for the relations between the three domains of life, are summarized. The RNAP of *Sulfolobus acidocaldarius* is considered in detail as an example. The nature of promoter motifs, their role in transcription initiation and details of their interaction with the RNAP are discussed on the basis of the archaeal promoter consensus and of mutation analysis. The corresponding knowledge of archaeal transcription terminators is comparatively modest.

In vitro transcription and its prerequisites in several archaeal systems are described. Several transcription controls, especially those operating in archaeal phage host systems but also those regulating the expression of cellular genes, are briefly summarized.

Note added in proof

Recent developments in understanding the transcription machinery of Archaea, the structure and function of RNAP components and transcription factors, and the homologies of components of the transcription apparatus in the three domains of life are reviewed by Lanzendörfer et al. [158] and Thomm et al. [159].

References

[1] Woese, C.R. and Fox, G.E. (1977) J. Mol. Evol. 10, 1–6.
[2] Fox, G.E., Stackebrandt, E., Hespell, R.B., Gibson, J., Maniloff, J., Dyer, T.A., Wolfe, R.S., Balch, W.E., Tanner, R.S., Magrum, L.J., Zablen, L.B., Blakemore, R., Gupta, R., Bonen, L., Lewis, B.J., Stahl, D.A., Luehrsen, K.R., Chen, K.N. and Woese, C.R. (1980) Science 209, 457–463.
[3] Woese, C.R., Kandler, O. and Wheelis, M.L. (1990) Proc. Natl. Acad. Sci. U.S.A. 87, 4576–4579.
[4] Leffers, H., Kjems, J., Ostergaard, L., Larsen, N. and Garrett, R.A. (1987) J. Mol. Biol. 195, 43–61.
[5] Zillig, W., Klenk, H.-P., Palm, P., Pühler, G., Gropp, F., Garrett, R.A. and Leffers, H. (1989) Can. J. Microbiol. 35, 73–80.
[6] Pühler, G., Leffers, H., Gropp, F., Palm, P., Klenk, H.-P., Lottspeich, F., Garrett, R.A. and Zillig, W. (1989) Proc. Natl. Acad. Sci. U.S.A. 86, 4569–4573.
[7] Auer, J., Spicker, G. and Böck, A. (1990) System. Appl. Microbiol. 13, 354–360.
[8] Hasegawa, M., Iwabe, N., Mukohata, Y. and Miyata, T. (1990) Jpn. J. Genet. 65, 109–114.
[9] Gogarten, J.P., Kibak, H., Dittrich, P., Taiz, L., Bowman, E.J., Bowman, B.J., Manolson, M.F., Poole, R.J., Date, T., Oshima, T., Konishi, J., Denda, K. and Yoshida, M. (1989) Proc. Natl. Acad. Sci. U.S.A. 86, 6661–6665.

[10] Iwabe, N., Kuma, K.-I., Hasegawa, M., Osawa, S. and Miyata, T. (1989) Proc. Natl. Acad. Sci. U.S.A. 86, 9355–9359.
[11] Konishi, J., Denda, K., Oshima, T., Wakagi, T., Uchida, E., Ohsumi, Y., Anraku, Y., Matsumoto, T., Wakabayashi, T., Mukohata, Y., Ihara, K., Inatomi, K., Kato, K., Ohta, T., Allison, W.S. and Yoshida, M. (1990) J. Biochem. 108, 554–559.
[12] Kates, M. and Kushwaha, S.C. (1978) In: Energetics and Structure of Halophilic Microorganisms (Caplan, S.R. and Ginzburg, M., Eds.), pp. 461–480, Elsevier, Amsterdam.
[13] Kates, M. and Moldoveanu, N. (1991) In: General and Applied Aspects of Halophilic Microorganisms (Rodriguez-Valera, F., Ed.), pp. 191–198, Plenum Press, New York.
[14] Thurl, S. and Schäfer, W. (1988) Biochim. Biophys. Acta 961, 253–261.
[15] Kalmokoff, M.L., Karnauchow, T.M. and Jarrell, K.F. (1990) Biochem. Biophys. Res. Commun. 167, 154–160.
[16] Langworthy, T.A. and Pond, J.L. (1986) System. Appl. Microbiol. 7, 253–257.
[17] Zillig, W., Palm, P., Klenk, H.-P., Pühler, G., Gropp, F. and Schleper, C. (1991) In: General and Applied Aspects of Halophilic Microorganisms (Rodriguez-Valera, F., Ed.), pp. 321–332, Plenum Press, New York.
[18] Klenk, H.-P., Palm, P., Lottspeich, F. and Zillig, W. (1992) Proc. Natl. Acad. Sci. U.S.A. 89, 407–410.
[19] Edmonds, C.G., Crain, P.F., Gupta, R., Hashizume, T., Hocart, C.H., Kowalak, J.A., Pomerantz, S.C., Stetter, K.O. and McCloskey, J.A. (1991) J. Bacteriol. 173, 3138–3148.
[20] Kjems, J. and Garrett, R.A. (1990) J. Mol. Evol. 31, 25–32.
[21] Kaine, B.P. (1990) Mol. Gen. Genet. 221, 315–321.
[22] Gehrmann, R., Henschen, A., Postulka, W. and Klink, F. (1986) System. Appl. Microbiol. 7, 115–122.
[23] Kohiyama, M., Nakayama, M. and Ben Mahrez, K. (1986) System. Appl. Microbiol. 7, 79–82.
[24] Zabel, H.-P., Holler, E. and Winter, J. (1987) Eur. J. Biochem. 165, 171–175.
[25] Reiter, W.-D., Palm, P. and Zillig, W. (1988) Nucl. Acids Res. 16, 1–19.
[26] Thomm, M. and Wich, G. (1988) Nucl. Acids Res. 16, 151–163.
[27] Reiter, W.-D., Hüdepohl, U. and Zillig, W. (1990) Proc. Natl. Acad. Sci. U.S.A. 87, 9509–9513.
[28] Lanzendörfer, M., Palm, P., Grampp, B., Peattie, D.A. and Zillig, W. (1992) Nucl. Acids Res. 20, 1145.
[29] Kaine, B.P. (1987) J. Mol. Evol. 25, 248–254.
[30] Wich, G., Leinfelder, W. and Böck, A. (1987) EMBO J. 6, 523–528.
[31] Daniels, C.J., Gupta, R. and Doolittle, W.F. (1985) J. Biol. Chem. 260, 3132–3134.
[32] Zillig, W., Palm, P., Reiter, W.-D., Gropp, F., Pühler, G. and Klenk, H.-P. (1988) Eur. J. Biochem. 173, 473–482.
[33] Lake, J.A., Anderson, E., Oakes, M. and Clark, M.W. (1984) Proc. Natl. Acad. Sci. U.S.A. 81, 3786–3790.
[34] Lake, J.A., Clark, M.W., Henderson, E., Fay, S.P., Oakes, M., Scheinman, A., Thornber, J.P. and Mah, R.A. (1985) Proc. Natl. Acad. Sci. U.S.A. 82, 3716–3720.
[35] Lake, J.A. (1987) Cold Spring Harbor Symposia on Quantitative Biology, Vol. 52, Cold Spring Harbor Laboratory, pp. 839–845.
[36] Gouy, M. and Li, W.-H. (1989) Nature 339, 145–147.
[37] Langworthy, T.A., Tornabene, T.G. and Holzer, G. (1982) Zbl. Bakt. Hyg., I. Abt. Orig. C 3, 228–244.
[38] De Rosa M. and Gambacorta, A. (1986) System. Appl. Microbiol. 7, 278–285.
[39] Danson, M.J. (1988) Advances in Microbial Physiology, 29, 165–231.
[40] Hensel, R., Zwickl, P., Fabry, S., Lang J. and Palm, P. (1989) Can. J. Microbiol. 35, 81–85.
[41] Honka, E., Fabry, S., Niermann, T., Palm, P., Hensel, R, (1990) Eur. J. Biochem. 188, 623–632.
[42] Fabry, S., Heppner, P., Dietmaier, W. and Hensel, R. (1990) Gene 91, 19–25.

[43] Prangishvilli, D., Zillig, W., Gierl, A., Biesert, L. and Holz, I. (1982) Eur. J. Biochem. 122, 471–477.
[44] Thomm, M., Madon, J. and Stetter, K.O. (1986) Biol. Chem. Hoppe-Seyler 367, 473–481.
[45] Schnabel, R., Thomm, M., Gerardy-Schahn, R., Zillig, W., Stetter, K.O. and Huet, J. (1982) EMBO J. 2, 751–755.
[46] Leffers, H., Gropp, F., Lottspeich, F., Zillig, W. and Garrett, R.A. (1989) J. Mol. Biol. 206, 1–17.
[47] Pühler, G., Lottspeich, F. and Zillig, W. (1989) Nucl. Acids Res. 17, 4517–4534.
[48] Berghöfer, B., Kröckel, L., Körtner, C., Truss, M., Schallenberg, J. and Klein, A. (1988) Nucl. Acids Res. 16, 8113–8128.
[49] Zillig, W., Holz, I., Janekovic, D., Schäfer, W. and Reiter, W.-D. (1983) System. Appl. Microbiol. 4, 88–94.
[50] Sturm, S., Schönefeld, U., Zillig, W., Janekovic, D. and Stetter, K.O. (1980) Zbl. Bakt. Hyg., I. Abt. Orig. C 1, 12–25.
[51] Stetter, K.O., Lauerer, G., Thomm, M. and Neuner, A. (1987) Science 236, 822–824.
[52] Huet, J., Schnabel, R., Sentenac, A. and Zillig, W. (1983) EMBO J. 2, 1291–1294.
[53] Sentenac, A. (1985) CRC Crit. Rev. Biochem. 18, 31–91.
[54] Woychik, N.A., Liao, S.-M., Kolodziej, P.A. and Young, R.A. (1990) Genes Dev. 4, 313–323.
[55] Auer, J., Spicker, G, Mayerhofer, L., Pühler, G. and Böck, A. (1991) System. Appl. Microbiol. 14, 14–22.
[56] Schröder, J. and Klink, F. (1991) Eur. J. Biochem. 195, 321–327.
[57] Tesch, A. and Klink, F. (1990) FEMS Lett. 71, 293–298.
[58] Itoh, T. (1989) Eur. J. Biochem. 186, 213–219.
[59] Auer, J., Lechner, K. and Böck, A. (1989) Can. J. Microbiol. 35, 200–204.
[60] Arndt, E. (1990) FEBS Lett. 267, 193–198.
[61] Arndt, E., Krömer, W. and Hatakeyama, T. (1990) J. Biol. Chem. 265, 3034–3039.
[62] Zillig, W., Holz, I., Klenk, H.-P., Trent, J., Wunderl, S., Janekovic, D., Imsel, E. and Haas, B. (1987) System. Appl. Microbiol. 9, 62–70.
[63] Dayhoff, M.O., Schwartz, R.M. and Orcutt, B.C. (1978) In: Atlas of Protein Sequence and Structure, (Dayhoff, M.O., Ed.), pp. 345–358, National Biomedical Research Foundation, Washington, DC.
[64] Ahearn, J.M., Bartolomei, M.S., West, M.L., Cisek, L.J. and Cordon, J.L. (1987) J. Biol. Chem. 262, 10695–10705.
[65] Allison, L.A., Moyle, M., Shales, M. and Ingles, C.J. (1985) Cell 42, 599–610.
[66] Bird, D.M. and Riddle, D.L. (1989) Mol. Cell Biol. 9, 4119–4130.
[67] Li, W.B., Bzik, D.J., Gu, H., Tanaka, M., Fox, B.A. and Inselburg, J. (1989) Nucl. Acids Res. 17, 9621–9636.
[68] Evers, R., Hammer, A., Köck, J., Jess, W., Borst, P., Mémet, S. and Cornelissen, A.W.C.A. (1989) Cell 56, 585–597.
[69] Evers, R., Hammer, A. and Cornelissen, A.W.C.A. (1989) Nucl. Acids Res. 17, 3403–3413.
[70] Mémet, S., Gouy, M., Marck, C., Sentenac, A. and Buhler, J.-M. (1988) J. Biol. Chem. 263, 2823–2839.
[71] Jess, W., Hammer, A. and Cornelissen, A.W.C.A. (1989) FEBS Lett. 248, 123–128.
[72] Jess, W., Hammer, A. and Cornelissen, A.W.C.A. (1989) FEBS Lett. 258, 180.
[73] Yamagishi, M., and Nomura, M. (1988) Cell 74, 503–515.
[74] Ovchinnikov, Y.A., Monastyrskaya, G.S., Gubanov, V.V., Gureyev, S.O., Salomatina, I.S., Shuvaeva, T.M., Lipkin, V.M. and Sverdlov, E.D. (1982) Nucl. Acids Res. 10, 4035–4044.
[75] Ohyama, K., Fukuzawa, H., Kohchi, T., Shirai, H., Sano, T., Sano, S., Umesono, K., Shiki, Y., Takeuchi, M., Chang, Z., Aota, S., Inokuchi, H. and Ozeki, H. (1986) Nature 322, 572–574.

[76] Hudson, G.S., Holton, T.A., Whitfeld, P.R. and Bottomley, W. (1988) J. Mol. Biol. 200, 639–654.
[77] Danilkovich, A.V., Borodin, A.M., Allikmets, R.L., Rostapshov, V.M., Chernov, I.P., Azkhikina, T.L., Monastyrskaya, G.S. and Sverdlov, E.D. (1988) Dokl. Biochem. 303, 241–245.
[78] Fitch, W.M. and Margoliash, E. (1967) Science 155, 279–284.
[79] Felsenstein, J. (1989) PHYLIP, version 3.21. Copyright 1986, University of Washington, and 1989, J. Felsenstein.
[80] Felsenstein, J. (1973) Syst. Zool. 22, 240–249.
[81] Felsenstein, J. (1981) J. Mol. Evol. 17, 368–376.
[82] Lake, J.A. (1987) J. Mol. Evol. 26, 59–73.
[83] Lake, J.A. (1987) Mol. Biol. Evol. 4, 167–197.
[84] Felsenstein, J. (1985) Evolution 39, 783–791.
[85] Murphy, S., Di Liegro, C. and Melli, M. (1987) Cell 51, 81–87.
[86] Mattaj, I.W., Dathan, N.A., Parry, H.D., Carbon, P. and Krol, A. (1988) Cell 55, 435–442.
[87] Mereschkowsky, C. (1910) Biol. Zentralbl. 30, 278–303, 321–347, 353–367.
[88] Whatley, J.M., Chapman, D.J. and Andresen, C. (1989) In: Handbook of Protista (Margulis, L., Corliss, J.O., Melkonian, M. and Chapman, D.J., Eds.), pp. 167–185, Jones and Bartlett Publishers, Boston.
[89] Iwabe, N., Kuma, K., Kishino, H., Hasegawa, M. and Miyata, M. (1991) J. Mol. Evol. 32, 70–78.
[90] Cendrin, F., Chroboczek, J., Zaccai, G., Eisenberg, H. and Mevarech, M. (1993) Biochemistry 32, 4308–4318.
[91] Schägger, H. and von Jagow, G. (1987) Anal. Biochem. 166, 368–379.
[92] Langer, D. (1991) Diploma thesis, Ludwig-Maximilians-Universität München.
[93] Heil, A. and Zillig, W. (1970) FEBS Lett. 11, 165–168.
[94] Wich, G., Hummel, H., Jarsch, M., Bär, U. and Böck, A. (1986) Nucl. Acids Res. 14, 2459–2479.
[95] Palm, P., Schleper, C., Grampp, B., Yeats, S., McWilliam, P., Reiter, W.-D. and Zillig, W. (1991) Virology 185, 242–250.
[96] Kjems, J. and Garrett, R.A. (1987) EMBO J. 6, 3521–3530.
[97] Dennis, P.P. (1985) J. Mol. Biol. 186, 457–461.
[98] Shimmin, L. and Dennis, P.P. (1989) EMBO J. 8, 1225–1235.
[99] May, B. and Dennis, B.P. (1990) J. Bacteriol. 172, 3725–3729.
[100] Blanck, A. and Oesterhelt, D. (1987) EMBO J. 6, 265–273.
[101] DasSarma, S., RajBhandary, U.L. and Khorana, H.G. (1984) Proc. Natl. Acad. Sci. U.S.A. 81, 125–129.
[102] Lechner, J. and Sumper, M. (1987) J. Biol. Chem. 262, 9724–9729.
[103] Gerl, L. and Sumper, M. (1988) J. Biol. Chem. 263, 13246–13251.
[104] Salin, M.L., Duke, M.V., Ma, D.-P. and Boyle, J.A. (1991) Free Rad. Res. Commun. 12–13, 443–449.
[105] Horne, M. and Pfeifer, F. (1989). Mol. Gen. Genet. 218, 437–444.
[106] Jones, J.G., Hackett, N.R., Halladay, J.T., Scothorn, D.J., Yang, C.-F., Ng, W.-L. and Dassarma, S. (1989) Nucl. Acids Res. 17, 7785–7793.
[107] Gropp, F., Palm, P. and Zillig, W. (1989) Can. J. Microbiol. 35, 182–188.
[108] Mevarech, M., Hirsch-Twizer, S., Goldman, S., Yakobson, E., Eisenberg, H. and Dennis, P.P. (1989) J. Bacteriol. 171, 3479–3485.
[109] Reeve, J.N., Beckler, G.S., Cram, D.S., Hamilton, P.T., Brown, J.W., Krzycki, J.A., Kolodziej, A.F., Alex, L., Orme-Johnson, W.H. and Walsh, C.T. (1989) Proc. Natl. Acad. Sci. U.S.A. 86, 3031–3035.

[110] Bokranz, M., Bäumer, G., Allmansberger, R., Ankel-Fuchs, D. and Klein, A. (1988) J. Bacteriol. 170, 568–577.
[111] Brown, J.W. and Reeve, J.N. (1989) FEMS Microbiol. Lett. 60, 131–136.
[112] Auer, J., Spicker, G. and Böck, A. (1989) J. Mol. Biol. 209, 21–36.
[113] Brown, J.W., Thomm, M., Beckler, G.S., Frey, G., Stetter, K.O. and Reeve, J.N. (1988) Nucl. Acids Res. 16, 135–150.
[114] Thomm, M., Sherf, B.A. and Reeve, J.N. (1988) J. Bacteriol. 170, 1958–1961.
[115] Lechner, K., Heller, G. and Böck, A. (1989) J. Mol. Evol. 29, 20–27.
[116] Ree, H. and Zimmermann, R.A. (1990) Nucl. Acids Res. 18, 4471–4478.
[117] Kjems, J., Leffers, H., Olesen, T., Holz, I. and Garrett, R.A. (1990) System. Appl. Microbiol. 13, 117–127.
[118] Wich, G., Leinfelder, W. and Böck, A. (1987) EMBO J. 6, 523–528.
[119] Betlach, M., Friedman, J., Boyer, H.W. and Pfeifer, F. (1984) Nucl. Acids Res. 12, 7949–7959.
[120] Englert, C., Horne, M. and Pfeifer, F. (1990) Mol. Gen. Genet. 222, 225–232.
[121] Bucher, P. and Trifonov, E.N. (1986) Nucl. Acids Res. 14, 10009–10026.
[122] Hain, J., Reiter, W.-D., Hüdepohl, U. and Zillig, W. (1992) Nucl. Acids Res. 20, 5423–5428.
[123] Hüdepohl, U., Gropp, F., Horne, M. and Zillig, W. (1991) FEBS L. 285, 257–259.
[124] Reiter, W.-D., Palm, P. and Zillig, W. (1988) Nucl. Acids Res. 16, 2445–2459.
[125] Wiest, D.K. and Hawley, D.K. (1990) Mol. Cell. Biol. 10, 5782–5795.
[126] Kerppola, T.K. and Kane, C.M. (1990) Biochemistry 29, 269–278.
[127] Müller, B., Allmansberger, R. and Klein, A. (1985) Nucl. Acids Res. 13, 6439–6445.
[128] Wich, G., Sibold, L. and Böck, A. (1986) System. Appl. Microbiol. 7, 18–25.
[129] Brown, J.W., Daniels, C.J. and Reeve, J.N. (1989) CRC Crit. Rev. Microbiol. 16, 287–338.
[130] Frey, G., Thomm, M., Brüdigam, B., Gohl, H.P. and Hausner, W. (1990) Nucl. Acids Res. 18, 1361–1367.
[131] Reiter, W.-D., Palm, P., Voos, W., Kaniecki, J., Grampp, B., Schulz, W. and Zillig, W. (1987) Nucl. Acids Res. 15, 5581–5595.
[132] Hüdepohl, U., Reiter, W.-D. and Zillig, W. (1990) Proc. Natl. Acad. Sci. U.S.A. 87, 5851–5855.
[133] Knaub, S. and Klein, A. (1990) Nucl. Acids Res. 18, 1441–1446.
[134] Wasylyk, B. (1988) CRC Crit. Rev. Biochem. 23, 77–120.
[135] Saltzman, A.G. and Weinmann, R. (1989) FASEB J. 3, 1723–1733.
[136] Zillig, W., Reiter, W.-D., Palm, P., Gropp, F., Neumann, H. and Rettenberger, M. (1988) In: The Bacteriophages (R. Calendar, Ed.), Vol. 1 of The Viruses (Fränkel-Conrat, H. and Wagner, R.R., series Eds.), pp. 517–558, Plenum Press, New York.
[137] Reiter, W.-D., Zillig, W. and Palm, P. (1988) In: Advances in Virus Research, Vol. 34 (Maramorosch, K., Murphy, F.A. and Shatkin, A.J., Eds.), pp. 143–188, Academic Press, Orlando, FL.
[138] Ken, R. and Hackett, N.R. (1991) J. Bacteriol. 173, 955–960.
[139] Martin, A., Yeats, S., Janekovic, D., Reiter, W.D., Aicher, W. and Zillig, W. (1984) EMBO J. 3, 2165–2168.
[140] Reiter, W.-D., Palm, P., Yeats, S. and Zillig, W. (1987) Mol. Gen. Genet. 209, 270–275.
[141] Reiter, W.-D. (1988) Doctor's thesis. Ludwig-Maximilians-Universität München.
[142] Zillig, W., Yeats, S., Holz, I., Böck, A., Gropp, F., Rettenberger, M. and Lutz, S. (1985) Nature 313, 789–791.
[143] Zillig, W., Yeats, S., Holz, I., Böck, A., Rettenberger, M., Gropp, F. and Simon, G. (1986) System. Appl. Microbiol. 8, 197–203.
[144] Kletzin, A. (1986) Diploma thesis. Christian-Albrechts-Universität Kiel.
[145] Kletzin, A. (1989) J. Bacteriol. 171, 1638–1643.
[146] Shand, R.F. and Betlach, M.C. (1991) J. Bacteriol. 173, 4692–4699.
[147] May, B.P., Tam, P. and Dennis, P.P. (1989) Can. J. Microbiol. 35, 171–175.

[148] Rospert, S., Linder, D., Ellermann, J. and Thauer, R.K. (1990) Eur. J. Biochem. 194, 871–877.
[149] Reiter, W.-D. and Palm, P. (1990) Mol. Gen. Genet. 221, 65–71.
[150] Klenk, H.-P., Schwass, V., Lottspeich, F. and Zillig, W. (1992) Nucl. Acids Res. 20, 4659.
[151] Klenk, H.-P., Renner, O., Schwass, V. and Zillig, W. (1992) Nucl. Acids Res. 20, 5226.
[152] Kletzin, A. (1992) Nucl. Acids Res. 20, 5389–5396.
[153] Kletzin, A. (1992) J. Bacteriol. 174, 5854–5859.
[154] Gropp, F., Grampp, B., Stolt, P., Palm, P. and Zillig, W. (1992) Virology 190, 45–54.
[155] Stolt, P. and Zillig, W. (1992) Mol. Gen. Genet. 235, 197–204.
[156] Klenk, H.-P., Schleper, C., Schwass, V. and Brudler, R. (1993) Biochim. Biophys. Acta 1174, 95–98.
[157] Klenk, H.-P., Schwass, V. and Zillig, W. (1993) Nucl. Acids Res. 19, 6047.
[158] Lanzendörfer, M., Langer, D., Hain, J., Klenk, H.-P., Holz, I., Arnold-Ammer, I. and Zillig, W. (1993) System. Appl. Microbiol., in press.
[159] Thomm, M., Hausner, W. and Hethke, C. (1993) System. Appl. Microbiol., in press.

CHAPTER 13

Translation in archaea

Ricardo AMILS[1], Piero CAMMARANO[2] and Paola LONDEI[2]

[1]*Centro De Biologia Molecular, CSIC-UAM, Cantoblanco, Madrid 28049, Spain*, [2]*Istituto Pasteur–Fondazione Cenci-Bolognetti; Dipt. Biopatologia Umana, Sez. Biologia Cellulare, Università di Roma "La Sapienza", Policlinico Umberto I, Roma, Italy*

1. Introduction

Macromolecules (proteins, RNAs and their genes) involved in translation processes possess the ubiquity and the high degree of evolutionary stability required for inferring the deepest genealogical relationships among (extant) life forms. The ribosome itself, that centerpiece of the translation machinery, has provided the Rosetta stone for unveiling the hidden dichotomy of the microbial world and the distinctiveness of the three domains of life [1,2].

The inventory of phylogenetically relevant components offered by ribosomes comprises no less than three RNA molecules and about fifty proteins. Aminoacyl-tRNA synthetases and protein factors assisting the initiation, elongation and termination reactions of polypeptide synthesis constitute further probes of potential usefulness to delineate the unfolding of the early lineages.

Although the mechanism whereby proteins are synthesized is the same in all living forms, classical distinctions exist between eucaryal (post-transcriptional) and bacterial (co-transcriptional) translation. In-frame read-out of (usually polycistronic) bacterial mRNAs is established via Shine–Dalgarno mRNA 16S-rRNA recognition mechanisms; polypeptide synthesis is initiated by a formylated methionine and the initiation reactions are assisted by a limited number of protein factors (three in *Escherichia coli*) that primarily influence kinetic parameters [3].

By contrast, eucaryal mRNAs are translated after extensive modifications of the primary transcripts that yield mature (generally "capped" and polyadenylated) monocistronic mRNAs. Recognition of translation start sites does not rely upon Shine–Dalgarno recognition; instead, the small ribosomal subunit (generally) binds to the "capped" 5′ end of mRNA and "scans" its nucleotide sequence until the initiator AUG codon is encountered. The polypeptide chains are initiated by a non-formylated methionine and the initiation reactions are aided by as many as 8–10 protein factors, some of which possess ATPase activity and perform functions not encountered in bacteria, such as "cap" recognition and mRNA unwinding (for a detailed review see ref. [4]).

Knowledge of translation in archaea has steadily increased over the past decade. In this chapter, we compare the archaeal translation apparatus with the eucaryal and the bacterial versions thereof, and review the phylogenetic information deduced from a comparative consideration of the archaeal translational components.

2. Structure of translational components

2.1. Transfer RNAs and aminoacyl–tRNA synthetases

The sequences of nearly one hundred archaeal tRNAs have been deduced from either cloned genes or the purified tRNA species (for a review see ref. [5]). Compared to their bacterial and eucaryal homologs, archaeal tRNAs systematically lack ribothymidine and 7-methylguanosine. In addition to modified nucleotides (pseudouridine, $2'$-O-methylcytidine, 1-methylguanosine and N^2, N^2-dimethylguanosine) that are also found in bacteria and eucarya, archaeal tRNAs contain unique modified bases not encountered in either of the two other domains [6]. Also, the CCA stems of initiator tRNAs from distant archaea (*Thermoplasma acidophilum, Halococcus morrhuae, Sulfolobus acidocaldarius*) contain an identical stretch of five base pairs that has no counterpart in eucaryal and bacterial initiator tRNAs [7].

As in eucarya, the CCA-terminus of archaeal tRNAs is not encoded in the tRNA genes (with the exception of certain methanogens) and many archaeal tRNA genes resemble their eucaryal homologs in harbouring introns that are often located in the proximity of the $3'$ position of the anticodon [8–12].

Little is known about archaeal aminoacyl–tRNA synthetases, except that the phenylalanyl tRNA synthetases from the archaea *Methanosarcina barkeri* and *S. acidocaldarius* resemble their bacterial and eucaryal counterparts in being tetrameric proteins with an aggregate mass of 270 kDda. The archaeal enzymes, however, do not share antigenic determinants with the bacterial and eucaryal enzymes and are functionally restricted to tRNAs of their own lineage [13,14]. Thus, they appear to constitute a third class of tRNA charging enzymes, evolutionarily distinct from those of the other domains.

2.2. Messenger RNAs

Archaeal genomes contain clusters of cooperating genes often resembling those of bacteria in both gene composition and gene order (reviewed in refs. [15,16]). The archaeal clusters, however, are not strictly comparable to bacterial operons. First, they contain internal promoters and terminators. Secondly, transcripts of varying lengths are often produced from the same package of physically linked genes.

This situation is exemplified by the *Sulfolobus* virus-like particle genes V1, V2 and V3 and by the *Sulfolobus* RNA polymerase large subunit genes termed rpoB, rpoA and rpoC. The three viral genes are tightly linked and cotranscribed, but mono-cistronic mRNAs encoding the VP1 gene alone are produced along with the polycistronic ones [17]. Likewise, the three *Sulfolobus* RNA polymerase subunit genes are transcriptionally linked; however, shorter transcripts encoding only rpoC and certain ribosomal proteins whose genes are located downstream from rpoC are produced along with the polycistronic transcripts [18]. Interestingly, when shorter transcripts are produced, these are often more abundant than the longer ones [16–18]. Most probably, this multiplicity of mRNAs arises from transcription initiation at different internal promoter sites, although post-transcriptional processing of polycistronic transcripts cannot be ruled out.

It is not clear, however, whether this situation can be generalized to all archaeal gene clusters and to all archaeal lineages. Physically linked genes encoding enzymes involved in methane formation appear to be transcribed in a single polycistronic mRNA in *Methanobacterium formicicum* [16]. The same also holds true for the *Methanococcus voltae* genes encoding enzymes of the histidine biosynthesis pathway [16]. Amongst the halophiles, tight clustering of cooperating genes is exemplified by the genes encoding the RNA polymerase large subunit of *Halobacterium halobium* [19] and by several ribosomal protein genes of *Halobacterium marismortui* [20]. It is not known, however, whether these clusters are transcriptionally linked.

It is also unclear whether archaeal mRNAs can be cotranscriptionally or post-transcriptionally modified. Polyadenylated RNAs have been identified in several archaea [21–23], the poly(A) tails being quite short in methanogens [22] and rather long in the halophiles and the crenarchaeota [21–23]. However, evidence that they are indeed mRNAs is lacking. In any event Poly(A)+ RNAs are not "capped" in *Methanococcus vannielii* [22].

2.3. Messenger RNA–ribosome interaction

Judging from the nucleotide sequences of cloned protein coding genes, Shine–Dalgarno mRNA–ribosome recognition appears to be a frequent although not generalized feature of archaea.

Within the euryarchaeota, potential Shine–Dalgarno motifs, preceding the initiation codon, and complementary to the 3' end of 16S rRNA, exist in the elongation factor 1α (EF-1α) gene from *Thermococcus celer*, *Pyrococcus woesei*, *Thermoplasma acidophilum* and *H. marismortui* [24–27]. Extended purine-rich Shine–Dalgarno motifs are also generally present immediately upstream of the putative initiator ATG in methanogenic archaea (for a detailed review before 1989 see ref. [16]). Similarity of ribosome–mRNA recognition mechanisms in archaea and bacteria is also inferred from the ability of cloned methanogen genes to complement auxotrophic mutations of *Escherichia coli* (see ref. [16]).

The situation is less straightforward in the halophiles, because relatively few halophile genes have been sequenced and because most cloned genes are obtained from the single species *Halobacterium halobium*. Among these, a potential Shine–Dalgarno motif has only been identified in the ribosomal A protein gene [28]. Interestingly, Shine–Dalgarno motifs are uniquely located immediately downstream from the putative initiator ATG in certain halobacterial genes (namely the *bop*, *brp* and *hop* genes) having very short (or lacking) untranslated 5' leader sequences. It has been surmised that these mRNAs form stem–loop structures amenable to expose the putative Shine–Dalgarno sequence to a conformation favorable for interaction with the 16S rRNA [29–32].

In the crenarchaeota (exemplified by only *Sulfolobus*), Shine–Dalgarno sequences are not an obligatory feature. Certain *Sulfolobus* genes (notably the EF-1α [33], EF-2 [34] and RNA polymerase rpoC genes [18]) exhibit potential Shine–Dalgarno motifs located immediately upstream of the initiator ATG, whereas others (notably, the ribosomal protein S12 and RNA polymerase rpoB genes [18]) have, at best, very weak Shine–Dalgarno sequences.

2.4. Polypeptide chain initiation

The synthesis of most archaeal proteins appears to start at AUG codons. Translation initiation at a GUG codon has been reported for the VP1 protein of the *Sulfolobus* virus-like particle SSV1 [17]. Interestingly, overexpression of the *Sulfolobus* L12 gene in early-log phase *E. coli* cells results in the almost exclusive production of a short form of L12 arising from translational initiation at an AUA codon located 60 bases downstream of the regular AUG [35]. Anomalous initiation at AUA codons appears to depend on the stage of cell growth, as late-log phase *E. coli* cells predominantly produce the normal "long" form of L12 protein. It is not known whether a short form of L12 is also synthesized in the natural host.

Initiation of protein synthesis in archaea seems to occur with methionyl–tRNA as in eucarya [36]. There is no information, however, as to the number and properties of the archaeal initiation factors (IF). The interesting possibility that archaea possess eucarya-like initiation factors is suggested by the finding that total proteins from phylogenetically disparate archaea contain the unusual amino acid hypusine which is unique to eucaryal initiation factor 4D (eIF-4D) [37]. Hypusine seems to be more abundant amongst the crenarcheota whereas euryarcheota contain predominantly deoxyhypusine, a putative precursor of hypusine [38a]. The molecular mass (17 kDa) of the hypusine-containing protein of *Sulfolobus* is very similar to that of eIF-4D protein [38a]. Recently, the gene encoding *Sulfolobus* hypusine-containing protein has been cloned and sequenced; the derived amino-acid sequence has been found to bear considerable homology to eukaryotic IF-4D [38b]. However, there is no evidence that the archaeal hypusine-containing protein plays a role in translation.

2.5. Elongation factors

Elongation factors (EF) catalyzing the binding of incoming aminoacyl–tRNA to ribosomes (EF-Tu for bacteria, EF-1α for eucarya and archaea) and the translocation of peptidyl–tRNA from the ribosomal A site to the P site concomitant with the ejection of deacylated tRNA (EF-2 for eucarya and archaea, EF-G for bacteria) are universal, highly conserved GTP-binding proteins coded for by paralogous genes [39]. Both factors consist of a single polypeptide chain, the length of the translocating factor (700–860 residues) being about twice that of aminoacyl–tRNA binding factor (390–460 residues).

2.5.1. Elongation factor sequences
Several archaeal EF-1α and EF-2 genes have been cloned and sequenced over the past five years. Derived amino-acid sequences are available for *H. marismortui* EF-1α [27] *H. halobium* EF-2 [40] *M. vannielii* EF-1α [41] and EF-2 [42], *T. acidophilum* EF-1α [26] and EF-2 [25], *P. woesei* EF-1α [24] and EF-2 [Creti, R., unpublished results], *T. celer* EF-1α [43] and *S. acidocaldarius* EF-1α [33] and EF-2 [34].

TABLE 1

Comparison of the amino acid sequences around the ADP-ribosylatable site in archaeal and eucaryal EF-2 sequences[a]

Organism	Amino acid position																		
	1	2	3	4	5	6	7	8	9	10*11	12	13	14	15	16	17	18	19	
S. cerevisiae	D	V	T	L	H	A	D	A	I	H	R								
D. discoideum	D	V	T	L	H	T	D	A	I	H	R	G	G	G	Q	I	I	P	T
D. melanogaster	D	V	T	L	H	A	D	A	I	H	R	G	G	G	Q	I	I	P	T
Hamster	D	V	T	L	H	A	D	A	I	H	R	G	G	G	Q	I	I	P	T
M. vannielii	D	A	T	F	H	E	D	A	I	H	R	G	P	S	Q	I	I	P	A
H. halobium	D	A	R	L	H	E	D	A	I	H	R	G	P	A	Q	V	I	P	A
R. acidophilum	D	A	K	L	H	E	D	S	I	H	R	G	P	A	Q	V	I	P	A
R. celer[b]	D	A	K	I	H	E	D	N	V	H	R	G	P	A	Q	I	Y	P	A
P. woesei[b]	D	A	Q	V	H	E	D	N	V	H	R	G	P	A	Q	I	Y	P	A
D. mobilis[b]	D	A	V	V	H	E	D	P	A	H	R	G	P	A	Q	I	F	P	A
S. acidocaldarius	D	A	V	V	H	E	D	P	A	H	R	G	P	A	Q	L	Y	P	A
S. solfataricus[c]	D	A	V	I	H	E	D	P	A	H	R	G	P	A	Q	I	Y	P	A
E. coli	D	V	D	S	S	E	L	A							F	K	L	A	
T. thermophilus	E	V	D	S	S	E	M	A							F	K	I	A	
Consensus sequences																			
Eucarya	D	V	–	L	H	–	D	A	I	H	R	G	G	G	Q	I	I	P	T
Archaea	D	A	–	–	H	E	D	–	–	H	R	G	P	–	Q	–	–	P	A

[a] Updated from Schroder and Klink [34]. The hystidine which is post-translationally converted to diphthamide is marked by an asterisk.
[b] Data for *Pyrococcus woesei* are from Cammarano et al. (unpublished);); data for *Desulfurococcus mobilis* are from Ceccarelli, E., et al. (unpublished); data for *T. celer* are from Auer, J. (unpublished).
[c] *Sulfolobus solfataricus* strain formerly called *Caldariella acidophila* MT4 strain. (Data from Raimo, G., et al. (1992) Biochim. Biophys. Acta 1132, 127–132.)

Both archaeal factors share with the equivalent eucaryal and bacterial factors, three conserved motifs [(A/G)HXDXGK(T/S), DXXGH, NKXD] [44] spanning residues 20–144 (*E. coli* numbering) of EF-1α [26] and residues 17–142 (*E. coli* numbering) of EF-2 [25]. The three motifs are correlated with specific functions such as GTP binding and GTP hydrolysis and represent specific variants of the consensus A, B and C motifs [(A/G)XXXXGK(T/S), DXXG, NKXD)] of nucleotide-binding proteins. In addition, archaeal EF-2 shares with eucaryal EF-2 the C-terminal sequence which is required for ADP-ribosylation by diphtheria toxin [34] (section 4.4) (Table 1).

As might be expected, elongation factors from the extremely halophilic archaea possess features that enable them to withstand the hypersaline intracellular environments. Halotolerance is exemplified by the EF-1α of *H. marismortui* which remains soluble and functions in 3.5 M salt while denaturing slowly at salt concentrations lower than about 2.0 M [45]. Both tolerance to extreme K^+ and Na^+ levels, and instability in low salt, appear to be accounted for by a greater proportion of amino acids with negatively charged side chains than exists in non-halophilic factors; these are grouped in patches on the protein surface [27] and compete with salt for available water, thereby retaining an appropriate hydration layer even at high ionic strength; the underlying principle is that glutamic and aspartic acids bind about twice as many water molecules (6.0–7.5 mol water per mol residue) than positively charged polar residues [46,47].

A more difficult task is to identify features conferring heat stability to factors from the hyperthermophilic archaea (and in fact to factors from all hyperthermophilic organisms, including bacteria such as *Thermotoga maritima* and *Aquifex pyrophylus*). Contrary to original considerations and expectations, thermophilic and mesophilic enzymes do not exhibit extensive and pronounced structural differences (for recent reviews see refs. [48–50]) and it is often difficult to discriminate "adaptive" changes from primary structural changes accumulated during the separate evolutionary course of the lineages. Recent data, however, indicate that the GTP-binding domain of *T. maritima* EF-Tu possesses a more compact hydrophobic core than the corresponding region of mesophile (*E. coli*) EF-Tu; this appears to be principally accounted for by the substitution of small hydrophobic residues in *E. coli* protein with bulkier ones in the *Thermotoga* one (Creti, R., personal communication). A similar mechanism may apply to the EF-1α protein of the hyperthermophilic archaea.

Average sequence identities between homologous factors from the three domains are given in matrix form in Tables 2 and 3. The archaeal sequences are clearly segregated from those of eucarya and bacteria. The distinctness of archaeal factors is also borne from the immunochemical cross-reactivities of EF-1α/EF-Tu proteins from phylogenetically disparate archaea, eucarya and bacteria [51]. However, the most impressive result from comparing the sequences is the evidence that the similarity of archaeal EF-1α to eucaryal EF-1α (49.0–57.3% identical residues) is far greater than that (32.3–40.7% identity) between the archaeal and the bacterial factors. A similar situation also obtains for EF-2; compared to EF-1α, however, the similarity between archaeal and eucaryal factors is less pronounced, the archaeal EF-2 sharing 33–40% identical residues with eucaryal EF-2 and 30–34% identical residues with bacterial EF-G.

Phylogenies inferred from the amino-acid identities also bear this out (Fig. 1), the root of the archaeal subtree being closer to the (present) eucaryal root than to the bacterial root, and more evidently so in the case of the EF-1α trees. Furthermore, both the EF-1α/EF-Tu and the EF-2/EF-G trees confirm the deep subdivision of the archaea into two major branches constituting the kingdoms of Crenarchaeota and Euryarchaeota [52].

2.5.2. Elongation factor gene order
The chromosomal organization of the archaeal EF-1α and EF-2 genes is illustrated in Chapter 12 by Zillig et al. in this volume (Fig. 2).

Bacterial (*Escherichia coli*) genes encoding elongation factors Tu and G are typically clustered with ribosomal protein genes in a transcriptional unit – the streptomycin (*str*) operon – which comprises (from the 5' end) the genes for ribosomal proteins S12 and S7, the EF-G (*fus*) gene and the EF-Tu (*tuf*) gene, in that order [53]. An identical organization of the *str* operon genes is found in methanogens (*M. vannielii*), except that the *Methanococcus* "*str* operon" equivalent gene cluster contains two additional open reading frames (ORF1 and ORF2) at the promoter proximal side of the cluster, and the S10 ribosomal protein gene at the promoter-distal site [54]. Unlike *M. vannielii*, all other archaea investigated exhibit *str* operon equivalent gene clusters having incomplete and scrambled sets of "streptomycin operon" genes (see

TABLE 2
EF-1α/(EF-Tu) percent identities from amino acid sequences[a,b]

Asa																		
Dme	87.9																	
Hsa	83.6	85.2																
Sce	79.6	79.3	82.0															
Egr	77.0	78.4	80.4	75.7														
Ddi	75.7	76.1	78.9	75.1	74.1													
Sso	51.0	52.3	52.5	50.7	50.5	52.1												
Tce	54.5	56.0	55.4	54.7	54.5	55.6	60.2											
Pwo	55.3	56.4	56.3	54.8	55.0	57.3	59.7	88.8										
Mva	54.2	53.8	52.4	52.9	51.2	53.6	54.8	63.2	62.4									
Hma	49.3	49.9	50.1	49.0	49.8	50.0	52.3	60.3	59.3	61.7								
Tac	54.3	55.3	54.7	54.2	52.8	56.0	59.7	63.6	64.5	64.0	59.4							
Tma	36.1	35.2	35.4	35.8	34.8	36.6	35.8	39.0	40.4	35.4	37.4	40.1						
Tth	35.4	34.8	34.5	35.3	33.8	35.9	35.1	39.0	40.1	36.2	37.7	38.5	74.0					
Eco	34.5	32.9	33.7	34.5	32.9	35.0	34.1	39.2	40.7	36.8	39.1	37.5	71.8	72.7				
Mlu	33.7	33.3	35.7	33.4	33.6	35.1	34.0	38.6	39.2	35.7	37.1	37.3	67.3	69.6	71.4			
Spl	31.4	31.1	31.4	31.2	29.5	32.0	33.8	36.5	37.9	32.8	33.9	38.6	65.1	70.8	73.4	66.5		
Ech	33.0	31.7	32.5	31.2	30.6	32.4	34.3	39.2	39.6	33.5	34.9	36.6	64.8	69.3	71.4	65.1	77.3	
Smt	35.4	34.2	34.2	32.9	32.8	33.9	32.3	37.8	39.2	34.9	37.0	38.0	62.8	67.1	64.1	61.8	60.5	62.3
	Asa	Dme	Hsa	Sce	Egr	Ddi	**Sso**	**Tce**	**Pwo**	**Mva**	**Hma**	**Tac**	Tma	Tth	Eco	Mlu	Spl	Ech	Smt

[a] Abbreviations for archaea are in boldface type.
[b] Abbreviations: Asa, *Artemia salina*; Dme, *Drosophila melanogaster*; Hsa, Homo sapiens; Egr, *Euglena gracilis*, Ddi, *Dictyostelium discoideum*; Sso, *S. solfataricus;* Tce, *Thermococcus celer*; Pwo, *Pyrococcus woesei*; Mva, *M. vannielii*; Hma, *Halobacterium marismortui*; Tac, *Thermoplasma acidophilum*; Tma, *Thermotoga maritima*; Tth, *Thermus thermophilus*; Eco, *E. coli*; Mlu, *Micrococcus luteus*; Spl, *Spirulina platensis*; Ech, *E. gracilis* chloroplast; Smt, *Saccharomyces cerevisiae* mitochondria.

Chapter 12 by Zillig et al., this volume, Fig. 2). The *str* operon equivalent cluster of *H. halobium* lacks the EF-1α gene and contains an additional short ORF[55]. The *str* operon equivalent cluster of *Sulfolobus* lacks both the EF-2 gene, which is separately transcribed[33,34] and the S12 gene, which is affiliated to an upstream (RNA polymerase) operon; also, a tRNA$_{ser}$ gene is located immediately downstream of the 3' end of the *Sulfolobus* S10 gene [33]. Interestingly, the EF-1α and EF-2 genes have different chromosomal locations in the two closely related, early branching euryarchaeota *P. woesei* and *T. celer*. In *P. woesei*, a region comprising the EF-1α, S10 and tRNA$_{ser}$ gene cluster (in that order) has been sequenced. No EF-2 gene sequences have been found over a DNA region spanning 600 nucleotides upstream of the EF-1α gene, and no EF-1α gene sequences occur about 800 nucleotides downstream of the terminating codon of the EF-2 gene [Cammarano, P., unpublished data]. This is in contrast to the situation in *T. celer* in which the 3' end of the EF-2 coding gene is situated 460 nucleotides upstream of the EF-1α-S10 gene cluster [Auer, J., personal communication]. Lastly, in *T. acidophilum* the EF-1α and EF-2 genes follow each other in reversed order[56].

The striking conservation of the *str* operon gene order across domain boundaries (i.e., in the bacterium *E. coli* and the archaeon *M. vannielii*) suggests that the *(str* operon)-

TABLE 3
EG–2/(EF–G) percent identities from amino acid sequences[a,b]

	Ham	Dme	Ddi	Sso	Pwo	Mva	Hha	Tac	Tma	Tth	Eco	Mlu
Ham												
Dme	79.5											
Ddi	64.3	61.6										
Sso	37.7	36.8	37.3									
Pwo	36.9	36.6	35.4	49.9								
Mva	35.8	35.8	35.4	46.0	59.6							
Hha	36.5	36.2	35.4	45.8	55.5	53.3						
Tac	33.8	33.7	33.2	43.4	58.5	55.3	51.1					
Tma	27.8	27.4	27.4	32.4	32.6	32.9	31.6	31.3				
Tth	26.3	26.0	26.6	31.8	36.1	33.8	32.6	32.9	61.4			
Eco	26.4	25.3	25.6	31.3	33.6	31.2	31.2	29.8	58.0	60.3		
Mlu	25.7	25.0	23.6	29.3	31.2	30.2	29.7	28.3	55.9	60.1	61.3	
	Ham	Dme	Ddi	**Sso**	**Pwo**	**Mva**	**Hha**	**Tac**	Tma	Tth	Eco	Mlu

[a] Abbreviations for archaea are in boldface type.
[b] Abbreviations: Ham, hamster; Dme, *Drosophila melanogaster*; Ddi, *Dyctiostelium discoideum*; Sso, *Sulfolobus solfataricus*; Pwo, *Pyrococcus woesei*; Mva, *Methanococcus vannielii*; Hha, *Halobacterium halobium*; Tac, *Thermoplasma acidophilum*; Tma, *Thermotoga maritima*; Tth, *Thermus thermophilus*; Eco, *Escherichia coli*; Mlu, *Micrococcus luteus*.

like gene organization is primitive. Within bacteria that primitive organization has been maintained across the vast evolutionary distance separating *E. coli* and *T. maritima* [57]. In contrast, more or less extensive rearrangements of the primitive order appear to have been experienced by different archaea during the separate evolution of the individual lineages. This is exemplified by the loss of the EF-2 and EF-1α gene linkage in *P. woesei* and the maintenance of their proximity in the closely related organism *T. celer*.

2.6. Archaeal ribosomes. Halotolerance and heat-stability

Archaeal ribosomes resemble those of bacteria in the number and the sizes of their rRNAs (23S, 16S and 5S, but no separate 5.8S as found in eucarya except Microsporidia [58]) and in the sedimentation behaviour of their two subunits in sucrose density gradients. The organization of the rRNA genes (reviewed in ref. [16]) also follows the general pattern seen in bacteria (see ref. [56]), although exceptions have been found [59].

However, ribosomes from most archaea must possess unique features that enable them to secure acceptable levels of efficiency and accuracy in code translation under conditions (temperatures higher than the boiling point of water, saturating salt concentrations) that are lethal for most known bacteria. The physicochemical properties of ribosomes from extremely thermophilic (*Sulfolobus solfataricus*) and halophilic (*Halobacterium cutirubrum*) archaea have been studied in some detail. The melting temperature of *S. solfataricus* ribosomal subunits (T_m = 92–94°C) is considerably higher (about 20°C) than that observed for mesophile (*E. coli*) ribosomes, and is also 15°C higher than that of subunits from the bacterium *Bacillus acidocaldarius* (maximum growth temperature ∼70°C). The great stability of the *Sulfolobus* 50S and 30S subunits is principally accounted for by more extensive rRNA–protein and protein–protein interactions than exist in mesophile (*E. coli*) ribosomes [60,61]. This is argued from (i) the difference in

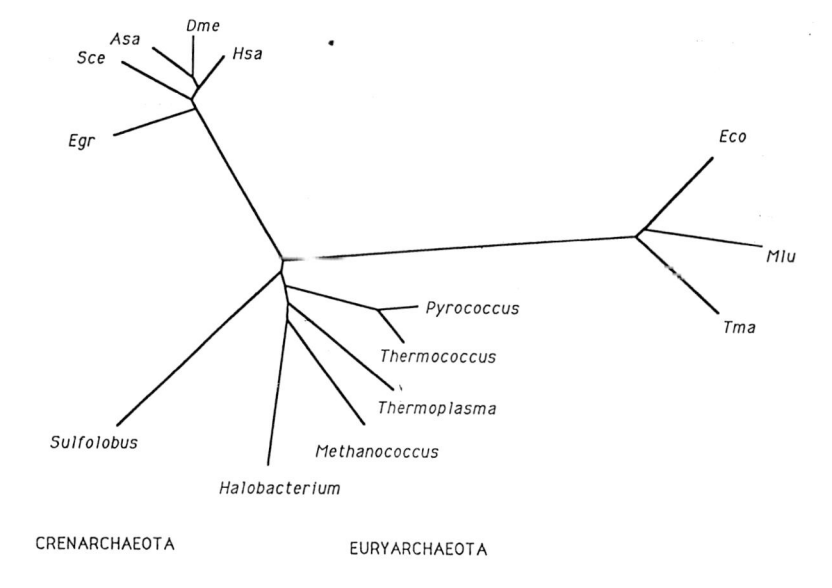

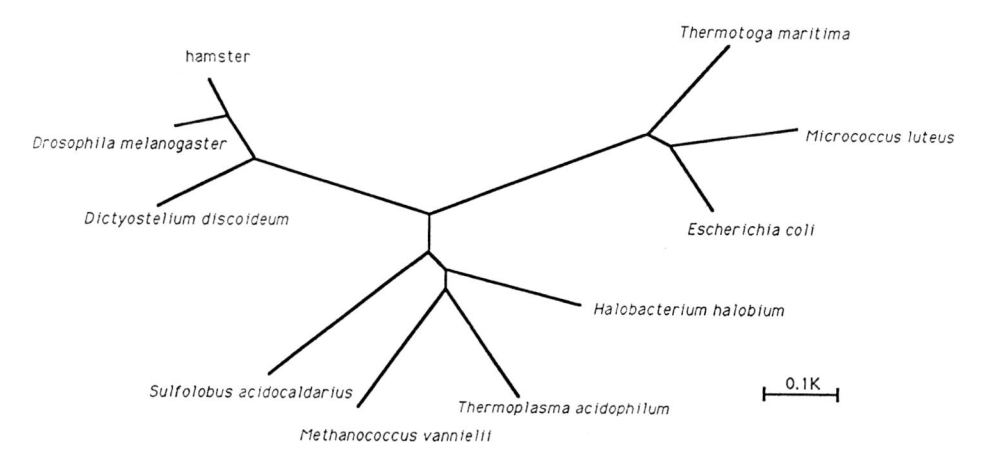

Fig. 1. Elongation factor distance-matrix phylogenetic trees inferred from amino-acid sequence identities. Top: phylogenetic tree inferred from EF-1α/(EF-Tu) sequences. Bottom: phylogenetic tree inferred from EF-2/(EF-G) sequences. Abbreviations: Egr, *Euglena gracilis*; Sce, *Saccharomyces cerevisiae*: Asa, *Artemia salina*; Dme, *Drosophila melanogaster*; Hsa, *Homo sapiens*; Eco, *Escherichia coli*; Mlu, *Micrococcus luteus*; Tma, *Thermotoga maritima*.

melting temperature (ΔT_m) between free rRNA and rRNA stabilized by proteins within the intact subunit, the T_m being markedly larger for thermophile (*S. solfataricus*) than for mesophile (*E. coli*) ribosomes; (ii) the unusual resistance of the *Sulfolobus* ribosomal subunits to urea and LiCl concentrations about fourfold higher than those resulting in complete disassembly of mesophile (*E. coli*) ribosomes; and (iii) the low susceptibility of the thermophile subunits to treatments designed to promote unfolding of the quaternary packing through depletion of Mg^{2+} ions [60,61]. Presumably, a similar situation applies to ribosomes from hyperthermophilic bacteria such as *T. maritima* and *A. pyrophylus*.

An even more striking case is offered by halophile ribosomes. Ribosomal subunits from the extreme halophiles are uniquely stable at K^+ and Mg^{2+} levels (3.5 M and 100 mM, respectively) that disrupt ribosomes of all other sources; lowering the K^+ and Mg^{2+} concentrations causes unfolding of the halophile particles and loss of their proteins to an extent depending on the final K^+/Mg^{2+} ratio [62]. A possible contribution to the high degree of ribosome stability in high salt is given by the fact that halophile r-proteins (*H. cutirubrum*) are higher in acidic residues than those of non-halophile ribosomes [63]. As mentioned previously (section 2.5) this allows the halophile proteins to maintain an appropriate hydration volume at near to saturating salt concentrations.

2.6.1. Ribosomal subunit interaction
Ribosomes extracted in a standard "low-salt" medium (10 mM Mg^+ and 40 mM NH_4^+) from all archaea, except halophiles, are generally found as synthetically active 50S and 30S subunits free of adhering factors following centrifugation in "high-salt" sucrose density gradients (10 mM Mg^+, 1.0 M NH_4Cl).

This may generate the false impression that archaeal ribosomes generally exist as 70S monomers in cell extracts but dissociate into subunits during zone-velocity centrifugation due to a high ratio of monovalent (NH_4^+) to divalent cations (Mg^{2+}), or to high hydrostatic pressure, or both. On the contrary, ribosomes from *Sulfolobus*, *Thermoproteus tenax* and *Desulfurococcus mobilis* (crenarchaeota) differ from those of *Thermococcus*, *Thermoplasma* and methanogens (euryarchaeota) in the tightness of the subunit interaction [66]. Ribosomes from the former group of organisms are in fact composed of weakly interacting subunits and exist almost exclusively as free 50S and 30S particles in the cell extracts, regardless of the stage of cell growth (*Sulfolobus*), or Mg^{2+} (and/or polyamine) concentrations used during cell fractionation. In contrast, ribosomes from organisms belonging to the latter group generally consist of tightly bonded subunit couples whose dissociation requires exposure to high salt; certain methanogen ribosomes are in fact partially (*Methanococcus thermolithotrophicus*) or completely resistant (*Methanobacterium thermoautotrophicum*) to salt-induced dissociation. As will be illustrated in later sections (3.5–3.5.4), the two ribosome types also exhibit opposite behaviours in their ionic (NH_4^+) and polyamine (spermine, thermine) requirements for polyphenylalanine synthesis (cf. Table 5, below).

Halobacterial ribosomes are a special case: they exist as 70 monomers in the presence of 100 mM Mg^{2+} and near to saturating concentrations (3.1–4.0 M) of K^+ ions [67,68] and dissociate progressively upon lowering the Mg^{2+} concentration [68]; complete dissociation of 70S monomers into subunits that are synthetically active [69] occurs only upon exposure to a tenfold lower concentration of Mg^{2+} ions in the presence of a stabilizing (3.1 M) concentration of monovalent (K^+) cations [67,68].

2.6.2. Ribosome mass and composition
Since archaeal ribosomes cosediment with those of eubacteria (*E. coli*) in sucrose density gradients, it was tacitly implied, and generally assumed (until 1986) that they conformed

to the prokaryotic (*E. coli*) paradigm (1.46×10^6 Da large subunits and 0.84×10^6 Da small subunits containing, respectively, 67% and 60% protein by weight)[70]. However, the hydrodynamic behaviour of ribosomes as measured by S_0^{20} does not provide a reliable estimate of the particle mass (M). Lack of correlation between S and M is dramatically illustrated by the protein-rich ribosomes of mammalian mitochondria, which exhibit an S value of 55S although they are no smaller in mass than *E. coli* 70 ribosomes [71].

To circumvent these pitfalls, the gross physicochemical properties (aggregate mass and protein content) of archaeal ribosomal subunits have been calculated directly as the ratio of rRNA chemical weight (estimated from the nucleotide sequences) to rRNA weight fraction; the latter parameter being accurately inferred from the linear relationship between rRNA weight fraction and the subunits' buoyancies in CsCl gradients after fixation with formaldehyde. Since rRNA molecular weights are nearly invariant among prokaryotes, changes in the ribosomes' buoyant densities reflect only the relative abundance of the protein moiety. In general, a transition from high buoyancy to low buoyancy indicates accretion of ribosomal protein, because the buoyant density of protein in CsCl (1.33 g cm^{-3}) is much less than that of free RNA in the same salt (1.87 g cm^{-3}). The ribosome weights determined in this way are "formula weights", free of contributions from bound water molecules, polyamines and salt).

An updated list of ribosome weights and compositions obtained from the buoyancies of "high-salt washed" factor-free subunits is given in Table 4.

In bacteria, the mass and composition of the ribosomal subunits coincide almost exactly with those calculated from the primary sequences of all of the individual rRNA and protein components of *E. coli* ribosomes; typically, 50S and 30S particles from the recent bacterial phyla contain $0.46-0.48 \times 10^6$ and $0.33-0.38 \times 10^6$ Da protein, respectively. Until now, a dramatic increase in protein mass has been found only in the extremely thermophilic bacterium *A. pyrophylus* [Cammarano, P., unpublished results] representing an extremely deep branching in the bacterial tree [Stetter, K.O. personal communication].

Archaea exhibit intralineage variations of their ribosome type. To generalize, all archaea contain ribosomes (especially their small subunits) that are considerably richer in protein than those of bacteria, with the exception of the Halobacteriales and of methanogens belonging to the Methanomicrobiales and Methanobacteriales. These differ from other archaea in harbouring ribosomes having the same protein mass and the same (3:2) RNA to protein mass ratio as those of typical bacteria. Because the size and the secondary structure of the archaeal rRNAs do not differ from those of bacterial rRNAs, it is likely that the accreted proteins do not interact directly with the rRNA moiety.

The accretion in protein mass of (most) archaeal ribosomes is substantiated by gel-electrophoretic and sequence analysis of the subunit proteins. Thus, Methanobacteriales [72,73] and halophiles [74] contain a number of proteins which lie in the bacterial range (20–22 for the small subunit and 31–32 for the large subunit). In contrast, ribosomes from the Methanococcales (*M. vannielii*) [74] and Sulfolobales [64,73,75] possess proteins that are more numerous and also heavier, on average, than those found in the typical bacteria. For instance, *S. solfataricus* 30S and 50S subunit proteins have average molecular weights of 20.7×10^3 and 19.3×10^3, respectively, compared to values of 16.6×10^3 and 14.9×10^3 for the corresponding *E. coli* proteins [64]. This situation is more precisely documented by sequence comparisons of ribosomal proteins from the three domains. Böck and coworkers [54,76]

TABLE 4
Particle weight and protein composition of archaeal ribosomal units[a,b,c]

	30S mass ×10⁻⁶	30S protein mass ×10⁻⁶	50 S mass ×10⁻⁶	50S protein mass ×10⁻⁶
Bacteria				
E. coli[a]	0.84	0.33	1.46	0.46
C. aurantiacus[b]	0.895	0.39	1.60	0.60
T. maritima[b]	0.92	0.41	1.59	0.595
F. islandicum[b]	0.88	0.37	1.60	0.60
T. africanus[b]	0.895	0.39	1.64	0.65
A. pyrophylus[b]	1.17	0.66	1.87	0.876
Archaea				
Euryarchaeota				
H. halobium[a]	0.80	0.31	1.52	0.51
M. barkeri[a]	0.80	0.31	1.55	0.54
A. fulgidus[b]	1.14	0.64	1.84	0.81
M. tindarius[a]	0.80	0.32	1.52	0.51
M. formicicum[a]	0.80	0.31	1.58	0.57
T. acidophilum[a]	1.10	0.605	1.81	0.78
M. vannielii[a]	1.00	0.52	1.81	0.85
M. thermolithotrophicus[a]	1.00	0.52	1.81	0.85
T. celer[a]	1.14	0.645	1.90	0.87
P. woesei[b]	1.14	0.645	1.84	0.81
P. furiosus[b]	1.16	0.67	1.90	0.87
Crenarchaeota				
D. mobilis[a]	1.16	0.66	1.92	0.86
T. tenax[a]	1.15	0.66	1.97	0.94
P. occultum[a]	1.16	0.665	2.00	0.97
S. marinus[b]	1.16	0.665	1.90	0.87
S. solfataricus[a]	1.14	0.645	1.80	0.77
Eucarya	1.50		2.80	

[a] Data from ref. [65].
[b] Data from Cammarano, P., et al. (unpublished).
[c] Species not mentioned in the text: *Chloroflexus aurantiacus*, *Fervidobacterium islandicum*, *Thermosypho africanus*, *Methanosarcina barkeri*, *Archaeoglobus fulgidus*, *Fervidobacterium islandicum*, *Thermosypho africanus*, *Methanosarcina barkeri*, *Archaeoglobus fulgidus*, *Methanolobus tindarius*, *Methanococcus thermolithotrophicus*, *Pyrococcus furiosus*, *Pyrodictium occultum*, *Staphylothermus marinus*.

have cloned and sequenced the *M. vannielii* ribosomal protein gene clusters equivalent to the *E. coli* streptomycin (*str*) and spectinomycin (*spc*) operons. These two clusters encode (in addition to elongation factor genes) the proteins equivalent to *E. coli* S12,

S7, S10 (*str*-operon) S17, L14, L24, L5, S14, S8, L6, L18, S5, L30, L15 (*spc*-operon), L22, S3, L29 (which in *M. vannielii* belong to a separate gene cluster), and also proteins equivalent to eucaryal ribosomal proteins having no counterpart in bacteria. From a comparison of the deduced amino-acid sequences, the methanococcal ribosomal proteins appear to be generally longer than their bacterial (*E. coli*) homologs. In cases where a related eucaryal sequence exists, this has the same size as, or is larger than, the archaeal sequence. Furthermore, open reading frames are found in the ribosomal protein gene clusters of *Methanococcus* that may encode additional proteins having no counterpart in bacteria [54].

Because ribosomes that are heavier and richer in protein than those of (recent) bacterial phyla span both archaeal kingdoms and can be traced back to the deepest node of the archaeal tree (the *Thermococcus–Pyrococcus* lineage), they seem – for this reason – to represent the aboriginal archaeal particle from which the leaner ribosomes of Methanobacteriales, Methanomicrobiales and halophiles arose, presumably by streamlining of the protein moieties. The primitive character might have been retained in the extremely thermophilic members of the Euryarchaeota and lost by the mesophilic ones. This is exemplified by *Archaeoglobus fulgidus*, a sulfur-reducing hyperthermophilic member of the phylogenetic unit comprising Methanomicrobiales and extreme halophiles [77] which is endowed with protein-rich "heavy" ribosomes, unlike the other (methanogenic and halophilic) members of the same grouping which typically harbour bacteria-like, protein-poor, ribosomes.

Interestingly, in both prokaryotic domains the deepest offsprings are extreme thermophiles endowed with heavy-sized, protein-rich ribosomes. By inference, protein-rich ribosomes of most archaea and of deep branching bacteria such as *A. pyrophylus* could be vestiges of the prototype particle harbored by the primitive thermophilic ancestor.

2.6.3. Ribosome shape
The overall shape, and the distinguishing architectural details, of eucaryal and bacterial ribosomes exhibit a high degree of intralineage constancy [78] although the full spectrum of diversity (e.g. ribosomes from Microsporidia) [79,90] has not been explored. In contrast, archaeal ribosomes appear to display intralineage diversity in sharing with eucarya all or few of several structural attributes that typically distinguish the ribosomes of eucarya from those of bacteria [81].

Unlike bacterial 50S subunits, eucaryal 60S particles display a bulging of the so-called L1 protuberance, an incision below it and a protrusion at the base of the subunit [81] (Fig.2A); the 40S subunits differ from bacterial 30S particles in having a beak-like structure of the head of the particle, a bifurcated lower pole of the particle body, and a bifurcation, or split, of so-called subunit platform [81] (Fig. 2B). All of the above ribosome features are systematically lacking [81,82], or are rudimentary [83–87], in bacteria but are differentially represented amongst archaeal phenotypes, so that three ribosome types occur [89,90].

According to Lake and coworkers [89], ribosomes from the sulfur-dependent thermophiles (exemplified by *S. acidocaldarius, T. tenax, Desulfurococcus mucosus* and *Thermofilum pendens*) share a special relationship with those of eucarya in possessing a full complement of the above (eucaryal) attributes; in contrast, halobacterial ribosomes

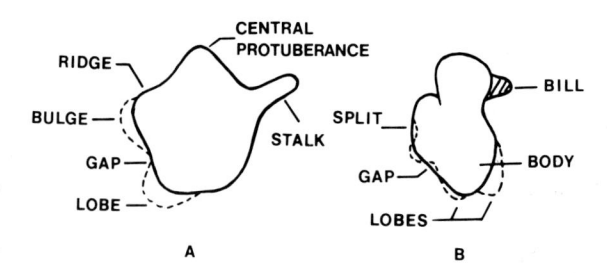

Fig. 2. Distinguishing morphological features of archaeal and bacterial ribosomal subunits. The solid curves show the shapes of bacterial particles; additional features of archaeal subunits are hatched, or drawn as dashed lines. (A) large subunits; (B) small subunits.

are indistinguishable from those of bacteria in lacking all of the above features except the beak of the 30S subunit [81,82]; methanogen ribosomes (exemplified by *M. thermoautotrophicum* and *M. barkeri*) illustrate an intermediate situation in displaying a significant split of the 30S subunit platform and an evident 30S subunit bill, but essentially none of the remaining attributes [81,82].

These similarities led to the proposition that sulfur-dependent thermophiles (represented by Sulfolobales and Thermoproteales) and methanogens constitute two different kingdoms (the eocytes and the archaebacteria proper) while halophiles are to be lumped with (eu)bacteria to form the kingdom of photocytes [82]. To generalize, different ribosome structures would define four higher-order taxa: eocytes, eukaryotes, archaebacteria and photocytes, the implication being that archaea, as originally defined by Woese and Fox [1], constitute an invalid (phylogenetically incoherent) taxon and should be split into three higher-order taxa, each having a kingdom status.

However, doubt on the validity of these conclusions has been cast by the finding that the same ribosome features used to isolate sulfur-dependent thermophiles from the other archaea (and to raise them to the kingdom rank) are also present in *Methanococcus vannielii*, a typical methanogenic archaeon representative of the order Methanococcales [87,90]. Following the same logic, that ribosome features define higher-order taxa, Methanococcales should be extracted from the remaining methanogens and lumped into eocytes, which is an obvious paradox. The use of details of ribosome structure to delineate the unfolding of the early lineages has been questioned on the grounds that the parameters involved are too elusive and ill-defined [90].

Apart from that, presence of eucarya-like bulges and protuberances on archaeal ribosomes clearly correlates with the relative abundance of the protein moiety inferred from buoyancy data, from SDS gel-electrophoretic comparisons and from sequencing of ribosomal protein genes. Reference to Table 4 shows that within the archaea all of the eucaryal attributes are in fact present in the protein-rich ribosomes of Methanococcales and sulfur-dependent thermophiles (eocytes) but are absent, or nearly so, in the protein-poor particles of the remaining methanogens, the halophiles and most bacteria.

TABLE 5

Optimal conditions for in vitro reconstitution of *Sulfolobus solfataricus* and *Haloferax mediterranei* 50S subunits

	K$^+$(M)	Mg^{2+} (mM)	Spermine (mM)	T (°C)	Time (min)
S. solfataricus[a]					
1st step	0.3	20	10–12	65	45
2nd step	0.3	40	10–12	80	60
H. mediterranei[b]	3.0	40–60	–	42	120

[a] Data from ref.[99].
[b] Data from ref.[100].

2.7. In vitro reconstruction of ribosomal subunits

The reconstitution of bacterial ribosomal subunits from the separated rRNAs and proteins, first announced in 1968, provides a potent tool to investigate such essential aspects of ribosome structure and evolution as the subunits' assembly pathway [92,93], the locations and neighbourhoods of the subunit proteins [94,95], the roles of the individual proteins in both assembly and function [92,93,96], and the degree of exchangeability of ribosomal components both within and across domain boundaries [97,98].

Recently, both crenarchaeal (*S. solfataricus*) [99] and euryarchaeal (*Haloferax mediterranei*) [100] 50S ribosomes have been reconstituted.

2.7.1. Reconstruction of Sulfolobus 50S subunits

The *Sulfolobus* 50S assembly protocol superficially resembles the Nierhaus and Dohme protocol [101] for the reconstruction of *E. coli* 50S subunits in entailing two incubation steps, the second of which requires a higher temperature and a higher concentration of Mg^{2+} ions than the first one.

For *Sulfolobus* 50S reconstitution, 23S + 5S S rRNAs and total 50S subunit proteins (TP 50) are first incubated at 60–65°C in the presence of 20 mM Mg^{2+}. The incubation temperature is then raised to 75–80°C while the Mg^{2+} concentration is raised to 40 mM. The corresponding conditions for the reconstruction of *E. coli* 50S particles are 44°C and 4 mM Mg^{2+} in the first step; 50°C and 20 mM Mg^{2+} in the second step. Unlike the situation in *E. coli*, however, the reassembly of *Sulfolobus* 50S subunits is obligatorily dependent on a concentration (10–12 mM) of spermine (or thermine) considerably higher than that (3–4 mM) required for polyphenylalanine synthesis on *Sulfolobus* ribosomes (section 3.3). Spermidine (1,8-diamino-4-azaoctane) is about half as effective as spermine while putrescine and cadaverine are totally ineffective. *Sulfolobus* 50S subunits reconstructed by the protocol illustrated in Table 5 (cf. ref. [99]) are indistinguishable from authentic 50S particles in sedimentation behaviour, thermal-melting temperature ($T_m = 90$–92°C in the presence of 0.1 mM Mg^{2+}), compactness of the quaternary packing and degree of susceptibility to the ribosome-inactivating protein α-sarcin [99]. The synthetic capacity of the reconstructed particles is about 70% that of the native 50S subunits [99].

The in vitro assembly of *Sulfolobus* large subunits differs in certain aspects from that of *E. coli* 50S particles. Typically, the reconstitution of *E. coli* 50S subunits occurs via an inactive 48S assembly intermediate which is formed at a temperature (44°C) about 10°C higher than the physiological optimum for cell growth; the conversion of the 48S intermediate into a synthetically active 50S particle necessitates a further heat treatment at 50°C [102]. In contrast, the *Sulfolobus* 50S particle formed during the first incubation step at 65°C is structurally indistinguishable from native subunits and already possesses appreciable synthetic capacity [99]; in fact, the subsequent incubation at 80°C only improves the particle activity, without attendant changes of either the S value or the thermal-melting behaviour.

Presumably, the incubation at 80°C acts by ultimately shaping the assembly product formed at 65°C into a spatial design capable of efficiently interacting with the other translational components. This conformational adjustment, in fact a fine tuning of the particle shape, is absolutely dependent on spermine. This observation, and the evidence that spermine is essential to activate the peptidyltransferase center of crenarchaeal ribosomes (see section 3.4), suggest that spermine-like polyamines (notably thermine) are an integral and essential constituent of the large subunits of crenarchaeal ribosomes.

Recently, *Sulfolobus* 23S + 5S RNA and total large subunit proteins (TP 50) have been found to recombine into a low-sedimenting (42S), protein-deficient particle at low temperature (0–20°C) and in the absence of spermine [103]. The 42S particle, which contains about one half of the subunit proteins, appears to represent a genuine reconstitution intermediate as it is converted into an active 50S subunit following further incubation at 75–80°C in the presence of spermine [103]. Therefore, the correct initiation of *Sulfolobus* ribosome assembly is independent of both heat and spermine. The identification and the isolation of the "low-temperature" precursor particles has provided insight into the order in which the various ribosomal proteins are sequentially assembled in the nascent particle.

In fact, data by Altamura et al. [103] show that the proteins constituting the 42S particle (Fig. 3) occupy an internal position within the mature subunit and comprise all of the primary RNA-binding proteins, namely those proteins (about 9) that are capable of strongly and independently interacting with the cognate 23S RNA (Fig. 3). Moreover, the proteins of the 42S particle are capable of specifically interacting with 23S RNA from evolutionarily distant organisms (section 5.3, below). Therefore, they are likely to be highly conserved molecules suitable, in principle, for inferring deep genealogical relationships among organisms.

2.7.2. Reconstitution of Haloferax 50S subunits

Total reconstitution of halophile ribosomes [100] has been accomplished from rRNA and proteins of *Haloferax mediterranei*, a solar saltern species thriving in 20% NaCl and maintaining an internal KCl concentration of about 2 M. Reconstitution of *H. mediterranei* 50S particles (which contain about 30 proteins) takes place at 40°C in the presence of high concentrations of monovalent (3 M KCl) and divalent (60 mM Mg^{2+}) cations (Table 5). Unlike *Sulfolobus* and *E. coli,* a single incubation step is necessary and sufficient for the formation of a complete and synthetically active particle. Notably, no raise in temperature above physiological values (40°C) nor polyamines are required. Therefore, a complex procedure involving at least one incubation step at relatively high temperatures (> 50°C) is not, as was previously postulated [102], a prerequisite for the in vitro reconstitution of very large macromolecular assemblies.

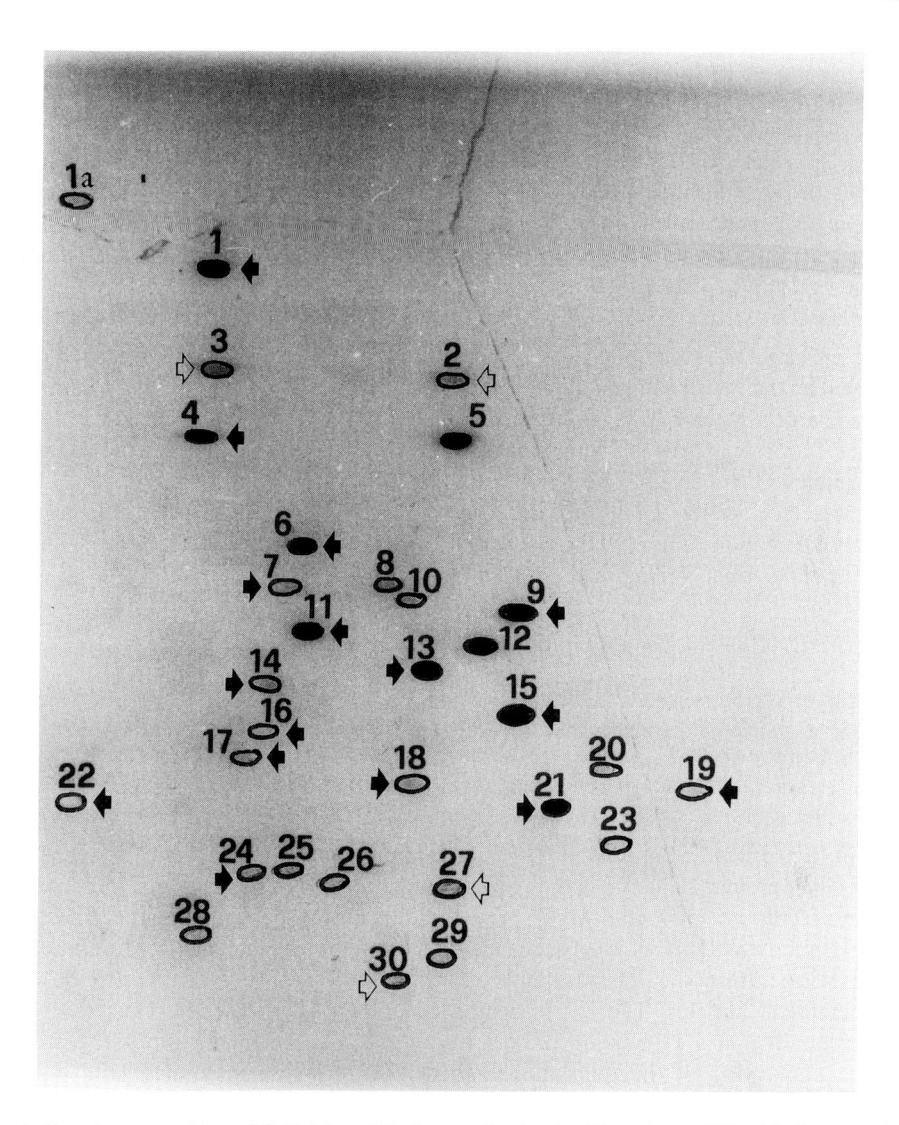

Fig. 3. Protein composition of *Sulfolobus* 50S ribosomal subunits. The primary RNA-binding proteins are evidenced as black spots. The solid arrows indicate proteins found in the "low-temperature reconstitution intermediate" *and* in the 42S hybrid particles obtained from the reaction between *Sulfolobus* TP 50 and either *E. coli* or *H. mediterranei* 23S RNA. The open arrows indicate proteins found in the 42S hybrid particles but not in the low-temperature reconstitution intermediate.

The in vitro assembly of *Haloferax* 50S subunits appears to be primarily governed by the concentration of monovalent cations. If the reconstitution assay is carried out in the presence of KCl, which is the predominant salt within the living cells, a concentration of at least 2.5 M is required for formation of fully active particles. It is somewhat intriguing, however, that the "physiological" salt (KCl) can be entirely replaced by ammonium sulphate, even though NH_4^+ ions are not especially abundant within the living halobacterial cells. Even more intriguing, and as yet unexplained, is the fact that assembly does not take place in the presence of ammonium chloride, regardless of concentration [100].

An important aspect of the *Haloferax* 50S subunit reconstitution is the fact that near-to-saturating salt concentrations are neither essential to attain a functional conformation of the individual ribosomal components, nor are they needed for a correct RNA/protein recognition. Rather, high salt appears to be required for adjusting the ribosome into an efficient working shape. This is inferred from the fact that particles having significant activity are formed at salt concentrations (0.5–1 M KCl) considerably lower than that (3 M KCl) required for full functional reconstitution. The particles reconstructed in the presence of suboptimal salt concentrations exist in a partially unfolded state, as indicated by their low sedimentation coefficients (30–40S); however, they can refold into a fully active 50S subunit upon raising salt concentration to the optimum value of 3 M [100].

Thus, the role played by high salt during the assembly of halophile ribosomes is comparable to that played by heat and spermine during the reconstitution of thermophilic ribosomes (see section 2.7.1); in fact, none of the above factors directly influences the primary interactions amongst the ribosomal components, but all of them participate in assisting the correct folding of the nascent particles.

2.8. Protein targeting and signal recognition in archaea

In eucarya, secretory and membrane proteins are targeted to their respective locations through a transport pathway that involves the recognition, by a signal recognition particle (SRP), of a hydrophobic sequence, encoded in the N-terminus of the nascent protein [104]. The eucaryal SRP consists of six polypeptide chains closely associated with a 7S RNA molecule. As soon as the nascent protein emerges from the exit domain of the large ribosomal subunit, the hydrophobic signal sequence is recognized by the SRP; protein synthesis is then arrested until the SRP–ribosome complex interacts with a specific receptor or docking protein (DP) located on the cytosolic side of the endoplasmic reticulum. As a consequence of this interaction, translation is resumed, the SRP is released from the ribosome, and cotranslational transport of the polypeptide across the membrane initiates. The DP is composed of two subunits [105]. One, designated DPα subunit, is a GTP-binding protein responsible for the interaction with SRP and for the unblocking of the translational arrest [106]. The other, termed DPβ subunit, has no known function [105]. In bacteria, the targeting of secretory proteins occurs by similar mechanisms except that the bacterial SRP differs from eucaryal SRP in having a much smaller (4.5S) RNA moiety [107]. In addition, bacteria (*E. coli*) contain a gene (*ftsY*) encoding a GTP-binding protein that shares significant sequence similarity with the eucaryal DPα subunit [108].

There are hints that a protein targeting system resembling that of eucarya may exist in archaea. Archaeal genes for integral membrane proteins and for secreted proteins have been found to encode N-terminal signal sequences similar to those found in secreted eucaryal and bacterial proteins [32,109–111]. Importantly, however, archaea resemble eucarya, and differ from bacteria, in having a 7S RNA molecule. Genes encoding 7S RNA have been isolated from such phylogenetically disparate archaea as *H. halobium* [112], *M. voltae* [113], *Pyrodictium occultum* [114], *A. fulgidus, Methanosarcina acetivorans,*

S. solfataricus, and *T. celer* [115]. The archaeal 7S RNA, however, has not yet been identified as a component of a ribonucleoprotein complex comparable to eucaryal SRP, and its functional role in the cell remains uncertain. In spite of absence of identification of a SRP-like particle, *S. solfataricus* has been found to contain a gene coding for a GTP-binding protein that shares significant sequence similarity with both the eucaryal DPα subunit and the product of the *E. coli ftsY* gene [116].

Interestingly, the 7S RNA region whose sequence is conserved between archaea and eucarya is also specifically related to regions of the bacterial 4.5S RNA [107,113]. Conservation of this primary structural element across domain boundaries suggests that archaeal and eucaryal 7S RNAs, and bacterial 4.5S RNAs, are the descendants of a unique subcellular structure that existed before the earliest radiation of the lineages.

3. In vitro translation systems

Information on archaeal translation is essentially based on poly(U)- and poly(UG)-programmed cell-free systems and on peptidyltransferase assay systems. Poly(U)-directed systems have been used to monitor the reconstruction of archaeal ribosomal subunits and the susceptibility of archaea to protein synthesis inhibitors.

A Poly(U)-programmed cell-free system from *H. cutirubrum* requiring saturating salt concentrations was developed by Bayley and Griffiths [117] ten years before extreme halophiles were assigned to the newly discovered archaea. Poly(U)- and poly(UG)-directed systems using purified ribosomes and post-ribosomal supernatant fractions have been subsequently developed from most known archaea.

The salient features of in vitro translation systems from halophiles, methanogens, and from sulfur-dependent thermophiles belonging to both archaeal kingdoms are summarized below.

3.1. Translation systems from halophilic archaea

All halophile poly(U)-programmed cell-free systems are descendants from the original *H. cutirubrum* system of Bayley and Griffiths [117]. Starting from that system, a progressive improvement of the synthetic capacity has been attained over the past decade through a fine tuning of the relative K^+, Na^+, NH_4Cl and $(NH_4)_2SO_4$ concentrations while keeping constant the Mg^{2+} levels (50–72 mM). The original (*H. cutirubrum*) system, which was poorly active, contained $>3.8\,M\ K^+$, $1.0\,M\ Na^+$ and $0.4\,M\ NH_4Cl$, the Na^+ and NH_4^+ concentrations being much higher than the physiological ones. The Bayley–Griffiths system was subsequently modified by Kessel and Klink [118] (*H. cutirubrum*) through removal of Na^+ ions and through the addition of moderate concentrations (0.085–0.35 M) of $(NH_4)_2SO_4$ while keeping the NH_4Cl concentration at 0.4 M. However, the Kessel–Klink system was no more active than the initial system of Bayley and Griffiths. More substantial changes were implemented by Saruyama and Nierhaus [119] (*H. halobium*) who decreased the concentration of K^+ ions (from 3.8 M to 2.0 M) while considerably increasing that of $(NH_4)_2SO_4$ (from 0.35 M to 2 M); a further improvement consisted in the addition of yeast $tRNA^{Phe}$. More recently, Sanz et al. [120] developed a halophile poly(U)-directed system which is virtually independent of K^+ ions and is generally applicable to all of the six halophile genera known today (namely, *Halobacterium*, *Haloferax*, *Haloarcula*, *Halococcus*, *Natronobacterium* and *Natronococcus*). In the presence of added *Saccharomyces cerevisiae* or *S. solfataricus* tRNA the system exhibits an activity optimum with as low as 1.0 M K^+ (higher concentrations being inhibitory), with a high concentration (1.5 M) of $(NH_4)_2SO_4$ and with a moderate concentration (0.4 M) of NH_4Cl; in these conditions the synthetic capacity of the halophile ribosomes is not strictly dependent on a specific Mg^{2+} concentration but is active over a broad range (20–120 mM). An intriguing aspect of the halophile system is the fact that the ionic conditions which allow for maximal in vitro activity differ more or less extensively from those existing in the intracellular environment of the different halophiles, in which K^+ is by far the predominant cationic species (1.8 M for *H. mediterranei*, 2.6 M for *Natronobacter faraonis* and 3.8 M for *H. halobium*) while NH_4^+ is always present in extremely low concentrations (10–30 mM) [120]. Evidently, the translation machinery of the halophilic archaea is not regulated by the compatible solute (K^+) and is largely insensitive to its variations. However, the striking dependence of the halophile in vitro systems on a concentration of NH_4^+ ions tenfold higher than the physiological one remains unexplained.

3.2. Translation systems from methanogenic archaea

Cell-free systems for methanogen ribosomes are based on the poly(U)- and poly(UG)-programmed system originally developed by Elhardt and Böck [121] for *M. vannielii*. Aside from the tedious procedures involved in carrying out the initial cell fractionations under anaerobic conditions, the methanogen systems do not differ substantially from the classical *E. coli* cell-free system in their ionic requirements (10 mM Mg^{2+}, 10 mM K^+ and 100 mM NH_4^+); polyamines (spermidine or spermine) are not absolutely required although they improve synthetic capacity. The *M. vannielii* protocol has been successfully used with Methanobacteriales (*M. thermoautotrophicum*, *M. formicicum*, *Methanobacterium bryantii*), Methanosarcinaceae (*M. barkeri*) and Methanospirillaceae (*Methanospirillum hungatei*) [73,121,122].

3.3. Translation systems from sulfur-dependent archaea

Poly(U)-programmed systems from the sulfur-dependent thermophilic members of the two archaeal kingdoms are descendants from a parent system from *Caldariella acidophila* (now *S. solfataricus*, probably identical to DSM 1616 strain) reconstituted from purified ribosomes and from an $(NH_4)_2SO_4$-fractionated post-ribosomal cell extract [123]. Perhaps, the most conspicuous feature of the *Sulfolobus* system is its dependence upon spermine. At relatively high (18–20 mM) concentration of Mg^{2+} ions the synthetic capacity of *Sulfolobus* ribosomes is in fact absolutely dependent upon the presence of a relatively high (2–4 mM) concentration of spermine. Only the spermine-related polyamine thermine (1, 11-diamino-4, 8-diazaundecane), a physiological component of *Sulfolobus* cells, can effectively substitute for spermine (1, 12-diamino-4, 9-diazadodecane). The spermine effect cannot be duplicated by increased concentrations of Mg^{2+} ions, and other polyamines are much less effective (spermidine) [123] or even inhibitory (putrescine) [124]. Furthermore, the capacity of *Sulfolobus* ribosomes to perform poly(phe) synthesis is strongly counteracted by NH_4^+ ions and is also inhibited by concentrations of TRIS buffer in excess of 25 mM.

Under the optimized conditions a stable plateau level of maximum activity is attained between 75°C and 80°C; below this optimum the rates of peptide bond formation display a stringent Arrhenius dependence upon temperature (20–75°C range) with an Arrhenius activation energy of $20\,\text{kcal}\,\text{mol}^{-1}$; an identical value has been inferred for E. coli ribosomes over the 10°–48°C temperature range (Fig. 4).

At 75–80°C the synthetic capacity of Sulfolobus ribosomes is critically dependent upon the stage of cell growth. Preparations from cells harvested in the mid-log phase of growth statistically polymerize about 40 phenylalanine residues per ribosome in 30 min [66] compared to only 5–7 residues found in cells harvested in the late phase of exponential growth [123].

Under optimal conditions for poly(phe) synthesis Sulfolobus ribosomes are highly accurate, the incorporation of a non-cognate amino acid (leucine) being approximately 0.3% that of the cognate one; also, the error frequency in tRNA selection does not increase at temperatures (55°C) well below the temperature optimum of the system [66,125]. A less refined spermine- (thermine)-dependent Sulfolobus cell-free system has been described by Friedman [126] who used an unfractionated cell extract (S30).

Starting from the composition of the Sulfolobus system, *reconstituted* poly(U)-programmed systems have been developed from other members of the crenarchaeota (T. tenax, D. mobilis) [66] Acidianus infernus (DSM 3191), Acidianus brierleyi (formerly called Sulfolobus brierleyi), Methallosphaera sedula (Sulfolobaceae) [Sanz, J.L., unpublished), and from the euryarchaeota T. acidophilum [66,127], T. celer [66] and P. woesei [128].

Strikingly, the ionic and polyamine requirements of the two euryarchaeal systems (T. acidophilum and T. celer) deviate drastically from that for the Sulfolobus, D. mobilis and T. tenax systems in being independent of (T. acidophilum), or inhibited by (T. celer), spermine while being absolutely dependent upon a high concentration (80–100 mM) of NH_4^+ ions. Aside from that, all systems display maximal synthetic rates at temperatures equal, or close, to the physiological optimum for cell growth (58°C for T. acidophilum and 80°C for all others) and all are highly accurate in codon reading, although somewhat less efficient synthetically than the Sulfolobus system; the synthetic capacity (Phe residues polymerized per ribosome in 40 min) being 4 for T. tenax, 10 for T. celer and 20–25 for D. mobilis and T. acidophilum [66].

3.4. Peptidyltransferase assay systems

The peptidyltransferase activity of ribosomes can be segregated from other translation reactions by the so-called puromycin reaction [129] which monitors the formation of [acetyl-aminoacyl]-puromycin or [peptidyl]-puromycin from puromycin and either [acetyl-aminoacyl]-tRNA or [peptidyl]-tRNA. In its simplest form (termed "uncoupled" or 30S-subunit-independent peptidyltransferase) the reaction requires large ribosomal subunits and an organic solvent (ethanol or methanol) which is presumably needed to promote the binding of tRNA and puromycin to the 50S subunit. In the absence of organic solvents, however, the reaction (then termed coupled or 30S-subunit-dependent

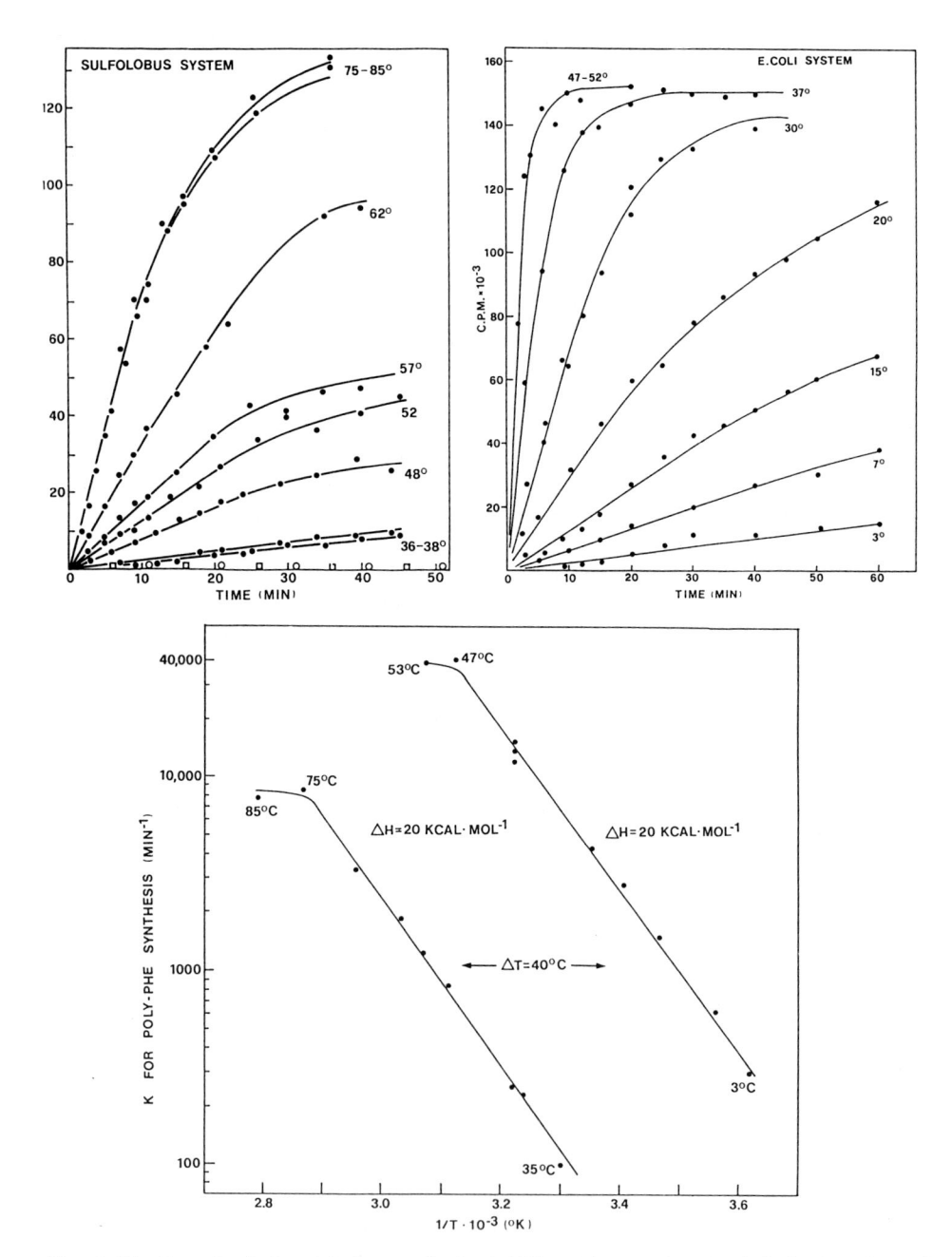

Fig. 4. Kinetics of polyphenylalanine synthesis at different temperatures, and Arrhenius plots of poly(phe) synthesis in *S. solfataricus* and *E. coli* poly-U programmed cell-free systems. Rates of polyphenylalanine synthesis were calculated from the initial slopes of the incorporation kinetics at different temperatures [Cammarano, P., unpublished results].

peptidyltransferase) is obligatorily dependent on the presence of small ribosomal subunits and mRNA. Both coupled and uncoupled peptidyltransferase assay systems have been developed from *S. solfataricus, D. mobilis, T. tenax, T. celer* and *T. acidophilum* [66].

Perhaps the most interesting conclusion from the peptidyltransferase assays is the evidence that ribosomes from extremely thermophilic archaea display significant peptidyltransferase activity (both uncoupled and coupled) at temperatures (37°C) well below that required for poly(U)-directed polypeptide synthesis. Therefore, in the thermophile systems high temperature appears to activate reactions other than peptide bond formation.

Peptidyltransferase assays have also provided insight into the mechanisms whereby spermine promotes, and NH_4^+ ions inhibit, polypeptide synthesis on the ribosomes of *S. solfataricus, T. tenax* and *D. mobilis* (see section 3.3). First, the 30S uncoupled peptidyltransferase activity is absolutely dependent on spermine while being totally unaffected by monovalent cations. Secondly, monovalent cations strongly inhibit the 30S subunit coupled reaction [66]. Thus, polyamines appear to be obligatorily required to convert the catalytic center of the spermine-dependent ribosomes into an active conformation, whereas monovalent cations inhibit polypeptide synthesis by preventing 30S subunits from interacting with the cognate 50S particles (ref. [66], see below).

Two considerations argue against the idea [124,126] that spermine activates polypeptide synthesis in the thermophile (*Sulfolobus*) systems by protecting ribosomes against thermal inactivation. First, spermine is absolutely required for the functioning of the sperminedependent 50S subunits at low temperature (37°C). Secondly, polyamines are not required for (and in fact inhibit) the peptidyltransferase activity of 50S subunits from the hyperthermophilic archaeon *T. celer* [66].

3.5. Distinctness of euryarchaeal and crenarchaeal translation systems

The ionic and polyamine requirements of the archaeal cell-free systems, and the tightness of the ribosomal subunit interactions in different archaea, are summarized in Table 6. The interesting result is the evidence that according to requirements for polypeptide synthesis and to stability of the ribosomal subunit association, the archaea fall into two rather sharply defined classes, corresponding to the kingdoms of Euryarchaeota and Crenarchaeota. Clearly, the two classes exhibit opposite behaviours in their requirements for spermine and for NH_4^+ ions and in the tightness of the association of their subunits. Namely, the Crenarchaeota (*S. solfataricus, T. tenax, D. mobilis*) harbor 70S ribosomes composed of weakly associated subunits, whose synthetic capacity is absolutely dependent upon relatively high concentrations of spermine while being drastically inhibited by NH_4^+ ions. Conversely, the Euryarchaeota (*T. celer, T. acidophilum*, halophiles and methanogens) contain 70S particles composed of tightly bonded subunits, whose synthetic capacity is independent of spermine while being totally dependent on a relatively high concentration (80–100 mM) of NH_4^+ ions. Clearly, the dependence upon spermine is not related to thermophily. The opposite behaviour of the Thermococcales (thriving at 97–105°C) and the Sulfolobales–Thermoproteales (optimum growth temperature 87–95°C) in their dependence on spermine (or thermine) can only be interpreted as indicating a phylogenetic, rather than adaptive, distinction of the 50S subunits from the two archaeal groupings. The differences between the two groups are in fact so sharp that one might anticipate the phylogenetic affiliation of new archaea

TABLE 6

Optimal conditions for polyphenylalanine synthesis and tightness of subunit interaction in archaea[a]

	Crenarchaeota				Euryarchaeota						
	Sso	Mse	Dmo	Tte	Tce	Pwo	Tac	H	Mva	Mfo	Mth
Tris-HCl pH 7.4 (mM)	16	16	20	20	20	20	–	30	20	20	20
Hepes pH 6.5 (mM)	–	–	–	–	–	–	20	–	–	–	–
Mg acetate (mM)	18	18	18	10	18	10	15	30	20	20	20
KCl (mM)	–	–	–	–	–	–	–	1000	40	40	25
NH$_4$Cl (mM)	6	25	10	10	120	100	80	400	100	100	100
(NH$_4$)$_2$SO$_4$ (mM)	–	–	–	–	–	–	–	1500	–	–	–
Spermine (mM)	3	3	3	1	–	0.5	1	–	1	1	1
PEG 6000 (mM)	–	–	–	–	–	–	–	–	50	50	50
ATP (mM)	2.5	2.5	2.5	2.5	2.5	3	1	2	1	1	1
GTP (mM)	1.5	1.5	1.5	1.5	1.5	1	1	0.5	0.5	0.5	0.5
Phoshoenolpyruvate (mM)	–	–	–	–	–	–	–	5	6	6	6
Pyruvate kinase (μg/ml)	–	–	–	–	–	–	–	50	50	50	50
Poly(U) (μg/ml)	160	160	160	160	160	160	200	800	480	480	480
E. coli tRNA (μg/ml)	–	–	–	–	–	–	–	–	–	1200	–
Yeast tRNA (μg/ml)	–	–	–	–	–	–	–	1000	–	–	–
Sulfolobus tRNA (μg/ml)	80	80	80	80	80	80	–	–	–	–	–
M. vannielii tRNA (μg/ml)	–	–	–	–	–	–	–	–	240	–	240
Source of S-100[b]	Sso	Sso	Dmo	Tte	Tce	Pwo	Tac	H	Mva	Mfo	Mva
T (°C)	75	65	75	75	75	80	58	40	37	37	37
Time (min)	30	30	30	30	30	30	30	60	45	90	75
Aggregation state of ribosomes	30+50	n.d.	30+50	30+50	70S	70S	70S	70S	70S	70S	70S

[a] Data are from the following sources: *S. solfataricus*: [66,23]; *D. mobilis, T. celer, T. tenax* and *T. acidophilum*: [66]; *P. woesei*: [128]; *Halobacterium*: [120]; *Methallosphaera sedula*: Amils, R. and Sanz, J.L., unpublished.

[b] Abbreviations used: Dmo, *D. mobilis*; H, halobacteria; Mfo, *M. formicicum*; Mse, *Metallosphaera sedula*; Mva, *M. vannielii*; Mth, *Methanobacterium thermoautotrophicum*; Pwo, *P. woesei*; Sso, *S. solfataricus*; Tac, *T. acidophilum*; Tce, *T. celer*; Tte, *T. tenax*.

to either one or the other of the two archaeal kingdoms by testing the requirements of their translation systems.

4. Sensitivity of archaea to protein synthesis inhibitors

Bacteria and eucarya are differentially affected by antibiotics that impair ribosome function (and/or fidelity) by directly interacting with, or by perturbing, functionally essential domains of ribosomes and factors. On the basis of their specificity, ribosome-directed and elongation factor-directed antibiotics have been classically subdivided into (i) anti-70S-directed inhibitors (*Group I*) which selectively affect (eu)bacterial translation, (ii) anti-80S-directed inhibitors (*Group II*) having a stringent specificity for eucaryal (cytosolic) translation systems, and (iii) compounds having a dual (anti-70S *plus* anti-80S) action (*Group III*). For a comprehensive survey of the inhibitory actions of *Group I–III* compounds known until 1979 see ref. [130]. The potential of ribosome-targeted drugs as phylogenetic probes is exemplified by their ability to discriminate mitochondrial and chloroplast ribosomes (which are susceptible to a variety of *Group I* inhibitors and refractory to *Group II* drugs) from their cytosolic–80S counterparts [131].

Following the discovery that methanogens, halophiles and sulfur-dependent extreme thermophiles constitute a third domain of life, the sensitivity of archaeal ribosomes and elongation factors to a host of *Group I–III* inhibitors was systematically surveyed. Although the principal aim of the initial studies was to probe the uniqueness of archaea relative to bacteria and eucarya, the wealth of antibiotic sensitivity spectra accumulated over the past decade has far-reaching implications that extend beyond the initial goal; in fact, they highlight the great phylogenetic depth and the degree of evolutionary stability of the functionally essential (tertiary and/or primary structural) traits that act as the drug binding sites within each of the three domains.

4.1. Ribosome-targeted inhibitors: in vivo assays

Hilpert et al. [132] and Pecher and Böck [133] were the first to show that archaea display a unique – although not uniform – pattern of sensitivity to a host of *Group I–III* inhibitors. Namely, Halobacteriales and methanogens of different orders (Methanobacteriales, Methanococcales, Methanomicrobiales) were found to be refractory to many classical anti-70S-directed drugs (e.g. streptomycin) while being sensitive to certain anti-80S-directed inhibitors (e.g. anisomycin). An additional unexpected outcome of the early studies was the evidence that, unlike the situation in bacteria and eucarya, archaea (notably methanogens of different orders) display a remarkable heterogeneity in their response to compounds classically considered to be selective inhibitors of bacterial or eucaryal translation, as well as non-selective inhibitors of both. The lack of susceptibility to many 70S-targeted drugs was further supported by the evidence that archaeal ribosomes lack binding sites for streptomycin and chloramphenicol [73].

In vivo response to ribosome-directed drugs, however, does not provide decisive information on the absence or presence of specific ribosomal binding sites. Lack of susceptibility may result from impermeability of the cell envelopes, from the absence of a membrane transport system, or from the ability of the organisms to inactivate the drug. Conversely, in vivo sensitivity may reflect secondary effects of the antibiotics, not related to ribosome function. For instance, chloramphenicol inhibits the growth of methanogenic and halophilic archaea although ribosomes from these organisms lack a chloramphenicol-binding site [73].

4.2. Ribosome-targeted inhibitors: in vitro assays

Reliable antibiotic sensitivity spectra have been obtained by monitoring the inhibition of poly(Phe) synthesis on poly(U)-programmed ribosomes from methanogens representative of the Methanobacteriales, Methanococcales [121] Methanosarcinaceae [121,134] and Methanospirillaceae [134]; from sulfur-dependent thermophiles belonging to both the Crenarchaeota (*S. solfataricus* [135], *D. mobilis, T. tenax* [136]) and the Euryarchaeota

(*T. celer* [136], *P. woesei* [128], *T. acidophilum* [137]), and from several halophiles (*Halobacterium, Halococcus, Natronococcus*) [138]. When aminoglycoside antibiotics (which affect both synthetic capacity and translational fidelity) were assayed, the misincorporation of non-cognate amino acids (isoleucine, leucine, serine) was also measured. Cell-free systems from thermophilic bacteria (*Bacillus stearothermophilus, Thermus thermophilus*), and from halophilic bacteria (*Vibrio costicola*) were used as the control systems. Fortunately, all of the *Group I–III* inhibitors assayed, including the peptide (thiostrepton) and polypeptide inhibitors (e.g. ricin, α-sarcin, mitogillin, restrictocin), retain full activity after prolonged incubation at temperatures (75°C) close to the physiological optimum for cell growth and for maximum in vitro activity of the hyperthermophile species [135].

The response of archaeal ribosomes to compounds that affect reactions of the elongation cycle and/or promote misreading is summarized in Table 7 (elongation inhibitors) and Table 8 (misreading inducers belonging to the streptomycin group, and the mono- and disubstituted 2-deoxystreptamine compounds). The cut-off point to discriminate true inhibition from spurious effects of the drug has been set at 10^{-3} M. The in vitro assays allow the following conclusions:

(i) Archaea, as a whole, have unique susceptibility patterns both to compounds classically considered to be selective inhibitors of (eu)bacterial (*Group I*) or eukaryotic (*Group II*) ribosomes, as well as to non-selective inhibitors of both (*Group III*). They display unique mosaics of eucaryal and bacterial attributes, although on the whole they are sensitive to few (only about one-half) of the compounds belonging to Groups I and II. For instance, amongst the Group I compounds archaeal ribosomes are characteristically refractory to the classical bacterial inhibitors streptomycin and chloramphenicol while being sensitive (although not always) to virginiamycin and thiostrepton; amongst the Group II inhibitors they are refractory to the classical 80S-directed drug cycloheximide while being susceptible (although not uniformly) to other 80S inhibitors, such as anisomycin.

(ii) Unlike the situation in bacteria and eucarya, archaea exhibit a dramatic intralineage diversity in their response to ribosome-directed drugs, regardless of mode of action of the drugs, or selectivity to bacterial or eucaryal translation. A trend can be recognized only in that sulfur-dependent thermophiles belonging to both the Crenarchaeota and the Euryarchaeota (notably *Sulfolobus* and *Thermoplasma*) are refractory to most ribosome-targeted inhibitors, while methanogens (notably *Methanobacterium* and *Methanospirillum*) are susceptible to about one-half of all *Group I–III* inhibitors tested. Halophiles exhibit an intermediate situation; however, results obtained using halophile cell-free systems should be judiciously considered because control systems from the halophilic bacterium *Vibrio costicola* suggest that insensitivity of halobacterial ribosomes to certain drugs (notably aminoglycosides) may result from competition between antibiotics and monovalent cations [138]. Aside from that, diversified antibiotic-sensitivity spectra are also found between organisms that are closely related phylogenetically, such as different Thermoproteales (*Thermoproteus* and *Desulfurococcus*), methanogens of different orders (e.g. *Methanococcus* and *Methanobacterium*), halophiles belonging to different cell lines, and between the methanogens and the closely related organism *T. acidophilum* [2].

TABLE 7
Sensitivity of archaea to ribosome-targeted antibiotics[a]

Antibiotic	Tp[b]	Grp[c]	SS	TA	DM	TC	TT	MF	MV	HM	EC	SC
Althiomycin	I	B	−	−	−	−	−	−	−	−	+	−
Amicetin	III	B	−	−	−	−	±	−	−	−	+	−
Anisomycin	II	B	−	−	−	−	−	+	−	+	−	+
Anthelmycin	III	B	±	±	−	−	±	+	−	±	+	+
Blasticidin-S	III	B	−	−	−	−	−	++	±	+	++	++
Bruceantine	II	B	−	−	−	−	−	+	−	−	−	++
Carbomycin A	I	B	−	−	−	−	−	+	+	+	+	−
Cycloheximide	II	C	−	−	−	−	−	−	−	−	−	+
Edeine A1	III		+	+	−	+	++	+	±	−	−	++
Gentamycin	III	A	−	±	±	±	±	++	±	−	+	−
Griseoviridin	I	B	−	−	−	−	−	±	±	−	++	±
Haemantamine	II	B	−	−	−	−	−	−	−	−	±	+
Harringtonine	II	A+B	−	−	−	−	−	−	−	−	−	+
Hygromycin B	III	A+C	−	±	−	−	±	+	−	−	++	+
Kanamycin	III	A	−	±	±	−	±	−	−	−	+	±
Mitogillin	III	A	±	+	+	±	±	−	−	−	+	+
Narciclasine	II	B	−	−	±	−	±	++	−	−	−	++
Neomycin	III	A	+	+	+	+	+	++	+	−	+	+
Paromomycin	III	A	±	±	±	±	+	−	−	−	+	+
Puromycin	III	B	±	±	±	±	+	±	+	±	+	+
Restrictocine	II	A	−	+	+	±	±	±	±	−	+	+
Ribostamycin	I	A	−	−	−	−	±	+	−	−	+	−
α-Sarcin	III	A	±	+	+	−	±	+	+	−	+	++
Sparsomycin	III	B	+	±	−	−	±	+	+	++	+	+
Streptimidone	II	C	−	−	−	−	−	−	−	−	−	+
Streptomycin	I	A	−	−	−	−	−	−	−	−	+	−
Strepeovitacin A	II	C	−	−	−	−	−	−	−	−	−	+
Tetracycline	III	A	−	±	−	±	−	+	−	−	+	+
Thiostrepton	I	A	−	±	+	++	−	+	++	+	++	−
Tobramycin	III	A	−	±	±	+	+	±	−	−	+	+
Toxin T-2	II	B	−	−	−	−	−	−	−	−	−	++
Tubulosine	II	C	−	−	−	−	−	−	−	−	−	+
Tylophorine	II	C	−	−	−	−	−	−	−	−	−	+
Tylosine	I	B	−	−	−	−	−	+	±	±	++	−
Viomycin	I	B	±	±	±	±	−	±	±	−	++	±
Virginiamycin M	I	B	±	±	±	±	−	±	±	−	++	±

[a] Abbreviations used: SS, *Sulfolobus solfataricus*; TA, *Thermoplasma acidophilum*; DM, *Desulfurococcus mobilis*; TC, *Thermococcus celer*; TT, *Thermoproteus tenax*; MV, *Methanococcus vannielii*; MF, *Methanobacterium formicicum*; HF, *Haloferax mediterranei*.
Inhibition values are rated from ++ to − according to the algorithm described in ref. [171], as follows: −, 0–200; ±, 200–500; +, 500–1000; ++ 1500–2000.
[b] Tp, Type: antibiotic domain specificity: I, eubacteria-specific inhibitors; II, eucarya-specific inhibitors; III, non-specific inhibitors.
[c] Grp, Group: antibiotic group specificity: A, inhibitors of aminoacyl–tRNA binding; B, inhibitors of peptide bond formation; C, inhibitors of translocation.

TABLE 8

Effect of aminoglycoside antibiotics on the fidelity and efficiency of polyphenylalanine synthesis in archaea, bacteria and eucarya [125]

	Strepto-mycin	4-5 disubstituted 2-deoxystreptamine		4-6 disubstituted 2-deoxy-streptamine[a]	Monosubstituted 2-deoxy-streptamine (Hygromycin)
		Neomycin	Paromomycin		
Archaea					
M. vannielii	−	+	+	−	−
M. barkeri	−	+	+	−	n.d.
M. hungatei	−	+	+	−	n.d.
T. acidophilum	−	+	+	+	±
T. celer	−	+	−	−	−
D. mobilis	−	−	−	−	−
S. solfataricus	−	−	+	−	∓
P. woesei	−	+	+	+	+
Bacteria[b]	+	+	+	+	±
Eucarya[c]	−	±	+	+	n.d.

[a] Kanamycin, gentamicin, tobramycin.
[b] *E. coli* and *Bacillus stearothermophilus*.
[c] Exemplified by *Tetrahymena thermophila*.

This situation is in sharp contrast with the constancy of the antibiotic sensitivity spectra within each of the two classically recognized domains [134]; the only known exception being the deep-branching bacterium *T. maritima* [2,24,57] whose ribosomes differ from those of the recent bacterial phyla in being totally refractory to the miscoding-inducing action of all known groups of aminoglycoside antibiotics [139].

Clearly, numerous homologous sites of action are shared by archaea and bacteria (e.g. thiostrepton), by eucarya and archaea (e.g. anisomycin), by eucarya and bacteria (e.g. tetracycline) and by all three (neomycin, edeine). Why these "shared" sites are phylogenetically unstable within archaea while being invariant within eucarya and in most known bacterial phyla (except the *Thermotoga* lineage) is unclear (see ref. [136]).

The sensitivity of the Methanobacteriales (a late offspring of the euryarchaeal division) to many *Group I–III* compounds, and the lack of sensitivity to the same drugs in the deeper offshoots of the same division (notably the *Thermococcus–Pyrococcus lineage*) indicates that (i) ribosomal binding sites for drugs affecting methanobacteria were present in the primitive archaeon and (ii) these were lost in a seemingly erratic manner during the separate evolution of individual archaeal lineages.

4.3. Structural correlates of sensitivity to ribosome-targeted drugs

In both bacteria and eucarya the susceptibility to certain ribosome-targeted inhibitors correlates with specific primary structural features of rRNA, or with features of the ribosomal domain that acts as the antibiotic binding sites. Firstly, resistance to specific ribosome-directed drugs is conferred by single-base changes within phylogenetically

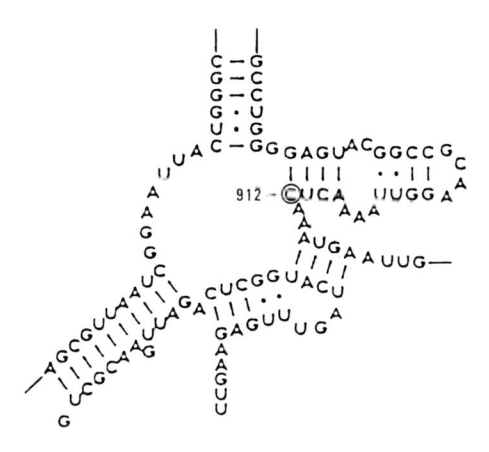

Fig. 5. Secondary structure of the *E. coli* 16S rRNA region involved in streptomycin sensitivity. The encircled C at position 912 is mutated to U in streptomycin-resistant chloroplast ribosomes and in archaeal ribosomes, with the exception of those of *Desulfurococcus mobilis*.

invariant regions of rRNA [140–144]. Secondly, the rRNA sites whose mutation confers resistance are juxtaposed to sites that are protected from chemical attack by the bound antibiotic [145].

A satisfactory correlation has been found to exist between sensitivity of archaeal ribosomes to certain antibiotics (streptomycin, erythromycin, α-sarcin) and possession of the specific structural motifs that are involved in antibiotic action.

Streptomycin (a *Group I* inhibitor) is ineffective in all of the archaeal poly(U)-directed cell-free systems studied until now. This correlates well with the evidence that the resistance to streptomycin of chloroplast (*Euglena gracilis*) ribosomes is conferred by a C to U base change at the invariant residue C 912 of 16S rRNA (*E. coli* numbering scheme) (Fig. 5) [144–146]. In accord with this evidence, archaea contain a U, instead of a C, at the 912-equivalent position of their S-rRNA, similarly to their eucaryal counterparts [147,148]. There is one exception however: *Desulfurococcus* ribosomes are insensitive to streptomycin [136] despite the fact that they resemble bacteria and organelles in possessing the "streptomycin-sensitive" C residue at position 912 of their 16S rRNA [149]. Evidently, the C 912 is a necessary although not sufficient determinant of streptomycin action.

Erythromycin (a macrolide *Group I* inhibitor ineffective in the archaea) impairs the functioning of bacterial 50S subunits by interacting with the central loop of domain IV of the 23S rRNA (Fig. 6). In bacteria (*E. coli*), erythromycin resistance is conferred by an A to U base change, or by the dimethylation of the A residue [142]. In erythromycin-resistant yeast (*S. cerevisiae*) mitochondria, the equivalent A residue is replaced by a G [141]. In accord with these observations, all of the archaeal 23S rRNAs sequenced until now resemble the large subunit rRNA of eucarya and of the erythromycin-resistant yeast mitochondrial ribosomes in having a G instead of the critical A residue [150]. Since the remaining structure of the peptidyltransferase loop is conserved in all of the

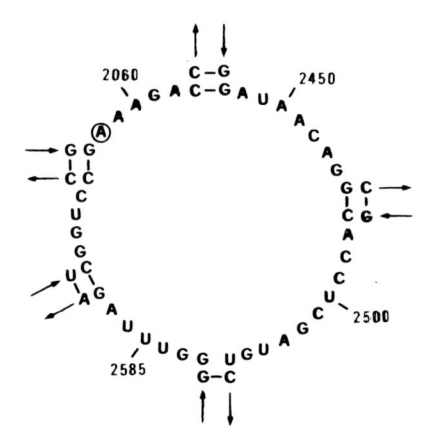

Fig. 6. The "central loop" of E. coli 23S RNA involved in erythromycin sensitivity. The encircled A residue is methylated in erythromycin-resistant ribosomes and is substituted by a G in all of the archaeal and eucaryal ribosomes.

large subunit rRNAs, the conclusion is almost inescapable that erythromycin insensitivity of archaeal ribosomes resides in this specific sequence feature.

There are several instances, however, in which the correlation between antibiotic sensitivity and possession of the prerequisite elements for antibiotic action is less straightforward. This is exemplified by thiostrepton, monosubstituted and disubstituted 2-deoxystreptamines and α-sarcin.

Thiostrepton (a *Group I* inhibitor which also affects archaeal ribosomes, albeit not systematically) binds with high affinity to the GTPase domain of bacterial ribosomes. The target domain comprises the 23S rRNA region (spanning nucleotides 1052–1110 of E. coli numbering) which interacts with ribosomal protein L11 (Fig. 7). Susceptibility to thiostrepton is primarily determined by a highly conserved adenylic residue at position 1067 [151]. First, the A 1067 is specifically methylated in the thiostrepton-producing bacterium *Streptomyces azureus* [151]. Secondly, substitution of A 1067 with either U or C by site-directed mutagenesis renders *E. coli* ribosomes resistant to the drug [152]. However, the L11 protein also appears to play a role in modulating thiostrepton sensitivity, as *Bacillus megatherium* mutants lacking L11 bind the antibiotic with low affinity [153].

The varying thiostrepton sensitivity of different archaea appears to be accounted for by the presence or the absence of an L11-like protein, rather than by primary structural features of the archaeal 23S rRNA [153]. Firstly, all of the sequenced archaeal 23S RNAs possess the prerequisite A residue at the 1067-equivalent position [154]. Secondly, hybrid complexes formed by 23S rRNA of (the thiostrepton-insensitive) *S. solfataricus* and L11 proteins of (the thiostrepton-sensitive) *E. coli* bind the drug with high affinity [153]. Thirdly, the thiostrepton-binding capacity of ribosomes from *M. formicicum*, which are sensitive to thiostrepton, but less so than either *E. coli* or *M. vannielii* ribosomes, can be significantly enhanced by the addition of *E. coli* protein L11; in contrast, no enhancement

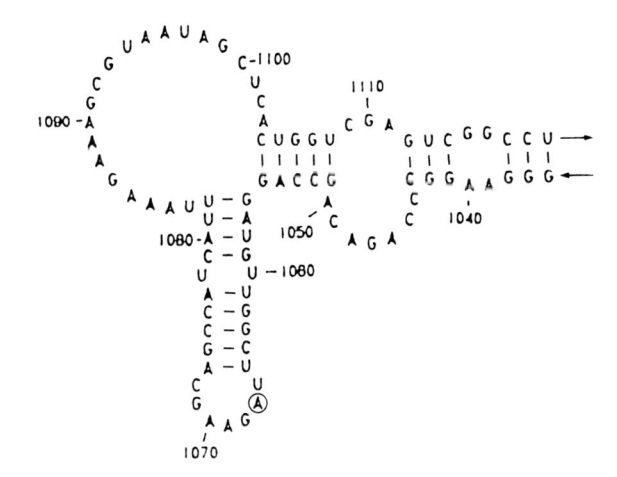

Fig. 7. The region of *E. coli* 23S RNA involved in thiostrepton sensitivity. The encircled A residue at position 1067 is methylated, or mutated, in thiostrepton-resistant ribosomes. The region spanning nucleotides 1052–1110 constitutes the binding site for protein L11.

of thiostrepton binding occurs when *Methanococcus* ribosomes (which are as sensitive as *E. coli* ribosomes to thiostrepton) are supplemented with *E. coli* L11 [153].

An additional possibility is that sensitivity to thiostrepton is influenced by the stability of the double-helical stem of the A1067-containing loop [155].

The monosubstituted (hygromycin) and the disubstituted 2-deoxystreptamine compounds (gentamicin, kanamycin, paromomycin, and neomycin) (*Group III* antibiotics) interact with the small subunit rRNA of both bacteria and eucarya (*S. cerevisiae, Tetrahymena thermophila*). The 16S/18S rRNA regions which confer sensitivity to both classes of aminoglycoside antibiotics span two nucleotide stretches (1405–1412 and 1490–1499 of *E. coli* numbering scheme) of the small subunit rRNA that are close in rRNA secondary structure [140,143,145] (Fig. 8). These sequences are invariant within the archaea, and no base changes comparable to those conferring resistance to both the monosubstituted and the disubstituted 2-deoxystreptamine aminoglycosides in eucarya (*S. cerevisiae, T. thermophila*) are found in archaeal 16S rRNAs (see ref. [136]). However, archaeal ribosomes, except those of Methanobacteriales, are insensitive to hygromycin and to most disubstituted 2-deoxystreptamines [136] despite the fact that they possess the primary structural elements that are required for antibiotic action.

α-sarcin, traditionally considered to be a Group II inhibitor, impairs eucaryal protein synthesis by cleaving a single phosphodiester bond within a looped-out region of the large subunit RNA [156] (Fig. 9). The α-sarcin cleavable site lies within a fourteen-nucleotide stretch whose sequence is identical in eucarya and archaea. Despite the identity of the α-sarcin cleavable site, however, archaeal ribosomes (methanogens and sulfur-dependent thermophiles) are two to three orders of magnitude less sensitive to α-sarcin than their eucaryal counterparts [135,136,138]. In fact, the same holds true in the case of bacterial ribosomes which are scarcely affected by α-sarcin although their target loop is also cleavable by the inhibitor [157].

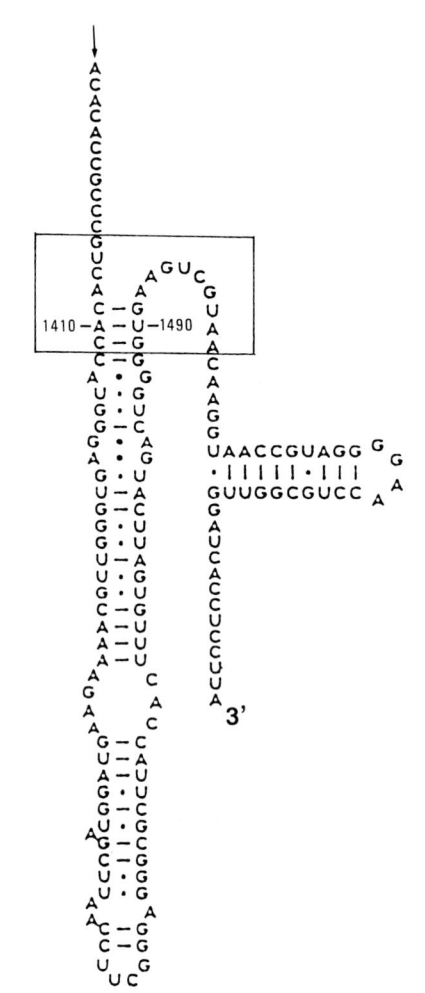

Fig. 8. The region of E. coli 16S rRNA involved in the response to monosubstituted and disubstituted 2-deoxystreptamine compounds. Mutations conferring resistance to antibiotics of these classes are included within the boxed region; most of them destroy the terminal G.C base pair of the long imperfect stem.

Presumably, the insensitivity to aminoglycosides, and weak susceptibility to α-sarcin, are accounted for by unique three-dimensional features of the interacting site.

```
         ↓       ⎛A G U A⎞
         G    2650 U       C
         C U G C U C C     G
         ¦ ·· ¦ ¦ ¦ ¦      A
         G G U G A G G     G
         A      ¦      ⎝C A G G A⎠
         ↓     2670
```

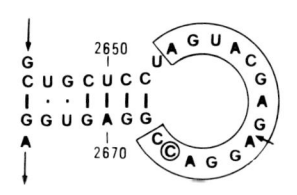

Fig. 9. The α-sarcin target loop of large ribosomal subunit RNA. The α-sarcin cleaving site is indicated by the arrow (right). The boxed 14-nucleotide stretch is universally conserved, except for the encircled residue which is a cytosine in all bacteria and an adenine in all archaea and eucarya.

4.4. Elongation factor-targeted inhibitors

Certain inhibitors – kirromycin, pulvomycin, fusidic acid and diphtheria toxin – which block protein synthesis by interacting with either elongation factor constitute potent probes to reveal eucaryal and/or bacterial traits on archaeal factors.

The structurally related antibiotics kirromycin and pulvomycin both act upon EF-Tu of most bacteria (and chloroplasts) although not upon its eucaryal (EF-1α) counterpart [158,159]. The steroid antibiotic fusidic acid interacts systematically with both the eucaryal (EF-2) and the bacterial (EF-G) translocating factors, including chloroplasts of higher plants. Diphtheria toxin (fragment A) discriminates between bacterial-mitochondrial and eucaryal translocating factors by selectively and irreversibly impairing the eucaryal EF-2 factors [160,161].

The four factor-targeted inhibitors act by different mechanisms. Pulvomycin inhibits the formation of a ternary [EF-Tu · GTP · aminoacyl–tRNA] complex [159]. Kirromycin prevents the [EF-Tu · GDP] complex from leaving the ribosomal A site [158]. Fusidic acid acts by stabilizing the [ribosome · GDP · EF-G(EF-2)] ternary complex [162]. Diphtheria toxin (fragment A) inactivates EF-2 by catalyzing the covalent bonding of the adenosine diphosphate-ribose moiety of NAD to a unique post-translationally modified histidine designated diphthamide [2-(3-carboxyamido-3-[trimethylammonio]propyl)histidine] [163], which is not found in bacteria [164]. Whereas kirromycin, pulvomycin and diphtheria toxin all interact with the susceptible factors in the absence of ribosomes, fusidic acid only binds to a ribosome-bound [EF-G(EF-2) · GDP] complex [162]; the specificity of the response to fusidic acid, however, appears to be determined by the factor rather than the ribosome itself [165]. Therefore, pulvomycin and kirromycin reveal structural homologies to bacterial EF-Tu. Fusidic acid reveals homologies to both EF-G (bacterial) and EF-2 (eucaryal) proteins. Diphtheria toxin reveals the presence of traits (the diphthamide residue which is ADP-ribosylatable and the histidine modifying enzymes) that are present in the eucaryal EF-2 but not in the bacterial EF-G.

The results of in vitro assays, summarized in Table 9, show that archaea exhibit a uniform response to two of the factor-targeted inhibitors (diphtheria toxin and kirromycin) while being heterogeneous in their response to the other two (fusidic acid and pulvomycin).

TABLE 9
Sensitivity of archaea to elongation factor-targeted drugs

Organisms	Kirromycin	Pulvomycin	Fusidic acid	Diphtheria toxin
Euryarchaeota				
Methanogens[b]	−	−	+	+
Halophiles[b]	−	−	+	+
T. acidophilum	−	−	−	+
T. celer	−	+	+	+
Crenarchaeota				
T. tenax	−	−	+	+
S. solfataricus	−	−	−	+
D. mobilis	−	−	−	+
E. coli	+	+	+	−
S. cerevisiae	−	−	+	+

[a] Data from refs. [128,165].
[b] Methanogens, M. formicicum, M. thermoautotrophicum, M. barkeri, M. vannielii. Halophiles, Halobacterium mediterranei, Halobacterium marismortui.

4.4.1. EF-2- and EF-G-targeted inhibitors

Archaea are systematically susceptible to ADP-ribosylation from NAD by the diphtheria toxin [118,166–168] although the rate of the ADP-ribosylation reaction is three orders of magnitude slower than that typically observed with eucaryal EF-2 [164,168]. Furthermore, in accordance with the sensitivity data, diphthamide occurs in archaea (H. halobium) but not in bacteria (E. coli) [164]. The archaeal EF-G-equivalent factor, therefore, resembles the eucaryal factor in having both the diphthamide that acts as the receptor for the ADP-ribose moiety of NAD, and the enzyme system which assists the post-translational conversion of histidine to diphthamide.

The EF-2 region comprising the histidine residue which is post-translationally converted to diphthamide, and renders the protein sensitive to diphtheria toxin, is the most highly conserved one between the two domains [34]. A consensus sequence which is likely to constitute the recognition site for both the diphtheria toxin and the histidine modifying enzymes has been drawn from aligned archaeal and eucaryal EF-2 sequences (Table 1, above). The consensus archaeal and eucaryal sequences (comprising 9 residues N-terminally, and 9 residues C-terminally around the histidine) are highly congruent but not identical: three positions of this region (2, 13 and 19 in Table 1) are invariant within each kingdom but differ between archaea and eucarya; these substitutions may account for the differing ADP ribosylation rates of the eucaryal and the archaeal EF-2 factors. Correspondingly, the histidine-modifying systems of archaea and eucarya appear to be specifically adapted to the respective consensus sequences since the archaeal (M. vannielii) EF-2 expressed in a recombinant strain of S. cerevisiae is consistently diphthamide free [42].

It should be stressed that the simplicity of the diphtheria toxin assay system provides an unequivocal and rapid diagnostic test for a preliminary assignment of new isolates.

The systematic susceptibility of the archaeal EF-2 to diphtheria toxin contrasts with the heterogeneity in the factor response to fusidic acid. In this latter case, the results of in vitro assays [165] indicate a sharp distinction between methanogenic–halophilic archaea (which are systematically susceptible to fusidic acid) and sulfur-dependent thermophiles (including *P. woesei* [128] and *T. acidophilum* [165]) which are insensitive to fusidic acid with the exception of *T. tenax*. Thus, there is only a vague relationship between the EF-2 sensitivity to fusidic acid and the affiliation of the organisms to the kingdom of Euryarchaeota (which includes *T. acidophilum*) and to the kingdom of Crenarchaeota (which includes *T. tenax*).

4.4.2. EF-1α- and EF-Tu-targeted inhibitors

In bacteria, insensitivity to kirromycin and pulvomycin is restricted to certain *Lactobacillus* strains [169]. Unlike bacteria, the archaea are systematically insensitive to kirromycin [128,165] but exhibit a distribution in their response to pulvomycin [165] (Table 9). To summarize, two features, insensitivity of EF-1α to kirromycin and sensitivity of EF-2 to ADP ribosylation by diphtheria toxin, are systematic of archaea. Two others (the sensitivity of EF-1α to pulvomycin and the sensitivity of EF-2 to fusidic acid) are unevenly distributed amongst the archaea in a manner that reflects only vaguely the phylogenetic placements of the organisms inferred from similarities in 16S rRNA sequences.

4.5. Phylogeny inference from antibiotic sensitivity spectra

The sensitivities of ribosomes and elongation factors to selected Group I–III inhibitors have been used by Amils and coworkers [170,171] to infer genealogical relationships between the archaea and the other two domains of life. The underlying assumption is that susceptibility to a given antibiotic reveals the presence or the absence of specific details of the (functionally essential) domain that acts as the antibiotic binding site.

A significant statistical segregation of three archaeal groupings (the sulfur-dependent thermophiles of all known orders and genera, the methanogens and the halophiles) was obtained initially by scoring the antibiotic sensitivities as "all or none" effects [170]. However, unlike phylogenies inferred from molecular sequences [2,24,33,172,173] all of the treeing methods used (cluster analysis, compatibility, parsimony) indicated the archaea as a being a paraphyletic, rather than a holophyletic grouping (bacteria originating *from within* the methanogen–halophile cluster).

In order to overcome the simplification implicit in scoring antibiotic action as an "all or none" effect, on a subsequent approach antibiotic sensitivities were quantified by using an algorithm designed to convert the dose–response relationships into dimensionless values, so as to incorporate the concentration dependence of antibiotic action [171]. Using these new values in phylogeny treeing resulted in the segregation of three monophyletic–holophyletic clusters (eucarya, archaea, bacteria), in accord with phylogenies inferred from similarities in rRNA [2] and protein sequences [2,24,33,172,173] (Fig.10); unlike phylogenies based on sequence data, however, all of the sulfur-dependent thermophiles (Sulfolobales, Thermoproteales, Thermococcales, Thermoplasmales) were treed together, rather than spanning both divisions of the archaeal subtree.

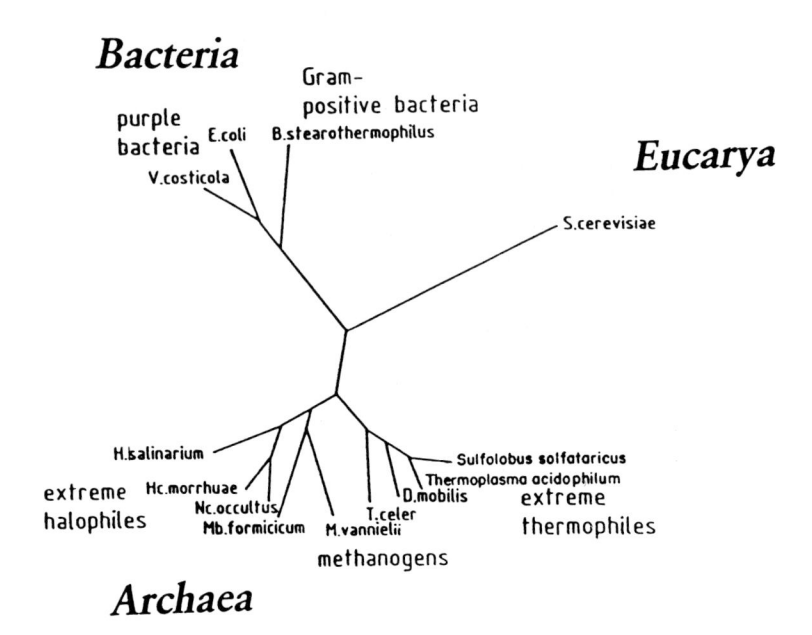

Fig. 10. Unrooted phylogenetic tree inferred from analysis of antibiotic sensitivity curves (ref. [155]).

5. Interchangeability of translational components

5.1. Interchangeability of ribosomal subunits

Within each of the two classical domains, bacteria and eucarya, reciprocal combinations of ribosomal subunits from distantly related organisms yield synthetically active hybrid ribosomes [174–176]. In contrast, formation of hybrid particles from eucaryal and bacterial ribosomal subunits appears to be subject to severe constraints [174]. One case of hybrid monosome formation from subunits of bacterial (*E. coli*) and eucaryal ribosomes (*Artemia salina*) has been reported by Boublik et al. [177]; hybrid ribosomes (73S), however, were assembled only from *Artemia* 40S subunits and *E. coli* 50S subunits, and only at Mg^{2+} levels (30 mM) considerably higher –about twofold – than those (15–18 mM) normally required for poly(U)-directed poly(Phe) synthesis in both bacterial and eucaryal cell-free systems.

Altamura et al. [178] and Londei et al. [179] investigated the ability of 50S and 30S subunits from phylogenetically disparate archaea to form synthetically active hybrid ribosomes with subunits from bacteria and eucarya, in the presence of Mg^{2+} concentrations (15–18 mM) which are optimal for polyphenylalanine synthesis. With poly(U) as the template and Phe-tRNA (or [*N*-acetyl-Phe]-puromycin) as the substrate, 50S and 30S subunits from Euryarchaeota (*M. vannielii*) and Crenarchaeota (*S. solfataricus*) could be assembled into hybrid active monosomes in all reciprocal combinations; surprisingly, however, both reciprocal combinations of archaeal (*S. solfataricus*, *M. vannielii*) and eucaryal (*S. cerevisiae*) ribosomal subunits gave rise to a prominent peak of hybrid monosomes. Subunits from other archaea (*T. celer, T. acidophilum*) gave similar results, except that only one of the two combinations (archaeal 30S *plus* eucaryal 60S) was found to be productive. Under no circumstances were hybrid couples formed between the bacterial subunits and the cognate subunits from both archaea or eucarya.

Although the structural basis for these affinities is not understood, the mutual compatibilities and exclusions indicate that eucaryal and archaeal ribosomes are structurally closer to one another than either is to those of bacteria. It is an attractive possibility that the ribosomes of modern eucarya, having rRNA "expansion segments" [180] and a high (1:1) protein/RNA mass ratio, resemble more closely the prototype particle, and that archaeal and bacterial ribosomes arose from this less parsimoniously-built antecedent by amputation of RNA "expansion segments", and/or trimming or deletions of the primitive ribosomal proteins (see refs. [65,68,76,181]). The "miniaturization" process [76] would have been less severe amongst the archaea (most of which still possess a more abundant protein complement than do their bacterial counterparts), thus explaining the observed compatibilities between archaeal and eucaryal ribosomal subunits.

5.2. Interchangeability of elongation factors

Hybrid poly(U)-programmed cell-free systems containing heterologous combinations of factors and (factor-free) ribosomes have been used to assess the ability of archaeal factors to cooperate with heterologous ribosomes [127,182]. In accordance with the phylogenetic status of the archaea, archaeal EF-2 and EF-1α factors (*T. acidophilum, M. vannielii*) do not cooperate with eucaryal (mammalian) and bacterial (*E. coli*) ribosomes. The reciprocal condition however, in which archaeal ribosomes are tested with either eucaryal or bacterial factors, is less clear-cut in that bacterial (*T. thermophilus*) EF-Tu appears to significantly cooperate with archaeal (*Thermoplasma*) ribosomes [127]. With this limitation, the functional restrictions define a third class of elongation-factor specificity [127].

Unlike the situation in eucarya and bacteria, however, restrictions in ribosome specificity of elongation factors which are similar to, or even more severe than those existing between different domains have been found to exist *within* the archaea. Three groups, the Methanococcaceae/Thermoplasmales, the Sulfolobales and the Thermoproteales are separated from one another by these restrictions. Thus, ribosomes and factors from different archaea do not freely cooperate in polypeptide synthesis, in striking contrast with the uniform compatibility of ribosomes and elongation factors within each of the two classically recognized domains [182].

5.3. Interchangeability of ribosomal RNAs and proteins

The development of protocols for the reconstitution of archaeal ribosomal subunits has disclosed the possibility of determining the degree of compatibility between ribosomal components from evolutionarily disparate organisms. The evidence accumulated so far is reviewed below.

5.3.1. Interchangeability of 5S RNAs

Archaeal and bacterial 5S rRNAs are freely exchangeable in ribosome reconstitution assays. *B. stearothermophilus* 50S subunits reconstituted from archaeal (*T. acidophilum, H. halobium*) 5S rRNAs are as active (in poly(U) translation) as subunits reconstituted from the homologous 5S rRNA [98]. Conversely, bacterial (*E. coli*) 5S RNA can be integrated into *Sulfolobus* large subunits with no loss of synthetic capacity at 75–80°C [183] despite the fact that *E. coli* 5S rRNA is considerably less heat-stable than *Sulfolobus* 5S rRNA [184]; importantly, the heterologous (*E. coli*) 5S RNA appears to interact with the same *Sulfolobus* ribosomal proteins that normally interact with the homologous 5S RNA species [183]. In contrast to the interchangeability of archaeal and bacterial 5S RNAs, exchangeability between eucaryal and bacterial 5S rRNAs appears to be subject to more severe constraints, as *B. stearothermophilus* 50S subunits reconstituted from eucaryal 5S rRNA are essentially inactive in poly(U) translation. To make a complete case, however, it remains to be established whether active archaeal ribosomes can be reconstituted from eucaryal 5S RNA, or whether 5S RNA from the Crenarchaeota can be integrated into synthetically active bacterial ribosomes.

5.3.2. Interchangeability of 50S subunit proteins

Within the bacteria, 16S rRNA and proteins from evolutionarily diverse organisms can be reconstituted into synthetically active hybrid particles in spite of extensive primary structural changes undergone by the interacting components [97]. In addition, it is known that the structure of some rRNA/r-protein interaction sites has been stringently conserved throughout evolution; this is demonstrated from the ability of certain ribosomal proteins from *E. coli* [185] and yeast [186] to recognize their specific binding domains on evolutionarily distant RNA molecules.

Hybrid reconstitution assays using archaeal and bacterial ribosomal constituents [103] suggest that the large ribosomal subunits from the two prokaryotic domains possess a structurally conserved common "core". This is inferred from the fact that the proteins involved in the early assembly of *Sulfolobus* 50S subunit (including all of the primary RNA-binding proteins) are also able to specifically and cooperatively interact with 23S rRNAs from *H. mediterranei* and *E. coli* [103], and to generate a discrete particle sedimentally similar (40S) to the *Sulfolobus* low-temperature reconstitution intermediate (see section 2.7.1).

Unlike the early assembly proteins, the proteins and/or the rRNA domains involved in the late assembly reactions of bacterial and archaeal 50S ribosomes appear to share little structural homology; this is indicated by the inability of the hybrid particles to bind additional proteins and to complete the assembly process, regardless of the environmental conditions employed [103].

6. Conclusions

Both genotypic (primary structural) and functional properties of the archaeal translational machinery confirm the distinctness of the archaea originally inferred from comparative analysis of rRNA sequences.

Phylogenies reconstructed from polypeptide elongation-factor sequences support the monophyly–holophyly of the archaea, in accord with phylogenies based on sequence analysis of other proteins such as RNA polymerase [172] and H^+-ATPase subunits [187,188] (see section 2.5.1). In perspective, other translational components (notably, aminoacyl–tRNA synthethases, and RNA-binding proteins involved in the initial steps of ribosome assembly) may provide new themes to delineate the unfolding of the early lineages and to investigate further such vexing question as the monophyly–holophyly of the archaeal phenotypes.

Importantly, the distinctness of archaea is also inferred from functional data. First, segregation of archaea as a monophyletic–holophyletic cluster, similar to that inferred from analysis of molecular sequences, is obtained by using antibiotic sensitivity spectra in phylogeny treeing (see section 4.5). Secondly, archaeal elongation factors are functionally restricted to the cognate components of their own lineage (see section 5.2). A functional restriction of archaeal aminoacyl–tRNA synthetases to tRNAs of the same lineage is also indicated by studies on a single archaeal tRNA charging enzyme (phenylalanyl-tRNA synthetase) (see section 2.1).

Components of the archaeal translation apparatus, however, display an intra-domain diversity, not encountered within the other two domains. This situation is exemplified by (i) the lack of a uniform compatibility between ribosomes and elongation factors from different archaeal lineages (see section 5.2) (ii) the impressive diversity of archaeal ribosomes and factors in their response to a host of protein synthesis inhibitors (see section 4) (iii) the heterogeneity in shape, mass and composition of archaeal ribosomal subunits (see sections 2.6.2, 2.6.3). Importantly, within-domain diversity is also exemplified by the different complexities of the RNA polymerase subunit patterns from the sulfur-dependent and the methanogenic–halophilic archaea (see Zillig et al., Chapter 12 of this volume).

The origin of this diversity is open to speculation. As regards eucarya, however, it must be pointed out that the full spectrum of diversity, from the deep-branching Diplomonads to the recently evolved higher animals [189], has not been explored. For instance, certain protists, such as *Tetrahymena thermophila*, differ from higher eukaryotes in being highly susceptible to the miscoding-inducing action of some aminoglycoside antibiotics (see refs. [125,136]).

An important implication of the phylogenetic reconstruction of Woese and Fox [1] is that it enables the nature of the most recent common ancestor to be inferred from properties of present-day organisms. Since all extant life forms can be traced back to three lines of cell descent, properties shared by all three domains of life must pre-date the radiation of the early lineages. Following that logic, the most recent common ancestor of extant life forms must have possessed a rather sophisticated translation machinery, not overly different from the modern versions thereof. Evidently, all three domains of life harbour similarly designed (albeit compositionally distinct) ribosomes sharing numerous antibiotic binding sites; polypeptide chain elongation is ubiquitously assisted by two factors, coded for by paralogous genes, which universally bind to the large stalk of the large ribosomal subunit. Even a rudimentary protein-targeting system based on the recognition of hydrophobic signal sequences might well pre-date the earliest radiation of the lineages (see section 2.8).

If this is so, the most recent common ancestor of archaea, bacteria and eucarya was a genote, rather than a progenote: a rather advanced cellular entity that had already perfected the genotype–phenotype link, and whose genome was probably organized in packages of cooperating genes (see section 2.5.2). Since the earliest offshoots of both bacteria and archaea are hyperthermophiles, by inference the common ancestor was also a hyperthermophile. Judging from the ribosomes of the deepest bacterial (notably *Aquifex pyrophylus*) and archaeal offsprings (the *Thermococcus–Pyrococcus* lineage), that hyperthermophilic ancestor possessed ribosomes considerably richer in protein than those typically found in the recent bacterial and archaeal phyla (section 2.6.2). However, the relationship between the protein-rich ribosomes of most archaea (and of deep branching bacteria) and those of eucarya is still a matter of speculation. The possibility that prokaryotic ribosomes at large arose by streamlining from a less parsimoniously built antecedent, compositionally similar to the 80S ribosomes of present-day eucarya, has been proposed more than once (see sections 2.6.1 and 5.1). In that proposition, the protein-rich ribosomes of most archaea are assumed to have arisen from incomplete streamlining of the prototype particle, a view supported by the analysis of *M. vannielii* and *Sulfolobus* ribosomal proteins (see section 2.6.2). Clearly, however, a great number of homologous ribosomal proteins from the three domains must be identified and sequenced if genealogical relationships between extant ribosome types are to be inferred.

On the whole, our present understanding of archaeal translation is far from being complete. There is a considerable dearth of information on such essential aspects of archaeal biochemistry as the structure and sequences of the aminoacyl–tRNA synthethases, the mechanism of polypeptide chain termination and release, the number and complexity of the initiation factors; the possibility that the archaea may resemble eucarya in having a more complex set of initiation factors than exists in bacteria is, in fact, suggested by the identification in archaea of hypusine-containing proteins (see section 2.4). The development of efficient and accurate cell-free systems using natural messenger RNAs is an obvious priority in order to elucidate these points.

References

[1] Woese, C.R. and Fox, G.E. (1977) Proc. Natl. Acad. Sci. U.S.A. 74, 5088–5090.
[2] Woese, C.R. (1987) Microbiol. Rev. 51, 221–271.
[3] Gualerzi, C.O., Pon, C.L., Pawlik, R.T., Canonaco, M.A., Paci, M. and Wintermeyer, W. (1986) In: Structure, Function and Genetics of Ribosomes (Hardesty, B. and Kramer, G., Eds.), pp. 621–641, Springer, Berlin.
[4] Kozak, M. (1983) Microbiol. Rev. 47, 1–45.
[5] Gupta, R. (1985) In: The Bacteria, Vol. 8: Archaebacteria (Woese, C.R. and Wolfe, R.S., Eds.), pp. 311–344, Academic Press, New York.
[6] McCloskey, J.A. (1986) Syst. Appl. Microbiol. 7, 246–252.
[7] Kuchino, Y., Ihara, M., Yabusaki, Y. and Nishimura, S. (1982) Nature 298, 684–685.
[8] Kaine, B.P., Gupta, R. and Woese, C.R. (1983) Proc. Natl. Acad. Sci. U.S.A. 80, 3309–3312.
[9] Kaine, B.P. (1987) J. Mol. Evol. 25, 248–255.
[10] Wich, G., Sibold, L. and Böck, A. (1986) Syst. Appl. Microbiol. 7, 18–25.
[11] Daniels, C.J., Douglas, S.E. and Doolittle, W.F. (1986) Syst. Appl. Microbiol. 7, 26–29.
[12] Wich, G., Leinfelder, W. and Böck, A. (1987) EMBO J. 6, 523–527.

[13] Rauhut, R., Gabius, H.-J., Kühn, W. and Cramer, F. (1984) J. Biol. Chem. 259, 6340–6345.
[14] Rauhut, R., Gabius, H.-J. and Cramer, F. (1985) Biosystems 19, 173–183.
[15] Zillig, W., Palm, P., Reiter, W.D., Gropp, F., Pühler, G. and Klenk, H.-P. (1988) Eur. J. Biochem. 173, 473–482.
[16] Brown, J.W., Daniels, C.J. and Reeve, J.N. (1989) Crit. Rev. Microbiol. 16, 287–338.
[17] Reiter, W.D., Palm, P., Henschen, A., Lottspeich, F., Zillig, W. and Grampp, B. (1987) Mol. Gen. Genet. 206, 144–153.
[18] Pühler, G., Lottspeich, F. and Zillig, W. (1989) Nucl. Acids Res. 17, 4517–4534.
[19] Leffers, H., Gropp, F., Lottspeich, F., Zillig, W. and Garrett, R.A. (1989) J. Mol. Biol. 206, 1–17.
[20] Arndt, E., Krömer, W. and Hatakeyama, T. (1990) J. Biol. Chem. 265, 3034–3039.
[21] Oshima, T., Obha, M. and Wakagi, T. (1984) Origins Life 14, 665–669.
[22] Brown, J.W. and Reeve, J.N. (1985) J. Bacteriol. 162, 909–917.
[23] Brown, J.W. and Reeve, J.N. (1986) J. Bacteriol. 166, 686–688.
[24] Creti, R., Citarella, F., Tiboni, O., Sanangelantoni, A.M., Palm, P. and Cammarano, P. (1991) J. Mol. Evol. 33, 332–342.
[25] Pechmann, H., Tesch, A. and Klink, F. (1991) FEMS Microbiol. Lett. 79, 51–56.
[26] Tesch, A. and Klink, F. (1990) FEMS Microbiol. Lett. 71, 293–298.
[27] Baldacci, G., Guinet, F., Tillit, J., Zaccai, G., de Recondo, A.M. (1990) Nucl. Acids Res. 18, 507–511.
[28] Itoh, T., Kumazaki, T., Sugiyami, M. and Otaka, E. (1988) Biochim. Biophys. Acta 940, 110–116.
[29] Betlach, M.C., Friedman, J., Boyer, H.W. and Pfeifer, F. (1984) Nucl. Acids Res. 12, 7949–7959.
[30] Betlach, M.C., Leong, D. and Boyer, H.W. (1986) Syst. Appl. Microbiol. 7, 83–89.
[31] Blanck, A. and Oesterhelt, D. (1987) EMBO J. 6, 265–271.
[32] Dunn, R., McCoy, J., Simsek, M., Majundar, A., Chang, S.H., RajBhandhary, U.L. and Khorana, H.G. (1981) Proc. Natl. Acad. Sci. U.S.A. 78, 6744–6748.
[33] Auer, J., Spicker, J., Mayerhofer, L., Pühler, G. and Böck, A. (1991) Syst. Appl. Microbiol. 14, 14–22.
[34] Schröder, J. and Klink, F. (1991) Eur. J. Biochem. 195, 321–327.
[35] Köpke, A.K.E. and Leggatt, P.A. (1991) Nucl. Acids Res. 19, 5169–5172.
[36] Gupta, R. (1985) In: The Bacteria, Vol. 8: Archaebacteria (Woese, C.R. and Wolfe, R.S., Eds.), pp. 311–343, Academic Press, New York.
[37] Schümann, H. and Klink, F. (1989) Syst. Appl. Microbiol. 11, 103–107.
[38a] Bartig, D., Schümann, H. and Klink, F. (1990) Syst. Appl. Microbiol. 13, 112–116.
[38b] Bartig, D., Lemkemeier, K., Frank, J., Lottspeich, F. and Klink, F. (1992) Eur. J. Biochem. 204, 751–758.
[39] Iwabe, N., Kuma, K.I., Hasegawa, M., Osawa, S. and Miyata, T. (1989) Proc. Natl. Acad. Sci. U.S.A. 86, 9355–9359.
[40] Itoh, T. (1989) Eur. J. Biochem. 186, 213–219.
[41] Lechner, K. and Böck, A. (1987) Mol. Gen. Genet. 208, 523–528.
[42] Lechner, K., Heller, G. and Böck, A. (1988) Nucl. Acids Res. 16, 7817–7826.
[43] Auer, J., Spicker, G. and Böck, A. (1990) Nucl. Acids Res. 18, 3989–3998.
[44] Walker, J.E., Saraste, M., Runswick, M. and Gay, N. (1982) EMBO J. 1, 945–951.
[45] Guinet, F., Rainer, F. and Leberman, R. (1983) Eur. J. Biochem. 172, 687–694.
[46] Kuntz, I.D. (1971) J. Am. Chem. Soc. 93, 514–525.
[47] Zaccai, G., Cendrin, F., Haik, Y., Borochov, N. and Eisenberg, H. (1989) J. Mol. Biol. 208, 491–500.
[48] Brock, T.D. (1985) Science 230, 132–136.
[49] Jaenicke, R. (1988) Naturwissenschaften 75, 604–610.

[50] Jaenicke, R. and Zavodszky, P. (1990) FEBS Lett. 268, 344–349.
[51] Cammarano, P., Tiboni, O. and Sanangelantoni, A.M. (1989) Can. J. Microbiol. 35, 2–10.
[52] Woese, C.R., Kandler, O. and Wheelis, M. (1990) Proc. Natl. Acad. Sci. U.S.A. 87, 4576–4579.
[53] Post, L.E. and Nomura, M. (1980) J. Biol. Chem. 255, 4660–4666.
[54] Auer, J., Lechner, K. and Böck, A. (1989) Can. J. Microbiol. 35, 200–204.
[55] Zillig, W., Palm, P., Klenk, H.-P., Pühler, G., Gropp, F. and Schleper, C. (1991) In: General and Applied Aspects of Halophilic Microorganisms (Rodriguez-Valera, F., Ed.), pp. 321–332, Plenum Press, New York.
[56] Zillig, W., Palm, P. and Klenk, H.-P. (1992) In: Frontiers of Life (Trân Thanh Vân, J. and K., Mounolou, C., Schneider, J. and McKay, C., Eds.), pp. 181–193, Editions Frontières, Gif-sur-Yvette, France. in press.
[57] Tiboni, O., Cantoni, R., Creti, R., Cammarano, P. and Sanangelantoni, A.M. (1991) J. Mol. Evol. 33, 142–151.
[58] Vossbrinck, C.R. and Woese, C.R. (1986) Nature 320, 287–288.
[59] Ree, K.H. and Zimmermann, R.A. (1990) Nucl. Acids Res. 18, 4471–4478.
[60] Cammarano, P., Mazzei, F., Londei, P., De Rosa, M. and Gambacorta, A. (1982) Biochim. Biophys. Acta 699, 1–14.
[61] Cammarano, P., Mazzei, F., Londei, P., Teichner, A., De Rosa, M. and Gambacorta, A. (1983) Biochim. Biophys. Acta 740, 300–312.
[62] Strom, A.R., Hasnain, S., Smith, N., Matheson, A.T. and Visentin, L.P. (1975) Biochim. Biophys. Acta 383, 325–337.
[63] Bayley, S.T. (1966) J. Mol. Biol. 15, 420–427.
[64] Londei, P., Teichner, A., Cammarano, P., De Rosa, M. and Gambacorta, A. (1983) Biochem. J. 209, 461–470.
[65] Cammarano, P., Teichner, A. and Londei, P. (1986) Syst. Appl. Microbiol. 7, 137–146.
[66] Londei, P., Altamura, S., Cammarano, P. and Petrucci, L. (1986) Eur. J. Biochem. 157, 455–462.
[67] Visentin, L.P., Chow, C., Matheson, A.T., Yaguchi, M. and Rollin, F. (1972) Biochem. J. 130, 103–110.
[68] Teichner, A., Londei, P. and Cammarano, P. (1986) J. Mol. Evol. 23, 343–353.
[69] Shevack, A., Gewitz, H.S., Hennemann, B., Yonath, A. and Wittmann, H.G. (1985) FEBS Lett. 184, 68–71.
[70] Wittmann, H.G. (1982) Annu. Rev. Biochem. 51, 155–183.
[71] Sacchi, A., Ferrini, U., Londei, P., Cammarano, P. and Maraldi, N. (1977) Biochem. J. 168, 245–259.
[72] Douglas, C., Achatz, F. and Böck, A. (1980) Zentralbl. Bakteriol. Parasitenk. Infektionskr. Hyg. Abt. 1 Orig. C 1, 1–11.
[73] Schmid, G., Pecher, T. and Böck, A. (1982) Zbl. Bakt. Hyg., I. Abt. Orig. C 3, 209–217.
[74] Strom, A.R. and Visentin, L.P. (1973) FEBS Lett. 37, 274–280.
[75] Schmid, G. and Böck, A. (1982) Mol. Gen. Genet. 185, 498–501.
[76] Auer, J., Spicker, G. and Böck, A. (1989) J. Mol. Biol. 209, 21–36.
[77] Woese, C.R., Achenbach, L., Rouviere, P. and Mandelco, L. (1991) Syst. Appl. Microbiol. 14, 364–371.
[78] Lake, J.A., Henderson, E., Clark, M.W. and Matheson, A.T. (1982) Proc. Natl. Acad. Sci. U.S.A. 79, 5948–5932.
[79] Ishikara, R. and Hayashi, Y (1968) J. Invertebr. Pathobiol. 11, 377–385.
[80] Curgy, J., Vavra, J. and Vivares, C. (1980) Biol. Cellulaire 38, 49–52.
[81] Lake, J.A., Henderson, E., Clark, M.V., Scheinman, A. and Oakes, M.J. (1986) Syst. Appl. Microbiol. 7, 131–136.
[82] Lake, J.A., Clark, M.W., Henderson, E., Fay, S., Oakes, M., Scheinman, A., Thornber, J.P. and Mah, R.A. (1985) Proc. Natl. Acad. Sci. U.S.A. 82, 3716–3720.

[83] Kiselev, N.A., Sel'mansschuk, V.Ya., Orlava, E.V., Vasiliev, V.D. and Selivanova, O.M. (1982) Mol. Biol. Rep. 8, 191–197.
[84] Verschoor, A., Frank, J., Rademacher, W., Wagenknecht, T. and Boublik, M. (1984) J. Mol. Biol. 178, 677–698.
[85] Meisenberger, O., Pilz, I., Stöffler-Meilicke, M. and Stöffler, G. (1984) Biochim. Biophys. Acta 781, 225–233.
[86] van Heel, M. and Stöffler-Meilicke, M. (1985) EMBO J. 4, 2389–2395.
[87] Stöffler-Meilicke, M., Böhme, C.V., Strobel, O., Böck, A. and Stöffler, G. (1986) Science 231, 1306–1308.
[88] Lake, J.A., Henderson, E., Oakes, M. and Clark, M.W. (1984) Proc. Natl. Acad. Sci. U.S.A. 81, 3786–3790.
[89] Henderson, E., Oakes, M., Clark, M.V., Lake, J.A., Matheson, A.T. and Zillig, W. (1984) Science 225, 510–512.
[90] Stöffler, G. and Stöffler-Meilicke, M. (1986) Syst. Appl. Microbiol. 7, 123–130.
[91] Traub, P. and Nomura, M. (1968) Proc. Natl. Acad. Sci. U.S.A. 59, 777–782.
[92] Mizushima, S. and Nomura, M. (1970) Nature 266, 1214–1216.
[93] Röhl, R. and Nierhaus, K.H. (1982) Proc. Natl. Acad. Sci. U.S.A. 79, 729–733.
[94] Moore, P.B., Capel, M., Kjeldgaard, M. and Engelman, D.M. (1986) In: Structure, Function and Genetics of Ribosomes (Hardesty, B. and Kramer, G., Eds.), pp. 87–100, Springer, Berlin.
[95] Nowotny, V., May, R.P. and Nierhaus, K.H. (1986) In: Structure, Function and Genetics of Ribosomes (Hardesty, B. and Kramer, G., Eds.), pp. 101–111, Springer, Berlin.
[96] Nowotny, V. and Nierhaus, K.H. (1982) Proc. Natl. Acad. Sci. U.S.A. 79, 7238–7242.
[97] Nomura, M., Traub, P. and Bechmann, H. (1968) Nature 219, 793–799.
[98] Hartmann, R.K., Vogel, D.W., Walker, R.T. and Erdmann, V.A. (1988) Nucl. Acids Res. 16, 3511–3524.
[99] Londei, P., Teixidò, J., Acca, M., Cammarano, P. and Amils, R. (1986) Nucl. Acids Res. 14, 2269–2285.
[100] Sanchez, M.E., Ureña, D., Amils, R. and Londei, P. (1990) Biochemistry 29, 9256–9261.
[101] Nierhaus, K.H. and Döhme, F. (1974) Proc. Natl. Acad. Sci. U.S.A. 71, 4713–4717.
[102] Nierhaus, K.H. (1980) In: Ribosomes: Structure, Function and Genetics (Chambliss, G., Craven, G.R., Davies, J., Davis, K., Kahan, L. and Nomura, M., Eds.), pp. 267–294, University Park Press, Baltimore, MD.
[103] Altamura, S., Caprini, E., Sanchez, M.E. and Londei, P. (1991) J. Biol. Chem. 266, 6195–6200.
[104] Saier, M.H., Werner, P.K. and Mueller, M. (1989) Microbiol. Rev. 53, 333–336.
[105] Tajima, S., Lauffer, L., Rath, V.L. and Walter, P. (1986) J. Cell Biol. 103, 1167–1178.
[106] Gilmore, R., Blöbel, G. and Walter, P. (1982) J. Cell Biol. 95, 463–499.
[107] Poritz, M.A., Bernstein, H.D., Strub, K., Zopf, D., Wilhelm, H. and Walter, P. (1990) Science 250, 1111–1117.
[108] Römisch, K., Webb, J., Herz, J. Prehn, S., Frank, R., Vingron, M. and Dobberstein, B. (1989) Nature 340, 478–482.
[109] Dellweg, H.G. and Sumper, M. (1980) FEBS Lett. 116, 303–306.
[110] Lechner, J. and Sumper, M. (1987) J. Biol. Chem. 262, 9724–9729.
[111] Lin, X. and Tang, J. (1990) J. Biol. Chem. 265, 1490–1495.
[112] Moritz, A. and Goebel, W. (1985) Nucl. Acids Res. 13, 6969–6980.
[113] Kaine, B.P. and Merkel, U.L. (1989) J. Bacteriol. 171, 4261–4266.
[114] Kaine, B.P., Schurke, C.M. and Stetter, K.O. (1989) Syst. Appl. Microbiol. 12, 8–14.
[115] Kaine, B.P. (1990) Mol. Gen. Genet. 221, 315–321.
[116] Ramirez, C. and Matheson, A.T. (1991) Mol. Microbiol. 5, 1687–1693.
[117] Bayley, S.T. and Griffiths, E. (1968) Biochemistry 7, 2249–2256.
[118] Kessel, M. and Klink, F. (1981) Eur. J. Biochem. 114, 481–486.

[119] Saruyama, H. and Nierhaus, K.H. (1985) FEBS Lett. 183, 390–394.
[120] Sanz, J.L., Marin, I., Balboa, M.A., Ureña, D. and Amils, R. (1988) Biochemistry 27, 8194–8199.
[121] Elhardt, D. and Böck, A. (1982) Mol. Gen. Genet. 188, 128–134.
[122] Böck, A., Bär, U., Schmid, G. and Hummel, H. (1983) FEMS Microbiol. Lett. 20, 435–438.
[123] Cammarano, P., Teichner, A., Chinali, G., Londei, P., De Rosa, M., Gambacorta, A. and Nicolaus, B. (1982) FEBS Lett. 148, 255–259.
[124] Friedman, S.M. and Oshima, T. (1983) J. Biochem. 105, 1030–1033.
[125] Londei, P., Altamura, S., Sanz, J.L. and Amils, R. (1988) Mol. Gen. Genet. 214, 48–54.
[126] Friedman, S.M. (1986) Syst. Appl. Microbiol. 7, 325–329.
[127] Klink, F., Schümann, H. and Thomsen, A. (1983) FEBS Lett. 115, 173–177.
[128] Catani, M.V., Altamura, S. and Londei, P. (1990) FEMS Microbiol. Lett. 70, 285–290.
[129] Monro, R.E. (1974) Methods Enzymology XX, 472–481.
[130] Vazquez, D. (1979) Inhibitors of Protein Biosynthesis, Molecular Biology, Biochemistry and Biophysics Series, Vol. 30, Springer, Berlin.
[131] Boynton, J., Gillham, N.W. and Lambowitz, A.M. (1980) In: Ribosomes: Structure, Function and Genetics (Chambliss, G., Craven, R., Davies, J., Davis, K., Kahan, L. and Nomura, M., Eds.), pp. 903–950, University Park Press, Baltimore, MD.
[132] Hilpert, R., Winter, J., Hammes, W. and Kandler, O. (1981) Zbl. Bakt. Hyg., I. Abt. Orig. C 2, 11–20.
[133] Pecher, R. and Böck, A. (1981) FEMS Microbiol. Lett. 10, 295–297.
[134] Hummel, H., Bär, U., Heller, G. and Böck, A. (1985) Syst. Appl. Microbiol. 6, 125–131.
[135] Cammarano, P., Teichner, A., Londei, P., Acca, M., Nicolaus, B., Sanz, J.L. and Amils, R. (1985) EMBO J. 4, 811–816.
[136] Altamura, S., Sanz, J.L., Amils, R., Cammarano, P. and Londei, P. (1988) Syst. Appl. Microbiol. 10, 218–255.
[137] Sanz, J.L., Altamura, S., Mazziotti, I., Amils, R., Cammarano, P. and Londei, P. (1987) Mol. Gen. Genet. 207, 385–394.
[138] Amils, R. and Sanz, J.L. (1986) In: Structure, Function and Genetics of Ribosomes (Hardesty, B. and Kramer, G., Eds.), pp. 605–620, Springer, Berlin.
[139] Londei, P., Altamura, S., Huber, R., Stetter, K.O. and Cammarano, P. (1988) J. Bacteriol. 170, 4353–4360.
[140] Li, M., Tzagoloff, A., Underbrink-Lyon, K. and Martin, N.C. (1982) J. Biol. Chem. 257, 5921–5928.
[141] Sor, F. and Fukuhara, H. (1982) Nucl. Acids Res. 10, 6571–6577.
[142] Sigmund, C.D., Ettayebi, M. and Morgan, E.A. (1984) Nucl. Acids Res. 12, 4653–4663.
[143] Spangler, E.A. and Blackburn, E.H. (1985) J. Biol. Chem. 260, 6334–6340.
[144] Montandon, P.E., Nicolas, P., Schürmann, P. and Stutz, E. (1985) Nucl. Acids Res. 13, 4299–4310.
[145] Moazed, D. and Noller, H.F. (1987) Nature 327, 389–394.
[146] Montandon, P., Wagner, R. and Stutz, E. (1986) EMBO J. 5, 3705–3708.
[147] McCarroll, R., Olsen, G.J., Stahl, U.D., Woese, C.R. and Sogin, M.L. (1983) Biochemistry 22, 5828–5868.
[148] Eckenrode, V., Arnold, J. and Meager, R. (1985) J. Mol. Evol. 21, 259–269.
[149] Kjems, J., Garrett, R.A. and Ansorge, W. (1987) Syst. Appl. Microbiol. 9, 22–28.
[150] Leffers, H., Kjems, J., Ostergaard, L., Larsen, N. and Garrett, R.A. (1987) J. Mol. Biol. 195, 43–61.
[151] Thompson, J., Schmidt, F. and Cundliffe, E. (1982) J. Biol. Chem. 257, 7915–7917.
[152] Thompson, J., Cundliffe, E. and Dahlberg, A.E. (1988) J. Mol. Biol. 203, 457–465.

[153] Beauclerk, A.A.D., Hummel, H., Holmes, D.J., Böck, A. and Cundliffe, E. (1985) Eur. J. Biochem. 151, 245–255.
[154] Leffers, H., Kjems, J., Oestergaard, L., Larsen, N. and Garrett, R.A. (1987) J. Mol. Biol. 195, 43–61.
[155] Amils, R. Ramirez L,, Sanz, J.L., Marin, I., Pisabarro, A.G., Sanchez, E. and Ureña, D. (1991) In: Structure, Function and Genetics of Ribosomes (Hill, W. Ed.), pp. 645–654, American Society of Microbiology, Washington, D.C.
[156] Schindler, D.G. and Davies, J.E. (1977) Nucl. Acids Res. 4, 1097–1110.
[157] Wool, I.G. (1984) Trends Biochem. Sci. 9, 14–17.
[158] Wolf, H.G., Chinali, G. and Parmeggiani, A. (1974) Proc. Natl. Acad. Sci. U.S.A. 71, 4910–4914.
[159] Wolf, H.D., Assman, D. and Fischer, E. (1978) Proc. Natl. Acad. Sci. U.S.A. 75, 5324–5328.
[160] Johnson, W.R., Kuchler, R.J. and Solotorovsky, M. (1968) J. Bacteriol. 96, 1089–1098.
[161] Richter, D. and Lippman, F. (1970) Biochemistry 9, 5065–5070.
[162] Tanaka, N., Kinoshita, T. and Masukawa, H. (1968) Biochem. Biophys. Res. Comm. 30, 278–283.
[163] Van Ness, B.G., Howard, J.B. and Bodley, J.W. (1980) J. Biol. Chem. 255, 10710–10716.
[164] Pappenheimer, A.M., Dunlop, P.C., Adolph, K.W. and Bodley, J.W. (1983) J. Bacteriol. 153, 1342–1347.
[165] Londei, P., Sanz, J.L., Altamura, S., Hummel, H., Cammarano, P., Amils, R. and Böck, A. (1986) J. Bacteriol. 167, 265–271.
[166] Kessel, M. and Klink, F. (1980) Nature 287, 250–251.
[167] Kessel, M. and Klink, F. (1982) Zbl. Bakt. Hyg. I. Abt. Orig. C 3, 140–148.
[168] Klink, F. (1985) In: The Bacteria, Vol. 8: Archaebacteria (Woese, C.R. and Wolf, R.S., Eds.), pp.379–410, Academic Press, New York.
[169] Worner, W., Glockner, C., Mierzowski, M. and Wolf, H. (1983) FEMS Microbiol. Lett. 18, 69–73.
[170] Oliver, J.L., Sanz, J.L., Amils, R. and Marin, A. (1987) J. Mol. Evol. 24, 281–288.
[171] Amils, R., Ramirez, L., Sanz, J.L., Marin, I., Pisabarro, A.G. and Ureña, D. (1989) Can. J. Microbiol. 35, 141–147.
[172] Pühler, G., Leffers, H., Gropp, F., Palm, P., Klenk, H., Lottspeich, H.P., Garrett, R.A. and Zillig, W. (1989) Proc. Natl. Acad. Sci. U.S.A. 86, 4569–4573.
[173] Cammarano, P., Palm, P., Creti, R., Ceccarelli, E., Sanangelantoni, A.M. and Tiboni, O. (1992) J. Mol. Evol. 34, 396–405.
[174] Takeda, M. and Lippman, F. (1966) Proc. Natl. Acad. Sci. U.S.A. 56, 1875–1879.
[175] Martin, T.E, and Wool, I.G. (1969) J. Mol. Biol. 43, 151–161.
[176] Cammarano, P., Felsani, A., Gentile, M., Gualerzi, C., Romeo, A. and Wolf, G. (1972) Biochim. Biophys. Acta 281, 625–642.
[177] Boublik, M., Wydro, R., Hellmann, W. and Jenkins, F. (1979) J. Supramol. Struc. 10, 397–404.
[178] Altamura, S., Cammarano, P. and Londei, P. (1986) FEBS Lett. 204, 129–133.
[179] Londei, P., Altamura, S., Tiboni, O. and Cammarano, P. (1988) In: Genetics of Translation: New Approaches (Tuite, M.F., Picard, M. and Bolotin-Fukuhara, M., Eds.), pp. 181–194, Springer, Berlin.
[180] Clark, C.G., Tague, B.W., Ware, V.C. and Gerbi, S.A. (1984) Nucl. Acids Res. 12, 6197–6220.
[181] Clark, C.G. (1987) J. Mol. Evol. 25, 343–350.
[182] Gehrmann, R., Henschen, A., Postulka, W. and Klink, F.(1986) Syst. Appl. Microbiol. 7, 115–122.
[183] Teixidò, J., Altamura, S., Londei, P. and Amils, R. (1989) Nucl. Acids Res. 17, 845–851.
[184] Dams, E., Londei, P., Cammarano, P., Vandenberghe, A. and De Wachter, R. (1983) Nucl. Acids Res. 11, 4667–4676.

[185] Zimmermann, R.A. (1980) In: Ribosomes: Structure, Function and Genetics (Chambliss, G., Craven, R., Davies, J., Davis, K., Kahan, L. and Nomura, M., Eds.), pp. 135–169, University Park Press, Baltimore, MD.
[186] El Baradi, T.A.L., Raué, H.A., de Regt, V.C., Verbree, E.C. and Planta, R.J. (1985) EMBO J. 4, 2101–2107.
[187] Gogarten, J.P., Rausch, T., Bernasconi, P., Kibak, H. and Taiz, L. (1989) Z. Naturforschung 44, 97–105.
[188] Gogarten, J.P., Kibak, H., Dittrich, P., Lincoln, T., Bowman, E., Bowman, H., Manolson, M.F., Poole, R.J., Date, T., Oshima, T., Knoishi, J., Danda, K. and Yoshida, M. (1989) Proc. Natl. Acad. Sci. U.S.A. 86, 6661–6665.
[189] Sogin, M.L. (1991) Curr. Op. Gen. Develop. 1, 457–463.

CHAPTER 14

The structure, function and evolution of archaeal ribosomes

C. RAMÍREZ*, A.K.E. KÖPKE*, D-C. YANG*, T. BOECKH* and A.T. MATHESON

Department of Biochemistry and Microbiology, University of Victoria, Victoria, B.C. V8W 3P6, Canada

1. Introduction

Ribosomes are present in all living cells and play a central role in the synthesis of cellular protein. Because of their ubiquitous nature they have proven to be a valuable phylogenetic probe.

During the past decade a great deal of information has been obtained on the structure of archaeal ribosomes and their constituent rRNA and r-protein components. Although these molecules share many properties with their counterparts in ribosomes from other organisms, they also show unique features which have proven to be of great value in the elucidation of the evolution of archaea in relation to bacteria and eucarya.

In this chapter we review the current knowledge of the structure and evolution of archaeal ribosomes with special emphasis on the rRNA and r-protein molecules that form complex macromolecules. In a separate chapter, Amils, Cammarano and Londei (Chapter 13) describe the structural and functional aspects of the archaeal translational apparatus including the structural aspects of the archaeal ribosome.

2. The archaeal ribosomes

The 70S archaeal ribosome consists of two subunits, a large 50S subunit and a smaller 30S subunit. In this regard these subunits are similar in size to the bacterial ribosomes

---

* Present addresses:
CR, Department of Microbiology, University of British Columbia, Vancouver, B.C. V6T 1Z3, Canada.
AK, Max-Planck-Institute for Experimental Medicine, Department of Molecular Neuroendocrinology, Hermann Rein Strasse 3, 3400 Goettingen, Germany.
DCY, Research Center of the University of British Columbia, 950 West 20th Ave., Vancouver, B.C. V5Z 4H4, Canada
TB, Max-Planck-Institut für Molekulare Genetik, Abt. Wittmann, D-1000 Berlin 33 (Dahlem), Germany.

and significantly smaller than the 80S eucaryal ribosome with its 60S and 40S subunits. Although ribosomes in extreme halophiles, methanogens and sulfur-dependent extreme thermophiles (the three groups of organisms that make up the archaea domain [1,2]) have similar sedimentation values, studies by Cammarano et al.[3] indicate that the archaea contain two classes of ribosomes based on the number and size of the r-proteins in these particles. The extreme halophiles [4] and most of the methanogens [5] were thought to contain a similar number of r-proteins to those found in bacterial ribosomes, whereas the sulfur-dependent extreme thermophiles [6,7] and *Methanococcus* species [5] appear to contain a larger number of proteins with higher average molecular weights, primarily in the 30S subunit, than bacterial ribosomes. From phylogenetic data it was suggested [3,8,9] that these larger archaeal ribosomes may more closely resemble the 'ancestral ribosome' while the smaller bacterial and the other archaeal ribosomes have undergone additional streamlining [8–10].

However, the exact number of proteins present in the archaeal ribosome is still unknown. The number quoted by various investigators, based on 2D polyacrylamide gel electrophoresis, depends on the resolving power of the gel system used. Recent studies by Casiano et al. [11] suggest that the *Sulfolobus* 50S ribosomal subunit could contain as many as 43 r-proteins, and work from our laboratory [12] on the structure of the *Sulfolobus* r-proteins and from Wittmann's laboratory [13] on *Haloarcula* (formerly *Halobacterium*) *marismortui* r-proteins indicates an increasing number of proteins present in the archaeal ribosome that are not present in the bacterial ribosome. It may well be that the archaeal ribosome contains more proteins than are present in the bacterial (*E. coli*) ribosome, or the bacterial ribosome contains proteins not found in the archaeal ribosome (see section 6).

Early results from Lake's laboratory [14] on the morphology of the archaeal ribosomes, using electron microscopy, suggested that the 30S and 50S ribosomal subunits of the archaea contained unique features that could be used to establish phylogenetic relationships. The archaeal ribosomes appeared to contain features not present in bacterial ribosomal subunits but which, in turn, were less complex than the eucaryal ribosomes. Results from other laboratories [15,16] confirmed the increasing complexity of the ribosomal morphology as one moved from bacteria to archaea and eucarya. However, the use of electron-microscopic data to postulate new phylogenetic relationships has been seriously questioned since the fine structural details used for such classifications are at the border of resolution of the systems, and features that initially were thought to be unique to one group of organisms (domain) were often seen in other organisms [16]. Computer-imaging techniques [17] have confirmed the structural differences between bacterial and archaeal ribosomes. The availability of three-dimensional crystals of archaeal ribosomal subunits [18] and the improvement in X-ray crystallography techniques should provide more detailed information on the fine structure of the ribosome particle in the future.

A further example of the diversity of the archaeal ribosome comes from the studies of Londei and her co-workers [19], which show that the sulfur-dependent extreme thermophiles can be divided into two groups, based on the stability of the interaction between the ribosomal subunits. One group, which contains *Thermococcus* and *Thermoplasma*, has highly stable 70S ribosomes while another group, which includes *Sulfolobus, Thermoproteus* and *Desulfurococcus*, contains easily dissociated 70S ribosomes. The cell extracts from this latter group of organisms contain only the dissociated 50S and 30S subunits. It is interesting to note that the ribosomes of this group also require spermine for in vitro protein synthesis.

It is now possible to specifically modify the individual rRNA and r-protein molecules that form the archaeal ribosome, enabling one to study the structural/functional relationships of these molecules. Recent studies on the reconstitution of the archaeal 50S ribosomal subunit from the extreme halophile *Haloferax mediterranei* [20] and the extreme thermophile *Sulfolobus solfataricus* [21] open the way for the identification of the individual functions of the r-proteins in these particles. Experiments on site-specific changes in the r-protein L12 from *Sulfolobus* and their effect on the structure and function of the ribosome are described later in this chapter.

3. Archaeal ribosomal RNA

The rRNA components of the archaeal ribosome are similar in size to the rRNA molecules in bacteria (23S, 16S, 5S). The archaea are also similar to the bacteria in that, unlike the eucarya, no separate 5.8S rRNA molecule is present in the ribosome but rather the 23S rRNA contains a sequence homologous to 5.8S rRNA at its 5' end [22]. The structure of the rRNA in archaea is discussed in the Introduction to this volume, by Woese.

3.1. Gene organization

Fig. 1 shows a comparison of the different types of rRNA gene organization present in members of the three domains [23–32]. The rRNA gene organization of the archaea is very variable and includes examples of organisms where the three rRNA genes are closely linked [24] and cases where the three genes are unlinked [32]. The number of rRNA operons is also variable. There are four operons in *Methanococcus vannielii* [33], two in *Methanobacterium thermoautotrophicum* [34], *Methanothermus fervidus* [35] and *Haloarcula marismortui* [36], and one in all the other archaea so far examined [37]. A gene for alanine tRNA has been detected between the 16S and 23S rRNA genes in all the methanogens and halophiles studied, while in the extreme thermophiles, this gene is present only in *Thermococcus celer* [37].

Although the number of rRNA operons varies within the cell the sequence of the rRNA in these multicopy genes is usually identical or nearly so [38]. One exception is the rRNA genes in *H. marismortui*. The two 16S rRNA genes differ significantly [39] and these differences are mainly clustered within the central domain of the 16S rRNA. Of even greater interest is the observation that both genes are expressed and the 16S rRNA molecules assembled into active 70S ribosomes.

Analysis of the sequences located upstream of archaeal rRNA genes has shown that archaeal rRNA promoters consist of the sequence TTTA(A/T)A located 20–30 nucleotides upstream of the transcription initiation site and a weakly conserved sequence, (A/T)TG(A/C) around the transcription initiation site [40]. Transcription of the rRNA genes terminates within pyrimidine-rich regions in the extreme thermophiles [30,41], in pyrimidine-rich regions followed by a short hairpin loop in the methanogens [27,34] and in AT-rich regions preceded by a GC-rich region in the extreme

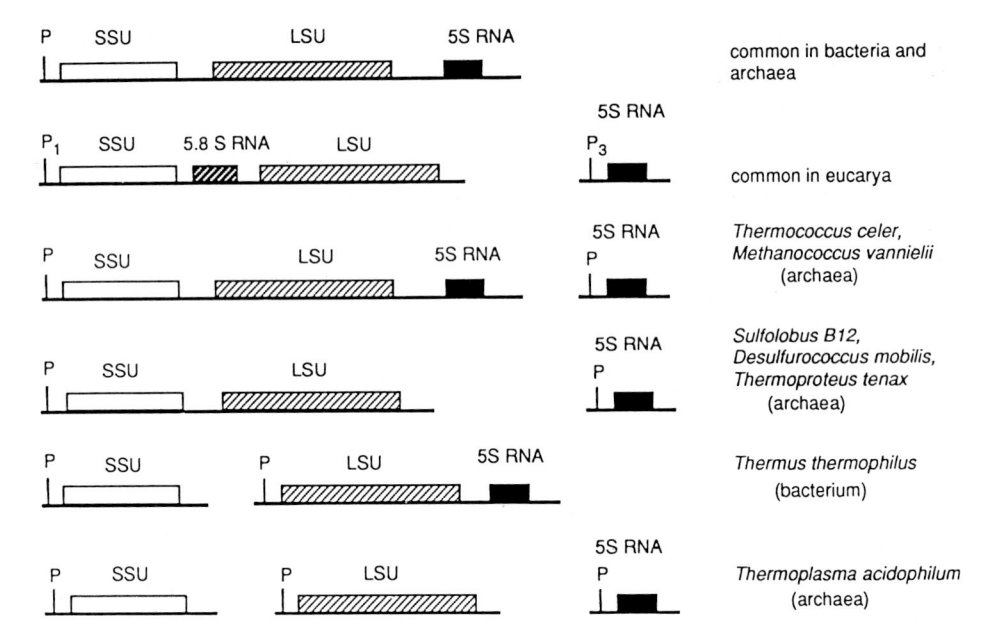

Fig. 1. rRNA gene organization in the three domains. Abbreviations: SSU, small subunit or 16S-like rRNA; LSU, large subunit or 23S-like rRNA; P, promoters; P_1, promoter for RNA polymerase I; P_3, promoter for RNA polymerase III. In *T. celer*, the unlinked 5S RNA gene is part of an operon that includes an aspartic acid tRNA [26] while in *M. vannielii*, the unlinked 5S RNA gene is part of an operon that includes 7 tRNA genes [27]. In *Sulfolobus* B12, *D. mobilis* and *T. tenax,* the unlinked 5S rRNA is not part of an operon [28–31]. Data from: bacteria [23], archaea [24], eucarya [25], *T. celer* [26], *M. vannielii* [24, 27], *Sulfolobus* B12 [28], *D. mobilis* [29], *T. tenax* [30], *T. thermophilus* [31], and *T. acidophilum* [32].

halophiles [42,43]. In some cases, multiple promoters have been identified in front of the rRNA genes [43].

Introns have been detected in the 23S rRNA from *Desulfurococcus mobilis* [44] and *Staphylothermus marinus* [45]. The splicing mechanism which has been studied in *D. mobilis* indicates that the intron is excised by an endonuclease producing a 3' phosphate which serves as a substrate for an RNA ligase, which in turn generates the mature 23S rRNA and a full-length circular intron [46]. Since crude extracts from related organisms, lacking an rRNA intron, can splice the *D. mobilis* intron, it is probable that the splicing enzymes have other cellular functions [46].

Internal transcribed spacers (ITS) [47] or intervening sequences (IVS) [48], which in bacteria and eucarya are responsible for the generation of fragmented 23S-like rRNA molecules, have not been detected in the archaea.

Very little is known about the processing of rRNA transcripts to generate mature rRNAs in the archaea, except that one of the enzymes involved recognizes a structure that consists of bulges that are staggered on opposite sides of the chain and that the processing cuts occur in the bulges [27,29,43].

3.2. rRNA structure and function

rRNA has had a major role in the history of archaeal studies since it was on the basis of the sequence of the 16S rRNA that Woese and his co-workers [1,49] discovered the existence of the archaeal domain. In the last decade, the comparison of the sequences of the three rRNA molecules from the three domains has not only allowed us to establish the phylogenetic relationships among the different types of organisms (refs. [50–56], and the Introduction to this volume) but has also radically changed our views on the structure and function of rRNA. Models of the secondary [57–62] and even tertiary structure [63–70] of the three rRNA molecules have been proposed and the old notion that rRNA is simply a scaffold for the assembly of r-proteins [71] has given way to the view that rRNA plays a major role in the function of the ribosome [72–74] and that the most primitive ribosome was probably composed solely of rRNA [75].

Based on the rRNA data available from the three domains, the following general conclusions can be drawn:

1. rRNA is one of the most highly conserved molecules found in nature [61,76].

2. Comparison of the structures of the two large rRNA (16S-like or small subunit [SSU], and 23S-like or large subunit [LSU]) from the three domains, reveals in both cases, the existence of a conserved common structural core to which variable extension segments are attached at specific sites [61,62,73,76]. There are ten variable extension segments in the case of the SSU rRNA and 18 in the case of the LSU rRNA [73].

3. Almost all of the sites that have been implicated in translation are located within these conserved cores [72,73,76–78]. This suggests that these cores represent the basic functional unit of each rRNA molecule and probably the regions of contemporary rRNA molecules that first arose during evolution [47,61,73,76].

4. Structurally, the SSU rRNA can be divided into three major domains and a 3' minor domain [61,77]. The 3' minor domain is involved in translational initiation and decoding, the 3' major domain in elongation and termination, the central domain in subunit association and the 5' domain in translational accuracy [72,73,77].

5. The LSU rRNA can be divided into six domains [62,73,76,77]. Domain II contains the GTPase center, certain regions of Domains IV and V are part of the peptidyltransferase center and Domain VI is involved in elongation [73,77].

6. Within the variable extension segments of each of the rRNA molecules, certain structural elements can be identified which are only present in one lineage and can be used to identify the members of that lineage [50,61,62].

7. Reconstitution experiments in which archaeal 5S rRNA was incorporated into bacterial large subunits [79] and bacterial 5S rRNA into archaeal large subunits [80] to produce active ribosomes, indicate that the tertiary structure of this rRNA has also been conserved.

8. Comparison of the binding sites on rRNA for r-proteins shows that structurally many of these binding sites have been conserved in the three lineages [73,76]. However, very few experiments have been done to show that a r-protein from one lineage can bind to the rRNA from the other two lineages. Examples are the L11 proteins from *E. coli* and *S. cerevisiae* (see subsection 3.2.1.2 on the GTPase center) [81,82], protein L1 from *E. coli* [83] and *M. vannielii* [84] which can bind to eucaryal LSU rRNA, L23 from *E. coli* and its eucaryal counterpart, L25 from yeast, which can bind to the heterologous

LSU rRNA [73,85], and the L10 protein from *E. coli* and its archaeal equivalent from *M. vannielii* which can be incorporated into the heterologous large subunit [86].

3.2.1. Functional domains

Although most of the studies concerning the function of rRNA have been carried out using the rRNA from *E. coli* (see ref. [72]), there are some functional regions for which there is information available from the archaeal domain [29,30,81,87,88]. Furthermore, since most of the functional centers are located in the regions of rRNA that are phylogenetically conserved, most of the results obtained from *E. coli* can be extended to the archaea and eucarya.

3.2.1.1. The peptidyltransferase center. The central loop of Domain V of the LSU rRNA constitutes the major element of the peptidyltransferase center of the ribosome and has been highly conserved in the three lineages (see Fig. 2) [72,73,77,89]. Evidence of its role in translation was first suggested by the mapping of mutations in this loop that give resistance to antibiotics that affect peptidyl transfer in bacterial and chloroplast ribosomes (see Fig. 2) [89]. Recently, some of these sites have also been shown to be protected against chemical modification in the presence of these antibiotics, indicating that the drug is interacting with the regions containing these nucleotides [91]. Other evidence for its role in translation comes from the crosslinking of azidopuromycin [92] and Phe-tRNA [93] to this region, as well as protection against chemical modification by tRNA bound at the A, P, and E sites [94].

Studies on archaeal ribosomes have shown that mutations that confer resistance to the peptidyl-transfer inhibitor anisomycin (an antibiotic that also affects eucaryal ribosomes) also map to this central loop [87]. Sequencing of several archaeal LSU rRNA has also revealed that they have either a G or a U at a position equivalent to A2058, which in bacteria produces sensitivity to erythromycin. As a consequence of this change, archaea are insensitive to this drug [29,30,87,88,95].

3.2.1.2. The GTPase center. The GTPase center is located in Domain II of the LSU rRNA and contains the binding site for protein L11 [62,76,96]. This region is one of the most highly conserved structural features of the LSU rRNA, being clearly distinguishable even in the highly reduced LSU rRNA from trypanosomal mitochondria [62,76]. Studies with protein L11 from *E. coli* have shown that it can bind to the equivalent region of both archaeal [81] and eucaryal [82] LSU rRNA. The eucaryal equivalent protein has also been shown to bind to bacterial LSU rRNA [82]. However, although archaeal L11 proteins have been identified, they have not been tested for binding [97]. Recently, it has also been shown that the *S. cerevisiae* GTPase center can be entirely replaced by the equivalent region from *E. coli* without loss of function [73].

The GTPase center is involved in the interaction of the elongation factors with the ribosome and is responsible for triggering their GTPase activity [98]. Evidence for the function of this region initially came from studies on the interaction of thiostrepton with bacterial ribosomes [98]. This antibiotic, which inhibits elongation-factor-G-dependent GTP hydrolysis, was found bound to a complex containing 23S rRNA and protein L11 [99]. Analysis of *Streptomyces azureus,* the organism that produces this antibiotic, showed that this organism is insensitive due to the methylation of an adenine at position 1067 [100]. Bacteria, which are sensitive to this antibiotic, lack this modification [98,101]. Archaea also have an unmodified adenine at this position and are

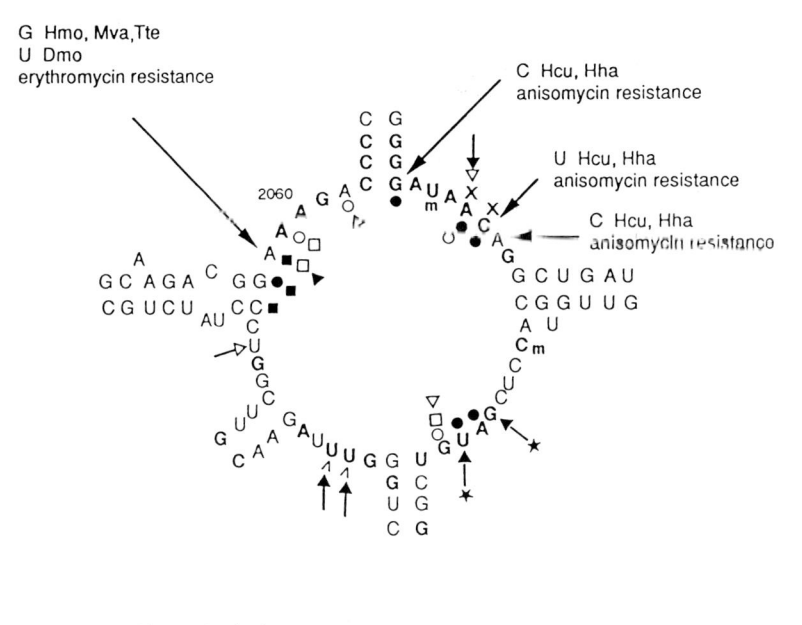

Fig. 2. The peptidyltransferase center. The structure of the central loop of Domain V of
E. coli 23S rRNA is shown. Nucleotides involved in resistance against different inhibitors are indicated.
Closed symbols indicate resistance and open symbols protection against chemical modification
by bound antibiotic. Mutations that confer resistance to anisomycin in archaea are indicated [87]
(Hcu, *Halobacterium cutirubrum*; Hha, *H. halobium*). The presence of either a G or U at
position 2058 in archaea is also indicated. As a consequence of this change archaea are resistant
to erythromycin (Hmo, *Halococcus morrhuae*; Mva, *Methanococcus vannielii*; Tte, *Thermoproteus
tenax*, Dmo, *Desulfurococcus mobilis*) [29,30,88,90]. Positions where crosslinking to photoreactive
derivatives of Phe-tRNA and puromycin have been observed as well as nucleotides protected by
bound tRNA are also indicated. Modified from ref. [73].

sensitive, while eucarya have a guanine residue and are insensitive [29,98,101]. Further evidence of the function of this region has been provided by analysis of bacterial and archaeal mutants resistant to thiostrepton [98,101,102], site-directed mutagenesis of this site [103], chemical protection experiments using thiostrepton [101] and EF-G [104], and by crosslinking of EF-G to this region [95].

4. Structure of archaeal ribosomal proteins

The r-proteins, like the rRNA, are important phylogenetic probes in the study of molecular evolution. In addition, comparative studies on the amino-acid sequence of these proteins often reveal regions that are highly conserved, suggesting these regions may have important functional roles. The sequences of a large number of archaeal r-proteins from *Haloarcula marismortui, Methanococcus vannielii* and *Sulfolobus* are now available, either from the sequencing of the purified protein or from the structure of the r-protein genes. In addition, the ability to overexpress archaeal r-protein genes in *E. coli* has allowed studies into the functional interchangeability of diverse ribosomal proteins [105]. (see section 4.5).

4.1. Nomenclature of ribosomal proteins

Since all the r-proteins from both the large (L) and the small (S) ribosomal subunits of *E. coli* are sequenced and well characterized [106], homologous proteins from other organisms are named according to their *E. coli* counterpart. For example, proteins homologous to L23 of *E. coli* (Eco L23) will be called Sac L23 in *Sulfolobus acidocaldarius,* Hma L23 in *Haloarcula marismortui* and Sce L23 in *Saccharomyces cerevisiae*. In Table 2 (below), we have used this nomenclature to identify any r-protein that, from its sequence similarity, was equivalent to a known *E. coli* r-protein. Proteins which have no counterparts in *E. coli* (or where the sequence similarity is too weak to identify the relationship) are listed by the numbers described in the original publications and a trivial name. These numbers usually refer to the mobility of the protein in a 2D polyacrylamide gel. For example, HS15 is a protein from the small ribosomal subunit of *H. marismortui* which has no counterpart in *E. coli* [107].

4.2. Comparison of the archaeal r-proteins with those from bacteria and eucarya

Protein sequence comparisons are a powerful tool to determine the evolutionary relationship between homologous proteins in the various organisms. The sequence of these proteins, even from distantly related organisms, is often very similar. For example, the primary sequence of r-protein L2, believed to be part of the peptidyltransferase center [108], is identical in humans and hamsters. On the other hand some r-proteins, such as L4, appear to be of more structural importance and are so distantly related between eucarya and bacteria, that a relationship could only be established with the aid of the archaeal protein sequences [107].

In Table 1 the archaeal r-proteins are listed whose sequences were available at the time this chapter was prepared. These data were obtained from the RIBO database of the Max-Planck-Institut für Molekulare Genetik, Berlin, which contains over 860 published and unpublished r-protein sequences as well as partial protein sequences (December, 1991). Using this database, the similarity (identical amino acids at comparable locations) of the r-proteins from the three domains (archaea, bacteria and eucarya) was calculated and the results are shown in Table 2.

For some archaeal r-proteins a homologous protein was found in the eucarya but no bacterial counterpart could be detected in the complete set of *E. coli* r-proteins. This

TABLE 1

Archaeal ribosomal protein sequences in the RIBO Database (Max-Planck-Institut für Molekulare Genetik, Berlin; Boeckh, Köpke, Beck, Dzionara, Wittmann-Liebold) up to December, 1991.

	Proteins	N-terminal sequences
Euryarchaeota		
Halophilic archaea		
Haloarcula marismortui	44	6
Halobacterium cutirubrum	5	20
Halobacterium halobium	11	2
Halobacterium mediterranei	–	2
Halobacterium salinarium	–	2
Halobacterium volcanii	–	1
Haloanaerobium praevalens	1	–
Methanogenic archaea		
Methanococcus vannielii	34	46
Methanobacterium thermoautotrophicum	–	1
Crenarchaeota		
Sulfur-dependent thermophilic archaea		
Sulfolobus acidocaldarius	22	40
Sulfolobus solfataricus	–	3

means either that no bacterial r-proteins homologous to these sequences are present, or that the protein sequences are so diverged that no significant similarity can be detected. An example in which the primary structure of a protein in bacteria and its counterpart in the other two domains are so diverged that no unambiguous alignment could be found is r-protein L12 [109]. Other archaeal r-proteins have equivalents in the bacteria but none as yet in the eucarya, which is likely due to the incomplete set of eucaryal r-proteins available for sequence comparison at this time.

The following general conclusions can be made from the r-protein data available from the three domains (Table 2):

1. The sequences of the various r-protein families have diverged to a different extent, and this is likely dependent on the function of the individual r-protein in the ribosome.

2. The r-proteins show, in general, much more sequence similarity within a domain than they do to sequences of the equivalent proteins from the other domains.

3. In general, the archaeal r-proteins are more related to those of the eucarya than to those of the bacteria, and vice versa.

4. The sequence similarity of the archaeal r-proteins to the equivalent proteins in eucarya increases from the extreme halophiles to the methanogens (in the Euryarchaeota) to the sulfur-dependent extreme thermophiles (in the Crenarchaeota), while the similarity to the bacterial r-proteins increases in the reverse order.

In summary, the r-protein sequence data (Table 2) show that the bacteria, archaea and eucarya are well-defined and separate domains. It would also appear from the r-protein sequence data that the archaea are more diverged as a group of organisms than are the bacteria and eucarya, suggesting a more ancient origin of these organisms.

TABLE 2
Similarity of ribosomal proteins from the different domains[a,b]

Ribosomal Protein Family	Bacteria	Archaea	Eucarya	$\frac{Bac}{Arch}$ (%)	$\frac{Bac}{Euc}$ (%)	$\frac{Arch}{Euc}$ (%)
S3	EcoS3 BstS3 (59%) McaS3 (47%)	HmaS3 MvaS3 (44%) HhaS3 (76%)	RnoS3	18	15	27
S5	EcoS5 BstS5 (55%) McaS5 (42%)	HmaS5 MvaS5 (49%) SacS5*(46%)	SceS5 MmuS5 (59%) HsaS5 (58%)	29	26	37
S7	EcoS7 BstS7 (57%) TaqS7 (52%) CpdS7 (50%) SplS7 (55%) AniS7 (54%) MluS7 (53%)	HmaS7 HhaS7 (84%) MvaS7 (47%) SacS7 (50%)	–	27	–	–
S8	EcoS8 BstS8 (49%) McaS8 (43%) TaqS8 (49%)	HmaS8 MvaS8 (53%) SacS8*(38%)	SceS8	25	24	44
S9	EcoS9 BstS9 (54%)	HmaS9	RnoS9 MmuS9 (99%) LpoS9 (66%)	29	29	42
S10	EcoS10 CphS10 (52%) CpdS10 (51%) McaS10 (49%)	HmaS10 MvaS10 (44%) SacS10 (54%)	RnoS10 XlaS10 (97%)	33	26	34
S11	EcoS11 BstS11 (68%) BsuS11 (80%)	HmaS11	SceS11 CgrS11 (79%)	38	30	47
S15	EcoS15 BstS15 (58%)	HmaS15	RnoS15	28	20	45
S17	EcoS17 BstS17 (48%) BsuS17 (48%) McaS17 (51%) TqaS17 (44%)	HmaS17 MvaS17 (44%)	RnoS17	32	36	25
S19	EcoS19 YpsS19 (96%) CpdS19 (69%) BstS19 (66%) McaS19 (57%)	HmaS19 HhaS19 (81%)	–	34	–	–
–	–	HS15	XlaS19 RnoS19 (98%)	–	–	44

continued on next page

TABLE 2, continued

Ribosomal Protein Family	Bacteria	Archaea	Eucarya	Bac/Arch (%)	Bac/Euc (%)	Arch/Euc (%)
–	–	HS12	YeaS19	–	–	33
–	–	HS13	YeaS10	–	–	27
L1	EcoL1 BstL1 (52%) PvuL1 (89%) SmaL1 (95%)	HmaL1 Hha/HcuL1(79%) MvaL1 (42%) SsoL1 (30%)	–	28	–	–
L2	EcoL2 BstL2 (57%) YpsL2 (96%) McaL2 (52%)	HmaL2 MvaL2 (59%)	SpoL2 DdiL2 (61%)	37	27	39
L3	EcoL3 BstL3 (49%) CpdL3 (44%) YpsL3 (93%)	HmaL3 MvaL3* (50%)	SceL3 AthL3 (64%) RnoL3 (19%) HsaL3 (21%)	24	34	21
L4	EcoL4 BstL4 (43%) McaL4 (36%) YpsL4 (94%)	HmaL4 MvaL4* (48%)	SceL4 XlaL4 (56%) DmeL4 (53%)	22	19	31
L5	EcoL5 BstL5 (60%) BsuL5 (59%) McaL5 (55%) TaqL5 (56%) TflL5 (56%)	HmaL5 MvaL5 (44%) SacL5 (41%)	SceL5 DdiL5 (80%)	28	28	41
L6	EcoL6 BstL6 (48%) McaL6 (42%) TaqL6 (35%)	HmaL6 MvaL6 (41%) SacL6* (34%)	RnoL6	15	14	34
L10	EcoL10	HhaL10 MvaL10 (36%) SsoL10 (27%)	SceL10 RnoL10 (54%) HsaL10 (54%)	nd	nd	26
L11	EcoL11 SmaL11 (94%) PvuL11 (89%) BstL11* (68%)	HcuL11 HmaL11 (78%) SsoL11 (40%)	SceL11	34	18	23
L12	EcoL12 BstL12 (64%) Bsu12 (52%)	Hha/HcuL12 HmaL12 (68%) Mva12 (43%) SsoL12 (45%)	SceL12 AsaL12 (50%) DmeL12 (50%) RnoL12 (45%) HsaL12 (56%)	nd	nd	31
L14	EcoL14 BstL14 (69%) BsuL14 (64%) McaL14 (53%) TaqL14 (70%)	HmaL14 MvaL14 (52%)	SceL14	36	31	44

continued on next page

TABLE 2, continued

Ribosomal Protein Family	Bacteria	Archaea	Eucarya	$\frac{Bac}{Arch}$ (%)	$\frac{Bac}{Euc}$ (%)	$\frac{Arch}{Euc}$ (%)
L15	EcoL15 BstL15 (48%) McaL15 (49%)	HmaL15 MvaL15 (40%) SacL15* (22%)	SceL15 MmuL15 (61%) NcrL15 (70%)	15	16	27
L18	EcoL18 BstL18 (53%) McaL18 (40%)	HmaL18 MvaL18 (49%) SacL18* (42%)	RnoL18 XlaL18 (91%)	21	19	33
L22	EcoL22 BstL22 (52%)	HmaL22 HhaL22 (65%) MvaL22 (42%)	HsaL22	29	27	33
L23	EcoL23 BstL23 (33%) McaL23 (39%) YpsL23 (92%)	HmaL23 MvaL23 (36%) SacL23 (36%)	SceL23 CutL23 (82%)	32	28	38
L24	EcoL24 BstL24 (47%) BsuL24 (46%) TaqL24 (44%) McaL24 (37%) PsaL24 (34%)	HmaL24 MvaL24 (47%)	RnoL24	23	22	33
L30	EcoL30 BstL30 (53%)	HmaL30 MvaL30 (40%) SacL30* (30%)	SpoL30 DdiL30 (46%) MmuL30 (48%) RnoL30 (48%) HsaL30 (49%)	24	21	22
–	–	HL21 HL24 MvaO1* (42%) SacOE* (43%)	YeaL9 RnoL19	–	–	26 35
–	–	HL30	YeaL34 RnoL31 (58%)	–	–	25
–	–	HL46 SsoL46 (45%)	SceL46 RnoL39 (62%)	–	–	40

[a] Within the domain the numbers following a protein name are the percentages of identical residues at matched positions (gaps not included) of this protein sequence to the first protein sequence in the list. These numbers were calculated using the GAP program of the GCG6 program package (employing the new default values). The similarity between the domains is the calculated average of all protein similarity values between these two domains utilizing a multiple alignment (LINEUP) of these proteins and the DISTANCES (GCG program package) values for identical residues at matched positions. The protein names were changed according to the standard nomenclature that ribosomal proteins similar to *E. coli* ribosomal proteins get a similar name. The proteins in this table are contained in the RIBO database and/or the PIR database. Proteins marked with an asterisk were unpublished at the time of preparation of this table. References to the individual proteins are available from the databases.

continued on next page

TABLE 2, continued

[b] Abbreviations: Ani, *Anacystis nidulans*; Asa, *Artemia salina*; Ath, *Arabidopsis thaliana*; Bst, *Bacillus stearothermophilus*; Bsu, *Bacillus subtilis*; Cgr, *Cricetulus griseus*; Cpd, *Cyanophora paradoxa*; Cph, *Cryptomonas phi*; Cut, *Candita utilis*; Ddi, *Dyctyostelium discoideum*; Dme, *Drosophila melanogaster*; Eco, *Escherichia coli*; Hcu, *Halobacterium cutirubrum*; Hha, *Halobacterium halobium*; Hma (or H), *Haloarcula marismortui*; Hmo, *Halococcus morrhuae*; Hsa, *Homo sapiens*; Lpo, *Lupinus polyphyllus*; Mca, *Mycoplasma capricolum*; Mlu, *Micrococcus luteus*; Mmu, *Mus musculus*; Mva, *Methanococcus vannielii*; Ncr, *Neurospora crassa*; Psa, *Pisum sativum*; Pvu, *Proteus vulgaris*; Rno (or Rat), *Rattus norwegicus*; Sac, *Sulfolobus acidocaldarius*; Sce (or Yea), *Saccharomyces cerevisiae*; Sma, *Serratia marcescens*; Spl, *Spirulina platensis*; Spo, *Schizosaccharomyces pombe*; Sso, *Sulfolobus solfataricus*; Taq, *Thermus aquaticus*; Tfl, *Thermus flavus*; Xla, *Xenopus laevis*; Yps, *Yersina pseudotuberculosis*.

4.3. The L2 r-protein family

Some r-proteins show a great deal of sequence similarity in all three domains. One such protein is L2, which is thought to be part of the peptidyltransferase center[108]. The peptidyltransferase center, containing the central loop of Domain V of the LSU rRNA (see section 3.2.1.1), is present in the upper central region of the large ribosomal subunit[110], and of the r-proteins located in this region (L3, L4, L16 and L2), only the L2 r-proteins from the three domains show a striking degree of structural similarity (see Fig. 3). It is thought, therefore, that this protein may play a key role in the peptidyl-transfer reaction, especially a region in the C-terminal portion of the protein molecule which is very conserved. Site-directed mutagenesis of residues in this protein[111] should provide an insight into its possible role in translation.

4.4. The stalk protuberance in the large ribosomal subunit (r-proteins L12/L10)

The stalk region, present in all large ribosomal subunits, consists of at least one dimer of r-protein L12. It has been shown that two dimers of the L12 protein are bound to one molecule of L10 protein, which in turn binds to the large subunit r-RNA[117]. In *E. coli*, the second dimer of L12 is located in the body of the large ribosomal subunit at the base of the stalk[118]. The L12 proteins can be removed from the *E. coli* ribosome by salt/ethanol washes[119] either as the two dimers or in a pentameric complex with L10.

When the amino-acid sequences of the L12 protein from the three domains are compared, it is obvious that the L12 protein has changed dramatically during evolution. The sequences of the bacterial L12 protein cannot be unambiguously aligned to the equivalent proteins from archaea and eucarya[109,120] while the multiple alignment of the L12 proteins from the latter two domains shows significant similarities (Fig. 4), as first suggested by the results of Amons et al.[121] when they compared the N-terminal regions of the *H. cutirubrum* and *A. salina* L12 proteins. In addition, the C-terminal region of the archaeal and eucaryal L12 and L10 proteins, which contain a very hydrophilic and highly charged region of alternating pairs of glutamic acids and lysine residues, is highly conserved[97,131,133]. In *Sulfolobus*, for example, the last 33 amino-acid residues of the L10 and L12 proteins are identical[97]. Interestingly, the conservation of this region is higher between the L10 and L12 proteins of the same organism than is the overall

```
          40         50         60         70         80         90        100        110        120       130
DdiL2   MGRIIRAQRKGKAGSVFGAHTHHRKGTPRFRALDYAERQGYVKGVVKEIIHDPGRGAPLARVVFKGLTQFKLDKQLFIAPEGMHTGFQ
         *               **                  *          **         *     *   **    *            *
SpoL2   MGRVIRAQRKS..GGIFQAHTRLRKGAAQLRTLDFAERHGYIRGVVOKIIHDPGRGAPLAKVAFRNPYHYRTDVETFVATEGMYTGQF
         *               **                  *          ***        *     *   *  **  *            *
MvaL2   MGKRLISQNRGRGTPKYRSPSHKRRGEVKYRSYDEMEKVGKVLGTVIDVLHDPGRSAPVAKVRFANG.....EERLVLIPEGISVGEQ
                              *                         *****        *                           *
HmaL2   GRRIQGQRRGRGTSTFRAPSHRYKADLEHRKV...EDGDVIAGTVVDIEHDPARSAPVAAVEFEDG.....DRLLILAPEGVGVGDE
                               *                        *  ***        *      *   ****            *
EcoL2   SKSGGRN...NNGRITTRHIGGGHKQAY....RIVDFKRNKDGIPAVVERLEYDPNRSANIALVLYKDG.....ERRYILAPKGLKAGDQ
         *                              *                **  *     **   *    *    **    *          *
BstL2   KKRAGRN...NQGKITVRHQGGGHKRQY....RIIDFKRDKDGIPGRVATIEYDPNRSANIALINYADG.....EKRYIIAPKNLKVGME
         *                              *                **   *    **   *    *    **              *

          131        140        150        160        170        180        190        200        210       220
DdiL2   FVAGKKATLTIRHILPIGKLPEGTIICNVEEKLGDCGAVARCSGNYATIVSHNPDEGVTLYQTIRFKEERSSLARAMIGIVAGGRIDK
         **            **        * ***    *            * **                   *                 ****
SpoL2   VYCGKNAALTVGNVLPVGEMPEGTIISNVEEKAGDRGALGRSSGNYVISVGHDVDTGKTRVKLPSGAKKVVPSSARGVVGIVAGGRIDK
         **            **        * ***    *            * **                                     ****
MvaL2   IECGISAEIKPGNVLPLGEIPEGIPVNIETIPGDGGKLIVEAGGCYAHVVAH..DIGKTIVKLPSGYAKVLNPACRATIGVVAGGRKEK
         **            **       ** ***    *            *  *                 *                  ****
HmaL2   LQVGVDAEIAPGNTLPLAEIPEGVPVCNVESSPGDGGKRARASGVNAQLLTH..DRNVAVVKLPSGEMKRLDPQCRATIGVVGGGRTDK
         **            **       ** ***    *            *  *                 *                 * ****
EcoL2   IQSGVDAAIKPGNTLPMRNIPVGSTVHNVEMKPGKGGQLARSAGTYVQIVARD..GAYVTLRLRSGEMRKVEADCRATIGEVGNAEHMLR
         **            **        * ***    *            *  *                 *
BstL2   IMSGPDADIKIGNALPLENIPVGTLVHNIELKPGRGGQLVRAAGTSAQVIGKE..GKYVIVRLASGEVRMILGKCRATVGEVGNEQHELV
         **            **        * ***    *            *  *                 *

          221        230        240        250        260        270        280        290        300
DdiL2   PMLKAGRAFHKYRVKKNNWPKVRGVAMNPVEHHTVVIINMLV...M....PLQP............RETIQLVRKLV
         ***                                             *
SpoL2   PLLKAGRAFHKYRVKRNCWPRTRGVAMNPVDHPHGGGNHQHVG...HSTTVPRQSAPGQKVG.....LIAARRTGLLRGAAAVEN
         ***   *     **                           ***   *
MvaL2   PFVKAGKKHHSLSAKAVAWPKVRGVAMNAVDHPYGGGRHQHLG...KPSSVSRNTSPGRKVG......HIASRRT
         ***   *     **    ******                 ***   *
HmaL2   PFVKAGNKHHKMKARGTKWPNVRGVAMNAVDHPFGGGGRQHPG...KPKSISRNAPPGRKVG.......DIASKRTGRGGNE
         ***   *     **    ******                 ***   *
EcoL2   VLGKAGAARW......RGVRPTVRGTAMNPVDHPHGGGEGRNFG..KHPVTPWGVQTKGKKTRSNKRT.DKFIVRRRSK
         ***   *                ***                  ***   *
BstL2   NIGKAGRARW......LGVRPTVRGSVMNPVDHPHGGGEGKAPIGRKSPMTPWGKPTLGYKTRKKKNKSDKFIIRRKK
         ***   *                ***                  ***   *
```

Fig. 3. Comparison of the amino-acid sequence of ribosomal protein L2 from eucarya (Ddi[112] and Spo[113]), archaea (Mva[109] and Hma[114]) and bacteria (Eco[115] and Bst[116]). See Table 2 for abbreviations. An asterisk indicates amino acids conserved in all L2 proteins while a dot indicates amino acids conserved in at least two of the three domains.

```
            1        10        20        30        40        50        60
Hma      MEYVYAALILN.EADEEINEDNLTDVLDAAGVDVEESR.VKALVAALEDVD.IEEAVDQA
         * ** * *  •                ••     *          *      • ••
Hha      MEYVYAALILN.EADEELTEDNITGVLEAAGVDVEESR.AKALVAALEDVD.IEEAVEEA
         * ** * *  •                ••     *          *      • •••
Sso      MEYIYASLLLH.AAKKEISEENIKNVLSAAGITVDEVR.LKAVAAALKEVN.IDEILKTA
         * *  *  ••• •              •• ••    *  •         *  •  • ••
Mva      MEYIYAALLLN.SANKEVTEEAVKAVLVAGGIEANDAR.VKALVAALEGVD.IAEAIAKA
         * *   * ••  •               ••  *••  •   •  *  • • • •  ••
Spo      MKYLAAYLLLTVGGKDSPSASDIESVLSTVGIEAESER.IETLINELNGKD.IDELIAAG
         * *   * ••• •              *  •        *   •     •  •  • ••
Sce      MKYLAAYLLLVQGGNAAPSAADIKAVVESVGAEVDEAR.INELLSSLEGKGSLEEIIAEG
         * *   * •••                *  ••         •           ••  ••
Asa      MRYVAAYLLAALSGNADPSTADIEKILSSVGIECNPSQ.LQKVMNELKGKD.LEALIAEG
         * ••  * *•                    *••      •  •     •    •  • ••
Dme      MRYVAAYLLAVLGGKDSPANSDLEKILSSVGVEVDAER.LTKVIKELAGKS.IDDLIKEG
         * ••  * * *                *             •    •      •  • ••
Rno      MRYVAAYLLAALGGNSNPSAKDIAKILDSVGIEADDERKLKNVISELNGKN.IEDVIAQG
         * ••  * *                  *•• • • •      *   •       ••  ••
Hsa      MRYVAAYLLAALGGNSSPSAKGIKKILDSVGIEADDDR.LKNVISELNGKN.IEDVIAQG
         * ••  * *                  *•• • • •      *   •       ••  ••

            60       70        80        90       100        110      118
Hma      AAAPVPASGGAAAPAEGDADEADEADEEAEEEAADDGGDDDDDEDDEASGEGLGELFG
           •                 •     •         •           • •        **
Hha      AAAPA.AAPAASGSDDEAAADDGDDDEEADADEAAEAEDAGDDDDEEPSGEGLGDLFG
          •           ••        •      •               ••     **
Sso      TAMPV.AA........VAAPAGQQTQQAAEKKEEKKEEEKKGPSEEEIGGGLSSLFG
          •          •           •   • •••  • • •• • •  ••      **
Mva      AIAPV.AA.........AAPVAAAAAPAEVKKEEKKEDTTAAA.....AAGLGALFM
                                  • •• •••   • •                **
Spo      NEKLA.TVPTGGA...ASAAPAAAAGGAAPAAEEAAKEEAK.EEEESDECDMGFGLFD
              •                             • • • •             **
Sce      QKKFA.TVPTGGAS..SAGPASAGAAAGGGDAAEEKEEEAK....EESDDDMGFGLFD
              •                 •• •       • • • ••            **
Asa      QTKLA.SMPTGGAPAAAAGGAAAAPAAE....AKEAKKEEKKEESSEEDEDMGFGLFD
              •              • •             • •  ••            **
Dme      REKLS.SMPVGGGGAVAAA..DAAPAAAAGGDKKEAKKEEKKEESESEDDDMGFGLFD
              •        •      ••  •           • •• • •          **
Rno      VGKLA.SVPAGGAVAVSAAPGAAAPAAGSAPAAAEEK.....EESEEKKDDMGFGLFD
              •       •  • ••• ••   • • •                ••••   **
Hsa      IGKLA.SVPAGGAVAVSAAPGSAAPAAGSAPAAAEEKKDEKKEESEESDDDMGFGLFD
              •       •  • ••• ••   • • •    •• ••••             **
```

Fig. 4. Comparison of the amino-acid sequence of ribosomal protein L12 from archaea (Hha [122], Hma [123], Sso [124], Mva [125]) and eucarya (Spo [126], Sce [127], Asa [128], Dme [129], Rno [130], Hsa [131]). See Table 2 for abbreviations. Hcu [124] is identical to Hha, while Sac [132] is identical to Sso except that a lys at position 46 in Sso is replaced by a glu in Sac. An asterisk indicates amino acids conserved in all archaeal and eucaryal L12 proteins while a dot indicates amino acids conserved in both domains in at least half of the sequences shown.

conservation of this region between the homologous proteins of the different archaeal and eucaryal organisms, suggesting the coevolution of this region in these two proteins in order to conserve an important functional property [133]. In bacteria there is no evidence of sequence similarity between L12 and L10 in the C-terminal region [106].

The overall structure of the bacterial L12 protein consists of a globular C-terminal region connected by a flexible hinge region to a rod-like N-terminal region [134]. The N-terminal region binds to the L10 protein and is also responsible for the dimerization of this molecule [135]. The C-terminal region, however, interacts with the different factors involved in translation [136].

Recent experiments on the archaeal L12 protein from *Sulfolobus* indicate that the ribosome-binding site and the dimerization site of this protein involve the N-terminal region as was the case in the bacterial protein [137]. The hinge region can easily be detected by its special amino-acid composition. Since the C-terminal region in the archaeal protein is not involved in ribosome binding, it is likely implicated in factor binding. These results indicate that earlier models [138], suggesting the N-terminal region of the bacterial L12 is equivalent to the C-terminal region of archaeal and eucaryal L12, are incorrect.

4.5. Interchangeability of ribosomal components from different organisms

A valuable test for tertiary structure similarity of ribosomal constituents from different domains is their interchangeability into other organisms. For example, *Sulfolobus* 5S rRNA can be incorporated into *E. coli* ribosomes or into yeast ribosomes, and the hybrid ribosomes are fully active [80]. The same is true with the incorporation of protein L1 from *M. vannielii* [84]. In vitro experiments indicated the L12 protein from *H. marismortui* can be replaced by *M. vannielii* L12 to give an active translation system in the high-salt environment needed for protein synthesis in extreme halophiles. However, attempts to incorporate the L12 protein from *E. coli* or *B. stearothermophilus* into this halophilic ribosome were unsuccessful [105], confirming a similar observation with the L12 r-protein from *H. cutirubrum* [139]. These experiments indicate therefore that certain ribosomal components can be interchanged while others are so diverse in structure that they can only be interchanged within a similar domain (kingdom). What is surprising, however, is the ability of the L12 protein from a non-halophilic organism (*M. vannielii*) to be fully active in the in vitro halophilic translation system. The r-proteins from extreme halophiles appear to be much more acidic than the equivalent proteins from non-halophilic sources and one might have expected denaturation of the non-halophilic L12 protein in the above experiment. It will be of interest to see whether other non-halophilic r-proteins from archaea are active in chimeric halophilic ribosomes or whether halophilic r-proteins are active in chimeric *E. coli* ribosomes.

5. Archaeal ribosomal protein genes

5.1. Gene organization

The organization of archaeal r-protein genes [8,9,12,84,107,109,114,120,122,123,133, 140–152] resembles that found in the bacteria, where the r-protein genes are clustered in operons [153–155] (Fig. 5), and differs therefore from the eucarya where the genes are dispersed throughout the genome and are transcribed as single units [156]. In addition, the order of the r-protein genes in the archaea is very similar to that found in bacteria. However, even though the organization of the r-protein genes is similar in these two domains, some important differences have been observed:

1. The position of promoters and terminators in the archaeal r-protein genes is different from that in *E. coli,* and even varies within the archaea [8,9,12,84,109,114,120,122,123, 140–146,148–152,155,157–159]. As a consequence of this, the transcriptional units are different in the two domains (see Fig. 5).

2. Some archaeal r-protein operons contain open reading frames (ORFs) that are not present in the equivalent bacterial operon. In many cases, these ORFs code for proteins which show sequence similarity to eucaryal r-proteins. For example, ORF c, d and e in the 'spc' operon of *Methanococcus vannielu* [100] and *Sulfolobus acidocaldarius* [Kusser et al., unpublished] code for r-proteins equivalent to the eucaryal r-proteins S4 (*Rattus norwegicus*), L32 (*Homo sapiens*) and L19 (*R. norwegicus*) respectively.

3. Some archaeal r-protein genes have been located in operons that have no equivalents in *E. coli*. Examples are the genes for L46 [149] and LX [12] in *Sulfolobus solfataricus*.

4. Some genes that form part of a certain operon in *E. coli,* are not part of that operon in the archaea. For example, the S10 gene, which is the first gene of the S10 operon in the bacteria [159] is part of the 'str' operon in the archaea [141,143–145,148,160] (see Fig. 5), while S17, which is the last gene in the S10 operon in bacteria, is the first gene in the 'spc' operon in archaea [9]. The 'str', S10 and 'spc' operons are much closer together than they are in bacteria and it is possible that they were once connected and part of a single large operon.

5.2. Transcription

Archaeal promoters resemble the RNA II promoters of the eucarya [40,161] and are different from those in the bacteria [162]. Analysis of the promoters of archaeal r-protein genes shows the presence of two conserved regions: a hexanucleotide sequence (TTTAAA/T) located 20–30 nucleotides upstream of the transcription initiation site and two conserved nucleotides around the transcription initiation site (shown in lower case (Ta/gT/G)) [9,114,120,141,144,145,148,150]. Transcription in the archaea usually starts at a purine residue as is the case in the other two domains [161,162]. The presence of several initiation sites is common among the extreme thermophiles [120,141,148], a situation similar to that found for yeast r-proteins genes [156].

Transcription usually ends within or after pyrimidine-rich regions [120,141,145,150]. Among the methanogens, inverted repeats that might form stem–loop structures, reminiscent of the rho-independent terminators of bacteria, have been identified close to the termination sites [100]. Usually, there are multiple transcription stops in the archaea.

5.3. Translation signals

In the bacteria, the recognition of the initiator codon is mediated by the interaction of the $3'$ end of the 16S rRNA and a purine-rich region (the Shine–Dalgarno or SD sequence) located upstream of the initiation codon [148]. Putative SD sequences have been found upstream of many archaeal r-protein genes [8,9,12,84,109,114,120,122,123,133,140,143–146,148–152]. However, in no case have these sequences been shown experimentally to interact with the 16S rRNA. In *E. coli,* the distance between the SD sequence and the initiation codon is critical, greater than 5 nucleotides and less than 13 [163]. Fig. 6 shows

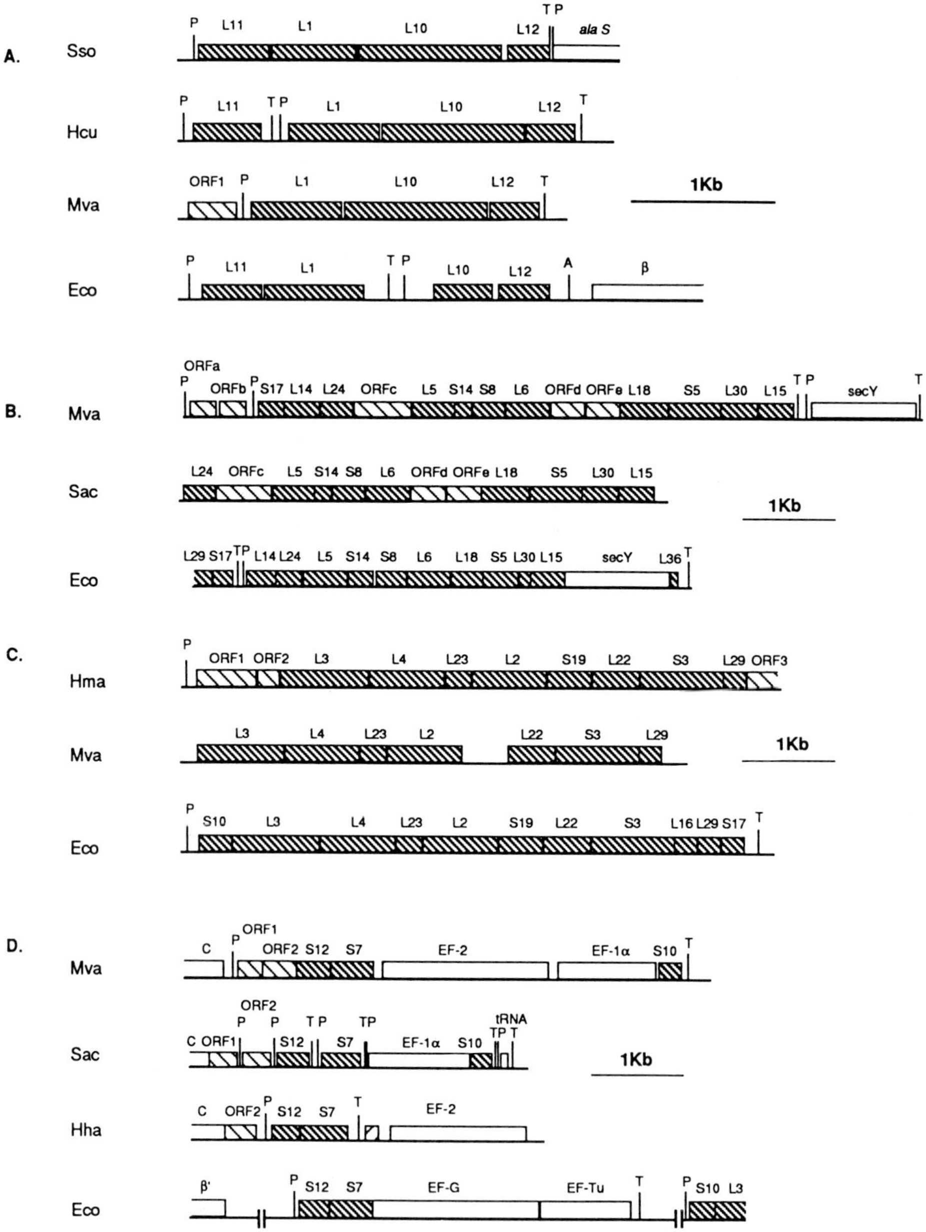
A.
Sso
P L11 L1 L10 L12 T P ala S
Hcu
P L11 T P L1 L10 L12 T
Mva
ORF1 P L1 L10 L12 T
1Kb
Eco
P L11 L1 T P L10 L12 A β
B.
Mva
ORFa
P ORFb P S17 L14 L24 ORFc L5 S14 S8 L6 ORFd ORFe L18 S5 L30 L15 T P secY T
Sac
L24 ORFc L5 S14 S8 L6 ORFd ORFe L18 S5 L30 L15
1Kb
Eco
L29 S17 T P L14 L24 L5 S14 S8 L6 L18 S5 L30 L15 secY L36 T
C.
Hma
P ORF1 ORF2 L3 L4 L23 L2 S19 L22 S3 L29 ORF3
Mva
L3 L4 L23 L2 L22 S3 L29
1Kb
Eco
P S10 L3 L4 L23 L2 S19 L22 S3 L16 L29 S17 T
D.
Mva
C P ORF1 ORF2 S12 S7 EF-2 EF-1α S10 T
Sac
ORF2
C ORF1 P P S12 T P S7 TP EF-1α S10 tRNA TP T
1Kb
Hha
C ORF2 P S12 S7 T EF-2
Eco
β' P S12 S7 EF-G EF-Tu T P S10 L3

Fig. 5. Comparison of the gene organization of the "L11", "L10", "spc", "S10" and "str" operons in the archaea with the Related Transcriptional Units in *Escherichia coli*. Gene sizes are drawn to scale. P and T indicate the location of promoters and terminators respectively. An A indicates a transcriptional attenuator. ORF indicates an open reading frame. (A) "L11" and "L10" operons: *ala* S, alanine–tRNA synthetase gene; β, gene for the β subunit of the *E. coli* RNA polymerase. Data: Sso (ref. [120], and Ramírez et al., in preparation), Hcu [150], Mva [84], Eco [158]. (B) "spc" operon: *sec* Y, protein export gene. Data: Mva [9], Sac [Kusser et al., unpublished], Eco [156]. (C) "S10" operon. Data: Hma [114], Mva [8,107,109], Eco [159]. (D) "str" operon: EF, elongation factor; C, gene for the C subunit of archaeal RNA polymerase; β', gene for the β' subunit of the *E. coli* RNA polymerase. Data: Mva [144], Sac [141,148], Hha [143,145], Eco [157]. Recent data [160] indicate that the "str" operon in the archaeate *Pyrococcus woesei* has a gene order EF-1α–S10–tRNA$_{ser}$ as in *Sulfolobus*.

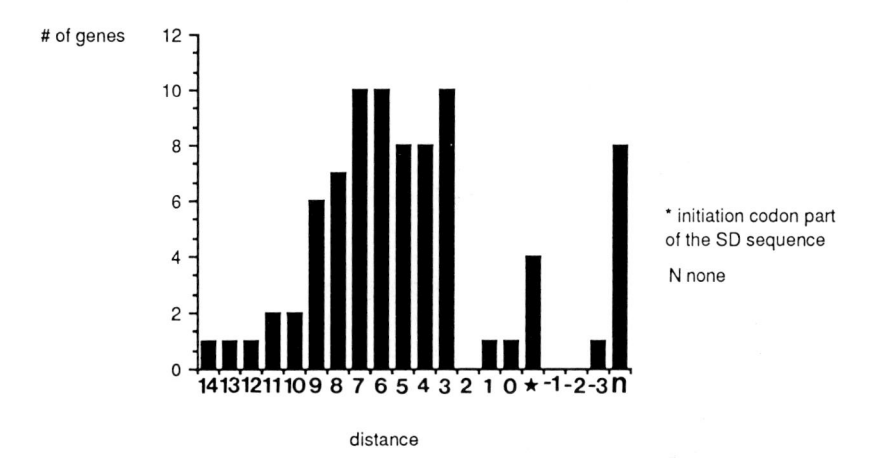

Fig. 6. Distance from the Shine–Dalgarno sequence to the Initiation Codon in archaeal genes. Data for this graph were obtained from refs. [8,9,12,84,114,120,122,123,133,140,142–146,148–152].

the distances of the putative SD sequences from the initiation codon of the different archaeal r-protein genes. Although in most cases the distance is similar to that found in the bacteria, there is a large number of genes in which the distance is less than 5 nucleotides. In addition, there are several genes that appear to lack a SD sequence or have a putative SD structure which includes the initiation codon or is located downstream from it (see Fig. 6), suggesting that the archaea might have other ways to identify the initiation codon.

5.4. Regulation

In bacteria, the synthesis of r-proteins is autogenously regulated at the level of translation [154,162,164]. When r-proteins are synthesized in excess, one of the r-proteins in the operon acts as a repressor of translation by binding to the mRNA. The binding of a single protein is able to block the translation of the whole operon because the translation of all the genes in the operon is coupled [154,162,164]. Although we still lack experimental evidence regarding the control of r-protein synthesis in the archaea,

L1 binding site:

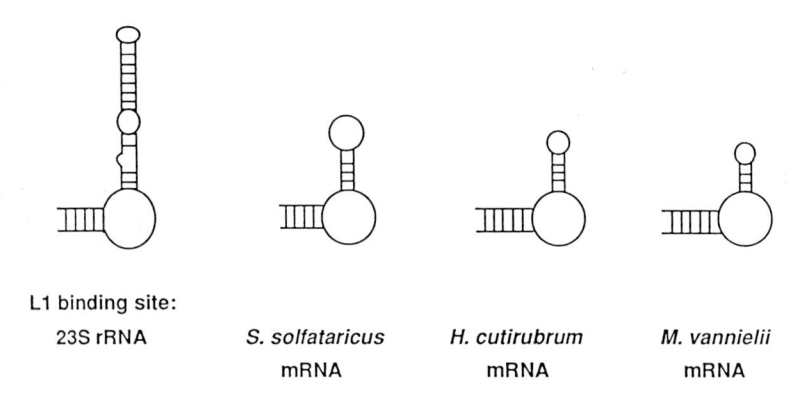

| 23S rRNA | S. solfataricus mRNA | H. cutirubrum mRNA | M. vannielii mRNA |

Fig. 7. Comparison of the secondary structure of the L1 binding site in archaeal 23S rRNA and similar structures in the mRNA from three different archaea. Structure of the L1 binding site in the archaeal 23S rRNA taken from ref. [145], *S. solfataricus* L11–L1–L10–L12 mRNA from [Ramírez et al. in preparation], *H. cutirubrum* L1–L10–L12 mRNA from ref. [150], and *M. vannielii* L1–L10–L12 mRNA from ref. [84].

certain structural features observed in the different operons suggest a control mechanism similar to that found in bacteria. A large number of overlapping stop/start codons has been observed in different operons [114,120,145,151]. In bacteria, overlapping stop/start codons represent one of the mechanisms by which the cell couples the translation of two genes in order to obtain equimolar amounts of their products [165,166]. Thus, the presence of overlapping stop/start codons in the archaea suggests the existence of translational coupling in this domain. It has also been found in the bacteria that the binding site of the repressor r-protein on the mRNA has a similar secondary structure to the binding site of the protein on the rRNA [154,162,164]. Structures that resemble the L1 binding site on the 23S rRNA have been identified in the mRNA from the L1–L10–L12 operons in *M. vannielii* [84] and *H. cutirubrum* [150], and in the L11–L1–L10–L12 operon in *S. solfataricus* [Ramírez et al., in preparation] as shown in Fig. 7.

6. Evolution of the ribosome

Comparison of the structures and gene organization of the rRNAs and r-proteins from the three domains has provided us with valuable information regarding the evolution of the ribosome.

The discovery that rRNA plays an active role in all the functions of the ribosome, as well as the observation that there is a core in the rRNA molecules that has been conserved in all organisms, has given strong support to Woese's proposal [75] that the most primitive ribosomes were composed solely of RNA. On the genetic level, the identification of internal transcribed spacers (ITS) [47] or intervening sequences (IVS) [48] in the rRNA genes of bacteria and eucarya has provided us with a model with which to explain how the present-day rRNA molecules could have evolved from small fragments [47,54].

Perhaps the most remarkable result to arise from the study of archaeal r-proteins and their genes, is the observation that the genes are clustered into operons as in bacteria and the order of the genes is very similar in the two domains [8,9,12,84,109,114,120,122,123,133,140–152]. However, the archaeal r-proteins, as a group, are considerably closer in primary structure to the equivalent r-proteins in eucarya than to their bacterial counterparts [107,147]. These observations would suggest that the ancestral cell that gave rise to the present-day domains had its r-protein genes in an operon-like structure with the ribosomal genes organized in a distinct sequence. The bacterial domain is thought to have separated early [2,167], retaining the operon structure but its r-proteins evolved away from those in the archaeal–eucaryal branch. This branch eventually divided to give rise to the archaea, which retained the operon gene structure, and the eucarya which have lost the operon structure. However, because the archaea diverged from the eucarya after the bacteria separated, the archaeal r-proteins are more similar to those of the eucarya than are the equivalent proteins in bacteria [169].

Nierhaus and his colleagues [168] have suggested that in *E. coli* the r-proteins which form part of an operon constitute assembly units. Thus, the order of the r-protein genes seems to be important for the assembly of the ribosome in those organisms in which the synthesis of r-proteins and the assembly of the ribosome take place in a single compartment. Since this is also the case with the archaea, it is perhaps for this reason that the gene order has been conserved in both domains for efficient ribosome production. In the eucarya, however, where synthesis and assembly take place in different cell compartments, the order in which the r-proteins are synthesized might be of less importance and the operon structure has not been conserved.

In most of the archaeal operons sequenced thus far, ORF's are present which appear to code for proteins which are absent in the bacterial ribosome but in many cases are equivalent to eucaryal r-proteins [9,114,141,143,148]. Since the early reports on the number of r-proteins from the extreme halophiles indicate that they might contain a similar number of r-proteins to those found in the bacterial ribosome [3,4], it is possible that the bacterial ribosome might contain r-proteins which are absent in the archaea but present in the eucarya. As yet, such proteins have not been reported. However, it appears more likely, as discussed in section 2, that the early published reports on the total number of r-proteins in the different archaeal groups may be in error and the archaeal ribosome contains more r-proteins than the bacterial ribosome.

The ancestral ribosome is thought to have contained more proteins than are now found in the archaeal and bacterial ribosomes, and may have lost these extra r-proteins by a streamlining process [8,10,147]. During this process, the function of various r-proteins may have been combined and the number of r-proteins in the ribosome thereby reduced. Since this streamlining probably occurred independently in the archaea and bacteria, different proteins may have resulted from the combination of these functional components [169]. This hypothesis would be supported by the observation that some r-proteins found in archaea do not seem to have an equivalent in bacteria.

Phylogenetic trees based on either rRNA [2,50] or r-protein [170] sequence data support a monophyletic origin for the archaea. However, the evolutionary distance of the three domains relative to each other varies when different macromolecules are used [170,171].

Trees based on SSU and LSU rRNA data give a larger distance between archaea and eucarya than trees based on 5S rRNA or r-proteins [170,171]. These differences are partly due to the difficulty of determining the position of the ancestral root [50] and the uncertainty introduced by sampling error [171]. Recently, however, Iwabe and his colleagues [167] have developed a method to root phylogenetic trees by using paralogous genes; i.e., genes which were duplicated before the three domains separated. Their results suggest that the root of the universal tree is located close to the bacteria, indicating that this domain was the first to separate and that the archaea and eucarya share a common ancestor. The position of the root has now been incorporated into trees based on rRNA [2] or r-protein sequence data [170] and the two types of trees show a similar trend. In contrast, Hensel and his co-workers [172] have found that the sequences of some metabolic enzymes, like glyceraldehyde-3-phosphate dehydrogenase, are closer between eucarya and bacteria than between eucarya and archaea. These authors believe that a possible explanation of these results is lateral gene transfer [172]. Other possibilities are unequal rates of evolutionary change in different lineages [170], mistaken homology or evolutionary convergence [171] or, as Zillig and co-workers [173] have suggested, that the eucaryal nucleocytoplasm is the result of a fusion between an archaeal and a bacterial cell.

A great deal of work is currently underway in many laboratories on the structure and function of the archaeal ribosome. These data, coupled with equivalent data forthcoming from the bacterial and eucaryal ribosomes, should answer many of the questions raised in this chapter. Of special interest will be the information obtained on the three-dimensional structure of the ribosome in the three domains.

Acknowledgements

We wish to thank those colleagues who provided unpublished information for this review. The authors are grateful to the Natural Sciences and Engineering Council of Canada for funding.

References

[1] Woese, C.R. and Fox, G.E. (1977) Proc. Natl. Acad. Sci. U.S.A. 74, 5088–5090.
[2] Woese, C.R., Kandler, O., and Wheelis, M.L. (1990) Proc. Natl. Acad. Sci. U.S.A. 87, 4576–4579.
[3] Cammarano, P., Teichner, A. and Londei, P. (1986) Syst. Appl. Microbiol. 7, 137–146.
[4] Ström, A.R. and Visentin, L.P. (1973) FEBS Lett. 37, 274–280.
[5] Schmid, G. and Böck, A. (1982) Zentralbl. Bakteriol. Parasitenk. Infektionskr. Hyg. Abt. 1 Orig. C 3, 347–353.
[6] Londei, P., Teichner, A. and Cammarano, P. (1983) Biochem. J. 209, 461–470.
[7] Schmid, G. and Böck, A. (1982) Mol. Gen. Genet. 185, 498–501.
[8] Auer, J., Lechner, K. and Böck, A. (1989) Can. J. Microbiol. 35, 200–204.
[9] Auer, J., Spicker, G. and Böck, A. (1989) J. Mol. Biol. 209, 21–36.

[10] Wool, I.G. (1980) In: The Ribosomes: Structure, Function and Genetics (Chambliss, G.R., Craven, G.R., Davies, J., Davis, K., Kahan, L. and Nomura, M., Eds.), pp. 797–824, University Park Press, Baltimore, MD.
[11] Casiano, C., Matheson, A.T. and Traut, R.R. (1990) J. Biol. Chem. 265, 18757–18761.
[12] Ramírez, C., Louie, K.A. and Matheson, A.T. (1991) FEBS Lett. 284, 39–41.
[13] Arndt, E., Scholzen, T., Krömer, W., Hatakeyama, T. and Kimura, M. (1991) Biochimie 73, 657–668.
[14] Lake, J.A., Clark, M.W., Henderson, E., Fay, S., Oakes, M., Scheiman, A., Thornber, P. and Mah, R.A. (1985) Proc. Natl. Acad. Sci. U.S.A. 82, 3716–3720.
[15] Stöffler, G. and Stöffler-Meilicke, M. (1986) System. Appl. Microbiol. 7, 123–130.
[16] Harauz, G., Stöffler-Meilicke, M. and van Heel, M. (1987) J. Mol. Evol. 26, 347–357.
[17] Yonath, A., Leonard, K.R., Weinstein, S. and Wittmann, H.G. (1987) Cold Spring Harbor Symp. Quant. Biol. 52, 729–741.
[18] Yonath, A., Bennett, W., Weinstein, S. and Wittmann, H.G. (1990) In: The Ribosomes: Structure, Function and Evolution (Hill, W.E., Dahlberg, A., Garrett, R.A., Moore, P.B., Schlessinger, D. and Warner, J.R., Eds.), pp. 134–147, American Society for Microbiology, Washington, D.C.
[19] Londei, P., Altamura, S., Cammarano, P. and Petrucci, L. (1986) Eur. J. Biochem. 157, 455–462.
[20] Sánchez, M.E., Urena, D., Amils, R. and Londei, P. (1990) Biochemistry 29, 9256–9261.
[21] Altamura, S., Caprini, E. and Londei, P. (1991) J. Biol. Chem. 266, 6195–6200.
[22] Woese, C.R., Gutell, R.R., Gupta, R., and Noller, H.F. (1983) Microbiol. Rev. 47, 621–669.
[23] Nomura, M., and Post, L.E. (1980) In: Ribosomes: Structure, Function and Genetics (Chambliss, G., Craven, G.R., Davies, J., Davis, K., Kahan, L. and Nomura, M., Eds.), pp. 761–791, University Park Press, Baltimore, MD.
[24] Neumann, H., Gierl, A., Tu, J., Leibrock, J., Staiger, D. and Zillig, W. (1983) Mol. Gen. Genet. 192, 66–72.
[25] Planta, R.J., and Meyerink, J.H. (1980) In: Ribosomes: Structure, Function and Genetics (Chambliss, G., Craven, G.R., Davies, J., Davis, K., Kahan, L. and Nomura, M., Eds.), pp. 871–887, University Park Press, Baltimore, MD.
[26] Culham, D.E. and Nazar, R.E. (1988) Mol. Gen. Genet. 212, 382–385.
[27] Wich, G., Jarsch, M. and Böck, A. (1984) Mol. Gen. Genet. 196, 146–151.
[28] Reiter, W.D., Palm, P., Voos, W., Kaniecki, J., Grampp, B., Schulz, W. and Zillig, W. (1987) Nucl. Acids Res. 15, 5581–5595.
[29] Leffers, H., Kjems, J., Østergaard, L., Larsen, N. and Garrett, R.A. (1987) J. Mol. Biol. 195, 43–61.
[30] Kjems, J., Leffers, H., Garrett, R.A., Wich, G., Leifelder, W. and Böck, A. (1987) Nucl. Acids Res. 15, 4821–4835.
[31] Hartmann, R.K. and Erdmann, V.A. (1989) J. Bacteriol. 171, 2933–2941.
[32] Ree, H.K., Cao, K., Thurlow, D.L. and Zimmermann, R.A. (1989) Can. J. Microbiol. 35, 124–133.
[33] Jarsch, M., Altenbuchner, J. and Böck, A. (1983) Mol. Gen. Genet. 189, 41–47.
[34] Østergaard, L., Larsen, N., Leffers, H., Kjems, J. and Garrett, R. (1987) System. Appl. Microbiol. 9, 199–209.
[35] Brown, J.W., Daniels, C.J. and Reeve, J.N. (1989) CRC Crit. Rev. Microbiol. 16, 287–338.
[36] Mevarech, M., Hirsh-Twizer, S., Goldman, S., Yakobson, E., Eisenberg, H. and Dennis, P.P. (1989) J. Bacteriol. 171, 3479–3485.
[37] Achenbach-Richter, L., Gupta, R., Zillig, W. and Woese, C.R. (1988) System. Appl. Microbiol. 10, 231–240.
[38] Maden, B.E.H., Dent, C.L., Farrell, T.E., Garde, J., McCallum, F.S. and Wakeman, J.A. (1987) Biochem. J. 246, 519–527.
[39] Mylvaganam, S. and Dennis, P.P. (1992) Genetics 130, 399–410.

[40] Zillig, W., Palm, P., Reiter, W., Gropp, F., Pühler, G. and Klenk, H. (1988) Eur. J. Biochem. 173, 473–482.
[41] Kjems, J. and Garrett, R.A. (1987) EMBO J. 6, 3521–3530.
[42] Chant, J. and Dennis, P.P. (1986) EMBO J. 5, 1091–1097.
[43] Chant, J., Hui, I., DeJong-Wong, D., Shimmin, L.C. and Dennis, P.P. (1986) System. Appl. Microbiol. 7, 106–114.
[44] Kjems, J., and Garrett, R.A. (1985) Nature 318, 675–677.
[45] Kjems, J. and Garrett, R. (1991) Proc. Natl. Acad. Sci. U.S.A. 88, 439–443.
[46] Kjems, J., Jensen, J., Olesen, T., and Garrett, R.A. (1989) Can. J. Microbiol. 35, 210–214.
[47] Gray, M.W. and Schnare, M.N. (1990) In: The Ribosomes: Structure, Function and Evolution (Hill, W.E., Dahlberg, A., Garrett, R.A., Moore, P.B., Schlessinger, D. and Warner, J.R., Eds.), pp. 589–597, American Society for Microbiology, Washington, D.C.
[48] Pace, N.R. and Burgin, A.B. (1990) In: The Ribosomes: Structure, Function and Evolution (Hill, W.E., Dahlberg, A., Garrett, R.A., Moore, P.B., Schlessinger, D. and Warner, J.R., Eds.), pp. 417–425, American Society for Microbiology, Washington, D.C.
[49] Woese, C.R. and Fox, G.E. (1977) J. Mol. Evol. 10, 1–6.
[50] Woese, C.R. (1987) Microbiol. Rev. 51, 221–271.
[51] Gouy, M. and Li, W.H. (1989) Nature 339, 145–147.
[52] Olsen, G.J. (1987) Cold Spring Harbor Symp. Quant. Biol. L11, 825–839.
[53] Cedergren, R., Gray, M.W., Abel, Y. and Sankoff, D. (1988) J. Mol. Evol. 28, 98–112.
[54] Clark, C.G. (1987) J. Mol. Evol. 25, 343–350.
[55] Gray, M.W., Sankoff, D. and Cedergren, R.J. (1984) Nucl. Acids Res. 12, 5837–5852.
[56] Hori, H. and Osawa, S. (1987) Mol. Biol. Evol. 4, 445–472.
[57] Wolters, J. and Erdmann, V.A. (1988) Nucl. Acids Res. 16 (Suppl.) r1–r70.
[58] Delihas, N., Andersen, J. and Singhal, R.P. (1984) Prog. Nucl. Acid Res. Mol. Biol. 31, 161–190.
[59] Dams, E., Hendricks, L., Van de Peer, Y., Neefs, J.M., Smits, G., Vandenbempt, I. and De Wachter, R. (1988) Nucl. Acids Res. 16 (Suppl.), r87-r173.
[60] Neefs, J., Van de Peer, Y., De Rijk, P., Goris, A. and DeWachter, R. (1991) Nucl. Acids Res. 19 (Suppl.), 1987–2015.
[61] Gutell, R.R., Weiser, B., Woese, C.R. and Noller, H.F. (1985) Prog. Nucl. Acid Res. Mol. Biol. 32, 155–216.
[62] Gutell, R.R. and Fox, G.E. (1988) Nucl. Acids Res. 16, r175-r269.
[63] Romby, P., Westhof, E., Toukifimpa, R., Mache, R., Ebel, J.P., Ehresmann, C. and Ehresmann, B. (1988) Biochemistry 27, 4721–4730.
[64] Westhof, E., Romby, P. Romaniuk, P.J., Ebel, J.P., Ehresmann, C. and Ehresmann, B. (1989) J. Mol. Biol. 207, 417–431.
[65] Expert-Bezançon, A. and Wollenzien, P.L. (1985) J. Mol. Biol. 184, 53–66.
[66] Stern, S., Powers, T., Changchien, L.M. and Noller, H.F. (1989) Science 244, 783–790.
[67] Stern, S., Weiser, B. and Noller, H.F. (1988) J. Mol. Biol. 204, 447–481.
[68] Woese, C.R. and Gutell, R.R. (1989) Proc. Natl. Acad. Sci. U.S.A. 86, 3119–3122.
[69] Gutell, R.R., Noller, H.F. and Woese, C.R. (1986) EMBO J. 5, 1111–1113.
[70] Brimacombe, R., Atmadja, J., Stiege, W. and Schüler, D. (1988) J. Mol. Biol. 199, 115–136.
[71] Fellner, P. (1974) In: Ribosomes (Nomura, M., Tissieres, A. and Lengyel, P., Eds.), pp. 169–191, Cold Spring Harbor Laboratory, Cold Spring Harbor, NY.
[72] Noller, H.F., Moazed, D., Stern, S., Powers, T., Allen, P.N., Robertson, J.M., Weiser, B. and Triman, K. (1990) In: the Ribosomes: Structure, Function and Evolution (Hill, W.E., Dahlberg, A., Garrett, R.A., Moore, P.B., Schlessinger, D. and Warner, J.R., Eds.), pp. 73–91, American Society for Microbiology, Washington, D.C.

[73] Ravé, H.A., Musters, W., Rugers, C.A., Van't Riet, J. and Planta, R.J. (1990) In: The Ribosomes: Structure, Function and Evolution (Hill, W.E., Dahlberg, A., Garrett, R.A., Moore, P.B., Schlessinger, D. and Warner, J.R., Eds.), pp. 217–235, American Society for Microbiology, Washington, D.C.
[74] Noller, H.F. (1991) Annu. Rev. Biochem. 60, 191–227.
[75] Woese, C.R. (1980) In: Ribosomes: Structure, Function and Genetics (Chambliss, G., Craven, G.R., Davies, J., Davis, K., Kahan, L. and Nomura, M., Eds.), pp. 357–373, University Park Press, Baltimore, MD.
[76] Ravé, H.A., Klootwijk, J. and Musters, W. (1988) Prog. Biophys. Mol. Biol. 51, 77–129.
[77] Zimmermann, R.A., Thomas, C.L. and Wower, J. (1990) In: The Ribosomes: Structure, Function and Evolution (Hill, W.E., Dahlberg, A., Garrett, R.A., Moore, P.B., Schlessinger, D. and Warner, J.R., Eds.), pp. 331–347, American Society for Microbiology, Washington, D.C.
[78] Dahlberg, A.E. (1989) Cell 57, 525–529.
[79] Erdmann, V.A., Pieler, T., Wolters, J., Digweed, M., Vogel, D. and Hartmann, R. (1986) In: Structure, Function and Genetics of Ribosomes (Hardesty, B. and Kramer, G., Eds.), pp. 164–183, Springer, New York.
[80] Teixido, J., Altamura, S., Londei, P. and Amils, R. (1989) Nucl. Acids Res. 17, 845–851.
[81] Beauclerk, A.A.D., Hummel, H., Holmes, D.J., Böck, A. and Cundiffe, E. (1985) Eur. J. Biochem. 151, 245–255.
[82] El-Baradi, T.T.A.L., de Regt, V.C.H.F., Einerhand, S.W.C., Teixido, J., Planta, R.J., Ballesta, J.P.G. and Raue, H.A. (1987) J. Mol. Biol. 195, 909–917.
[83] Gourse, R.L., Thurlow, D.L., Gerbi, S.A. and Zimmermann, R.A. (1981) Proc. Natl. Acad. Sci. U.S.A. 78, 2722–2726.
[84] Baier, G., Piendl, W., Redl, B. and Stöffler, G. (1990) Nucl. Acids Res. 18, 719–724.
[85] El-Baradi, T.T.A.L., de Regt, V.C.H.F., Planta, R.J., Nierhaus, K.H. and Raué, H.A. (1987) Biochimie 69, 939–948.
[86] Stöffler-Meilicke, M. and Stöffler, G. (1991) Biochimie 73, 797–804.
[87] Hummel, H. and Böck, A. (1987) Nucl. Acids Res. 15, 2431–2443.
[88] Jarsch, M. and Böck, A. (1985) Mol. Gen. Genet. 200, 305–312.
[89] Vester, B. and Garrett, R.A. (1988) EMBO J. 7, 3577–3587.
[90] Cammarano, P., Teichner, A., Londei, P., Acca, M., Nicolaus, B., Sanz, J.L. and Amils, R. (1985) EMBO J. 4, 811–816.
[91] Moazed, D. and Noller, H.F. (1987) Biochimie 69, 879–884.
[92] Hall, C.C., Johnson, D. and Cooperman, B.S. (1988) Biochemistry 27, 3983–3990.
[93] Steiner, G., Kuechler, E. and Barta, A. (1988) EMBO J. 7, 3949–3955.
[94] Moazed, D. and Noller, H.F. (1989) Cell 57, 585–597.
[95] Sköld, S.K. (1983) Nucl. Acids Res. 11, 4923–4932.
[96] Egebjerg, J., Douthwaite, S.R., Liljas, A. and Garrett, R.A. (1990) J. Mol. Biol. 213, 275–288.
[97] Ramírez, C., Shimmin, L.C., Newton, C.H., Matheson, A.T. and Dennis, P.P. (1989) Can. J. Microbiol. 35, 234–244.
[98] Cundiffe, E. (1986) In: Structure, Function, and Genetics of Ribosomes (Hardesty, B. and Kramer, G., Eds.), pp. 586–604, Springer, New York.
[99] Thompson, J., Cundiffe, E. and Stark, M. (1979) Eur. J. Biochem. 98, 261–265.
[100] Thompson, J., Schmidt, F. and Cundiffe, E. (1982) J. Biol. Chem. 257, 7915–7917.
[101] Egebjerg, J., Douthwaite, S. and Garrett, R.A. (1989) EMBO J. 8, 607–611.
[102] Hummel, H. and Böck, A., (1987) Biochimie 69, 857–861.
[103] Thompson, J., Cundiffe, E. and Dahlberg, A.E. (1988) J. Mol. Biol. 203, 457–468.
[104] Moazed, D., Robertson, J.M. and Noller, H.F. (1988) Nature 334, 362–364.
[105] Köpke, A.K.E., Paulke, C. and Gewitz, H.S. (1990) J. Biol. Chem. 265, 6436–6440.

[106] Wittmann-Liebold, B. (1986) In: Structure, Function and Genetics of Ribosomes (Hardesty, B. and Kramer, G., Eds.); pp. 326–361, Springer, New York.
[107] Wittman-Liebold, B., Köpke, A.K.E., Arndt, E., Krömer, W., Hatakeyama, T. and Wittmann, H.G. (1990) In: The Ribosomes: Structure, Function and Evolution (Hill, W.E., Dahlberg, A., Garrett, R.A., Moore, P.B., Schlessinger, D. and Warner, J.R., Eds.), pp. 598–616, American Society for Microbiology, Washington, D.C.
[108] Hampl, H., Schulze, H. and Nierhaus, K.H. (1981) J. Biol. Chem. 256, 2284–2288.
[109] Köpke, A.K.E. and Wittmann-Liebold, B. (1989) Can. J. Microbiol. 35, 11–20.
[110] Walleczek, J., Schüler, D., Stöffler-Meilicke, M., Brimacombe, R. and Stöffler, G. (1988) EMBO J. 7, 3571–3576.
[111] Romero, D.P., Arredondo, J.A. and Traut, R.R. (1990) J. Biol. Chem. 265, 18185–18191.
[112] Singleton, C.K. (1989) Nucl. Acids Res. 17, 7989.
[113] Gatermann, K.B., Teletski, C., Gross, T. and Kaufer, N.F. (1989) Curr. Genet. 16, 361–367.
[114] Arndt, E., Krömer, W. and Hatakeyama, T. (1990) J. Biol. Chem. 265, 3034–3039.
[115] Kimura, M., Mende, L. and Wittman-Liebold, B. (1982) FEBS Lett. 149, 304–312.
[116] Kimura, M., Kimura, J. and Watanabe, K. (1985) Eur. J. Biochem. 153, 289–297.
[117] Österberg, R., Sjöberg, B., Pettersson, I., Liljas, A. and Kurland, C.G. (1977) FEBS Lett. 73, 22–24.
[118] Olson, H.M., Sommer, A., Tewari, D.S., Traut, R.R. and Glitz, D.G. (1986) J. Biol. Chem. 261, 6924–6932.
[119] Hamel, E., Koka, M. and Nakamoto, T. (1972) J. Biol. Chem. 247, 805–814.
[120] Ramírez, C. (1990) Ph.D. Thesis, University of Victoria.
[121] Amons, R., Van Agthoven, A., Pluijms, W., Möller, W., Higo, K., Itoh, T. and Osawa, S. (1977) FEBS Lett. 81, 308–310.
[122] Itoh, T. (1988) Eur. J. Biochem. 176, 297–303.
[123] Arndt, E. and Weigel, C. (1990) Nucl. Acids Res. 18, 1285.
[124] Ramírez, C., Shimmin, L.C., Newton, C.H., Matheson, A.T. and Dennis, P.P. (1989) Can. J. Microbiol. 35, 234–244.
[125] Strobel, O., Köpke, A.K.E., Kamp, R.M., Böck, A. and Wittmann-Liebold, B. (1988) J. Biol. Chem. 263, 6538–6546.
[126] Beltrame, M. and Bianchi, M.E. (1987) Nucl. Acids Res. 15, 9089.
[127] Remacha, M., Sáenz-Robles, M.T., Vilella, M. and Ballesta, J.P.G. (1988) J. Biol. Chem 263, 9094–9101.
[128] Maassen, J.A., Schop, E.N., Brands, J.H.G.M., Van Hemert, F.J., Lenstra, J.A. and Möller, W. (1985) Eur. J. Biochem. 149, 609–616.
[129] Qian, S., Zhang, J.Y., Kay, M.A. and Jacobs-Lorena, M. (1987) Nucl. Acids Res. 15, 987–1003.
[130] Lin, A., Wittmann-Liebold, B., McNally, J. and Wool, I.G. (1982) J. Biol. Chem. 257, 9189–9197.
[131] Rich, B.E. and Steitz, J.A. (1987) Mol. Cell Biol. 7, 4065–4074.
[132] Matheson, A.T., Louie, K.A. and Böck, A. (1988) FEBS Lett. 231, 331–335.
[133] Köpke, A.K.E., Baier, G. and Wittmann-Liebold, B. (1989) FEBS Lett. 247, 167–172.
[134] Leijonmarck, M. and Liljas, A. (1987) J. Mol. Biol. 195, 555–580.
[135] Gudkov, A.T., Tumanova, L.G., Gongadze, G.M. and Bushuev, V.N. (1980) FEBS Lett. 109, 34–38.
[136] Möller, W. and Maassen, J.A. (1986) In: Structure, Function, and Genetics of Ribosomes (Hardesty, B. and Kramer, G., Eds.), pp. 309–326, Springer, New York.
[137] Köpke, A.K.E., Leggatt, P. and Matheson, A.T. (1992) J. Biol. Chem. 267, 1382–1390.
[138] Liljas, A., Thirup, S. and Matheson, A.T. (1986) Chem. Scripta 26B, 109–119.
[139] Boublik, M., Visentin, L.P., Weissbach, H. and Brot, N. (1979) Arch. Biophys. Biochem. 198, 53–59.

[140] Arndt, E. and Kimura, M. (1988) J. Biol. Chem. 263, 16063–16068.
[141] Auer, J., Spicker, G., Mayerhofer, L., Pühler, G. and Böck, A. (1991) System. Appl. Microbiol. 14, 14–22.
[142] Baldacci, G., Guinet, F., Tillit, J., Zaccai, G. and de Recondo, A. (1989) Nucl. Acids Res. 18, 507–511.
[143] Itoh, T. (1989) Eur. J. Biochem. 186, 213–219.
[144] Lechner, K., Heller, G. and Böck, A. (1989) J. Mol. Evol. 29, 20–27.
[145] Leffers, H., Gropp, F., Lottspeich, F., Zillig, W. and Garrett, R.A. (1989) J. Mol. Biol. 206, 1–17.
[146] Mankin, A.S. (1989) FEBS Lett. 246, 13–16.
[147] Matheson, A.T., Auer, J., Ramirez, C. and Böck, A. (1990) In: The Ribosomes: Structure, Function and Evolution (Hill, W.E., Dahlberg, A., Garrett, R.A., Moore, P.B., Schlessinger, D. and Warner, J.R., Eds.), pp. 617–635, American Society for Microbiology, Washington, D.C.
[148] Pühler, G., Lottspeich, F. and Zillig, W. (1989) Nucl. Acids Res. 17, 4517–4534.
[149] Ramirez, C., Louie, K.A. and Matheson, A.T. (1989) FEBS Lett. 250, 416–418.
[150] Shimmin, L.C. and Dennis, P.P. (1989) EMBO J. 8, 1225–1235.
[151] Spiridonova, V.A., Akhmanova, A.S., Kagramanova, V.K., Köpke, A.K.E. and Mankin, A.S. (1989) Can. J. Microbiol. 35, 153–159.
[152] Köpke, A.K.E. (1989) Ph.D. Thesis, Freie Universität Berlin.
[153] Mager, W.H. (1988) Biochim. Biophys. Acta 949, 1–15.
[154] Nomura, M., Gourse, R. and Baughman, G. (1984) Annu. Rev. Biochem. 53, 75–117.
[155] Cerretti, D.F., Dean, D., Davis, G.R., Bedwell, D.M. and Nomura, M. (1983) Nucl. Acids Res. 11, 2599–2616.
[156] Planta, R.J., Mager, W.H., Leer, R.J., Woudt, L.P., Raue, H.A. and El-Baradi, T.T.A.L. (1986) In: Structure, Function and Genetics of Ribosomes (Hardesty, B. and Kramer, G., Eds.), pp. 699–718, Springer, New York.
[157] Post, L.E. and Nomura, M. (1980) J. Biol. Chem. 255, 4660–4666.
[158] Post, L.E., Strycharz, G.D., Nomura, M., Lewis, H. and Dennis, P.P. (1979) Proc. Natl. Acad. Sci. U.S.A. 76, 1697–1701.
[159] Zurawski, G. and Zurawski, S.M. (1985) Nucl. Acids Res. 13, 4521–4526.
[160] Creti, R., Citarella, F., Tiboni, O., Sanangelantoni, A., Palm, P. and Cammarano, P. (1991) J. Mol. Evol. 33, 332–342.
[161] Gluzman, Y., Ed. (1985) Eukaryotic Transcription, Cold Spring Harbor Laboratory, Cold Spring Harbor, NY.
[162] Lindahl, L. and Zengel, J.M. (1986) Annu. Rev. Genet. 20, 297–326.
[163] Gold, L. (1988) Annu. Rev. Biochem. 57, 199–233.
[164] Nomura, M. (1986) In: Symp. Society for General Microbiology: Regulation of Gene Expression (Booth, I. and Higgins, C., Eds.), pp. 199–220, Cambridge University Press, Cambridge.
[165] Askoy, S., Squires, C.L. and Squires, C. (1984) J. Bacteriol. 157, 363–367.
[166] Yanofsky, C., Platt, T., Crawford, I., Nichols, B., Christie, G., Horowitz, H., Van Cleemput, M. and Wu, A. (1981) Nucl. Acids Res. 9, 6647–6668.
[167] Iwabe, N., Kuma, K.I., Hasegawa, M., Osawa, S. and Miyata, T. (1989) Proc. Natl. Acad. Sci. U.S.A. 86, 9355–9359.
[168] Röhl, R. and Nierhaus, K.H. (1982) Proc. Natl. Acad. Sci. U.S.A. 79, 729–733.
[169] Matheson, A.T. (1992) Biochem. Soc. Symp. 58, 89–98.
[170] Auer, J., Spicker, G. and Böck, A. (1990) System Appl. Microbiol. 13, 354–360.
[171] Sneath, P.H.A. (1989) System. Appl. Microbiol. 12, 15–31.
[172] Hensel, R., Zwickl, P., Fabry, S., Lang, J. and Palm, P. (1989) Can. J. Microbiol. 35, 81–85.

[173] Zillig, W., Klenk, H.P., Palm, P., Leffers, H., Pühler, G., Gropp, F. and Garrett, R.A. (1989) Endocyt. C. Res. 6, 1–25.

CHAPTER 15

Halobacterial genes and genomes

Leonard C. SCHALKWYK*

Department of Biochemistry, Dalhousie University, Halifax, N.S. B3H 4H, Canada

Abbreviations

CoA	Coenzyme A	IS(H)	insertion sequence (halobacterial)
HMG–CoA	3-hydroxy-3-methylglutaryl–coenzyme A	EF	elongation factor
		ORF	open reading frame
(k, M)bp	(kilo, mega) base pairs		

1. Introduction

Work on the structure of archaeal genes and genomes, most advanced in halobacteria, has led to a general and surprising consensus that these are largely bacterial in character. Halobacterial genomes consist of circular DNAs similar in size to those of bacteria, apparently without introns in protein-coding genes, and with genes in operons or clusters sometimes closely resembling their bacterial counterparts. Archaea, however, are not closely related to bacteria, but according to the rooting of the universal tree by Iwabe et al. (ref. [1], see also the Introduction to the present volume, by Woese), are instead a sister group of the eucarya. This indicates that this size and shape of genome and perhaps some snatches of gene order were features of the last common ancestor of all life.

There are six recognized genera of halobacteria (*Halobacterium, Halococcus, Haloarcula, Haloferax, Natronococcus* and *Natronobacterium*) and evidence of the need for another four [2,3]. Sequences of 16S RNA genes of representatives of the six genera have been used to produce a phylogenetic tree of the halobacteria (Fig. 1). Most of these genera contain species currently or previously referred to as *Halobacterium* species. Furthermore, strains now known to belong in different genera were originally given the same name, so it can be important to pay attention to strain designations. Most of the work reviewed here has been done on two species of halobacteria and some of their close relatives. The first is *Halobacterium salinarium*, an extreme halophile of which some strains are known as *Halobacterium halobium* NRC1, *Halobacterium halobium* NRC817, *Halobacterium halobium* NRC34020 (strains R_1 and S9 are derived from 34020) and *Halobacterium cutirubrum* NRC34001 [2]. The other is

---

* Current address: Genome Analysis Department, Imperial Cancer Research Fund, P.O. Box 123, Lincoln's Inn Fields, London WC2A 3PX, UK.

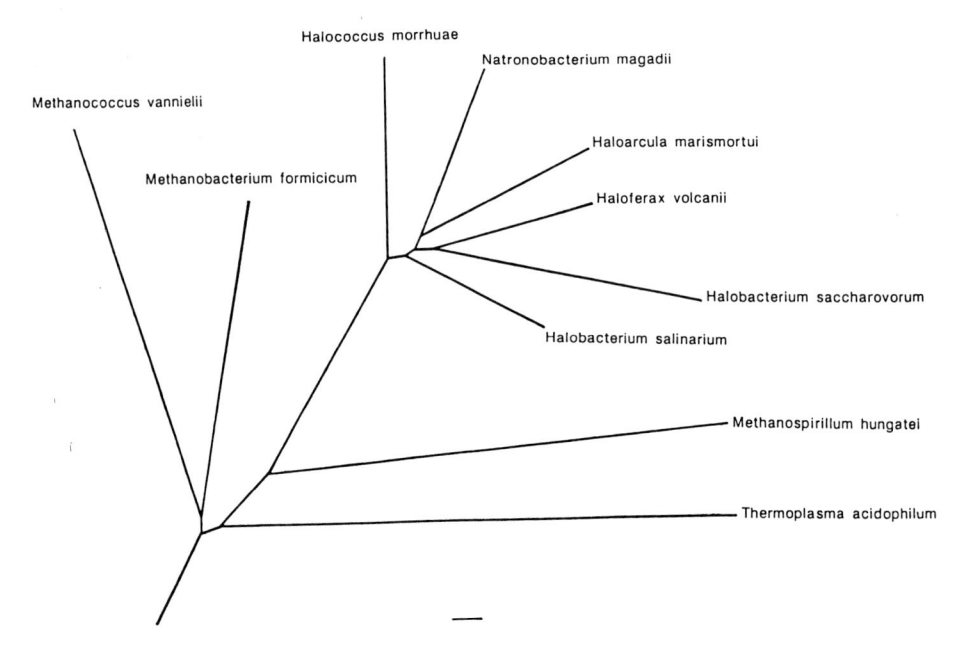

Fig. 1. Phylogenetic (distance) tree for the 16S rRNAs of the halobacteria, reproduced from Lodwick et al. [3]. The bar represents 0.01 substitutions per site.

Haloferax (formerly *Halobacterium*) *volcanii*, a moderately halophilic species. A close relative of *Haloferax volcanii* is *Haloferax* phenon K isolate Aa 2.2 [4]. *Halobacterium marismortui*, the subject of extensive work on ribosomal proteins, is not closely related to *Halobacterium salinarium*, instead, it belongs in the genus *Haloarcula* [2].

Reviews covering many of the topics treated here [5–9] have previously appeared.

2. Halobacterial genomes

2.1. Size

Genome sizes of a variety of archaea have been determined by renaturation ($Cot_{0.5}$) analysis and more recently by determination of large restriction fragment sizes using pulsed-field gel electrophoresis (see Table 1). The sizes fall within the bacterial range indicated by the last three entries in the table. The $Cot_{0.5}$-derived sizes are based on the *Escherichia coli* chromosome as a standard, and have been adjusted using a more recent estimate of the *Escherichia coli* genome size (4.7 Mbp [10,11]).

Archaeal chromosomes have been shown to be circular in three cases. For *Thermococcus celer* [12], restriction mapping of the entire chromosome with three rare-cutting enzymes clearly demonstrated circularity. *Sulfolobus solfataricus* DNA gives two

TABLE 1
Genome sizes of halobacteria and other prokaryotes

Species	Size (Mbp)	Method	Reference
Methanosarcina barkeri	2.0	C_0t	[177]
Halobacterium halobium	4.3	C_0t	[29]
Haloferax volcanii	2.9	cloning	[13]
Halococcus morrhuae	4.3	C_0t	[29]
Thermoplasma acidophilum	0.81	C_0t	[178]
Thermococcus celer	1.89	PFG	[10]
Sulfolobus acidocaldarius	2.1	PFG	[11]
Ureaplasma urealyticum	0.9	PFG	[179]
Escherichia coli	4.7	cloning	[9]
Myxococcus xanthus	10	PFG	[180]

fragments when digested with NotI. Cloned DNAs containing NotI sites indicate that each end of one fragment links an end of the other, so this organism also most likely has a circular chromosome [13]. The genome of *Haloferax volcanii* has been mapped in detail [14,15] and consists of a circular chromosome (2.9 Mbp) and four plasmids of 690, 442, 89, and 6.4 kbp. The other halobacteria listed in Table 1 are likely to have similarly modest-sized genomes and several large plasmids included in the sizes given. *Halobacterium halobium* NRC1 has a chromosome of 2 000 kbp and plasmids of 400 and 200 kbp, as determined by mapping using pulsed-field electrophoresis (Hackett, N., personal communication).

2.2. Plasmids

Circular plasmids are widely distributed among the archaea. They have been detected in most of the halobacterial species studied (Table 2) and are of a wide range of sizes, and are cryptic except for pHH1 (150 kbp) of *Halobacterium halobium* NRC817 [16] which bears one [17] of the two [18] gas-vesicle genes in this organism (see section 3.4.8). The gas-vesicle gene previously thought to be borne by the chromosome is apparently on the newly discovered 400 kbp plasmid in *Halobacterium halobium* NRC1 (Hackett, N., personal communication). The *Halobacterium halobium* phage ΦH1 is a plasmid in its lysogenic form, so to this extent it is also a plasmid with known phenotype [19].

Using an in situ lysis method, Gutierrez et al. [20] electrophoretically surveyed sixty-five halobacterial strains for the presence of large plasmids. Three quarters of the strains had plasmids visible by this method, the majority containing three or four [20]. Due to the limitations of the electrophoretic technique used, this analysis no doubt missed some of the largest plasmids (such as the largest *Haloferax volcanii* plasmid [15]).

Different strains of *Halobacterium halobium* contain different relatives of pHH1, such as pNRC100 in *Halobacterium halobium* NRC1. Rearranged or deleted versions of pHH1 are readily isolated [21]. These events involve insertions of IS elements and deletions flanking them. pNRC100 has an inverted repeat of 35–38 kbp bounded by insertion

TABLE 2

Halobacterial plasmids. Only relatively small, well characterized plasmids mentioned in the text are listed

Species	Name	Size	Data	Derivatives	Ref.
Halobacterium halobium NRC1	pNRC100	100	map		[22]
H. halobium NRC817	pHH1	150	map	pHH4...9	[16,196]
				pUBP2	[60]
H. halobium R_1L	pΦHL	12	map		[49]
			transcripts		[167]
Halobacterium sp. GRB	pGRB1	1.781	seq	pMPK29	[23,50]
				pMPK52	[63]
Halobacterium sp. SB3	pHSB1	1.736	seq		[25,62]
Halobacterium sp. GN101	pHGN1	1.765	seq		[24]
Haloferax volcanii WR11	pHV11	3	map, compatibility		[28]
Hf. volcanii WR12	pHV12	5	map		[28]
Hf. volcanii WR13	pHV13	44	map		[28]
Hf. volcanii DS2	pHV2	6.354	seq	p455, pWL102	[55]
	pHV1	86	map cloned		[15]
Haloferax p. K Aa 2.2	pHK	10.5	map	pMDS2	[4,58]

sequences, ISH2 at one end and ISH3 at the other, and exists as an approximately equal mixture of two inversion isomers [22].

Several of the purple-membrane-containing extreme halophiles (i.e., *Halobacterium salinarium* and its close relatives) contain small, high-copy-number plasmids, whose DNA sequences have been determined. The sequences are quite similar, and all have an open reading frame of about 1 kbp and a set of hexanucleotide repeats that may be involved in the maintenance of the plasmid [23–25].

The supercoiled fraction obtained by CsCl-ethidium bromide equilibrium ultracentrifugation of *Halobacterium halobium* DNA contains, in addition to the characterized plasmids, a heterogeneous population of "minor ccc DNAs". It is supposed that these arise by intramolecular recombination of the chromosome and it is not known whether they are replicated [26,27].

An indication of the diversity of plasmids that might be found in wild populations of halobacteria is given by the characterization of three new isolates of *Haloferax volcanii*, each of which contains a different plasmid (3, 5, and 44 kbp). These plasmids do not hybridize with each other or with the 6.4 kbp plasmid pHV2 from the original *Haloferax volcanii* isolate DS2. One of the plasmids was shown to be compatible with pHV2 [28].

2.3. Inhomogeneity of composition

As first reported by Joshi et al. [29] in *Halobacterium salinarium* and *Halobacterium cutirubrum*, the DNA of halobacteria can be separated into two fractions on the basis of composition. This observation was confirmed and extended by Moore and McCarthy [30] who demonstrated a minor more AT-rich "satellite" fraction in the DNA of *Halobacterium halobium, Halobacterium salinarium, Halobacterium cutirubrum* and *Halococcus morrhuae*. Renaturation studies [31] indicated that the minor fraction was neither a simple repeated sequence nor multiple copies of a small episomal element. The minor fraction (also known as FII DNA) is approximately 10 mol% lower in G+C content than the bulk of the DNA (FI) and varies from about 10% to roughly 30% of the total in different species. It may be absent from *Halobacterium trapanicum* [27].

Although Moore and McCarthy concluded that the satellite fraction was not a plasmid, it has since been found that halobacteria do contain plasmids, and that these are more AT-rich than the bulk of the DNA [30], but also that some satellite DNA is interspersed in the genome [33].

2.4. Repeated sequences and instability

Halobacteria, especially the extreme halophiles, show a variety of visibly different phenotypes when plated from a single colony, at frequencies (in *Halobacterium halobium*) of 10^{-2} for loss of the gas vesicle and 10^{-4} for loss of bacterioruberin and bacteriorhodopsin [34]. Vesicle loss is correlated with loss of [35], insertions into [32], or rearrangements of [34] the large plasmid pHH1, from which the gas-vesicle gene was eventually cloned [17]. Complex and varied rearrangements of the plasmid also accompanied bacterioruberin and bacteriorhodopsin mutations [34], and it was clear that an extraordinarily active mechanism for DNA rearrangement was present. Insertion, inversion and deletion variants of the *Halobacterium halobium* phage ΦH1 could also be readily observed [36].

Supposing that the observed instability might be due to transposable elements, Sapienza and Doolittle [37] demonstrated by hybridization that *Halobacterium halobium* NRC1 contains repeated sequences, and then screened randomly cloned EcoRI and EcoRI/BamHI fragments of about 3 kbp length for hybridization to multiple EcoRI fragments. Thirty-one of 35 EcoRI/BamHI fragments from *Halobacterium halobium* R1 (a gas-vesicle mutant strain) and 27 of 28 EcoRI fragments from NRC1 (wild type) contained repeated sequences. A similar experiment using PstI fragments gave only unique sequences, leading to the conclusion that this organism contains many repeated sequences, perhaps 50 families of 2–20 copies each, which are clustered in that part of the genome frequently cut by EcoRI, but not PstI ("Pst-poor regions"). Some of the repeated elements hybridized with repeated elements in the genome of *Haloferax volcanii*. Comparison of the patterns revealed by hybridization of repeated sequences to independent but not visibly different single-colony isolates of *Halobacterium halobium* indicated that these elements are highly mobile [38]. A quantitative study produced an estimate of $>4\times10^{-3}$ per family per generation for changes detectable on Southern transfers probed with repeated sequences. No changes were seen in the two unique sequences tested. The analysis also suggested that the events occur in bursts, where a single isolate was affected by events involving several repeat families [39].

Sequencing has shown that the repeated sequences are insertion sequences similar to those widely distributed in the bacteria and in eucarya. The first of these characterized was ISH1, an insertion in the bacteriorhodopsin gene of *Halobacterium halobium* S9. The element is 1118 bp in length, with an 8 bp terminal inverted repeat. Eight base pairs of the target are duplicated. The target sequence is itself flanked by a 9 bp inverted repeat which is similar to the terminal repeat. Within the element are found an additional 7 inverted repeats of over 8 bp in length. There are two overlapping reading frames on opposite strands, of which one could be shown to correspond to a transcript by Northern analysis. Twenty-one independent insertions of ISH1 into the bacteriorhodopsin gene have been characterized, all at the same position, but in either orientation [40].

ISH2, an unusually short element, was also isolated as an insertion in the bacteriorhodopsin gene. It is 520 bp in length and has a 19 bp terminal inverted repeat. Unlike ISH1, the target-site duplication varies in length. ISH2 has little insertion specificity and has been found at several sites in the bacteriorhodopsin gene and at one site 102 bp upstream of the gene. The upstream insertion also prevents bacteriorhodopsin expression [41].

Five other, less commonly inserted ISH elements have been found in this way: ISH23, ISH24 [40], ISH26, ISH27 and ISH28 [43,44]. ISH23 is very similar in sequence to ISH50 [43], which was isolated from the 50 kbp plasmid in *Halobacterium halobium* strain R1. ISH27 resembles in sequence ISH51 of *Haloferax volcanii* [21,46]. In *Haloferax volcanii*, ISH51 makes up a large family of degenerate repeated sequences. Different copies (of which there are 20–30) have sequence similarity of 85%, on average, and are present in both FI and FII regions of the genome [46]. A spontaneous insertion of ISH51 into the plasmid pHV2 has been detected [47].

Pfeifer and Blaseio [21] have characterized numerous deletion events in the *Halobacterium halobium* NRC817 plasmid pHH1 and its derivative pHH4, each flanked by a copy of ISH2 or ISH27 [21]. They have also observed a burst of transposition in the strain carrying pHH4 after long storage at 4°C, in which 20% of colonies contained a detectable transposition. In this strain, but not in the wild type, a transcript corresponding to an open reading frame in the sequence of ISH27 could be detected. On subsequent continuous growth at 37°C, further transposition events were not seen, indicating that the high transposition activity was likely to be due to the environment rather than to a regulatory mutation in the IS element. Ten ISH27 elements were sequenced. These fell into three groups on the basis of sequence similarity, one of which is more similar to ISH51 of *Haloferax volcanii* (91%) than it is to the other two groups (82–83%) [48].

Insertions into the DNA of the phage ΦH are readily isolated. These are due to a second copy of the resident element ISH1.8, of which there are two copies in the genome of the host. This element does not have terminal inverted repeats, nor does it apparently cause a duplication of the target sequence [36,49].

It is likely that further *Halobacterium halobium* IS elements will be discovered if additional substrates for insertion are assayed. Krebs, RajBhandary and Khorana [50] screened 212 transformants of *Halobacterium halobium* R_1 containing the plasmid pGRB and discovered one containing a new IS element, called ISH11, of 1068 bp, most of which is composed of a 384-codon open reading frame [50].

Starting with a chromosomal copy of ISH1 in *Halobacterium halobium* NRC817, Pfeifer and Betlach [33] carried out a cosmid walk in both directions and succeeded in cloning an entire "island" of FII DNA, that is to say a region of more AT-rich sequence embedded in the chromosome. A total of 160 kbp was cloned, of which 70 kbp was FII, as judged by probing isolated fragments to fractionated genomic DNA. One copy of

ISH1, two each of ISH2 and ISH26 and ten or more other, uncharacterized repeated sequences were found in the cloned region, mostly in the FII part [33]. *Halobacterium* sp GRB is an extreme halophile with purple membrane, which does not have sequences hybridizing with pHH1, and which thus must lack many of the ISH elements of *Halobacterium halobium*. As might be expected from this finding, the frequency of easily scored mutations resulting in loss of gas-vesicle or ruberin synthesis is appreciably lower than in *Halobacterium halobium* [51,52].

3. Genetics

Haloferax volcanii is particularly attractive as a halobacterial experimental system because it is a relatively fast grower and is capable of growing on a simple minimal medium [53], so that there is the possibility of isolating auxotrophic strains and thus identifying genes for many biosynthetic functions. Auxotrophic strains and methods for mating [53] and protoplast fusion [28] of *Haloferax volcanii* have been developed.

Transformation was first demonstrated in halobacteria using DNA of the bacteriophage ΦH of *Halobacterium halobium* [54]. The DNA of this phage could be introduced into *Halobacterium halobium* cells by spheroplast formation and treatment with polyethylene glycol 600. Transfected cells could be detected as infective centers. Further work concentrated on *Haloferax volcanii* which can be transfected with ΦH DNA by a similar method. Although *Haloferax volcanii* is not a natural host it can produce a burst of phage which can be detected using a lawn of *Halobacterium halobium*. A *Haloferax volcanii* strain was cured of the native, cryptic plasmid pHV2, whose sequence had been determined. The plasmid could be reintroduced by transformation (scored by colony hybridization), as could a deletion variant constructed in vitro [55,56].

The availability of transformation and a plasmid paved the way for the development of *E. coli*–*Haloferax volcanii* shuttle vectors. Required for this were a selectable marker and knowledge of which parts of the plasmid might be dispensable. The latter was serendipitously supplied (along with useful restriction sites) by the discovery of a spontaneous insertion of ISH51 into pHV2. The selectable marker was obtained by selection for mutants resistant to the HMG–CoA-reductase inhibitor mevinolin [57] and shotgun cloning of the resistance determinant into the ISH51-bearing derivative of pHV2. This was combined with the *E. coli* cloning vector pAT153 to produce, after further tailoring, the 10.5 kbp shuttle vector pWL102 [47].

Another shuttle vector with a very useful second selectable marker has been constructed from the plasmid pHK2 from *Haloferax* phenon K isolate Aa 2.2 and a mutant novobiocin-resistant DNA gyrase B subunit gene from the same strain [4,58]. A *Halobacterium halobium* shuttle vector, pUBP2, has been constructed from the minimal replicating fragment of the plasmid pHH1 (4.3 kbp), the *Haloferax volcanii* mevinolin resistance determinant, and the *E. coli* vector pIBI31. This construct can be maintained by both *Halobacterium halobium* and *Haloferax volcanii*, as can pWL102 [60]. *Haloferax mediterranei*, *Haloarcula hispanica* and *Haloarcula vallismortis* [61] can also be transformed with pWL102. The small, high-copy-number plasmid pHSB1 from *Halobacterium* SB3 has been sequenced [25,62], and has been transformed into a derivative of *Halobacterium halobium* NRC1, and is thus another candidate for vector construction [25]. The related plasmid pGRB1 has (in an insertion-sequence-bearing form) been used as a vector for the reintroduction of the bacteriorhodopsin gene to a mutant strain of *Halobacterium halobium*. Transformants could be scored by their purple color, though not selected [63].

Haloferax volcanii auxotrophs can be stably transformed to prototrophy with genomic DNA, at an efficiency depending strongly upon fragment size [64]. Other, presumably nonreplicating, circular or linear DNAs such as cosmids, DNA fragments recovered from gels, and single or double stranded M13 clones can also be used, allowing precise localization and isolation of genes [65].

For these reasons, *Haloferax volcanii* seemed to be a good subject for a genome mapping project. Physical mapping of the genomes of two other halobacteria is now also underway (*Halobacterium* sp. GRB [Charlebois, R.L., personal communication], *Halobacterium halobium* NRC1 [Hackett, N., personal communication].

3.1. Physical mapping: Introduction

In the study of gene structure by sequencing, the comparison of genes from the same species and of similar genes from different species has been a very powerful method of predicting which sequence features are functionally important. Sequencing has also allowed inferences to be made about the history of the gene in question. In a similar fashion, comparison of maps should be informative about aspects of gene organization dictated by function and history. The depth to which we can see into the history of genomes will be determined by the extent to which gene order is maintained over geological time. Comparison of the detailed genetic maps of several bacteria whose phylogeny has been independently inferred by comparison of their 16S ribosomal RNAs [66] gives a basis on which to make predictions. In this connection it is necessary to point out that genetic maps are only available for species representing a small part of the diversity of the bacteria. All of the existing genetic maps are of gamma-subdivision purple bacteria (proteobacteria [66,67]) or of gram-positive bacteria. There are likely some surprises left in more deeply branching groups, such as green non-sulfur bacteria, deinococci, and spirochetes.

3.2. Clues from comparison of bacterial genetic maps

The most highly developed genetic map is that of *Escherichia coli* K12, and it has recently been augmented with coarse [10] and fine [11] restriction maps. As the most nearly complete genetic map it can give us clues to structural features which may be functionally important, especially since some data exist on the effects of chromosomal rearrangements. This makes it a natural reference to which other maps of similar-sized genomes can be compared.

The genetic maps of *Escherichia coli* K12 and the closely related *Salmonella typhimurium* LT2 can be aligned, but differ by an inversion of about 10% of their length symmetrically disposed about the terminus of replication [68] and by 14 (in *Escherichia coli* K12) and 15 (in *Salmonella typhimurium* LT2) loops of sequence not present in the other species [69,70].

The *Bacillus subtilis* 168 genome is about 20 percent larger than that of *Escherichia coli*, and also circular, but it is difficult to see any alignment of its map [71] with the *Escherichia coli* map. This impression agrees with the much more rigorous approach of Sankoff, Cedergren and Abel [72], who have developed a measure of the number of rearrangements of one map order relative to another and compare this to what one would obtain by comparing with a selection of randomly shuffled maps. Comparison of the *Escherichia coli* and *Salmonella typhimurium* maps gives 5% as many intersections (the measure of rearrangement) as comparison with shuffled genomes, and the comparison of *Escherichia coli* and *Bacillus subtilis* gives 96%, or no detectable similarity [72].

Having said that there is no detectable similarity between gene order in *Escherichia coli* and *Bacillus subtilis*, it is clear that it is a question of at what level of structure there is conservation. At the nucleotide sequence level, many genes are similar and can be aligned. Some operons also have genes in the same or similar order, for one example the rRNA operons in which the order is 16S–23S–5S throughout the bacteria. Up to what level is order then conserved? As an example, in comparing the sequence of the region of the origin of replication of *Bacillus subtilis* with the *rpn*A–*rpm*H–*dna*A–*dna*N–*rec*F–*gyr*B region of *Escherichia coli,* Ogasawara et al. [73,74] found similar genes in similar relative positions. In this remarkable 10 kbp alignment, the order and orientation of 6 replication-related genes, though not the details of transcription or the size of the spacer sequences, are the same [73,74].

Some simple ideas of general principles that govern bacterial genome organization are beginning to be formulated, and it will be interesting to see them expanded and to determine whether they apply also to archaeal genomes. These include a preferred orientation for genes (such that they are transcribed in the same direction they are replicated) [75–77] and a preference of highly expressed genes for the region nearest the origin of replication due to higher effective copy number [78].

3.3. The Haloferax volcanii map

Restriction maps of entire genomes can be obtained by two approaches. In the top-down approach, infrequently-cutting restriction enzymes and pulsed-field gel electrophoresis are used to map a chromosome much as one would a plasmid. In the bottom-up approach, analysis of many clones in parallel leads to the accumulation of a set of clones which represent the whole genome. Restriction maps of the clones can then be merged to make the genome map. The clones themselves are also a valuable resource. The earliest complete and nearly complete physical mapping projects, both top-down (*E. coli* [10], *Schizosaccharomyces pombe* [79]) and bottom-up (*E. coli* [11], *Saccharomyces cerevisiae* [82], *Caenorhabditis elegans* [80,81]), were on genomes which were already extensively mapped genetically. In cases in which we haven't the luxury of a pre-existing map, and these are after all the most interesting ones, bottom-up and top-down approaches have to be integrated in order to produce both continuity and detail.

After obtaining estimates of genome size and restriction-site frequency with pulsed-field electrophoresis, we analyzed 1000 randomly chosen cosmid clones using infrequent restriction sites as landmarks to efficiently determine overlaps, and further clones chosen by cosmid walking. A minimal set of cosmids was restriction-mapped using six enzymes, and where necessary a seventh enzyme was used to determine the exact extent of overlap.

Inconsistent areas of the map were clarified, gaps mapped, and the overall continuity of the map was checked by hybridizing whole cosmids or end probes to genomic pulsed-field Southern blots. The final minimal set contains 151 cosmids and covers 96% of the genome with 7 gaps. The completed map (Fig. 2) shows five circles: a 2920 kilobase-pair circular chromosome and plasmids of 690 (pHV4), 442 (pHV3), 86 (pHV1) and 6.4 kbp (pHV2) [14,15].

The distribution of sites for the six restriction enzymes used for mapping the *Haloferax volcanii* genome is far from random. Two small (50 and 36 kbp) regions of the chromosome, two larger (119 and 128 kbp) sections of pHV4 and all of pHV1 (86 kbp) are particularly rich in sites for the six enzymes. These "oases" are highly significant deviations from what one would expect of random sequence, and correspond to segments of FII DNA, also called "islands". As discussed in section 2.4, in *Halobacterium halobium* these islands are often the homes of insertion sequences of many types. We mapped the locations of the most common *Haloferax volcanii* insertion sequence, ISH51, and the related element D by hybridization to dot blots and Southern transfers of the minimal set and found that copies of these elements are concentrated in the oases and in their surrounding neighborhoods, but are by no means restricted to these areas. The oasis definition is based on a 40 kbp window and finer resolution is not possible because of the small numbers of sites involved, but it seems likely from the data that there are smaller interspersions of oasis-like sequence, and that the oasis sequence of pHV4 has a different character from that on the chromosome [15].

Hybridization to dot blots of the minimal set was also used to map a number of genes. Among protein-coding genes mapped with specific probes from *Haloferax volcanii* or other halobacteria were those for the cell surface glycoprotein, dihydrofolate reductase, gyrase B, histidinol-phosphate aminotransferase, HMG–CoA reductase, RNA polymerase subunits B″, B′, A, and C, superoxide dismutase (two genes), enzymes of tryptophan biosynthesis (in two clusters), and several ribosomal proteins. Ten previously cloned and sequenced tRNA genes (refs. [83,84], and Gupta, R., personal communication) were mapped, and an additional 39 fragments likely encoding tRNA were found by using end-labelled tRNA as probe. The two ribosomal RNA (*rrn*) operons were located with the 5S-containing clone pVT6 [81] and oriented using internal restriction sites. The 7S RNA gene was located using 3′ end-labelled, gel-purified 7S RNA [14].

Transformation with cosmid DNA from the *Haloferax volcanii* minimal set has been used to map 140 ethylmethanesulfonate-induced mutants requiring one of 14 amino acids, uracil, adenine, or guanine. This has added an additional 35 loci to the map, most represented by several mutants in the collection. Some indicate operons, such as the 19 *arg* mutations mapped to the overlap between cosmids 21 and 247, as well as the neighboring part of cosmid 21. Other loci are likely to contain single genes, such as some of the four unlinked *his* loci [86].

The *Haloferax volcanii* map is becoming quite detailed, and a few general observations can be made about its structure. One striking feature is the high density of IS elements on two of the plasmids, pHV1 and pHV4. All of pHV1 and more than a third of pHV4 are composed of oasis (FII) DNA. pHV3 on the other hand, is not especially rich in IS elements, and in restriction-site distribution resembles typical (non-oasis) *Haloferax volcanii* chromosomal DNA. An attractive hypothesis is that pHV1 and pHV4 are mobile plasmids, recent immigrants to *Haloferax volcanii* from a more heavily IS-infested host.

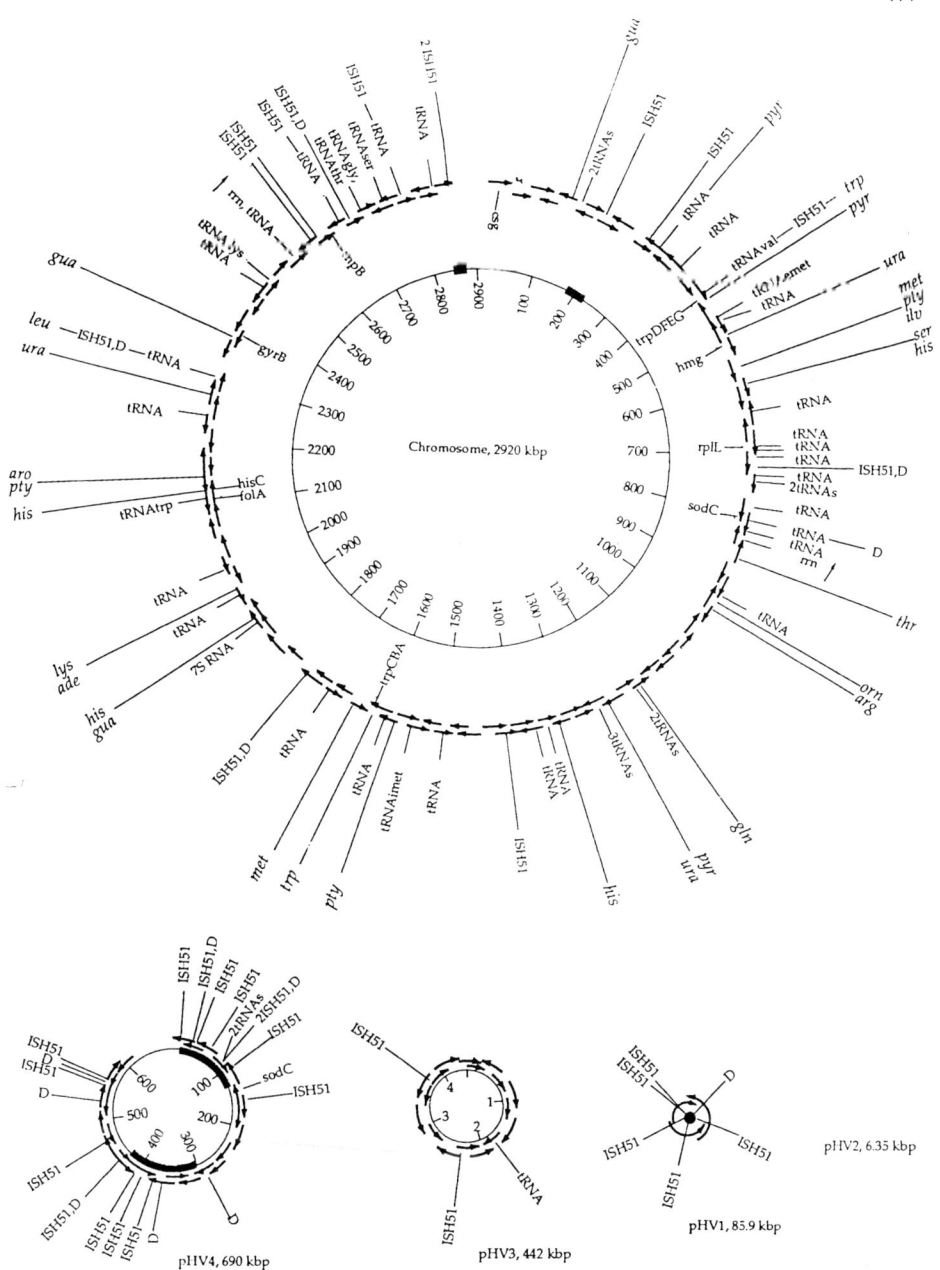

Fig. 2. Map of the *Haloferax volcanii* genome: (top) the chromosome; (bottom) the plasmids. Cosmids are represented by arrows, site-rich regions (oases) by heavy line, and markers placed by hybridization with unique probes are shown inside the circle (all from ref. [15]). Outside the circle are the results of probing with repeated sequences (ISH51, D), with labelled tRNA, 7S and ribosomal RNAs and (outermost) auxotrophic markers mapped by complementation. From ref. [86].

This is attractive in particular because of the high degree of similarity of ISH51 to ISH27 of *Halobacterium halobium* [48]. The large plasmids pHH1 and pNRC100 from *Halobacterium halobium* NRC817 and NRC1 have also been restriction mapped and probed for IS elements. These are also rich in IS elements, including ISH27 (ISH51) in the case of pHH1 (pNRC100 has not been assayed for this element) [8,22].

The chromosomal region from 2600–2900 kbp is rich in ISH51 elements, though less so than pHV1 and pHV4 (Fig. 2), and a small part of the region is oasis DNA. None of the auxotrophies mapped to the region, though it includes several tRNA genes, including the sequenced tRNA ser, gly and thr genes. The map is not yet detailed enough to shed light on the question of the conservation of map order, for which we have so far some information from the sequencing of operons. We should soon be able to determine whether this organism has some of the general features of genome structure recently discerned in bacteria, in particular a preferred orientation for genes [75–77] and a preference of highly expressed genes for the region nearest the origin of replication [78]. Other halobacterial maps now being developed will soon allow more detailed comparisons.

3.4. Genes and operons

The number of halobacterial genes characterized to date is still modest enough to allow me to catalogue them, but genes are becoming more accessible, and both sequences and information about expression are beginning to accumulate rapidly. Protein coding genes were initially isolated by probing with oligonucleotides designed on the basis of peptide sequences, antibody screening of expression libraries, or in a few exceptional cases hybridization with bacterial genes. Sequencing of genes obtained in these ways has also allowed nearby related genes to be harvested and demonstrated that halobacteria contain operon-like clusters of genes, some of which have been shown to give rise to polycistronic transcripts.

Few details are so far known about regulation of gene expression in halobacteria, though there are several characterized halobacterial genes that are known to be regulated. Convincing consensus sequences for archaeal promotors have been produced by sequence comparison [87,88], reviewed by Zillig et al. in this volume (Chapter 12) and supported and extended by in vitro studies: footprinting using purified *Methanococcus vannielii* RNA polymerase [89,90] and in vitro transcription of insertion, deletion, and linker-scanning mutants of the 16S/23S rRNA from *Sulfolobus* B12 [91]. The significance of matches to the consensus in individual cases are hard to evaluate without functional evidence. So far there is not enough information to tell whether there are different types of promotors for different classes of genes, or whether promotors differ between the sulfur/thermophile branch (Crenarchaeota [92]) and the halobacteria/methanogen branch (Euryarchaeota [92]) of the archaea. The situation with transcription termination is less clear, with perhaps more than one type of termination, as in *E. coli*. Potential ribosome-binding (Shine–Dalgarno, 16S rRNA-complementary) sites can be found in many of the sequences, sometimes upstream of, sometimes within the open reading frame, but there is no direct evidence for their importance. I will not dwell on the putative promotors, terminators and ribosome binding sites in the genes catalogued here (see Zillig et al.,

Chapter 12 of this volume), but comment on how they were isolated, and implications of their structure and organization.

Sequencing of halobacterial protein-coding genes has helped to add detail and robustness to our understanding of the phylogenetic relationship between archaea, bacteria and eucarya first lined out by ribosomal RNA oligonucleotide catalogues and sequences. Further impetus and direction has been given to this work by the method of Iwabe et al. [1] for rooting of the universal tree. The method uses pairs of related genes whose origin is an ancient gene duplication. One gene can thus serve as the outgroup for a tree of the other gene. The analysis of F1 ATPase α and β subunits as well as Ef-Tu and EF-G in this way indicates that the root of the universal tree is between the archaea and the eucarya on the one branch and the bacteria on the other [1]. Initiator and elongator methionine tRNAs do not contradict this, but do not provide enough information to give significant support. Sequences of more gene pairs, as well as more examples of the pairs already analyzed will allow the conclusion to be tested further.

Although the overall genome maps of diverse prokaryotes are unlikely to resemble one another [72], operons or clusters of archaeal genes are in several cases similar to their bacterial counterparts (S10 and *rplKAJL* ribosomal protein operons, for example), though details of transcription and regulation may not be. Although it is hard to make an objective assessment of the importance of these fragmentary similarities, it seems likely that they are due to some advantage in the linkage of the genes, maintained by selection. One possibility is that these groups of genes whose products participate in complexes are coevolved to an extent that makes it disadvantageous for them to be separated by gene-exchange processes. The prevalence of the kind of interspecific gene exchange that this would imply is unknown both for the present and the past. Another possibility is that when transcription is coupled with translation, gene products needed in a particular process or complex could be produced in proximity if the genes were linked, and this might procure a kinetic advantage. This effect would disappear when transcription and translation become uncoupled, as must have happened in the ancestor of the eucarya. The conclusion of Röhl and Nierhaus [93] that in *Escherichia coli*, groups of ribosomal protein genes whose products are interdependent for assembly are grouped together in operons is suggestive of this. Some features of the ribosome assembly process might thus favor the maintenance of a particular operon structure.

3.4.1. Ribosomal RNA genes

Archaeal ribosomal RNAs resemble their bacterial counterparts in size, number, and nucleotide sequence. Those of the halophiles are also transcribed from genes organized in operons similar to those of bacteria, in the order 5′–16S–tRNA (ala where known)–23S–5S–(tRNA cys). Although the unprocessed precursor has not been detected, archaeal rRNA operons have flanking and spacer regions which can base pair to form structures similar to those important in bacterial rRNA transcript processing [94–102]. In *Halobacterium cutirubrum* 34001, there is a specific endonucleolytic cleavage site in each stem [103], the characteristic bulge–helix–bulge structure of which is also found in *Thermoproteus tenax* [94], *Desulfurococcus mobilis* [94], *Thermofilum pendens* [104], and *Methanobacterium thermoautotrophicum* [105]. This processing occurs at such a rate that a full-length precursor is probably not formed [103]. See ref. [9] for another recent review of archaeal rRNA operons.

The number of rRNA operons in archaea is always small. *Haloferax volcanii, Methanobacterium thermoautotrophicum* and *Methanothermus fervidus* have two, and *Methanobacterium vannielii* has four (reviewed by Brown et al. [5]). On the basis of Southern transfers of pulsed field gels, Sanz et al. [106] report that the halophilic archaea have from one to four rRNA operons: *Haloarcula californiae*, 4; *Haloferax gibbonsii*, 4; *Halobacterium halobium* NCMB 777, 3; *Halobacterium marismortui*, 3; *Halococcus morrhuae*, 2; and *Halobacterium salinarium*, 1. Bacteria have from one to eleven copies (e.g., *Bacillus subtilis*, 11 [107]; *Escherichia coli*, 7 [108], and *Mycoplasma pneumoniae*, 1 [109]). In *Halobacterium halobium*, the presence of only one rRNA operon facilitated the isolation of strains with mutated 23S rRNA genes, which are resistant to thiostrepton [110], anisomycin [111] or chloramphenicol [112].

3.4.2. Transfer RNA genes
Ten *Haloferax volcanii* tRNA genes have been cloned and sequenced (refs. [83,84], and Gupta, R., personal communication), as have a number from other archaea. None have the 3' CCA encoded; it is added after transcription, as it generally is in eucarya. The sequenced tRNA genes not in rRNA operons are found as single genes, except for the tandemly repeated tRNAval^{GAC}. This contrasts with the situation in several methanogens in which tRNA operons have been found [5]. Introns have been found in the tRNAtrp genes of *Haloferax volcanii* [1113], *Haloferax mediterranei* and *Halobacterium cutirubrum* [83], and in the elongator methionine tRNA of *Haloferax volcanii* [84]. These short introns are placed in exactly the same position (1 nucleotide 3' of the anticodon) as the introns of eucaryal nuclear tRNA genes, but the mechanism of identifying cleavage sites is different [114–116]. These introns have in common the possibility of forming a characteristic base-paired stem with the splice sites in staggered bulges. These sites resemble the exonuclease III sites involved in 16S and 23S rRNA processing, and cleavage could be achieved by the same enzyme [114]. Cleavage at splice sites can be achieved in vitro, and the use of various synthetic substrates has demonstrated the importance of the bulge–helix–bulge structure [104,114,115,117,118]. Forty-one *Haloferax volcanii* tRNAs, as well as several other archaeal tRNAs have been sequenced (reviewed in [119]). These fit the generalized structure known chiefly from *Escherichia coli* and yeast tRNAs, but in the sequence and pattern of modification there is a mixture of eucaryal, bacterial and unique features, including that, like in eucarya, leucine and serine but not tyrosine tRNAs are of class II (large extra arm).

3.4.3. 7S RNA
Archaea contain a stable 7S RNA species of unknown function, which is not associated with the ribosome. The 7S RNA gene from *Halobacterium halobium* NRC817 has been cloned and sequenced [120]. The sequence resembles the 7SL RNA of eucarya [85] and the 4.5S (*Escherichia coli*) or sc (*Bacillus subtilis*) RNAs of bacteria [121] in potential secondary structure. The 7SL RNA forms part of the signal-recognition particle, involved in protein translocation. The function of the 4.5S RNA is not known exactly, but it is known to be essential [122].

3.4.4. RNase P RNA
RNase P, which removes the 5′ leader from tRNA transcripts, contains an RNA moiety in bacteria, eucarya, and archaea. The enzyme from *Haloferax volcanii* has been purified and the 345-nucleotide RNA portion used to generate a probe for cloning of the corresponding gene [123]. The sequence can be folded into a structure similar to that of bacterial RNase P RNAs. S1 nuclease and primer extension localize the 5′ end of the transcript adjacent to four potential archaeal promotor sequences. An in vitro transcript corresponding to the native RNA plus twenty 5′ and nine 3′ flanking nucleotides did not by itself exhibit RNase P activity (as bacterial RNase P RNAs do), under a variety of conditions. It was, however, able to reconstitute an active enzyme in combination with the protein moiety from *Bacillus subtilis*. This indicates that in *Haloferax volcanii* the RNase P RNA is the catalytic part of the enzyme but it may require some structural help [123].

3.4.5. Bacteriorhodopsin
Halobacterium halobium contains four retinylidene proteins: bacteriorhodopsin [124], halorhodopsin [125] and two sensory rhodopsins [126]. Purple membrane was first recognized microscopically in preparations of fractionated *Halobacterium halobium* cells and bacteriorhodopsin isolated as its sole protein component. In vitro studies subsequently revealed its role as a light-driven proton pump [127]. Bacteriorhodopsin is of extraordinary interest because it contains everything necessary for active proton translocation in a single small protein. The protein has been studied in great detail, including the complete peptide sequence [128,129] which was used to design a dodecanucleotide primer to make a specific cDNA [130], which was then used as a probe to obtain the entire gene, the first halobacterial protein-coding gene cloned [130]. Comparison of the peptide and DNA sequences demonstrated that halobacteria use the universal genetic code. S1 nuclease protection and capping analysis reveal that the transcript begins only 2 bp upstream from the start codon and ends at two sites 45 bp (major) and 170 bp (minor) beyond the stop codon [131]. Numerous *bop*⁻ strains have been isolated. These contain insertion sequences within or upstream of the *bop* gene [131,132]. Investigation of the insertions upstream of the *bop* gene brought to light two additional genes which are involved in *bop* expression: *brp* and *bat* [132–134]. They are located upstream of the *bop* gene, and are opposite to it in orientation. *brp* is 526 bp from the *bop* ORF, and *bat* is in turn downstream of *brp*. The start codon of *bat* overlaps the stop codon of *brp*. Protein products of these two open reading frames have not yet been characterized, but judging from Northern transfers, both are expressed at low levels, and are transcribed separately even though the reading frames overlap [135]. *bat* appears to be an activator of *bop* expression, and ISH insertions into *brp* drastically reduce *bat* transcript levels. The predicted *brp* protein has the potential to form six or seven hydrophobic helices, so it may be a membrane protein. Yet another open reading frame, for which a transcript can be detected, extends downstream of *bat*, in the opposite direction [134].

The bacteriorhodopsin gene, like the gas-vesicle protein gene (section 3.4.16), is more highly expressed under conditions of low oxygen tension, a response blocked by the DNA-gyrase inhibitor novobiocin. Bacteriorhodopsin message levels increase 20-fold and *gvpA* message levels five-fold between mid-logarithmic and stationary phase. Levels of both are lower in cultures grown with increased aeration, and it is not known whether the variation over the culture cycle is purely a matter of oxygen availability. The connection between supercoiling and expression that the novobiocin effect indicates may have to do with Z-DNA formation in $(YR)_n$ tracts found upstream of *bop, gvpA* and *sod* [153,155]. This might turn out to be a mechanism of regulation that is common to bacteria and archaea, because the *bat* gene involved in *bop* expression is recognizably similar to *nifL,* the oxygen sensor in nitrogen fixation in *Klebsiella pneumoniae* [153].

Measurements of *bop, brp* and *bat* transcript levels in cultures grown under low and high oxygen tension with and without illumination suggest a regulatory scheme in which *bat* is involved in activating *bop* at low oxygen tension, and this is modulated by *brp* in response to light [154].

The structure and function of bacteriorhodopsin has been studied in great detail, both because it is an accessible example of a light-transducing protein, and because it is an useful model for membrane proteins in general. Much of this work has involved in vitro mutagenesis of a synthetic *bop* gene designed for expression in *E. coli,* followed by purification of the altered apoprotein and in vitro regeneration of altered bacteriorhodopsin. That approach was necessitated by the initial unavailability of a method of reintroducing altered genes into halobacteria, and the difficulty of finding point mutations in *Halobacterium halobium* against the high background of insertion mutants. Point mutations in *bop* have been obtained from *Halobacterium* species GRB after selection (using 5-bromo-2'-deoxyuracil) against cells able to grow phototrophically [136,137]. Derivatives of the *bop* gene can now be reintroduced by transformation. This has been done with *bop* cloned in p455, a precursor of pWL102, which could restore a *bop*$^-$ insertion mutant to *bop*$^+$, though it is unclear whether this construct was maintained as a plasmid [138]. The *bop* gene has been cloned into a derivative of pGRB1 and reintroduced into a *bop*$^-$ strain, where its presence can be scored by purple color. The bacteriorhodopsin produced is similar in its physical properties to that of the wild type, though lower in amount. Bacteriorhodopsin expression responds to oxygen tension in this strain as it does in the wild type, and the lower overall level of expression is attributed to the instability of the construct, which does not include a selectable marker [63].

3.4.6. Halorhodopsin

Halorhodopsin is a light-driven chloride pump, whose activity was detected in membrane vesicles prepared from *bop*$^-$ *Halobacterium halobium* strains [125]. Genes encoding the halorhodopsins of *Halobacterium halobium* and *Natronobacterium pharaonis* have been cloned and sequenced [139–141]. The *Halobacterium halobium* halorhodopsin (*hop*) gene was cloned from a cosmid library using probes designed on the basis of partial peptide sequence of the protein, which was purified from a *bop*$^-$ strain. The N-terminus of the mature product is apparently heterogeneous, but about 25% begins with predicted residue 22. Two aspartic-acid residues are processed from the C-terminus. Seven transmembrane helices can be predicted from the predicted amino-acid sequence, just as in bacteriorhodopsin, and the sequence resembles that of bacteriorhodopsin in these regions (identical at 36% of positions), but much less in the intervening loops (19%) [139].

Natronobacterium pharaonis, an alkaliphilic halobacterium, contains a halorhodopsin as well, though at only 10% of the level of halobium *hop*. The activity of pharaonis halorhodopsin in vitro is not more alkali tolerant than is halobium *hop*: their pH dependence is similar. Unlike halobium halorhodopsin, the pharadopsin *hop* transports nitrate as well as it does chloride[141]. The pharaonis *hop* gene was cloned by probing with the gene from *Halobacterium halobium*, and its identity was confirmed by comparison with partial peptide sequence obtained from the purified protein[142].

3.4.7. Sensory rhodopsins
Halobacterium halobium is motile and its swimming is modulated by light, which attracts (in the orange-red) or repels (in the ultraviolet) the cells. Sensory rhodopsin I was first detected in a mutant lacking bacteriorhodopsin and halorhodopsin, whose presence had previously made the studies of the spectral properties of the phototaxis receptor(s) difficult[126]. Further study of the repellent effect of blue-green light led to the discovery of sensory rhodopsin II. The gene for sensory rhodopsin I was isolated from *Halobacterium halobium* by the use of a degenerate oligonucleotide probe derived from N-terminal peptide sequence. The gene encodes a C-terminal aspartic acid that is not present in the mature protein, but the N-terminus is not processed. Judging from Northern hybridization, the transcript is monocistronic. The amino-acid sequence is detectably similar to those of bacteriorhodopsin and halorhodopsin, but not animal rhodopsins[143].

3.4.8. Gas-vesicle proteins
Halobacterium halobium and its relatives contain gas vesicles, which are composed of a single protein, and are similar to those found in cyanobacteria[144,145]. That a gene involved in vesicle synthesis was plasmid borne was suspected because of a high frequency of loss of gas vesicles and its correlation with plasmid loss[35] or rearrangement[32,34]. A gas-vesicle protein gene was located and cloned in *Halobacterium halobium* NRC1 using a cyanobacterial probe. The gene, *gvpA*, proved to be located on the plasmid pNRC100[17]. The major transcript begins 20 bp upstream of the ATG. The corresponding gene from *Halobacterium halobium* NRC817 plasmid pHH1 has also been isolated and sequenced, along with a second gas-vesicle protein gene found in the chromosomal DNA[18,146]. These two genes (called p-vac and c-vac by Pfeifer and collaborators and *gvpA* and *gvpB* by DasSarma) are nearly identical in amino-acid sequence, differing by two substitutions and a gap[18]. *gvpB* is expressed in stationary phase, unlike *gvpA*, which is expressed throughout the culture cycle at a higher level. This explains the presence of two proteins differing in size in gas-vesicle preparations[145]. Extremely halophilic strains *Halobacterium* spp SB3, YC1819-9, and GN101 which, like *H. halobium gvpA* mutant strains, have gas vesicles only in stationary phase, have only *gvpB* [147]. *Halobacterium halobium* NRC1 has recently been found to contain a 400 kbp plasmid which contains *gvpB* (Hackett, N., personal communication).

A minor *gvpA* transcript continues downstream to include another open reading frame of unknown function, dubbed *gvpC* [148]. Two additional genes necessary for *gvpA* expression were found by analysis of insertion mutants. Insertions both within and well upstream of *gvpA* gave rise to the Vac⁻ phenotype. Two open reading frames, called *gvpD* and *gvpE*, oriented oppositely to *gvpA*, are thus also required for *gvpA* expression. Southern hybridization results indicate that the *gvpB* locus is also accompanied by sequences similar to *gvpD* and *gvpE* [149]. An additional seven open reading frames thought to be related to gas-vesicle synthesis (*gvpFGHIJKLM*) have been identified by sequencing downstream of *gvpE* [150].

Haloferax mediterranei has gas vesicles whose electron-microscopic appearance is like that of the *gvpB* product of *Halobacterium halobium*. Like *gvpB*, the *H. mediterranei* gene is expressed in stationary phase. Expression of the *Haloferax mediterranei gvp* is also modulated by salt: vesicles are only seen at salt concentrations above 17% [151]. A DNA-modification activity in *Haloferax mediterranei* has also been reported to be modulated by salt concentration [152]. A construction in pWL102 containing an 11 kbp fragment including the *Haloferax mediterranei gvp* gene could transform *Haloferax volcanii* (which does not naturally contain gas vesicles) but not a *gvpA* deletion mutant of *Halobacterium halobium* to Vac$^+$. A 4.5 kbp *gvpA* fragment (which would also include *gvpD* and *gvpE*) was also unable to restore Vac$^+$, indicating that yet more sequences are required for expression of the Vac$^+$ phenotype.

3.4.9. Cell surface glycoprotein

The surface of halobacterial cells is covered by a hexagonally packed surface layer [156–159]. The protein of which the *Halobacterium halobium* surface layer is composed was the first glycoprotein isolated from a prokaryote, though there is now evidence than an bacterial S-layer protein may also be glycosylated [160]. A fragment of the *Halobacterium halobium* cell-surface glycoprotein gene was isolated by screening an expression library with an antibody raised to the purified protein, and the complete gene was isolated using the fragment as a probe. Comparison of the gene sequence with N-terminal peptide sequence reveals that the primary translation product contains either 18 or 34 amino acids which are subsequently removed. The transcript was shown by primer extension to begin 346 bp upstream of the beginning of the mature protein's amino-acid sequence. The sequence at the processing site is (ala)$_4$, with the last ala residue as the mature N-terminus. This resembles a consensus for bacterial and eucaryal signal peptides which the *bop* "signal peptide" does not share [157].

The *Haloferax volcanii* cell-surface glycoprotein (*csg*) gene was cloned using a probe produced by PCR using primers designed on the basis of peptide sequences. The protein is synthesized as a precursor with a 34-residue N-terminal extension, cleaved, similar to *Halobacterium halobium*, at (ala)$_3$. The amino-acid sequence has a hydrophobic C-terminus presumed to be the membrane anchor. The remainder of the protein contains largely polar amino acids, with, as in most halobacterial proteins, an excess of acidic groups. Next to the hydrophobic region are clusters of threonine residues, most or all of which in the corresponding position in *Halobacterium halobium* are involved with *O*-glycosidic linkages. Similarity with the *Halobacterium csg* is spread in short stretches over the length of the sequence [156].

3.4.10. Flagellins

Halobacterial flagella differ from those typical of bacteria in that they are right-handed helices, and reversing the direction of rotation reverses direction of swimming; the bundle of polar flagella remains together [174]. The flagella can be purified and are composed of at least three heterogeneous sulfated glycoproteins. Antibodies raised to purified flagellins were used to isolate a fragment of a flagellin gene from an expression library. This fragment was in turn used as a probe to isolate a clone which proved to contain two similar open reading frames separated by 11 bp. The sequence did not match what was expected from the fragment used as probe, and another locus containing three additional genes was subsequently found. All five genes are quite similar to each other in predicted amino-acid sequence, except for well-defined variable regions. They do not bear a detectable resemblance to bacterial flagellins. The two spacers separating the three genes of the second locus are the identical 11 bp sequence as found in the first. Transcripts of both loci are detectable on Northern blots, and their sizes as well as the transcription-initiation sites defined by primer extension show that both transcripts are polycistronic.

3.4.11. Superoxide dismutase

Superoxide dismutases in the bacterial and eucaryal lineages are of different types, containing Fe or Mn in the former case and Cu and Zn in the latter. Archaeal superoxide dismutases contain Fe (*Methanobacterium bryantii* [161], *Thermoplasma acidophilum* [162]) or Mn (*Halobacterium cutirubrum* [163], *Halobacterium halobium* [164]). The *Halobacterium cutirubrum* 34001 *sod* gene was cloned using an oligonucleotide probe designed on the basis of N-terminal amino-acid sequence. The amino-acid sequence predicted from the sequence of the gene is quite similar to those of bacterial superoxide dismutases of which it is clearly a homologue. The N-terminal methionine is not present in the mature protein. Primer extension, S1 mapping and capping analysis show that the major transcript begins only 3 bp upstream of the start codon. In this, and in the fact that there are only poor matches to the archaeal promotor consensus sequence, this gene resembles *brp*. SOD activity is present even when cells are grown anaerobically and is inducible by about fivefold by paraquat (which generates superoxide in vivo). The induction can also be seen at the mRNA level, but is then less pronounced. The inducible and constitutive *sod* transcripts are identical [155,165]. There is another sequence in the *Halobacterium cutirubrum* genome which hybridizes with the *sod* gene and proves to be an expressed open reading frame resembling *sod*, with flanking sequences that are not detectably related. The *sod*-like gene (*slg*) transcript is at a level similar to that of *sod*, but is not paraquat-inducible. May and Dennis [165] argue that the pattern of differences between the *sod* and *slg* genes indicates that the latter is not a pseudogene, or responsible for a second SOD activity, but a copy of the *sod* gene that has been recruited for another function. There is only one detectable SOD activity, corresponding to the *sod* gene, and although differences between the two sequences are spread through all three reading frames, transversions outnumber transitions. That would seem to be evidence for selection of some sort, but if it were selection for conserved function, changes would be concentrated in silent or conservative-substitution positions, which is not the case here. Hybridizations with the *Halobacterium cutirubrum* 34001 *sod* gene have also revealed two *sod*-like sequences in *Haloferax volcanii* [15].

3.4.12. Dihydrofolate reductase

Haloferax volcanii mutants resistant to the competitive dihydrofolate-reductase inhibitor trimethoprim contain amplifications of various lengths of DNA from a particular region and overproduce a particular protein [166], which proved to be dihydrofolate reductase. The *dhf* gene was cloned from one of the amplification mutants, and its sequence can be aligned with those of bacterial and eucaryal enzymes, though it is identical to them at less than 30% of positions. Structures and function of both eucaryal and bacterial dihydrofolate reductases have been studied in detail, and most of the positions which are conserved or thought to be functionally important are present in the halobacterial enzyme. The halobacterial enzyme is more acidic than its counterparts from bacteria or eucarya, with the excess negative charges distributed throughout the molecule. S1-nuclease analysis indicates that transcription is initiated more than 220 nucleotides upstream of the start codon, so this gene may be part of a polycistronic message [166].

3.4.13. DNA gyrase

The DNA gyrase subunit B gene from *Haloferax* phenon K isolate Aa 2.2 was isolated as a novobiocin-resistance determinant, by shotgun cloning into the endogenous plasmid pHK2 as a 6.7 kbp fragment. The location of the determinant was further narrowed by subcloning in *E. coli* and testing of subclones for their ability to confer novobiocin resistance when transformed into *Haloferax* (relying on recombination with the chromosome). The 1.4 kbp fragment identified in this way, along with flanking fragments, were sequenced and a 1920 bp ORF was found, highly similar to the *E. coli gyrB* gene. Only 1 bp downstream the gyrase subunit A ORF was found. Primer extension revealed a transcript beginning 115 bp upstream from the ATG, and preliminary Northern data indicate that *gyrA* and *gyrB* are cotranscribed. Compared to the wild-type sequence, the nov^r *gyrB* sequence differs in 3 amino-acid replacements in a highly conserved region in eucaryal and bacterial gyrases [4,58].

3.4.14. Photolyase

Photolyase, the enzyme responsible for photoreactivation of UV damage to DNA from four eucaryal and bacterial species, has been characterized at the sequence level, and these sequences are highly similar in their C-terminal halves, even though two use folate as a prosthetic group (*E. coli* and yeast), and the other two use 7,8-didemethyl-8-hydroxy-5-deazaflavin (DHF, *Anacystis nidulans, Streptomyces griseus*). It was possible to use a probe from *Streptomyces* to isolate the *Halobacterium halobium* photolyase gene [184]. The amino-acid sequence predicted is similar to other sequenced examples, more so with the DHF-containing enzymes, as had been predicted from the action spectrum of reactivating light. As in dihydrofolate reductase and superoxide dismutase, the halobacterial enzyme is more acidic than its non-halobacterial homologue, which may be an indication of the adaptation to halophily. When expressed in *E. coli* as a fusion protein, the gene confers increased UV resistance to a photoreactivation mutant.

The photolyase (*phr*) gene proves to be immediately upstream of *slg* (see section 3.4.11) and is flanked on the upstream side by a 151-codon ORF of unknown function. The 151 and *slg* ORFs are preceded by recognizable promotor sequences, and Northern analysis indicates that the *phr* transcript includes 151, *phr* and *slg*, and that its level does not respond to UV light or O_2, but is repressed by visible light. *slg* is transcribed separately at a much higher level, in addition to the longer transcript. Although *slg* is not paraquat-inducible [165], the shorter *slg* transcript is seen at much higher levels in aerobically grown cells, and is also induced by high levels of ultraviolet light [184].

3.4.15. Bacteriophage ΦH
The temperate bacteriophage ΦH infects *Halobacterium halobium*. In its lysogenic form ΦH DNA is a 57 kbp circle. A defective lysogen, containing a 12 kb portion of the phage genome (L-segment) as a plasmid confers resistance to the wild-type phage, but not to an insertion mutant of the phage called ΦHL1 [19,36,49]. The principal bacteriophage transcripts have been mapped over the course of a single-burst experiment by using pulse-labelled RNA as a probe on Southern transfers of phage DNA digests. Distinct early, middle, and late patterns are seen [167]. Acting on the hypothesis that the L segment encodes a phage repressor similar to those of bacterial lambdoid phages, and that the site of the insertion in the more virulent phage ΦHL1 defines an operator, Ken and Hackett [168] set out to find and isolate these parts of an archaeal regulatory system. Using a mobility shift assay, it was possible to find a fragment in the proposed operator region which is bound by a factor present in extracts of lysogenic but not uninfected cells. Even though it was necessary to do the assay under quite unphysiological conditions, the binding could be shown to be specific; it is abolished by the addition of excess specific competitor DNA. Methylation footprinting further narrowed the binding site to a set of two nested imperfect inverted repeats, with G residues susceptible to methylation interference spaced 11 nucleotides apart on both strands. Sequencing in the operator region led to the identification of an 89-codon ORF corresponding to a previously mapped transcript [167], which has a weak resemblance to the helix–turn–helix family of DNA-binding proteins. When expressed in *E. coli*, the product of this ORF could be shown to bind the operator sequence. This is thus the first example of archaeal gene regulation studied at the level of mechanism, and it is much like what one would see in a bacterium.

3.4.16. H^+ ATPase
The soluble "headpiece" of the membrane-associated H^+ ATPase of *Halobacterium halobium* r_1m_1 has been isolated, and partial peptide sequence was used to design oligonucleotide probes for cloning the genes for the two subunits, α and β [172]. As in *Sulfolobus acidocaldarius* [173], the genes (*atp*A and *atp*B) are closely linked in the order AB, with open reading frames separated by 0 bp (*Sulfolobus*) and 5 bp (*Halobacterium*). The amino-acid sequences of archaeal ATPases are more closely related to eucaryal vacuolar ATPases than to the bacterial (or mitochondrial) F_0F_1-type ATPases used for ATP synthesis by bacteria and eucarya. The α and β subunits are detectably related to each other and must have arisen from a gene duplication pre-dating the divergence of all existing forms of life. Such pairs (including the ATPase α and β subunits from *Sulfolobus*) were used by Iwabe et al. [1] to assign a root to the universal tree.

3.4.17. Histidinol-phosphate aminotransferase
A homologue of the *E. coli hisC* gene was isolated by transformation of the his^-, arg^- *Haloferax volcanii* strain WR256 with cosmid DNAs from the *Haloferax volcanii* minimal set. The deduced amino-acid sequence can be aligned with corresponding bacterial and eucaryal enzymes, with which it is identical at approximately 20 (yeast) or 30 (*E. coli* or *Bacillus subtilis*) percent of positions. *E. coli* and *B. subtilis* are in turn only 30% identical [65]. The strain had previously been shown to be transformable

to wild type with genomic DNA and to have a low rate of spontaneous reversion [61]. This study demonstrated the usefulness of transformation with non-replicating *E. coli*-grown DNAs despite a restriction barrier which is estimated to reduce transformation efficiency by 10^4-fold [47]. More recently, use of *dam*⁻ *E. coli* hosts has been shown to improve transformation efficiency, apparently circumventing the restriction system [59].

3.4.18. 3-Hydroxy-3-methylglutaryl–coenzyme A reductase
The gene for 3-hydroxy-3-methylglutaryl–coenzyme A reductase from *Haloferax volcanii* was isolated as the mevinolin-resistance determinant used in the construction of the shuttle vector pWL102 [47]. Mevinolin is a competitive inhibitor of the enzyme in mammals [169] and in *Halobacterium halobium* [55], which produces mevalonate, precursor of the isoprenoid lipids of the halobacterial membrane. A survey of mevinolin-resistant *Haloferax volcanii* isolates revealed two classes. In one, amplification of the *hmg* gene was seen, and in the other, a particular MluI site was destroyed. Sequencing in the latter case revealed a single base change with respect to the wild-type sequence, which simultaneously abolished the MluI site and made the sequence more similar to the proposed TATA-like distal promotor element [88,89]. The deduced amino-acid sequence of the gene can be aligned with those of the soluble HMG–CoA reductases of plants and with the C-terminal parts of the mammalian and yeast enzymes, which are anchored by membrane-spanning domains in the N-terminal half. The *Haloferax volcanii* enzyme is less similar to the only sequenced bacterial example, from *Pseudomonas mevalonii*, which has a catabolic role and uses NADH instead of NADPH.

3.4.19. Tryptophan biosynthesis
Haloferax volcanii genes similar to all seven *E. coli* structural *trp* genes have been isolated by complementation of tryptophan auxotrophs with DNA shotgun cloned into pHV51 (an ISH51-bearing variant of pHV2) or pWL102. The mutations could be mapped to two locations on the *Haloferax volcanii* cosmid map. On sequencing, the first of these proved to contain three open reading frames similar to the *trp C, B* and *A* genes of *E. coli*, in that order [170], and the other *trp D, F, E* and *G* [171]. The adjacent open reading frames all overlap by 1 or 4 nucleotides, and upstream of each cluster is a sequence resembling the archaeal promotor consensus. Each open reading frame is preceded by what may be a ribosome-binding site. There is no direct information on transcription or regulation, but the *CBA* cluster has an upstream structure capable of forming alternative stem–loop structures reminiscent of an attenuator, and in the *trpE* ORF there is an amino-acid sequence believed to be conserved in bacteria as a site of feedback inhibition.

3.4.20. Ribosomal proteins
The ribosome is easily the most complicated enzyme known, consisting (in *E. coli* for example) of three RNAs and 52 proteins. The structures of the ribosomal proteins of diverse organisms are being investigated for the light the comparative approach will shed on the functional significance of different parts of the ribosome, and they can also be phylogenetically informative. Sequencing of ribosomal-protein genes builds on a large body of work in characterizing and sequencing of the proteins themselves. Cloning of ribosomal-protein genes using oligonucleotide probes designed from peptide sequences and sequencing has yielded many additional genes in the flanking sequences. This work

has been exciting because many of the halobacterial and other archaeal ribosomal-protein genes proved to be in clusters similar in order to those found in bacteria. This despite the fact that the degree of sequence similarity with eucaryal homologues is greater in most cases, and the presence of some archaeal ribosomal proteins with eucaryal but no bacterial counterparts. Ten *Halobacterium marismortui* proteins so far are without bacterial counterparts: HS3, HS6, HS12, HS13, HL6, HS15, HL29, HL21/22, HL24 and HL30. (The numbering corresponds to positions on two-dimensional electrophoresis of the protein; where homologous proteins have been identified in *E. coli*, the archaeal proteins are named after them, either as, e.g., L12e [175] or as, e.g., HmaL12 [176]. In *E. coli,* the gene names rpsA, B, C... and rplA, B, C... parallel the protein names S1, 2, 3... and L1, 2, 3.... For convenience I will refer to the halobacterial genes by the names of their products). The eucaryal ribosomal proteins are less extensively characterized than those of *E. coli,* so that some archaeal ribosomal proteins that appear to be unique or have only bacterial counterparts will likely prove to have eucaryal counterparts. The genes without bacterial counterparts are grouped both with each other (5'–HL46e–HL30–3', separated by 1 bp [177]), or within otherwise "bacterial" groupings (HL6 occupies what would be the place of *rplD* in the *E. coli* S10 operon [178]; HS3, HL5 and HL24 are inserted into the equivalent of the *E. coli* spectinomycin operon [179]). The *Halobacterium marismortui* S10 operon also lacks the equivalent of the S10 gene at the 5' end, where it has instead two unidentified ORFs, and the L16 gene and S17 gene at the 3' end. The same arrangement has been found in *Halobacterium halobium* [180,181].

The main emphasis of the work has been to find the sequences of the proteins, so in most cases the details of transcription are unknown, and the regulation completely unknown, but where known, the details of transcription are different from the bacteria. For example, *Halobacterium marismortui* and *Halobacterium cutirubrum* NRC34001 have the genes for the large-subunit proteins adjacent to each other in the order and orientation 5'–L11–L1–L10–L12–3', just as in the *E. coli rplKAJL* region, but in the halobacteria there are two major transcripts, one of L11 (*rplK*) and the other of the other three (*rplAJL*) [175,176,182] (in *Halobacterium halobium* S9 the *rplAJL* operon has so far been characterized, but not *rplJ* [197]). In *Methanococcus vannielii, rplAJL* are adjacent but all transcribed separately, and upstream there is an unidentified ORF instead of *rplK* [183a], while in *E. coli* there are *rplKA* and *JL* transcripts. The *E. coli rplJL* transcript also continues downstream to include *rpoBC*.

3.4.21. Elongation factors
The elongation factor Tu was purified from *Halobacterium marismortui* and N-terminal peptide sequence was used to design an oligonucleotide probe. The clone obtained using the oligonucleotide probe was sequenced, and the gene has 61% identity with its homologue from *Methanococcus vannielii,* and it can also be unambiguously aligned with EF1α from yeast and EfTu from *E. coli,* from which it differs most notably in the number of acidic residues [183b].

3.4.22. RNA polymerase
Archaea, like bacteria, apparently have a single, all-purpose DNA-dependent RNA polymerase core enzyme, and the genes of the four largest subunits have been cloned and sequenced from halobacteria [185] as well as a methanogen [186] and

Sulfolobus acidocaldarius [187,188]. Relatedness between polymerases had previously been investigated immunochemically [189]. The genes for the largest subunits of the *Halobacterium halobium* RNA polymerase were cloned by screening of a *Halobacterium halobium* expression library with antibodies raised against the purified C subunit. The fragment obtained was then used as a probe to isolate a larger fragment bearing seven open reading frames, including four corresponding to the subunits B″, B′, A and C of the polymerase, all but A confirmed by comparison with N-terminal peptide sequence, and two ribosomal-protein genes. The cluster is composed of a 75-codon ORF of unknown function, but which has counterparts in *Sulfolobus* and *Methanobacterium*, the B″ ORF immediately downstream, followed after 1 bp by the B′ ORF, then A overlapping by 1 bp and C overlapping in turn by 5 bp and another unknown ORF (139 codons) overlapping by 1 bp. Northern, S1 nuclease protection, and primer extension analyses show that this unit is cotranscribed in a transcript of over 8 kbp starting 42 bp upstream of the 75 ORF. Partial, ragged termination is seen downstream of the 139 ORF, as well as a small amount after the subunit A gene, and there is some readthrough into the S12 and S7 ribosomal-protein genes which begin 163 nucleotides downstream of the end of the 139 ORF and are separated by 3 bp. The ribosomal-protein genes are separately transcribed, at 5–10 times the level of the RNA-polymerase operon, starting at the first nucleotide of the S12 ORF.

The *rpo* genes are large and conservative enough to be useful for phylogenetic comparison at the deepest level, and as such are a welcome complement to ribosomal RNA comparisons. Analyses of these sequences [190,191] support the conclusion [66] that the archaea are monophyletic. It seems likely that they diversified from a single all-purpose ancestor after the divergence of the ancestral eucarya [191], but the divergence between the three specialized polymerases of the eucarya is very great and the history may be more complicated [190].

4. Future directions

The study of halobacterial genes has entered a phase in which, as in the best-studied eucaryal and bacterial systems, genes are readily isolated, modified, and reintroduced. This allows emphasis to begin to shift from sequence comparison (not by any means finished) to a study of processes such as transcription and its regulation, translation, replication and recombination.

The proposed rooting of the universal tree [1] as well as the similarities in structure between archaeal and bacterial genomes, make it seem that the most common ancestor of all modern life had a genome something like those of present-day prokaryotes. This means that very large changes in genome organization must have occurred early in the history of the eucaryal lineage, and information on this process may be available by finding and comparing very early-diverging eucarya.

The similarities of gene organization between archaea and bacteria that are already apparent will become easier to interpret as more data become available. Comparative information within both domains will give some power to discriminate between functionally important and chance juxtapositions.

References

[1] Iwabe, N., Kuma, K.-I., Hasegawa, M., Osawa, S. and Miyata, T. (1989) Proc. Natl. Acad. Sci. U.S.A. 86, 9355–9359.
[2] Grant, W.D. and Larsen, H. (1989) In: Bergey's Manual of Systematic Bacteriology (Staley, J.T., Bryant, M.P., Pfennig, N. and Holt, J.G., Eds.), pp. 2216–2219, Williams and Wilkins, Baltimore.
[3] Lodwick, D., Ross, H.N., Walker, J.A., Almond, J.W., and Grant, W.D.(1991) System. Appl. Microbiol. 14, 352–357.
[4] Holmes, M. and Dyall-Smith, M. (1990) J. Bacteriol. 172, 756–761.
[5] Brown, J.W., Daniels, C.J. and Reeve, J.N. (1989) CRC Crit. Rev. Microbiol. 16, 287–338.
[6] Doolittle, W.F. (1985) In: The Bacteria: A Treatise on Structure and Function (Woese, C.R. and Wolfe, R.S., Eds.), pp. 545–560, Academic Press, Orlando, FL.
[7] Charlebois, R.L. and Doolittle, W.F. (1989) In: Mobile DNA (Berg, D.E. and Howe, M.M., Eds.), pp. 297–308, American Society for Microbiology, Washington, D.C.
[8] Pfeifer, F. (1988) In: Halophilic Bacteria (Rodriguez-Valera, F., Ed.), pp. 105–133, CRC Press, Boca Raton, FL.
[9] Garrett, R.A., Dalgaard, J., Larsen, N., Kjems, J. and Mankin, A.S. (1991) Trends Biochem. Sci. 16, 22–26.
[10] Smith, C.L., Econome, J.G., Schutt, A., Klco, S. and Cantor, C.R. (1987) Science 236, 1448–1453.
[11] Kohara, Y., Akiyama, K. and Isono, K. (1987) Cell 50, 495–508.
[12] Noll, K.M. (1989) J. Bacteriol. 171, 6720–6725.
[13] Yamagishi, A. and Oshima, T. (1990) Nucl. Acids Res. 18, 1133–1136.
[14] Charlebois, R.L., Hofman, J.D., Schalkwyk, L.C., Lam, W.L. and Doolittle, W.F. (1989) Can. J. Microbiol. 35, 21–29.
[15] Charlebois, R.L., Schalkwyk, L.C., Hofman, J.D. and Doolittle, W.F. (1991) J. Mol. Biol. 222, 509–524.
[16] Pfeifer, F., Weidinger, G. and Goebel, W. (1981) J. Bacteriol. 145, 369–374.
[17] DasSarma, S., Damerval, T., Jones, J.G. and Tandeau de Marsac, N. (1987) Mol. Microbiol. 1, 365–370.
[18] Horne, M., Englert, C. and Pfeifer, F. (1988) Mol. Gen. Genet. 213, 459–464.
[19] Schnabel, H. and Zillig, W. (1984) Mol. Gen. Genet. 193, 422–426.
[20] Gutierrez, M.C., Ventosa, A. and Ruiz-Berraquero, F. (1990) Biochem. Cell Biol. 68, 106–110.
[21] Pfeifer, F. and Blaseio, U. (1989) J. Bacteriol. 171, 5135–5140.
[22] Ng, W.L., Kothakota, S. and DasSarma, S. (1991) J. Bacteriol. 173, 1958–1964.
[23] Hackett, N.R., Krebs, M.P., DasSarma, S., Goebel, W., RajBhandary, U.L. and Khorana, H.G. (1990) Nucl. Acids Res. 18, 3408.
[24] Hall, M.J. and Hackett, N.R. (1989) Nucl. Acids Res. 17, 10501.
[25] Hackett, N.R. and DasSarma, S. (1989) Can. J. Microbiol. 35, 86–91.
[26] Ebert, K. and Goebel, W. (1985) Mol. Gen. Genet. 200, 96–102.
[27] Pfeifer, F., Ebert, K., Weidinger, G. and Goebel, W. (1982) Zbl. Bakt. Hyg., I. Abt. Orig. C 3, 110–119.
[28] Rosenshine, I. and Mevarech, M. (1989) Can. J. Microbiol. 35, 92–95.
[29] Joshi, J.G., Guil, W.R. and Handler, P. (1963) J. Mol. Biol. 6, 34–38.
[30] Moore, R.L. and McCarthy, B.J. (1969) J. Bacteriol. 99, 248.
[31] Moore, R.L. and McCarthy, B.J. (1969) J. Bacteriol. 99, 255.
[32] Weidinger, G., Klotz, G. and Goebel, W. (1979) Plasmid 2, 377–386.
[33] Pfeifer, F. and Betlach, M. (1985) Mol. Gen. Genet. 198, 449–455.
[34] Pfeifer, F., Weidinger, G. and Goebel, W. (1981) J. Bacteriol. 145, 375–381.

[35] Simon, R.D. (1978) Nature 273, 314–317.
[36] Schnabel, H., Schramm, R., Schnabel, R. and Zillig, W. (1982) Mol. Gen. Genet. 188, 370–377.
[37] Sapienza, C. and Doolittle, W.F. (1982) Zbl. Bakt. Hyg. I. Abt. C3, 120–127.
[38] Sapienza, C. and Doolittle, W.F. (1982) Nature 295, 384–389.
[39] Sapienza, C., Rose, M.R. and Doolittle, W.F. (1982) Nature 299, 182–185.
[40] Simsek, M., DasSarma, S., RajBhandary, U.L. and Khorana, H.G. (1982) Proc. Natl. Acad. Sci. U.S.A. 79, 7268–7272.
[41] DasSarma, S., RajBhandary, U.L. and Khorana, H.G. (1983) Proc. Natl. Acad. Sci. U.S.A. 80, 2201–2205.
[42] Pfeifer, F., Friedman, J., Boyer, H.W. and Betlach, M. (1984) Nucl. Acids Res. 12, 2489–2497.
[43] Pfeifer, F., Betlach, M., Martienssen, R., Friedman, J. and Boyer, H. (1983) Nucl. Acids Res. 12, 2489–2497.
[44] Pfeifer, F. and Ghahraman, P. (1991) Nucl. Acids Res. 19, 5788.
[45] Xu, W.-L. and Doolittle, W.F. (1983) Nucl. Acids Res. 11, 4195–4199.
[46] Hofman, J.D., Schalkwyk, L.C. and Doolittle, W.F. (1986) Nucl. Acids Res. 14, 6983–7000.
[47] Lam, W.L. and Doolittle, W.F. (1989) Proc. Natl. Acad. Sci. U.S.A. 86, 5478–5482.
[48] Pfeifer, F. and Blaseio, U. (1990) Nucl. Acids Res. 18, 6921–6925.
[49] Schnabel, H. (1984) Proc. Natl. Acad. Sci. U.S.A. 81, 1017–1020.
[50] Krebs, M.P., RajBhandary, U.L. and Khorana, H.G. (1990) Nucl. Acids Res. 18, 6699.
[51] Ebert, K., Goebel, W. and Pfeifer, F. (1984) Mol. Gen. Genet. 194, 94–97.
[52] Ebert, K., Goebel, W., Rdest, U. and Surek, B. (1986) System. Appl. Microbiol. 7, 30–35.
[53] Mevarech, M. and Werczberger, R. (1985) J. Bacteriol. 162, 461–462.
[54] Cline, S.W. and Doolittle, W.F. (1987) J. Bacteriol. 169, 1341–1344.
[55] Charlebois, R.L., Lam, W.L., Cline, S.W. and Doolittle, W.F. (1987) Proc. Natl. Acad. Sci. U.S.A. 84, 8530–8534.
[56] Cline, S.W., Lam, W.L., Charlebois, R.L., Schalkwyk, L.C. and Doolittle, W.F. (1989) Can. J. Microbiol. 35, 148–152.
[57] Cabrera, J.A., Bolds, J., Shields, P.E., Havel, C.M. and Watson, J.A. (1986) J. Biol. Chem. 261, 3578–3583.
[58] Holmes, M.L. and Dyall-Smith, M.L. (1991) J. Bacteriol. 173, 642–648.
[59] Holmes, M.L., Nuttall, S.D. and Dyall-Smith, M.L. (1991) J. Bacteriol. 173, 3807–3813.
[60] Blaseio, U. and Pfeifer, F. (1990) Proc. Natl. Acad. Sci. U.S.A. 87, 6772–6776.
[61] Cline, S.W. and Doolittle, W.F. (1991) J. Bacteriol. 174, 1076–1080.
[62] Kagramanova, V.K., Derckacheva, N.I. and Mankin, A.S. (1989) Can. J. Microbiol. 35, 160–163.
[63] Krebs, M.P., Hauss, T., Heyn, M.P., RajBhandary, U.L. and Khorana, H.G. (1991) Proc. Natl. Acad. Sci. U.S.A. 88, 859–863.
[64] Cline, S.W., Schalkwyk, L.C. and Doolittle, W.F. (1989) J. Bacteriol. 171, 4987–4991.
[65] Conover, R.K. and Doolittle, W.F. (1990) J. Bacteriol. 172, 3244–3249.
[66] Woese, C.R. (1987) Microbiol. Rev. 51, 221–271.
[67] Stackebrant, E., Murray, R.G.E. and Trüper, H.G. (1988) Int. J. System. Bacteriol. 38, 321–325.
[68] Sanderson, K.E. and Hall, C.A. (1970) Genetics 64, 215–228.
[69] Riley, M. and Anilionis, A. (1978) Annu. Rev. Microbiol. 32, 519–560.
[70] Riley, M. and Sanderson, K.E. (1990) In: The Bacterial Chromosome (Drlica, K. and Riley, M., Eds.), pp. 85–95, American Society for Microbiology, Washington, D.C.
[71] Piggot, P.J. (1990) In: The Bacterial Chromosome (Drlica, K.and Riley, M., Eds.), pp. 107–146, American Society for Microbiology, Washington, D.C.
[72] Sankoff, D., Cedergren, R. and Abel, Y. (1990) Methods Enzymol. 183, 428–438.
[73] Moriya, S., Ogasawara, N. and Yoshikawa, H. (1985) Nucl. Acids Res. 13, 2251–2265.

[74] Ogasawara, N., Moriya, S., von Meyenburg, K., Hansen, F.G. and Yoshikawa, H. (1985) EMBO J. 4, 3345–3350.
[75] Brewer, B. (1990) In: The Bacterial Chromosome (Drlica, K. and Riley, M., Eds.), pp. 61–84, American Society for Microbiology, Washington, D.C.
[76] Brewer, B. (1988) Cell 53, 679–686.
[77] Zeigler, D.R. and Dean, D.H. (1990) Genetics 125, 703–708.
[78] Schmid, M. and Roth, J.R. (1987) J. Bacteriol. 169, 2872–2875.
[79] Fan, J.B., Chikashige, Y., Smith, C.L., Niwa, O., Yanagida, M. and Cantor, C.R. (1989) Nucl. Acids Res. 17, 2801–2818.
[80] Coulson, A., Sulston, J., Brenner, S. and Karn, J. (1986) Proc. Natl. Acad. Sci. U.S.A. 83, 7821–7825.
[81] Coulson, A., Waterston, R., Kiff, J., Sulston, J. and Kohara, Y. (1988) Nature 335, 184–186.
[82] Olson, M.V., Dutchik, J.E., Graham, M.Y., Brodeur, G.M., Helms, C., Frank, M., MacCollin, M., Scheinman, R. and Frank, T. (1986) Proc. Natl. Acad. Sci. U.S.A. 83, 7826–7830.
[83] Daniels, C.J., Douglas, S.E., McKee, A.H.Z. and Doolittle, W.F. (1986) In: Microbiology-1986 (Silver, S.D., Ed.), pp. 349–355, American Society for Microbiology, Washington, D.C.
[84] Datta, P.K., Hawkins, L.K. and Gupta, R. (1989) Can. J. Microbiol. 35, 189–194.
[85] Daniels, C.J., Hofman, J.D., MacWilliam, J.G., Doolittle, W.F., Woese, C.R., Luehrsen, K.R. and Fox, G.E. (1985) Mol. Gen. Genet. 198, 270–274.
[86] Cohen, A., Lam, W.L., Charlebois, R.L., Doolittle, W.F. and Schalkwyk, L.C. (1991) Proc. Natl. Acad. Sci. U.S.A. 89, 1602–1606.
[87] Reiter, W.D., Palm, P. and Zillig, W. (1988) Nucl. Acids Res. 16, 1–19.
[88] Zillig, W., Palm, P., Reiter, W.D., Gropp, F., Pühler, G. and Klenk, H.P. (1988) Eur. J. Biochem. 173, 473–482.
[89] Thomm, M. and Wich, G. (1988) Nucl. Acids Res. 16, 151–163.
[90] Thomm, M., Wich, G., Brown, J.W., Frey, G., Sherf, B.A. and Beckler, G.S. (1989) Can. J. Microbiol. 35, 30–35.
[91] Reiter, W.D., Hüdepohl, U. and Zillig, W. (1990) Proc. Natl. Acad. Sci. U.S.A. 87, 9509–9513.
[92] Woese, C.R., Kandler, O. and Wheelis, M.L. (1990) Proc. Natl. Acad. Sci. U.S.A. 87, 4576–4579.
[93] Röhl, R. and Nierhaus, K.H. (1982) Proc. Natl. Acad. Sci. U.S.A. 79, 729–733.
[94] Kjems, J., Garrett, R.A. and Ansorge, W.A. (1987) System. Appl. Microbiol. 9, 22–28.
[95] Kjems, J., Leffers, H., Garrett, R.A., Wich, G., Leinfelder, W. and Böck, A. (1987) Nucl. Acids Res. 15, 4821–4835.
[96] Hui, I. and Dennis, P.P. (1985) J. Biol. Chem. 260, 899–906.
[97] Mankin, A.S. and Kagramanova, V.K. (1986) Mol. Gen. Genet. 202, 152–161.
[98] Lechner, K., Wich, G. and Böck, A. (1985) System. Appl. Microbiol. 6, 157–163.
[99] Larsen, N., Leffers, H., Kjems, J. and Garrett, R. (1986) System. Appl. Microbiol. 7, 49–57.
[100] Chant, J. and Dennis, P.P. (1986) EMBO J. 5, 1091–1097.
[101] Chant, J., Hui, I., De Jong-Wong, D., Shimmin, L. and Dennis, P.P. (1986) System. Appl. Microbiol. 7, 106–114.
[102] Achenbach-Richter, L. and Woese, C.R. (1988) System. Appl. Microbiol. 10, 211–214.
[103] Dennis, P.P. (1985) J. Mol. Biol. 186, 457–461.
[104] Kjems, J. and Garrett, R.A. (1990) J. Mol. Evol. 31, 25–32.
[105] Østergaard, L., Larsen, N., Leffers, H., Kjems, J. and Garrett, R. (1987) System. Appl. Microbiol. 9, 199–209.
[106] Sanz, J.L., Marin, I., Ramirez, L., Abad, J.P., Smith, C.L. and Amils, R. (1988) Nucl. Acids Res. 16, 7827–7832.
[107] Widom, R.L., Jarvis, E.D., LaFauci, G. and Rudner, R. (1988) J. Bacteriol. 170, 605–610.
[108] Bachmann, B.J. (1990) Microbiol. Rev. 54, 130–197.

[109] Wenzel, R. and Herrmann, R. (1988) Nucl. Acids. Res. 16, 8223–8336.
[110] Hummel, H. and Bčk, A. (1987) Biochimie 69, 857–861.
[111] Hummel, H. and Bčk, A. (1987) Nucl. Acids Res. 15, 2431–2443.
[112] Mankin, A.S. and Garrett, R.A. (1991) J. Bacteriol. 173, 3559–3563.
[113] Daniels, C.J., Gupta, R. and Doolittle, W.F. (1985) J. Biol. Chem. 260, 3132–3134.
[114] Thompson, L.D. and Daniels, C.J. (1988) J. Biol. Chem. 263, 17951–17959.
[115] Thompson, L. and Daniels, C.J. (1990) J. Biol. Chem. 265, 18104–18111.
[116] Perlman, P.S., Peebles, C.L. and Daniels, C. (1990) In: Intervening Sequences in Evolution and Development (Stone, E.M. and Schwartz, R.J., Eds.), pp. 112–161, Oxford University Press, New York.
[117] Kjems, J. and Garrett, R.A. (1988) Cell 54, 693–703.
[118] Kjems, J., Jensen, J., Olesen, T. and Garrett, R.A. (1989) Can. J. Microbiol. 35, 210–214.
[119] Gupta, R. (1985) In: Archaebacteria (Woese, C.R. and Wolfe, R.S., Eds.), pp. 311–343, Academic Press, Orlando, FL.
[120] Moritz, A., Lankat-Buttgereit, B., Gross, H.J. and Goebel, W. (1985) Nucl. Acids Res. 13, 31–43.
[121] Struck, J.C., Toschka, H.Y., Specht, T. and Erdmann, V.A. (1988) Nucl. Acids Res. 16, 7740.
[122] Brown, S. (1987) Cell 49, 825–833.
[123] Nieuwlandt, D.T., Haas, E.S. and Daniels, C.D. (1991) J. Biol. Chem. 266, 5689–5695.
[124] Stoeckenius, W. (1979) Biochim. Biophys. Acta 505, 215–278.
[125] Lanyi, J.K. (1986) Annu. Rev. Biophys. Biophys. Chem. 15, 11–28.
[126] Spudich, J.L. and Bogomolni, R.A. (1988) Annu. Rev. Biophys. Biophys. Chem. 17, 193–215.
[127] Stoeckenius, W., Lozier, R. and Bogomolni, R. (1979) Biochim. Biophys. Acta 505, 215–278.
[128] Khorana, H.G., Gerber, G.E., Herlihy, W.C., Gray, C.P., Anderegg, R.J., Nihei, K. and Biemann, K. (1979) Proc. Natl. Acad. Sci. 76, 5046–5050.
[129] Ovchinnikov, Y.A., Abdulaev, N.G., Feigina, M.Y., Kiselev, A.V. and Lobanov, N.A. (1979) FEBS Lett. 100, 219–224.
[130] Dunn, R., McCoy, J., Simsek, M., Majumdar, A., Chang, S.H., RajBhandary, U.L. and Khorana, H.G. (1981) Proc. Natl. Acad. Sci. U.S.A. 78, 6744–6749.
[131] DasSarma, S., RajBhandary, U.L. and Khorana, H.G. (1984) Proc. Natl. Acad. Sci. U.S.A. 81, 125–129.
[132] Betlach, M., Friedman, J., Boyer, H.W. and Pfeifer, F. (1984) Nucl. Acids Res. 12, 7949–7959.
[133] Leong, D., Pfeifer, F., Boyer, H. and Betlach, M. (1988) J. Bacteriol. 170, 4903–4909.
[134] Betlach, M.C., Shand, R.F. and Leong, D.M. (1989) Can. J. Microbiol. 35, 134–140.
[135] Leong, D., Boyer, H. and Betlach, M. (1988) J. Bacteriol. 170, 4910–4915.
[136] Soppa, J. and Oesterhelt, D. (1989) J. Biol. Chem. 264, 13043–13048.
[137] Soppa, J., Otomo, J., Straub, J., Tittor, J., Meessen, S. and Oesterhelt, D. (1989) J. Biol. Chem. 264, 13049–13056.
[138] Ni, B.F., Chang, M., Duschl, A., Lanyi, J. and Needleman, R. (1990) Gene 90, 169–172.
[139] Blanck, A. and Oesterhelt, D. (1987) EMBO J. 6, 265–273.
[140] Hegemann, P., Blanck, A., Vogelsang-Wenke, H., Lottspeich, F. and Oesterhelt, D. (1987) EMBO J. 6, 259–264.
[141] Duschl, A., Lanyi, J.K. and Zimanyi, L. (1990) J. Biol. Chem. 265, 1261–1267.
[142] Lanyi, J.K., Duschl, A., Hatfield, G.W., May, K. and Oesterhelt, D. (1990) J. Biol. Chem. 265, 1253–60.
[143] Blanck, A., Oesterhelt, D., Ferrando, E., Schegk, E.S. and Lottspeich, F. (1989) EMBO J. 8, 3963–71.
[144] Walker, J.E., Hayse, P.K. and Walsby, A.E. (1984) J. Gen. Microbiol. 28, 327–358.
[145] Walsby, A.E. (1978) Symp. Soc. Gen. Microbiol. 27, 327.

[146] Surek, B., Pillay, B., Rdest, U., Beyreuther, K. and Goebel, W. (1988) J. Bacteriol. 170, 1746–1751.
[147] Horne, M. and Pfeifer, F. (1989) Mol. Gen. Genet. 218, 437–444.
[148] DasSarma, S. (1989) Can. J. Microbiol. 35, 65–72.
[149] Jones, J.G., Hackett, N.R., Halladay, J.T., Scothorn, D.J., Yang, C.F., Ng, W.L. and DasSarma, S. (1989) Nucl. Acids Res. 17, 7785–7793.
[150] Jones, J.G., Young, D.C., and DasSarma, S. (1991) Gene 102, 117–122.
[151] Englert, C., Horne, M. and Pfeifer, F. (1990) Mol. Gen. Genet. 222, 225–322.
[152] Juez, G., Rodriguez, V.F., Herrero, N. and Mojica, F.J. (1990) J. Bacteriol. 172, 7278–7281.
[153] Yang, C.F. and DasSarma, S. (1990) J. Bacteriol. 172, 4118–4121.
[154] Shand, R.F. and Betlach, M. (1991) J. Bacteriol. 173, 4692–4699.
[155] May, B.P., Tam, P. and Dennis, P.P. (1989) Can. J. Microbiol. 35, 171–175.
[156] Sumper, M., Berg, E., Mengele, R. and Strobel, I. (1990) J. Bacteriol. 172, 7111–7118.
[157] Lechner, J. and Sumper, M. (1987) J. Biol. Chem. 262, 9724–9729.
[158] Mescher, M.F. and Strominger, J.L. (1976) J. Biol. Chem. 251, 2005–2014.
[159] Kessel, M., Wildhaber, I., Cohen, S. and Baumeister, W. (1988) EMBO J. 7, 1549–1554.
[160] Peters, J., Peters, M., Lottspeich, F. and Baumeister, W. (1989) J. Bacteriol. 171, 6307–6315.
[161] Kirby, T.W., Lancaster, J.R. and Fridovich, I. (1981) Arch. Biochem. Biophys. 210, 140–148.
[162] Searcy, K.B. and Searcy, D.G. (1981) Biochim. Biophys. Acta 670, 39–46.
[163] May, B.P. and Dennis, P.P. (1987) J. Bacteriol. 169, 1417–1422.
[164] Salin, M.L. and Oesterhelt, D. (1988) Arch. Biochem. Biophys. 260, 806–810.
[165] May, B.P. and Dennis, P.P. (1989) J. Biol. Chem. 264, 12253–12258.
[166] Zusman, T., Rosenshine, I., Boehm, G., Jaenicke, R., Leskiw, B. and Mevarech, M. (1989) J. Biol. Chem. 264, 18878–18883.
[167] Gropp, F., Palm, P. and Zillig, W. (1989) Can. J. Microbiol. 35, 182–188.
[168] Ken, R. and Hackett, N.R. (1991) J. Bacteriol. 173, 955–960.
[169] Alberts, A.W., Chen, J., Kuron, G., Hunt, V., Huff, J., Hoffman, C., Rothrock, J., Lopez, M., Joshua, H., Harris, E., Patchett, A., Monaghan, R., Currie, S., Stapley, E., Albers-Schoenberg, G., Hensens, O., Hirschfield, J., Hoogsteen, K., Liesc, J. and Springer, J. (1980) Proc. Natl. Acad. Sci. U.S.A. 77, 3957–3961.
[170] Lam, W.L., Cohen, A., Tsouluhas, D. and Doolittle, W.F. (1990) Proc. Natl. Acad. Sci. U.S.A. 87, 6614–6618.
[171] Lam, W.L., Logan, S.E. and Doolittle, W.F. (1991) J. Bacteriol. 174, 1694–1697.
[172] Ihara, K. and Mukohata, Y. (1991) Arch. Biochem. 286, 111–116.
[173] Denda, K., Konishi, J., Oshima, T., Date, T. and Yoshida, M. (1988) J. Biol. Chem. 263, 17251–17254.
[174] Alam, M. and Oesterhelt, D. (1984) J. Mol. Biol. 176, 459–475.
[175] Shimmin, L.C. and Dennis, P.P. (1989) EMBO J. 8, 1225–1235.
[176] Arndt, E. and Weigel, C. (1990) Nucl. Acids Res. 18, 1285.
[177] Bergmann, U. and Arndt, E. (1990) Biochim. Biophys. Acta 1050, 56–60.
[178] Arndt, E., Kromer, W. and Hatakeyama, T. (1990) J. Biol. Chem. 265, 3034–3039.
[179] Scholtzen, T. and Arndt, E. (1991) Mol. Gen. Genet. 228, 70–80.
[180] Mankin, A.S. (1989) FEBS Lett. 246, 13–16.
[181] Spiridonova, V.A., Akhmanova, A.S., Kagramanova, V.K., Kopke, A.K. and Mankin, A.S. (1989) Can. J. Microbiol. 35, 153–159.
[182] Shimmin, L.C., Newton, C.H., Ramirez, C., Yee, J., Downing, W.L., Louie, A., Matheson, A.T. and Dennis, P.P. (1989) Can. J. Microbiol. 35, 164–170.
[183a] Baier, G., Piendl, W., Redl, B. and Stoffler, G. (1990) Nucl. Acids Res. 18, 719–724.
[183b] Baldacci, G., Guinet, F., Tillit, J., Zaccai, G. and de Recondo, A.M. (1990) Nucl. Acids Res. 18, 507–511.

[184] Tako, M., Kobayashi, T., Oikawa, A. and Yasui, A. (1989) J. Bacteriol. 171, 6323–6329.
[185] Leffers, H., Gropp, F., Lottspeich, F., Zillig, W. and Garrett, R.A. (1989) J. Mol. Biol. 206, 1–17.
[186] Schallenberg, J., Moes, M., Truss, M., Reiser, W., Thomm, M., Stetter, K.O. and Klein, A. (1988) J. Bacteriol. 170, 2247–2253.
[187] Zillig, W., Klenk, H.P., Palm, P., Pühler, G., Gropp, F., Garrett, R.A. and Leffers, H. (1989) Can. J. Microbiol. 35, 73–80.
[188] Pühler, G., Lottspeich, F. and Zillig, W. (1989) Nucl. Acids Res. 17, 4517–4534.
[189] Huet, J., Schnabel, R., Sentenac, A. and Zillig, W. (1983) EMBO J. 2, 1291–1294.
[190] Pühler, G., Leffers, H., Gropp, F., Palm, P., Klenk, H.P., Lottspeich, F., Garrett, R.A. and Zillig, W. (1989) Proc. Natl. Acad. Sci. U.S.A. 86, 4569–4573.
[191] Iwabe, N., Kuma, K., Kishino, H., Hasegawa, M. and Miyata, T. (1991) J. Mol. Evol. 32, 70–78.
[192] Klein, A. and Schnorr, M. (1984) J. Bacteriol. 158, 628.
[193] Searcy, D.G. and Doyle, E.K. (1975) Int. J. System. Bacteriol. 25, 286.
[194] Pyle, L.R., Corcoran, L.N., Cocks, B.G., Bergemann, A.D., Whitley, J.C. and Finch, L.R. (1988) Nucl. Acids Res. 16, 6015–6025.
[195] Chen, H., Keseler, I.M. and Shimkets, L.J. (1990) J. Bacteriol. 172, 4206–4213.
[196] Pfeifer, F., Blaseio, U. and Ghahraman, P. (1988) J. Bacteriol. 170, 3718–3724.
[197] Itoh, T. (1988) Eur. J. Biochem. 176, 297–303.

M. Kates et al. (Eds.), *The Biochemistry of Archaea (Archaebacteria)*

CHAPTER 16

Structure and function of methanogen genes

J.R. PALMER and J.N. REEVE

Department of Microbiology, The Ohio State University, Columbus, OH 43210, U.S.A.

Abbreviations

ACS	acetyl–coenzyme A synthetase
CODH	carbon monoxide dehydrogenase
FDH	formate dehydrogenase
FTR	formylmethanofuran:tetra-hydromethanopterin formyltransferase
FRH	cofactor F_{420}-reducing hydrogenase
GAPDH	glyceraldehyde-3-phosphate dehydrogenase
GS	glutamine synthetase
gp	gene product
HMf	histone from *Methanothermus fervidus*
ISM1	insertion sequence methanogen #1
MDH	L-malate dehydrogenase
MRI	methyl–coenzyme M reductase isoenzyme 1
MRII	methyl–coenzyme M reductase isoenzyme 2
MVH	methylviologen-reducing hydrogenase
NNC/U	codons containing any base in positions 1 and 2 and C or U in the wobble position
ORF	open reading frame
PGK	3-phosphoglycerate kinase
RBS	ribosome-binding site
RNAP	DNA-dependent RNA-polymerase
rRNA	ribosomal RNA
r-protein	ribosomal protein
SOD	superoxide dismutase
SUMT	*S*-adenosyl-L-methionine:uroporphyrinogen III methyltransferase
tRNA	transfer RNA

1. Introduction

Methanogens are defined, and unified as a group, by their common energy-conserving metabolism which generates methane [1,2], but they are otherwise a very diverse group of archaea. Methanogens can be isolated from virtually every anaerobic environment, growing at temperatures from below 30°C to above 100°C [1,3,4]. There are rod-shaped, coccal and spiral methanogens and their genomic DNAs range from 26 to 68 mol% G+C [5]. Despite this diversity all methanogens, until very recently, were placed in one of three phylogenetic Orders: the Methanococcales, the Methanobacteriales or the Methanomicrobiales [1,5]. It has, however, now been established that the novel hyperthermophile, *Methanopyrus kandleri* [3,4], while undoubtedly a methanogen with standard methanogenesis biochemistry, does not belong to any of the above groups but represents a separate lineage which originates near the root of the archaeal tree [6]. A phylogenetic tree of the methanogens is provided in the Introduction to this volume, by Woese.

Shortly after the recognition of the archaea it was demonstrated by DNA renaturation studies that the genome of *Methanobacterium thermoautotrophicum*, although relatively small ($\sim 10^9$ daltons), nevertheless had typically prokaryotic physical features [7]. This genomic DNA was comprised essentially of unique sequences with a uniform distribution of bases averaging 49.7 mol% G+C. A more recent study confirmed these results for *M. thermoautotrophicum* and revealed that the genomes of *Methanococcus thermolithotrophicus*, *Methanobrevibacter arboriphilicus*, *Methanosarcina barkeri* and *Methanococcus voltae* all appeared to be approximately the same size as the genome of *M. thermoautotrophicum*, ranging from 1.0 to 1.8×10^9 daltons [8]. The melting profiles of the genomic DNAs from the latter three methanogens were, however, biphasic, indicating the presence of regions with A+T contents substantially higher than the average mol% A+T for these genomes. It is now known from sequencing that intergenic regions in methanogens are frequently very A+T-rich [9–11] and it is possible, although not proven, that these could be responsible for the biphasic melting curves. The recently established physical map of the *M. voltae* genome is somewhat smaller (1900 kbp, equivalent to $\sim 1.25 \times 10^9$ daltons) [12] than the size estimated for this molecule from the renaturation studies (1.8×10^9 daltons) [8]. It does, however, appear to be typically prokaryotic; being a single, circular DNA molecule of predominantly unique sequences. Genes are located at positions all around the molecule and cross-hybridization experiments do not indicate large numbers of repetitive elements, as found in the genomes of some halophilic archaea [13–15]. The only methanogen repetitive element so far documented is the insertion sequence ISM1, present in approximately 10 copies per genome, in *Methanobrevibacter smithii* [16]. As the sequence 5'GATC is underrepresented in the genome of *M. voltae* [17], the restriction enzymes *BamHI*, *BglII*, *BclI* and *PvuI*, which have 5'-NGATCN as recognition and cleavage sites, were particularly useful in establishing the *M. voltae* physical map. These enzymes cut the *M. voltae* genome 11, 6, 6 and 4 times, respectively [12].

Extrachromosomal genetic elements, superficially very similar to extrachromosomal elements found in bacteria, have been described in methanogens. Cryptic plasmids have been isolated from several different methanococci [18,19], *Methanosarcina acetivorans* [20], *Methanolobus vulcani* [21,22], three strains of *Methanobacterium thermoformicicum* [23] and *M. thermoautotrophicum* strain Marburg [24]. As *M. thermoautotrophicum* strains are used widely in research, plasmid pME2001, isolated from *M. thermoautotrophicum* strain Marburg, has received much attention [25–27]. Its complete sequence (4439 bp) has been determined revealing the presence of four open reading frames (ORFs), three of which are preceded by strong ribosome-binding sites (RBS) [27], and the template region for a short transcript, synthesized in vivo from pME2001 has been mapped [26]. The biological function(s) encoded by pME2001 remain unknown and a derivative of *M. thermoautotrophicum* strain Marburg cured of pME2001 is fully viable. Several potential shuttle vectors and cloning vehicles have been constructed based on the pME2001 replicon [25].

A virulent phage, designated ψM1, has been isolated which infects *M. thermoautotrophicum* strain Marburg producing ~ 5 phage per infected cell [28,29]. The double-stranded (ds) DNA genome of this phage is ~ 30 kbp in length, circularly permuted and terminally redundant. The genome is packaged into a polyhedral head attached to a flexible, probably non-contractile tail. Phage ψM1 does not productively infect other strains of *M. thermoautotrophicum* or *Methanobacterium wolfei* although there are sequences in the genome of *M. wolfei* that hybridize to ψM1 DNA. Interesting features of ψM1 which are potentially very important for its practical use, are its ability to package multimers of pME2001 and to transduce fragments of chromosomal DNA between different auxotrophic strains of *M. thermoautotrophicum* strain Marburg. There are also preliminary reports of virulent phages that infect two *Methanobrevibacter* species [Baresi, L. and Bertani, G. (1984) Annu. Meeting Am. Soc. Microbiol., Abstr. I74; Knox, M.R. and Harris, J.E. (1986) XIV Int. Congr. Microbiol., Abstr. PG 3–8] and more detailed studies of an unusual virus-like particle (VLP) which accumulates in the medium during growth of cultures of *Methanococcus voltae* strain A3 [30]. A circular dsDNA molecule, 23 kbp in length, has been isolated from these VLPs and designated pURB600. Identical DNA molecules can be isolated as plasmid DNA from the cytoplasm of *M. voltae* A3 cells and hybridization experiments have shown that a copy of pURB600 is integrated into the chromosome of *M. voltae* A3. This element therefore has properties consistent with it being a temperate phage although infectivity has not been demonstrated. Plasmid pURB600 DNA also hybridizes to DNA sequences in the genome of the widely-used laboratory strain *M. voltae* PS, but not to genomic DNAs from *Methanococcus vannielii*, *Methanococcus maripaludis*, *Methanococcus deltae* or *Methanococcus thermolithotrophicus*.

Overall, the organization and mechanisms of expression of genes in methanogens [10, 31–34] appear to follow patterns well established in bacteria. Tightly-linked clusters of genes are frequently found organized into single transcriptional units, conventionally designated as operons, although operator sequences per se have not been documented. Immediately preceding virtually every polypeptide-encoding gene is a sequence complementary to the sequence at the 3' terminus of the methanogen's 16S rRNA. These are presumably ribosome-binding sites (RBS) which, as in bacteria, direct the binding of mRNAs to ribosomes to facilitate translation initiation. Highly expressed genes tend to have RBS with more potential for 16S rRNA base-pairing (strong RBS) than do genes which encode polypeptides synthesized in lesser amounts [33,34]. To date, introns have not been reported in methanogens and although polyA sequences are attached to the 3' termini of some methanogen RNAs, these average only 12 bases in length, typical of the short polyA sequences found in bacteria [35]. Gene and protein sequencing has revealed coding sequences, in some genes, for amino-terminal amino acids that are not present in the mature protein [36–38]. In some cases [36] these resemble signal peptides which direct sub-cellular protein localization or protein secretion in bacteria [39]. The demonstration that a *sec*Y-related gene cloned from *Methanococcus vannielii* can complement *sec*Y mutants of *E. coli* [40], together with the conserved structure of some putative methanogen signal sequences [36], argues for the presence of bacterial-like protein maturation and localization mechanisms in these methanogens.

Promoter structure is the one feature of methanogen gene organization that is conspicuously different from its bacterial counterpart. Methanogens, in common with all archaea, have complex, multi-subunit DNA-dependent RNA-polymerases (RNAP) [31,32,41,42] which recognize and bind to sequences which have very little in common with the canonical −10 and −35 promoter sequences established in bacteria (see Chapter 12 of this volume). Methanogen promoters have a TATA-box, conventionally designated as Box A, with a consensus sequence of 5′TTTAAATA, centered ~−25 bp relative to the site of transcription initiation. A second conserved motif, designated as Box B, with the consensus sequence 5′ATGC is frequently found at the site of transcription initiation [31,32,43,50]. These sequences were proposed as promoter elements, initially, because of their conservation upstream of many sequenced genes. Subsequently, footprinting experiments demonstrated that purified methanogen RNAPs bound to Box A regions in vitro protecting ~50 bp, including the Box A sequence, from nuclease digestion [44–46,48,49]. The functionally important components of the Box A and Box B sequences and the importance of the spacing between them, have now been demonstrated directly by transcription in vitro using wild-type and mutated DNA template sequences [50–53]. The primary sequences and spacing of promoter elements appear to be surprisingly conserved in all archaea [31,32]. The same consensus sequence is not only applicable to both stable RNA and protein-encoding genes in methanogens but there is substantial overlap with the consensus sequences established for promoters in non-methanogenic archaea. In addition to the RNAP holoenzyme and the primary sequence of the promoter, transcription factors are also needed to regulate transcription initiation in methanogens [51].

Most sequences postulated to be transcription terminators have been identified on the basis of their location and potential, when transcribed, to form hairpin-loop structures, [31,32,54–64] analogous to rho-independent bacterial transcription terminators. There are, however, also examples of transcription being terminated in vivo in regions lacking an inverted repeat but containing several oligo-T sequences [36,40,65–70]. Transcription termination in the non-methanogenic, thermophilic archaea has also been shown to occur predominantly in regions containing several oligo-T sequences [31,32,71,72].

The first report of cloning of methanogen DNAs was published approximately a decade ago [73]. Described below are the results obtained by gene cloning since that date and their analyses in terms of the structure and mechanisms of expression of genes in methanogens.

2. Genes encoding enzymes involved in methanogenesis

2.1. Methyl–coenzyme M reductase (MR)

The CH_4-generating reaction in methanogenesis is catalyzed by methyl–coenzyme M reductase (MR) [1,2,5,74–77] (see Chapters 3 and 4 of this volume for detailed reviews of the biochemistry of methanogens). The CH_3-group of methyl–coenzyme M (CH_3–S–CoM) is reduced to CH_4 by reductant from the methanogen-specific cofactor, mercaptoheptanoylthreonine phosphate (HS–HTP) [78–83]. This MR-catalyzed reaction also generates the heterodisulphide, CoM–S–S–HTP, which must then be reduced by a specific heterodisulphide reductase to regenerate the cofactors, HS–CoM and HS–HTP, needed for continued methanogenesis [84–86]. As MR constitutes up to 10% of the total cellular protein in methanogens [87] and plays this central, conserved role in methanogenesis, much attention has been directed towards understanding its structure and the regulation of its activity. Regions of genomic DNA encoding MR, together with their flanking sequences, have been cloned and sequenced from the two mesophilic methanococci, *Methanococcus vannielii* [56] and *Methanococcus voltae* [88]; a mesophilic methanosarcina, *Methanosarcina barkeri* [89]; a thermophilic methanobacterium, *Methanobacterium thermoautotrophicum* strain Marburg [57]; and a hyperthermophilic methanobacterium, *Methanothermus fervidus* [67]. In all cases a cluster of five tightly linked genes in the order *mcrBDCGA*, termed the *mcr* operon, has been found [61]. The *mcrA*, *mcrB* and *mcrG* genes encode the α, β, and γ subunits, respectively, of the MR holoenzyme which, with stoichiometries of $\alpha_2\beta_2\gamma_2$, have native molecular masses of $\sim$300 kDa. Functions have not yet been assigned to the gene products of *mcrC* or *mcrD* (gp*mcrC* and gp*mcrD*) although their presence and conservation in all five sequenced *mcr* operons implies a significant and homologous role in these methanogens. Preliminary evidence indicates that the *mcr* operon, including the *mcrC* and *mcrD* genes, is also conserved in *Methanopyrus kandleri* [77] [Palmer, J.R., Steigerwald, V.J., Daniels, C.J. and Reeve, J.N., unpublished results]. Antibodies raised against either the MR holoenzyme from *M. vannielii* or the product of a *mcrD–lacZ* gene fusion (gp*mcrD–lacZ*) synthesized in *E. coli* coprecipitate both MR and gp*mcrD* from extracts of *M. vannielii* cells [90] and partially inhibit methanogenesis in vitro [Stroup, D. and Reeve, J.N., unpublished results]. Electrophoresis, through nondenaturing polyacrylamide gels, abolishes the association of MR and gp*mcrD*. The DNA sequences responsible for expression of the *mcr* operons are consistent with high levels of both transcription and translation. Consensus Box A and Box B promoter sequences are located $\sim$25 bp upstream from, and at the sites of *mcr* transcription initiation, respectively. The nontranslated leader sequences transcribed upstream of the *mcrB* genes range in length from 32 bases in *M. vannielii* to 107 bases in *M. thermoautotrophicum* strain Marburg. Transcription termination in *M. vannielii*, *M. voltae* and *M. thermoautotrophicum* strain Marburg occurs downstream of *mcrA* following sequences which, when transcribed, could form stem–loop structures. The region downstream of *mcrA* in *M. fervidus* lacks such a G+C-rich inverted repeat sequence and transcription terminates following several oligo-T sequences [67]. Every *mcr* gene is preceded by a RBS sequence, however the RBS for the genes encoding the α, β and γ subunits appear to be much stronger than those preceding the *mcrC* and *mcrD* genes. This predicts that in vivo much smaller amounts of gp*mcrC* and gp*mcrD* are synthesized than of MR and this has been confirmed directly in *M. vannielii* [Stroup, D. and Reeve, J.N., unpublished results]. Differences in codon usage in the different *mcr* genes are also consistent with this observation. The NNC codon is used preferentially in NNC/U pairs of synonymous codons in the *mcrA*, *mcrB* and *mcrG* genes but not in the *mcrC* or *mcrD* genes [56,61]. These codon pairs are translated by the same tRNA, with a G in the first anticodon position. Codons with C in the wobble position are therefore expected to enhance the stability of codon–anticodon interactions by forming a C:G base pair at this location which may increase the rate of translation. NNC codons have been shown to direct the formation of complexes between mRNA, ribosomes and cognizant tRNAs in vitro at nearly twice the rate of the synonymous NNU codons [91]. This bias for NNC codons is least evident in the *mcrA*, *mcrB* and *mcrG* genes of the hyperthermophile *M. fervidus*, suggesting that little advantage is afforded by NNC codons over NNU codons in codon–anticodon interactions in cells growing above 80°C.

All five fully sequenced *mcr* operons, and the partially sequenced *M. kandleri mcr* operon, must have evolved from a common ancestral sequence. Quantitative comparisons of the sequences of the *mcr* genes and gene products should therefore provide valid estimates of evolutionary divergence. The polypeptides encoded by the *mcrBCGA* genes are conserved to approximately the same extent and somewhat more than the gp*mcrD*s [61]. The sequences of the corresponding MR polypeptides from methanogens placed in the same Order are from 74 to 94% identical whereas polypeptides from different Orders are only 50–60% identical. These results do, therefore, support the phylogenetic relationships of methanogens established originally on the basis of their 16S rRNA sequences (see the Introduction to this volume and refs. [1,5,92,93]). Changing from growth at mesophilic temperatures (*M. voltae*, *M. vannielii* and *M. barkeri*) to thermophilic temperatures (*M. thermoautotrophicum* and *M. fervidus*) does not result in the same pattern of amino-acid substitutions in MR that has been correlated with such a change in other proteins [67]. In view of the antiquity of methanogenesis, most amino-acid residues still conserved in all MRs are very likely to play essential roles, constituting active sites or involved in subunit assembly or coenzyme binding. The α subunits are predicted to possess a hydrophobic pocket, bounded by two exposed hydrophilic regions, forming a structure that resembles a substrate- or cofactor-binding site and it has been suggested that coenzyme M may bind to this subunit [61,94]. A hydrophobic domain in the α subunits, containing a block of 23 highly conserved amino-acid residues, could be a region which interacts, via hydrophobic bonds, with the other subunit polypeptides or possibly with gp*mcrD* [61,88].

Two similar but distinct MR isoenzymes (MRI and MRII) have been purified from *M. thermoautotrophicum* strain Marburg [95]. Southern blot analyses have also demonstrated two separate *mcr* operons in the genome of this methanogen and in *M. thermoautotrophicum* strain ΔH, *M. wolfei* and *M. fervidus*, although there was no evidence for a second *mcr* operon in the genomes of *M. vannielii*, *M. voltae*, *Methanococcus thermolithotrophicus* or *M. kandleri* [Hennigan, A., Stroup, D., Palmer, J.R. and Reeve, J.N. (1991) Annu. Meeting Am. Soc. Microbiol., Abstr I118, and unpublished results]. Changes in the relative amounts of MRI and MRII in *M. thermoautotrophicum* strain Marburg cells, during growth in batch culture [95], indicates that growth rate may differentially regulate the expression of the two *mcr* operons. The MR isolated from cells growing exponentially under substrate sufficient conditions (MRII) has an approximately fourfold greater specific activity than MRI isolated from cells entering and in stationary phase. The *mcr* operons first cloned and sequenced from *M. thermoautotrophicum* strain Marburg [57] and *M. fervidus* [67] encode MRI, the isoenzymes synthesized preferentially at the end of exponential growth, however MRII encoding genes have now also been cloned and partially sequenced. They are located immediately downstream from the *mvh* operon (see section 2.2), predicting coordinated synthesis of MRI with the methylviologen-reducing hydrogenase and the polyferredoxin [112,385].

2.2. Hydrogenases and ferredoxins

Hydrogen-dependent methanogenesis requires hydrogenase activity to obtain reductant and mechanisms to transfer this reductant from H_2 to reaction centers. Two different nickel- and iron-containing hydrogenases (NiFe-hydrogenases) have been detected in most methanogens [96–111]; a higher-molecular-mass, cofactor F_{420}-reducing hydrogenase (FRH) and a lower-molecular-mass hydrogenase, which is incapable of reducing the cofactor F_{420} and is conventionally known as the methylviologen-reducing hydrogenase (MVH). Some FRHs also contain selenium (NiFeSe-hydrogenase) [100]. *M. thermoautotrophicum*, *M. formicicum*, *M. barkeri*, *M. voltae* and *Methanococcus jannaschii* possess both FRH and MVH whereas *Methanospirillum hungatei* appears to contain only MVH. FRH has been shown to be membrane-associated in *M. voltae* [103], *M. formicicum* [106] and *M. barkeri* [111]. The natural electron acceptor(s) from MVH and the precise biochemical roles of the two hydrogenases in vivo have yet to be determined. The clusters of genes encoding these two hydrogenases are located relatively close to each other on the *M. voltae* genome but separated by at least 60 kbp from the *M. voltae mcr* operon [12].

The FRH- and MVH-encoding genes from *M. thermoautotrophicum* strain ΔH [69, 112] and part of the MVH-encoding region from *M. fervidus* [113] have been cloned and sequenced. The α, β and γ subunits of FRH, encoded by genes *frhA*, *frhB* and *frhG*, respectively, are arranged within a tightly linked cluster of genes, in the order *frhADGB* [69]. This forms a single transcriptional unit, the *frh* operon. The product of the *frhD* gene, which appears to share a common evolutionary ancestry with the *hydD* gene of *E. coli* [114], does not co-purify with the active enzyme and its function is unknown. All four *frh* genes are preceded by RBS, the promoter conforms to the consensus for methanogen promoters and the sequence of nucleotides downstream of *frhB* is consistent with it being a transcription terminator. The MVH-encoding genes also appear to form a single transcriptional unit arranged in the order *mvhDGAB* [112,113]. The *mvhD*, *mvhG* and *mvhA* genes, which are preceded by strong RBS, encode the δ, γ and α subunits of MVH, respectively. Transcription in *M. thermoautotrophicum* strain ΔH is initiated 42 bp upstream of *mvhD*. Sequences upstream of the *mvh* operons in two additional strains of *M. thermoautotrophicum* have also been determined, revealing a highly conserved Box A-containing sequence over 100 bp in length, in all three cases [109]. This is presumably the length of sequence essential for regulation and expression of these *mvh* operons. A comparison of the large subunits of both FRH and MVH (gp*frhA* and gp*mvhA*) with the corresponding large subunits of NiFe-hydrogenases from several phylogenetically very diverse prokaryotes has revealed blocks of conserved amino-acid residues at both the carboxyl termini and the amino termini [69,109,112,115–118]. These regions each contain a pair of cysteine residues which, together with several invariant histidine residues, could form the Ni-binding site [69,109]. Conserved regions, including four conserved cysteine residues, are also found in the amino termini of the small subunits of these prokaryotic NiFe-hydrogenases, arguing for a common evolutionary ancestry for all these enzymes. Reduction of cofactor F_{420} by FRH requires a complex of the α, β and γ subunits, but an αγ complex alone can reduce one-electron acceptor dyes [69]. The β subunit of FRH, or a multi-subunit structure involving β, is therefore likely to be the site of cofactor-F_{420} reduction. The fourth ORF in the *mvh* operon, namely *mvhB*, has been cloned and sequenced from both *M. thermoautotrophicum* strain ΔH[112] and *M. fervidus* [113]. The encoded gp*mvhB*s are predicted to have amino-acid sequences that are 56% identical and contain six tandemly repeated, bacterial ferredoxin-like domains. This polymeric structure suggests that these proteins, designated polyferredoxins, could function as electron conduits. Electrons might transfer through the different [4Fe–4S] centers into the subcellular structures, designated methanoreductosomes, which have been implicated as the sites of methanogenesis [119,120]. Different domains of the polyferredoxins might supply electrons to different reactions during the reduction of methanogenic substrates to CH_4, including the reduction of the heterodisulphide, CoM–S–S–HTP, the reaction now considered to be at the center of energy conservation during methanogenesis [83,86]. Polyferredoxins could also function in a protective or storage capacity. The availability of electrons stored in these molecules might allow the cell to survive transient exposure to an oxidizing environment or sudden starvation for reductant [112]. The domains of the *M. thermoautotrophicum* polyferredoxin are similar to the ferredoxin-like sequences in the carboxyl-terminal region of the α subunit of FRH and in the γ subunit of carbon-monoxide dehydrogenase (CODH) from *Methanothrix soehngenii* [121]. The pentapeptide PTAAI is conserved at the same location in domain 1 of this polyferredoxin and in the ferredoxin-like region in the large subunit of the Fe-hydrogenase from *Desulfovibrio vulgaris* [122]. Domains 2, 3, 4 and 5 of the polyferredoxins have amino-acid sequences consistent with their being classified as archaeal ferredoxins, although domain 5 has an additional conserved sequence in both cases [123]. In addition, a block of 30 amino-acid residues in domain 3 of the *M. thermoautotrophicum* polyferredoxin includes 18 that are identical to a sequence in a mono-ferredoxin isolated from *M. barkeri* MS [124]. The conservation of a heptapeptide, PKGALSL, following the CysIV residue in both domains 5 and 6 of the *M. thermoautotrophicum* polyferredoxin, suggests that these domains were formed by a relatively recent sequence duplication [112].

Monomeric ferredoxins have been isolated and characterized from three methanosarcinaceae [124–127] and one methanococcal species [128,129]. The methanosarcinal ferredoxins have primary sequences similar to clostridial 2×[4Fe–4S] ferredoxins but differ in the configuration of their [Fe–S] centers. Although the ferredoxins isolated from *M. barkeri* MS and *M. barkeri* Fusaro both have molecular masses of approximately 6 kDa they contain a [3Fe–3S] cluster and 2×[4Fe–4S] clusters, respectively. The ferredoxin from *Methanosarcina thermophila* has a molecular mass of only 4.9 kDa and contains either a [3Fe–3S] or a [4Fe–4S] center. These ferredoxins appear to be involved in electron transport from pyruvate dehydrogenase and from CODH [125,130]. A ferredoxin purified from *Methanococcus thermolithotrophicus* has 2×[4Fe–4S] centers, a molecular mass of 7.3 kDa and also appears to accept electrons from CODH [128,129].

2.3. Formate dehydrogenase (FDH)

Approximately 50% of methanogens so far characterized can dissimilate formate [131, 132] and under these conditions formate dehydrogenase (FDH) may account for 2–3% of the total soluble protein [133]. During methanogenesis from formate, FDH catalyzes a two-electron oxidation of formate generating CO_2 and the reducing equivalents needed to reduce CO_2 to CH_4. In *Methanobacterium formicicum*, the *fdhA* and *fdhB* genes which encode the α and β subunits of FDH, respectively, overlap by 1 bp and have strong RBS [134–135a]. Molybdenum starvation causes a decline in the activity and synthesis of FDH and an increase in transcription of the *fdhAB* genes [133]. Replacement of molybdenum with tungsten results in the synthesis of an inactive enzyme containing the same polypeptide subunits as active FDH but with a metal-free molybdopterin cofactor [133]. In contrast, *M. vannielii* requires tungsten for optimum growth on formate [136,137].

In addition to the molybdopterin cofactor, *M. formicicum* FDH contains a FAD molecule which is involved in hydride transfer to cofactor F_{420} and several [Fe–S] centers. The amino-acid sequence of the β subunit of FDH is 26% identical (54% similar allowing for conservative amino-acid substitutions) to the amino-acid sequence of the β subunit of FRH from *M. thermoautotrophicum* strain ΔH [69]. As FRH also catalyzes hydride transfer to cofactor F_{420}, these conserved β-subunit sequences are likely candidates for regions involved in this common activity. There are also two clusters of four cysteine residues in the β subunit of FDH which appear to be arranged in a manner similar to the 2×[4Fe–4S] clusters found in bacterial and archaeal ferredoxins [123,134].

Exposure to H_2 causes a decline in FDH activity in cells of *M. thermolithotrophicus* [139]. As the enzyme itself is not sensitive to H_2, this reduction in FDH activity appears to reflect decreased FDH synthesis in response to the presence of H_2. *M. thermolithotrophicus* cells previously grown on CO_2 and H_2 require a very extended period of exposure to formate before they synthesize sufficient amounts of FDH to grow on formate [139].

2.4. Formylmethanofuran:tetrahydromethanopterinformyltransferase (FTR)

During methanogenesis from CO_2 and formate, methyl groups are generated by a series of reductive steps. Two methanogen-specific cofactors, methanofuran (MFR) and tetrahydromethanopterin (H_4MPT) carry the C_1 moiety through the formyl, methenyl and methylene levels to the methyl level of reduction [1,2,140]. The enzyme formylmethanofuran:tetrahydromethanopterinformyltransferase (FTR) catalyzes the transfer of the formyl group from formyl–MFR to H_4MPT yielding formyl–H_4MPT [141]. The FTR-encoding gene (*ftr*) plus its flanking sequences have been cloned and sequenced from *M. thermoautotrophicum* strain ΔH [62]. This gene encodes an acidic protein with a calculated molecular mass of 31 401 Da which, with some differences in K, N and D codons, has a codon-usage pattern generally similar to that of the *mcrA*, *mcrB* and *mcrG* genes in *M. thermoautotrophicum* strain Marburg. A RBS is located from position −11 to position −6 relative to the translation initiation codon. A sequence consisting of two inverted repeats, one of which is flanked by oligo-T sequences, is located immediately downstream of the *ftr* gene and is likely to be the transcription terminator. The presence of an ORF terminating only 61 bp upstream of the *ftr* gene and the absence of a promoter-like sequence within this upstream intergenic region suggest that the *ftr* gene is part of a polycistronic transcriptional unit. Although *M. thermoautotrophicum* strain ΔH is an anaerobe, with an optimum growth temperature of 65°C, synthesis of a functional *ftr* gene product was obtained in *E. coli* growing aerobically at 37°C. The primary sequence of this FTR has no obvious similarity to functionally analogous formyltetrahydrofolate synthetases in non-methanogens nor to other pterin-binding proteins.

2.5. Carbon-monoxide dehydrogenase (CODH) and acetyl–coenzyme A synthetase (ACS)

Acetate is activated to acetyl–CoA by acetyl–coenzyme A synthetase (ACS) during methanogenesis from acetate in *Methanothrix* species [142–144] whereas, in *Methanosarcina* species, activation occurs by the sequential actions of acetate kinase and phosphotransacetylase [145–149]. Carbon-monoxide dehydrogenase (CODH) then catalyzes the cleavage of the C–C bond of the acetyl residue generating a methyl group and low-potential electrons used to reduce the CH_3 group of CH_3–S–CoM to CH_4 [150–158]. The amount of CODH in *M. barkeri* cells growing on acetate is much higher than in cells growing on CH_3OH [153,154] indicating that the expression of the CODH-encoding genes is substrate regulated. As CODH is also required for carbon assimilation during growth by methanogens there is considerable interest in determining how the expression of the CODH-encoding genes is regulated and how the anabolic and catabolic activities of the enzyme are coordinated.

The genes (*cdhA* and *cdhB*) encoding the large (79.4 kDa) and small (19.4 kDa) subunits of CODH have been cloned and sequenced from *Methanothrix soehngenii* [159]. A large subunit (89 kDa) encoding *cdhA* gene has also been cloned and sequenced from *Methanosarcina thermophila* [160]. Sequences conforming to the consensus for methanogen promoters are located upstream of both *cdhA* genes. The absence of a promoter sequence or transcription terminator in the 19 bp intergenic region separating *cdhA* and *cdhB* in *M. soehngenii* strongly suggests that these genes are cotranscribed, in the direction *cdhAB*, as part of the same polycistronic mRNA. The site of transcription termination is uncertain as there is no obvious transcription terminator downstream of *cdhB*. Both *cdh* genes in *M. soehngenii* are preceded by strong RBSs. Consistent with the observed substrate regulation of CODH levels [153,154], transcription of *cdhA* in *M. thermophila* is stimulated by the presence of acetate and is much reduced in methanol- and methylamine-containing media [160].

The small *M. soehngenii* CODH subunit (gp*cdhB*) is not obviously related to any known protein whereas the large subunit (gp*cdhA*) contains an archaeal ferredoxin-like domain [123] and a region with substantial similarity to two eucaryal acyl–CoA oxidases [161,162]. These relationships are consistent with CODH functioning as an electron carrier and also binding acetyl–CoA and catalyzing its dissimilation to CH_3–S–CoM and CO_2. The *M. soehngenii cdhAB* genes have been expressed in *E. coli* but the polypeptides synthesized were not catalytically active [159].

An acetyl–coenzyme A synthetase (ACS) encoding gene (*acs*) has been cloned and sequenced from *M. soehngenii*; it appears to form a monocistronic transcriptional unit. Two ORFS immediately downstream of this gene if joined by a frameshift, would also form an *acs* gene [70]. The intact *acs* gene, which is preceded by a typical TATA-box and RBS, encodes a 73 kDa polypeptide that has been synthesized in *E. coli*. The amounts and specific activity of the ACS synthesized in *E. coli* were similar to those of the ACS synthesized in *M. soehngenii* [70]. Comparison of the amino-acid sequence deduced for the ACS with the sequences of other proteins which bind coenzyme A, ATP or both, revealed conserved regions which might therefore bind these common cofactors.

3. Amino-acid and purine biosynthetic genes

Complementation of auxotrophic mutations in *E. coli*, by expression of cloned fragments of methanogen genomic DNAs, has facilitated the isolation and characterization of methanogen genes encoding enzymes which participate in amino-acid and purine biosynthetic pathways [16,58,60,163–170]. As methanogen promoters [32,50] do not resemble *E. coli* promoters, transcription of these randomly cloned methanogen genes in *E. coli* is unlikely to be directed by their natural promoters. The A+T rich intergenic sequences that are common in methanogen DNAs [9–11] apparently contain sequences that, in *E. coli*, function fortuitously as promoters. This has been demonstrated directly for transcription of the *hisA* gene from *M. vannielii* in *E. coli* [46]. As the nucleotide sequences at the 3' termini of the 16S rRNAs in methanogens, *E. coli* and *B. subtilis* are very similar, it is very likely that methanogen RBS in mRNAs can function correctly in translation initiation in these bacteria.

3.1. Histidine

HisA genes and their flanking regions have been cloned and sequenced from *M. vannielii* [165], *M. voltae* [163,165] and *M. thermolithotrophicus* [168]. These genes appear to have diverged from each other to an equal extent, which is consistent with their phylogenetic relationships based on their 16S rRNA sequences. Although all three *hisA* genes are embedded in multigene transcriptional units, with A+T-rich intergenic sequences, the flanking regions appear to have undergone different genomic rearrangements. An increased amount of the *hisA* transcript has been detected in *M. voltae* cells exposed to aminotriazole, an inhibitor of histidine biosynthesis, suggesting that histidine starvation causes increased transcription of this *hisA* gene [171].

The *hisI* gene from *M. vannielii*, which is separated by at least 10 kbp of genomic DNA from its *hisA* gene, also appears to be part of an operon. The flanking ORFs are not, however, related to other known *his* genes [167]. The *hisI* gene of *M. vannielii* encodes a polypeptide with a single enzymatic activity whereas the functionally and evolutionary homologous DNA sequences in *E. coli* and *Saccharomyces cerevisiae* are now fused within larger genes, which encode polypeptides with two and four different enzymatic activities, respectively [172].

3.2. Arginine

Argininosuccinate synthetase (AS) catalyzes the penultimate step in arginine biosynthesis, and AS-encoding *argG* genes have been cloned and sequenced from *M. barkeri* and *M. voltae* [58]. These two *argG* genes are located downstream from unrelated ORFs and both are followed by inverted repeat sequences likely to be transcription terminators. The upstream ORF in *M. barkeri* has been designated *carB* as it appears to encode carbamyl-phosphate synthetase, an enzyme also involved in arginine biosynthesis. The intergenic region separating the *carB* and *argG* genes is, however, 389 bp in length and contains six tandemly repeated copies of a 14 bp sequence, an inverted 9 bp repeat sequence, and three tandem copies of a 29 bp sequence. It seems unlikely therefore that the *carB* and *argG* genes in *M. barkeri* are cotranscribed. All *carB* genes sequenced to date from bacteria and eucarya contain a tandem duplication but, as only part of the *M. barkeri carB* gene was originally cloned [58], it was uncertain whether this archaeal gene also contained the duplication. Additional cloning and sequencing has now demonstrated the duplication within the *M. barkeri carB* gene [Schofield, P. (1991) personal communication] indicating that the duplication event most probably occurred before the separation of the Archaea, Bacteria and Eucarya. Comparisons of the methanogen *argG* and *carB* gene products with the functionally homologous enzymes in bacteria and eucarya indicate that the archaeal enzymes are more closely related to their eucaryal equivalents than to their bacterial equivalents [58,173]. The human AS-encoding gene contains at least nine introns [174] which are not present in the archaeal or bacterial *argG* genes. The *M. barkeri argG* gene also complements mutations in the *argA* gene of *B. subtilis* [164].

3.3. Proline and ISM1

A gene complementing the *E. coli proC* mutation, and therefore likely to encode 1-pyrroline-5-carboxylate reductase, has been cloned and sequenced from *Methanobrevibacter smithii* [16]. A copy of the methanogen insertion sequence ISM1, located directly adjacent to the *Mb. smithii proC* gene, was cloned and sequenced fortuitously as part of the *proC* analysis. ISM1 is 1183 bp in length, has 29 bp terminal inverted repeats and is mobile. It duplicates 8 bp at the sites of its insertion and is present in ~10 copies per *Mb. smithii* genome. One ORF, which presumably encodes a transposase, occupies 87% of the ISM1 sequence.

3.4. Tryptophan

The *trpEGCFBAD* genes from *M. thermoautotrophicum* strain Marburg [170] and the *trpDFBA* genes from *M. voltae* [169] have been cloned and sequenced. In *M. thermoautotrophicum* strain Marburg, the promoter proximal *trpE* and *trpG* genes overlap by 2 bp and the remainder are separated by intergenic regions ranging in length from 5 to 56 bp. As in many other prokaryotic species, the methanogen *trp* genes are clustered, but the functionally equivalent genes are arranged *trpEGDCFBA* in the enteric bacteria, *trpEDCFBA* plus an unlinked *trpG* gene in *B. subtilis* [175], and *trpCBA* in *Haloferax volcanii* [176]. Only the order and close linkage of the *trpBA* genes, which encode the α and β subunits of tryptophan synthetase, respectively, seem to be conserved in all *trp* gene clusters so far studied. In *S. cerevisiae* and *Neurospora crassa,* the *trpA* and *trpB* functions are combined in the product of the *TRP5* gene [177]. The absence of promoter-like sequences between the *trp* genes in *M. thermoautotrophicum* strain Marburg and the presence, upstream of *trpE,* of the octanucleotide 5′TTTAAATA, which conforms precisely to the consensus sequence for methanogen promoters, strongly suggest that these *trp* genes are transcribed into a single mRNA. A region of dyad symmetry, which resembles the *E. coli trp* operator [178], is located immediately downstream of this promoter, indicating that repressor binding may also regulate expression of the methanogen *trp* operon. Sequences likely to be transcription terminators are not obvious downstream of either *trpD* in *M. thermoautotrophicum* strain Marburg or *trpA* in *M. voltae.*

Comparisons of the *trp* proteins from *M. thermoautotrophicum* strain Marburg with the corresponding archaeal proteins from *M. voltae* (gp*trpB* and gp*trpA*) and *H. volcanii* (gp*trpC*, gp*trpB* and gp*trpA*), and with the functionally equivalent bacterial and yeast sequences, indicate similar overall extents of divergence [170]. Extant gp*trpB*s are 51–62% similar whereas gp*trpA*s and gp*trpC*s are 32–42% similar. These results are consistent with a common ancestral origin for all *trp* genes followed by divergence, at similar rates, within the different biological domains. In *Brevibacterium lactofermentum* [179] and *Salmonella typhimurium* [175], tryptophan biosynthesis is subject to feedback inhibition. Tryptophan binds to an amino-acid sequence, LLES, located in the amino-terminal region of the α subunit of anthranilate synthetase (gp*trpE*) causing this regulation. As this amino-acid sequence is also present at a similar location in the anthranilate synthetase of *M. thermoautotrophicum* strain Marburg, tryptophan biosynthesis may be similarly regulated in this methanogen.

3.5. Glutamine

The *glnA* gene in *M. voltae*, encoding glutamine synthetase (GS), forms a monocistronic transcriptional unit flanked by sequences unrelated to sequences surrounding *glnA* genes in other prokaryotes [60]. The intergenic region upstream of *glnA* contains Box A and Box B promoter elements separated by the palindromic sequence, 5'TATCGGAAATATATTTCCGATA, which is similar to a palindrome found in the promoter region of the *nifH1* gene of *Methanococcus thermolithotrophicus* [180]. Immediately downstream of *glnA*, in *M. voltae*, are several oligo-T sequences and a sequence with the potential to form three hairpin-loop structures which presumably play a role in transcription termination. The codon usage in *glnA* generally reflects the overall 31 mol% G+C of the *M. voltae* genome. There is a strong bias for A or U in the wobble position in L, V, K, E, P, A and R codons although Y, H, N and D codons have predominantly C in the wobble position as is also found in the highly expressed MR-encoding genes (*mcrBGA*) of *M. voltae* [88]. The *M. voltae* GS is similar in size to bacterial GSs ($\sim$50 kDa) and has sequence identities ranging from 33% with the GSs from *Anabaena* [181] and *Thiobacillus ferrooxidans* [182] to 51% with GSs from *B. subtilis* [183] and *Clostridium acetobutylicum* [184]. Eucaryal GSs are smaller and the amino-acid sequences of GSs from alfalfa [185], *Phaseolus vulgaris* [186] and hamsters [187] are less than 12% identical to the sequence of *M. voltae* GS. The most highly conserved regions, thought to be involved in ATP-binding, glutamate-binding and Mn^{2+}-binding are, however, present in all GSs. The activities of GS from both *M. voltae* [60] and *Methanobacterium ivanovii* [188] are repressed by high ammonia concentrations. In the enterobacteria, GS activity is regulated by adenylation of a tyrosine residue within a highly conserved amino-acid sequence. In the *M. voltae* GS this tyrosine residue is replaced by a phenylalanine residue and there is only limited conservation in the flanking sequences [60]. Regulation of GS activity in *M. voltae* cells may not therefore be by the same adenylation mechanism.

3.6. Adenine

Restriction fragments which complement both the *purE* and *purK* genes of *E. coli* have been cloned and sequenced from both the thermophile *M. thermoautotrophicum* strain ΔH [166] and the mesophile *Methanobrevibacter smithii* [16]. Both methanogen-derived *purE* genes appear to be part of longer transcriptional units. They encode polypeptides that are 45% identical to each other and 38% identical to the *E. coli purE* gene product [189]. Differences in codon usage in the two methanogen *purE* genes reflect the different overall G+C contents of the two methanogen genomes. Most changes occur in the wobble position and are A or T to G or C in the direction from *Mb. smithii* (38 mol% G+C) to *M. thermoautotrophicum* (50 mol% G+C) [166]. As the methanogen *purE* genes complement mutations in both the *purE* and *purK* genes of *E. coli*, they must encode proteins with both 5'-phosphoribosyl-5-aminoimidazole carboxylase activity (*E. coli purE* function) and CO_2-binding activity (*E. coli purK* function) [189]. Transposon mutagenesis has located the *purK*-complementing activity within the 75-amino-acid residues at the carboxyl terminus of the 37 kDa gp*purE* from *Mb. smithii* [16,189].

Sequencing the cloned regions flanking the gene (*ileS*) which encodes isoleucyl–tRNA synthetase, in *M. thermoautotrophicum* strain Marburg [190], revealed a truncated ORF beginning 2 bp downstream of *ileS*. This ORF could encode a polypeptide 57% identical in sequence to the amino-acid sequence at the amino terminus of the enzyme formylglycineaminidine ribonucleotide synthetase, encoded by *purL* in *B. subtilis* [191]. This truncated methanogen ORF, which appears to be cotranscribed with *ileS*, has therefore also been designated *purL*.

4. Transcription and translation machinery genes

4.1. Stable RNA genes

4.1.1. tRNA genes
Genes encoding tRNAs in methanogens are arranged in linked clusters some of which also contain embedded 5S rRNA genes, or are found as individual transcriptional units, or are linked to ribosomal operons [192–203]. (For a review of archaeal tRNA genes, see Chapter 13 of this volume). Nineteen tRNA genes, arranged in seven transcriptional units with specificities for a total of 15 amino acids, have been cloned and sequenced from *M. vannielii* [66,192–194]. Twelve of these are in two tRNA clusters, one of which contains seven tRNA genes linked to a single 5S gene [194]. The second cluster encodes five tRNA molecules, and a single $tRNA^{Phe}$-encoding gene is located 115 bp from this cluster but on the opposite DNA strand [66]. Related groups of linked tRNA genes have also been characterized from *M. voltae* [199] and *Methanothermus fervidus* [68]. The $tRNA^{Met}_{CAU}$ gene sequenced from all three methanogens possesses only limited homology to other $tRNA^{Met}$ genes but is very similar to $tRNA^{Ile}_{CAU}$ sequences. This gene may therefore encode a $tRNA^{Ile}$ that, with a modified nucleotide in the wobble position, recognizes AUA codons [68]. Eight of the nine tRNA genes sequenced from *M. fervidus* translate abundant codons based on the codon usage in

the *M. fervidus mcrA, mcrB* and *mcrG* genes. The exception is a tRNA^{Lys} gene, the only class-II tRNA gene sequenced [68]. These tRNAs, from the hyperthermophile, are predicted to possess increased numbers of G:C base pairs when compared to the functionally homologous tRNAs from the mesophiles *M. vannielii, M. voltae* and *M. formicicum* and from the thermophile *M. thermoautotrophicum*. The additional base pairing presumably enhances the resistance of the *M. fervidus* tRNAs to thermal denaturation. Most methanogen tRNA genes do not encode the 3′ terminal–CCA sequence, a feature they share with most other archaeal and eucaryal tRNA genes. This trinucleotide sequence is, however, encoded in the tRNA^{Pro}, tRNA^{Asn}, and tRNA^{His} genes of *M. vannielii* [66,194], a tRNA^{Pro} gene in *M. voltae* [199] and in the four tRNA genes sequenced from *Methanopyrus kandleri* [203a]. Introns have been detected in tRNA-encoding genes in halophilic [204,205] and sulfur-dependent [206–208] archaea but not, so far, in tRNA genes from methanogens. Although tRNAs in the archaea exhibit both bacterial and eucaryal post-transcriptional modifications these modifications are, overall, more eucaryal than bacterial in nature [209]. Modified nucleotides, unique to the archaea, have been discovered in tRNAs from *M. thermoautotrophicum, M. fervidus, Methanolobus tindarius* and *M. vannielii*. Several mature tRNAs in *M. vannielii* contain 2-selenouridine [210].

4.1.2. rRNA genes
Most methanogen rRNA genes are arranged in bacterial-like 5′–16S–23S–5S transcriptional units [195–198,201–203,211]. (For reviews of archaeal ribosomal RNAs see Chapters 13 and 14 of this volume). The 16S and 23S rRNA genes are flanked by inverted repeat sequences that, when transcribed, could form secondary structures similar to bacterial RNase III cleavage sites. There is one rRNA operon in the genome of *M. voltae* [12,199], two in *M. thermoautotrophicum* [198], *M. formicicum* [197], *M. fervidus* [203] and *M. soehngenii* [201], and four in *M. vannielii* [192]. Most of these operons contain a tRNA^{Ala} gene in the spacer region between the 16S and 23S rRNA-encoding sequences. A tRNA^{Ala} gene is also found at this location in the rRNA operons of the extreme halophiles and *Archaeoglobus* and *Thermococcus celer*, non-methanogenic archaea that cluster phylogenetically with the methanogens in the Euryarchaeota [200]. Preliminary evidence indicates that there is only one 16S and one 23S rRNA-encoding gene in the phylogenetically remote methanogen *Methanopyrus kandleri* and that these genes are not closely linked to each other nor associated with a tRNA^{Ala} gene [Palmer, J.R., Daniels, C.J. and Reeve, J.N. (1991) unpublished results].

In *M. thermoautotrophicum* strain Marburg and in *M. fervidus*, 7S RNA- and tRNA^{Ser}-encoding genes are located immediately upstream of one of their two rRNA operons [203]. In *M. fervidus*, there is a methanogen promoter sequence upstream of the 7S RNA gene but not between the tRNA^{Ser} and 16S rRNA genes, suggesting that 7S RNA–tRNA^{Ser}–16S rRNA–tRNA^{Ala}–23S rRNA–5S rRNA–3′ might be a single transcriptional unit [203]. When compared with other 16S rRNAs the secondary structure predicted for the 16S rRNA from the hyperthermophile *M. fervidus* has additional base pairing presumably to enhance its resistance to heat denaturation. The secondary structure predicted for the *M. fervidus* 5S rRNA molecule [68] also conforms to this pattern. It resembles, most closely, the 5S rRNA from *T. celer* which, while not a methanogen, is also a thermophilic archaeon.

The methanococci *M. vannielii* [194] and *M. voltae* [199] contain one and two additional 5S rRNA-encoding genes, respectively, not located in rRNA operons. These 5S rRNA genes are, instead, clustered with tRNA-encoding sequences. The operon-associated 5S rRNA genes in the two different methanococci are more similar to each other than are the operon-linked and operon-unlinked 5S rRNA genes in the same methanogen [199].

4.1.3. 7S RNA genes
Archaeal species contain large amounts of a RNA molecule, ~300 nucleotides in length, known generically as the 7S RNA [212]. Secondary structures predicted for these archaeal RNAs are similar to those predicted for the 7S RNA component of eucaryal signal-recognition particles [203]. There is, however, only limited conservation at the primary sequence level. Archaeal 7S RNAs also have similarities in both their primary sequences and secondary structures to regions of the small cytoplasmic scRNA of *B. subtilis* and the 4.5S RNA of *E. coli* [203]. The function of the archaeal 7S RNA is unknown, although a ribosome-associated activity is suggested by the discovery that the 7S RNA-encoding gene in *M. fervidus* and in *M. thermoautotrophicum* strain Marburg is adjacent to, and possibly co-transcribed with, one of their two rRNA operons [203]. Genes encoding 7S RNAs and their flanking regions have also been cloned and sequenced from *M. voltae* [213] and *M. acetivorans* [214] but linkage to rRNA operons was not reported.

4.2. Genes encoding RNA polymerases, ribosomal proteins and elongation factors

Archaea, in common with Bacteria, appear to contain only one DNA-dependent RNA-polymerase (RNAP) [31,41,42]. Archaeal RNAPs, however, in contrast to the relatively simple $\alpha_2\beta\beta'\sigma$ subunit composition of bacterial RNAPs, resemble eucaryal RNAPs in containing eight to ten different polypeptide subunits [31,41]. Immunological studies and comparisons of the sequences of RNAP-encoding genes have confirmed that archaeal RNAPs are more closely related to eucaryal RNAPs than to bacterial RNAPs [41,42,215,219,386]. (Archaeal RNAPs are reviewed in Chapter 12 of this volume.) As the different eucaryal RNAPs all appear more closely related to archaeal enzymes than to each other, contemporary archaeal RNAPs are likely to be the enzymes now most similar to the common ancestor of all RNAPs [216]. The genes encoding the four largest subunits of RNAP in *M. thermoautotrophicum* strain Marburg are arranged in the order B″–B′–A–C [42]. The B′ and C subunit-encoding genes have GTG translation initiation codons and the A-encoding gene uses TTG as the start codon. Initiation codons other than ATG are relatively common in methanogens. GTG initiation codons have been reported for *mcrB* [89] and *atpA* [220] in *M. barkeri*, *sod* [63] and *trpG* [170] in *M. thermoautotrophicum*, *acs* in *M. soehngenii* [70] and *slgA* in *M. fervidus* and *Methanothermus sociabilis* [36], and TTG initiation codons for *mdh* [221] and *pgk* [222] in *M. fervidus* and for *nifH2* in *M. thermolithotrophicus* [223].

The combined A+C and B″+B′ subunits of RNAP in *M. thermoautotrophicum* strain Marburg are phylogenetically homologous to the largest and second-largest subunits, respectively, of eucaryal and bacterial RNAPs [42]. There is substantial sequence conservation in the B″+B′ subunits and the corresponding regions of the eucaryal enzymes. These subunits may contain the active center of the *M. thermoautotrophicum* RNAP as they bind radioactively-labeled ribonucleoside triphosphate analogs [217,224]. Labeling of the B' subunit occurs in a region that is highly conserved in all RNAPs. This region could form an α helix, with a high density of basic amino-acid residues on one side, which could bind negatively charged nucleotide residues [217]. RNAPs must also recognize and bind to DNA, and amino-acid sequences which could form zinc-finger structures, similar to those found in the largest subunit of eucaryal RNAPs, are present in the A subunit of the *M. thermoautotrophicum* RNAP [42]. In *E. coli*, the genes encoding the β and β' subunits of RNAP are adjacent to genes encoding ribosomal proteins (r-proteins), in the order L11–L1–L10–L12–β–β', forming two transcriptional units, L11–L1 and L10–12–β–β' [225]. The same r-protein genes are also linked in the same order in *Halobacterium halobium* [226], *Halobacterium cutirubrum* [227] and *Sulfolobus solfataricus* [228]. In *M. vannielii*, an ORF of unknown function is located 440 bp upstream of the L1 gene, the L11 gene is absent and the L1–L10–L12 genes form a transcriptional unit [229]. Two sequences resembling methanogen promoters are located 22 and 64 bp upstream of the site of initiation of transcription of this L1–L10–L12 operon. Two similar Box A TATA-motifs have also been demonstrated upstream of the equivalent operon in *H. cutirubrum* [230]. In its primary sequence and predicted secondary structure the 5' leader region of the *M. vannielii* L1–L10–L12 transcript is similar to the L1-binding site on its 23S rRNA [229]. It also resembles the leader sequence of the L11–L1 transcript in *E. coli* that binds L1 protein and thereby regulates the synthesis of these r-proteins in *E. coli* [231,232]. Regulation of r-protein synthesis by L1 binding to its own mRNA may therefore also occur in *M. vannielii*.

The genes encoding the *M. vannielii* equivalent of the β' subunit of *E. coli* RNAP and the H subunit of *M. vannielii* RNAP have been located immediately upstream of the *M. vannielii* equivalent of the *E. coli* streptomycin operon [233–235,386]. Close linkage of the streptomycin operon and RNAP-encoding genes has also been found in other archaea, whereas these two regions are separated by $\sim$480 kbp in the *E. coli* genome. The *M. vannielii* streptomycin operon encodes r-proteins S7, S10 and S12, elongation factors EF-2 (equivalent to bacterial EF-G) and EF-1α (equivalent to bacterial EF-Tu) and two additional proteins in the sequence ORF1–ORF2–S12–S7–EF-2–S2–EF-1α–S10. ORF1 encodes a protein with an amino-acid sequence that is 37% identical (58% similar allowing for conservative substitutions) to that of the r-protein L30 from rat liver [233,234]. The *M. vannielii* elongation factors have sequences which are much more similar to their eucaryal than to their bacterial counterparts, and consistent with this, this methanococcal EF-2 is ADP-ribosylated and inhibited by diphtheria toxin [233–236].

Located 30 kbp from the *M. vannielii* streptomycin operon are the *M. vannielii* equivalents of the *E. coli* S10 and spectinomycin operons [59,234]. The arrangement of genes in the *M. vannielii* spectinomycin operon is similar to that in *E. coli*, however it contains a S17-encoding gene and additional ORFs, in the order ORFa–ORFb–S17–L14–L24–ORFc–L5–S14–S8–L6–ORFd–ORFe–L18–S5–L30–L15. Although ORFs c, d and e do not have bacterial counterparts, they do appear to encode r-proteins as they are clearly related to r-proteins L32 and L19 from mouse and rat liver, and to yeast r-protein S6, respectively. Two promoters direct the synthesis of three differently sized transcripts from the *M. vannielii* spectinomycin operon [59]. A promoter located upstream of ORFa directs the synthesis of two transcripts, 0.8 kbp and 8.0 kbp in length. The shorter transcript terminates between ORFb and the S17 gene and the longer terminates downstream of the L15 gene. The second promoter, located upstream of the S17 gene, directs the synthesis of a 7 kbp transcript that also terminates downstream of the L15 gene. The *M. vannielii* equivalent of the *E. coli* S10 operon is located immediately upstream of its spectinomycin operon. It contains r-protein genes in the order L22–S3–L29 [59,234], which is the same order as in the S10 operon of *E. coli* but which lacks several r-protein genes, including the S10 gene. Based on the sequenced genes, most *M. vannielii* proteins directly involved in transcription and translation appear more similar overall to their eucaryal than to their bacterial counterparts [59,237–240,386].

4.3. Aminoacyl–tRNA synthetase

A gene (*ileS*) encoding isoleucyl–tRNA synthetase has been cloned and sequenced from the wild-type *M. thermoautotrophicum* strain Marburg and from a pseudomonic acid-resistant mutant, designated strain MBT10 [190]. This *ileS* gene encodes 1045 amino-acid residues resulting in a protein with a calculated molecular mass of 121 244. A single base change in codon 590 in the *ileS* gene from MBT10 replaces a glycine residue with an aspartic-acid residue. This substantially reduces the enzyme's affinity for pseudomonic acid and confers pseudomonic acid resistance on strain MBT10. Both the wild-type and mutated *ileS* genes have been expressed in *E. coli*, but the mutated gene did not provide *E. coli* with increased resistance to pseudomonic acid [190]. Three possible RBS are found upstream of *ileS*, two close to the ATG translation initiation codon (-10 to -7 and -13 to -10) and a third located relatively far upstream (-21 to -15). Sequences resembling Box A and Box B promoter elements have been identified 420 and 401 bp upstream of the *ileS* coding sequence, within an ORF transcribed in the same direction as *ileS*. The sequenced part of this ORF encodes a product unrelated to protein X encoded by the gene upstream of *ileS* in *E. coli* [241] or to any other known protein sequence. An ORF preceded by two RBS begins 2 bp downstream of the *ileS* termination codon. The sequenced part of this gene encodes a protein with an amino-acid sequence that is 57% identical to the corresponding region of formylglycineaminidine ribonucleotide synthetase II, encoded by *purL* in *Bacillus subtilis* [191]. There is no obvious transcription terminator in the 25 bp intergenic region upstream of the *ileS* gene, which may therefore be part of a polycistronic transcriptional unit containing at least the upstream ORF, *ileS* and *purL*. For comparison, in *E. coli* [242], *Enterobacter aerogenes* [243] and *Pseudomonas aeruginosa* [244], *ileS* is within a polycistronic transcriptional unit arranged gene X–*ileS*–*lsp*–ORF149–ORF316. *lsp* encodes a prolipoprotein signal peptidase but the functions of the gene X, ORF149 and ORF316 products are unknown. In *S. cerevisiae* a gene encoding a heat shock protein is located immediately upstream of *ileS* [245].

The amino-acid sequence predicted for the isoleucyl–tRNA synthetase from *M. thermoautotrophicum* strain Marburg is 36% and 32% identical to the sequences of the functionally equivalent enzymes from the *S. cerevisiae* cytoplasm [245] and from *E. coli* [246], respectively. There are four conserved regions found only in the archaeal and eucaryal enzymes. The yeast and *E. coli* enzymes are themselves 27% identical. The primary sequence of the isoleucyl–tRNA synthetase from *M. thermoautotrophicum* strain Marburg conforms well to the consensus sequence derived for all tRNA synthetases in the Class I isoleucyl family. Nine such enzymes have so far been sequenced, representing all three biological domains, and all these enzymes must have a common ancestor [245–248]. The methanogen enzyme contains the KMSKS sequence thought to be essential for tRNA binding [249–251], and a variant of the common peptide sequence implicated in ATP binding [246,248].

5. Nitrogen fixation genes

The ability to fix dinitrogen (N_2) is found in all three phylogenetic Orders of methanogens [252–258]. Similarly, N_2 fixation is widely distributed in the bacterial domain but has yet to be demonstrated in eucarya. Southern hybridizations, using the *nifH* gene from *Anabaena* as the probe, revealed related nucleotide sequences in the genomes of 14 methanogens [259–261] and probes containing the *nifD* and *nifK* genes from *Klebsiella pneumoniae* hybridized specifically to genomic DNA from *M. ivanovii* and *M. voltae* [259]. Hybridization was not, however, detected to genomic DNAs from *Thermoplasma acidophilum* or to DNA from four other thermophilic, sulfur-dependent, non-methanogenic archaea [260].

Two separate *nifH* genes (*nifH1* and *nifH2*) plus related flanking regions have been cloned and sequenced from *M. ivanovii* [223,262], *M. thermolithotrophicus* [180,223] and *M. barkeri* [262] and one *nifH* gene from *M. voltae* [263]. Downstream of *nifH1* in *M. barkeri* and *M. ivanovii* are related genes arranged ORF105–ORF122/123–*nifD* [262]. In *M. thermolithotrophicus* the arrangement is *nifH1*–ORF105–ORF128–*nifD*–*nifK*, in which *nifD* and *nifK* overlap by 8 bp [180]. Downstream of the *nifH2* gene in *M. barkeri* is the sequence ORF105–ORF125 [262]. All the downstream ORFs appear to be related. A gene-duplication event apparently gave rise to the common ancestors of the current ORF105 group and the current ORF122/123/125/128 group [262]. The sequences of polypeptides encoded by these ORFs are all $\sim$50% identical to the amino-acid sequences of the P_{II} regulatory proteins encoded by *glnB* genes in bacteria [264–269]. They may therefore function similarly in the methanogens, regulating N_2 fixation in response to changes in the availability of fixed nitrogen. A methanogen Box A promoter element is present 21 bp upstream of the site of initiation of transcription of *nifH1* in *M. thermolithotrophicus*. Transcription in vivo occurs only under N_2-fixing conditions, resulting in a transcript containing *nifH1*–ORF105–ORF128 sequences, however this regulation of *nif* gene expression could not be duplicated in vitro [387]. A separate transcript of *nifD*–*nifK* is initiated 85 bp upstream of the ATG translation-initiation codon of *nifD* [180]. Transcription of the *nifH2* gene in *M. thermolithotrophicus* could not be detected under diazotrophic growth conditions nor in ammonium-supplemented media [180]. A partially sequenced ORF upstream of this *nifH2* gene appears to be part of the same transcriptional unit [223]. A similar gene organization also appears to exist for the second *nifH* gene in *M. ivanovii* [223]. Approximately 60 bp upstream of the translation-initiation codon of the single *nifH* gene cloned from *M. voltae* is the sequence 5'ATGGCATA [263] which resembles the sequence 5'CTGGPyAPyPu found in the promoter region of some bacterial *nif* genes and genes under *ntrC*/*nifA* control [270].

Approximately 25% of the amino-acid residues in all gp*nifH*s, the Fe-protein components of MoFe nitrogenases, are conserved [271]. The amino-acid sequences of the methanogen gp*nifH*s, which contain from 263 to 292 residues, are themselves only

47–55% conserved. It appears that more divergence has occurred in these methanogen gp*nifH*s than between all pairs of bacterial gp*nifH*s [271], which indicates a very ancient origin for this protein in the methanogens. Despite this extensive divergence, some common features remain in almost all gp*nifH*s. Five cysteine residues have conserved locations [180], and with the exception of the gp*nifH* from *Rhizobium trifoli* [272], they all lack tryptophan residues. The amino-acid sequence GKGGIGKS, which conforms to a consensus sequence found in ATP-binding proteins, is present in the amino-terminal region of all gp*nifH*s, except the gp*nifH2* from *M. barkeri* [262]. Although the sequences of the *M. thermolithotrophicus* gp*nifH1* and gp*nifH2* are only 47% similar to each other, the sequence of this gp*nifH1* is ~70% similar to the sequence of the gp*nifH3*s from *Azotobacter vinelandii* and *Clostridium pasteurianum* [180]. In *A. vinelandii* this *nifH3* gene has been shown to encode the Fe-protein of a recently discovered and novel Fe-nitrogenase [273]. The *M. thermolithotrophicus nifH1*, *C. pasteurianum nifH3* and *A. vinelandii nifH3* genes may therefore encode a family of enzymes distinct from the well-established Mo and V nitrogenases [180].

The α and β subunits of nitrogenase, encoded by the *M. thermolithotrophicus nifD* and *nifK* genes, are also predicted to contain amino-acid sequences conserved in the α and β subunits of the MoFe nitrogenases encoded by *nifD* and *nifK* genes in bacteria [180].

6. Genes encoding metabolic enzymes

6.1. Glyceraldehyde-3-phosphate dehydrogenase (GAPDH)

The nucleotide sequences, plus flanking regions, of GAPDH encoding genes (*gap*) from the mesophiles *M. bryantii* and *M. formicicum* [274], and from the hyperthermophile *M. fervidus* have been cloned and sequenced [275]. The GAPDH-encoding sequences are 91% identical in the two mesophiles and these sequences are 71% identical to the *M. fervidus gap* gene. An ORF identified immediately downstream of the *M. fervidus gap* gene is not present at this location in either of the mesophiles. Sequences conserved immediately downstream of the *gap* genes in the mesophiles are limited to 35 bp regions enriched for pyrimidine bases, indicative of transcription terminators. An ORF, conserved upstream of the *gap* gene in both mesophiles, terminates 1 bp into the *gap* genes.

Amino-acid sequence comparisons indicate that bacterial and eucaryal GAPDHs are more closely related to each other than to these methanogen GAPDHs [274,275]. Within the archaea, GAPDH primary sequences range from 95% identical in the two mesophilic methanogens [274] to ~50% identity between the GAPDHs from *Pyrococcus woesei* and the methanogens [276]. The sequences assumed to be catalytic sites in the methanogen GAPDHs are considerably different from those identified in other GAPDHs. Although the nucleotide-binding region and the region surrounding the active cysteine are well conserved, a histidine residue previously thought essential for catalysis, and a substrate-binding arginine are absent at the corresponding positions of the methanogen GAPDHs [274]. Despite this substantial divergence in their primary sequences, methanogen, eucaryal and bacterial GAPDHs do have similar secondary structures based on hydropathy, chain flexibility and amphipathy analyses [274].

The patterns of amino-acid substitutions observed in the GAPDHs from the mesophilic methanogens, when compared with the GAPDH from the hyperthermophile, mostly conform to the pattern established previously in comparisons of other functionally identical enzymes from mesophiles and thermophiles. In the *M. fervidus* GAPDH, most glycine and serine residues are replaced, particularly in loop structures, by bulkier and/or more hydrophobic residues. These changes are expected to increase the thermostability of the enzyme by increasing its hydrophobicity and primary-chain rigidity and by improving the packing of internal amino-acid residues [274]. The predominant substitution of isoleucine for leucine residues and the lack of arginines in the *M. fervidus* GAPDH are, however, deviations from the established pattern of amino-acid substitutions associated with thermophilic growth [274]. Expression of the *M. fervidus gap* gene in *E. coli* results in the synthesis of an enzyme with properties identical to the GAPDH synthesized in *M. fervidus* [277]. Construction of recombinant *gap* genes, containing regions of both the *M. bryantii* and *M. fervidus gap* genes, followed by their expression in *E. coli*, has revealed that a short region in the carboxyl terminus of the *M. fervidus* protein provides most of the thermostability of this enzyme [278].

6.2. L-Malate dehydrogenase (MDH)

The L-malate dehydrogenase-encoding gene (*mdh*) cloned from *M. fervidus* directs the synthesis of a polypeptide containing 339 amino-acid residues. The MDH holoenzyme is a homodimer and therefore its native molecular mass is ~70 kDa [221]. The *mdh* gene has a TTG translation-initiation codon immediately preceded by a RBS. The amino-acid sequence deduced for the *M. fervidus* MDH has only limited similarity to the amino-acid sequences of bacterial and eucaryal MDHs. These enzymes are more similar to their NADH-dependent L-lactate dehydrogenases than to this methanogen MDH. The two eucaryal enzymes must have diverged from each other after the ancestor of the *M. fervidus* MDH-encoding gene separated from their common ancestral sequence. Despite this extensive divergence, all MDHs have conserved active site residues and similar secondary structures. The *M. fervidus* MDH has a high ratio of isoleucine to leucine residues and no preference for arginine residues, features it shares with the *M. fervidus* GAPDH [274]. All archaeal MDHs can use either NADH or NADPH and catalyze oxaloacetate reduction to malate more efficiently than the reverse of this reaction [279–282]. These are, nevertheless, very different enzymes as demonstrated by the very different amino-acid sequences of the MDHs from *Sulfolobus acidocaldarius* and *M. fervidus* [221,283].

6.3. 3-Phosphoglycerate kinase (PGK)

Genes (*pgk*) encoding 3-phosphoglycerate kinase (PGK) have been cloned and sequenced from *Methanobacterium bryantii* and *Methanothermus fervidus*, and the deduced amino-acid sequences have been analyzed for thermophily-conferring features in the *M. fervidus* protein [222]. The *M. bryantii* and *M. fervidus* PGK molecules are predicted to contain 409 and 410 amino-acid residues, respectively, giving calculated molecular masses of 44.8 kDa and 45.9 kDa. Their sequences are 61% identical and 32–36% identical to bacterial and eucaryal PGKs. Lysine and arginine residues are used frequently in the *M. fervidus* PGK, in contrast to the situation in the *M. fervidus* GAPDH and MDH, and arginine for lysine is the most frequent substitution in this PGK when it is compared to mesophilic PGKs. Although the relatedness of the methanogen PGK enzymes to the sequences of bacterial and eucaryal PGKs is lower than between PGKs from bacteria and eucarya, there is significantly more conservation than is found in the equivalent comparisons of GAPDHs and MDHs. The *M. fervidus pgk* gene has a TTG translation-initiation codon whereas ATG is found in the translation-initiation codon for *pgk* in *M. bryantii* [222].

6.4. ATPases

The *atpA* and *atpB* genes which encode the α (578 amino-acid residues) and β (459 amino-acid residues) subunits of a membrane-bound ATPase in *M. barkeri* have been cloned and sequenced. The *atpA–atpB* intergenic region is only 2 bp in length, indicating that these genes are co-transcribed into a single mRNA. RBS precede both genes and the translation initiation codons for *atpA* and *atpB* are GTG and ATG, respectively [220]. The encoded polypeptides are very similar, having identical amino-acid residues at 126 positions and, allowing for conservative amino-acid substitutions, they have 227 residues in common. The *atpA* and *atpB* genes appear therefore to have evolved from a common ancestral gene following its duplication. The sequences of the α and β subunits of this methanogen ATPase are from 52% to 60% conserved in the corresponding subunits of vacuolar H^+-ATPases from carrot [284], *Neurospora crassa* [285,286], *Saccharomyces cerevisiae* [287] and *Arabidopsis thaliana* [288] and 57% and 53% identical to the subunits of a *Sulfolobus acidocaldarius* ATPase [289,290]. Unlike the *M. barkeri* ATPase, the *S. acidocaldarius* ATPase is, however, insensitive to DCCD [291,292]. The sequences of the α and β subunits of this methanogen ATPase are also 22% and 24% identical to the amino-acid sequences of the β and α subunits of *E. coli* F_1 H^+-ATPase, respectively [293–296]. This conservation of sequences in the bacterial F_1, eucaryal vacuolar and archaeal ATPases strongly implies a common ancestry for all these enzymes. Two peptide sequences in the α subunit [220,297] and one in the β subunit of the *M. barkeri* ATPase, that conform to the consensus sequence for nucleotide-binding sites [220] are also present in the ATPase subunits from *S. acidocaldarius* [284,285,289]. An unrelated vanadate-sensitive ATPase has been purified and characterized from *M. voltae* [299,300]

6.5. Superoxide dismutase

A gene (*sod*) encoding a Fe-superoxide dismutase (Fe-SOD), together with its flanking regions, has been cloned and sequenced from *M. thermoautotrophicum* strain Marburg and functionally expressed in *E. coli* [63]. Translation starts at a GTG codon, preceded by a RBS, and results in the synthesis of a polypeptide containing 205 amino-acid residues. In common with the archaeal SODs from *Thermoplasma acidophilum* [301] and *Methanobacterium bryantii* [302], the methanogen enzyme is a homo-tetramer, giving it a calculated molecular mass of 105 kDa. A Box A promoter sequence is located 72 bp upstream from the start codon, and an inverted repeat sequence, likely to be a transcription terminator, is located within the SOD-encoding sequence immediately preceding the translation termination codon [63]. The upstream intergenic region contains an inverted repeat sequence which is conserved, at the same location, upstream of the *Halobacterium halobium* SOD-encoding gene [303] but not upstream of the *H. cutirubrum* SOD-encoding gene [304]. Downstream of the *M. thermoautotrophicum sod* gene is a 45 bp direct repeat, inside an ORF which, if expressed, would be transcribed from the opposite DNA strand.

The amino-acid sequence deduced for the *M. thermoautotrophicum* SOD most closely resembles bacterial, archaeal and eucaryal Mn-SODs, however atomic absorption spectroscopy has established that this is a Fe-SOD [305]. In contrast to other Fe-SODs, including the Fe-SOD from *Methanobrevibacter smithii* [302], this enzyme is resistant to azide and hydrogen peroxide, features it has in common with Mn-SODs [305]. Amino-acid residues identified as metal ligands or active sites are, however, conserved in both Mn-SODs and Fe-SODs, implying that all these enzymes have a common ancestry and similar enzyme mechanisms [63].

6.6. S-adenosyl-L-methionine:uroporphyrinogen III methyltransferase (SUMT)

A gene (*corA*), implicated in corrinoid biosynthesis, which encodes *S*-adenosyl-L-methionine:uroporphyrinogen III methyltransferase (SUMT) activity has been cloned and sequenced from *Methanobacterium ivanovii* [306]. Synthesis of the *corA* gene product has been demonstrated in *E. coli* and *Pseudomonas denitrificans* [306]. The *corA* gene is preceded by a RBS, has codon usage similar to the *nifH* gene from *M. ivanovii* [223], and encodes a polypeptide containing 231 amino-acid residues giving a calculated molecular mass of 24.9 kDa. The amino-acid sequence predicted for the *M. ivanovii* SUMT is 40.4% and 47.0% identical to the sequences of *cobA*-encoded SUMTs from *P. denitrificans* [307] and *Bacillus megaterium* [308], respectively, and 47% identical to a sequence in the carboxyl-terminal region of the *E. coli* gpcysG, a protein which also possesses SUMT activity [309,310]. These SUMTs all contain highly conserved sequences, most of which are within conserved domains of known (*CobA, CobI*) and proposed (*CobF, CobJ, CobL* and *CobM*) SAM methyltransferases which participate in cobalamin biosynthesis in *P. denitrificans* [311]. These conserved amino-acid sequences are therefore likely to be involved in methyltransferase activity.

Compared to the corresponding *P. denitrificans* enzyme [312], the *M. ivanovii* SUMT has a 20-fold lower K_M for its substrate, urogen III, and does not display substrate inhibition. Substrate inhibition of the SUMT from *P. denitrificans* is consistent with the proposed very low requirement of this bacterium for cobalamin [306]. In contrast, methanogens require and synthesize large amounts of corrinoids [313,314]. Cofactor F_{430} of MR, for example, contains corrin and porphyrin-based structures which are thought to be synthesized from precorrin-2 [315], the product of SUMT activity. The *M. ivanovii* SUMT has properties that indicate participation in the synthesis of very large amounts of corrinoids.

7. Chromosomal proteins

Historically, the separation of prokaryotes and eukaryotes was based on visible differences in the organization of the nuclear material. This division was further strengthened by the discovery that in eukaryotes, histones compact the nuclear DNA into nucleosomes whereas nucleosomes have not been routinely demonstrated in bacteria [316]. The possibility that archaea might contain nucleosomes led to the isolation and characterization of histone-like DNA-binding proteins from a range of archaeal species [316]. Many archaea contain several small and closely related DNA-binding proteins, some of which have been sequenced directly without resorting to gene cloning [316–324]. Proteins of this type isolated from two *Methanosarcinae* [317–323] and from *Methanothrix soehngenii* [324] have been shown to be related to each other, but are not closely related to eucaryal histones. In contrast, DNA-binding proteins isolated from *Methanothermus fervidus* and *M. thermoautotrophicum* strain ΔH, designated HMf and HMt respectively, are closely related to histones [325–327a]. Over 30% of the 69 amino-acid residues in HMf-2, encoded by *hmfB* in *M. fervidus*, are conserved in the consensus sequence for eucaryal core histones [325]. The intergenic region upstream of the cloned *hmfB* contains two directly repeated promoter-contained copies of a 73 bp sequence, however there is only one copy of this sequence in the genome [388]. Transcription of *hmfB* appears to be terminated by the presence of several oligo-T sequences located immediately downstream from the coding sequence [325]. *M. fervidus* and *M. thermoautotrophicum* strain ΔH are both thermophiles, and HMf and HMt may play a role in protecting their genomic DNAs from heat denaturation [326–328]. DNA binding by HMf and HMt in vitro does result in the formation of nucleosome-like structures but the DNA molecules in these structures are wrapped in positive toroidal supercoils and not in a negative supercoil, as in eucaryal nucleosomes [326,328].

8. Surface-layer glycoproteins

The surfaces of *Methanothermus fervidus* and *Methanothermus sociabilis*, two closely related hyperthermophilic methanogens, are covered by a layer of glycoprotein. The genes (*slgA*) encoding the protein components of these surface-layer (S-layer) glycoproteins have been cloned and sequenced, revealing only nine nucleotide differences, corresponding to three amino-acid differences [36]. These *slgA* genes encode polypeptides containing 593 amino-acid residues which possess amino-terminal sequences containing 22 amino-acid residues that are not present in the mature proteins. The predicted structures and probable cleavage sites for the removal of these signal peptides appear to be typically bacterial [39,298], indicating that translocation of these proteins to the cell surface may proceed via a pathway very similar to that established in bacteria.

Both *slgA* genes are preceded by readily identifiable Box A and Box B promoter elements that direct transcription initiation 33 bp upstream of GTG translation-initiation codons. In the intergenic regions upstream of the two *slgA* genes, 111 of 214 nucleotides are conserved, including conserved RBS, however there is virtually no sequence conservation downstream of the *slgA* genes. In *M. fervidus* the *slgA* gene appears to form a monocistronic transcriptional unit terminated by three oligo-T sequences located immediately downstream of the gene. This putative transcription terminator is not, however, present in the DNA sequence immediately downstream of the *slgA* gene in *M. sociabilis*.

The 571 amino-acid residues of the mature glycoproteins contribute a molecular mass of ~65 kDa, and with their carbohydrate moieties, composed of mannose, 3-*O*-methylglucose, galactose, *N*-acetylglucosamine and *N*-acetylgalactosamine [329], these glycoproteins have a total molecular mass of 76 kDa.

9. Flagellins

Although archaeal flagella contain several different flagellin proteins and may be glycosylated, they are much thinner than bacterial flagella [330–332]. Four related flagellin-encoding genes (*flaA, flaB1, flaB2* and *flaB3*) have been cloned and sequenced from *M. voltae* [38]. They are physically adjacent and appear to form two transcription units, *flaA* and *flaB1–flaB2–flaB3*. The encoded flagellins are predicted to be synthesized as precursors with short amino-terminal sequences that are not found in the mature proteins but which do not resemble bacterial signal peptides [39,298]. Although several potential sites for glycosylation are present in the sequences of these flagellins, their glycosylation has not been demonstrated directly [38].

10. Gene regulation and genetics

Practical difficulties inherent in growing and handling methanogens have severely limited studies of genetics and gene regulation. To obtain growth, O_2 must be scrupulously excluded and CO_2 and H_2 supplied, optimally, at above atmospheric pressure [333–337]. Alternatively, if a metabolically more versatile methanogen is studied, CO_2 can be replaced by methanol, methylamines or acetate, but as cells of these more versatile methanogens grow in clumps they have, until recently, been avoided by geneticists. Media have now been reported in which these methanogens do grow as single cells [338,339], and clump-disaggregating enzymes are now available [340], so that clumping may no longer be a limiting problem. Undertaking genetic experiments with *Methanobacterium thermoautotrophicum* strains, the methanogens most thoroughly studied biochemically, adds the practical problem that these are thermophiles. Standard microbial procedures become very difficult when attempted in the total absence of O_2, in the presence of pressurized CO_2 and H_2 and at temperatures above 55°C. Nevertheless, substantial progress has recently been made both in identifying genetically regulated systems and in developing genetic procedures [341–350].

10.1. Regulated systems of gene expression

M. voltae responds classically to heat shock by synthesizing, transiently, a limited number of heat-shock proteins [351], and a gene homologous to the heat-shock *dnaK* gene of *E. coli* has been cloned and sequenced from *Methanosarcina mazei* [389].

Opportunities therefore exist to study and exploit this well-established regulatory system in methanogens. Exposure of *M. thermolithotrophicus* to elevated hydrostatic pressures causes the synthesis of several novel proteins [352] which therefore also provides an environmentally regulated system to study gene expression. Several Methanobacteriales are now known to have two *mcr* operons which are expressed differently in response to growth conditions [95]. Investigating this regulation of MR synthesis and its coordination with the synthesis of the methylviologen-reducing hydrogenase [385], are now feasible and important projects. Similarly, determining how the synthesis of substrate-specific catabolic methyltransferases [353–355] and secondary alcohol dehydrogenases [356–359] are regulated and how acetate stimulates transcription of *cdhA* in *M. thermophila* [160] are attractive and tractable projects. The observation that *nif* gene expression is regulated in methanogens, as in bacteria, by the availability of fixed nitrogen [180,387] indicates that studies of this regulation might progress rapidly, especially if the large body of information already available from studies of bacterial *nif* systems is directly applicable to *nif* regulation in methanogens.

10.2. Transformation systems

Establishing transformation procedures, and selectable phenotypes to facilitate the introduction of DNA into methanogens, has been very time consuming. Methanogens are sensitive to only a limited number of antibiotics [360–368], some analogs of natural metabolites [369–375] and to heavy metals [376,377]. Mutants, spontaneously resistant to some of these inhibitors, have however been isolated and chromosomal DNA from these mutants has been used as the donor DNA to provide selectable phenotypes in the development of transformation protocols [343]. Auxotrophs have also been isolated [378–383], and chromosomal DNA from the wild-type strain then used as the donor DNA to transform the auxotrophic strain back to prototrophy [383]. Techniques to enrich for auxotrophs, analogous to the classical penicillin-enrichment technique used with bacteria, have been developed using bacitracin for *Methanobacterium* species [363,380] and base analogs for methanococcal species [382,383]. Chromosomal transformation has been demonstrated by adding DNA directly to cells of *Methanococcus voltae* PS [341] and *Methanococcus maripaludis* [350] in suspension and to cells of *Methanobacterium thermoautotrophicum* strain Marburg grown on the surface of a gellan gum solidified medium [345]. Electroporation has been used to increase the efficiency of chromosomal transformation of *M. voltae* [383]. Transformation resulting in the introduction and stable replication of an extrachromosomal element has not yet been reported. A puromycin acetyltransferase-encoding gene (*pac*), from *Streptomyces alboniger*, has been identified that confers puromycin resistance when introduced into *M. voltae* [347,384] and *M. maripaludis* [350]. Directed insertion of this gene into the genome, by homologous recombination using flanking sequences, has been developed as a procedure for insertional mutagenesis, but resistance plasmids, carrying this *pac* gene, have yet to be established in the cytoplasm of these methanogens.

11. Summary

As described in this chapter many methanogen genes have now been cloned and are therefore available for in vitro manipulation. With improved techniques to facilitate their reintroduction and expression in their native environments, investigating the consequences of such in vitro manipulations on methanogen gene expression in vivo should soon be possible. As methanogens play a central role in biotechnologies of such global importance as anaerobic waste treatment and biogas production [346], their continued study and "improvement" by genetic engineering is inevitable.

References

[1] Jones, W.J., Nagle Jr., D.P. and Whitman, W.B. (1987) Microbiol. Rev. 51, 135–177.
[2] Rouvière, P.E. and Wolfe, R.S. (1988) J. Biol. Chem. 263, 7913–7916.
[3] Huber, R., Kurr, M., Jannasch, H.W. and Stetter, K.O. (1989) Nature 342, 833–834.
[4] Kurr, M., Huber, R., Konig, H., Jannasch, H.W., Fricke, H., Trincone, A., Kristjansson, J.K. and Stetter, K.O. (1991) Arch. Microbiol. 156, 239–247.
[5] Balch, W.E., Fox, G.E., Magrum, L.J., Woese, C.R. and Wolfe, R.S. (1979) Microbiol. Rev. 43, 260–296.
[6] Burggraf, S., Stetter, K.O., Rouvière, P. and Woese, C.R. (1991) System. Appl. Microbiol. 14, 346–351.
[7] Mitchell, R.M., Loeblich, L.A., Klotz, L.C. and Loeblich III, A.R. (1979) Science 204, 1082–1084.
[8] Klein, A. and Schnorr, M. (1984) J. Bacteriol. 158, 628–631.
[9] Bollschweiler, C., Kühn, R. and Klein, A. (1985) EMBO J. 4, 805–809.
[10] Reeve, J.N., Hamilton, P.T., Beckler, G.S., Morris, C.J. and Clarke, C.H. (1986) System. Appl. Microbiol. 7, 5–12.
[11] Allmansberger, R., Knaub, S. and Klein, A. (1988) Nucl. Acids Res. 16, 7419–7436.
[12] Sitzmann, J. and Klein, A. (1991) Mol. Microbiol. 5, 505–513.
[13] Sapienza, C. and Doolittle, W.F. (1982) Nature 295, 384–389.
[14] Sapienza, C. and Doolittle, W.F. (1982) Zentralbl. Bakteriol. Parasitenkd. Infectionskr. Hyg. Abt. 1, Orig., C 3, 120–126.
[15] Sapienza, C., Rose, M.R. and Doolittle, W.F. (1982) Nature 299, 182–185.
[16] Hamilton, P.T. and Reeve, J.N. (1985) Mol. Gen. Genet. 200, 47–59.
[17] Jarrell, K.F., Julseth, C., Pearson, B. and Kuzio, J. (1987) Mol. Gen. Genet. 208, 191–194.
[18] Wood, A.G., Whitman, W.B. and Konisky, J. (1985) Arch. Microbiol. 142, 259–261.
[19] Zhao, H., Wood, A.G., Widdel, F. and Bryant, M.P. (1988) Arch. Microbiol. 150, 178–183.
[20] Sowers, K.R. and Gunsalus, R.P. (1988) J. Bacteriol. 170, 4979–4982.
[21] Thomm, M., Altenbuchner, J. and Stetter, K.O. (1983) J. Bacteriol. 153, 1060–1062.
[22] Wilharm, T., Thomm, M. and Stetter, K.O. (1986) System. Appl. Microbiol. 7, 401.
[23] Nölling, J., Frijlink, M. and DeVos, W.M. (1991) J. Gen. Microbiol. 137, 1981–1986.
[24] Meile, L., Keiner, A. and Leisinger, T. (1983) Mol. Gen. Genet. 191, 480–484.
[25] Meile, L. and Reeve, J.N. (1985) Bio/Technology 3, 69–72.
[26] Meile, A., Madon, J. and Leisinger, T. (1988) J. Bacteriol. 170, 478–481.
[27] Bokranz, M., Klein, A. and Meile, L. (1990) Nucl. Acids Res. 18, 363.
[28] Meile, L., Jenal, U., Studer, D., Jordan, M. and Leisinger, T. (1989) Arch. Microbiol. 152, 105–110.
[29] Jordan, M., Meile, L. and Leisinger, T. (1989) Mol. Gen. Genet. 220, 161–164.

[30] Wood, A.G., Whitman, W.B. and Konisky, J. (1989) J. Bacteriol. 171, 93–98.
[31] Zillig, W., Palm, P., Reiter, W.-D., Gropp, F., Pühler, G. and Klenk, H.-P. (1988) Eur. J. Biochem. 173, 473–482.
[32] Brown, J.W., Daniels, C.J. and Reeve, J.N. (1989) Crit. Rev. Microbiol. 16, 287–338.
[33] Reeve, J.N. (1989) In: Genetics and Molecular Biology of Industrial Microorganisms (Hershberger, C.L., Queener, S.W. and Hegeman, G., Eds.), pp. 207–214, American Society for Microbiology, Washington, D.C.
[34] Palmer, J.R. and Reeve, J.N. (1992) In: Genetics and Molecular Biology of Anaerobic Bacteria (Sebald, M., Ed.), pp. 13–35, Springer Verlag, New York.
[35] Brown, J.W. and Reeve, J.N. (1985) J. Bacteriol. 162, 909–917.
[36] Bröckl, G., Behr, M., Fabry, S., Hensel, R., Kaudewitz, H. and Biendl, E. (1991) Eur. J. Biochem. 199, 147–152.
[37] Dharmavaram, R., Gillevet, P. and Konisky, J. (1991) J. Bacteriol. 173, 2131–2133.
[38] Kalmokoff, M.L. and Jarrell, K.F. (1991) J. Bacteriol. 173, 7113–7125.
[39] Pollitt, S. and Inouye, M. (1987) In: Bacterial Outer Membranes as Model Systems (Inouye, M., Ed.), pp. 117–139, Wiley, New York.
[40] Auer, J., Spicker, G. and Böck, A. (1991) Biochimie (Paris) 73, 683–688.
[41] Thomm, M., Madon, J. and Stetter, K.O. (1986) Biol. Chem. Hoppe-Seyler 367, 473–481.
[42] Berghöfer, B., Kröckel, L., Körtner, C., Truss, M., Schallenberg, J. and Klein, A. (1988) Nucl. Acids Res. 16, 8113–8128.
[43] Wich, G., Leinfelder, W. and Böck, A. (1987) EMBO J. 6, 523–528.
[44] Thomm, M. and Wich, G. (1988) Nucl. Acids Res. 16, 151–163.
[45] Thomm, M., Sherf, B.A. and Reeve, J.N. (1988) J. Bacteriol. 170, 1958–1961.
[46] Brown, J.W., Thomm, M., Beckler, G.S., Frey, G., Stetter, K.O. and Reeve, J.N. (1988) Nucl. Acids Res. 16, 135–150.
[47] Allmansberger, R., Knaub, S. and Klein, A. (1988) Nucl. Acids Res. 16, 7419–7436.
[48] Thomm, M., Wich, G., Brown, J.W., Frey, G., Sherf, B.A. and Beckler, G.S. (1989) Can. J. Microbiol. 35, 30–35.
[49] Brown, J.W. and Reeve, J.N. (1989) FEMS Microbiol. Lett. 60, 131–136.
[50] Hausner, W., Frey, G. and Thomm, M. (1991) J. Mol. Biol. 222, 495–508.
[51] Frey, G., Thomm, M., Brüdigam, B., Gohl, H.P. and Hausner, W. (1991) Nucl. Acids Res. 18, 1361–1367.
[52] Knaub, S. and Klein, A. (1990) Nucl. Acids Res. 18, 1441–1446.
[53] Thomm, M., Frey, G., Hausner, W. and Brüdigam, B. (1990) In: Microbiology and Biochemistry of Strict Anaerobes Involved in Interspecies Transfer (Belaich, J.-P., Bruschi, M. and Garcia, J.L., Eds.), pp. 305–312, Plenum Press, New York.
[54] Müller, B., Allmansberger, R. and Klein, A. (1985) Nucl. Acids Res. 13, 6439–6445.
[55] Allmansberger, R., Bollschweiler, C., Konheiser, U., Müller, B., Muth, E., Pasti, G. and Klein, A. (1986) System. Appl. Microbiol. 7, 13–17.
[56] Cram, D.S., Sherf, B.A., Libby, R.T., Mattaliano, R.J. Ramachandran, K.L. and Reeve, J.N. (1987) Proc. Natl. Acad. Sci. U.S.A. 84, 3992–3996.
[57] Bokranz, M, Bäumner, G., Allmansberger, R., Ankel-Fuchs, D. and Klein, A. (1988) J. Bacteriol. 170, 568–577.
[58] Morris, C.J. and Reeve, J.N. (1988) J. Bacteriol. 170, 3125–3130.
[59] Auer, J., Spicker, G. and Böck, A. (1989) J. Mol. Biol. 209, 21–36.
[60] Possot O., Sibold, L. and Aubert, J.-P. (1989) Res. Microbiol. 140, 335–371.
[61] Weil, C.F., Sherf, B.A. and Reeve, J.N. (1989) Can. J. Microbiol. 35, 101–108.
[62] DiMarco, A.A., Sment, K.A., Konisky, J. and Wolfe, R.S. (1990) J. Biol. Chem. 265, 472–476.
[63] Takao, M., Oikawa, A. and Yasui, A. (1990) Arch. Biochem. Biophys. 283, 210–216.
[64] Dharmavaram, R., Gillevet, P. and Konisky, J. (1991) J. Bacteriol. 173, 2131–2133.

[65] Wich, G., Hummel, H., Jarsch, M., Bär, U. and Böck, A. (1986) Nucl. Acids Res. 14, 2459–2479.
[66] Wich, G., Sibold, L. and Böck, A. (1986) System. Appl. Microbiol. 7, 18–25.
[67] Weil, C.F., Cram, D.S., Sherf, B.A. and Reeve, J.N. (1988) J. Bacteriol. 170, 4718–4726.
[68] Haas, E.S., Daniels, C.J. and Reeve, J.N. (1989) Gene 77, 253–263.
[69] Alex, L.A., Reeve, J.N., Orme-Johnson, W.H. and Walsh, C.T. (1990) Biochem. 29, 7237–7244.
[70] Eggen, R.I.L., Geerling, A.C.M., Boshoven, A.B.P. and DeVos, W.M. (1991) J. Bacteriol. 173, 6383–6389.
[71] Kjems, J., Leffers, H., Garrett, R.A., Wich, G., Leinfelder, W. and Böck, A. (1987) Nucl. Acids Res. 15, 4821–4835.
[72] Kjems, J. and Garrett, R.A. (1988) EMBO J. 6, 3521–3530.
[73] Reeve, J.N., Trun, N.J. and Hamilton, P.T. (1982) In: Genetic Engineering of Microorganisms for Chemicals (Hollaender, A., DeMoss, R.D., Kaplan, S., Konisky, J., Savage, D. and Wolfe, R.S., Eds.), pp. 233–244, Plenum, New York.
[74] Ellerman, J., Hedderich, R., Böcher, R. and Thauer, R.K. (1988) Eur. J. Biochem. 172, 669–677.
[75] Rouvière, P.E., Bobik, T.A. and Wolfe, R.S. (1988) J. Bacteriol. 170, 3946–3952.
[76] Ellermann, J., Rospert, S., Thauer, R.K., Bokranz, M., Klein, A., Voges, M. and Berkessel, A. (1989) Eur. J. Biochem. 184, 63–68.
[77] Rospert, S., Breitung, J., Ma, K., Schwörer, B., Zirngibl, C., Thauer, R.K., Linder, D., Huber, R. and Stetter, K.O. (1991) Arch. Microbiol. 156, 49–55.
[78] Noll, K.M., Rinehart Jr., K.L., Tanner, R.S. and Wolfe, R.S. (1986) Proc. Natl. Acad. Sci. U.S.A. 83, 4238–4242.
[79] Bobik, T.A., Olson, K.D., Noll, K.M. and Wolfe, R.S. (1987) Biochem. Biophys. Res. Commun. 149, 455–460.
[80] Ellermann, J., Kobelt, A., Pfaltz, A. and Thauer, R.K. (1987) FEBS Lett. 220, 358–362.
[81] Noll, K.M., Donnelly, M.I. and Wolfe, R.S. (1987) J. Biol. Chem. 262, 513–515.
[82] Bobik, T.A. and Wolfe, R.S. (1988) Proc. Natl. Acad. Sci. U.S.A. 85, 60–63.
[83] Hauska, G. (1988) Trends Biochem. Sci. 13, 2–4.
[84] Hedderich, R. and Thauer, R.K. (1988) FEBS Lett. 234, 223–227.
[85] Hedderich, R., Berkessel, A. and Thauer, R.K. (1989) FEBS Lett. 255, 67–71.
[86] Deppenmeier, U., Blaut, M., Mahlmann, A. and Gottschalk, G. (1990) Proc. Natl. Acad. Sci. U.S.A. 87, 9449–9453.
[87] Ellefson, W.L. and Wolfe, R.S. (1981) J. Biol. Chem. 256, 4259–4262.
[88] Klein, A., Allmansberger, R., Bokranz, M., Knaub, S., Müller, B. and Muth, E. (1988) Mol. Gen. Genet. 213, 409–420.
[89] Bokranz, M. and Klein, A. (1987) Nucl. Acids Res. 15, 4350–4351.
[90] Sherf, B.A. and Reeve, J.N. (1990) J. Bacteriol. 172, 1828–1833.
[91] Thomas, L.K., Dix, D.B. and Thompson, R.C. (1988) Proc. Natl. Acad. Sci. U.S.A. 85, 4242–4246.
[92] Woese, C.R. and Fox, G.E. (1977) Proc. Natl. Acad. Sci. U.S.A. 74, 5088–5090.
[93] Fox, G.E., Stackebrandt, E., Hespell, R.B., Gibson, J., Maniloff, J., Dyer, T.A., Wolfe, R.S., Balch, W.E., Tanner, R.S., Magrum, L.J., Zablen, L.B., Blakemore, R., Gupta, R., Bonen, L., Lewis, B.J., Stahl, D.A., Luehrsen, K.R., Chen, K.N. and Woese, C.R. (1980) Science 209, 457–463.
[94] Hartzell, P.L. and Wolfe, R.S. (1986) System. Appl. Microbiol. 7, 376–382.
[95] Rospert, S., Linder, D., Ellermann, J. and Thauer, R.K. (1990) Eur. J. Biochem. 194, 871–877.
[96] Jacobson, F.S., Daniels, L., Fox, J.A., Walsh, C.T. and Orme-Johnson, W.H. (1982) J. Biol. Chem. 257, 3385–3388.
[97] Yamazaki, S. (1982) J. Biol. Chem. 257, 7926–7929.
[98] Catherine Jin, S.-L., Blanchard, D.K. and Chen, J.-S. (1983) Biochim. Biophys. Acta 748, 8–20.

[99] Kojima, N., Fox, J.A., Hausinger, R.P., Daniels, L., OrmeJohnson, W.H. and Walsh, C. (1983) Proc. Natl. Acad. Sci. U.S.A. 80, 378–382.
[100] Fauque, G. (1989) In: Microbiology of Extreme Environments and Its Potential for Biotechnology (Costa, M.S., Da Duarte, J.C. and Williams, R.A.D., Eds.), pp. 216–236, Elsevier, Amsterdam.
[101] Fox, J.A., Livingston, D.J., Orme-Johnson, W.H. and Walsh, C.T. (1987) Biochem. 26, 4219–4227.
[102] Hausinger, R.P. (1987) Microbiol. Rev. 51, 22–42.
[103] Muth, E., Mörschel, E. and Klein, A. (1987) Eur. J. Biochem. 169, 571–577.
[104] Sprott, G.D., Shaw, K.N. and Beveridge, T.J. (1987) Can. J. Microbiol. 33, 896–904.
[105] Wackett, L.P., Hartwieg, E.A., King, J.A., Orme-Johnson, W.H. and Walsh, C.T. (1987) J. Bacteriol. 169, 718–727.
[106] Baron, S.F., Williams, D.S., May, H.D. Patel, P.S., Aldrich, H.C. and Ferry, J.G. (1989) Arch. Microbiol. 151, 307–313.
[107] Fiebig, K. and Friedrich, B. (1989) Eur. J. Biochem. 184, 79–88.
[108] Bhosale, S.B., Yeole, T.Y. and Kshirsagar, D.C. (1990) FEMS Microbiol. Lett. 70, 241–248.
[109] Reeve, J.N. and Beckler, G.S. (1990) FEMS Microbiol. Rev. 87, 419–424.
[110] Shah, N.N. and Clark, D.S. (1990) Appl. Environ. Microbiol. 56, 858–863.
[111] Lünsdorf, H., Niedrig, M. and Fiebig, K. (1991) J. Bacteriol. 173, 978–984.
[112] Reeve, J.N., Beckler, G.S., Cram, D.S., Hamilton, P.T., Brown, J.W., Krzycki, J.A., Kolodziej, A.F., Alex, L., Orme-Johnson, W.H. and Walsh, C.T. (1989) Proc. Natl. Acad. Sci. U.S.A. 86, 3031–3035.
[113] Steigerwald, V.J., Beckler, G.S. and Reeve, J.N. (1990) J. Bacteriol. 172, 4715–4718.
[114] Menon, N.K., Robbins, J., Peck, H.D., Chatelus, C.Y., Choi, E.S. and Przybyla, A.E. (1990) J. Bacteriol. 172, 1969–1977.
[115] Prickril, B.C., Czechowski, M.H., Przybyla, A.E., Peck Jr., H.D. and LeGall, J. (1986) J. Bacteriol. 167, 722–725.
[116] Menon, N.K., Peck Jr., H.D., LeGall, J. and Przybyla, A.E. (1987) J. Bacteriol. 169, 5401–5407. Correction (1988) J. Bacteriol. 170, 4429.
[117] Li, C., Peck Jr., H.D., LeGall, J. and Przybyla, A.E. (1987) DNA 6, 539–551.
[118] Leclerc, A., Colbeau, A., Cauvin, B. and Vignais, P.M. (1988) Mol. Gen. Genet. 214, 97–107.
[119] Mayer, F., Rohde, M., Salzmann, M., Jussofie, A. and Gottschalk, G. (1988) J. Bacteriol. 170, 1438–1444.
[120] Hoppert, M. and Mayer, F. (1990) FEBS Lett. 267, 33–37.
[121] Rik, I.L.E., Geerling, A.C.M., Jetten, M.S.M. and de Vos, W.M. (1991) J. Biol. Chem. 266, 1–5.
[122] Voordouw, G. and Brenner, S. (1985) Eur. J. Biochem. 148, 515–520.
[123] Otaka, E. and Ooi, T. (1987) J. Mol. Evol. 26, 257–267.
[124] Hausinger, R.P., Moura, I., Moura, J.J.G., Xavier, A.V., Santos, M.H., LeGall, J. and Howard, J.B. (1982) J. Biol. Chem. 257, 14192–14197.
[125] Hatchikian, E.C., Bruschi, M., Forget, N. and Scandellari, M. (1982) Biochem. Biophys. Res. Commun. 109, 1316–1323.
[126] Moura, J., Moura, J.J.G., Huynh, B.H., Santos, H., LeGall, J. and Xavier, A.V. (1982) Eur. J. Biochem. 126, 95–101.
[127] Terlesky, K.C. and Ferry, J.G. (1988) J. Biol. Chem. 263, 4080–4082.
[128] Hatchikian, E.C., Fardeau, M.L., Bruschi, M., Belaich, J.P., Chapman, A. and Cammack, R. (1989) J. Bacteriol. 171, 2384–2390.
[129] Bruschi, M., Bonicel, J., Hatchikian, E.C., Fardeau, M.L., Belaich, J.P. and Frey, M. (1991) Biochim. Biophys. Acta 1076, 79–85.
[130] Terlesky, K.C. and Ferry, J.G. (1988) J. Biol. Chem. 263, 4075–4079.

[131] Jarrell, K.F. and Kalmokoff, M.L. (1988) Can. J. Microbiol. 34, 557–576.
[132] Benstead, J., Archer, D.B. and Lloyd D. (1991) Arch. Microbiol. 156, 34–37.
[133] May, H.D., Patel, P.S. and Ferry, J.G. (1988) J. Bacteriol. 170, 3384–3389.
[134] Shuber, A.P., Orr, E.C., Recny, M.A., Schendel, P.F., May, H.D., Schauer, N.L. and Ferry, J.G. (1986) J. Biol. Chem. 261, 12942–12947.
[135] Patel, P.S. and Ferry, J.G. (1988) J. Bacteriol. 170, 3390–3395.
[135a] White, W.B. and Ferry, J.G. (1992) J. Bacteriol. 174, 4997–5004.
[136] Jones, J.B. and Stadtman T.C. (1977) J. Bacteriol. 130, 1404–1406.
[137] Jones, J.B. and Stadtman, T.C. (1981) J. Biol. Chem. 256, 656–663.
[138] Birkmann, A., Zinoni, F., Sawers, G. and Böck, A. (1987) Arch. Microbiol. 148, 44–51.
[139] Sparling, R. and Daniels, L. (1990) J. Bacteriol. 172, 1464–1469.
[140] DiMarco, A.A., Bobik, T.A. and Wolfe, R.S. (1990) Annu. Rev. Biochem. 59, 355–394.
[141] Donnely, M.I. and Wolfe, R.S. (1986) J. Biol. Chem. 261, 16653–16659.
[142] Kohler, H.-P.E. and Zehnder, A.J.B. (1984) FEMS Microbiol. Lett. 21, 287–292.
[143] Pellerin, P., Bruson, B., Prensier, G., Albagnac, G. and Debeire, P. (1987) J. Bacteriol. 146, 377–381.
[144] Jetten, M.S.M., Stams, A.J.M. and Zehnder, A.J.B. (1989) J. Bacteriol. 171, 5430–5435.
[145] Laufer, K., Eikmanns, B., Frimmer, U. and Thauer, R.K. (1987) Z. Naturforsch. 42c, 360–372.
[146] Terlesky, K.C., Barber, M.J., Aceti, D.J. and Ferry, J.G. (1987) J. Biol. Chem. 262, 15392–15395.
[147] Aceti, D.J. and Ferry, J.G. (1988) J. Biol. Chem. 263, 15444–15448.
[148] Fischer, R. and Thauer, R.K. (1988) FEBS Lett. 228, 249–253.
[149] Lundie Jr., L.L. and Ferry, J.G. (1989) J. Biol. Chem. 264, 18392–18396.
[150] Vogels, G.D. and Visser, C.M. (1983) FEMS Microbiol. Lett. 20, 291–297.
[151] Eikmanns, B. and Thauer, R.K. (1984) Arch. Microbiol. 138, 365–370.
[152] Eikmanns, B. and Thauer, R.K. (1985) Arch. Microbiol. 142, 175–179.
[153] Krzycki, J.A., Lehman, L.J. and Zeikus, J.G. (1985) J. Bacteriol. 163, 1000–1006.
[154] Terlesky, K.C., Nelson, M.J.K. and Ferry, J.G. (1986) J. Bacteriol. 168, 1053–1058.
[155] Bhatnagar, L., Krzycki, J.A. and Zeikus, J.G. (1987) FEMS Microbiol. Lett. 337–343.
[156] Krzycki, J.A. and Prince, R.C. (1990) Biochim. Biophys. Acta 1015, 53–60.
[157] Abbanat, D.R. and Ferry, J.G. (1990) J. Bacteriol. 172, 7145–7150.
[158] Raybuck, S.A., Ramer, S.E., Abbanat, D.R., Peters, J.W., Orme-Johnson, W.H., Ferry, J.G. and Walsh, C.T. (1991) J. Bacteriol. 173, 929–932.
[159] Eggen, R.I.L., Geerling, A.C.M., Jetten, M.S.M. and DeVos, W.M. (1991) J. Biol. Chem. 266, 6883–6887.
[160] Sowers, K.R., Thai, T.T. and Gunsalus, R.P. (1992) J. Biol. Chem., submitted for publication.
[161] Okazaki, K., Takechi, T., Kambara, N., Fukui, S., Kuboto, I. and Kamiryo, T. (1986) Proc. Natl. Acad. Sci. U.S.A. 83, 1232–1236.
[162] Murray, W.W. and Rachubinski, R.A. (1987) Gene 51, 119–128.
[163] Wood, A.G., Redborg, A.H., Cue, D.R., Whitman, W.B. and Konisky, J. (1983) J. Bacteriol. 156, 19–29.
[164] Morris, C.J. and Reeve, J.N. (1984) In: Microbial Growth on C1 Compounds (Crawford, R.L. and Hanson, R.S., Eds.), pp. 205–209, American Society for Microbiology, Washington, D.C.
[165] Cue, D., Beckler, G.S., Reeve, J.N. and Konisky, J. (1985) Proc. Natl. Acad. Sci. U.S.A. 82, 4207–4211.
[166] Hamilton, P.T. and Reeve, J.N. (1985) J. Mol. Evol. 22, 351–360.
[167] Beckler, G.S. and Reeve, J.N. (1986) Mol. Gen. Genet. 204, 133–140.
[168] Weil, C.F., Beckler, G.S. and Reeve, J.N. (1987) J. Bacteriol. 169, 4857–4860.
[169] Sibold, L. and Henriquet, M. (1988) Mol. Gen. Genet. 214, 439–450.
[170] Meile, L., Stettler, R., Banholzer, R., Kotik, M. and Leisinger, T. (1991) J. Bacteriol. 173, 5017–5023.

[171] Sment, K.A. and Konisky, J. (1986) System. Appl. Microbiol. 7, 90–94.
[172] Bruni, C.B., Carlomagno, M.S., Formisano, S., Paolella, G. and Fink, G.R. (1986) Mol. Gen. Genet. 203, 389–396.
[173] Van Vliet, F., Crabeel, M., Boyen, A., Tricot, C., Stalon, V., Falmagne, P., Nakamura, Y., Baumberg, S. and Glansdorff, N. (1990) Gene 95, 99–104.
[174] Freytag, S.O., Beaudet, A.L., Böck, H.G.O. and O'Brien, W.F. (1984) Mol. Cell. Biol. 4, 1978–1984.
[175] Crawford, I.P. (1989) Annu. Rev. Microbiol. 43, 567–600.
[176] Lam, W.L., Cohen, A., Tsouluhas, D. and Doolittle, W.F. (1990) Proc. Natl. Acad. Sci. U.S.A. 87, 6614–6618.
[177] Zalkin, H. and Yanofsky, C. (1982) J. Biol. Chem. 257, 1491–1500.
[178] Yanofsky, C., Platt, T., Crawford, I.P., Nichols, B.P., Christie, G.E., Horowitz, H., Van Cleeput, M. and Wu, A.M. (1981) Nucl. Acids Res. 9, 6647–6668.
[179] Matsui, K., Miwa, K. and Sano, K. (1987) J. Bacteriol. 169, 5330–5332.
[180] Souillard, N. and Sibold, L. (1989) Mol. Microbiol. 3, 541–551.
[181] Tumer, N.E., Robison, S.J. and Haselkorn, R. (1983) Nature 306, 337–342.
[182] Rawlings, D.E., Jones, W.A., O'Neill, E.G. and Woods, D.R. (1987) Gene 53, 211–217.
[183] Strauch, M.A., Aronson, A.L., Brown, S.W., Schreier, H.J. and Sonenshein, A.L. (1988) Gene 71, 257–265.
[184] Janssen, P.J., Jones, W.A., Jones, D.T. and Woods, D.R. (1988) J. Bacteriol. 170, 400–408.
[185] Tischer, E., DasSarma, S. and Goodman, H.M. (1986) Mol. Gen. Genet. 203, 221–229.
[186] Gebhardt, C., Oliver, J.E., Forde, B.G., Saarelainen, R. and Miflin, B.J. (1986) EMBO J. 5, 1429–1435.
[187] Hayward, B.E., Hussain, A., Wilson, R.H., Lyons, A., Woodcock, V., McIntosh, B. and Harris, T.J.R. (1986) Nucl. Acids Res. 14, 999–1008.
[188] Bhatnagar, L., Jain, M.K., Zeikus, J.G. and Aubert, J.-P. (1986) Arch. Microbiol. 144, 350–354.
[189] Tiedeman, A.A., Keyhani, J., Kamholz, J., Daum III, M.A., Gots, J.S. and Smith, J.M. (1989) J. Bacteriol. 171, 205–212.
[190] Jenal, U., Rechsteiner, T., Tan, P.-Y., Bühlmann, E., Meile, L. and Leisinger, T. (1991) J. Biol. Chem. 266, 10570–10577.
[191] Ebbole, D.J. and Zalkin, H. (1987) J. Biol. Chem. 262, 8274–8287.
[192] Jarsch, M., Altenbuchner, J. and Böck, A. (1983) Mol. Gen. Genet. 189, 41–47.
[193] Jarsch, M. and Böck, A. (1983) Nucl. Acids Res. 11, 7537–7544.
[194] Wich, G., Jarsch, M. and Böck, A. (1984) Mol. Gen. Genet. 196, 146–151.
[195] Jarsch, M. and Böck, A. (1985) Mol. Gen. Genet. 200, 305–312.
[196] Jarsch, M. and Böck, A. (1985) System. Appl. Microbiol. 6, 54–59.
[197] Lechner, K., Wich, G. and Böck, A. (1985) System. Appl. Microbiol. 6, 157–163.
[198] Østergaard, L., Larsen, N., Leffers, H., Kjems, J. and Garrett, R. (1987) System. Appl. Microbiol. 9, 199–209.
[199] Wich, G., Sibold, L. and Böck, A. (1987) Z. Naturforsch. 42c, 373–380.
[200] Achenbach-Richter, L. and Woese, C.R. (1988) System. Appl. Microbiol. 10, 211–214.
[201] Eggen, R., Harmsen, H., Geerling, A. and de Vos, W.M. (1989) Nucl. Acids Res. 17, 9469.
[202] Eggen, R., Harmsen, H. and de Vos, W.M. (1990) Nucl. Acids Res. 18, 1306.
[203] Haas, E.S., Brown, J.W., Daniels, C.J. and Reeve, J.N. (1990) Gene 90, 51–59.
[203a] Palmer, J.R., Baltrus, T., Reeve, J.N. and Daniels, C.J. (1992) Biochim. Biophys. Acta 1132, 315–318.
[204] Daniels, C.J., Gupta, R. and Doolittle, W.F. (1985) J. Biol. Chem. 260, 3132–3134.
[205] Thompson, L.D., Brandon, L.D., Nieuwlandt, D.T. and Daniels, C.J. (1989) Can. J. Microbiol. 35, 36–42.
[206] Kjems, J. and Garrett, R.A. (1985) Nature 318, 675–677.

[207] Kaine, B.P. (1987) J. Mol. Evol. 25, 248–256.
[208] Kjems, J. and Garrett, R.A. (1991) Proc. Natl. Acad. Sci. U.S.A. 88, 439–443.
[209] Edmonds, C.G., Crain, P.F., Gupta, R., Hashizume, T., Hocart, C.H., Kowalak, J.A., Pomerantz, S.C., Stetter, K.O. and McCloskey, J.A. (1991) J. Bacteriol. 173, 3138–3148.
[210] Politino, M., Tsai, L., Veres, Z. and Stadtman, T.C. (1990) Proc. Natl. Acad. Sci. U.S.A. 87, 6345–6348.
[211] Garrett, R.A., Dalgaard, J., Larsen, N., Kjems, J. and Mankin, A.S. (1991) Trends Biochem. Sci. 16, 22–26.
[212] Luehrsen, K.R., Nicholson Jr., D.E. and Fox, G.E. (1985) Curr. Microbiol. 12, 69–72.
[213] Kaine, B.P. and Merkel, V.L. (1989) J. Bacteriol. 171, 4261–4266.
[214] Kaine, B.P. (1990) Mol. Gen. Genet. 221, 315–321.
[215] Zabel, H.P., Fischer, H., Holler, E. and Winter, J. (1985) System. Appl. Microbiol. 6, 111–118.
[216] Gropp, F., Reiter, W.D., Sentenac, A., Zillig, W., Schnabel, R., Thomm, M. and Stetter, K.O. (1986) System. Appl. Microbiol. 7, 95–101.
[217] Thomm, M., Lindner, A.J., Hartmann, G.R. and Stetter, K.O. (1988) System. Appl. Microbiol. 10, 101–105.
[218] Zillig, W., Klenk, H.-P., Palm, P., Pühler, G., Gropp, F., Garrett, R.A. and Leffers, H. (1989) Can. J. Microbiol. 35, 73–80.
[219] Iwabe, N., Kuma, K., Kishino, H., Hasegawa, M. and Miyata, T. (1991) J. Mol. Evol. 32, 70–78.
[220] Inatomi, K.-I., Eya, S., Maeda, M. and Futai, M. (1989) J. Biol. Chem. 264, 10954–10959.
[221] Honka, E., Fabry, S., Niermann, T., Palm, P. and Hensel, R. (1990) Eur. J. Biochem. 188, 623–632.
[222] Fabry, S., Heppner, P., Dietmaier, W. and Hensel, R. (1990) Gene, 91, 19–25.
[223] Souillard, N., Magot, M., Possot, O. and Sibold, L. (1988) J. Mol. Evol. 27, 65–76.
[224] Zaychikov, E.F., Mustaev, A.A., Glaser, S.J., Thomm, M., Grachev, M.A. and Hartmann, G.R. (1990) System. Appl. Microbiol. 13, 248–254.
[225] Post, L.E., Strycharz, G.D., Nomura, M., Lewis, H. and Dennis, P.P. (1979) Proc. Natl. Acad. Sci. U.S.A. 76, 1697–1701.
[226] Itoh, T. (1988) Eur. J. Biochem. 176, 297–303.
[227] Shimmin, C., Newton, C.H., Ramirez, C., Yee, J., Downing, W.L., Louie, A., Matheson, A.T. and Dennis, P.P. (1989) Can. J. Microbiol. 35, 164–170.
[228] Ramirez, C., Shimmin, C., Newton, C.H., Matheson, A.T. and Dennis, P.P. (1989) Can. J. Microbiol. 35, 234–244.
[229] Baier, G., Piendl, W., Redl, B. and Stöffler, G. (1990) Nucl. Acids Res. 18, 719–724.
[230] Shimmin, C. and Dennis, P.P. (1989) EMBO J. 8, 1225–1235.
[231] Dean, D. and Nomura, M. (1980) Proc. Natl. Acad. Sci. U.S.A. 77, 3590–3594.
[232] Yates, J.L. and Nomura, M. (1981) Cell 24, 243–249.
[233] Lechner, K., Heller, G. and Böck, A. (1989) J. Mol. Evol. 29, 20–27.
[234] Auer, J., Lechner, K. and Böck, A. (1989) Can. J. Microbiol. 35, 200–204.
[235] Lechner, K., Heller, G. and Böck, A. (1988) Nucl. Acids Res. 16, 7817–7826.
[236] Lechner, K. and Böck, A. (1987) Mol. Gen. Genet. 208, 523–528.
[237] Strobel, O., Köpke, A.K.E., Kamp, R.M., Böck, A. and Wittmann-Liebold, B. (1988) J. Biol. Chem. 263, 6538–6546.
[238] Köpke, A.K.E. and Wittman-Liebold, B. (1989) Can. J. Microbiol. 35, 11–20.
[239] Auer, J., Spicker, G. and Böck, A. (1990) System. Appl. Microbiol. 13, 354–360.
[240] Köpke, A.K.E., Paulke, C. and Gewitz, H.S. (1990) J. Biol. Chem. 265, 6436–6440.
[241] Kamio, Y., Lin, C.-K., Regue, M. and Wu, M.C. (1985) J. Biol. Chem. 260, 5616–5620.
[242] Miller, K.W., Bouvier, J., Stragier, P. and Wu, H.C. (1987) J. Biol. Chem. 262, 7391–7397.
[243] Isaki, L., Kawakami, M., Beers, R., Hom, R. and Wu, H.C. (1990) J. Bacteriol. 172, 469–472.
[244] Isaki, L., Beers, R. and Wu, H.C. (1990) J. Bacteriol. 172, 6512–6517.

[245] Martindale, D.W., Gu, Z.M. and Csank, C. (1989) Curr. Genet. 15, 99–106.
[246] Burbaum, J.J. and Schimmel, P. (1991) J. Biol. Chem. 266, 16956–16968.
[247] Heck, J.D. and Hatfield, G.W. (1988) J. Biol. Chem. 263, 868–877.
[248] Schimmel, P. (1987) Annu. Rev. Biochem. 56, 125–158.
[249] Hountondji, C., Dessen, P. and Blanquet, S. (1986) Biochimie (Paris) 68, 1071–1078.
[250] Hountondji, C., Schmitter, J.-M., Beauvallet, C. and Blanquet, S. (1990) Biochemistry 29, 8190–8198.
[251] Myers, A.M. and Tzagloff, A. (1985) J. Biol. Chem. 260, 15371–15377.
[252] Belay, N., Sparling, R. and Daniels, L. (1984) Nature 312, 286–288.
[253] Murray, P.A. and Zinder, S.H. (1984) Nature 312, 284–286.
[254] Bomar, M., Knoll, K. and Widdel, F. (1985) FEMS Microbiol. Ecol. 31, 47–55.
[255] Fardeau, M.-L., Peillex, J.-P. and Belaich, J.-P. (1987) Arch. Microbiol. 148, 128–131.
[256] Lobo, A.L. and Zinder, S.H. (1988) Appl. Environ. Microbiol. 54, 1656–1661.
[257] Lobo, A.L. and Zinder, S.H. (1990) J. Bacteriol. 172, 6789–6796.
[258] Magingo, F.S.S. and Stumm, C.K. (1991) FEMS Microbiol. Lett. 81, 273–278.
[259] Sibold, L., Pariot, D., Bhatnagar, L., Henriquet, M. and Aubert, J.-P. (1985) Mol. Gen. Genet. 200, 40–46.
[260] Possot, O., Henry, M. and Sibold, L. (1986) FEMS Microbiol. Lett. 34, 173–177.
[261] Magot, M., Possot, O., Souillard, N., Henriquet, M., and Sibold, L. (1986) In: Biology of Anaerobic Bacteria (DuBourguier, H.C., Albagnac, G., Montreuil, J., Romond, C., Sautière, P. and Guillaume, J., Eds.), pp. 193–199, Elsevier, Amsterdam.
[262] Sibold, L., Henriquet, M., Possot, O. and Aubert, J.-P. (1991) Res. Microbiol. 142, 5–12.
[263] Souillard, N. and Sibold, L. (1986) Mol. Gen. Genet. 203, 21–28.
[264] Colonna-Romano, S., Riccio, A., Guida, M., Defez, R., Lamberti, A., Iaccarino, M., Arnold, W., Priefer, U. and Pühler, A. (1987) Nucl. Acids Res. 15, 1951–1964.
[265] Son, H.S. and Rhee, S.G. (1987) J. Biol. Chem. 262, 8690–8695.
[266] Holtel, A. and Merrick, M. (1988) Mol. Gen. Genet. 215, 134–138.
[267] Martin, G.B., Tomashow, M.F. and Chelm, B.K. (1989) J. Bacteriol. 171, 5638–5645.
[268] Kranz, R.G., Pace, V.M. and Caldicott, I.M. (1990) J. Bacteriol. 172, 53–62.
[269] De Zamaroczy, M., Delorme, F. and Elmerich, C. (1990) Mol. Gen. Genet. 224, 421–430.
[270] Dixon, R. (1984) J. Gen. Microbiol. 130, 2745–2755.
[271] Sibold, L. and Souillard, N. (1988) In: Nitrogen Fixation: Hundred Years After (Bothe, H., de Bruijn, F.J. and Newton, W.E., Eds.), pp. 705–710, Gustav-Fischer Verlag, Stuttgart.
[272] Scott, K.F., Rolfe, G.B. and Shine, J. (1983) DNA 2, 149–155.
[273] Bishop, P.E., Premakumar, R., Joerger, R.D., Jacobson, M.R., Dalton, D.A., Chisnell, J.R. and Wolfinger, E.D. (1988) In: Nitrogen Fixation: Hundred Years After (Bothe, H., de Bruijn, F.J. and Newton, W.E., Eds.), pp. 71–79, Gustav-Fischer Verlag, Stuttgart.
[274] Fabry, S., Lang, J., Niermann, T. Vingron, M. and Hensel, R. (1989) Eur. J. Biochem. 179, 405–413.
[275] Fabry, S. and Hensel, R. (1988) Gene 64, 189–197.
[276] Zwickl, P., Fabry, S., Bogedain, C., Haas, A. and Hensel, R. (1990) J. Bacteriol. 172, 4329–4338.
[277] Fabry, S., Lehmacher, A., Bode, W. and Hensel, R. (1988) FEBS Lett. 237, 213–217.
[278] Biro, J., Fabry, S., Dietmaier, W., Bogedain, C. and Hensel, R. (1990) FEBS Lett. 275, 130–134.
[279] Mevarech, M., Eisenberg, H. and Neumann, E. (1977) Biochem. 16, 3781–3785.
[280] Sprott, G.D., McKellar, R.C., Shaw, K.M., Giroux, J. and Martin, W.G. (1979) Can. J. Microbiol. 25, 192–200.
[281] Grossebüter, W., Hartl, T., Görisch, H. and Stezowski, K. (1986) Biol. Chem. Hoppe-Seyler 367, 457–463.

[282] Hartl, T., Grossebüter, W., Görisch, H. and Stezowski, K. (1987) Biol. Chem. Hoppe-Seyler 368, 259–267.
[283] Görisch, H. and Jany, K.-D. (1989) FEBS Lett. 247, 259–262.
[284] Zimniak, L., Dittrich, P., Gogarten, J.P., Kibak, H. and Taiz, L. (1988) J. Biol. Chem. 263, 9102–9112.
[285] Bowmann, E.J., Tenney, K. and Bowman, B.J. (1988) J. Biol. Chem. 263, 13994–14001.
[286] Bowman, B.J., Allen, R., Wechser, M.A. and Bowman, E.J. (1988) J. Biol. Chem. 263, 14002–14007.
[287] Nelson, H., Mandiyan, S. and Nelson, N. (1989) J. Biol. Chem. 264, 1775–1778. Correction (1989) J. Biol. Chem. 264, 5313.
[288] Manolson, M.F., Ouellette, B.F.F., Filion, M. and Poole, R.J. (1988) J. Biol. Chem. 263, 17987–17994.
[289] Denda, K., Konishi, J., Oshima, T., Date, T. and Yoshida, M. (1988) J. Biol. Chem. 263, 6012–6015.
[290] Denda, K., Kohishi, J., Oshima, T., Date, T. and Yoshida, M. (1988) J. Biol. Chem. 263, 17251–17254.
[291] Lübben, M. and Schäfer, G. (1987) Eur. J. Biochem. 164, 533–540.
[292] Kohishi, J., Wakagi, T., Oshima, T. and Yoshida, M. (1987) J. Biochem. (Tokyo) 102, 1379–1387.
[293] Gay, N.J. and Walker, J.E. (1981) Nucl. Acids Res. 9, 3919–3926.
[294] Kanazawa, H., Kayano, T., Mabuchi, K. and Futai, M. (1981) Biochem. Biophys. Res. Commun. 103, 604–612.
[295] Saraste, M., Gay, N.J., Eberle, A., Runswick, M.J. and Walker, J.E. (1981) Nucl. Acids Res. 9, 5287–5296.
[296] Kanazawa, H., Kayano, T., Kiyasu, T. and Futai, M. (1982) Biochem. Biophys. Res. Commun. 105, 1257–1264.
[297] Inatomi, K. and Maeda, M. (1988) J. Bacteriol. 170, 5960–5962.
[298] Pollitt, S. and Inouye, M. (1987) In: Bacterial Outer Membranes as Model Systems (Inouye, M., Ed.), pp. 117–139, Wiley, New York.
[299] Dharmavaram, R.M. and Konisky, J. (1987) J. Bacteriol. 169, 3921–3925.
[300] Dharmavaram, R.M. and Konisky, J. (1989) J. Biol. Chem. 264, 14085–14089.
[301] Searcy, K.B. and Searcy, D.G. (1981) Biochim. Biophys. Acta 670, 39–46.
[302] Kirby, T.W., Lancaster Jr., J.R. and Fridovich, I. (1981) Arch. Biochem. Biophys. 210, 140–148.
[303] Takao, M., Kobayashi, T., Oikawa, A. and Yasui, A. (1989) J. Bacteriol. 171, 6323–6329.
[304] May, B.P. and Dennis, P.P. (1989) J. Biol. Chem. 264, 12253–12258.
[305] Takao, M., Yasui, A. and Oikawa, A. (1991) J. Biol. Chem. 266, 14151–14154.
[306] Blanche, F., Robin, C., Couder, M., Faucher, D., Cauchois, L., Cameron, B. and Crouzet, J. (1991) J. Bacteriol. 173, 4637–4645.
[307] Cameron, B., Briggs, K., Pridmore, S., Brefort, G. and Crouzet, J. (1989) J. Bacteriol. 171, 547–557.
[308] Robin, C., Blanche, F., Cauchois, L., Cameron, B., Couder, M., and Crouzet, J. (1991) J. Bacteriol. 173, 4893–4896.
[309] Peakman, T., Crouzet, J., Mayaux, J.-F., Busby, S., Mohan, S., Harborne, N., Wooton, J., Nicholson, R. and Cole, J. (1990) Eur. J. Biochem. 191, 315–323.
[310] Warren, M.J., Roessner, C.A., Santander, P.J. and Scott, A.I. (1990) Biochem. J. 265, 725–729.
[311] Crouzet, J., Cameron, B., Cauchois, L., Rigault, S., Rouyez, C.-M., Blanche, F., Thibaut, D. and Debussche, L. (1990) J. Bacteriol. 172, 5980–5990.
[312] Blanche, F., Debussche, L., Thibaut, D., Crouzet, J. and Cameron, B. (1989) J. Bacteriol. 171, 4222–4231.
[313] Krzycki, J. and Zeikus, J.G. (1980) Curr. Microbiol. 3, 243–245.

[314] Stupperich, E. and Kräutler, B. (1988) Arch. Microbiol. 149, 268–271.
[315] Gilles, H. and Thauer, R.K. (1983) Eur. J. Biochem. 135, 109–112.
[316] Drlica, K. and Rouviere-Yaniv, J. (1987) Microbiol. Rev. 51, 301–319.
[317] Chartier, F., Laine, B., Sautière, P., Touzel, J.-P. and Albagnac, G. (1985) FEBS Lett. 183, 119–123.
[318] Chartier, F., Laine, B. and Sautière, P. (1988) Biochim. Biophys. Acta 951, 149–156.
[319] Imbert, M., Laine, B., Prensier, G., Touzel, J.-P., and Sautière, P. (1988) Can. J. Microbiol. 34, 931–937.
[320] Chartier, F., Laine, B., Bélaïche, D., Touzel, J.-P., and Sautière, P. (1989) Biochim. Biophys. Acta 1008, 309–314.
[321] Imbert, M., Laine, B., Helbecque, N., Mornon, J.-P., Hénichart, J.-P. and Sautière, P. (1990) Biochim. Biophys. Acta 1038, 346–354.
[322] Laine, B., Culard, F., Maurizot, J.-C. and Sautière, P. (1991) Nucl. Acids Res. 19, 3041–3045.
[323] Katouzian-Safadi, M., Laine, B., Chartier, F., Cremet, J.Y., Bélaïche, D., Sautière, P. and Charlier, M. (1991) Nucl. Acids Res. 19, 4937–4941.
[324] Chartier, F., Laine, B., Bélaïche, D. and Sautière, P. (1989) J. Biol. Chem. 464, 17006–17015.
[325] Sandman, K., Krzycki, J.A., Dobrinski, B., Lurz, R. and Reeve, J.N. (1990) Proc. Natl. Acad. Sci. U.S.A. 87, 5788–5791.
[326] Musgrave, D.R., Sandman, K.M. and Reeve, J.N. (1991) Proc. Natl. Acad. Sci. U.S.A. 88, 10397–10401.
[327] Krzycki, J.A., Sandman, K.M. and Reeve, J.N. (1990) In: Proc. 6th Int. Symp. on Genetics of Industrial Microorganisms (Heslot, H., Davies, J., Florent, J., Bobichon, L., Durand, G. and Penasse, L., Eds.), pp. 603–610, Societé Française de Microbiologie, Strasborg.
[327a] Tabassum, R., Sandman, K.M. and Reeve, J.N. (1992) J. Bacteriol. 174, 7890–7895.
[328] Musgrave, D.R., Sandman, K.M., Stroup, D. and Reeve, J.N. (1992) In: Biocatalysis at Extreme Temperatures: Enzyme Systems near and above 100°C (Adams, M.W.W. and Kelly, R.M., Eds.), pp. 174–188, American Chemical Society, Washington, D.C.
[329] Hartmann, E. and König, H. (1989) Arch. Microbiol. 151, 274–281.
[330] Nusser, E., Hartmann, E., Allmeier, H., König, H., Paul, G. and Stetter, K.O. (1988) In: Crystalline Bacterial Cell Surface Layers (Sleytr, U.B., Messner, P., Pum, D. and Sara, M., Eds.), pp. 71–95, Springer, Berlin.
[331] Kalmokoff, M.L. Koral, S.F. and Jarrell, K.F. (1988) J. Bacteriol. 170, 1752–1758.
[332] Southam, G., Kalmokoff, M.L., Jarrell, K.F., Koral, S.F. and Beveridge, T.J. (1990) J. Bacteriol. 172, 3221–3228.
[333] Kiener, A. and Leisinger, T. (1983) System. Appl. Microbiol. 4, 305–312.
[334] Winter, J. (1983) System. Appl. Microbiol. 4, 558–563.
[335] Bhatnagar, L., Henriquet, M. and Longin, R. (1983) Biotechnol. Lett. 5, 39–42.
[336] Jones, W.J., Whitman, W.B., Fields, R.D. and Wolfe, R.S. (1983) Appl. Environ. Microbiol. 46, 220–226.
[337] Scherer, P., Lippert, H. and Wolff, G. (1983) Biolog. Trace Element Res. 5, 149–163.
[338] Harris, J.E. (1987) Appl. Environ. Microbiol. 53, 2500–2504.
[339] Sowers, K.R. and Gunsalus, R.P. (1988) J. Bacteriol. 170, 998–1002.
[340] Xun, L., Mah, R.A. and Boone, D.R. (1990) Appl. Environ. Microbiol. 56, 3693–3698.
[341] Bertani, G. and Baresi, L. (1987) J. Bacteriol. 169, 2730–2738.
[342] Kiener, A., König, H., Winter, J. and Leisinger, T. (1987) J. Bacteriol. 169, 1010–1016.
[343] Meile, L., Rechsteiner, T., Jenal, U., Jordan, M., and Leisinger, T. (1990) In: Microbiology of Extreme Environments and its Potential for Biotechnology (DaRosta, M.S., Duarte, J.C. and Williams, R.A.D., Eds.), pp. 253–257, Elsevier Applied Science Publishers, Barking, UK.
[344] Nagle, D.P. Jr., (1989) Dev. Ind. Microbiol. 30, 43–51.
[345] Worrell, V.E., Nagle, D.P., McCarthy, D. and Eisenbraun, A. (1988) J. Bacteriol. 170, 653–656.

[346] Konisky, J. (1989) Trends Biotechnol. 7, 88–92.
[347] Possot, O., Gernharot, P., Foglino, M., Klein, A. and Sibold, L. (1989) In: Microbiology and Biochemistry of Strict Anaerobes Involved in Interspecies Hydrogen Transfer (Belaich, J.P., Bruschi, M. and Garcia, J.L., Eds.), pp. 527–529, Marseille, France.
[348] Jarrell, K.F., Faguy, D., Herbert, A.M. and M.L. Kalmokoff (1992) Can. J. Microbiol. 38, 65–68.
[349] Micheletti, P.A., Sment, K.A. and Konisky, J. (1991) J. Bacteriol. 173, 3414–1318.
[350] Sandbeck, K.A. and Leigh, J.A. (1991) Appl. Environ. Microbiol. 57, 2762–2763.
[351] Hebert, A.M., Kropinski, A.M. and Jarrell, K.F. (1991) J. Bacteriol. 173, 3224–3227.
[352] Jaenicke, R., Bernhardt, G., Lüdemann, H.D. and Stetter, K.O. (1988) Appl. Environ. Microbiol. 54, 2375–2380.
[353] Hippe, H., Caspari, D., Fiebig, K. and Gottschalk, G. (1979) Proc. Natl. Acad. Sci. U.S.A. 76, 494–498.
[354] Nauman, E., Fahlbusch, K. and Gottschalk, G. (1984) Arch. Microbiol. 138, 79–83.
[355] Zinder, S.H. and Elias, A.F. (1985) J. Bacteriol. 163, 317–323.
[356] Widdel, F. and Wolfe, R.S. (1989) Arch. Microbiol. 152, 322–328.
[357] Widdel, F. (1986) Appl. Environ. Microbiol. 51, 1056–1062.
[358] Widdel, F., Rouvière, P.E. and Wolfe, R.S. (1988) Arch. Microbiol. 150, 477–481.
[359] Frimmer, U. and Widdel, F. (1989) Arch. Microbiol. 152, 479–483.
[360] Pecher, T. and Böck, A. (1981) FEMS Microbiol. Lett. 10, 295–297.
[361] Elhardt, D. and Böck, A. (1982) Mol. Gen. Genet. 188, 128–134.
[362] Böck, A., Bär, U., Schmid, G. and Hummel, H. (1983) FEMS Microbiol. Lett. 20, 435–438.
[363] Harris, J.E. and Pinn, P.A. (1985) Arch. Microbiol. 143, 151–153.
[364] Hummell, H., Bär, U., Heller, G. and Böck, A. (1985) System. Appl. Microbiol. 6, 125–131.
[365] Hummel, H. and Böck, A. (1985) Mol. Gen. Genet. 198, 529–533.
[366] Haas, E.S., Hook, L.A. and Reeve, J.N. (1986) FEMS Microbiol. Lett. 33, 185–188.
[367] Kiener, A., Rechsteiner, T. and Leisinger, T. (1986) FEMS Microbiol. Lett. 33, 15–18.
[368] Possot, O., Gernhardt, P., Klein, A. and Sibold, L. (1988) Appl. Environ. Microbiol. 54, 734–740.
[369] Smith, M.R. and Mah, R.A. (1981) Curr. Microbiol. 6, 321–326.
[370] Kiener, A., Holliger, C. and Leisinger, T. (1984) Arch. Microbiol. 139, 87–90.
[371] Teal, R. and Nagle Jr., D.P. (1986) Curr. Microbiol. 14, 227–230.
[372] Nagle Jr., D.P., Teal, R. and Eisenbraun, A. (1987) J. Bacteriol. 169, 4119–4123.
[373] Santoro, N. and Konisky, J. (1987) J. Bacteriol. 169, 660–665.
[374] Grindbergs, A., Müller, V., Gottschalk, G. and Thauer, R.K. (1988) FEMS Microbiol. Lett. 49, 43–47.
[375] Knox, M.R. and Harris, J.E. (1988) Arch. Microbiol. 149, 557–560.
[376] Pankhania, J.P. and Robinson, J.P. (1984) FEMS Microbiol. Lett. 22, 277–281.
[377] Ahring, B.K. and Westermann, P. (1985) Curr. Microbiol. 12, 273–276.
[378] Bhatnagar, L., Jain, M.K., Zeikus, J.G. and Aubert, J.-P. (1986) Arch. Microbiol. 144, 350–354.
[379] Rechsteiner, T., Kiener, A. and Leisinger, T. (1986) System. Appl. Microbiol. 7, 1–4.
[380] Jain, M.K. and Zeikus, J.G. (1987) Appl. Environ. Microbiol. 53, 1387–1390.
[381] Tanner, R.S., McInerney M.J. and Nagle Jr., D.P. (1989) J. Bacteriol. 171, 6534–6538.
[382] Ladapo, J. and Whitman, W.B. (1990) Proc. Natl. Acad. Sci. U.S.A. 87, 5598–5602.
[383] Micheletti, P.A., Sment, K.A. and Konisky, J. (1991) J. Bacteriol. 173, 3414–3418.
[384] Gernhardt, P., Possot, O., Foglino, M., Sibold, L. and Klein, A. (1990) Mol. Gen. Genet. 221, 273–279.
[385] Steigerwald, V.J., Pihl, T.D. and Reeve, J.N. (1992) Proc. Natl. Acad. Sci. U.S.A. 89, 6929–6933.
[386] Klenk, H.-P., Palm, P., Lottspeich, F. and Zillig, W. (1992) Proc. Natl. Acad. Sci. U.S.A. 89, 407–410.
[387] Gohl, H.P, Hausner, W. and Thomm, M. (1992) Mol. Gen. Genet. 231, 286–295.

[388] Thomm, M., Sandman, K., Frey, G., Koller, G. and Reeve, J.N. (1992) J. Bacteriol. 174, 3508–3513.
[389] Macario, A.J.V., Dugan, C.B. and Conway de Macario, E. (1991) Gene 108, 133–137.

CHAPTER 17

Archaeal hyperthermophile genes

Jacob Z. DALGAARD and Roger A. GARRETT

Institute of Molecular Biology, Copenhagen University, Sølvgade 83, DK-1307 Copenhagen K, Denmark

1. Introduction

We present a survey of the currently available gene sequences from the sulfothermophiles (defined as hyperthermophiles which can metabolize sulfur compounds and do not produce methane) and the hyperthermophilic methanogens. The hyperthermophiles grow optimally at temperatures between about 70°C and 105°C. A few moderate thermophiles, growing optimally between 50°C and 70°C, are also included mainly for comparative purposes. The aims of the survey were threefold. First, to list all the thermophilic genes which have been sequenced; they are arranged in order of organisms together with Accession Numbers to the nucleotide-sequence databases. Second, to characterize the genes and their flanking regions for transcriptional signals, translational signals, codon usage, etc., and, third, to establish whether any of these features could be considered specific for archaeal hyperthermophiles.

Recently, comprehensive reviews have appeared on archaea in general by Brown et al.[1] and on the sulfothermophiles by Forterre[2]; the latter review also includes an excellent discussion of general factors influencing thermostability at the DNA level. Moreover, many new isolates of hyperthermophiles have been characterized during the past decade and they have been described in two recent reviews by Stetter and colleagues[3,4]; in Table 1 we present a recent view of the phylogeny of the hyperthermophiles taken from one of these[4]. The list of organisms is not complete since several isolates have recently been characterized which remain to be classified.

2. Gene sequences

A complete list of all the gene sequences present in the EMBL-GenBank databases in October 1991 is given in Table 2 together with a few which are not yet in the database. They are arranged alphabetically according to the name of the organism, and Accession Numbers and literature references are given. Many of the sequences contain flanking non-coding regions, some of which have been analysed for transcriptional and translational signals; other gene sequences are partial. Several putative open reading frames (ORFs) of unknown identity are included, and for each of these we searched the

TABLE 1
Taxonomy of archaeal hyperthermophiles[a]

Order	Genus	Species
Sulfolobales	*Sulfolobus*	*S. acidocaldarius*
		S. shibatae
		S. solfataricus
	Metallosphaera	*Ms. sedula*
	Acidianus	*Aa. infernus*
		Aa. brierleyi
	Desulfurolobus	*Dl. ambivalens*
Thermoproteales	*Thermoproteus*	*Tp. tenax*
		Tp. neutrophilus
	Pyrobaculum	*Pb. islandicum*
		Pb. organotrophum
	Thermofilum	*Tf. pendens*
		Tf. librum
	Desulfurococcus	*Dc. mobilis*
		Dc. mucosus
		Dc. saccharovorans
	Staphylothermus	*St. marinus*
Pyrodictiales	*Pyrodictium*	*Pd. occultum*
		Pd. brockii
		Pd. abyssum
	Thermodiscus	*Td. maritimus*
Thermococcales	*Thermococcus*	*Tc. celer*
		Tc. stetteri
	Pyrococcus	*Pc. furiosus*
		Pc. woesei
Archaeoglobales	*Archaeoglobus*	*Ag. fulgidus*
		Ag. profundus
Thermoplasmales	*Thermoplasma*	*Tp. acidophilum*
		Tp. volcanium
Methanobacteriales	*Methanobacterium*	*M. thermoautotrophicum*
	Methanothermus	*Mt. fervidus*
		Mt. sociabilis
Methanococcales	*Methanococcus*	*Mc. thermolithotrophicus*
		Mc. jannaschii

[a] The classification is from Stetter et al. [4].

DNA/protein databases for sequence similarities; potential homologs that were found are included in Table 2.

3. Nucleotide composition and optimal growth temperature

The results summarized in Table 3 show that there is no simple relationship between optimal growth temperature and the G+C content of the genomic DNA. Although

TABLE 2
Compilation of sequences of archaeal thermophile genes[a]

Gene product	Gene	Accession no.	Ref.
Archaeoglobus fulgidus			
16S rRNA		X05567, Y00275	5
7S rRNA		X17237	6
tRNA-Ala		M19340	7
RNA polymerase subunit A	*rpoT*	X72728	8
RNA polymerase subunit C	*rpoX*	X72728	8
Desulfurococcus mobilis			
16S rRNA		M36474	9
23S rRNA		X05480	10
23S intron		X03263	11
5S rRNA		X07545	12
tRNA-Met (+ intron)			13
	ORF 1	X06190	12
	ORF 1	X07545	12
	ORF 2	X06190	12
	ORF 2	X07545	12
Methanobacterium thermoautotrophicum			
16S rRNA (M)		X15364, X05482	14
23S rRNA (M)		X15364, X05482	14
5S rRNA (M)		M36186, M36187	15
7S rRNA (M)		X15364, X05482	14
tRNA-Ala (M)		X15364, X05482	14
tRNA-Asp		X06788	16
tRNA-Gly		X06787	16
tRNA-Ser (M)		X15364, X05482	14
RNA polymerase subunit A (M)	*rpoT*	M02391	17
RNA polymerase subunit B' (M)	*rpoU*	M02391	17
RNA polymerase subunit B'' (M)	*rpoV*	M02391	17
RNA polymerase subunit C (M)	*rpoX*	M02391	17
5'-phosphoribosyl-5-aminoimidazole carboxylase (ΔH)	*purE*	X03250	18
8-hydroxy-5-deazaflavin-reducing hydrogenase subunit α (ΔH)	*frhA*	J02914	19
8-hydroxy-5-deazaflavin-reducing hydrogenase subunit β (ΔH)	*frhB*	J02914	19
8-hydroxy-5-deazaflavin-reducing hydrogenase subunit δ (ΔH)	*frhD*	J02914	19
8-hydroxy-5-deazaflavin-reducing hydrogenase subunit γ (ΔH)	*frhG*	J02914	19
Isoleucyl–tRNA synthetase (M)			20
Methylviologen-reducing hydrogenase small subunit (ΔH)	*mvhG*	J04540	21
Methylviologen-reducing hydrogenase protein D (ΔH)	*mvhD*	J04540	21
Methylviologen-reducing hydrogenase large subunit (ΔH)	*mvhA*	J04540	21
Methyl–CoM reductase subunit α (M)	*mcrA*	X07794, M18969	22
Methyl–CoM reductase subunit β (M)	*mcrB*	X07794, M18969	22
Methyl–CoM reductase subunit γ (M)	*mcrG*	X07794, M18969	22
Methyl–CoM reductase operon protein C (M)	*mcrC*	X15364, X05482	14
Methyl–CoM reductase operon protein D (M)	*mcrD*	X15364, X05482	14

continued on next page

TABLE 2, continued

Gene product	Gene	Accession no.	Ref.
Polyferredoxin (ΔH)	mvhB	J04540	21
Superoxide dismutase	sod	D00614	23
Tetrahydromethanopterin formyltransferase (ΔH)	ftr	J05173	24
(M)	ORF 1	X07794, M18969	22
(M)	ORF 2	X15364 X05482	14
(M)	ORF 3	X15364, X05482	14
(ΔH)	ORF-B(t)	X03250	18
pME2001 (plasmid) (M)		M19040, X17205	25,26
Methanococcus thermolithotrophicus			
ATPase (H$^+$) – ATP regulatory subunit	atp60	X56519	27
ATPase – ATP bindinq site	atp70	X56520	28
Formimino-5-aminoimidazole phosphoribosyl carboxamideribotide isomerase	hisA	M17742	29
Nitrogenase subunit D	nifD	X13830	30
Nitrogenase subunit H	nifH	X13830	30
Nitrogenase subunit K	nifK	X13830	30
	ORF 96	X13830	30
* Nitrogen regulating PII	ORF 105	X13830	30
* Nitrogen regulating PII	ORF 128	X13830	30
Methanothermus fervidus			
16S rRNA		M32222	31
5S rRNA		M26976	32
7S rRNA		M32222	31
tRNA-Ala		M32222	31
tRNA-Asn		M26978	32
tRNA-Asp		M26677	32
tRNA-Glu		M26978	32
tRNA-His		M26978	32
tRNA-Leu		M26978	32
tRNA-Lys		M26677	32
tRNA-Met		M26978	32
tRNA-Pro		M26677	32
tRNA-Ser		M32222	31
tRNA-Thr		M26677	32
3-Phosphoglycerate kinase	pgk	M55529	33
DNA binding protein HMfB	hmfb	M34778	34
Glyceraldehyde-3-phosphate dehydrogenase	gap	M19980	35
L-malate dehydrogenase		X51714	36
Methyl–CoM reductase subunit α	mcrA	J03375	37
Methyl–CoM reductase subunit β	mcrB	J03375	37
Methyl–CoM reductase subunit γ	mcrG	J03375	37
Methyl–CoM reductase subunit δ	mcrD	J03375	37
Methylviologen-reducing hydrogenase large subunit	mvhA	M34016	38
Methylviologen-reducing hydrogenase small subunit	mvhG	M34016	38

continued on next page

TABLE 2, continued

Gene product	Gene	Accession no.	Ref.
Polyferredoxin	mvhB	M34016	38
Surface-layer glycoprotein	slgA		39
	ORF	M26978	32
	ORF 260	M32222	33
* Agmatinase	ORF 285	M34778	33
Pyrobaculum islandicum			
23S rRNA (partial)		M86622	40
Pyrobaculum organotrophum			
23S rRNA (partial)			40
23S intron 1			41
23S intron 2			41
Pyrococcus furiosus			
23S rRNA (partial)		M86627	40
Pyrococcus woesei			
5S rRNA		X15329	42
Glyceraldehyde-3-phosphate dehydrogenase	gapdh		43
	ORF a		43
	ORF b		43
	ORF x		43
* Xenopus L32, Rat/Human L35A	ORF y		43
	ORF z		43
Pyrodictium occultum			
16S rRNA		M21087	44
23S rRNA (partial)		M86622	40
5S rRNA		M21086	44
7S rRNA		M21085	44
Staphylothermus marinus			
23S rRNA (+ 2 introns, partial)		M38363, M86623	45
Sulfolobus acidocaldarius			
5S rRNA		V01286	46
tRNA-Met		X01223	47
tRNA-Ser		X52382	48
RNA polymerase subunit A	rpoT	X14818	49
RNA polymerase subunit B	rpoU	X14818	49
RNA polymerase subunit C	rpoX	X14818	49
Elongation factor EF-1α	tuf	X52382	48
Elongation factor EF-2	fus	X54972	50
S7		X52382	48
S10		X52382	48
S12		X14818	49
ATPase subunit α	atpA	J03218	51
ATPase subunit β	atpB	M22403	52

continued on next page

TABLE 2, continued

Gene product	Gene	Accession no.	Ref.
ATPase subunit γ	atpG	M57238, J05671	53
ATPase subunit δ	atpD	M57236, J05671	53
ATPase subunit ε	atpE	M57238, J05671	53
Thermopsin		J05184	54
	ORF	X14818	49
* RNA polymerase subunit	ORF 88	X14818	49
* Yeast L32, Rat L30	ORF 104	X14818	49
* (b)	ORF 130	X14818	49
Sulfolobus solfataricus			
16S rRNA		X03235	55
7S rRNA		X17239	6
tRNA-Gly (+ intron)		X06053	56
tRNA-Leu (+ intron)		V01548	57
tRNA-Met (+ intron)		X06054	56
tRNA-Phe (+ intron)		X06053	56
tRNA-Ser (+ intron)		V01548	57
tRNA-Val		X06054	56
LX			58
L46e		X16161	59
Aspartate aminotransferase	AspATS	X16505	60
β-galactosidase	bgaS	X15950, X15372	61
β-D-galactosidase	lacS	M34696	62
***Sulfolobus* sp.**			
5S rRNA		M16530	63
Sulfolobus shibatae			
16S rRNA		X05869, M02729	64,65
5S rRNA		X05870	64
Thermococcus celer			
16S rRNA		M21529	66
5S rRNA		X07692, M12711	67–69
7S rRNA		X17240	6
tRNA-Ala		M19344	70
tRNA-Asp		X07692	67
Elongation factor EF-1α	tuf	X52383	71
Thermofilum pendens			
23S rRNA		X14835	72
16S rRNA		X14835	72
tRNA-Gly (+ intron)		X14835, J05061	72
tRNA-Met		X14835	72
	ORF 1	X14835	72
	ORF 2	X14835	72

continued on next page

TABLE 2, continued

Gene product	Gene	Accession no.	Ref.
Thermoplasma acidophilum			
16S rRNA		M38637, M20822	73
23S rRNA		M32298, M20822	74
5S rRNA		X02709, M32297	75,73
tRNA Met		V01382	76
tRNA-Met (initiator)		X01221	53
Elongation factor EF-1α	tuf	X53866	77
Elongation factor EF-2			78
Citrate synthase		X55282	79
Proteasome large (α-) subunit			80
Thermoproteus tenax			
16S rRNA		X05073	81
23S rRNA		X05074	82
tRNA-Ala (+ intron)		X05070	83
tRNA-Leu (+ intron)		X05071	83
tRNA-Leu		X05072	83
tRNA-Met (+ intron)		X05069	83
	ORF	X06157	81
SSV1 (Sulfolobus shibatae)			
Coat protein VP1		X07234	84
Coat protein VP2		X07234	84
Coat protein VP3		X07234	84
TTV1 (Thermoproteus tenax)			
Coat protein TP1		X14855	85
Coat protein TP2		X14855	85
Coat protein TP3		X14855	85
Protein TPX		X14855	86,87

[a] Almost all of the sequences were present in the EMBL-GenBank databases for nucleotide sequences in October, 1991 and their Accession Numbers are given; sequences for which no numbers are given are not yet submitted to the database; some are from our own laboratory. For *M. thermoautotrophicum* we indicate whether the dissimilar Marburg (M) or ΔH strains were used. Numbers preceded by S or L correspond to ribosomal Small or Large subunit proteins; LX refers to an unidentified large subunit protein. A larger variant of the TTV1 virus protein TPX (from *Tp. tenax*) has been described [87]. ORFs of unknown identity are all listed. For some we identified possible homologs (> 30% sequence similarity) and these are indicated with asterisks in the first column; for most of the remaining ORFs no evidence has been provided for their being expressed.
[b] For ORF 130 of *S. acidocaldarius*, similar amino-acid sequences have been detected in other archaea, *Halobacterium halobium* and *Halococcus morrhuae*, although the protein product has not yet been identified [88].

Pd. occultum, with the highest optimal growth temperature (> 100°C), exhibits the highest G+C content (62%), for the next group of organisms, growing at 95–100°C, the G+C contents lie in the range of 38–46%. Thus, one can infer that factors other than nucleotide composition of the DNA determine its thermal stability. Examples of

TABLE 3
Comparison of optimal growth temperatures and genomic G–C contents

Organism	Optimal growth temperature (°C)	Genomic G–C content (%)	Reference
Pd. occultum	100–110	62	89
Pc. woesei	95–100	38	90
Pc. furiosus	95–100	38	91
Pb. islandicum	95–100	46	92
Pb. organotrophum	95–100	46	92
St. marinus	90–95	35	93
Tc. celer	85–95	56	94
Tp. tenax	85–95	56	95
Tf. pendens	85–90	57	96
Dc. mobilis	80–90	51	97
S. solfataricus	80–90	36	98
Ag. fulgidus	80–90	46	99
S. acidocaldarius	70–85	37	23
S. shibatae	70–85	35	65
Tp. acidophilum	55–65	46	100
Mt. fervidus	75–85	33	101
M. thermoautotrophicum	65–70	50	14
Mc. vannielii	35–40	32	83

[a] The names of the organisms are abbreviated as in Table 1.

these factors, including the presence of polyamines and histone-like proteins, have been summarized by Forterre [2].

In contrast to the genomic DNA, the G+C contents of the large rRNAs show a direct correlation with the optimal growth temperature. The results are illustrated for the four nucleotides in Fig. 1 and reveal a substantial increase in both G and C content (with a corresponding decrease in both A and U content) at growth temperatures above 70°C. At the highest temperatures, cytidine increases more than guanosine, suggesting that G–U pairs may convert to G–C pairs to enhance thermostability. Additional stability of the rRNA structures is probably provided by chemical modification since many modified nucleotides have been detected in the large rRNAs of *Sulfolobus* [102]. The same may be true for the tRNAs since several novel modified nucleosides have been characterized in McCloskey's laboratory, some of which are exclusive to the sulfothermophiles [103-106]; moreover, often the 2'-OH group which directly contributes to thermal instability of RNA is methylated.

For the open reading frames, the nucleotide compositions were also correlated with the optimal growth temperatures. No direct relationship was observed between G+C content and temperature. However, as illustrated in Fig. 2, the G+C content of the protein-coding

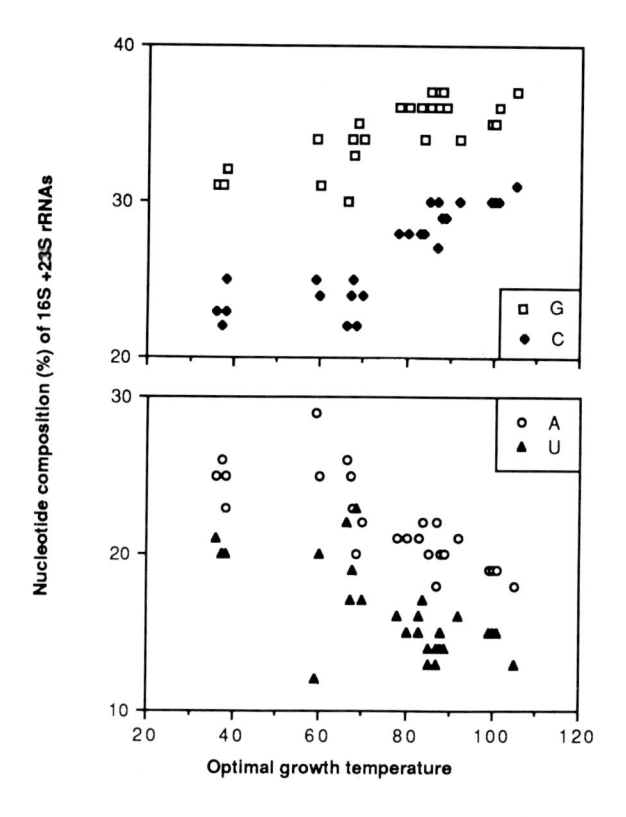

Fig. 1. Dependence of the nucleotide compositions of the large rRNAs on the optimal growth temperature of the organisms listed in Table 3. The bacterial rRNAs from *Eschericia coli*, *Bacillus stearothermophilus* and *Fervidobacterium nodosum* were used to cover the lower temperature range.

regions increases proportionally to the G+C content of the genome and, moreover, the relative increase in C content is greater than that of G. We did not investigate the probability (see ref.[107] and references therein) that the general usage of the three codon positions responds differently to alterations in the genomic G+C content.

4. Gene organization

The 16S and 23S rRNA genes of the hyperthermophiles are generally coupled physically and transcriptionally except in *Tp. acidophilum* where they are transcribed separately [73,74]. However, whereas the 5S rRNA genes of the thermophilic methanogens are coupled to the 16S and 23S rRNA genes, in the sulfothermophiles they are located distantly on the genome and transcribed separately [12,64,82] as are the tRNA genes [83]. Only one example of a transcriptionally coupled 5S rRNA and tRNA^{Asp} gene has been described for *Tc. celer* [67-69]. Transcriptional units containing clusters of tRNA genes 5'-tRNA^{Thr}-tRNA^{Pro}-tRNA^{Asp}-tRNA^{Lys}-3' and 5'-tRNA^{Asn}-tRNA^{Met}-tRNA^{Glu}-tRNA^{Leu}-tRNA^{His}-3' have been characterized for *Mt. fervidus* [32].

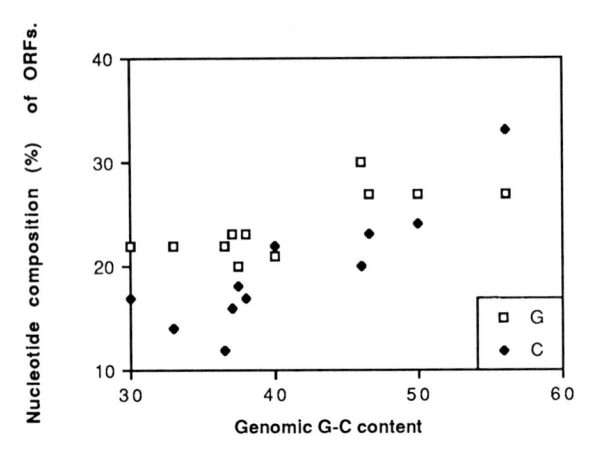

Fig. 2. Dependence of the nucleotide composition of the protein-coding regions on the G–C contents of the genomes (listed in Table 3) for *S. acidocaldarius, S. solfataricus,* SSV1 plasmid, *Tc. celer, Tp. acidophilum,* TTV1 virus, *Ag. fulgidus, Mt. fervidus, M. thermoautotrophicum* and *Mc. vannielii.*

In sulfothermophile transcripts, the 16S-23S rRNA spacers are short compared with those of other archaea and bacteria; they lack tRNAs and, although they contribute to the processing stems of the 16S and 23S rRNAs, the stems are truncated and, at least for *Tp. tenax,* a single, bifurcated processing stem is generated for the two rRNAs [82]. However, despite differences between the organization and structure of the sulfothermophile rRNA transcripts and those of the other archaea, secondary structural motifs can be discerned in the 16S rRNA leader sequences and in the 16S–23S rRNA spacer regions which are common and exclusive to the archaea [108].

Several introns have been located in both tRNA and rRNA genes of the sulfothermophiles (Table 2); they have also been located in tRNA genes of the extreme halophiles (ref. [109] and references therein). The locations of the tRNA introns are more variable than those of the eukaryotic nuclear tRNAs which generally occur after the nucleotide following the anticodon. Most of the archaeal introns occur at different positions in the anticodon loop and one, in *Tf. pendens,* was located in the extra arm [110]. The rRNA introns, like most of those found in eukaryotic nuclei and organelles, occur in domains IV and V of the 23S-like rRNAs which have been strongly implicated in tRNA binding and peptidyltransferase [111]. This led to a recent suggestion that a structural similarity between rRNA introns and the aminoacyl stem of tRNAs could provide a basis for reverse splicing and, hence, mobility of introns at the rRNA level [112].

All of the archaeal introns appear to be spliced by a protein-containing enzyme which recognizes a secondary structural motif ("bulge–helix–bulge") at the intron–exon junction; in this respect, they are quite distinct from the introns found in eukaryotes and bacterial phages [45,109]. However, some of the rRNA introns contain open reading frames [11,113] which show some amino-acid sequence similarity to group I-encoded RNA maturases, and/or DNA endonucleases, suggesting that the open reading frames, at least, may have a common origin [41].

Like the rRNA genes, the open reading frames are generally efficiently packed in the genome, consistent with the latter's relatively small size in sulfothermophiles [114].

Often, little space occurs between the protein genes, and occasionally there are small overlaps. As for the rRNA genes, the organization is sometimes exceptional. Thus, the RNA-polymerase genes of *S. acidocaldarius* are ordered B–A, as in eukaryotes [49], while for the methanogens and extreme halophiles, although the order is maintained, the B gene is split into B′–B′. The A gene is split into A–C in all archaea [17,88]. Nevertheless, a high degree of sequence similarity exists between the corresponding archaeal subunits [88,115].

5. Transcriptional signals

The region upstream from transcriptional start sites of archaeal genes contains a core promoter region which has been estimated as lying between positions −2 and −38 [116]. Alignment of the upstream sequences reveals a partial similarity to eukaryotic RNA polymerase II promoters but not to those of bacterial promoters [12,117,118]. A partially conserved A–T-rich box is centered approximately −26 nucleotides from the initiation site which both in position and in A/T composition resembles the eukaryotic TATA box. This similarity correlates with the observation that the amino-acid sequences of the archaeal RNA polymerases are more closely related to eukaryotic RNA polymerase II (and to a lesser degree to RNA polymerase III) than to the bacterial polymerases [49,88,116,119]. Moreover, as for RNA polymerase III-directed transcription in eukaryotes, transcriptional initiation in sulfothermophiles generally occurs at, or close to, the 5′ end of a mature 5S rRNA or tRNA [12,64,74,83].

Termination often occurs inefficiently at T-rich polypyrimidine sequences and, so far, no potential secondary structural features have been detected, either in the transcript or in the corresponding DNA strand, which are common amongst archaeal termination regions [1].

5.1. Promoter regions

In this section approximately aligned sequences from the promoter core region are presented and analyzed. The alignments are based on the A–T-rich box centered about 26 nucleotides upstream from the transcriptional start site and they are presented in Table 4. Sequence regions extending approximately from positions −50 to +10 are analyzed.

With few exceptions, all the promoter regions contain the consensus sequence TTTA$^A/_T$A. There are some indications that minor variations in this sequence motif occur in different orders of the hyperthermophiles but it is difficult to assess how significant this is given the relatively small number of sequences available. This is complicated further by recent analyses of the phylogeny of the hyperthermophiles based on partial 23S rRNA sequences [40] which suggest that the taxonomic classification in Table 1 needs revision (see section 7). For many promoters the A–T-rich motif is longer than 5–6 base pairs; indeed, for the *Mt. fervidus* genes it extends to 14–15 consecutive A–T pairs (Table 4).

The location of the TTTA$^A/_T$A motifs relative to the transcriptional start sites is so consistent that it can probably be used as a criterion for an active promoter region. Thus,

TABLE 4
Alignment of promoter regions directly upstream from coding regions[a]

	Sequence
Dc. mobilis	
rRNA operon	
p1	TCATGTGGCATAAAGTATATAAACCCTTATCGCATAGAGTAAGATTCCAGACGCTTAC AGCGGACAA
p2	CCTACACGGGAGGCAACCCGTCATGA<u>TTAATA</u>CCCTTGGAGCAAATAGATTCATCA AGCCCGCGGCA
p3	CGGAGCCCGATGTCATAGTGAACGC<u>TTTGAA</u>AGCAGCTGGTGTTCCACGGAGT GAAGCACTCTACGT
p4	GCTTGTCCAAGAGAACTGGTTCAAACACGTCAGGCTTTTCCCCGACGTCATCCCCGT GCTCAGGCAG
p5	--------GAATTCATCCTTGAGGCAGTGGTGGGAACCGGGTTGAGCAGGGAGGAT GCCGCCAGGTT
5S rRNA p1	TCTACGCTGTCGAGAAGAGTAAGG<u>TTTAAA</u>ACCCCAGTAATAGATTATGGGACT AcGGTGCCCGAC
p2	CAGCCGTGAACCTAGACCCTAACACACTATACACCCTGGGCCCAGTAGCCGGGATC ACGATGTAGT-(7n)-atg
orf1	TGGATGGATTGGTCAAGGTTAAAATAATAGCTACCCTGGGGCCCAGTAGCCGGGATC ACGATGTAGT-(7n)-atg
orf1 (rRNA operon)	TACTACACCCTATAAATTTTTATACGTGTAGCAGTGTCAAGTGAGATAAACAACCCA GCTCCTCGAGG-(7n)-atg
Pc. woesei	
gapdh	GTGAAAACAACCCTTAAAATTTCAT<u>TTTATA</u>ATCTAAATCTGGTGAGGTAAAgtgAAAATAAAGGTCG
orfA	AAGATTAATTCCCAAGCAAAAAACA<u>TTTATA</u>GGTGACCATTCACTCATGCAAACatgAAAGATTCTatg
S. solfataricus	
tRNA^{Gly}	TGCGACCCTCTTTTCATCATTAAAGG<u>TTAAA</u>TAGGCTTGAAAAAGATATTAATATTgCGGCCGTCGTC
tRNA^{Leu}	GTTCAAATCCCGCCCCCGGCGTTAGT<u>TTTAAA</u>ATTCTCCAATTTATATCATATATTGATTgCGGGGGTG
tRNA^{Met}	TCATAACCTTATCACGGCATAACT<u>TTTAAA</u>AGGTAACTATTTTATTATGTTATAGTgGGCCCGTAGC
tRNA^{Phe}	AAGATAAAATAATAACCAATAAACC<u>TATAAA</u>GTCATATGTAAATAATAATAATgCCGCCGTAGCTCAG
tRNA^{Val}	TCATTCATGTATTTTACACGAAGAGT<u>TTAAA</u>AACGGGTAAAGATTAAACTATTAGAGAgGGCCCGTCG
AspATS	ATCACTGATGCGAAACCCAGAAAAGA<u>TTAAA</u>GAATTTATACTATAGCGTTCTTCATATACACTTgtgG
bgas	CTTAGTAATATCCATGACAGAGTCCA<u>TTAAAA</u>TAATAAAGAGCCTTGGTCAACTTATATCTGTCTatg
L46e	----------------AGAAAAGA<u>TTATAA</u>GATTACGATTAAAGAGAGAGGATGGAAatgAGCAAG
lacS	GACCATAAAAGATACTCGCTCAAAGC<u>TTAAA</u>TAATATTAATCATAAATAAAGTCatgTACTCATTTCC
S. acidocaldarius	
tRNA^{Ser}	ATTGAGCTAATTTGATTTGATTATT<u>TTTATA</u>AGCATCTCGTATAATCTTATTAATgCCGGGGTGCCCG
ATPase	AGAAAAAATAAGATTACAGAAGAA<u>TTAAA</u>AAAATTTTATCTGAAATGAATCAAATAATAGATGAGG-(121n)-atg
EF-1α	CTCTAGGTAAATTGTAGCTAAACT<u>TTTAAT</u>GCTAGTCTAGTTATAAAAATGT AAGAGGTAATTAGTatg
EF-2	GTATATCAATCTAAATCCAACACAC<u>TTTAAT</u>ACTTAAAAGTTAACTTACTATTC GTAAACTGGTGAT-(7n)-ttg
rpoB, orf88	TAGAAATTCGTGTATCTAACATAAT<u>TTTATT</u>AAGCTTAAATGAGTAGGATGATATTACT GTACGCTACatg
rpoC[b]	TAGAATATGATGGAACTGTGAGAAG<u>TTTATA</u>TGGAGATATTGTTCAAGTAGTATAT GGTGATGATGC-(86n)-atg
S7	AATAAATTATTACTACACAAACTCT<u>ATTAAA</u>CCCTT QTAATATATTTAAGGT AGCAGTAAAGTAAA-(13n)-atg
S12[c]	GAAATGTTGAAAGGGCTAAGTTAATA<u>TTAAAA</u>AAGGTATTTGGATATTGATTCT GTTGTTGTAGTAtaa-(112n)-atg
thermopsin	TTTTTCTTTATTTTTAAAAGAAAGC<u>TTATA</u>TACATGAAAATTATTTAAAAAGTGatgAATTTTAAAT
orf130[b]	ATTAGGGACATTATGCGGTAAACCT<u>TTTAT</u>GGTATCTGTTATAGGTATAGTT GATGAAGGGGAATCA-(75n)-atg
S. shibatae	
16S tRNA	TTACACGGAATATATAGAAGTTAGA<u>TTTATA</u>TGGGATTTCAGAACAATATGTATAAT GCGGATGCCC
5S tRNA	TGAGGAAAAGAAGGGTAGTTAATT<u>TTTTATA</u>TGTGTTATGAGTACTTAATTTT gCCCACCCGGCCAC
SSV1	
T5	TAGATAGAGTATAGATAGAGTAAAG<u>TTTAAA</u>TACTTATATAGATAGAGTATAGAT AGAGGGTTCAAA-(2n)-atg
T6	TAGATAGAGTATAGATAGAGTAAGAC<u>TTAAAT</u>ACTAATTTATACATAGAGTATAGAT AGAGTGGGAT-(6n)-atg

continued on next page

TABLE 4, continued

	Sequence
T3	TAATATCTGCGTTAGTTAGGCTCTT<u>TTTAAAG</u>TCTACCTTCTTTTTCGCTTAC AatgAGGAAGTCCC
T4	AACTTAAATTTAGAAGATAGCCCT<u>TTTAAAG</u>CCATAAATTTTTTATCGCTT AatgAAGTGGGGACT
T2	GATTCTGAATTCAGAACTGGAGGGG<u>TTTAAAA</u>ACGTAAGCGGGAAGCCGATAGTT GACCAAGGATGA-(224n)-atg
T9	AAGCGGATGCACCGCAAGTAGGCCC<u>TTTAAAA</u>AGTCATATTCTTTTTCTTTCCCT GatgAGTGCGTT

Tp. acidophilum
16S tRNA	AGAAAATTGGAGTTCCGCTTCGAAAG<u>TTATA</u>TATACTGATTTGCTATTCTTTACTTT GCACATAACA
23S tRNA	ATAATTGGGCTTCCGGATCAAAATGC<u>TTATA</u>TCCCTCTTAATGATATAGTCCAT ACACGCTTACAAT
5S tRNA	GCTGAAGCCGTAAAATCACGAAAATC<u>TTATA</u>TAGATGTGTTCTATATAGTGT TCgGCAACGGTCATA
citrate synt.	TTAAAATATGTATACCGATCCCAAAA<u>TTATA</u>CGACGAAATTTCAAATAGCGTTAAAAGATATCATATA-(15n)-atg

Tc. celer
5S	TGTTGAGATGCTAATGACCGTAACC<u>TTTATA</u>AACCCAGGCCGGGAGTTCC GTACCGgTACGGCGGTC

Tf. pendens
rRNA operon	AGGACTTTTACAAGCATAATATTCA<u>TATAA</u>CCCCCCGTTTACTAACTAGATT GCCGCCATGGGCACC
tRNA^{Gly}	CTAGTCTTTGCACGCGGGAAAAGC<u>TTTTAAG</u>CATGCCTTTTTACTTCCTTCTAGAGGCTCA gCGGCC
tRNA^{Met}	CCGCCGGGGAACGCGATGCTAAAGG<u>TTTATT</u>ACCCAGGAAGTATTCCGGTCAT GGGGGGTTACGAAg
orf1	TGCTGTATCGGCATTCGACAAATATT<u>TTTAA</u>CCTCGTTCGAGGAGTATA GAGGTGGTTatgAGAGTCA
orf2	CCGGCTACTTCACCTTAAATAGCTT<u>TTTAA</u>CGTAGGAGGTGCTACGTCCAC GGGATCGTAatgGGGG

Tp. tenax
rRNA operon	TTTATGAATGCGGGAGCGAAAAATT<u>TTTAAA</u>TTTAGGGTGTTTTAGGATGGTC GCGCCTTAATTGTTT
tRNA^{Ala}	TCCGCGTGGCGCTCTAGCGAAAAAAT<u>TTAAAT</u>CGGTGAGTAAGTACGCTCG gGGGCCGGTAGTAGTC
tRNA^{Met}	TAGAGCGCCACGCGGACAAAAGCTT<u>TTTAAA</u>TTCGCGCAAAGCTTAGACCTA gCGGGGTAGGCCAGC
tRNA^{Ala}	GCGATGAGCTCGGAGGGGCTTAAAGC<u>TTAAA</u>ATATCCTGTCATATAACGAGTT gGGCCGGTAGTCTA
tRNA^{Leu}	CTGATcGAAcAAGGGGcTGAAAAAT<u>TTTAAA</u>CTGAGCAGTTTATATCAGAGAcGgcGGGGGTGcccGA
tRNA^{Leu}	AGGGATGTGCGTTGCGAGAAAACAT<u>TTTAAT</u>CCTGAGGAGAAAATACTGGACAG gCGGGGGTGCCCG

TTV1
T3^d	CGTGGTCAAGTGTGCCTATAAATCA<u>TATAAT</u>ATTACCATTTGGTACATCGGGAAGAGTAGCTTATTCA-(342n)-atg
T4^d	GCTCATAGGAAAGTAAATGAAAAAGT<u>TTTATA</u>AATAGATTGTTACTTATGCTTATGTCAGTAGTTGTTG-(850n)-atg
T1	TTATACCGCCATTTAGGGAGAAAGA<u>TTTAAA</u>TACCATGGATAGTAATTAGGAC AATGgttGAGATTA

Mt. fervidus
rRNA operon^{b,e}	TGAAGAAATATGTGAAATAGAAAAA<u>TTTATA</u>TTATCTCATAATTTGAAACCAATAATGATCAAATGTA
tRNA^{Thr} tRNA^{Pro}	ATTAAGAATTAAATTTTATTTATTA<u>TTTATA</u>AAATCGAAAAATATAAATATGGTTTTATCTAATCTAT
tRNA^{Asp} tRNA^{Lys}	TTAACACATTCATAAAACCGAAATA<u>TTTATA</u>TATTGATGGCTACCATACCAAACTATTTGGACACTAA
tRNA-operon	TCAAAAATTTTATTTGTCCGAAAAC<u>TTTATA</u>TATGAAAAATTCAAAGGTAAATTATAGCTAAGAAGAA
HMF p1	AAACATCAACGATGCAAACAATAAA<u>TTTATA</u>TAGGATAAATTTGATAATATTCTTTCGTAAGAATAGA-(12n)-atg
p2	AAACATCAACGATGCAAACAATAAA<u>TTTATA</u>TAGGATAAATTTGATAATATTCTTTCGTAAGAATAGA-(85n)-atg
p3^e	TTTGATACAATATCTATATTATCCA<u>TTTATA</u>TGGCTATCAAAGCACAATATAAACTTTTTATGGTTAA-(181n)-atg
mcr	AAAGAAAATTTCATTAAATATGTAA<u>TTTATA</u>CAATTATTCACAAAAAGATATTTAAAAAGGAGGTTAA-(3n)-atg
mvh	ATGTTGTAAAGAGACTATGATAATA<u>TTTATA</u>TGTGAGTAATCATCTACTGTGGATATATCATACTACA-(17n)-atg

M. thermoautotrophicum
frh	GAAAAACTTATAACATTTAAAAATAG<u>TTATA</u>ATATTGAAGGTTTTTTGAACTTTAAAAACAAGAGgtg

continued on next page

TABLE 4, continued

	Sequence
mcr	CGACTTGATTCAGAAAGAAATCTTAA<u>TTAAT</u>TATGATCAAGTTAAATATGCAC CGTATACTAAAA-(98n)-atg
purE	ATGGTCCCTGCGGGAGGTGCCATCTA<u>TTAAA</u>TATGACCTCCGATAGATTCTCTATATATAAAGTTAAA-(13n)-atg
vrh	TTTCGAAATTGTTTCATAAGTAACC<u>TTTATA</u>CTTACCACTCCTAAACCATATACCAAGTACAATTATA-(29n)-atg
pME2001	
orf	GCAGAAAATGTAGGGGTACACACCTTGCAGAATGGTACCCTGCATAAGGGTTGTAAA GCCCTGGCTT-(293n)-atg
Mc. thermolithotrophicus	
nif	AATTTAAATTTTTAAAATAAAAAAG<u>TTTATA</u>TATTATAAATACATATGTCTGCT TGTTGACAGGAAA-(87n)-atg

[a] DNA sequences corresponding to the transcript are aligned on the basis of the TTTA^A/$_T$A motif which is underlined and centred 25–27 nucleotides upstream from the putative transcriptional start sites. Putative transcriptional start sites are preceded by a space. The first nucleotide of mature 5S rRNAs or tRNAs is denoted by a single lower-case letter. Putative start codons are indicated by three lower-case letters. Accession Numbers to sequence databases, and literature references, are listed in Table 2. Names of organisms are abbreviated as in Table 1.
[b] The promotor is located within an ORF preceding the gene.
[c] The stop codon of the preceding ORF is indicated by lowercase letters.
[d] Initiation of transcription has been mapped close to this region.
[e] The underlined sequence is not that suggested in the original article.

some of the earlier putative motifs suggested for *Dc. mobilis* genes look doubtful despite supportive S_1 nuclease mapping data [12]; these include P4 and P5 of the rRNA operon and P2 of the 5S rRNA gene. Moreover, P1 of the *Dc. mobilis* rRNA operon and two neighboring ORFs exhibit motifs displaced 7–8 base pairs upstream (Table 4) which may, or may not, participate in transcriptional initiation. Finally, the degree of conservation of the location and sequence of these putative motifs correlates poorly with the multiple promoter motifs that precede the rRNA operons of the extreme halophile *Halobacterium halobium* [120,121]. This does not imply, however, that multiple promoters may be absent from the hyperthermophiles; the histone-like genes of *Mt. fervidus* also exhibit well-conserved promoter motifs repeated three times upstream from the transcriptional unit, although evidence for their involvement in transcriptional initiation is lacking [34]. A putative ORF on the methanogen plasmid pME2001 is not preceded by a characteristic promoter motif [25,26].

At this point, it is perhaps appropriate to emphasize that only a few 5' termini of transcripts have been assayed for a tri-phosphate group using the capping assay and, therefore, one cannot exclude that some of the transcriptional start sites determined by nuclease S_1 mapping or primer-directed reverse transcription (Table 4) correspond to processing sites.

Reiter et al. [116] investigated the importance of the sequence and composition of the upstream region of the rRNA operon of *Sulfolobus shibatae* (formerly *Sulfolobus* B12) for transcriptional initiation using an in vitro system employing *Sulfolobus* extracts. The approach of linker substitution mutagenesis was used to maintain a promoter region of constant length. They found that in addition to the TTTA^A/$_T$A motif centered on position −26, the region −11 to −2 (5'–CAATATGTATA–3'), which is always A–T-rich but not conserved in sequence (Table 4), is important. Moreover, they identified a region of negative control in the approximate region −93 to −38 which is also A–T-rich. Both lack any obvious sequence similarity with other archaeal promoter regions (Table 4).

With the above results in mind we undertook an analysis of how frequently the TTTA^A/$_T$A motif occurs generally in the available hyperthermophile sequences, and we could demonstrate that it is not confined to regions immediately upstream from genes. This indicated that other sequence regions must be involved in specific transcriptional initiation. This inference correlates with the characterization of a crude extract fraction from *S. shibatae* which, when added to the purified RNA polymerase, ensures specific recognition of the TTTA^A/$_T$A promoter motif, in an in vitro system, possibly as a result of factor(s) binding to the RNA polymerase and inducing specific initiation further downstream [122]. The latter results render it possible, as for RNA polymerase II activation in eukaryotes, that different transcription factors recognize, and bind to, short DNA regions and facilitate initiation.

We also surveyed the two control regions identified by Reiter et al. [116] for purine/pyrimidine distributions. Given the uncertainty of the alignments, the analysis was approximate. It was improved, however, first by aligning on the basis of the TTTA^A/$_T$A motif and second, for the −11 to −2 region, by aligning on the basis of the putative transcriptional starts. In the negative control region, runs of purines predominate in the regions −55 to −48 and −36 to −32, but in between the sequence shows a strong bias to alternating purines and pyrimidines. Similarly, in the positive control region (−11 to −2), there is a strong tendency to form alternating purines and pyrimidines in positions −10 to −4, in contrast to further upstream (−25 to −10) where stretches of purines or pyrimidines predominate. Regions containing alternating purine–pyrimidine sequences would produce relatively weak B-form double helices exhibiting a tendency to unwind and produce the Z-conformer [123]. For the downstream region (positions −11 to −2), at least, unwinding of the double helix could be concomitant with the start of transcription.

No significant sequence conservation was detected bordering the transcriptional start sites. The earlier box B which included the sequence TGCA located at the start site, where transcription starts at G [1], and later modified to ^{A}/$_T$TG^A/$_C$ [116], has not survived the accumulation of many new gene sequences (Table 4).

5.2. Terminators

Comparisons have been made between termination in archaea and rho-independent termination in bacteria where the 3′ termini of transcripts often map to polythymidine stretches in the non-coding DNA strand preceded by short inverted repeats (reviewed by Brown et al. [1]). It has also been proposed that inverted repeats following the polythymidine sequences participate in termination in mesophilic methanogens [124]. In the present study, we have investigated which features are common to sulfothermophile and thermophilic methanogen terminators, or putative terminators. An alignment of the downstream regions of thermophilic transcripts for which the 3′ ends have been mapped using S$_1$ nuclease is presented in Table 5. The following conclusions could be drawn.

The 3′ ends of transcripts are frequently ragged and map to one sequence region or to closely spaced regions. For a few of the stable RNA genes the termination site mapped to

TABLE 5
Transcriptional termination regions

Organism	Region	Sequence
Dc. mobilis	rRNA T1	TGGATTCGGGTATAAACCCTGTCTGCAGCGCTCGTAACGGTTTTTATTCTCTCGGGTTTTCATCAACTAGTAGTGTGTGGTATGGGTCTTAGTATATTTTTCA
	rRNA T2	TTCATCAACTAGTAGTGGTTGGGTATGGTCTGTCTTAGATTATTTTCAATGATGAAGATACCTACCTCCCTTTATGAGTGGTGTTGAGCAGTATTCAGGAACTAC
	5S	CCGGCCACGTCAGAACGGCCGTGAGGTCACGAGGCCTCGCACGCGTTCGAGCTGGGCCGGCACGCCTCTACCCGTGTCAAGGCATCCGTCTATTTAAGCCGGTGTGAGATA
Sl. acido-	tRNA-Ser	CCAGTGTGATCTTCACGCGCGGTTCAAATCCGCCCCGCGGTAGTGGGGGGTTTCCCCACAGCAGAAATCCAT
caldarius	ef-tu	TGGAGTTATTATAGATGTTAAACCAAGAAAGAGGTAAAGTAAATATTATTTGGGACGTTAAAGCAATTTTCGACTTTGATAAATACATTAGTTTCTTTGATGAAGTAGAGATAATGTCTCTG
	ef-2	AGATTTCATATCGTGATTTCTTTATGGAGGAGAATAATTTATTCGGGACGTTAAAGCAATTTTTCTCACAGTAGTCTAGTTATAAAAATGTAAGTAGTTAGTATGTCACAGA
	S7	AGCAATAAGGAAGAAGAAGAGATAGAGAGAATAGCGTTAAGCTCTAGGTAAAAATGTTAGCTAAATCTCTCTCAATGCTAGTCTAGTTATAAAAATGTAAGTAGTTAGTATGTCACAGA
	S10	GTCAGAGTACCTGACGATGTATACATAGAAATTGAGCTAATTTGATTTGAAATCCTATTATTTTTATAAGCACCTCSTATAATCTTATTAATGCCGGGTGCCCGAGTGGACTAAGGGCTGGCCT
SSV1	T1	TGATAAGATATACAAAGACTGAGGTGTGAGGGATGAAATAGCCCCTTTATAAAGTCATCATTTTTTGTAGGGATAGCACTATTCGGCCTATAAACAGTGTTGTA
	T2	TAGCGAAAAGGCTTAAGGCAGTGCACCGCAAGTAGCCTATATTAAGACCAAATTTCTTTTTCTGGAGTGGGTTAGGGGGTAGGTGATGTAATCTACATCTTGGGTTTCTCTTTCCGG
	T3	AGGAGGAAGTGCACAAAACTAACACATAACTGCATGCAGCCCCTTTATAAAGTCATAATGCAATATATGAAAAAACTGTTACGGTGTAGGTTCTATTT
	Tind-1	AGTGAAGAAAAATAGCTTTCGTAGAAGCAATAAATGATTTGTTCTAAACATATTTTTTGTTATACTCTAACCGTATCTATATATTATACATATATATAGATAGAGAGATAGTATGTAG
	Tind-2	AACAAAAAACTAACAAATCAACTCACCATTATACAAACTCAGAAAACATATTTTTGTTATACTCTAACCGTAAATGTTTGTTCGGTGATTTGGTGTCAATT
Tp. acido-	16S	CGTAGGGAACCTGCGATGATCACCTCCAAGTTAGAACACTGGCTGCATCTGCCAGACCATGAAATTTTTCTTTATTTTTGCCGTTGACTGTGAACTATGATGAAAATTA
philum	23S	GCTACTAAAGATCCGAAGGCACAATCCATGCTAAATTCTCGTCTATGAAGTCATAAGCGTGTTTGATTTTTTTTGCCGTTGACTGTGAACTATGCAATGATTAATCCTTGTC
	5S	CGTATTGCGTTGTACTGTATGCCGGAGGTACGGAAGCGCAATATGCTGTTACCACTTTTGAAATGGCAAAAGCACCTTCGCGAGTGTTTGCATTGTTAAAGACGCTTAAACT
Tc. celer	5S rRNA operon	CCCGTGACTCGGGTTCAAATCCCGACCCGGCCCAACAATTTTTCCCGAAAAGTCCCGTCTCCACCTCTGGGGGGACCACCCCCGTAGCCGGT
Tf. pendens	rRNA operon T1	AGATATTCCTCGACGAGGCTACGAAGACGATCTTCGGGGGTAGGGTCACCTTCTGATACTCGTACGTTTTTCTCTCTTTCCGGCGGAACGGCATGCTAAAGGTTATTACCCG
	rRNA operon T2	ACTCCTCAACGCGAACCCCCGTAGCCGGTGACTAGCGTGGACGAAACCCGGTTCAAATCCCGGCGGCCGCACTTCCCATCTCTTAGAAGAAGGCTACTCGTGCATAGTTCTCGCGCGGGTTCGCTAA
	tRNA-Gly	CAAGCCGAGCCTTAAGGCGCGGAAAACCGGTTCAAATCCCGGCGGCCGCACTTCCCATCTCTTAGAAGAAGGCTACTCGTGCATAGTTCTCGCGCGGGTTCGCTAA
	tRNA-Met T1	CTAAATGGCAAAAGCTTCGTGGGACTCGTCGTGGAACACGCTTGGGGTTATTGAGGTTCGCGTTTTTTAAAGTGAAAAGTCTGTGAAAATCGTTAAAAGCGTTG
	tRNA-Met T2	GTGGAAGCGTGATGGCTGAGAAATCTGTAGAAAGAAGAAAAGCCCCGGCGCGGTGGGGTACTCGTAATCCTTCTCTCTCGGTCCGAGGAGAGGCGGCTCCGCGCG
Tp. tenax	rRNA operon	CCCAATCGCCCGAGCGTGCGACGGCGGCGAAAGACCCGGCGCGGTGCGATCCCGGGTTCGAATCCCGGCCTCTCTGGGGCCTCCGTTTGCCTACAGTGTGTTCTGGGAGTTTATGCA
	tRNA-Ala	CGCCCAGGCTGTGCCGCCGTAGGCCTTCGCGCGCGGGATCCCGGCGCGGTGCGATCCCGGCGCGGTGCGATCCCCGGCCCATCAGAAAGTTTATTGAGACGT
	tRNA-Ala	CGCCCAGGCTGTGCCGCCGTAGGCCTTCGCGCGCGGGATCCCGGCGCGGTGCGATCCCGGCGCGGTGCGATCCCCGGCCCAACTACTATTCTCGACTGGGTTCAAATTTAGGTT
	tRNA-Leu	GTCAAAGGGGCAGGGCTCAGACCCTGTGGCGTAGGCCTTGCGTGGGTTGAATCCGGCGCCCACCCCGCACTATTCTCGACTGGGTTCAAATTTAGGTT
	tRNA-Met	GGGCTCATAAAGGGCTAAGGCCCGAGGAGCCCCGAGGTCCCGGTTCAAATCCCGGGCCCGCTAAGTGCACACGGCCTCCTACAT
Mb. thermo-	rRNA	TATAAACCTATTTTCATAGAATTAACTGTCATCTTAACGACTCAAGGTTCTCATTGAATGCCCAAGGTCCTCTATGACATGGCGTGATGCCCGGATGATTAGAAGAAACCCAGATGACGACAGACTGCC
autotrophicum	mcr	ATAACACCCGCAAAATAGATTAAGAAGGTGCATTACAAAACTTTTAAGTGTAATGCACCTATTTTTATGAAAAAATAAACAGGAAGAAATATGACCCTATGATAACA

[a] The terminator regions on the DNA strand corresponding to the transcript are aligned. Vertical arrows indicate the 3' termini of transcripts which have been mapped. The boxed sequences constitute polypyrimidine (T-rich) sequences which are probably required for the termination process. Underlined sequences correspond to regions which can generate double helices in the transcript. The horizontal arrows denote putative base-pairing regions, and when a square is attached this indicates either that the base-paired region occurs within a stable RNA structure or that it constitutes part of the processing stem of a large rRNA. Generally the mapping data for the terminator regions are published together with the sequence data (Table 2). The data for the SSV1 virus from *S. shibatae* are given in ref. [126]. Names of organisms are abbreviated as defined in Table 1.

a single nucleotide or dinucleotide, which could reflect either very specific termination of these transcripts or that some undetected processing has occurred. The 3′ ends of the transcripts generally map within polypyrimidine (T-rich) stretches ranging from 4 to 30 nucleotides with an average nucleotide length of about 15.

Although some kind of inverted repeat is detectable upstream from each termination site (Table 5), the secondary structures produced in the transcripts would have little in common. Thus, for the rRNA operons of *Dc. mobilis*, *Tf. pendens* and *Tp. tenax*, and for the large rRNA genes of *Tp. acidophilum*, termination occurred after the long processing stems of the rRNAs (Table 5). In contrast, transcripts from the 5S rRNA or tRNA genes generally terminate immediately after the long-range helix joining the two ends of the mature RNA (Table 5). For the remainder of the transcripts, potential stem–loop structures are long and irregular, containing bulges and internal loops. It is questionable, therefore, whether formation of these structures can provide a general signal for transcript release from the RNA polymerase.

We infer that it is unlikely that the thermophile terminators resemble bacterial rho-independent terminators. Moreover, this view is reinforced by examining the sequences for the bacterial terminators using a program in the GCG-software package from the University of Wisconsin. This is based on the observations of Brendel and Trifonov [125] that in addition to the inverted repeat preceding a polythymidine sequence many bacterial terminators share sequence features located both upstream and downstream from the termination site. Only one of the termination sites shown in Table 5 is predicted by this program.

Experience gained from S_1 nuclease mapping of stable RNA transcripts from both extreme halophile [88] and hyperthermophile genes [72] suggested that termination of archaeal transcripts might be relatively inefficient, with extensive read-through occurring between genes. One intergenic region, downstream from the rRNA operon of *Tf. pendens*, which exhibited multiple polypyrimidine (T-rich) sequences on both DNA strands, was considered to be an efficient terminating region in both directions [72]. The available sequence regions downstream from the known protein genes, which are illustrated in Table 6, reveal a similar distribution of multiple polythymidine sequences.

6. Translational signals

Here we summarize the signals carried by the mRNA which regulate initiation, elongation and termination of translation.

6.1. Initiation

The sequences immediately upstream from the open reading frames of both the sulfothermophiles and the thermophilic methanogens are presented in Table 7. Stretches of purines located 3–9 nucleotides upstream from the start codon which could interact with the polypyrimidine sequence present at the 3′ terminus of each archaeal 16S rRNA (Shine Dalgarno interaction) are underlined. For the thermophilic methanogen transcripts, as for those of the methanogen and extreme halophiles [1], a purine-rich sequence, of varying length, is present at the appropriate position. However, for the sulfothermophiles

TABLE 6

Sequences downstream from open reading frames – putative transcriptional termination signals[a]

	Sequence
S. acidocaldarius	
rpo*	CAGAtaaGCCTACATATATTAATGC<u>TTTTT</u>TAAATTTGAACTTGCTTCCATTCCTATAACTGCTATCTAAGT<u>TCTTTC</u>TCTG<u>CTCATCAACTAAC</u>TTCTTCTTATGAATATA
thermopsin	AATTtaaTTGTTGTCCAAGCCAAAATAAGGAAGAGAAAAAAGGAAAATGTGCATTCATTTTTTACTTTTCCACTG<u>TTCTTT</u>AATTCATTC<u>TT</u>AGCTGGCTCATTACTCTCTATAAATCAATATGAT-TAGGA
S. solfataricus	
AspATS	CCGAtagTACGATTGATGAGCTAAACTCCCATGACTGATTAGCAGTT<u>TTTTAATT</u>TCCACTTTT<u>TCTTC</u>ATTCGATTCTATTAGTTCAAGATTAGTTAC<u>TTT</u>AATGTCTTCATTACGTCA<u>TCTTT</u>CACTTCATTAATT
β-galactosidase	GCACTtaaACTTTCTCAAGTCTCACTATACCAAATGAGT<u>TTCTTT</u>TAATCTTATTCAATTCATT<u>TT</u>CATTAGATTGCAATACTTTCATACCT<u>TC</u>TATATTATTTTGTACC<u>TTT</u>GGGATCTACACTTAATGT
Tp. acidophilum	
citrate synthase	AAAGtgaAAAACTTTTCTCAAATTTTTATTAATATATTTATATACAACGGCGCTTGTTGTGGAAAT<u>TTCG</u>ATTAGCAGTAGCCATGGTCTGTG<u>TTTTC</u>GGATCC--------
Tf. pendens	
orf 1	CTTCtgtaTACTCGTACGTAT<u>TTTTTT</u>CCCGGGGGAGCCTTTCTACTCCTCAAACGCGAACCCCGTAGCCGTGACTAGCGTGGACGAACGCTAGACT<u>TTTTTT</u>CCGCCGGGAACGC
Mt. fervidus	
mvh*	AAAAtaa<u>TTTTATTTTCTT</u>TTTCTCCATGATTATTACTAATTA<u>TT</u>TGTATT<u>TTT</u>TATTAACACATT<u>TCTTTT</u>TTATT<u>TTTTT</u>TAGTAAAGGATATAAACAGT<u>TTT</u>TAACATAGCT<u>TC</u>AAAAACATCTGGAT
gab	TCAAtaaTAAGGAACCATGATCGAGTTGGACCAGCTGGCAACCCTATAGGTTATAAAGGAAGACTGTTGACGTAT<u>TT</u>GATT<u>ATT</u>TAAAAAATTAGGGCTTGATGCATATGAATATCAAGCTACATATGG<u>TTT</u>AAG
hmfb	GAAAtaaTAT<u>TTTTC</u>TT<u>TTTT</u>CCT<u>TTT</u>ATAT<u>TTTT</u>ATTGCAAATAAAAATCATACAAAAAT<u>TTTTT</u>GCTGCATT<u>CATCGTCG</u>CGGGTCGCCTATTCTATCTGT<u>TTGCAACTTC</u>AACAATGTCAATGCCAATAAT<u>TTTTTTT</u>
mcr*	TAAAtaaATTCT<u>TAA</u>CATAATTCAAAAT<u>TTTTT</u>GAT<u>TTTTT</u>TCT<u>TTCTT</u>CAAAACTTTTTCAAAACT<u>TTT</u>TCAAAAC<u>TTTT</u>TAAAGAATTAGTAAGTAATAACTTAGTTAATTAATAAGTTAAAAGGAAGAAT<u>GAT</u>ATGGACCCCGCAGTGCCTA
M. thermoautotrophicum	
sod	CCTCtcaaCAGTAATCACAGGAGATCATCCTCTCTTATCT<u>TTTTT</u>TTCAAATGATT<u>TTC</u>CTTGAGTCAGATCAGAGGCGGCCC<u>TTATTT</u>T<u>TTTT</u>TAAATCATCACC<u>TTGA</u>GTCAGAT<u>TT</u>AGAGGCGGCCC<u>TTATT</u>T<u>TTTTT</u>
frh*	TCTGtgaAAAAACTCGGATTAAACATGGAGCTCGTTGAGGAGATGGACATAGGTAAAGGAAAAT<u>TC</u>TGGG<u>TC</u>TACACCAGGACGATGTCTACACACTCCCCC<u>TCAAGGA</u>GACCCATGATACGATTAACAGGCAGGATGC
ftr	CTTCtga<u>TTTTTT</u>TATCCATGCCCTGATCCTATCAGGGTTGACTTT<u>TTTT</u>ATTC<u>TT</u>ATGCAGAATT<u>TTTT</u>CCATCATTT<u>CGCTTT</u>CAACTATT<u>TC</u>CCATATCT<u>CGT</u>CT<u>CG</u>CGGAGGTTAACACCGAGAGTATACCCC<u>TT</u>
mvh*	CAAAtaaCCCCCCTCCT<u>TTTT</u>GT<u>TTTT</u>CAGATGCAGGATTCCAGTTATCACAGACCATGTTATTATCACTGTTAATAT<u>TTTTA</u>TATT<u>TTT</u>CATATT<u>TAAA</u>ATT<u>TTTT</u>CAT<u>ATTT</u>AAAATTT<u>TTT</u>CAT<u>ATTT</u>AAAATT<u>TTT</u>CT<u>TTTT</u>GAAAACAT<u>TTT</u>

[a] Downstream regions of DNA strands corresponding to known mRNA transcripts. They are aligned at the stop codons which are indicated by lower case letters. Polypyrimidine sequences of five nucleotides or longer (and T-rich) are underlined; some of the longer underlined sequences contain occasional purines. Accession numbers and literature references for the genes are given in Table 2. Names of organisms are abbreviated as in Table 1. An asterisk indicates that the sequence lies downstream from an operon.

the results are less clear. Many transcripts lack such a purine stretch while for others only two or three consecutive purines are present. These results suggest that for many sulfothermophile transcripts, factors other than a Shine–Dalgarno interaction are important for ribosomal recognition, although no evidence has been found for 5' capping of mRNA [1]. The sequence between the purine-rich site and the initiation codon is generally A- and U-rich. AUG is the preferred initiation codon, but GUG (15%) and UUG (3%) are occasionally used.

6.2. Codon usage

Insufficient sequences are available to do a meaningful statistical analysis of codon usage in several sulfothermophile organisms. Instead, we selected those organisms for which many sequences are available, including *S. acidocaldarius,* and *Ag. fulgidus* which reduces sulphate and produces small amounts of methane and should probably be classified with the methanogens. These are compared with methanogens of decreasing thermophilicity (Table 3).

As mentioned earlier, the G+C content of the genome is, in general, reflected in the use of codons (Fig. 1) and, therefore, G+C-rich organisms favor codons which tend to end in G or C. Interestingly, *S. acidocaldarius* which has a relatively low G+C content (37%) strictly selects against the use of G or C in the third codon position (Table 8), even more so than the methanogens which exhibit even lower G+C contents (see Table 3). Another strong bias for the sulfothermophiles and methanogens, in general, is in the use of arginine codons; of the six possibilities, two, AGG and AGA, are strongly preferred and the other four are rarely used (see Table 8). Indeed, for one of the thermophilic methanogens, *Mb. thermolithotrophicum,* the other four were not found at all (data not shown). The low number of cysteine codons in the hyperthermophiles (Table 8) may reflect the instability of the amino acid at high temperatures [127].

The successful expression of sulfothermophile genes in *E. coli* [43,128], in good yield, reinforces the universality of the genetic code and, together with the development of a potential vector/transformation system for *Sulfolobus* [129], provides a solid basis for investigating the genetics of these organisms.

6.3. Termination

Analysis of termination signals amongst bacteria and eukaryotes have suggested that the nucleotide following the stop codon may also be involved in the release factor (RF) recognition, at least for highly expressed genes [130,131]. Thus, for RF-2-dependent termination at UAA or UGA codons in bacteria, there is a strong bias to a 3'-uridine while in eukaryotes these codons were generally followed by a purine. Therefore, we analysed this position for the thermophilic methanogens and sulfothermophiles. For the former, there is a similar bias of about 50% to uridine, following both UAA and UGA codons, as was inferred earlier from analyses of methanogens in general [130]. However, for the sulfothermophiles there is no strong bias to uridine although at least two-thirds of the examined sites are occupied by adenosine or uridine. While double stop codons are relatively common amongst the methanogens examined (17%), they are relatively rare amongst the sulphothermophiles (2%).

TABLE 7
Alignment of putative Shine–Dalgarno sequences of sulfothermophiles and thermophilic methanogens[a]

Sulfothermophiles			Thermophilic methanogens		
	16S	HO-UCCUCCACUAGG-5′		16S	HO-UCCUCCACUAGG-5′
Ag. fulgidus	rpoX	UCGGGGAGGUGAAGAAaug	Mt. fervidus	gab	UUUUGGAGGAAUGCUaug
Dc. mobilis	23S ivs	UCGAUGGGGUAGGAUUAaug		hmfb	AUAGAGAGGUGGUAAGUaug
Pb. organotrophum	23S ivs1	ATACGGGGGUGGCAAUUGUAUAgug		mcrA	AUUAGGAGGUGUUUAAAUaug
	23S ivs2	UAUACCCGUUUAUUUUUCUCCaug		mcrB	AAAAGGAGGUUAAUUAaug
Pc. woesei	gapdh	UAAAUCUGGUGAGGUAAAaug		mcrD	UCCUGGGGGCAGGAAaug
	orf y	AGGCUGGGGUCCGAGAaug		mcrG	AUUAGGAGGUGUUUAGAUaug
Tp. acidophilum	citr. syn.[a]	AUGUAGAGGUGUAUUAaug		mvhA	UAAAGGAGGGAAAUAAaug
Tp. tenax TTV1	tp1	UACAGGAGAGAAUCaug		mvhB	CUGUGGUGGUCUCaug
	tp2	UUGAUGAGUGCGUUAAUGCUaug		orf 285	UUUAAGAGUUAAAUAAUaug
	tp3	CCAUGGAUAGUAAUUAGGACAaug	M. th.-auto[a]	frhA	AACAAGAGGUGAUACAUuug
	tp4	ACCUAAAAUACUCUAUAUUgug		frhB	UAUAGGAGGCUGGAAAAaug
	tpx	AAUAUAUAGCUAUUGAGUAaug		frhD	GACUCUAGGGAAUAACaug
S. acidocaldarius	atpA	UUUAUUUGGUGAGUAAUaug		frhG	AAGCGGAGGUUUGAAAAgug
	atpB	UAAAAGAGGUGAGCUAAaug		ftr	UCAACUAGGUGAUAGUUaug
	atpD	AAUAAGAGACUUAUAAGUAAaug		mcrA	CGGGGGAGGUGUAGAUUaug
	atpG	UAAUUACCGUGGUAAGAaug		mcrB	GUUAGGAGGGAAAAAaug
	atpE	CAAAGGAGGAGAGGGGAAAaug		mcrG	UUUAGGAGGUGCAUGAUaug
	atpP	AUAUCACGGUGUCUUGAUaug		mvhA	AGAAGGAGGGUAAAUAGaug
	fus	-UAAACUGGUGAUUUAGACCuug		mvhB	CCAGAAAGGUGGUAAAaug
	rpoT	CUUAGGAGGUGCUUCAAaug		mvhD	GCAAGGAGGAAACUCUaug
	rpoU	AUAACAGGGUGAGUUUaug		mvhG	AUAAGGAGGAUUUCAAaug
	rpoX	UGGAAGAGGUGAAAAAaug		purE	UAAAGGUGAAUCUCCAGaug
	S7	AUGAGGAGAAAGAGaug		rpoV	UUUUGGAGGAAGUCCaug
	S10	UAAAUUAGGUGUUUUUaug		rpoU	ACCGGGAGGCUGUUACgug
	S12	AUCCUUUAAGAGAAGAaug		rpoT	CAAAGGAGAGAAUACCuug
	thermopsin	GAAAAUUAUUAAAAAGUGaug		rpoX	GGGGUAAGGUGAUUCUgug
	tuf	UGUAAGAGGUAAUUAGaug		sod	ACCGCAAAGUAGGGGgug
	orf 88	GAUGAUAUUACUGUACGCUACaug	Mc. th.-litho[a]	hisA	UACCUAGGGAAAGAaug
	orf 104	AAUAUGAGGUGAAAUGUAAaug		nifD	AGAAAAGGUGAGAAUAUAaug
	orf 130	CAGAAGAGUUAAAAUACaug		nifH	ACCAGGGGUGUUUGUAaug
S. solfataricus	AspATS	CGUUCUUCAUAUACACUUgug		nifK	UUUAAAGGGGAUUUAAaug
	bgaS	GUCAACUUAUAUCUGUCUaug		orf 105	UGAUAUGGUGACAUAaug
	lacS	UUAAUCUAAAAUAAAGCUaug		orf 128	AUAGUUGGUGAAGUUaug
	L46e	AGAGAGAGGAUGGAAaug			
S. shibatae SSV1	vp1	AAGUGCUAGAAGGUUCgug			
	vp2	AUAAGGAGGAGUGAUaug			
	vp3	AGACUGAGGUGUGAGGGaug			

[a] Names of organisms are abbreviated as in Table 1; also, *M. th.-auto, M. thermoautotrophicum; Mc. th.-litho, Mc. thermolithotrophicus;* citr. syn., citrate synthase.

TABLE 8
Comparison of codon usage in sulfothermophiles and in methanogens of varying thermophilicity[a]

Amino acid	Codon	Codon usage per 1000				
		S.a.	A.f.	M.f.	M.t.	M.v.
Ala	GCG	4.5	17.3	1.4	2.9	3.9
Ala	GCA	26.0	17.3	46.1	40.8	53.7
Ala	GCT	22.0	17.3	29.3	10.5	34.1
Ala	GCC	6.6	28.1	5.7	26.5	2.9
Arg	AGG	15.4	28.1	7.2	32.7	9.6
Arg	AGA	44.6	41.0	35.4	9.7	29.4
Arg	CGG	0.2	0.0	0.0	1.4	0.6
Arg	CGA	0.5	0.0	2.9	0.8	1.8
Arg	CGT	0.9	1.1	2.1	5.7	1.0
Arg	CGC	0.0	1.1	0.4	3.4	0.0
Asn	AAT	30.7	5.4	25.4	3.5	15.3
Asn	AAC	12.5	23.8	7.9	28.0	25.5
Asp	GAT	43.6	19.4	49.0	19.9	26.3
Asp	GAC	12.0	35.6	16.1	44.5	27.7
Cys	TGT	3.3	4.3	22.9	11.1	6.5
Cys	TGC	0.7	4.3	3.6	11.1	2.5
Gln	CAG	11.6	28.1	6.8	26.2	11.8
Gln	CAA	17.5	2.2	21.4	1.2	17.8
Glu	GAG	26.7	71.3	16.1	46.1	5.3
Glu	GAA	46.8	34.6	72.2	46.4	76.7
Gly	GGA	30.2	33.5	40.4	24.3	29.6
Gly	GGT	33.5	23.8	26.8	31.6	34.7
Gly	GGC	6.3	7.6	7.9	13.9	8.4
Gly	GGG	4.3	11.9	2.5	8.8	5.5
His	CAT	10.8	5.4	14.3	5.6	4.1
His	CAC	4.7	10.8	6.8	16.5	13.9
Ile	ATA	51.9	36.7	38.2	40.1	11.0
Ile	ATT	30.5	28.1	28.9	8.2	40.8
Ile	ATC	5.7	20.5	6.1	20.2	22.9
Leu	TTG	12.0	11.9	7.9	1.2	7.6
Leu	TTA	43.6	3.2	36.1	2.9	41.4
Leu	CTG	2.6	18.4	1.8	15.6	0.6
Leu	CTA	12.3	3.2	10.7	4.3	3.5

continued on next page

TABLE 8, continued

Amino acid	Codon	Codon usage per 1000				
		S.a.	A.f.	M.f.	M.t.	M.v.
Leu	CTT	13.2	24.8	13.9	20.3	16.1
Leu	CTC	2.8	20.5	1.8	33.6	6.7
Lys	AAG	26.0	52.9	11.1	36.8	13.9
Lys	AAA	42.5	16.2	61.1	19.7	70.6
Met	ATG	25.9	33.5	22.5	29.7	30.4
Phe	TTT	18.0	6.5	21.8	6.8	15.7
Phe	TTC	11.1	22.7	10.4	25.3	14.3
Pro	CCG	2.8	10.8	0.7	5.7	3.3
Pro	CCA	21.0	8.6	29.3	20.2	20.6
Pro	CCT	18.9	10.8	18.6	11.9	15.5
Pro	CCC	6.4	9.7	1.4	13.4	2.5
Ser	AGT	16.5	2.2	8.9	4.5	8.2
Ser	AGC	5.9	17.3	2.1	8.8	4.5
Ser	TCG	1.6	5.4	0.7	1.2	1.4
Ser	TCA	16.0	4.3	15.0	19.9	23.7
Ser	TCT	14.6	2.2	9.3	2.8	5.7
Ser	TCC	4.5	4.3	2.5	8.9	4.7
Thr	ACG	3.3	16.2	2.5	2.8	3.5
Thr	ACA	17.5	9.7	25.4	23.9	18.8
Thr	ACT	22.2	7.6	17.5	2.5	16.3
Thr	ACC	5.6	13.0	2.9	17.7	8.4
Trp	TGG	7.6	1.1	5.4	6.3	4.7
Tyr	TAT	28.6	3.2	24.7	4.3	8.2
Tyr	TAC	9.5	24.8	8.9	27.7	18.4
Val	GTG	9.4	18.4	7.9	19.3	3.5
Val	GTA	33.1	10.8	31.4	9.7	30.6
Val	GTT	31.8	32.4	33.6	24.8	47.9
Val	GTC	6.9	13.0	5.7	23.4	2.5
End	TGA	1.0	2.2	0.4	0.9	0.2
End	TAG	0.0	0.0	0.0	0.6	0.2
End	TAA	1.4	0.0	2.9	1.2	3.3
	Codons	5764	926	2799	6491	4902

[a] Names of organisms: S.a., *Sulfolobus acidocaldarius*; A.f., *Archaeoglobus fulgidus*; M.f., *Methanothermus fervidus*; M.t., *Methanobacterium thermoautotrophicum*; M.v., *Methanococcus vannielii*.

7. Phylogenetic considerations

The view that the archaea constitute a distinct, monophyletic, taxon [132] has received much support from experimental studies, for example from comparative analyses of both the transcriptional apparatus (ref. [115] and references therein) and the structures of the pre-rRNAs [108]. Tree building exercises based on archaeal gene sequences, and in particular those of the rRNAs, have reinforced the concept, although not without controversy. Lake [133] has argued that artefacts occur during the derivation of the archaeal tree which result from the slow evolution rates of the sulfothermophiles relative to methanogens and extreme halophiles. Using tree-building methods which minimize such potential errors, and applying them to sequences of both 16S RNA and elongation factor Tu, which both exhibit slow rates of evolution, Lake [133,134] provided evidence for a major division between these groups of organisms and classified the sulfothermophiles as a separate kingdom called the eocytes. Others have applied similar methods both to the rRNA [135] and RNA polymerase sequences [115,136,137] and have obtained results which, with certain reservations, favor monophyletic archaea. However, the controversy remains unresolved and has recently been complicated further by the demonstration that the tree topology can be influenced by the order in which pairs of aligned sequences are compared [134].

For trees derived for sulfothermophiles the problem of differing evolution rates is probably minimal. The tree presented in Fig. 3 was derived from alignments of partial 23S RNA sequences (inferred from gene sequences), deriving from the 3'-two thirds of the molecule, using a distance-matrix method [40]; similar topologies were obtained using maximum-parsimony and transversion-parsimony methods [40]. The tree defines the relationships amongst the sulfothermophiles. *Pc. furiosus* and *Tc. celer*, which are both classified as Thermococcales (ref. [3]; see Table 1), form a closely related cluster clearly separated from the other organisms. *S. solfataricus* of the order Sulfolobales is also fairly distant from the other organisms. The remaining organisms, classified as Thermoproteales (Table 1), show a more complex pattern. Thus, *Pb. islandicum*, *Pb. organotrophum* and *Tp. tenax* are closely clustered, while *Tf. pendens* is more distant and *Dc. mobilis*, *Dc. mucosus* and *St. marinus* cluster closely to *Pd. occultum*, a member of the Pyrodictiales. Thus, the current division of organisms between the orders Thermoproteales and Pyrodictiales (Table 1) is not compatible with the molecular sequence analysis.

This picture is complicated further by the observation that pairs of organisms with almost identical 23S rRNA sequences require classification as different species. Thus, the Desulfurococcus *Dc. mobilis* is flagellated, while *Dc. mucosus* lacks flagella and secretes a slimy polymer [97]. Moreover, of two Pyrobacula, *Pb. islandicum* is facultatively organotrophic while *Pb. organotrophum* is strictly organotrophic [138] and exhibits an unusual two-layered protein envelope [139]. The latter results reinforce the need for biological criteria as well as molecular ones for taxonomical classification.

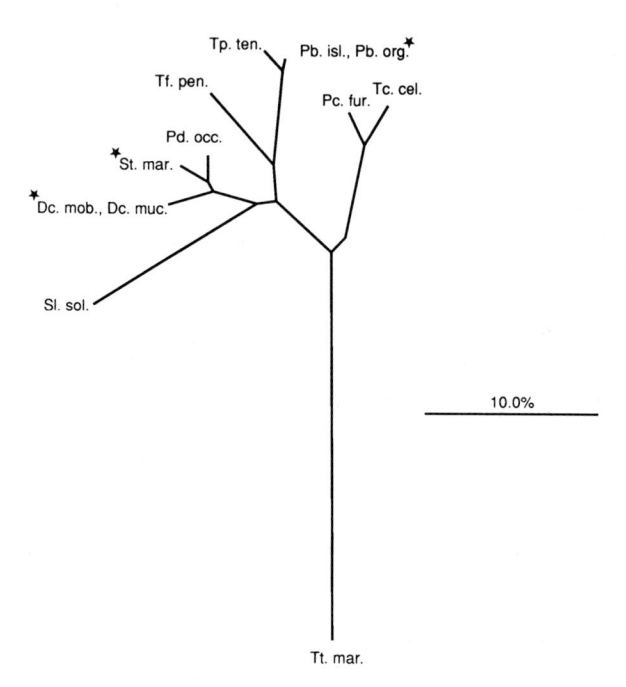

Fig. 3. A phylogenetic tree of the sulfothermophiles (crenarchaeota) derived from the available 23S ribosomal DNA sequences with the hyperthermophilic bacterium *Tt. maritima* as an outgroup. Full names of the organisms are given in Table 1. The distance measure corresponds to 10% mutational events per sequence position. Asterisks mark those organisms which contain introns within their 23S rRNA genes.

8. Summary

This survey defines our current knowledge of the primary structures of hyperthermophile genes and of sequence signals required for transcription and translation. The following conclusions are drawn:

1. No direct relationship exists between the optimal growth temperature of the archaeal thermophiles and their genomic G+C content. However, for the large rRNAs, the G+C content increases directly with optimal growth temperature and, at the highest temperatures, the C content increases relatively more than the G content, which suggests that G–U pairs may be converted to G–C pairs.

2. For the open reading frames the G+C content is directly proportional to the genomic G+C content but shows no direct relationship to the optimal growth temperature; at high growth temperatures the C content of the open reading frames increases more rapidly than the G content.

3. Consistent with the relatively small sizes of their genomes, the gene organization of the sulfothermophiles is simpler and more compact than that of the other archaea

including the hyperthermophilic methanogens. Characteristic features include low numbers of gene copies, short spacer regions between the large rRNA genes and between open reading frames.

4. Sulfothermophiles show a high incidence of archaeal-type introns relative to the extreme halophiles; none have yet been detected in either mesophilic or thermophilic methanogens.

5. All of the hyperthermophile promoter regions exhibit the archaeal $TTTA^A/_TA$ box centered about 26 nucleotides upstream from the transcriptional start site. Other promoter regions of negative control (-93 to -38) and positive control (-11 to -2) detected by Reiter et al. [116] contain, or overlap with, regions (-48 to -32 and -10 to -4, respectively) with a strong bias to alternating A–T/T–A pairs which generate weak B-form DNA.

6. Transcripts of 5S rRNAs and tRNAs generally start just before and terminate immediately after single genes amongst sulfothermophiles, while these stable RNA genes are often clustered in transcriptional units in the thermophilic methanogens and other archaea.

7. Transcripts from open reading frames and 16S-23S rRNA operons tend to terminate inefficiently at interspaced stretches of (T)-rich sequences. No common secondary structural motifs are associated with the terminator regions.

8. Putative Shine–Dalgarno polypurine sequences can be discerned just upstream from the start codons of the ORFs in all the archaea except the sulfothermophiles; the 3' ends of all of the 16S rRNAs exhibit a conserved polypyrimidine sequence.

9. The hyperthermophiles employ the universal genetic code but some biases are detectable, including preferential use of only two of the six arginine codons and rare usage of cysteine codons.

10. The sulfothermophiles cluster together in phylogenetic trees while the thermophilic methanogens branch with the extreme halophiles and methanogens. The results suggest that the preliminary classification of the orders Thermoproteales and Pyrodictiales (Table 1) needs revision.

Acknowledgements

All of the analyses presented in this review were performed using the Genetics Computer Group's (GCG) Sequence Analysis Software Package version 7.0, April 1991. The research from the authors' laboratory was supported by the Danish Science Research Council. J.Z. Dalgaard received a scholarship from the Carlsberg Foundation, and the NOVO Foundation supported our computer facilities. We thank Vicka Nissen for helping to prepare the manuscript.

References

[1] Brown, J.W., Daniels, C.J. and Reeve, J.N. (1989) CRC Crit. Rev. Microbiol. 16, 287–338.
[2] Forterre, P. (1993) In: Archaebacteria (Cowan, D., Ed.), Biotechnology Handbook, Plenum, in press.

[3] Stetter, K.O. (1986) In: Thermophiles: General, Molecular and Applied Microbiology (Brock, T.D., Ed.), pp. 39–74, Wiley, New York.
[4] Stetter, K.O., Fiala, G., Huber, G., Huber, R. and Segerer, A. (1990) FEMS Microbiol. Rev. 75, 117–124.
[5] Achenbach-Richter, L., Stetter, K.O. and Woese, C.R. (1987) Nature 327, 348–349.
[6] Kaine, B.P. (1990) Mol. Gen. Genet. 221, 315–321.
[7] Achenbach-Richter, L. and Woese, C.R. (1988) System. Appl. Microbiol. 10, 211–214.
[8] Leffers, H., Stetter, K.O. and Garrett, R.A. (1993) EMBL/GenBank databases. Accession Number X72728.
[9] Kjems, J., Garrett R.A., and Ansorge, W. (1987) System. Appl. Microbiol. 9, 22–28.
[10] Leffers, H., Kjems, J., Østergaard, L., Larsen, N., and Garrett, R.A. (1987) J. Mol. Biol. 195, 43–61.
[11] Kjems, J. and Garrett, R.A. (1985) Nature, 318, 675–677.
[12] Kjems, J. and Garrett, R.A. (1987) EMBO J. 6, 3521–3530.
[13] Kjems, J., Jensen, J., Olesen, T. and Garrett, R.A. (1989) Can. J. Microbiol. 35, 210–214.
[14] Østergaard, L., Larsen N., Leffers, H., Kjems, J., and Garrett, R.A. (1987) System. Appl. Microbiol. 9, 199–209.
[15] Willekens, P., Huysmans, E., Vandenberghe, A., and De Wachter, R. (1986) System. Appl. Microbiol. 7, 151–159.
[16] Gu, X.R., Nicoghosian, K., and Cedergren, R.J. (1984) FEBS Lett. 176, 462–466.
[17] Schallenberg, J., Moes, M., Truss, M., Reiser, W., Thomm, M., Stetter, K.O., and Klein, A. (1988) J. Bacteriol. 170, 2247–2253.
[18] Hamilton, P.T. and Reeve, J.N. (1985) J. Mol. Evol. 22, 351–360.
[19] Alex, L.A., Reeve, J.N., Orme-Johnson, W.H. and Walsh, C.T. (1990) Biochem. 29, 7237–7244.
[20] Jenal, U., Rechsteiner, T., Tan, P.Y., Buhlmann, E., Meile, L., and Leisinger, T. (1991) J. Biol. Chem. 266, 10570–10577.
[21] Reeve, J.N., Beckler, G.S., Cram, D.S., Hamilton, P.T., Brown, J.W., Krzycki, J.A., Kolodziej, A.F., Alex, L., Orme-Johnson, W.H. and Walsh, C.T. (1989) Proc. Natl. Acad. Sci. U.S.A. 86, 3031–3035.
[22] Bokranz, M., Bäumner, G., Allemansberger, R., Ankel-Fuchs, D. and Klein, A. (1988) J. Bacteriol. 170, 568–577.
[23] Takao, M., Oikawa, A. and Yasui, A. (1990) Arch. Biochem. Biophys. 283, 210–216.
[24] DiMarco, A.A., Sment, K. A, Konisky, J. and Wolfe, R.S. (1990) J. Biol. Chem. 265, 472–476.
[25] Bokranz, M., Klein, A. and Meile, L. (1990) Nucl. Acids Res. 18, 363.
[26] Meile, L., Madon, J., and Leisinger, T. (1988) J. Bacteriol. 170, 478–481.
[27] Bernaschoni, P., Rausch, T., Gogarten, J., Kibak, H. and Taiz, L. (1990) FEBS Lett. 259, 227–229.
[28] Rausch, T., Gogarten, J., Bernasconi, P., Kibak, H., and Taiz, L. (1990) Z. Naturforsch. C, Biosci. 44, 641–650.
[29] Weil, C.F., Beckler, G.S. and Reeve, J.N. (1987) J. Bacteriol. 169, 4857–4860.
[30] Souillard, N., and Sibold, L. (1990) EMBL/GenBank databases. Accession Number X13830.
[31] Haas, E.S., Brown, J.W., Daniels, C.J. and Reeve, J.N. (1990) Gene 90, 51–59.
[32] Haas, E.S., Daniels, C.J. and Reeve, J.N. (1989) Gene 77, 253–263.
[33] Fabry, S., Heppner, P., Dietmaier, W. and Hensel, R. (1990) Gene 91, 19–25.
[34] Sandman, K., Krzycki, J.A., Dobrinski, B., Lurz, R. and Reeve, J.N. (1990) Proc. Natl. Acad. Sci. U.S.A. 87, 5788–5791.
[35] Fabry, S. and Hensel, R. (1988) Gene 64, 189–197.
[36] Honka, E., Fabry, S., Niermann, T., Palm, P. and Hensel, R. (1990) Eur. J. Biochem. 188, 623–632.
[37] Weil, C.F., Cram, D.S., Sherf, B.A. and Reeve, J.N. (1988) J. Bacteriol. 170, 4718–4726.

[38] Steigerwald, V.J., Beckler, G.S. and Reeve, J.N. (1990) J. Bacteriol. 172, 1828.
[39] Bröckl, G., Behr, M., Fabry, S., Hensel, R., Kaudewitz, H., Biendl, E. and König, H. (1991) Eur. J. Biochem. 199, 147–52.
[40] Kjems, J., Larsen, N., Dalgaard, J.Z., Garrett, R.A., and Stetter, K.O. (1992) System. Appl. Microbiol. 15, 203–208.
[41] Dalgaard, J.Z., and Garrett, R.A. (1991) Gene 121, 103–110.
[42] De Wachter, R., Willekens, P., and Zillig, W. (1989) Nucl. Acids Res. 17, 5848.
[43] Zwickl, P., Fabry, S., Bogedain, C., Haas, A. and Hensel, R. (1990) J. Bacteriol. 172, 4329–4338.
[44] Kaine, B.P., Schurke, C. and Stetter, K.O. (1991) EMBL/GenBank databases. Accession Number M21087.
[45] Kjems, J. and Garrett, R.A. (1991) Proc. Natl. Acad. Sci. U.S.A. 88, 439–443.
[46] Stahl, D.A., Luehrsen, K.R., Woese, C.R. and Pace, N.R. (1981) Nucl. Acids Res. 9, 6129–6137.
[47] Kuchino, Y., Ihara, M., Yabusaki, Y. and Nishimura, S. (1982) Nature 298, 684–685.
[48] Auer, J., Spicker, G., Mayerhofer, L., Pühler, G. and Böck, A. (1991) System. Appl. Microbiol. 14, 14–22.
[49] Pühler, G., Lottspeich, F. and Zillig, W. (1989) Nucl. Acids Res. 17, 4517–4534.
[50] Schröder, J. and Klink, F. (1991) Eur. J. Biochem. 195, 321–327.
[51] Denda, K., Konishi, J., Oshima, T., Date, T. and Yoshida, M. (1988) J. Biol. Chem. 263, 6012–6015.
[52] Denda, K., Konishi, J., Oshima, T., Date, T. and Yoshida, M. (1988) J. Biol. Chem. 263, 17251–17254.
[53] Denda, K., Konishi, J., Hajiro, K., Oshima, T., Date, T. and Yoshida, M. (1990) J. Biol. Chem. 265, 21509–21513.
[54] Lin, X.L., and Tang, J. (1990) J. Biol. Chem. 265, 1490–1495.
[55] Olsen, G.J., Pace, N.R., Nuell, M., Kaine, B.P., Gupta, R. and Woese, C.R. (1985) J. Mol. Evol. 22, 301–307.
[56] Kaine, B.P. (1987) J. Mol. Evol. 25, 248–254.
[57] Kaine, B.P., Gupta, R. and Woese, C.R. (1983) Proc. Natl. Acad. Sci. U.S.A. 80, 3309–3312.
[58] Ramirez, C., Louie, K.A., and Matheson, A.T. (1991) FEBS Lett. 284, 39–41.
[59] Ramirez, C., Louie, K.A., and Matheson, A.T. (1989) FEBS Lett. 250, 416–418.
[60] Cubellis, M.V., Rozzo, C., Nitti, G., Arnone, M.I., Marino, G. and Sannia, G. (1989) Eur. J. Biochem. 186, 375–381.
[61] Little, S., Cartwright, P., Campbell, C., Prenneta, A., McChesney, J., Mountain, A., and Robinson, M. (1989) Nucl. Acids Res. 17, 7980.
[62] Cubellis, M.V., Rozzo, C., Montecucchi, P., and Rossi, M. (1990) Gene 94, 89–94.
[63] Stahl, D.A., Lane, D.J., Olsen, G.J. and Pace, N.R. (1985) Appl. Environ. Microbiol. 49, 1379–1384.
[64] Reiter, W.-D., Palm, P., Voos, W., Kaniecki, J., Grampp, B., Schulz, W. and Zillig, W. (1987) Nucl. Acids Res. 15, 5581–5595.
[65] Grogan, D.W., Palm, P. and Zillig, W. (1990) Arch. Microbiol. 154, 594–599.
[66] Achenbach-Richter, L., Gupta, R., Zillig, W. and Woese, C.R. (1988) System. Appl. Microbiol. 10, 231–240.
[67] Culham, D.E. and Nazar, R.N. (1988) Mol. Gen. Genet. 212, 382–385.
[68] Culham, D.E. and Nazar, R.N. (1989) Mol. Gen. Genet. 216, 412–416.
[69] McDougall, J. and Nazar, R.N. (1986) Biochem. Biophys. Res. Commun. 134, 1167–1174.
[70] Achenbach-Richter, L. and Woese, C.R. (1988) System. Appl. Microbiol. 10, 211–214.
[71] Auer, J., Spicker, G. and Böck, A. (1990) Nucl. Acids Res. 18, 3989.
[72] Kjems, J., Leffers, H., Olesen, T., Holz, I. and Garrett, R.A. (1990) System. Appl. Microbiol. 13, 117–127.

[73] Ree, H.K., Cao, K., Thurlow, D.L. and Zimmermann, R.A. (1989) Can. J. Microbiol. 35, 124–133.
[74] Ree, H.K. and Zimmermann, R.A. (1990) Nucl. Acids Res. 18, 4471–4478.
[75] Luehrsen, K.R., Fox, G.E. Kilpatrick, M.W., Walker, R.T., Domdey, H., Krupp, G., and Gross, H.J. (1981) Nucl. Acids Res. 9, 965–970.
[76] Kilpatrick, M.W. and Walker, R.T. (1981) Nucl. Acids Res. 9, 4387–4390.
[77] Tesch, A. and Klink, F. (1990) FEMS Microbiol. Lett. 71, 293–298.
[78] Pechmann, H., Tesch, A. and Klink, F. (1991) FEMS. Microbiol. Lett. 63, 51–56.
[79] Sutherland, K.J., Henneke, C.M., Towner, P., Hough, D.W. and Danson, M.J. (1990) Eur. J. Biochem. 194, 839–844.
[80] Zwickl, P., Lottspeich, F., Dahlmann, B., and Baumeister, D. (1991) FEBS Lett. 278, 217–221.
[81] Leinfelder, W., Jarsch, M. and Böck, A. (1985) System. Appl. Microbiol. 6, 164–170.
[82] Kjems, J., Leffers, H., Garrett, R.A., Wich, G., Leinfelder, W. and Böck, A. (1987) Nucl. Acids Res. 15, 4821–4835.
[83] Wich, G., Leinfelder, W. and Böck, A. (1987) EMBO J. 6, 523–528.
[84] Reiter, W.-D., Palm, P., Henschen, A., Lottspeich, F., Zillig, W. and Grampp, B. (1987) Mol. Gen. Genet. 206, 144–153.
[85] Neumann, H., Zillig, W., Schwass, V., and Eckerskorn, C. (1989) Mol. Gen. Genet. 217, 105–110.
[86] Neumann, H. and Zillig, W. (1990) Nucl. Acids Res. 18, 195.
[87] Neumann, H. and Zillig, W. (1990) Mol. Gen. Genet. 222, 435–437.
[88] Leffers, H., Gropp, F., Lottspeich, F., Zillig, W. and Garrett, R.A. (1988) J. Mol. Biol. 206, 1–17.
[89] Stetter, K.O., König, H. and Stackebrandt, E. (1983) System. Appl. Microbiol. 4, 535–551.
[90] Zillig, W., Holz, I., Klenk, H., Trent, J., Wunderl, S., Janekovic, D., Imsel, E. and Hass, B. (1987) System. Appl. Microbiol. 9, 62–70.
[91] Fiala, G. and Stetter, K.O. (1986) Arch. Microbiol. 145, 56–61.
[92] Huber, R., Kristjansson, J.K. and Stetter, K.O. (1987) Arch. Microbiol. 149, 95–101.
[93] Fiala, G., Stetter, K.O., Jannasch, H.W., Langworthy, T.A. and Madon, J. (1986) System. Appl. Microbiol. 8, 106–113.
[94] Zillig, W., Holz, I., Janekovic, D., Schäfer, W. and Reiter, W.D. (1983) System. Appl. Microbiol. 4, 88–94.
[95] Zillig, W., Stetter, K.O., Schäfer, W., Janekovic, D., Wunderl, S., Holz, I. and Palm, P. (1981) Zentralbl. Bakt. Hyg., I. Abt. Orig. C 62, 205–227.
[96] Zillig, W., Gierl, A. Schreiber, G., Wunderl, S., Janekovic, D., Stetter, K.O. and Klenk, H.P. (1983) System. Appl. Microbiol. 4, 79–87.
[97] Zillig, W., Stetter, K.O., Prangishvilli, D., Schäfer, W., Wunderl, S., Janekovic, D., Holz, I. and Palm, P. (1982) Zentralbl. Bakt. Hyg., I. Abt. Orig. C 2, 304–317.
[98] Zillig, W., Stetter, K.O., Wunderl, S., Schulz, W., Preiss, H. and Schulz, I. (1980) Arch. Microbiol. 125, 259–269.
[99] Stetter, K.O. (1988) System. Appl. Microbiol. 10, 172–173.
[100] Segerer, A., Langworthy, T.A. and Stetter, K.O. (1988) System. Appl. Microbiol. 10. 161–171.
[101] Stetter, K.O., Thomm, M., Winter, J., Wildgruber, G., Huber, H., Zillig, W., Janekovic, D., König, H., Palm, P. and Wunderl, S. (1981) Zentralbl. Bakt. Hyg., I. Abt. Orig. C 2, 166–178.
[102] Woese, C., Gupta, R., Hahn, C.M., Zillig, W. and Tu, J. (1984) Syst. Appl. Microbiol. 5, 97–105.
[103] McCloskey, J.A. (1986) System. Appl. Microbiol. 7, 246–252.
[104] Edmonds, C.G., Crain, P.F., Hashizume, T., Gupta, R., Stetter, K.O. and McCloskey, J.A. (1987) J. Chem. Soc. Chem. Commun., pp. 909–910.
[105] McCloskey, J.A., Edmonds, C.G., Gupta, R., Hashizume, T., Hocart, C.H. and Stetter, K.O. (1988) Nucl. Acids Res. Symp. 20, 45–46.

[106] Edmonds, C.G., Crain, P.F., Gupta, R., Hashizume, T., Hocart, C.H., Kowalak, J.A., Pomerantz, S.C., Stetter, K.O. and McCloskey, J.A. (1991) J. Bacteriol. 173, 3138–3148.
[107] Wada, A., Suyama, A. and Hanai, R. (1991) J. Mol. Evol. 32, 374–378.
[108] Kjems, J. and Garrett, R.A. (1990) J. Mol. Evol. 31, 25–32.
[109] Thompson, L.D. and Daniels, C.J. (1990) J. Biol. Chem. 265, 18104–18111.
[110] Kjems, J., Leffers, H., Olesen, T. and Garrett, R.A. (1989) J. Biol. Chem. 264, 17834–17837.
[111] Moazed, D. and Noller, H.F. (1989) Cell 57, 585–597.
[112] Garrett, R.A., Dalgaard, J., Larsen, N., Kjems, J. and Mankin, A.S. (1991) Trends Biochem. Sci. 16, 22–26.
[113] Kjems, J. and Garrett, R.A. (1988) Cell 54, 693–703.
[114] Woese, C.R. (1987) Microbiol. Rev. 51, 221–271.
[115] Pühler, G., Leffers, H., Gropp, F., Palm, P., Klenk, H.-P., Lottspeich, F., Garrett, R.A. and Zillig, W. (1989) Proc. Natl. Acad. Sci. U.S.A. 86, 4569–4573.
[116] Reiter, W.-D., Hüdepohl, U. and Zillig, W. (1990) Proc. Natl. Acad. Sci. U.S.A. 87, 9509–9513.
[117] Reiter, W.-D., Palm, P. and Zillig, W. (1988) Nucl. Acids Res. 16, 1–19.
[118] Thomm, M. and Wich, G. (1988) Nucl. Acids Res. 16, 151–163.
[119] Berghöfer, B., Kröckel, L., Körtner, C., Truss, M., Schallenberg, J. and Klein, A. (1988) Nucl. Acids Res. 16, 8113–8128.
[120] Dennis, P. (1985) J. Mol. Biol. 186, 457–461.
[121] Mankin, A.S. and Kagramanova, V.K. (1988) Nucl. Acids Res. 16, 4679–4692.
[122] Hüdepohl, U., Reiter, W.-D. and Zillig, W. (1990) Proc. Natl. Acad. Sci. U.S.A. 87, 5851–5855.
[123] Saenger, W. (1983) Principles of Nucleic Acid Structure, Springer, New York.
[124] Wich, G., Hummel, H., Jarsch, M., Bär, U. and Böck, A. (1986) Nucl. Acids Res. 14, 2459–2479.
[125] Brendel, V. and Trifonov, E.N. (1984) Nucl. Acids Res. 12, 4411–4427.
[126] Reiter, W.-D., Palm, P. and Zillig, W. (1988) Nucl. Acids Res. 16, 2445–2459.
[127] Volkin, D.B. and Klibanov, A.M. (1987) J. Biol. Chem. 262, 2945–2950.
[128] Cubellis, M.V., Rozzo, C., Montecucchi, P. and Rossi, M. (1990) Gene 94, 89–94.
[129] Martin, A., Yeats, S., Janekovic, D., Reiter, W.-D., Aicher, W. and Zillig, W. (1984) EMBO J. 3, 2165–2168.
[130] Brown, C.M., Stockwell, P.A., Trotman, C.N.A. and Tate, W.P. (1990) Nucl. Acids Res. 18, 6339–6345.
[131] Brown, C.M., Stockwell, P.A., Trotman, C.N.A. and Tate, W.P. (1990) Nucl. Acids Res. 18, 2079–2086.
[132] Fox, G.E., Stackebrandt, E., Hespell, R.B., Gibson, J., Maniloff, J., Dyer, T.A., Wolfe, R.S., Balch, W.E., Tanner, R.S., Magrum, L.J., Zablen, L.B., Blakemore, R., Gupta, R., Bonen, L., Lewis, B.J., Stahl, D.A., Luehrsen, K.R., Chen, K.N. and Woese, C.R. (1980) Science 209, 457–463.
[133] Lake, J. (1988) Nature 331, 184–186.
[134] Lake, J.A. (1991) Mol. Biol. Evol. 8, 328–335.
[135] Gouy, M. and Li, W-H. (1989) Nature 339, 145–147.
[136] Sidow, A. and Wilson, A.C. (1990) J. Mol. Evol. 31, 51–68.
[137] Iwabe, N., Kuma, K., Kishino, H. and Hasegawa, M. (1991) J. Mol. Evol. 32, 70–78.
[138] Huber, R., Kristjansson, J.K. and Stetter, K.O. (1987) Arch. Microbiol. 140, 95–101.
[139] Phipps, B.M., Huber, R. and Baumeister, W. (1991) Mol. Microbiol. 5, 253–265.

Epilogue

W. Ford DOOLITTLE

Department of Biochemistry, Dalhousie University, Halifax, N.S. B3H 4H7, Canada

Introduction

It has been little more than ten years since the first international meeting [1] bringing together molecular biologists and biochemists whose professional lives had been transformed by the "discovery" of the archaea. Many of the results of the last decade of intensive research are summarized in the chapters of this volume. In a very few words, I would like to position these results in the larger context of changing views on cellular evolution, and present some questions, cautions and prescriptions for the future.

1. Life's deepest branchings

Most of us have seen, in our lifetimes, three revolutions in the consensus view of what the natural lineages of organisms are, and how they are related. The first replaced the animal/plant dichotomy, still thought of as primary by non-biologists, with the prokaryote/eukaryote split described by Stanier and van Niel as "the largest and most profound single evolutionary discontinuity in the contemporary biological world" [2]. Both views still inform universal taxonomic–phylogenetic schemes such as that of Whittaker [3] presented in undergraduate biology texts. The second revolution, instigated by Lynn Margulis in 1967, gave new life to 19th-century European speculations about the prokaryotic origins of the cell-like organelles of eukaryotic respiration and photosynthesis [4]. Proving this *endosymbiont hypothesis* was the first serious and satisfactorily met challenge of the then infant discipline of molecular phylogenetics, now also textbook science. The third revolution, born of molecular phylogenetics, is that of the three kingdoms – Woese's revolution.

Woese surveys the field of battle in the Introduction to this volume. The war is clearly over, and we cannot return again to a simple prokaryote/eukaryote dichotomy of the kind drawn for us by Stanier and van Niel three decades ago. There may however be a deeper revolution ahead, one which challenges not our understanding of the relationships between evolutionary lineages but the very definition of lineage, and it is time for stock taking.

2. The coherence of the archaea

At that first international gathering of molecular biologists interested in archaea, there was a cheerful vagueness about the evolutionary position of these organisms [1]. In many ways, they were seen as primitive and/or as missing links between bacteria and eucarya, with properties of many of their cellular components or physiological processes arrayed in a spectrum from bacteria-like to eucarya-like, halophile–methanogen characteristics often on the first side, thermophilic features on the second. Fox, Luehrsen and Woese for instance ventured then that their "own prejudice is that the phylogenetic branching leading to the eukaryotic and eubacterial type ribosomes began among the archaebacteria" [5].

We all knew, however, that no organism can be other than metaphorically an evolutionary link between any other living organisms. If archaea *look* like universal ancestors that must be because they have evolved slowly overall – a dangerous notion because there is no clear reason why slow change in one organismal or molecular trait need imply slow change in others. One of the basic tenets of modern evolutionary biology is that the molecular clock keeps on ticking even when nothing happens to phenotype: horseshoe-crab hemoglobin has in fact evolved as fast as horse. And if some archaea more closely resemble bacteria while some look like eucarya, then the group itself is not coherent, but paraphyletic – as reptiles are, with some modern reptiles sisters to birds, some to mammals.

In fact almost all methods (including Lake's evolutionary parsimony) applied to almost all of the data – 16S ribosomal RNA sequences, 23S ribosomal RNA sequences, sequences of proteins and the genes encoding them – argue for the coherence of the archaea [6]. So does the strong resemblance in spacing and sequence of promoters from *Sulfolobus* [7] and *Methanococcus* [8]. So does the organization of genes for RNA-polymerase subunits in *Sulfolobus, Thermococcus, Methanococcus* and *Halobacterium,* which all show an *rpoH* gene encoding a homolog of a subunit of eucaryal RNA polymerases, not found in bacteria [9]. So do the very many features of the translation apparatus, lipid biochemistry and central metabolism reviewed in this volume. Few traits support a specific relationship between just some of the archaea and either bacteria or eucarya. It is most unlikely that we will soon retreat from the conclusion that archaea make up a monophyletic (single common ancestor, itself an archaeon), holophyletic (no descendants not archaea) group.

Each of these sorts of data also supports the deep split within the archaea, into what Woese, Kandler and Wheelis [10] call Euryarchaeota and Crenarchaeota. What the various data sets do not tell us unequivocally is whether, as a monophyletic assemblage, archaea are more recently diverged from eucarya or instead branch with the bacteria. As chapters in this volume reveal, ribosomal protein genes, those for RNA-polymerase subunits and translation elongation factors and some enzymes of metabolism (HMG–CoA reductase, for example [11]) show strong similarity specifically to their eukaryotic homologs, but others exhibit eubacterial affinities, or seem unique.

3. Rooting the universal tree

To resolve the question of branching order is to find the root of the universal tree – not in principle possible without an outgroup, and there can be no outgroup for all Life. Woese therefore rightly stresses the importance of the approach taken by Iwabe and collaborators[12], who used the products of gene duplications pre-dating the last common ancestor of all Life to provide paired trees, each serving as outgroup for the other. He also rightly stresses the fragility of this result. It rests on only two such pairs of paralogous genes, subunits of proton-pumping ATPases and translation factors EF-Tu and EF-G. One promising set of data might be provided by the aminoacyl–tRNA synthetases, whose use to this end I would like to encourage. G. Nagel and R. Doolittle[13] have recently shown that the available sequences for these enzymes form two (apparently nonhomologous) superfamilies. Within each superfamily, enzymes charging the same amino acid in different species are more alike than enzymes charging different amino acids in the same species. That is, each of the two superfamilies arose and diversified internally into its modern complement of families of amino-acid-specific enzymes before the appearance of the last common ancestor of all Life. We expect this already from the universality of the genetic code, but the fact that homology can still be shown between the several families within a superfamily means that each family can be used as an outgroup for a universal organismal tree constructed from sequences of members of each of the others, and the number of possible independently rooted independent trees is very large. So far only one archaeal aminoacyl–tRNA sequence, that of the *Methanobacterium thermoautotrophicum* isoleucine enzyme is available[14], and no tree including it has been published.

4. Implications of the root for eucarya

Accepting the rooting of Iwabe and coworkers (see Fig. 2 of the Introduction to this volume) leaves us, however, with another problem. Although perhaps the majority of primary gene sequence data is most consistent with a specific relationship between archaea and eucarya, gene organization in archaea often has a decidedly bacterial flavour. As the chapters in this volume on genes of halobacteria, methanogens and thermophiles show, archaea – like bacteria and unlike any known eucarya – have operons. Furthermore, many of these operons boast the very same genes in the very same order as known for *Escherichia coli*. The coincidences in order are far too extensive to be *just* coincidences. No reasonable scheme for convergence based on selection for coordinate regulation could possibly account for the similarity in gene order (but not in fact in transcriptional pattern) between archaeal and eubacterial ribosomal protein gene clusters, for instance. There is no way to avoid the conclusion that operons arose early – were present in the last common ancestor of all surviving organisms – and have been lost in eucarya, some time after they diverged from archaea.

In fact we don't know when after this divergence such loss occurred, or when many of the features of gene and genome organization and function we take as typically eukaryotic (division of labor and promoters between three classes of RNA polymerases, messenger binding by "scanning", interruption of coding sequences by introns) arose.

This is because eucaryal molecular biologists have limited their interests almost exclusively to animals, plants and fungi – all twigs at the tip of a single branch of the eucaryal tree (ref. [15]; see also Fig. 2 of the Introduction to this volume).

The little work that has been done on more early diverging eucarya has in fact produced major surprises. In trypanosomes, all messenger RNAs are transpliced to a short leader RNA, some genes are cotranscribed in an operon-like fashion, and coding sequences are not interrupted by introns. The diplomonad *Giardia* has a genome only a few times larger than *E. coli*'s, ribosomal RNA genes resembling those of archaea or bacteria in size and organization, and perhaps no splicing, either *trans* or *cis* [16]. It is possible to hope that this or some more early diverging archezoan will prove to look just like *Sulfolobus*, at the level of gene structure, function and organization, sporting only a nuclear membrane as its bid to membership in the eucarya.

5. Looking for "pre-adaptations" in archaea

In addition to thus pointing out that we should look for archaeal features in eucarya, the new rooting suggests that we might find, specifically in archaea, forerunners of the eucaryal state. Organisms of course do not truly "pre-adapt", do not prepare themselves for future evolutionary innovation by assembling precursors of no current use. But evolution *is* a tinkerer, and we expect that even complex structures and systems limited to eucarya were cobbled together from bits and pieces serving perhaps quite different functions in their immediate prokaryotic ancestors, bits and pieces we might detect by suitable tricks.

Searcy [17] has long claimed that *Thermoplasma acidophilum* produces true histones and proteins homologous to the actins and tubulins usually thought to be exclusively and always associated with the eucaryal cytoskeleton. Although there is, on the basis of the universal tree, no justification for proposing any single archaeal species as ancestral to eucarya, there is, as Woese argues in the Introduction to this volume and elsewhere [18], reason to suppose that the first archaeon, and the last common archaeal–eucaryal ancestor, was a thermophile. Searcy has recently [19] presented a more thoroughly articulated and testable hypothesis – that thermophiles maintain their irregular shape with the aid of a cytoskeleton (evolved as an adaptation for respiration on elemental sulfur), a pre-adaptation for the development of complex eucaryal cells.

More certain so far is the finding among thermophiles of the histone HMf, at the level of primary sequence and apparent DNA-compaction ability a much closer homolog of eucaryal histones than the "histone-like" proteins of eubacteria [20]. Although HMf effects positive supercoiling (while histones negatively supercoil), it is tempting to suggest that proteins stabilizing thermophilic archaeal DNA gave rise to eucaryal chromatin. More tantalizing still is the very recent finding, by Trent and collaborators [21] that a heat-shock protein of *Sulfolobus shibatae* is a chaperonin similar in structure but not primary sequence to bacterial chaperonins, and highly similar in sequence to the essential yeast protein TCP1, probably involved in mitotic spindle formation.

6. More courageous scenarios

There are other ways to reconcile parts of the data. Zillig – prompted by the observation that archaeal RNA-polymerase subunits are more similar in sequence to the subunits of eucaryal RNA polymerase II and III than these two are to those of RNA polymerase I, while this last most resembles bacterial RNA polymerases – suggests (ref. [22], and Chapter 12, this volume) that the eucaryal cell is more extensively chimaeric than Margulis imagined. By this view, nuclei were produced by the early fusion of archaeal and bacterial genomic lineages, and trees such as Woese's Fig. 2 (Introduction) reflect the history of only the first sort. Sogin's most recent suggestion [15] is still more radical – that the eukaryotic cell was produced through the engulfment of an archaebacterium by an RNA-genomed host, which provided both the machinery for translation and RNA processing and a cytoskeleton. This daring scheme is, as Sogin points out, testable in parts – the attempts to find archaeal pre-adaptations which could have given rise to the eucaryal cytoskeleton outlined above should fail resoundingly, for instance.

Such extensive reticulate evolution, or radical chimaerism, is of course an important feature of the "progenote" as first imagined by Fox and Woese. Woese wrote that "the universal ancestor progenote is not a single (well-defined) species. It is a varied collection of entities – cellular and subcellular – that exchange genetic information (and molecular structure) somewhat freely, with the result that in evolutionary perspective, progenotes appear to constitute a single species" [23]. There is no question that modern cells evolved from more primitive ancestors, and no reason not to suppose that genomes were much less cohesive units at the time of such ancestors. What is less clear is whether the last common ancestor of all surviving cells was such a primitive creature. If we are right in assuming that this ancestor already had operons, then such multigenic units may have been the independent contributors to Woese's "state of genetic communion" [23], as suggested by Zillig (Chapter 12, this volume, and personal communication).

7. The need for caution and more data

Most molecular biology/cell biology texts still discuss gene structure and function in terms of the prokaryote/eukaryote dichotomy and, what is worse, mean by this that a comparison between *E. coli* and mammalian cells in culture not only plumbs evolution's depth but gives a good idea of its direction and progress. We must avoid replacing this view with a similarly over-simplified trichotomy, although the temptation is strong because it is so difficult to assimilate the diversity of fragmentary data on gene structure and function within and between species, and so difficult to find general patterns once diversity is embraced.

We pretend for instance that bacteria have but a single kind of promoter, different from promoters in archaea and eucarya, when in fact there is within *E. coli* (among its many types of promoter) a minor (σ^{45}) class of RNA-polymerase recognition and binding sites more similar in structure to eukaryotic RNA polymerase II promoters than the usual (σ^{70}) class [24]. We contrast the histone-choked genes of "higher" eucarya with the presumed naked DNA of bacteria when speculating on the evolution of genetic regulation, although the bacterial nucleoid is highly structured and we have no clear understanding of the role of the various known prokaryotic DNA-binding proteins in gene expression, nor any real appreciation for the diversity of chromatin structures within most of the diverse eucarya, which are protists. We look for possible Shine–Dalgarno sequences upstream of archaeal genes and marvel when they are absent, though in reality some eubacterial messengers function without them [25].

We must deal with diversity by accepting it as part of evolution's pattern which will often put the lie to simple scenarios. We must face the possibility that no single gene tree may ever faithfully trace early organismal lineages – partly because there *has* been reticulate evolution (perhaps more of it earlier) and partly because no tree-building methods are perfect and no alignments are unequivocal. We must hope, with Woese, that some fraction of the funds expended on the junk-laden human genome will be turned to sequencing entire archaeal and bacterial genomes, as a well as a few of the smaller archezoal genomes [16]. With as few as ten such in hand (the equivalent of 1% of the human genome) we would have the basis for constructing several thousand gene trees, reconstructing scores of gene families and superfamilies, documenting many instances of the evolution of new enzymatic function, and assessing the roles of various sorts of genetic rearrangement and mechanisms of mutation. We could also then assess the extent of chimaerism within lineages – the degree to which any tree reflects the "true" evolutionary history of lineages of organisms. We could also then make sensible guesses about the timing and nature of major phenotypic innovations in the evolution of each of Life's three domains.

8. Archaea here and now

Questions about Life's earliest history may never be unequivocally answered. A more tangible product of the last ten years of archaeal research is a body of solid information about the biochemistry and physiology of these organisms and a growing set of molecular approaches and genetic tools for their deeper exploration and more facile exploitation. It seems unlikely that further undirected cloning and sequencing of genes will yield major surprises, equivalent in their evolutionary implications, say, to the discovery of introns. But we are in a position to understand specific adaptations of individual species to their chosen niches, and should pursue such questions with the same zeal as we have always shown for bacterial or eucaryal microbiology. We are also in position to design experimental systems to ask more general basic questions about protein structure and function at high ionic strength and extreme temperature, systems which can bring to bear the full power of mutation and selection in the laboratory. The biology of the archaea is now a mature scientific discipline.

References

[1] Kandler, O., Ed. (1982) Archaebacteria: Proc. First International Workshop on Archaebacteria, pp. 1–366, Gustav-Fischer Verlag, Stuttgart.
[2] Stanier, R.Y. and van Niel, C.B. (1962) Arch. Microbiol. 42, 17–35.
[3] Whittaker, R.H. (1969) Science 163, 150–162.

[4] Margulis, L. (1970) Origin of Eukaryotic Cells, pp. 1–349, Yale University Press, New Haven, CT.
[5] Fox, G.E., Luehrsen, K.R. and Woese, C.R. (1982) In: Archaebacteria: Proc. First International Workshop on Archaebacteria (Kandler, O., Ed.), pp. 330–345, Gustav-Fischer Verlag, Stuttgart.
[6] Gouy, M. and Li, W.-H. (1989) Nature 339, 145–147.
[7] Reiter, W.D., Hüdepohl, U. and Zillig, W. (1990) Proc. Natl. Acad. Sci. U.S.A. 87, 9509–9513.
[8] Hauser, W., Frey, G. and Thomm, M., (1991). J. Mol. Biol. 222, 495–508.
[9] Klenk, H. P., Palm, P., Lottspeich, F. and Zillig, W. (1991) Proc. Natl. Acad. Sci. U.S.A. 89, 407–410.
[10] Woese, C.R., Kandler, O. and Wheelis, M.L. (1990) Proc. Natl. Acad. Sci. U.S.A. 87, 4576–4579.
[11] Lam, W.L. and Doolittle, W.F. (1991) J. Biol. Chem. 267, 5829–5834.
[12] Iwabe, N., Kuma, K.I., Hasegawa, M., Osawa, S. and Miyata, T. (1989) Proc. Natl. Acad. Sci. U.S.A. 86, 9355–9359.
[13] Nagel, G.M. and Doolittle, R.F. (1991) Proc. Natl. Acad. Sci. U.S.A. 88, 8121–8125.
[14] Jenal, U., Rechsteiner, T., Tan, P.-Y., Bühlmann, E., Meile, L. and Leisinger, T. (1991) J. Biol. Chem. 266, 10570–10577.
[15] Sogin, M.L. (1991) Curr. Op. Genet. Dev. 1, 457–463.
[16] Adam, R.D. (1991) Microbiol. Rev. 55, 706–732.
[17] Searcy, D., Stein, D. and Green. G. (1978) BioSystems 10, 19–28.
[18] Woese, C.R. (1987) Microbiol. Rev. 51, 221–271.
[19] Searcy, D.G. and Hixon, W.G. (1991 BioSystems 25, 1–11.
[20] Sandman, K., Krzycki, J.A., Dobrinski, B., Lurz, R. and Reeve, J.N. (1990) Proc. Natl. Acad. Sci. U.S.A. 87, 5788–5791.
[21] Trent, J.D., Nimmesgern, E., Wall, J.S., Hartl, F.-U., and Horwich, A.L. (1991) Nature 354, 490–493.
[22] Zillig, W., Klenk, H.-P., Palm, P., Leffers, H., Pühler, G., Gropp, F. and Garrett, R.A. (1989) Endocytobiosis Cell Res. 6, 1–25.
[23] Woese, C.R. (1982) In: Archaebacteria: Proc. First International Workshop on Archaebacteria (Kandler, O., Ed.), pp. 1–17, Gustav-Fischer Verlag, Stuttgart.
[24] Gralla, J.D. (1991) Cell 66, 415–418.
[25] Gold, L. and Stormo, G. (1987) In: *Escherichia coli* and *Salmonella typhimurium*: Cellular and Molecular Biology (Neidhardt, F.C., Ed.), pp. 1303–1307, ASM Press, Washington, D.C.

Subject index

acetate
  methanogenesis from, 41, 61–65, 95–100, 118, 141–143, 147–153
  synthesis of, 141–143
    pyruvate as substrate for, 154–155, 162–164
  transport of, 156
acetate kinase, 61–62, 99–100, 147–148
  acetate formation from pyruvate and, 154–155
acetyl–CoA, 2, 6, 8–10, 11–13, 17
  acetate methanogenesis and, 61–65, 95–100, 147–153
  lactate oxidation and, 159–160
acetyl–CoA synthetase, 6, 13–14, 62, 99–100, 153
  ADP-forming, 162–164
  gene for, 506
acs gene, 506
adenine biosynthesis, 510
adenylate kinase, 62, 153
ADP-forming acetyl–CoA synthetase, 162–164
alcohol dehydrogenase, 57–58, 72–73, 141, 522
  of thermophiles, 217–218
alcohols, multiple-carbon, methanogenesis from, 41, 57–58, 72–73, 116–117, 139–141
aldol cleavage, 2, 4, 6
amino acid(s)
  metabolism of, 14
    genes for, 507–510
  Na^+, amino acid-symporters, halobacterial, 35–36
  transport of, 157
aminoacyl-tRNA synthetase(s), 394, 510, 514–515, 567
α-amylase, thermophilic, 217
anisomycin, 417, 418, 444
anthranilate synthetase, 509
antibiotic protein synthesis inhibitors, 416–427
  sensitivity to, 420–424, 427
aphidicolin, 351–357
Archaeoglobus fulgidus, xxii–xxiii
  central metabolism in, 14
  energetics of, 159–161

  ribosomes of, 405
  RNA polymerase of, 369, 373
  S-layers of, 247–248
archaeol (diphytanylglycerol diether), 261, 262–263, 265, 278
  biosynthesis of, 278, 279–285
    glycerol in, 279–282
    geranylgeranyl-PP in, 279–283
  membrane function and, 287–288
argG genes, 507–508
arginine
  biosynthesis of, 507–508
  fermentation of, 27, 33
argininosuccinate synthetase, 507–508
ATP (adenosine triphosphate), 2, 4, 6, 13–14
  halobacterial bioenergetics and, 27, 33–34, 37
  methanogen bioenergetics and, 113, 124–133, 144, 148–153
atpA/atpB genes, 487, 518
ATPase(s), 6, 297, 299–307, 318–319
  evolution of, 359, 377
  genes for, 487, 518
  of halobacteria, 27, 33–34, 37, 304–307, 487
  isolation of, 298–299
  of methanogens, 131–132, 300–302, 518
  of thermophiles, 216, 302–304

B_{12} cobamides (corrinoids), 51, 55–56, 58–59, 63–64, 93–94, 519
bacteriophage
  ΦH1, 469, 471, 472, 473, 487
  ψM1, 498–499
bacteriorhodopsin, 25–27, 32, 173–174, 189–202 passim
  energetics of, 201–202
  gene for, 471–472, 474, 481–482
  ion transport by, 27–30, 199–201
  photoreception by, 25, 30–32, 179, 196–199
  structure of, 190–196
bacterioruberin, 471
bacteropterins, 78
bat gene, 481–482
benzylviologen-reducing hydrogenase, 67

biotechnology, hyperthermophilic proteins and, 216–218
bop gene, 385, 395, 481–482
brp gene, 383, 385, 395, 481–482

caldarchaeol (dibiphytanyldiglycerol tetraether), 261, 264, 265
  biosynthesis of, 278, 285–286
  membrane function and, 288
carbamyl-phosphate synthetase, 507–508
carbon dioxide (CO_2)
  fixation of, 8, 12
  methanogenesis from, 53–58, 116–117, 119–143
carbon monoxide dehydrogenase (CODH)
  in *Archaeoglobus fulgidus*, 159–160
  in methanogens, 63, 64, 65, 95–98, 148, 506, 522
cdh genes, 506, 522
cell envelope(s), ix, 223–254
  see also S-layer proteins
  of gram-negative archaea, 239–252
  of gram-positive archaea, 223–239
central metabolism, 1–20
  see also individual enzymes
  comparative enzymology of, 18–20
  evolution of, 15–18
chloramphenicol, 417, 418
chloride ions (Cl^-), halobacterial transport of, 26, 30, 173, 174, 194–195, 198
chondroitin sulfate, 233
chromosome(s), 326–332, 468–469
  histones, x, xxvi, 327–331, 357–358, 568, 569–570
  genes for, 520, 568
  thermophile, 216, 568
  nucleosomes, 331, 520
citrate synthase, 11
  comparative enzymology, 18–20
citric acid cycle, 2, 11–13
  evolution of, 17–18
CO_2 fixation, 8, 12
CO_2 methanogenesis, 53–58, 116–117, 119–143
cob genes, 519
cobamides (corrinoids), 51, 55–56, 58–59, 63–64, 93–94
  C/Fe–sulfur protein, 63–64
  genes involved in, 519
coenzyme(s)
  see also individual coenzymes

  in methanogenesis, 43–53
coenzyme A (CoA), 6, 8, 10, 63, 141, 142, 149, 150, 152–154, 160, 163
coenzyme F_{420} (CoF_{420}), 45, 83, 93, 123, 124, 130, 141, 154, 159, 160, 503, 505
coenzyme F_{430}, 51–52, 91, 119, 121, 124
coenzyme M (CoM), 45, 49, 50–51, 55–56, 57, 58–60, 64–65, 87–94, 119, 121–124, 128, 144, 146, 150, 156, 159
coenzyme M (CoM)
  transport of, 156–157
corA gene, 519
corrinoids (cobamides), 51, 55–56, 58–59, 63–64, 93–94, 519
csg gene, 484
cyclic 2,3-diphosphoglycerate, 215
cyclic GMP, sensory transduction and, 182–183
cycloheximide, 418
2,3-cyclopyrophosphoglycerate, 7
cytochrome(s), 312–316
  of halobacteria, 33, 314–316
  of methanogens, 60, 65, 69–70, 129–130, 147, 148, 312
  of thermophiles, 312–314

demethylation, sensory transduction and, 181–182
denitrification, membrane-bound enzymes of, 317–318
dhf gene, 485–486
dibiphytanyldiglycerol tetraether (caldarchaeol), 261, 264, 265
  biosynthesis of, 278, 285–286
  membrane function and, 288
dibiphytanylglycerol nonitoltetraether (nonitol-caldarchaeol), 264, 265, 278, 286–287
dihydrofolate reductase gene, 476, 485–486
dihydrolipoamide dehydrogenase, 10, 17, 18
dihydroxyacetone, 15
diphthamide, 425, 426
Diphtheria toxin, 425–426, 427
diphytanylglycerol diether (archaeol), 261, 262–263, 265, 278
  biosynthesis of, 278, 279–285
  membrane function and, 287–288
discovery of archaea, vii–xi
DNA
  see also gene(s); genome
  replication of, 326, 351–357

stability of, 216, 331–332, 340, 342, 536, 541–543
topology of, 333–351
DNA gyrase (reverse gyrase), 216, 336–342, 349–350
genes for, 486
DNA polymerase(s), 351–357
of halobacteria, 356
of methanogens, 353, 355, 356
of thermophiles, 218, 353–355, 356
DNA topoisomerase(s), 333–351
evolution of, 347–349, 350
in halophiles, 346–349
in thermophiles, 343–346
type I, 333–335, 339, 343
reverse gyrase, 216, 336–342, 349–350, 476, 486
type II, 333–335, 342–343, 346–349
E. coli DNA topoisomerase IV, 335
DNA-binding proteins, see histones

ecology, microbial, xxv–xxvi
electron transport chain, membrane-bound enzymes of, 297, 308–316
elongation factors (EFs), 396–400
genes for, 398–400, 489, 513
inhibitors of, 416–417, 418, 425–427
interchangeability of, 429
Embden–Meyerhof pathway, 2, 4, 7, 15–17
endosymbiosis, xxvi, 565
origin of mitochondria and, 17
energy transduction
in extreme halophiles, 25–37
in methanogens, 113–156
in thermophiles, 161–164
Entner–Doudoroff pathway, 2–4, 5–6, 7, 16–17
epipodophyllotoxins, 346
erythromycin, 421–422, 444
esterase, thermophilic, 217
eukaryotes: prokaryote–eukaryote dichotomy, viii–xix, xxvi, 565
evolution, viii–xxvii, 565–570
of ATPase, 359, 377
of central metabolism, 15–18, 20
of DNA topoisomerases, 347–349, 350
of genome, 326, 358–360, 566–570
of membrane lipids, ix, 288, 289–291
prokaryote–eukaryote dichotomy, viii–xix, xxvi, 565
reticulate, 569, 570
of ribosomes, 458–460

of transcription, 373–377, 569
of translation, x–xi, xv–xix
extreme halophiles, see halobacteria
extreme thermophiles, see thermophiles

F_{420}, methanogenesis and, 45–47, 55, 56, 57–58, 60, 65, 72–75, 82–84, 93, 123, 503, 505
F_{420}-nonreducing hydrogenase, 56, 57, 65
F_{420}-reducing hydrogenase (FRH), 57, 65, 66, 68, 70–72, 124, 503, 505
gene for, 503
F_{430}, 51–52, 91, 119, 121, 124
fatty acid synthetase (FAS), 279, 289–291
fdh operon, 504–505
ferredoxin
methanogenesis and, 63, 64, 69, 148, 130–131, 504
polyferredoxin, 68–69, 130–131, 504
Pyrococcus furiosus and, 162
thermophile, 213
ferredoxin oxidoreductase, 8–10, 13, 17
flagella, 27, 36–37, 174, 175–183
flagellin(s), 176, 521
genes for, 484–485, 521
fluoroquinolones, 346
formaldehyde as methanogenic substrate, 119, 126–127, 135–137, 144–147
formate
methanogenesis from, 73–75, 116, 139
transport of, 156
formate dehydrogenase (FDH), 50, 73–75, 116, 139
genes for, 504–505
formyl-glycineamidinine ribonucleotide synthetase, 510
formylmethanofuran, 48, 53–54, 57, 75–78, 119, 122–123
formylmethanofuran dehydrogenase, 48, 49, 53–54, 57, 75–78, 119, 122–123, 136
formylmethanofuran:H_4MPT formyltransferase, 48, 49, 54, 78–79, 123
gene for, 505–506
frh operon, 503
fructose catabolism, 4
ftr gene, 505–506
fumarate, light-induced release of, 180–181
fumarate-binding protein, 181
N-furfurylformamide, 48, 78
fusidic acid, 425, 427

β-galactosidases, 385
gap genes, 516–517
gas vesicle genes/proteins, 383, 385, 469, 471, 483–484
gene(s)
  see also individual genes
  evolution of, x–xi, 566–570
  paralogous, xvii, xix, 359, 567
genetic code, evolution of, x–xi
genome, 326–332
  histones, x, xxvi, 327–331, 357–358, 568, 569–570
    genes for, 520, 568
    of thermophile, 216, 568
  nucleosomes, 331, 520
gentamicin, 423
geranylgeranyl-PP, 279–283
Giardia, 568
gln gene, 509
glucose
  catabolism, 2–7
    evolution of, 15–17
  gluconeogenesis, 7, 17
  Na^+, glucose-symporter, halobacterial, 35–36
glucose dehydrogenase, 18, 217
glutamate dehydrogenase(s)
  evolution of, 359
  of thermophiles, 213
glutamine biosynthesis, 509
glutaredoxin-like protein, 69
glyceraldehyde, 2, 4, 6, 7, 15
glyceraldehyde:ferredoxin oxidoreductase, 162
glyceraldehyde 3-phosphate dehydrogenase, 18
  of methanogens, 516–517
  of thermophiles, 164, 213–214, 516–517
glycerokinase, 279
glycerol, 15, 279–282
glycerol:$NADP^+$ oxidoreductase, 15
sn-1-glycerophosphate, 282, 283
sn-3-glycerophosphate, 275, 279–281, 283
glycerophosphate dehydrogenase, 279–282
glycolipids, 266–277, 291
  biosynthesis of, 284–285
β-glycosidase, thermophilic, 217
growth hormone gene, cloning of, vii
GTPase center, 444–445
gvp genes, 483–484
gyr genes, 486
gyrase, reverse, 216, 336–342, 349–350
  genes for, 476, 486

H_4MPT, *see* tetrahydromethanopterin
halobacteria
  cell envelopes of, 236–237, 243–246, 484
  central metabolism in, 2–4, 7, 8–9, 10, 11, 16–17, 18, 20
  energy transduction in, 25–37
  evolution of, 16–17, 289–291
  genes of, 476, 478–490
  genome of, 326, 331, 467, 468–473
    DNA polymerase, 356
    DNA topoisomerases, 346–349
    insertion sequences, 471–473, 476–478
    mapping of, 474–478
  lipids in, ix, 262, 263, 265–269, 289–291
    biosynthesis of, 15, 278, 279–282, 283, 284–285
    membrane function of, 287–288
  membrane-bound enzymes of, 297–299
    ATPases, 27, 33–34, 37, 304–307
    denitrification, 317–318
    electron transport chain, 308, 309–310, 311, 314–316
  motility of, 27, 36–37, 174, 175–183
  photoreception in, 25, 30–32, 173–174, 179–180, 183–185, 189–202
  phylogenetic relationships of, ix–x, xxi–xxiv, 267–269
  signal transduction in, 173–185
  transcription in
    control of, 385, 478
    genes for, 489–490
    in vitro systems, 380–383
    promoters, 380–383, 478
    RNA polymerase, 369, 371–373, 395, 489–490
    terminators, 478
  translation in, 395
    elongation factors, 397–398, 399
    genes for, 476, 479–481, 488–489
    in vitro systems, 411–412
    inhibitors of, 417, 418, 426, 427
    ribosomes, ix–x, xxii, xxiv, 402, 403–405, 406, 408–410, 440, 446, 447, 454, 476, 479–480
Halobacterium halobium phage ΦH1, 469, 471, 472, 473, 487
Haloferax volcanii, gene map of, 475–478
halophiles, *see* halobacteria
halorhodopsin, 26–27, 173–174, 189–202 passim
  energetics of, 201–202

gene for, 395, 482–483
ion transport by, 30, 199–201
photoreception by, 196–199
structure of, 190–196
heat-shock proteins, 521–522, 568
heterodisulfide reductase, 51, 53, 56–57, 92–93, 124, 143
hexose catabolism, 2–7
evolution of, 15–17
his genes, 507
histidine biosynthesis, 507
histidinol-phosphate aminotransferase gene, 476, 487–488
histones, x, xxvi, 327–331, 357–358, 568, 569–570
genes for, 520, 568
of thermophiles, 216, 327–329, 568
HMf protein, 329–331, 346, 358, 520, 568
hmfB gene, 520
hmg gene, 488
hop gene, 395, 482–483
HSHTP(7-mercaptoheptanoylthreonine phosphate), 53, 56 , 88–93, 124
HSNP protein, 327
HTa protein, 327–329, 358, 520
HU proteins, 327–329, 358
hydrogen ions (H^+)
halobacterial bioenergetics and, 25–27
bacteriorhodopsin and, 26, 27–30, 31–32, 173, 174, 199–202
motility and, 36–37
photoreception and, 31–32, 37
methanogens and: Na^+/H^+ antiporter, 60, 137–139, 145, 152, 155–156
hydrogenase(s), 316–317
in methanogens, 66–69, 124, 316
in *Pyrodictium brockii*, 316–317
3-hydroxy-3-methylglutaryl–coenzyme A reductase, gene for, 476, 488
hygromycin, 423
hyperthermophilic archaea, *see* thermophiles

ileS gene, 510, 514–515
initiation sequences, 396, 455, 457, 551, 553
insertion sequences
in halobacteria, 471–473, 476–478
in methanogens, 508
isoleucyl–tRNA synthetase, 510, 514–515, 567

kanamycin, 423
kirromycin, 425, 427

L_2 r-protein, 446, 451
L_{12} r-protein, 447, 451, 453–454, 513
lactate oxidation in *Archaeoglobus fulgidus*, 159–161
light, halobacteria and
H^+ pump, 25–30
motility, 174, 177–183
photoreception, 25, 30–32, 173–185, 189–202
lipid(s)
biosynthesis of, 14–15, 278–287
'core', 262–265, 279–283
evolution/taxonomic relations and, ix, 267–269, 273, 277, 289–291
membrane function of, 287–288
polar, 265–277, 283–286, 291
lipoic acid, 10

malate dehydrogenase, 18
gene for, 517
maltose fermentation in *Pyrococcus furiosus*, 161–162
MC1 protein, 329
mcr operon, 501–502, 522
mdh gene, 517
membrane(s)
enzymes bound to, 297–319
isolation of, 298–299
see also lipids
halobacterial bioenergetics and, 25–37
7-mercaptoheptanoylthreonine phosphate (HSHTP), 53, 56, 88–93, 124
messenger RNA, *see* mRNA
Methanobacterium thermoautotrophicum phage ψM1, 498–499
'methanochondrion' concept, 132–133
methanochondroitin, 232–236
methanofuran (MF; MFR), 47–48, 53–54, 119
methanogenesis, 41–100
from acetate, 41, 61–65, 95–100, 118, 141–143, 147–153
bioenergetics of, 113–156
from CO_2, 53–58, 116–117, 119–143
coenzymes for, 43–53
enzymes for, 66–100, 119–124, 143–144, 147–148
genes encoding, 500–506, 522
from methanol, 58–61, 93–94, 117–118, 143–147
from methylamines, 61, 118, 147

methanogenesis (cont'd)
  from non-methanol alcohols, 41, 57–58, 72–73, 116–117, 139–141
  substrates for, 41–42, 115–119
methanogens
  see also methanogenesis
  cell envelopes of, 223–236, 239–243, 520–521
  central metabolism in, 7, 8, 10, 13, 17, 20, 516–519
  discovery of, ix
  evolution/phylogenetic relationships of, xxi–xxiv, 114, 273, 289–291, 499–500, 508–521 passim
  genes of, 497–523
    amino acid biosynthesis, 507–510
    chromosomal proteins, 520
    expression of, 499–500, 521–522
    metabolic enzymes, 516–519
    methanogenesis, 500–506, 522
    nitrogen fixation, 515–516, 522
    organization of, 499
    phylogeny and, 499–500, 508–521 passim
    S-layer proteins, 520–521
    str operon, 398–400, 404–405, 513–514
    transcription/translation machinery, 510–515
    transformation systems, 522
  genome of, 326, 331, 498, 520
    DNA polymerase, 353, 355, 356
    histones, 329–331, 346, 520
    reverse gyrase, 339, 346
  growth rates of, 42
  lipids of, 262, 265, 269–273
    biosynthesis of, 15, 278, 282–283, 284, 285–286
    evolution of, 289–291
    membrane function of, 288
  membrane-bound enzymes of, 298–299
    ATPases, 131–132, 300–302
    cytochromes, 60, 65, 69–70, 129–130, 147, 148, 312
    hydrogenases, 316
  'methanochondrion' concept for, 132–133
  transcription in, 369–371, 395, 499–500
    genes for, 512–514
    in vitro systems, 384
    promoters, 380, 499–500
    RNA polymerase, 369, 371–373, 384, 512–513
    terminators, 384, 500
  translation in, 395
    elongation factors, 398, 513
    genes for, 510–515
    in vitro systems, 412
    inhibitors of, 417–418, 420, 427
    ribosomes, xxii, xxiv, 402, 403–405, 406, 428–429, 440, 446, 447, 454, 455, 499, 513–514
  transport in, 156–158
methanol, methanogenesis from, 58–61, 93–94, 117–118, 143–147
methanopterin, 48–50
Methanosarcina barkeri, pyruvate methanogenesis by, 118, 153–155
Methanothermus fervidus HMf protein, 329–331, 346, 358, 520, 568
methenyl–H_4MPT cyclohydrolase, 49, 54–55, 79–82, 123
methyl-accepting taxis proteins, 181–182
methylamine(s)
  methanogenesis from, 61, 118, 147
  transport of, 156
methylcobalamin:CoM methyltransferase, 56, 58–59, 64–65
methyl–CoM reductase, 51, 65, 88–92, 124, 143
  genes for, 500–502, 522
methylene–H_4MPT dehydrogenase, 49, 55, 82–84, 123, 124
methylene–H_4MPT reductase, 49, 55, 85–86, 123–124, 133, 142
methyl–H_4MPT:CoM methyltransferase, 49, 51, 87–88, 124, 134–135, 142, 145–146, 148, 151–153
methyltransferase(s)
  MT_1, 59, 93–94, 143
  MT_2, 55–56, 59, 93–94, 143
methylviologen-reducing hydrogenase (MVH), 66, 68–69
  genes for, 503–504, 522
mevinolin, resistance to, 473, 488
mitochondria, endosymbiotic origin of, 17
molybdopterin, 50, 78, 122–123, 505
motility, halobacterial, 27, 36–37, 174, 175–183
mRNA (messenger RNA), 394–395
  see also transcription; translation
  post-transcriptional modification of, 395, 511
  ribosome-mRNA interaction, 395–396, 499
mvh operon, 503–504

Na⁺, see sodium ions
NADH dehydrogenases, 309–310
NADH oxidases, 308–309
Natronobacterium pharaonis, halorhodopsin in, 30
neomycin, 423
nickel
   Ni/Fe–sulfur protein, 63–64
   transport of, 157
nif genes, 515–516, 522
nitrate pump, halobacterial, 30, 194–195, 198
nitrate reductase, 317–318
nitrogen fixation genes, methanogen, 515–516, 522
nucleosomes, 331, 520
nucleotide-activated oligosaccharides, 235–236

opsin shift, 193
oxidoreductase(s), 8–10, 13, 17, 162
2-oxoacid dehydrogenase complexes, 9–10, 17

paralogous genes/proteins, xvii, xix, 359, 567
paramomycin, 423
pentose-phosphate pathway, 2, 8
peptidyltransferase assay systems, 413, 415
peptidyltransferase center, 444
pgk gene, 517–518
pGRB, 472, 474
phage(s)
   ΦH1, 469, 471, 472, 473, 487
   ψM1, 498–499
pHH1, 469–470, 471, 472, 473
pHK2, 473
phosphate uptake
   in halobacteria, 36
   in methanogens, 158
phosphofructokinase, 2, 7, 16–17
phosphoglycerate kinase
   of methanogens, 517–518
   of thermophiles, 213–214
phosphohalopterin, 50
phospholipids, 265–277
   biosynthesis of, 283–284
5′-phosphoribosyl-5-aminoimidazole carboxylase, genes for, 510
phosphotransacetylase, 61–62, 99, 100, 147–148
   acetate formation from pyruvate and, 154–155
photolyase gene, 486
photomovement, 174, 177–183

photoreception, halobacterial, 25, 30–32, 173–174, 179–180, 183–185, 196–199
   motility and, 174, 175–183
phr gene, 486
plasmids
   in halobacteria, 469–470, 473–474
   in methanogens, 498, 499
pME2001, 498
pNRC100, 469–470
polyferredoxin, 68–69, 130–131
   genes for, 504
polymer-degrading enzymes, thermophilic, 217
poly(U)-programmed cell-free systems, 411–413, 416
post-transcriptional modification, 395, 511
potassium (K⁺)
   halobacterial bioenergetics and, 27, 34–35
   thermophile DNA stability and, 332
   thermophile proteins and, 215
   transport of, 157–158
proC gene, 508
progenote, xvii, xxvi–xxvii
prokaryote–eukaryote dichotomy, viii–xix, xxvi, 565
proline biosynthesis, 508
promoters, 379–383, 384, 478, 499–500, 545–549, 569
proteases: of thermophiles, 217
protein synthesis inhibitors, 416–427
   sensitivity to, 420–424, 427
protein ω, 334, 337–338
protometer, 31–32, 37
pseudomurein, 224–231
puc gene, 522
pulvomycin, 425, 427
pur genes, 510
pURB600, 499
puromycin, resistance to, 522
pWL102, 473
Pyrococcus furiosus, energetics of, 161–164
Pyrodictium brockii, hydrogenase of, 316–317
1-pyrroline-5-carboxylate reductase, gene for, 508
pyrophosphatase, methanogenesis and, 62, 99, 100, 153
pyruvate
   fermentation of, 13
   methanogenesis from, 118, 153–155
   oxidation of, 8–10, 14, 17
   production of, 2–7, 15–17
pyruvate oxidoreductase(s), 8–10, 13, 17, 162

quinol oxidase, 313–314
quinones, 308, 312–313

reduced nicotinamide dependent 2,2′-dithioethane-sulfonic acid reductase, 51
respiratory chain, energy transduction and, 25–27, 32–33
reticulate evolution, 569, 570
retinal, photoisomerization of, 27–30, 174, 184–185, 195
reverse gyrase, 216, 336–342, 349–350
  genes for, 476, 486
rhodopsin(s)
  ion transport, see bacteriorhodopsin; halorhodopsin
  sensory, see sensory rhodopsin
ribosome(s), 439–460
  evolution/phylogenetic relationships and, ix–x, xxiv–xxv, 405, 406, 415–416, 420, 427, 431–432, 440, 443–444, 446–447, 451, 453, 458–460
  hybrid, 428–429, 430
  in vitro translation systems, 411–416
  inhibitors targeted at, 416–427, 444–445
  mass of, 402–405
  mRNA interaction with, 395–396, 499
  proteins of, 403–405, 440, 446–454, 459–460
    genes for, 454–458, 476, 488–489, 513–514
  RNA of, see rRNA
  shape of, 405–406
  stability of, 400–402, 440–441
  subunits of, 402, 439–440
    interchangeability of, 428–429, 430, 454
    reconstruction in vitro of, 407–410, 441, 443
    stalk region, 451, 453–454
ribosome-binding sites, 499
RNA
  see also mRNA; rRNA
  post-transcriptional modification, 395, 511
  7S RNA molecule, 410–411, 476, 480, 511, 512
  RNaseP, 481
  thermophile, stability of, 332
RNA polymerase(s), 369–379, 383, 394–395
  evolution/phylogenetic relationships and, xxvi, 373–377, 569

  genes for, 371–373, 476, 489–490, 512–513, 545
  promoters, 379–383, 384, 478, 499–500, 545–549, 569
  terminators, 383–384, 478, 500
RNaseP RNA, 481
RPG effect, 57
rpo genes, 371–373, 476, 489–490, 512–513, 545
r-proteins, 403–405, 440, 446–454, 459–460
  genes for, 454–458, 476, 488–489, 513–514
rRNA (ribosomal RNA), 441–445, 458–459, 460
  evolution/phylogenetic relationships and, x, xi, xv–xvii, xxii–xxv
  genes for, 476, 479–480, 511–512, 557
  organization of, 441–442, 543–544
  thermostability and, 542
  promoters, 380–383, 384
  sequences of, x, xi, xv–xvii
  structure and function of, 443–445
rubidium uptake, 158

S-adenosyl-L-methionine:uroporphyrinogen III methyltransferase (SUMT), gene for, 519
α-sarcin, 423–424
sarcinapterin, 48–49, 64
Schiff base (bacterial rhodopsins), 27–29, 190, 192, 193, 194, 195–196, 199–202
sensory rhodopsin I, 30–32, 174, 179, 182, 183–185
  gene for, 483
sensory rhodopsin II, 30–32, 174, 179, 180–181, 183, 185
  gene for, 483
Shine–Dalgarno recognition, 393, 395–396, 455, 457, 551, 553
shuttle vectors
  in halobacteria, 473–474
  in methanogens, 498
S-layers (surface layers), 223, 254
  of halobacteria, 243–246, 476, 484
  of methanogens, 224, 231–232, 520–521
  of thermophiles, 213, 218, 248–252
slg gene, 485, 486, 520–521
sod gene, 485, 518–519
sodium ions (Na^+)
  halophilic archaea and
    bioenergetics and, 27, 31, 34–36
    motility and, 36–37
    Na^+-motive respiration, 37

methanogens and, 60, 113, 133–139, 155–156
  Na$^+$/H$^+$ antiporter, 60, 137–139, 145, 152, 155–156
solfapterin, 50
somatotropin gene, cloning of, vii
spc operon, 404–405, 455
spermine, 402, 407–408, 412–413, 415
str operon, 398–400, 404–405, 455, 513–514
streptomycin, 417, 418, 421
succinate dehydrogenases, 311
succinate thiokinase, 11
sugar
  catabolism of, 2–7
    evolution of, 15–17
    import of, halobacterial, 36
sulfate-reducing archaea, see thermophiles
superoxide dismutase gene(s), 385, 476, 485, 486, 518–519

"tatiopterin", 49
terminators, 383–384, 478, 500
tetrahydrofolate, 48, 54, 55, 63
tetrahydromethanopterin (H$_4$MPT), 48–50, 54–55, 59–60, 64 , 78–88, 119, 123–124
thermine, 402, 407–408, 412–413, 415
thermophiles
  cell envelopes of, 247–252
  central metabolism in, 5–7, 8, 10, 11–12, 13–14, 18–20
    citrate synthase, 19–20
    glyceraldehyde-3-phosphate dehydrogenase, 164, 213–214, 516–517
  energetics of, 161–164
  evolution/phylogenetic relationships of, ix, x, xxi–xxiv, 277, 289–291, 557, 568
  genes of, 516–517, 535–559
    nucleotide composition of, 536, 541–543, 553
    optimal growth temperature and, 536, 541–543
    organization of, 543–545
    phylogenetics and, 557
    sequences of, 535–536
    transcriptional signals, 380–384, 545–551
    translational signals, 551, 553
  genome of, 326, 331, 468–469
    DNA polymerase, 218, 353–355, 356
    DNA stability, 216, 331–332, 340, 342, 536, 541–543
    DNA topoisomerases, 343–346, 349–351
    histones, 216, 327–329, 568
    nucleotide composition of, 536, 541–543, 553
    RNA stability, 332
  lipids of, ix, 262, 265, 273–277
    biosynthesis of, 15, 278, 283, 286–287
    evolution of, 289–291
    membrane function of, 288
    membrane-bound enzymes of, 297–299
      ATPases, 216, 302–304
      electron transport chain, 308, 309, 311, 312–314
      hydrogenases, 316–317
  proteins of, 209–218
    biotechnological potential of, 216–218
    intracellular milieu and, 215
    structures of, 212–214
    thermoadaptive functions of, 215–216
  transcription in, 367–386
    control of, 385
    in vitro systems, 384
    promoters, 380–383, 384, 545–549
    RNA polymerase, 369, 371–373, 377–379, 383, 394–395, 545
    terminators, 383–384, 549–551
  translation in, 394, 395–396, 551, 553
    see also Archaeoglobus fulgidus
    elongation factors, 398, 399
    in vitro systems, 412–413, 415
    inhibitors of, 417–420, 421, 427
    ribosomes, xxii, xxiv, 400–402, 403, 405–406, 407–408, 428–430, 440–441, 446, 447, 451, 453–454, 455, 542, 557
Thermoplasma acidophilum HTa protein, 327–329, 358, 520
thiostrepton, 418, 422–423, 444–445
'third form of life', discovery of, vii–xi
TOP3 gene, 335
transcription, 367–386, 394–395, 478
  control of, 384–385
  evolution/phylogenetic relationships and, 367–369, 373–377, 569
  in vitro systems, 384
  post-transcriptional modification, 395, 511
  promoters, 379–383, 384, 394–395, 478, 499–500, 545–549, 569
  RNA polymerases, 369–379, 383, 394–395
    genes for, 371–373, 476, 489–490, 512–513, 545

transcription (cont'd)
  terminators, 383–384, 478, 500, 549–551
transfer RNA, see tRNA
translation, 393–432
  components of, 394–411
    see also ribosome; tRNA
  interchangeability of, 428–430, 454
  evolution of, x–xi, xv–xix, 398, 411,
    415–416, 420, 427, 431–432
  in vitro systems, 411–416
  inhibitors of, 416–427, 444–445
  initiation of, 396, 455, 457, 551, 553
  protein targeting in, 410–411
    7S RNA, 410–411, 476, 480, 511, 512
  termination of, 553
transport
  in halobacteria, 27–30, 199–201
  in methanogens, 156–158
tRNA (transfer RNA), 394
  genes for, xxiv–xxv, 380, 394, 476, 480,
    510–511, 543–544

trp genes, 488, 508–509
trypanosomes, 17, 568
tryptophan biosynthesis, genes for, 476, 488,
  508–509

universal ancestor, xvii, xix, xxvi–xxvii, 567,
  569
  genome and, 359–360, 567–568
    DNA topoisomerases, 350
  translation and, 431–432, 567

vac gene, 383
virginiamycin, 418
virus-like particle, 499

water, 53, 71, 73, 196, 201, 214, 316, 397–398

X-ray studies, 19, 91, 228, 379, 440

yeast, 99, 161, 329, 335, 349, 411, 421, 430,
  486, 509, 540, 568
Y:R (pyrimidine:purine), xvi, xix, xxiii